Mathematical theory of transport processes in gases

Mathematical theory of transport processes in gases

J. H. FERZIGER
Stanford University, Stanford, Calif., USA

and

H. G. KAPER
Argonne National Laboratory, Argonne, Ill., USA

1972

NORTH-HOLLAND PUBLISHING COMPANY – AMSTERDAM · LONDON
AMERICAN ELSEVIER PUBLISHING COMPANY, INC. – NEW YORK

Library of Congres Catalog Card Number 77–126501
ISBN North-Holland 0 7204 2046 6
ISBN American Elsevier 0 444 10352 X

Publishers:

NORTH-HOLLAND PUBLISHING COMPANY - AMSTERDAM

NORTH-HOLLAND PUBLISHING COMPANY, LTD. - LONDON

Sole Distributors for the U.S.A. and Canada:

AMERICAN ELSEVIER PUBLISHING COMPANY, INC.
52 VANDERBILT AVENUE
NEW YORK, N.Y. 10017

Printed in the Netherlands

To our children
Ruth, Shoshanah and Miriam
Tasso and Bertrand

Contents

Preface

During the past decade, research and development in the areas of engineering mathematics and computer technology have resulted in a substantial increase of both the sophistication and the quantity of scientific calculation. At times, advance has been so fast that it has proved impossible to provide exact values of fundamental parameters needed for the solution of specific technological problems. In such instances, it has been necessary to obtain these parameters from estimates, and various techniques have been used for this purpose – including interpolation (or even extrapolation) of experimental data and, perhaps more often than one would care to admit, pure guess.

Transport properties of gases certainly fit the above characterization. Their values are required in a large number of problems involving flowing gases. Such problems occur naturally in chemical engineering and aerodynamics, where one is dealing mostly with neutral gases at ordinary densities, and in astrophysics and certain areas of applied physics, where flows of ionized gases may play an important role. Measurement and tabulation of the various transport properties – thermal conductivity, viscosity, diffusion, etc. – for all gas mixtures and under all circumstances is not very well feasible. Thus, a means of predicting the needed properties from available data is necessary and, of course, a solid theory is the most desirable means of accomplishing this.

The present book offers a more or less complete theory for deriving transport properties of gases from first principles – that is, in terms of the intermolecular potentials which determine the forces between any two molecules in the gas. Both neutral gases and ionized gases are discussed, and, for the neutral gases, the theory of monatomic as well as polyatomic gases is covered. The principal results of these investigations are based on Boltzmann's equation for a dilute gas; however, extensions into the areas of dense gases and rarefied gases are also given.

This book is an outgrowth of a course offered somewhat irregularly at Stanford University and intended primarily for aeronautical and mechanical engineers and, to a lesser extent, for chemical engineers. In teaching this course we have found that the standard work on the kinetic theory of gases, "The Mathematical Theory of Non-uniform Gases", by S. Chapman and T. G. Cowling (Cambridge University Press, 1939, 1952) is rather difficult as a text and does not emphasize the calculation of transport properties sufficiently. On the other hand, "The Molecular Theory of Gases and Liquids" by J. O. Hirschfelder, C. F. Curtiss and R. B. Bird (J. Wiley & Sons, 1954), while very good in discussing the calculation of transport properties, compresses the theory far more than desirable. It was our aim to retain the best features of these works in a text geared to the typical needs of graduate students in the above mentioned disciplines. Our debt to the authors of these two works, however, will be particularly clear to anyone who is familiar with them.

More than a decade and a half have passed since the publication of Hirschfelder, Curtiss and Bird's book. We have attempted to assess the recent work and included what appeared to be the important advances. Of course, in those areas which are currently developing rapidly, such as polyatomic gases, we run the risk of being out of date very quickly. That is unavoidable; however, in such instances we have tried to point out what important developments may reasonably be anticipated in the next few years.

The collaboration that resulted in this book was made possible by a Fulbright grant to one of the authors (J. H. F.) for the year 1967–1968. That year, spent at the Mathematics Institute of the University of Groningen (The Netherlands), was extremely valuable in many ways and will be fondly remembered. Thanks are due to the United States Educational Foundation in The Netherlands and Mrs. H. J. van Doorne for making this possible. During the same year, 1967–1968, and the following year the second author (H. G. K.) was a faculty member at the host institute. Both authors wish to thank the Directors of the Mathematics Institute, Profs. J. C. H. Gerretsen and A. I. van de Vooren, for their superb hospitality and for their encouragement and support. Moreover, we wish to thank our present institutions, Stanford University and Argonne National Laboratory, for providing the facilities and the atmosphere necessary and conducive for the successful completion of this book.

It is a pleasure to acknowledge the assistance received in the preparation of the manuscript, which was a truly international endeavor. Miss Dineke Bakker of the Mathematics Institute of the University of Groningen, and Mrs. Ruth Korb of the Nuclear Engineering Division of the Mechanical Engineering Department of Stanford University, typed the first version of the manuscript. Mrs. Doris Haight of the Applied Mathematics Division of Argonne National Laboratory made final revisions and saw the manuscript to its final form, while the second author conducted a seminar on the subject of the kinetic theory of gases. The drawings for most of the figures were made by Mr. E. R. R. Timmerman of the Kapteyn Laboratory of the University of Groningen, and Mr. V. Grynczyk of the Applied Mathematics Division of Argonne National Laboratory. Library research was done for us by Miss Lydie Meijer of the Mathematics Institute of the University of Groningen. To all of these, our sincere thanks.

Lastly, the publisher, North-Holland Publishing Company, and their representative, Mr. W. H. Wimmers, deserve our very special thanks for an excellent job in producing the book and for their extremely pleasant manner of doing business.

Stanford, Cal. J. H. FERZIGER

Argonne, Ill. H. G. KAPER

November 1971

Introduction 1

1.1. Kinetic theory of transport processes

Statistical mechanics may be naturally divided into two branches, one dealing with equilibrium systems, the other with non-equilibrium systems. In this book attention will be focussed upon a segment of non-equilibrium statistical mechanics. In particular, we shall deal with dilute gases which are only slightly removed from equilibrium. This area of non-equilibrium statistical mechanics is customarily designated as *kinetic theory*.

The most common non-equilibrium situation is the one encountered, for example, in ordinary gas flows, in which the density, hydrodynamic velocity and temperature vary throughout the gas. In the approach to equilibrium these non-uniformities tend to be smoothed out through the transport of mass, momentum and energy from one part of the gas to another. Such processes are commonly and appropriately termed *transport processes*.

The purpose of this book is to give a mathematical account of the kinetic theory of transport processes. Of course, to render the problem mathematically tractable, certain assumptions and idealizations must be made, which will impose restrictions on the validity of the results. Although, at this stage, it would be premature to give a quantitative evaluation of these restrictions, it seems appropriate to discuss them at least qualitatively.

In kinetic theory, a gas is described in terms of a *distribution function* which contains information on both the distribution of molecules in the system under consideration and the distribution of molecular velocities. This distribution function, in general, will vary with time. If it is assumed that molecules can be treated as classical point centers of force surrounded by a force field, a nonlinear integro-differential equation, viz. *Boltzmann's equation*, may be derived for the distribution function. A detailed study of the hypotheses underlying its derivation reveals that Boltzmann's equation provides a valid description of the state of a gas as long as the density of the gas is "sufficiently" low and the gas is "sufficiently" uniform in space. Since Boltzmann's equation (or an appropriate modification thereof) will be the fundamental tool for the description of transport processes in gases in the present text, the restrictions which are necessary for Boltzmann's equation to yield a valid description of the physical behavior of gases will be assumed throughout much of the book.

The objective of kinetic theory is to explain phenomena that are macroscopically measurable in a gas at or near thermal equilibrium in terms of properties of the individual molecules, i.e., the intermolecular force law. The macroscopic observables are particular velocity moments of the distribution function mentioned earlier. Therefore, one must somehow solve for the distribution function from Boltzmann's equation. An *exact* solution to this equation can be found if and only if the gas is at thermal equilibrium. If the state of the gas is *near* thermal equilibrium, however, one can still give an *approximate* solution. Since the form of this solution is such that it is practically meaningful only as long as the density of the gas is "sufficiently" high, this will impose another restriction on the validity of our results.

Thus, it is evident that the range of densities for which a "clean" mathematical theory of transport processes in gases can be given with Boltzmann's equation, is rather limited. It is indeed fortunate that the gas densities encountered in many physical situations meet the limitations outlined above, the exceptions being dense gases and liquids on the one hand and rarefied gases on the other. For these latter cases separate discussions will have to be given.

From mathematical physics it is well known that a valid description of a gas is often given by fluid dynamics. The Navier-Stokes partial differential equations which describe the behavior of the macroscopic observables of a gas – density, hydrodynamic velocity and temperature – as a function of position and time, are indeed applied to gas flow problems with remarkable success. Therefore, it is reasonable to require that any solution to the equations of kinetic theory contain the fluid dynamic description in those cases in which the latter is expected to be valid. Accordingly, one of our objectives will be to elucidate the sense in which the fluid dynamic equations may be considered approximations to Boltzmann's equation and, thus, to emphasize the limitations of fluid dynamics. Fortunately enough, the Chapman-

Enskog method provides not only a solution to Boltzmann's equation but, at the same time, it reduces the kinetic description to the fluid dynamic description in a *constructive* manner. Thus, examination of the approximations required in this reduction will provide the desired exposition of the limitations of a fluid dynamic description of a gas.

There is further benefit in carrying out such a constructive procedure. In classical fluid dynamics, the basic equations are derived from conservation principles and from certain hypotheses which relate qualitatively the transport of mass, momentum and energy in a gas to gradients of macroscopic observables. Such hypotheses are, for example, Newton's law of viscosity and Fourier's law of heat conduction. These hypotheses introduce unknown, so-called *transport coefficients*, such as the coefficients of viscosity and thermal conductivity, into the equations of fluid dynamics. Now, if kinetic theory is self-contained, that is, if the laws of intermolecular interaction are specified, the fluid dynamic description which is constructed from the kinetic theory should contain no unknown coefficients. Thus, by comparison with the classical fluid dynamic equations, one obtains formulae that connect the transport coefficients with the laws of molecular interaction. These formulae are an especially valuable consequence of kinetic theory. They provide a means of obtaining transport coefficients in terms of intermolecular force laws or they may serve to establish models of molecular interaction from experimental data for the transport coefficients. Alternately, these formulae may be used to connect the various transport coefficients with each other.

The above remarks may serve to outline the domain of kinetic theory. The investigation of the various problems belonging to this domain will occupy the greater part of the book.

1.2. Historical summary

Kinetic theory of gases, as it is presently understood, has its historical origin in 1859, when Maxwell introduced the statistical approach in a paper read at a meeting of the British Association for the Advancement of Science. The assumption made by all previous workers, namely that all molecules of a gas move with the same speed, was thereby abandoned and the random character of molecular motion was recognized. In a series of two papers, Maxwell [1860] published his results on the law of distribution of molecular velocities for a uniform gas in equilibrium (the so-called *Maxwellian velocity distribution*) and on the law of *equipartition* of the mean molecular energy in a mixture of gases. These results were subsequently corrected and improved by Maxwell [1867] in a paper on the kinetic theory of non-uniform gases. In the latter paper, Maxwell derived the equations of transfer, which give the total rate of change of any mean molecular property, for a gas whose molecules are point centers of force repelling each other with forces that are inversely proportional to the fifth power of their mutual distance (so-called *Maxwell molecules*).

In an attempt to give a rigorous justification of Maxwell's assumption on the random character of molecular motion Boltzmann [1872] established the *H-theorem.* This theorem states the *irreversibility* of physical processes and shows how molecular collisions tend to increase entropy – any initial distribution of molecular positions and velocities will *almost certainly* evolve into an equilibrium state, in which the velocities are distributed according to Maxwell's law. In the same paper, Boltzmann derived an integro-differential equation to describe the evolution of the velocity distribution function in space and time and demonstrated that the formulae found by Maxwell for the various transport coefficients in a gas of Maxwell molecules could be obtained directly from the solution of the integro-differential equation. This integro-differential equation is now known as *Boltzmann's equation.* With the generalization of the H-theorem to the case of a gas in the presence of a conservative force field by Boltzmann [1875a] and Lorentz [1887], the formal framework of the kinetic theory of non-uniform gases was essentially completed.

The reader who wants a more detailed picture of this phase of the early history of kinetic theory may consult the monograph by Brush [1966] which contains in English translation many of the papers cited above.

From the investigations of Maxwell, Boltzmann, Stefan, Langevin and several others it became apparent that actually *solving* Boltzmann's equation was simple only in the case of a gas composed of Maxwell molecules. In all other cases, the mathematical complexities and uncertainties proved considerable. Lorentz [1905] succeeded in reducing Boltzmann's equation to a sufficiently simple form by considering a binary mixture of molecules, in which molecules of one type are of negligible mass compared to molecules of the other and encounters between the light molecules are ignored. His results, obtained in connection with the theory of electrons in a metal, were exact and of great value; however, the work gave no indication of a *general* method for solving Boltzmann's equation.

In 1910, the mathematician Hilbert published a study on the logical structure of Boltzmann's equation. Restricting himself to the case of a gas of rigid spherical molecules, Hilbert showed that Boltzmann's equation is equivalent to a certain linear integral equation of the second kind, for which a rigorous mathematical theory was available. Thus, he was able to prove the *existence* and *uniqueness* of a solution and to indicate some properties of the solution. Hilbert's results can be found in his treatise on the theory of integral equations (Hilbert [1912]). This purely mathematical approach to Boltzmann's equation was later followed by Lunn [1913], Carleman [1933], Wild [1951], Morgenstern [1954] and others. The importance of Hilbert's work is in the establishment of a firm logical structure for kinetic theory.

The problem of actually deriving a suitable formalism for solving Boltzmann's equation was solved independently by Chapman and Enskog shortly after publication of Hilbert's results. Chapman's work follows the method of Maxwell and is based upon the use of the equations of transfer, while Enskog's approach depends upon the solution of Boltzmann's equation for the velocity distribution function. Both methods lead to the same expressions for the transport coefficients. In two papers, Chapman [1916, 1917] deduced expressions for the coefficients of viscosity and thermal conductivity in a simple gas and in a gas mixture, assuming as did Maxwell, that the velocity distribution function, f, in a slightly non-uniform gas could be written as $f = f_0(1+\phi)$, where ϕ is supposed to vanish in a uniform gas. Enskog's theory, which was published in his doctoral dissertation – Enskog [1917] – was based upon a series solution of Boltzmann's equation. This method was inspired by Hilbert who, without success, had tried a similar technique of successive approximation.

A further improvement in the general theory of non-uniform gases at ordinary densities (i.e., normal temperature and pressure) is marked by two papers by Burnett [1935a, 1935b], who considered the second-order approximation to the velocity distribution function for simple gases and gas mixtures. He gave the method by which the velocity distribution function can be calculated to *any* order of approximation for a simple gas.

Once it had been shown that in a gas at normal temperature and pressure Boltzmann's equation could be solved to any degree of accuracy, attempts were made to overcome some of the limitations inherent to Boltzmann's equation. In particular, the assumption of binary encounters, which had been basic in the *heuristic* derivation of the integro-differential equation as first given by Boltzmann, prohibited application of the results of kinetic theory to dense gases or liquids. Of course, there is no *a priori* reason why a generalized Boltzmann equation should exist for dense gases. But the connection between kinetic theory and fluid dynamics – which had been clearly demonstrated by the methods of Chapman and Enskog – led one to *expect* the existence of such a generalized Boltzmann equation, although Boltzmann's derivation did not contain the slightest indication as to how such generalization could be obtained. The most successful attempt to generalize Boltzmann's equation to higher densities was due to Enskog [1922b]. However, this was purely an *ad hoc* generalization, which was valid for hard spheres only. It soon became apparent that a reevaluation of the very foundation of kinetic theory was necessary.

Bogoliubov [1946] was the first to succeed in this task. Starting with Liouville's equation which describes the temporal behavior of the state of a gas in phase space, Bogoliubov recognized the existence of *different time scales* on each of which the state of the gas needs to be described with an appropriate degree of fineness. He showed that a generalized Boltzmann equation is obtained if a description is sought on a rather coarse time scale and that this equation can be reduced to Boltzmann's original equation if a density expansion is performed and only the lowest-order terms are retained. This result cleared the way for a systematic extension to a kinetic theory of gases in which triple and higher-order collisions can be taken into account; such an extension subsequently was given by Choh and Uhlenbeck [1958] and further generalized by García-Colín, Green and Chaos [1966].

An approach which circumvents some of the logical gaps in Bogoliubov's was later worked out by Green [1956, 1958] and Cohen [1962a, 1962b], who adapted the method of *cluster expansions* developed for dense gases in equilibrium statistical mechanics. This approach, although presently far from complete, seems to be most promising for a better understanding of the fundamentals of kinetic theory (see Dorfman and Cohen [1967]).

As most gas flow problems at normal temperatures and pressures are adequately described by the equations of fluid dynamics, it is important to understand the relation between Boltzmann's equation and, say, the Euler and Navier-Stokes equations. Here, we must mention the work of Grad, who, in a series of papers, proved the equivalence of the equations of fluid dynamics to an *asymptotic* form of Boltzmann's equation. Again, the existence of different time scales plays a fundamental role, as the time scale appropriate for the fluid dynamic description of a gas is much coarser than the time scale of kinetic theory. Much work in this area remains to be done; particularly, a detailed study is required of the closely related question of the existence and uniqueness of solutions to initial and boundary value problems of kinetic theory.

Another limitation which is inherent to Boltzmann's equation and which severely restricts the applicability of the results obtained with it, is due to the fact that the equation is valid only for gases composed of monatomic molecules, while almost all gases, and especially those of physical interest, are composed of polyatomic molecules. Complications arise, firstly, because polyatomic molecules vary in shape and, thus, the intermolecular potential is no longer spherically symmetric, and secondly, because polyatomic molecules possess internal modes of motion with which energy is associated. To a large degree, the theory for monatomic gases could be modified to account for these complications, as was done, e.g., by Eucken [1913]. Such *a posteriori* theories were, in fact, sufficiently successful to delay the development of a rigorous theory of polyatomic gases for several decades. In the 1940's the need for a better theory became recognized and such a theory was put forward independently by De Boer and by Wang-Chang and Uhlenbeck (see Wang-Chang, Uhlenbeck and De Boer [1964]). Although not rigorous, it displayed some of the deficiencies in the theories based on modifying monatomic gas results and pointed the direction to a rigorous theory. The basis of the rigorous theory was finally laid down by Snider [1960] and used by McCourt and Snider [1964, 1965], among others. Simultaneously, experimental investigation of the transport properties of polyatomic gases was begun, particularly by Beenakker and co-workers at Leiden. The second half of the 1960's saw a great increase in the interest in this area and extensive development is currently taking place.

Two other areas of kinetic theory must be mentioned to complete this survey of current interests – viz., the study of ionized gas flows and rarefied gas dynamics. Interest in ionized gases has significantly increased over the last two decades. The impetus for this came from many diverse technical areas, each of which had its own peculiarities, and, in some of these areas, the theory had to be developed afresh from first principles. The subject – plasma physics – covers such a wide range of phenomena as to place the bulk of it outside the range of the present text. However, certain parts of this field, particularly those related to the flow of ionized gases, may properly be considered as belonging to kinetic theory. In fact, as long as the possibility of magnetic fields is ignored, the theory of neutral gases can be taken over with only a change in nomenclature. The admission of magnetic fields, on the other hand, poses some new problems, but these can be treated by a relatively straight-forward extension of the Chapman-Enskog method.

Rarefied gas dynamics is the study of phenomena taking place at an arbitrary ratio of the mean free path (time between two successive molecular encounters) to the characteristic dimension (time) of the phenomena. The range of problems includes, for example, problems of flow past aircraft flying at high altitudes, motion of gases in vacuum apparatus, propagation of sound, structure of shock waves, etc. These are situations for which Boltzmann's equation is appropriate, but the Chapman-Enskog method of finding approximate solutions to it becomes invalid. However, some problems of interest admit linearization and allow for an analytic solution with the aid of techniques used for solving the linear equations of, e.g., neutron transport theory and the theory of radiative transfer.

Properties of a gas 2

A number of different means of describing the state of a gas are possible and may be used. On the one extreme, one might consider a gas as a collection of molecules and follow the motion of the molecules by using any of a variety of formalisms of classical or quantum mechanics. This is clearly an impossible computational task and, furthermore, the surfeit of information produced is not required for any problem of interest. On the other extreme, one can adopt the crudest description of a gas that retains sufficient information to be of interest. For a gas at equilibrium, a thermodynamic description might suffice, while for a flowing gas one might use the standard equations of gas dynamics – the Navier-Stokes equations. The latter description is known to be a useful one and has been enormously successful. It has, in fact, been so successful that at one time it suppressed the development of theories based on the molecular nature of matter, viz., statistical mechanics and kinetic theory.

The well-known successes of the last named theories (some of which form the subject matter of this book) have rendered a defense of them unnecessary. There remains the question of how fundamental a theory one wishes to consider and what results one wishes the theory to produce. Thus one can begin with the study of the motions of a large number of particles and try to extract, by means of appropriate assumptions and/or approximations, the well-known macroscopic equations describing the behavior of the gas as a whole. While this can be done (and we shall do essentially just this later, cf. Sections 3.2–3.5) it is not clear that one need be quite *that* fundamental. An intermediate approach is, in fact, the historical one, and as it is quite a bit less formal and more easily understood we adopt it here and defer the more formal approach. Thus in this chapter we define the basic function that we will work with – the one-particle distribution function – and show how the various macroscopic properties of a gas may be related to it.

2.1. Basic definitions – Distribution functions and average properties

A gas molecule can be characterized by its *position* $\boldsymbol{r}$ in space and its *velocity* $\boldsymbol{c}$. For polyatomic molecules other coordinates are necessary to determine the state of a molecule; however, for the present, we treat only monatomic gases and so ignore internal degrees of freedom. If a gas contains molecules of more than one species, it is often convenient to append a species index to the velocity variable.

A volume element of physical space will be denoted by d^3r (i.e., in Cartesian coordinates $\mathrm{d}^3r = \mathrm{d}x\mathrm{d}y\mathrm{d}z$); the volume element d^3r is understood to contain the point $\boldsymbol{r}$. Likewise, a volume element in velocity space will be denoted by $\mathrm{d}^3c(= \mathrm{d}c_x\mathrm{d}c_y\mathrm{d}c_z)$ and the element d^3c extends about the velocity $\boldsymbol{c}$. Volume integration in physical space will be denoted by $\int \ldots \mathrm{d}^3r$, volume integration in velocity space by $\int \ldots \mathrm{d}^3c$. Unless expressly stated, these integrals extend over the entire physical system under consideration and over the whole of velocity space, respectively.

The variable t will represent time. An interval of time $\mathrm{d}t$, about time t, will be referred to simply as a time interval $\mathrm{d}t$.

Consider a simple gas, a gas composed of identical molecules of mass m. Since the number of molecules per unit volume is very large (approximately 3×10^{19} per cm^3 at standard conditions) we may expect that a statistical description, that is, a description based on the average behavior of large numbers of molecules, would be appropriate. Furthermore, such a description is certainly simpler than one which follows the detailed behavior of the large number of particles. Thus we assume that there exists a volume element d^3r whose dimension is small compared with the spatial scale on which macroscopic properties vary, but large enough to contain a sufficient number of molecules to allow a statistical description.*

* That such a choice of d^3r is possible in a normal gas can be seen by an example. Since there are about 3×10^{19} molecules per cm^3, if d^3r is a cube of edge 10^{-3} cm, there are of the order of 3×10^{10} molecules in d^3r, yet the edge of the cube is very small compared with the distance over which any of the macroscopic quantities may vary, except in shock waves.

Information is required not only about the spatial distribution of molecules, but about their velocity distribution as well. Only with the latter can we study the momentum and energy flows that play an important role in gas dynamics. The simplest function that contains the desired information is the *velocity distribution function*, $f(\boldsymbol{r}, \boldsymbol{c}, t)$, which we now introduce. It is defined in such a way that $f(\boldsymbol{r}, \boldsymbol{c}, t)\mathrm{d}^3r\mathrm{d}^3c$ is the *expected* number of molecules in the volume element $\mathrm{d}^3\boldsymbol{r}$ located at $\boldsymbol{r}$, whose velocities lie in d^3c about velocity $\boldsymbol{c}$ (i.e., whose x-velocities lie in $\mathrm{d}c_x$ about c_x, y-velocities in $\mathrm{d}c_y$ about c_y, and z-velocities in $\mathrm{d}c_z$ about c_z, all simultaneously) at time t. We emphasize again that this is an expectation value in the statistical sense and not the precise number of molecules with the stated coordinates. It is not obvious that the state of a gas is completely described by this function, and using it for this purpose requires justification. We will make the assumption that the justification exists and later present it (Sections 3.2–3.5).

Given the distribution function, it is an easy matter to obtain the macroscopic properties of the gas. To avoid the necessity of so stating for each property, we note that the values of all of these properties are expectation values in the statistical sense, but can be expected to coincide with the actual values to within narrow limits; we shall therefore conduct the discussion as if they are deterministic values. The simplest macroscopic property is the *number density* $n(\boldsymbol{r}, t)$, which is defined as the number of particles per unit volume at location $\boldsymbol{r}$ at time t. Thus $n(\boldsymbol{r}, t)\mathrm{d}^3\boldsymbol{r}$ is the total number of particles in $\mathrm{d}^3\boldsymbol{r}$ and n is clearly just the integral of f over velocity,

$$n(\boldsymbol{r}, t) = \int f(\boldsymbol{r}, \boldsymbol{c}, t)\mathrm{d}^3c. \qquad (2.1\text{-}1)$$

Since for a simple gas each particle has mass m, the *mass density* (or simply the *density*) at $\boldsymbol{r}$, at time t, is

$$\rho(\boldsymbol{r}, t) = mn(\boldsymbol{r}, t). \qquad (2.1\text{-}2)$$

Most equipment for measuring macroscopic gas velocity actually measures the associated momentum. The average velocity of a gas is therefore most properly defined in terms of average momentum. Now, since a particle of velocity $\boldsymbol{c}$ has momentum $m\boldsymbol{c}$ and there are $f\mathrm{d}^3r\mathrm{d}^3c$ particles with velocities in d^3c about $\boldsymbol{c}$ in $\mathrm{d}^3\boldsymbol{r}$ at time t, their total momentum is $m\boldsymbol{c}f\mathrm{d}^3r\mathrm{d}^3c$. Thus the total net momentum in d^3r is

$$d^3r \int mc\, f(r, c, t) d^3c.$$

Since the total mass in d^3r is $\rho(r, t)d^3r$, we have, for the *hydrodynamic velocity*,

$$v(r, t) = \frac{1}{\rho(r, t)} \int mc\, f(r, c, t)\, d^3c. \tag{2.1-3}$$

For a simple gas this is just also equal to the *average molecular velocity*. For gas mixtures, however, the average molecular velocity and the hydrodynamic velocity may differ by a small, but significant, amount.

The translational motion of an individual molecule may be specified either by its velocity c relative to a standard frame of reference, or by its velocity relative to a frame of reference moving with some given velocity. The molecular velocity in the frame of reference which moves with the local hydrodynamic velocity v will be called the *peculiar velocity* of the molecule and will be denoted by C. Obviously, $C = c - v$. From this definition it follows that in a simple gas the average peculiar velocity is always zero.

Next, consider the energy contained by the particles in d^3r. Since $c = C + v$, the kinetic energy of a particle may be written

$$\tfrac{1}{2}mc^2 = \tfrac{1}{2}mC^2 + mC \cdot v + \tfrac{1}{2}mv^2. \tag{2.1-4}$$

When this equation is averaged, the middle term will make no contribution because the average peculiar velocity is zero. On averaging, the last term yields $\frac{1}{2}\rho v^2$, which is clearly the kinetic energy associated with the flowing gas as a whole. It is then clear from thermodynamics that the average of $\frac{1}{2}mC^2$ must be the internal energy of the gas, so that

$$u(r, t) = \frac{1}{\rho(r, t)} \int \tfrac{1}{2}mC^2 f(r, c, t)\, d^3c, \tag{2.1-5}$$

where u is the *internal energy per unit mass* of the gas.

In the kinetic theory of gases the *temperature T* of a gas is defined in terms of u by the relation

$$\rho u = \tfrac{3}{2}nkT, \tag{2.1-6}$$

where k is a universal constant, known as Boltzmann's constant; its numerical value is $k = 1.380 \times 10^{-16}$ erg/°K. The quantity $\frac{3}{2}nkT$ will be recognized as the thermodynamic internal energy for a perfect gas, and the definition (2.1-6) is thus at least consistent with the thermodynamic definition of temperature. A rigorous proof of equivalence can be based on the proof of the *H*-theorem – see Section 4.2.

Finally, we note that, in general, the average of any function of velocity may be defined by

$$\bar{\varphi}(\boldsymbol{r}, t) = \frac{1}{n(\boldsymbol{r}, t)} \int \varphi(\boldsymbol{c}) f(\boldsymbol{r}, \boldsymbol{c}, t)\, \mathrm{d}^3 c. \tag{2.1-7}$$

2.2. Flux vectors

Except at a state of complete thermodynamic equilibrium there exist gradients in one or more of the macroscopic physical properties of the system: density, hydrodynamic velocity, and temperature. The gradients of these properties result in transport of mass, momentum, and kinetic energy through the gas. In this section we shall define the quantities that represent the transport of these properties in terms of the velocity distribution function.

Consider the passage of molecules across a small surface element d^2S, which is moving through the gas with the local average molecular velocity $\boldsymbol{v}$ so that a molecule's velocity with respect to the surface is its peculiar velocity. The orientation of d^2S is defined by a unit vector $\boldsymbol{n}$ normal to the surface. If a molecule whose peculiar velocity is $\boldsymbol{C}$ is to cross d^2S in a time interval $(t, t+\mathrm{d}t)$ so short that we may ignore the possibility of the molecule encountering another during it, then at the beginning of the interval the molecule must lie somewhere inside the cylinder with base d^2S and generators $-\boldsymbol{C}\,\mathrm{d}t$ (see Fig. 2.1). The volume of this cylinder is $(\boldsymbol{n} \cdot \boldsymbol{C})\mathrm{d}^2S\,\mathrm{d}t$. Since there are $f\mathrm{d}^3C$ molecules per unit volume which have their peculiar velocities in the range d^3C about $\boldsymbol{C}$, the number of such molecules in the cylinder at

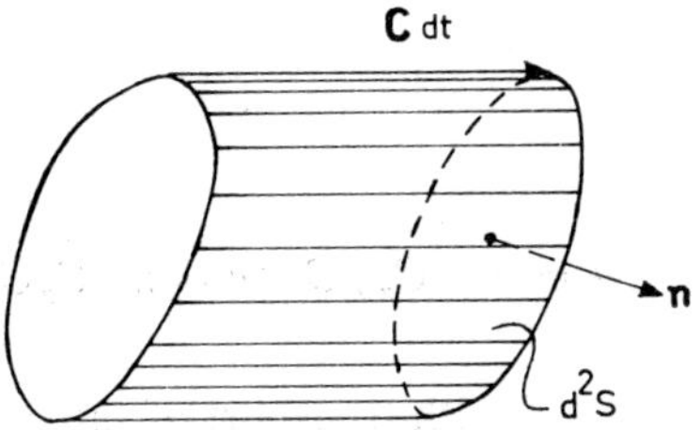

Fig. 2.1. Cylinder containing all molecules with peculiar velocity $\boldsymbol{C}$ which cross the surface element d^2S during the time interval $\mathrm{d}t$.

time t, i.e., the number which cross d^2S during a time interval $\mathrm{d}t$ is $(\boldsymbol{n}\cdot\boldsymbol{C})f\mathrm{d}^3C\mathrm{d}^2S\mathrm{d}t$.

Each molecule that crosses the surface carries with it its own mass, momentum, energy, and other properties. Letting $\varphi(\boldsymbol{C})$ represent any one property of interest, we see that the amount of property φ transported across d^2S during $\mathrm{d}t$ by the molecules in the velocity range d^3C is $(\boldsymbol{n}\cdot\boldsymbol{C})\varphi f\mathrm{d}^3C\mathrm{d}^2S\mathrm{d}t$, and the net flow of φ across d^2S during $\mathrm{d}t$ is $\mathrm{d}^2S\mathrm{d}t\int(\boldsymbol{n}\cdot\boldsymbol{C})\varphi f\mathrm{d}^3C$. This is equal to $\mathrm{d}^2S\mathrm{d}t\boldsymbol{n}\cdot\int\varphi f\boldsymbol{C}\mathrm{d}^3C$, which can also be written as $\mathrm{d}^2S\mathrm{d}t\boldsymbol{n}\cdot\int\varphi f\boldsymbol{C}\mathrm{d}^3c$, since integration over $\boldsymbol{C}$ is equivalent to integration over $\boldsymbol{c}$, as the two vectors differ only by a constant vector and the integration is performed over the whole velocity space. The vector

$$\boldsymbol{\Phi} = \int \varphi f\boldsymbol{C}\,\mathrm{d}^3c \tag{2.2-1}$$

is called the *flux vector* associated with the property φ. The physical significance of $\boldsymbol{\Phi}$ is that the rate of flow of property φ across a unit area of a surface which moves with the gas is the component of $\boldsymbol{\Phi}$ normal to the surface.

In particular, let us examine the flux vectors related to the transport of mass, momentum, and kinetic energy.

(i) *Transport of mass*

The mass flux vector is obtained by setting $\varphi(\boldsymbol{C}) = m$. Thus,

$$\boldsymbol{\Phi}(\boldsymbol{r}, t) = m\int f\boldsymbol{C}\,\mathrm{d}^3c = nm\overline{\boldsymbol{C}} = 0, \tag{2.2-2}$$

as a consequence of the definition of peculiar velocity and $\boldsymbol{v}$. Hence the mass flux vector vanishes in the case of a simple gas.

(ii) *Transport of momentum*

The flux vector for the α-component of momentum is obtained by setting $\varphi(\boldsymbol{C}) = mC_\alpha$. Thus,

$$\boldsymbol{\Phi} = m\int C_\alpha f\boldsymbol{C}\,\mathrm{d}^3c = nm\overline{C_\alpha\boldsymbol{C}} \tag{2.2-3}$$

is the flux vector associated with transport of the α-component of momentum. Since the index α may assume three different values, one has a total of three flux vectors associated with momentum transfer. The nine components of these vectors form a symmetric tensor of rank two, $\mathbf{P}$,

$$\mathbf{P} = m \int \mathbf{CC} f \, d^3c \tag{2.2-4a}$$

or, in component form,

$$P_{\alpha\beta} = m \int C_\alpha C_\beta f \, d^3c, \tag{2.2-4b}$$

which will be called the *pressure tensor*. The individual components have the following meaning. The diagonal elements are the *normal pressures*, $P_{\alpha\alpha}$ being the force per unit area in the α-direction exerted on a plane surface in the gas which is perpendicular to the α-direction. The non-diagonal elements are *shear stresses*, $P_{\alpha\beta}$ being the force per unit area in the α-direction exerted on a plane surface which is perpendicular to the β-direction. The vector with components $P_{\alpha x}, P_{\alpha y}, P_{\alpha z}$ represents the resultant force $\boldsymbol{P}_\alpha$ on a unit area perpendicular to the α-direction. All forces are those measured by an observer moving with the gas.

The sum of the normal pressures across three planes through any point parallel to the coordinate planes, i.e., the trace of the pressure tensor, is

$$P_{xx} + P_{yy} + P_{zz} = nm(\overline{C_x^2} + \overline{C_y^2} + \overline{C_z^2}) = \rho \overline{C^2}. \tag{2.2-5}$$

The *hydrostatic pressure*, or simply the *pressure*, is defined as the mean value of the normal pressures across any three orthogonal planes. It is thus one-third of the above and will be denoted by p,

$$p = \tfrac{1}{3}\rho\overline{C^2} = \tfrac{1}{3}\mathbf{P} : \mathbf{I}. \tag{2.2-6}$$

Here $\mathbf{I}$ is the unit tensor of rank two. In particular, if the shear stresses are zero and the normal pressures are equal, then

$$\mathbf{P} = p\mathbf{I}. \tag{2.2-7}$$

Since these conditions are satisfied in the hydrostatic case, such a pressure system is called *hydrostatic*.

Note that by combining Eqs. (2.1-5) and (2.1-6), we obtain

$$nkT = \frac{1}{3n} \int \rho C^2 f \, d^3c = \tfrac{1}{3}\rho\overline{C^2}. \tag{2.2-8}$$

Comparing this result with Eq. (2.2-6), we have

$$p = nkT, \tag{2.2-9}$$

which is the well-known *perfect gas law* or the *equation of state* for a perfect gas. It may be cast in the more usual form by considering one mole of gas. There are N ($N = 6.025 \times 10^{23}$, Avogadro's number) molecules in a mole and if we let V be the molar volume, $n = N/V$ and therefore

$$pV = RT, \tag{2.2-10}$$

where $R = Nk$ is the *universal gas constant*; numerically, $R = 8.314 \times 10^7$ erg/°K. Other forms of this equation of state are also in use.

In view of the choice of the definition of temperature via Eq. (2.1-6), a perfect gas relation, it is not surprising that the perfect gas equation of state is obtained. One must be careful, however, for in dense gases there are contributions to both the internal energy and the pressure tensor that are not accounted for here and, of course, the perfect gas equation of state is no longer valid. In fact, even the definition of temperature becomes ambiguous; see Chapter 13.

(iii) *Transport of kinetic energy*

The *heat flow vector* is obtained by setting $\varphi(\boldsymbol{C}) = \frac{1}{2}mC^2$ in Eq. (2.2-1),

$$\boldsymbol{q} = \tfrac{1}{2}m \int C^2 f \boldsymbol{C}\, \mathrm{d}^3c = \tfrac{1}{2}nm\overline{C^2 \boldsymbol{C}}. \tag{2.2-11}$$

Finally, we note that the net flow per unit time of any property φ across a surface d^2S moving with velocity $\boldsymbol{v}' \neq \boldsymbol{v}$ is given by $\boldsymbol{\Phi}' \cdot \boldsymbol{n}\,\mathrm{d}^2S$, where

$$\begin{aligned} \boldsymbol{\Phi}' &= \int \varphi f(\boldsymbol{c}-\boldsymbol{v}')\,\mathrm{d}^3c \\ &= \int \varphi f(\boldsymbol{C}+(\boldsymbol{v}-\boldsymbol{v}'))\,\mathrm{d}^3c \\ &= \boldsymbol{\Phi}+(\boldsymbol{v}-\boldsymbol{v}')n\overline{\varphi}, \end{aligned} \tag{2.2-12}$$

where $\overline{\varphi}$ is the local average of property φ. Occasionally, use will be made of this formula with $\boldsymbol{v}' = 0$.

2.3. Gas mixtures

The definitions given above for the case of a simple gas can be generalized to the case of a gas mixture.

Consider a mixture of K components. Let the expected number of molecules of species i which, at time t, are situated in the volume element d^3r with velocities in d^3c_i be $f_i(\boldsymbol{r}, \boldsymbol{c}_i, t)d^3r d^3c_i$. The function f_i is called the velocity distribution function of molecules of species i. The number density of the ith component of the mixture, n_i, is obtained from f_i by integration over velocity, cf. Eq. (2.1-1); its mass density, ρ_i, is defined as $\rho_i = m_i n_i$, where m_i is the mass of a molecule of species i. Then the *number density of the mixture*, n, is simply

$$n = \sum_{i=1}^{K} n_i \tag{2.3-1}$$

and the *mass density* or *density of the mixture*, ρ, is

$$\rho = \sum_{i=1}^{K} \rho_i = \sum_{i=1}^{K} n_i m_i, \tag{2.3-2}$$

where the sum extends over all components of the mixture. Of course, n and ρ are both functions of position and time.

The average value $\bar{\varphi}_i$ of a function φ of the molecular velocities for molecules of type i is given by

$$n_i \bar{\varphi}_i = \int f_i \varphi_i d^3c_i. \tag{2.3-3}$$

Then the average value $\bar{\varphi}$ for the mixture is defined by the relation

$$n\bar{\varphi} = \sum_i n_i \bar{\varphi}_i = \sum_i \int f_i \varphi_i d^3c_i. \tag{2.3-4}$$

In particular, for the average molecular velocity $\bar{\boldsymbol{c}}$ of a mixture one obtains

$$n\bar{\boldsymbol{c}} = \sum_i n_i \bar{\boldsymbol{c}}_i = \sum_i \int f_i \boldsymbol{c}_i d^3c_i. \tag{2.3-5}$$

This quantity is a simple numerical average of the velocities of all particles and should not be confused with the *hydrodynamic velocity* $\boldsymbol{v}$, which is defined as the ratio of average momentum density to average density,

$$\rho\boldsymbol{v} = \sum_i \rho_i \bar{\boldsymbol{c}}_i = \sum_i m_i \int f_i \boldsymbol{c}_i d^3c_i. \tag{2.3-6}$$

As discussed in Section 2.1, this is the velocity that is normally measured.

Since, in general, the average velocity of molecules of type i differs from both the hydrodynamic velocity and the average molecular velocity, a phenomenon which may be observed in a mixture is *diffusion* (relative motion) of one species with respect to another species. This diffusion can be characterized by the *diffusion velocity* of each species either with respect to $\boldsymbol{v}$ or with respect to $\bar{\boldsymbol{c}}$. In the former case we define the diffusion velocity of molecules of type i as

$$\boldsymbol{V}_i = \bar{\boldsymbol{c}}_i - \boldsymbol{v}, \tag{2.3-7}$$

in the latter as

$$\boldsymbol{W}_i = \bar{\boldsymbol{c}}_i - \bar{\boldsymbol{c}}. \tag{2.3-8}$$

Generally speaking, $\boldsymbol{V}_i$ is appropriate for the description of acceleration and friction phenomena, while $\boldsymbol{W}_i$ is sometimes more appropriate for the description of diffusion and heat conduction in slowly moving gases, when acceleration and friction phenomena are of minor importance.

Finally, the peculiar velocity $\boldsymbol{C}_i$ of a molecule of type i is defined in terms of the hydrodynamic velocity,

$$\boldsymbol{C}_i = \boldsymbol{c}_i - \boldsymbol{v}, \tag{2.3-9}$$

so that $\boldsymbol{V}_i = \overline{\boldsymbol{C}}_i$.

Similarly, the flux vectors, introduced in Section 2.2, may be generalized to the case of a mixture. If φ is any function of molecular velocity, the flux vector representing the transport of φ by molecules of the ith component, $\boldsymbol{\Phi}_i$, is defined by

$$\boldsymbol{\Phi}_i = \int \varphi_i f_i \boldsymbol{C}_i \, \mathrm{d}^3 c_i, \tag{2.3-10}$$

cf. Eq. (2.2-1). Then, the *flux vector for the mixture*, $\boldsymbol{\Phi}$, is defined as

$$\boldsymbol{\Phi} = \sum_i \boldsymbol{\Phi}_i. \tag{2.3-11}$$

In particular, for the pressure tensor and heat flow vector one has

$$\mathsf{P} = \sum_i n_i m_i \overline{\boldsymbol{C}_i \boldsymbol{C}_i} = \sum_i m_i \int \boldsymbol{C}_i \boldsymbol{C}_i f_i \, \mathrm{d}^3 c_i \tag{2.3-12}$$

and

$$\boldsymbol{q} = \sum_i \tfrac{1}{2} n_i m_i \overline{C_i^2 \boldsymbol{C}_i} = \sum_i \tfrac{1}{2} m_i \int C_i^2 \boldsymbol{C}_i f_i \, \mathrm{d}^3 c_i, \tag{2.3-13}$$

respectively. The hydrostatic pressure p of the gas at any point is again defined as

$$p = \tfrac{1}{3} \mathsf{P} : \boldsymbol{I} \tag{2.3-14}$$

and the temperature T of the gas by the relation

$$\rho u = \tfrac{3}{2}nkT, \tag{2.3-15}$$

where u is the local internal energy per unit mass,

$$\rho u = \sum_i \tfrac{1}{2}m_i \int C_i^2 f_i \, d^3c_i \tag{2.3-16}$$

and, again, the perfect gas law is valid.

2.4. Collision frequency, mean free path

When two molecules are within a few molecular diameters of each other, they interact in a rather complicated way (see Chapter 8 for a discussion of the intermolecular forces). Since there is a small residual interaction even at large distances, the definition of the beginning and end of a collision may be rather arbitrary. In order to avoid any ambiguity we shall assume for the moment that the molecules are smooth, rigid spheres. In that case an encounter between two molecules is instantaneous.

The average number of collisions undergone by a molecule per unit time is called the *collision frequency*. Thus, if N_{ij} represents the number of collisions occurring per unit volume and time between molecules of type i and molecules of type j, the collision frequency, ν_i, for a molecule of type i, is

$$\nu_i = (\sum_j N_{ij})/n_i, \tag{2.4-1}$$

where the sum extends over all constituents of the mixture. The *mean free time* between successive collisions, τ_i, is therefore

$$\tau_i = \nu_i^{-1}. \tag{2.4-2}$$

The mean distance, l_i, traveled by a molecule of type i between successive collisions is called its *mean free path.* In a time interval Δt, a molecule of type i travels, on the average, a distance $\bar{c}_i \Delta t$. Note that c_i is the average of the scalar speed c_i and not the magnitude of the vector average. The total distance traveled by all the particles of type i in a unit volume in time interval Δt is $n_i \bar{c}_i \Delta t$. The mean free path is obtained by dividing this total distance by the total number of collisions involving molecules of type i, $n_i \Delta t/\tau_i$. Thus we find for the mean free path

$$l_i = \bar{c}_i \tau_i = \bar{c}_i/\nu_i. \tag{2.4-3}$$

To obtain the values of ν_i and l_i, we begin by computing N_{ij}. To do this, let the diameters of molecules of type i and type j be denoted by σ_i and σ_j, respectively, and let $\sigma_{ij} = \frac{1}{2}(\sigma_i+\sigma_j)$. A collision will occur only when the centers of the molecules are separated by a distance σ_{ij}. Consider first only collisions of molecules of type j with velocities in d^3c_j about $\boldsymbol{c}_j$ and molecules of type i with velocity $\boldsymbol{c}_i$. The relative velocity is $\boldsymbol{g} = \boldsymbol{c}_j-\boldsymbol{c}_i$. Using a cylindrical coordinate system with the center of an i-molecule at the origin and polar axis parallel to $-\boldsymbol{g}$ one easily verifies that for an encounter between the two molecules during a short time interval dt, at the beginning of the interval, the center of molecule j must be in a right cylinder of base $\pi\sigma_{ij}^2$ and height $g dt$. Clearly, the volume of the cylinder is $\pi\sigma_{ij}^2 g\,dt$. The expected number of j-molecules in the velocity range under consideration that are in the cylinder is $f_j d^3c_j \pi\sigma_{ij}^2 g\,dt$, and this is also the probability that the i-molecule with velocity $\boldsymbol{c}_i$ will encounter a j-molecule with velocity in d^3c_j in time dt. As the number of i-molecules with velocities in d^3c_i is $f_i d^3c_i$, we have for the expected number of encounters in dt per unit volume between molecules in d^3c_i and d^3c_j

$$f_i f_j g \pi\sigma_{ij}^2 d^3c_i d^3c_j dt.$$

Integrating this result over all $\boldsymbol{c}_i$ and $\boldsymbol{c}_j$, and dividing by dt, we obtain N_{ij},

$$N_{ij} = \pi\sigma_{ij}^2 \iint f_i f_j g\, d^3c_i d^3c_j. \tag{2.4-4}$$

In particular, for a simple gas,

$$\nu = \frac{\pi\sigma^2}{n} \iint f(\boldsymbol{c}) f(\boldsymbol{c}_1) |\boldsymbol{c}_1-\boldsymbol{c}| d^3c\, d^3c_1. \tag{2.4-5}$$

The double integrals in these expressions can be interpreted as $n_i n_j \bar{g}$ and $n^2\bar{g}$, respectively, where $\bar{g}$ is the average relative speed of the colliding molecules. The problem is to find $\bar{g}$ in terms of the average molecular velocities. Of course, this problem cannot be solved unless the velocity distribution functions of the molecules have been specified. By using a "sneak preview" of Chapter 4 – where it is shown that, in the uniform steady state, f is the Maxwellian distribution function, i.e., $f = n(m/2\pi kT)^{\frac{3}{2}} \times \exp(-mc^2/2kT)$ – one finds for a simple gas

$$\nu = 4n\sigma^2(\pi kT/m)^{\frac{1}{2}} = \sqrt{2}\, n\pi\sigma^2 \bar{C}. \tag{2.4-6}$$

Consequently, the mean free path for a gas in equilibrium is

$$l = (\sqrt{2}\, n\pi\sigma^2)^{-1}, \tag{2.4-7}$$

i.e., the mean free path is inversely proportional to the density and to the square of the molecular diameter. The results (2.4-6) and (2.4-7) are also good estimates for a gas not far from equilibrium.

Boltzmann's equation 3

In this chapter we shall present a derivation of the fundamental equation of kinetic theory, Boltzmann's equation. As mentioned in the historical summary, Section 1.2, this equation was first derived by Boltzmann [1872] to describe the approach to equilibrium of a dilute gas. The basic assumptions in Boltzmann's derivation are (i) that only pairs of particles may interact simultaneously, i.e., collisions are short-duration events involving only two particles, and (ii) the so-called "Stosszahlansatz", sometimes known as the "molecular chaos" assumption, i.e., the particles are distributed in a statistical manner. The first assumption limits the applicability of the theory to gases of relatively low densities; at high densities, ternary and higher-order interactions become significant and one must expect deviations from the results obtained with Boltzmann's equation. The second assumption is of a statistical nature; it permits calculation of the *expected* number of pairs of molecules which collide during a given (short) time interval and its validity is much harder to discuss. It is well known that the latter statistical assumption makes Boltzmann's equation irreversible in time.

In an attempt to replace the above assumptions by more quantitative ones so that generalization of the theory to allow for higher densities would become possible, Bogoliubov [1946] presented a new approach to the problem of the derivation of a kinetic equation describing irreversible processes in a gas. His theory is based upon the (reversible) Liouville equation which describes the temporal behavior of the state of a gas in phase space. The fundamental idea is the introduction of different time scales on each of which the state of the gas needs to be described with an appropriate and different degree of fineness. If one seeks a description on a rather coarse time scale and assumes that the variables that describe the state of the gas can be written as power series with respect to the density, then it is possible to show that Boltzmann's equation is correct to the lowest order in the density. By systematic extension, higher-order collisions may be taken into account. Irreversibility is introduced, however, only when a particular type of initial condition is introduced into the theory.

A similar approach to the same problem was developed by Frieman [1963] and Sandri [1963]. These authors also start with the basic observation that, in a gas of not too high density, processes occur simultaneously on widely different time scales. The original problem is then *imbedded* into a larger class of problems by extending the notion of time. In the extended problem, considerable freedom exists in the prescription of initial conditions and Boltzmann's equation comes out if one looks for special solutions to the extended problem. The physical initial conditions that can be accommodated are more general than in Bogoliubov's approach and, therefore, this method of multiple time scales provides some insight into the problem of the relaxation of a gas in non-equilibrium toward a kinetic stage.

Despite the satisfying results of these approaches, there are several objections. The most serious objections are concerned with the basic assumption of Bogoliubov (his so-called functional assumption) and with the essential role played by non-physical functions in the theory of Frieman and Sandri (correlation functions in a non-physical time domain). Thus, neither of these theories seems completely satisfactory.

An approach which circumvents some of the disadvantages of the above methods is based on the use of cluster expansions, developed by Green [1956, 1958] and Cohen [1962a, b]. These cluster expansions are generalizations to non-equilibrium situations of the well-known cluster expansions for a dense gas in equilibrium, cf. Kahn [1938] or Uhlenbeck and Ford [1962]. At present this approach seems to be the most attractive, although problems also arise in this case. As, in the present text, the emphasis is on a discussion of transport coefficients in kinetic theory, we refer the reader to the literature for a complete survey of methods of deriving Boltzmann's equation. Besides the references cited above, the reader may wish to consult a paper by Cohen [1961] and a monograph by Wu [1966]. The derivation of Boltzmann's equation presented in this chapter is based entirely upon the method of cluster expansions. The modern derivation is preceded by a section in which Boltzmann's original derivation of the equation is given.

3.1. Classical derivation of Boltzmann's equation

In order to study the properties of a gas composed of molecules of one kind (simple gas) or of molecules of several kinds (gas mixture), we again introduce distribution functions $f_i(\boldsymbol{r}, \boldsymbol{c}_i, t)$, so defined that $f_i \mathrm{d}^3 r \mathrm{d}^3 c_i$ is the probable number of molecules of species i, the centers of which have, at time t, position coordinates in the range $\mathrm{d}^3 r$ about $\boldsymbol{r}$ and velocity coordinates in the range $\mathrm{d}^3 c_i$ about $\boldsymbol{c}_i$. The aim of kinetic theory is to find the distribution functions f_i given the form of the molecular interaction and initial and/or boundary conditions. In this section we shall derive the equation of evolution for the distribution functions using the classical arguments of Boltzmann.

Consider a gas in which each molecule of type i is subject to an external force $\boldsymbol{F}_i$ per unit mass, which may be a function of $\boldsymbol{r}$ and t but not of $\boldsymbol{c}_i$. In the interval between t and $t+\mathrm{d}t$ the velocity $\boldsymbol{c}_i$ of any molecule which does not collide with another molecule will change to $\boldsymbol{c}_i+\boldsymbol{F}_i\mathrm{d}t$, and its position vector $\boldsymbol{r}$ will change to $\boldsymbol{r}+\boldsymbol{c}_i\mathrm{d}t$. There are $f_i(\boldsymbol{r}, \boldsymbol{c}_i, t)\mathrm{d}^3 r \mathrm{d}^3 c_i$ molecules which at time t lie in the volume element $\mathrm{d}^3 r$ about $\boldsymbol{r}$ and have velocities in the range $\mathrm{d}^3 c_i$ about $\boldsymbol{c}_i$. After time interval $\mathrm{d}t$, if there were no molecular collisions, the same molecules, and no others, would be found in the volume element $\mathrm{d}^3 r'$ about $\boldsymbol{r}+\boldsymbol{c}_i\mathrm{d}t$ with velocities in the range $\mathrm{d}^3 c_i'$ about $\boldsymbol{c}_i+\boldsymbol{F}_i\mathrm{d}t$. The number in this set is $f_i(\boldsymbol{r}+\boldsymbol{c}_i\mathrm{d}t, \boldsymbol{c}_i+\boldsymbol{F}_i\mathrm{d}t, t+\mathrm{d}t)\mathrm{d}^3 r' \mathrm{d}^3 c_i'$. However, since we have assumed that $\boldsymbol{F}_i$ is independent of $\boldsymbol{c}_i$, we have $\mathrm{d}^3 r \mathrm{d}^3 c_i = \mathrm{d}^3 r' \mathrm{d}^3 c_i' + \mathrm{O}(\mathrm{d}t^2)$, so the number in the second set is also equal to $f_i(\boldsymbol{r}+\boldsymbol{c}_i\mathrm{d}t, \boldsymbol{c}_i+\boldsymbol{F}_i\mathrm{d}t, t+\mathrm{d}t)\mathrm{d}^3 r \mathrm{d}^3 c_i$. Hence, *in the absence of collisions* we have the equality

$$f_i(\boldsymbol{r}+\boldsymbol{c}_i\mathrm{d}t, \boldsymbol{c}_i+\boldsymbol{F}_i\mathrm{d}t, t+\mathrm{d}t) = f_i(\boldsymbol{r}, \boldsymbol{c}_i, t). \tag{3.1-1}$$

When there are collisions this equality must be modified. Then we write

$$f_i(\boldsymbol{r}+\boldsymbol{c}_i\mathrm{d}t, \boldsymbol{c}_i+\boldsymbol{F}_i\mathrm{d}t, t+\mathrm{d}t) = f_i(\boldsymbol{r}, \boldsymbol{c}_i, t) + \left(\frac{\partial f_i}{\partial t}\right)_{\mathrm{coll}} \mathrm{d}t \tag{3.1-2}$$

which defines $(\partial f_i/\partial t)_{\mathrm{coll}}$. On dividing by $\mathrm{d}t$ and letting $\mathrm{d}t$ tend to zero we obtain the equation

$$\left(\frac{\partial}{\partial t} + \boldsymbol{c}_i \cdot \boldsymbol{\nabla}_r + \boldsymbol{F}_i \cdot \boldsymbol{\nabla}_{c_i}\right) f_i(\boldsymbol{r}, \boldsymbol{c}_i, t) = \left(\frac{\partial f_i}{\partial t}\right)_{\mathrm{coll}}, \tag{3.1-3}$$

where $\boldsymbol{\nabla}_r$ and $\boldsymbol{\nabla}_{c_i}$ are, respectively, the gradient operators with respect to

$\boldsymbol{r}$ and $\boldsymbol{c}_i$. Of course, Eq. (3.1-3) is incomplete until we specify the term $(\partial f_i/\partial t)_{\text{coll}}$.

An expression for $(\partial f_i/\partial t)_{\text{coll}}$, i.e., the rate at which the velocity distribution function f_i is being altered by molecular collisions can be obtained by going back to its definition, Eq. (3.1-2). Let $\Gamma_i^+ \mathrm{d}^3 r \mathrm{d}^3 c_i \mathrm{d}t$ represent the expected number of collisions occurring during the time between t and $t+\mathrm{d}t$ in which the *final* state of one of the molecules is in $\mathrm{d}^3 r \mathrm{d}^3 c_i$ about $(\boldsymbol{r}, \boldsymbol{c}_i)$, and let $\Gamma_i^- \mathrm{d}^3 r \mathrm{d}^3 c_i \mathrm{d}t$ represent the expected number of collisions occurring during the time between t and $t+\mathrm{d}t$ in which the *initial* state of one of the molecules is in $\mathrm{d}^3 r \mathrm{d}^3 c_i$ about $(\boldsymbol{r}, \boldsymbol{c}_i)$. We assume that $\mathrm{d}^3 c_i$ is small enough that the probability both of the molecules involved in a collision are in $\mathrm{d}^3 c_i$ is negligibly small. Then, Eq. (3.1-2) is equivalent to

$$\left(\frac{\partial f_i}{\partial t}\right)_{\text{coll}} \mathrm{d}t = (\Gamma_i^+ - \Gamma_i^-)\,\mathrm{d}t. \tag{3.1-4}$$

Thus, the derivation of Eq. (3.1-3) must be supplemented by a further development of the quantities Γ_i^+ and Γ_i^-.

At this point we assume that the system is a dilute gas, so that only binary collisions need to be considered and the possibility that three or more molecules may interact simultaneously can be ignored. Then explicit expressions for the quantities Γ_i^+ and Γ_i^- can be derived in terms of the (unknown) functions f_i. We shall study the nature of binary collisions next.

Consider the collision of two molecules of masses m_1, m_2. The molecules are assumed smooth and spherically symmetric and can be treated as classical point centers of force. The force which either exerts on the other is directed along the line joining their centers and may be some function of the distance between their centers. It is supposed that any external forces which act on the molecules are so small compared with those involved in the collision that their effect on the dynamical outcome of a collision can be neglected.

Let the velocities before and after the collision (i.e., before the molecules have begun to influence each other appreciably and after the molecules have stopped influencing each other appreciably, respectively) be denoted by $\boldsymbol{c}_1$, $\boldsymbol{c}_2$ (before) and $\boldsymbol{c}_1'$, $\boldsymbol{c}_2'$ (after). The conservation equations of momentum and energy are

$$m_1 \boldsymbol{c}_1 + m_2 \boldsymbol{c}_2 = m_1 \boldsymbol{c}_1' + m_2 \boldsymbol{c}_2', \tag{3.1-5}$$

$$\tfrac{1}{2} m_1 c_1^2 + \tfrac{1}{2} m_2 c_2^2 = \tfrac{1}{2} m_1 c_1'^2 + \tfrac{1}{2} m_2 c_2'^2, \tag{3.1-6}$$

respectively. The *velocity of the center-of-mass*, which is constant throughout the collision, is given by

$$\boldsymbol{G} = \mu_1 \boldsymbol{c}_1 + \mu_2 \boldsymbol{c}_2 = \mu_1 \boldsymbol{c}_1' + \mu_2 \boldsymbol{c}_2', \tag{3.1-7}$$

where

$$\mu_1 = m_1/(m_1+m_2), \qquad \mu_2 = m_2/(m_1+m_2). \tag{3.1-8}$$

The *relative velocities* before and after the collision, are, respectively,

$$\boldsymbol{g} = \boldsymbol{c}_2 - \boldsymbol{c}_1, \quad \boldsymbol{g}' = \boldsymbol{c}_2' - \boldsymbol{c}_1'. \tag{3.1-9}$$

The equations inverse to Eqs. (3.1-7) and (3.1-9) are

$$\boldsymbol{c}_1 = \boldsymbol{G} - \mu_2 \boldsymbol{g}, \quad \boldsymbol{c}_2 = \boldsymbol{G} + \mu_1 \boldsymbol{g}, \tag{3.1-10}$$

$$\boldsymbol{c}_1' = \boldsymbol{G} - \mu_2 \boldsymbol{g}', \quad \boldsymbol{c}_2' = \boldsymbol{G} + \mu_1 \boldsymbol{g}'. \tag{3.1-11}$$

Substituting Eqs. (3.1-10) and (3.1-11) into Eq. (3.1-6), one readily finds that

$$g = g', \tag{3.1-12}$$

hence the collision merely rotates $\boldsymbol{g}$ into $\boldsymbol{g}'$ without changing its magnitude. Specifying $\boldsymbol{G}$, $\boldsymbol{g}$, and the angles χ and ε (see Fig. 3.1) completely determines $\boldsymbol{g}'$. Having $\boldsymbol{g}'$, $\boldsymbol{c}_1'$ and $\boldsymbol{c}_2'$ may be computed from Eq. (3.1-11) and hence the collision is completely determined by specifying $\boldsymbol{G}$, $\boldsymbol{g}$, χ and ε or $\boldsymbol{c}_1$, $\boldsymbol{c}_2$, χ and ε.

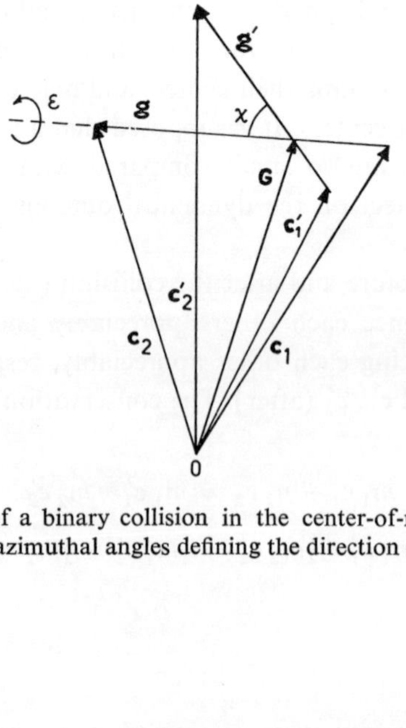

Fig. 3.1. The geometry of a binary collision in the center-of-mass coordinate system. χ and ε are the polar and azimuthal angles defining the direction of $\boldsymbol{g}'$ relative to that of $\boldsymbol{g}$.

In the following we shall need a relation between the differentials of the molecular velocities before and after the collision. This will be derived now. From Eqs. (3.1-7), (3.1-9) and (3.1-11) one obtains:

$$c_1' = \mu_1 c_1 + \mu_2 c_2 - \mu_2 g', \tag{3.1-13}$$

$$c_2' = \mu_1 c_1 + \mu_2 c_2 + \mu_1 g', \tag{3.1-14}$$

$$g = -c_1 + c_2. \tag{3.1-15}$$

Since these equations are linear and since the inverse transformations are obtained simply by interchanging primed and non-primed variables, it follows that the Jacobian of the transformation $(c_1, c_2, g') \to (c_1', c_2', g)$ and the Jacobian of the inverse transformation $(c_1', c_2', g) \to (c_1, c_2, g')$ must be the same:

$$\frac{\partial(c_1, c_2, g')}{\partial(c_1', c_2', g)} = \frac{\partial(c_1', c_2', g)}{\partial(c_1, c_2, g')}. \tag{3.1-16}$$

However, for any transformation one has

$$\frac{\partial(c_1, c_2, g')}{\partial(c_1', c_2', g)} \cdot \frac{\partial(c_1', c_2', g)}{\partial(c_1, c_2, g')} = 1, \tag{3.1-17}$$

so that

$$\left(\frac{\partial(c_1, c_2, g')}{\partial(c_1', c_2', g)}\right)^2 = 1. \tag{3.1-18}$$

Hence,

$$d^3c_1\, d^3c_2\, d^3g' = d^3c_1'\, d^3c_2'\, d^3g. \tag{3.1-19}$$

In particular, for an elastic collision, since $g = g'$, this reduces to

$$d^3c_1\, d^3c_2\, d^2\hat{g}' = d^3c_1'\, d^3c_2'\, d^2\hat{g}, \tag{3.1-20}$$

where $\hat{\boldsymbol{g}}$ and $\hat{\boldsymbol{g}}'$ are the unit vectors in the directions of $\boldsymbol{g}$ and $\boldsymbol{g}'$, respectively, and $d^2\hat{g}$ and $d^2\hat{g}'$ are the associated infinitesimal surface elements on the unit sphere. The relation (3.1-20) is known as *Liouville's law for elastic collisions.*

Consider the collision process in a coordinate system in which one of the molecules (mass m_1) is at rest with its center at the origin. The center of the other molecule (mass m_2) describes a trajectory in a plane through 0 as shown in Fig. 3.2; the velocities before and after the collision are $\boldsymbol{g}$ and $\boldsymbol{g}'$, respectively. The distance from 0 to either of the asymptotes of this tra-

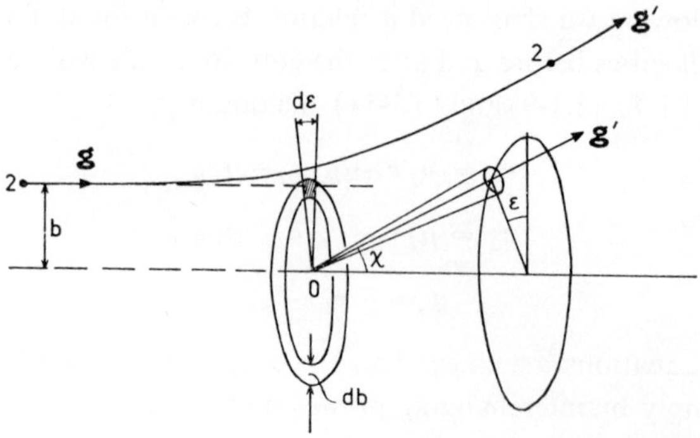

Fig. 3.2. The geometry of a binary encounter; molecule 1 is at rest with its center at the origin.

jectory is called the *impact parameter*, b. Let the frame of reference be chosen such that the polar axis is parallel to $\boldsymbol{g}$. Since $g = g'$, the final state is specified by the two scattering angles χ and ε, with χ the angle between $\boldsymbol{g}'$ and the polar axis and ε an azimuthal angle specifying the position of the plane in space. Of course, χ will depend on both g and b, the functional relation depending on the law of interaction of the molecules. Both for generality and for brevity, χ will be retained as an unspecified function of b and g as long as possible (see Chapter 9).

To derive explicit formulae for Γ^+ and Γ^-, cf. Eq. (3.1-4), we make the following assumptions: (i) only binary collisions are taken into account, (ii) the effects of external forces on the dynamical outcome of a collision are negligible, (iii) the expected number of collisions in a given volume element between molecules that belong to different velocity ranges can be computed statistically; this is known as the *Stosszahlansatz* (German for: collision number assumption). The validity of these assumptions will be discussed later.

Again, consider the motion of the center of a molecule 2 relative to the center of a molecule 1, as sketched in Fig. 3.2. If a collision between the two molecules in which the impact parameter is in the range $\mathrm{d}b$ about b and the azimuthal angle in the range $\mathrm{d}\varepsilon$ about ε is to occur, then the extension of the vector $\boldsymbol{g}$ must cut the plane through 0 perpendicular to the polar axis within the area bounded by circles of radii b, $b+\mathrm{d}b$ and center 0, and by radii from 0 including an angle $\mathrm{d}\varepsilon$. Let us first consider only those collisions which involve particles of relative velocity $\boldsymbol{g}$ and which have impact parameters in $\mathrm{d}b$ about b, and azimuths in $\mathrm{d}\varepsilon$ about ε. If such a collision is to occur in a time interval $\mathrm{d}t$ (which is large compared with the duration of a collision), then at the beginning of $\mathrm{d}t$ the center of molecule 2 must lie within a cylinder having an area $b\,\mathrm{d}b\,\mathrm{d}\varepsilon$ as base and generator equal to $-\boldsymbol{g}\,\mathrm{d}t$. This cylinder is called the *collision cylinder* associated with the particular collision under consideration; its volume is $gb\,\mathrm{d}b\,\mathrm{d}\varepsilon\,\mathrm{d}t$. According to the definition of the velocity distribution function, the total number of molecules of species 2 which have their velocities in the range d^3c_2 about $\boldsymbol{c}_2$ and lie in this collision cylinder is equal to $f_2(\boldsymbol{r}, \boldsymbol{c}_2, t)gb\,\mathrm{d}b\,\mathrm{d}\varepsilon\,\mathrm{d}^3c_2\,\mathrm{d}t$. We can imagine such a collision cylinder to be associated with each of the $f_1(\boldsymbol{r}, \boldsymbol{c}_1, t)\mathrm{d}^3c_1\,\mathrm{d}^3r$ molecules of the first kind within the specified velocity range d^3c_1 about $\boldsymbol{c}_1$, in the volume element d^3r about $\boldsymbol{r}$. Then the expected number of collisions in d^3r about $\boldsymbol{r}$, during a small interval $\mathrm{d}t$, between molecules in the velocity ranges d^3c_1, d^3c_2 about $\boldsymbol{c}_1$, $\boldsymbol{c}_2$ with geometrical collision variables in the range $\mathrm{d}b$, $\mathrm{d}\varepsilon$ about b, ε is

$$f_1(\boldsymbol{r}, \boldsymbol{c}_1, t)f_2(\boldsymbol{r}, \boldsymbol{c}_2, t)gb\,\mathrm{d}b\,\mathrm{d}\varepsilon\,\mathrm{d}^3c_1\,\mathrm{d}^3c_2\,\mathrm{d}^3r\,\mathrm{d}t. \tag{3.1-21}$$

Note that the derivation of this expression is based on a statistical argument as to the *expected* number of collisions. In counting the number of molecules in the collision cylinder we tacitly assumed a lack of correlation between the positions of the two colliding molecules, i.e., we assumed that the position of molecule 2 does not depend on where molecule 1 is located. The assumption is eminently reasonable but, on the other hand, there is no *a priori* reason for its truth. In fact, the assumption is certainly not true in solids, liquids or dense gases. This assumption is Boltzmann's *Stosszahlansatz* and is responsible for the introduction of irreversibility into the equations of motion. A further discussion of this point is given later in this chapter.

If there are several gases in a mixture, the quantities Γ_1^+ and Γ_1^- may be divided into the parts $\Gamma_{11}^+, \Gamma_{12}^+, \ldots$ and $\Gamma_{11}^-, \Gamma_{12}^-, \ldots$, respectively, due to the collisions of molecules of type 1 with molecules of type 1, 2, ... Of course, it suffices to derive expressions only for one set, $\Gamma_{12}^+, \Gamma_{12}^-$, say; the corresponding expressions for the other Γ_{ij}^+ and Γ_{ij}^- can be derived by changes of subscripts.

Consider the set of molecules of species 1, situated within d^3r about $\boldsymbol{r}$, which have velocities within the range d^3c_1 about $\boldsymbol{c}_1$. Every collision involving a molecule of this set results in a change of velocity and so involves the loss of a molecule to the set. Thus the number of molecules lost to the set during time $\mathrm{d}t$, due to collisions with molecules of type 2 that have velocities in the range d^3c_2 about $\boldsymbol{c}_2$ and such that the geometrical collision variables are in the range db, $d\varepsilon$ about b, ε is given by Eq. (3.1-21). The rate $\Gamma_{12}^{-}\,d^3r\,d^3c_1\,dt$ is obtained by integrating over all $\boldsymbol{c}_2$ and over all b and ε, so

$$\Gamma_{12}^{-} = \iiint f_1(\boldsymbol{r}, \boldsymbol{c}_1, t) f_2(\boldsymbol{r}, \boldsymbol{c}_2, t) g b \, db \, d\varepsilon \, d^3c_2 . \qquad (3.1\text{-}22)$$

In a similar fashion we calculate Γ_{12}^{+}. For this purpose we must consider those collisions which leave the velocity of a molecule of type 1 in the range d^3c_1 about $\boldsymbol{c}_1$ *after* the collision. This is most easily done by consideration of *inverse* collisions. According to the laws of classical mechanics, in a collision between a particle of mass m_1 with velocity $\boldsymbol{c}_1'$ and a particle of mass m_2 with velocity $\boldsymbol{c}_2'$ with impact parameter b and azimuthal angle $-\varepsilon$, particle 1 will be left with velocity $\boldsymbol{c}_1$ and particle 2 with velocity $\boldsymbol{c}_2$. Such a collision is called *inverse* to the *direct* collision discussed earlier.* The number of collisions in d^3r about $\boldsymbol{r}$ during the time interval dt, involving particles of species 1 with velocities in d^3c_1' about $\boldsymbol{c}_1'$ and particles of type 2 in d^3c_2' about $\boldsymbol{c}_2'$, such that the impact parameter is in the range b, $b+db$ and the azimuthal angle in the range $-\varepsilon$, $-\varepsilon-d\varepsilon$ is, by the previous argument:

$$f_1(\boldsymbol{r}, \boldsymbol{c}_1', t) f_2(\boldsymbol{r}, \boldsymbol{c}_2', t) g b \, db \, d\varepsilon \, d^3c_1' \, d^3c_2' \, d^3r \, dt. \qquad (3.1\text{-}23)$$

With the aid of Liouville's law for elastic collisions, Eq. (3.1-20), this is seen to be equal to

$$f_1(\boldsymbol{r}, \boldsymbol{c}_1', t) f_2(\boldsymbol{r}, \boldsymbol{c}_2', t) g b \, db \, d\varepsilon \, d^3c_1 \, d^3c_2 \, d^3r \, dt. \qquad (3.1\text{-}24)$$

The rate $\Gamma_{12}^{+}\,d^3r\,d^3c_1\,dt$ is obtained by summing over all $\boldsymbol{c}_2$ and over all b and ε, so

$$\Gamma_{12}^{+} = \iiint f_1(\boldsymbol{r}, \boldsymbol{c}_1', t) f_2(\boldsymbol{r}, \boldsymbol{c}_2', t) g b \, db \, d\varepsilon \, d^3c_2 . \qquad (3.1\text{-}25)$$

Combining the results (3.1-22) and (3.1-25), we find that the rate at which the velocity distribution function f_1 for molecules of species 1 is being

* This is not to be confused with a time reversed collision in which particles of velocities $-\boldsymbol{c}_1'$, $-\boldsymbol{c}_2'$ collide to produce particles with velocities $-\boldsymbol{c}_1$, $-\boldsymbol{c}_2$. This distinction is an important one because, while time reversibility is generally valid, the existence of inverse collisions is not generally guaranteed, e.g., inverse collisions do not necessarily exist for molecules with internal degrees of freedom.

altered by collisions with molecules of species 2, is given by

$$\left(\frac{\partial f_1}{\partial t}\right)_{2,\,\mathrm{coll}} = \Gamma_{12}^{+} - \Gamma_{12}^{-} = \iiint (f_1' f_2' - f_1 f_2) g b \, \mathrm{d}b \, \mathrm{d}\varepsilon \, \mathrm{d}^3 c_2 , \quad (3.1\text{-}26)$$

where the following abbreviations have been used:

$$\begin{aligned} f_1 &\equiv f_1(\boldsymbol{r}, \boldsymbol{c}_1, t), & f_2 &\equiv f_2(\boldsymbol{r}, \boldsymbol{c}_2, t), \\ f_1' &\equiv f_1(\boldsymbol{r}, \boldsymbol{c}_1', t), & f_2' &\equiv f_2(\boldsymbol{r}, \boldsymbol{c}_2', t). \end{aligned} \quad (3.1\text{-}27)$$

Having obtained this result, we can write down the equation for the evolution of the velocity distribution function in phase space. Thus we have

$$\left(\frac{\partial}{\partial t} + \boldsymbol{c}_i \cdot \nabla_r + \boldsymbol{F}_i \cdot \nabla_{c_i}\right) f_i(\boldsymbol{r}, \boldsymbol{c}_i, t) = \sum_j \iiint (f_i' f_j' - f_i f_j) g b \, \mathrm{d}b \, \mathrm{d}\varepsilon \, \mathrm{d}^3 c_j , \quad (3.1\text{-}28)$$

where the sum in the right member extends over all species in the mixture.* For the special case of a simple gas, this becomes

$$\left(\frac{\partial}{\partial t} + \boldsymbol{c} \cdot \nabla_r + \boldsymbol{F} \cdot \nabla_c\right) f(\boldsymbol{r}, \boldsymbol{c}, t) = \iiint (f' f_1' - f f_1) g b \, \mathrm{d}b \, \mathrm{d}\varepsilon \, \mathrm{d}^3 c_1 . \quad (3.1\text{-}29)$$

Eqs. (3.1-28) and (3.1-29) were first derived by Boltzmann and are known as *Boltzmann's equations*. Alternative forms of Eqs. (3.1-28) and (3.1-29) are

$$\left(\frac{\partial}{\partial t} + \boldsymbol{c}_i \cdot \nabla_r + \boldsymbol{F}_i \cdot \nabla_{c_i}\right) f_i(\boldsymbol{r}, \boldsymbol{c}_i, t) = \sum_j \iint (f_i' f_j' - f_i f_j) g \, \sigma_{ij}(g, \chi) \, \mathrm{d}^2\Omega \, \mathrm{d}^3 c_j \quad (3.1\text{-}30)$$

and

$$\left(\frac{\partial}{\partial t} + \boldsymbol{c} \cdot \nabla_r + \boldsymbol{F} \cdot \nabla_c\right) f(\boldsymbol{r}, \boldsymbol{c}, t) = \iint (f' f_1' - f f_1) g \, \sigma(g, \chi) \, \mathrm{d}^2\Omega \, \mathrm{d}^3 c_1 , \quad (3.1\text{-}31)$$

respectively, where σ is called the *differential collision cross-section*; σ_{ij} is defined in such a way that, when a stream of molecules of type i with velocity in the range $\mathrm{d}^3 c_i$ about $\boldsymbol{c}_i$ collides with a stream of molecules of type j with velocity in the range $\mathrm{d}^3 c_j$ about $\boldsymbol{c}_j$, the probable number of collisions per unit volume and time in which the direction of the relative velocity $\boldsymbol{g}'$ after collision (expressed by means of the unit vector $\boldsymbol{\Omega}$, say) lies in a solid

* It is tacitly assumed that the variable of integration $\boldsymbol{c}_j$ is replaced by $\boldsymbol{c}$ whenever $j = i$.

angle $d^2\Omega = \sin\chi\, d\chi\, d\varepsilon$, is given by $f_i f_j g \sigma_{ij}(g, \chi) d^2\Omega d^3 c_i d^3 c_j$.

In the next sections we shall rederive Eqs. (3.1-28) and (3.1-29) from first principles; a summary of the derivation may be found in Section 3.6.

3.2. The Liouville equation and the BBGKY-hierarchy

From the foregoing section one might conclude that kinetic theory can be formulated entirely in terms of the distribution function f. Clearly, the Boltzmann equation is closed in terms of this function and, presumably, it can be solved for this function if reasonable boundary and/or initial conditions are imposed on it. (This question is considered to some degree in Chapter 4.) Now the distribution function essentially describes the behavior of a "typical" particle. Interactions of the typical particle with others is, of course, accounted for, but only in an average, or statistical, manner; the other particles are assumed to behave in the same way as the one singled out for study. As was pointed out in the previous section, this is equivalent to ignoring all inter-particle correlations and is known to be incorrect for solids and liquids. Its validity for gases certainly deserves to be challenged, and a search for a more rigorous justification of the Boltzmann equation deserves considerable attention.

The second important assumption made in deriving the Boltzmann equation is that of accounting only for binary collisions. Intuitively, this seems quite reasonable at low densities where any sort of encounter is, in fact, rare. But one then must face the twofold problem of finding the density limitations of the Boltzmann equation and finding an improved or corrected equation valid at higher densities. Although Enskog succeeded in developing an equation valid at higher densities for hard sphere molecules (cf. Chapter 12) on the basis of physical reasoning, a general treatment appears to be attainable only via approximation of the rigorous equations. Derivation of the Boltzmann equation in such a manner, then, offers the possibility of being more readily generalized to higher densities.

For both of the reasons outlined above, a derivation of the Boltzmann equation from the rigorous laws of dynamics is important in any modern treatment of the subject of kinetic theory, and we now proceed with such a derivation.

Let the system consist of N identical molecules contained in a vessel of volume V. The molecules will be treated as classical point centers of force. The force which one molecule exerts on another is directed along the line joining their centers and may be some function of the distance between their centers. The internal state of the molecule is ignored, so the discussion is limited to monatomic gases.

In classical dynamics it is convenient to consider the motion of particles as trajectories in a six-dimensional space in which the coordinates are the three components of the position $\boldsymbol{r}$ and the three components of the momentum $\boldsymbol{p}$. This is called the *phase space* or *μ-space* of the particle. For a collection of N identical particles the state of the system may be specified by giving the coordinates of each particle in μ-space, i.e., the system as a whole is represented by a cloud of N points in μ-space. Alternatively, one can assign to each particle separate space and momentum coordinates $\boldsymbol{r}_i$, $\boldsymbol{p}_i$ $(i = 1, 2, \ldots, N)$ and consider the $6N$-dimensional phase space spanned by the coordinates $x_i \equiv (\boldsymbol{r}_i, \boldsymbol{p}_i)$, $i = 1, 2, \ldots, N$. In this space, called *Γ-space*, the state of the system is represented by a single point and its change in time by a single trajectory. This trajectory is in turn determined by the classical equations of motion; the most convenient choice of such equations for our purposes is the Hamiltonian set, see Goldstein [1950],

$$\dot{\boldsymbol{r}}_i = \frac{\partial H_N}{\partial \boldsymbol{p}_i}, \quad \dot{\boldsymbol{p}}_i = -\frac{\partial H_N}{\partial \boldsymbol{r}_i}, \tag{3.2-1}$$

where H_N is the Hamiltonian of the N particles,

$$H_N = \sum_{i=1}^{N} \left[\frac{\boldsymbol{p}_i^2}{2m} + \Phi(\boldsymbol{r}_i)\right] + \sum_{1 \leqq i < j \leqq N} \varphi(|\boldsymbol{r}_i - \boldsymbol{r}_j|). \tag{3.2-2}$$

In Eq. (3.2-2), Φ is the potential of external forces, including the wall potential, and φ is the molecular interaction potential. We shall write for brevity

$$\varphi_{ij} = \varphi(|\boldsymbol{r}_i - \boldsymbol{r}_j|). \tag{3.2-3}$$

Since Eqs. (3.2-1) are first-order differential equations in time, the path of the representative point in Γ-space is determined by the initial point. Following the Gibbs approach in statistical mechanics, we introduce the concept of an ensemble of many similar systems, all under the same external conditions (external forces, volume, total energy, etc.), but with different initial states. The collection of phase points of these systems forms a "gas" in Γ-space; it is customary to define a *distribution function*

$$D_N(x_1 \ldots x_N; t), \tag{3.2-4}$$

which we shall normalize according to

$$\int \ldots \int D_N(x_1 \ldots x_N; t)\,\mathrm{d}^6x_1 \ldots \mathrm{d}^6x_N = V^N, \tag{3.2-5}$$

so that $V^{-N}D_N(x_1 \ldots x_N; t)\mathrm{d}^6x_1 \ldots \mathrm{d}^6x_N$ is the probability that, at time t, a particular system will be located in the volume element $\mathrm{d}^6x_1, \ldots, \mathrm{d}^6x_N$ about $x_1, \ldots, x_N$ $(x_i \equiv (\boldsymbol{r}_i, \boldsymbol{p}_i))$. If the N particles are identical, one should require that D_N be a symmetric function of the phases x_i of the individual particles.

Since the systems comprising the ensemble are all identical, we may also view $V^{-N}D_N(x_1 \ldots x_N; t)\mathrm{d}^6x_1 \ldots \mathrm{d}^6x_N$ as the probability that in a particular system, particle 1 will be located in d^6x_1 about x_1, particle 2 in d^6x_2 about x_2, etc., all simultaneously at time t.

We now prove a theorem due to Liouville, which states that the ensemble of points in Γ-space moves like an incompressible fluid. Obviously, the total number of systems in an ensemble is conserved, so the rate of decrease of the number of representative points in any volume element in Γ-space equals the net number of such points that leave the same volume per unit time. Let Ω be an arbitrary volume element in Γ-space and let S be its surface. If $\dot{\boldsymbol{x}}$ denotes the vector whose components are the time derivatives of the components of $x_1, \ldots, x_N$, and if $\boldsymbol{n}$ is the vector locally normal to the surface S, then

$$\frac{\mathrm{d}}{\mathrm{d}t}\int_\Omega D_N \mathrm{d}^{6N}\Omega = -\int_S \boldsymbol{n} \cdot \dot{\boldsymbol{x}}\, D_N \mathrm{d}^{6N-1}S. \tag{3.2-6}$$

Since the volume Ω is thought to be fixed in Γ-space, we may use the divergence theorem to convert this equation into

$$\int_\Omega \left[\frac{\partial D_N}{\partial t} + \boldsymbol{\nabla} \cdot (\dot{\boldsymbol{x}} D_N)\right] \mathrm{d}^{6N}\Omega = 0, \tag{3.2-7}$$

where $\boldsymbol{\nabla}$ is the gradient operator in Γ-space. Eq. (3.2-7) holds for any volume Ω in Γ-space. Therefore one must have

$$\frac{\partial D_N}{\partial t} + \boldsymbol{\nabla} \cdot (\dot{\boldsymbol{x}} D_N) = 0, \tag{3.2-8}$$

or, equivalently,

$$\frac{\partial D_N}{\partial t} + \sum_{i=1}^{N} \left[\frac{\partial}{\partial \boldsymbol{r}_i} \cdot (\dot{\boldsymbol{r}}_i D_N) + \frac{\partial}{\partial \boldsymbol{p}_i} \cdot (\dot{\boldsymbol{p}}_i D_N)\right] = 0. \tag{3.2-9}$$

From Eqs. (3.2-1) one easily verifies that

$$\frac{\partial \dot{\boldsymbol{r}}_i}{\partial \boldsymbol{r}_i} + \frac{\partial \dot{\boldsymbol{p}}_i}{\partial \boldsymbol{p}_i} = 0, \tag{3.2-10}$$

so Eq. (3.2-9) is identical with

$$\frac{\partial D_N}{\partial t} + \sum_{i=1}^{N} \left(\frac{\partial D_N}{\partial \boldsymbol{r}_i} \cdot \dot{\boldsymbol{r}}_i + \frac{\partial D_N}{\partial \boldsymbol{p}_i} \cdot \dot{\boldsymbol{p}}_i \right) = 0. \tag{3.2-11}$$

This is *Liouville's theorem.* It may also be stated as

$$\mathrm{d}D_N/\mathrm{d}t = 0 \tag{3.2-12}$$

i.e., if we follow the motion of a representative point in Γ-space, we find that the density of representative points in its neighborhood is constant. Hence the ensemble of representative points moves in Γ-space like an *incompressible* fluid. By means of the canonical equations of motion (3.2-1), Eq. (3.2-11) can also be written as

$$\frac{\partial D_N}{\partial t} = \{H_N, D_N\}, \tag{3.2-13}$$

where $\{H_N, D_N\}$ is the *Poisson bracket* of H_N and D_N,

$$\{H_N, D_N\} = \sum_{i=1}^{N} \left(\frac{\partial H_N}{\partial \boldsymbol{r}_i} \cdot \frac{\partial D_N}{\partial \boldsymbol{p}_i} - \frac{\partial H_N}{\partial \boldsymbol{p}_i} \cdot \frac{\partial D_N}{\partial \boldsymbol{r}_i} \right). \tag{3.2-14}$$

An alternative form of Eq. (3.2-13) is

$$\partial D_N/\partial t = -\mathfrak{H}_N D_N, \tag{3.2-15}$$

where $\mathfrak{H}_N$ is the *Hamiltonian operator* for a system of N particles,

$$\mathfrak{H}_N \equiv \sum_{i=1}^{N} \left[\frac{\boldsymbol{p}_i}{m} \cdot \frac{\partial}{\partial \boldsymbol{r}_i} + \mathfrak{F}_i \cdot \frac{\partial}{\partial \boldsymbol{p}_i} \right] - \sum_{1 \leqq i < j \leqq N} \Theta_{ij}, \tag{3.2-16}$$

with the *molecular interaction operator* Θ_{ij} defined by

$$\Theta_{ij} \equiv \frac{\partial \varphi_{ij}}{\partial \boldsymbol{r}_i} \cdot \frac{\partial}{\partial \boldsymbol{p}_i} + \frac{\partial \varphi_{ij}}{\partial \boldsymbol{r}_j} \cdot \frac{\partial}{\partial \boldsymbol{p}_j}; \tag{3.2-17}$$

$\mathfrak{F}_i$ is the *external force* acting on particle *i*.

Solving Eq. (3.2-13) or, equivalently, Eq. (3.2-15) for D_N is completely equivalent to solving the equations of motion for N particles, so a direct study of the function D_N itself is impossible if N is very large. It is, therefore, expedient to seek a simpler representation of the state of the gas that still yields the information of real interest.

Let us consider the *reduced distribution functions* $F_s(x_1 \dots x_s; t)$, $s = 1, 2, \dots, N$, defined by

$$F_s(x_1 \dots x_s; t) = V^{-(N-s)} \int \dots \int D_N(x_1 \dots x_N; t) \mathrm{d}^6 x_{s+1} \dots \mathrm{d}^6 x_N, \qquad s = 1, 2, \dots, N-1. \quad (3.2\text{-}18)$$

From the assumed symmetry of D_N we conclude that F_s is a symmetric function of $x_1 \dots x_s$ and that the functional form of F_s is independent of any particular choice of s particles. The significance of the function F_s is that the expression $V^{-s} F_s(x_1 \dots x_s; t) \mathrm{d}^6 x_1 \dots \mathrm{d}^6 x_s$ gives the probability that the dynamic states of any s molecules of the system are located, respectively, in the volume elements $\mathrm{d}^6 x_1, \dots, \mathrm{d}^6 x_s$ about $x_1, \dots, x_s$ at time t. In particular we are interested in the equation for the *one-particle distribution function* F_1, which is related to the distribution function f of the previous section by

$$f(\boldsymbol{r}, \boldsymbol{c}, t)\, \mathrm{d}^3 c = (N/V)\, F_1(\boldsymbol{r}, \boldsymbol{p}; t)\, \mathrm{d}^3 p. \quad (3.2\text{-}19)$$

It is possible to derive, from Eq. (3.2-13), a system of equations for the functions F_s. This system, which was found independently by Bogoliubov [1946], Born and Green [1949], Kirkwood [1946, 1947] and Yvon [1935], is known as the BBGKY-*hierarchy of equations*. In order to derive the hierarchy, we write Eq. (3.2-13) in the form

$$\frac{\partial D_N}{\partial t} = \sum_{i=1}^{N} \left\{ \left(\frac{\boldsymbol{p}_i^2}{2m} + \Phi(\boldsymbol{r}_i) \right), D_N \right\} + \sum_{1 \leqq i < j \leqq N} \{\varphi_{ij}, D_N\}; \quad (3.2\text{-}20)$$

multiply this relation with $V^{-(N-s)}$ and integrate with respect to the variables $x_{s+1}, \dots, x_N$. Using the definition (3.2-18) we obtain

$$\frac{\partial F_s}{\partial t} = \sum_{i=1}^{N} V^{-(N-s)} \int \dots \int \left\{ \left(\frac{\boldsymbol{p}_i^2}{2m} + \Phi(\boldsymbol{r}_i) \right), D_N \right\} \mathrm{d}^6 x_{s+1} \dots \mathrm{d}^6 x_N$$
$$+ \sum_{1 \leqq i < j \leqq N} V^{-(N-s)} \int \dots \int \{\varphi_{ij}, D_N\}\, \mathrm{d}^6 x_{s+1} \dots \mathrm{d}^6 x_N. \quad (3.2\text{-}21)$$

If D_N is assumed to vanish for sufficiently large $\boldsymbol{r}_i$ and $\boldsymbol{p}_i$, we have, by use of the definition (3.2-14) and direct integration,

$$\int \left\{ \left(\frac{p_i^2}{2m} + \Phi(\boldsymbol{r}_i) \right), D_N \right\} \mathrm{d}^6 x_i = 0, \qquad i = 1, 2, \ldots, N \quad \text{(3.2-22a)}$$

and

$$\iint \{\varphi_{ij}, D_N\} \mathrm{d}^6 x_i \mathrm{d}^6 x_j = 0, \qquad i, j = 1, 2, \ldots, N. \quad \text{(3.2-22b)}$$

Further, changing the order of integration and differentiation in the terms with $i, j \leqq s$, we find

$$\sum_{i=1}^{s} V^{-(N-s)} \int \ldots \int \left\{ \left(\frac{p_i^2}{2m} + \Phi(\boldsymbol{r}_i) \right), D_N \right\} \mathrm{d}^6 x_{s+1} \ldots \mathrm{d}^6 x_N$$
$$+ \sum_{1 \leqq i < j \leqq s} V^{-(N-s)} \int \ldots \int \{\varphi_{ij}, D_N\} \mathrm{d}^6 x_{s+1} \ldots \mathrm{d}^6 x_N$$
$$= \left\{ \left[\sum_{i=1}^{s} \left(\frac{p_i^2}{2m} + \Phi(\boldsymbol{r}_i) \right) + \sum_{1 \leqq i < j \leqq s} \varphi_{ij} \right], F_s \right\} \equiv \{H_s, F_s\}, \quad \text{(3.2-23)}$$

where H_s is the Hamiltonian for a system of s particles. Finally, because of the symmetry of D_N with respect to its arguments,

$$\sum_{i=1}^{s} \sum_{j=s+1}^{N} V^{-(N-s)} \int \ldots \int \{\varphi_{ij}, D_N\} \mathrm{d}^6 x_{s+1} \ldots \mathrm{d}^6 x_N$$
$$= (N-s) \sum_{i=1}^{s} V^{-(N-s)} \int \ldots \int \{\varphi_{i,s+1}, D_N\} \mathrm{d}^6 x_{s+1} \ldots \mathrm{d}^6 x_N$$
$$= \frac{N-s}{V} \int \{ \sum_{i=1}^{s} \varphi_{i,s+1}, F_{s+1}\} \mathrm{d}^6 x_{s+1}. \quad \text{(3.2-24)}$$

When these results are substituted into Eq. (3.2-21), the latter becomes

$$\frac{\partial F_s}{\partial t} = \{H_s, F_s\} + \frac{N-s}{V} \int \{ \sum_{i=1}^{s} \varphi_{i,s+1}, F_{s+1}\} \mathrm{d}^6 x_{s+1}. \quad \text{(3.2-25)}$$

So far, we have not taken into account the fact that, usually, only systems consisting of an extremely large number of molecules are studied. In constructing the expressions for the distribution functions of interest we shall restrict our attention to the *asymptotic case* $N \to \infty$. Also, to concentrate our attention on the volume effects and to eliminate the wall effects we shall assume that, when $N \to \infty$, the boundary surfaces recede to infinity, so that

$V \to \infty$. However, the *density* $n = N/V$ shall remain *constant*. This way of taking the limit $N \to \infty$ will be called the *thermodynamic limit*. After this limiting process we find for the hierarchy of equations,

$$\frac{\partial F_s}{\partial t} = \{H_s, F_s\} + n \int \{\sum_{i=1}^{s} \varphi_{i,s+1}, F_{s+1}\} \mathrm{d}^6 x_{s+1}, \quad s = 1, 2, \ldots . \tag{3.2-26}$$

An alternative form of Eq. (3.2-26) is

$$\frac{\partial F_s}{\partial t} = -\mathfrak{H}_s F_s + n \mathfrak{L}_s F_{s+1}, \quad s = 1, 2, \ldots, \tag{3.2-27}$$

where $\mathfrak{H}_s$ is the *Hamiltonian operator* associated with a system of s particles,

$$\mathfrak{H}_s \equiv \sum_{i=1}^{s} \left[\frac{\boldsymbol{p}_i}{m} \cdot \frac{\partial}{\partial \boldsymbol{r}_i} + \mathfrak{F}_i \cdot \frac{\partial}{\partial \boldsymbol{p}_i} \right] - \sum_{1 \leq i < j \leq s} \Theta_{ij} \tag{3.2-28}$$

and $\mathfrak{L}_s$ is the *phase mixing operator*,

$$\mathfrak{L}_s \equiv \sum_{i=1}^{s} \int \mathrm{d}^6 x_{s+1} \, \Theta_{i,s+1} \, . \tag{3.2-29}$$

In these definitions the operator Θ_{ij} is the molecular interaction operator, defined in Eq. (3.2-17); $\mathfrak{F}_i$ is the external force acting on particle i.

The equations (3.2-27) form a *chain*: for every s, the equation for F_s contains the unknown F_{s+1}. The origin of this chain structure is obvious: the equation for F_s in the hierarchy is an s-particle Liouville equation, supplemented by a term that represents the interaction of the s particles with the remaining particles; since all of the remaining particles are identical these terms can be written together simply as the product of the interaction due to a single representative particle times the number of remaining particles.

The equation governing the evolution of the one-particle distribution function is obtained from the BBGKY-hierarchy as

$$\left(\frac{\partial}{\partial t} + \frac{\boldsymbol{p}_1}{m} \cdot \nabla_{r_1} + \mathfrak{F}_1 \cdot \nabla_{p_1} \right) F_1(x_1; t) = n \int \Theta_{12} F_2(x_1 x_2; t) \mathrm{d}^6 x_2 \, . \tag{3.2-30}$$

The left member of this equation has exactly the same form as the left member of the Boltzmann equation (3.1-29). On the other hand, the right member of Eq. (3.2-30) gives an *exact* expression for the change of F_1 due to interactions and it is clear that, if we could somehow express F_2 as a time-independent functional of F_1, then we would have a (generalized) Boltzmann equation. For then, Eq. (3.2-30) would be a closed equation for F_1 alone. This idea will be further worked out in the following sections.

3.3. Cluster expansion of the distribution functions

We now come to the problem of deducing, from Eq. (3.2-30), if possible, a closed equation for the single particle distribution. Before proceeding we note that Eq. (3.2-30) is an exact consequence of the laws of motion and therefore exact itself. More importantly, the laws of motion are time-reversible (if all velocities are reversed the system will trace its trajectory in phase space backwards), and Eq. (3.2-30) is therefore also time-reversible. On the other hand, the Boltzmann equation, which we hope to derive from this equation, is time-irreversible (the distribution function will not retrace its path if the velocities are reversed) since, as will be shown in Chapter 4, it leads to entropy production. This point will be discussed more fully later in this chapter, but it is clear that some approximation will need to be introduced if we are to obtain the Boltzmann equation.

It is also worthy of note that Eq. (3.2-30) is exact for *any* physical system of monatomic molecules and it is therefore clear that some information to the effect that a *dilute* gas result is sought must be added before Boltzmann's equation can be obtained. As the distinguishing feature of a dilute gas is its low density, it is reasonable to suppose that an expansion in powers of the density is in order. This idea is further supported by experience with the case of a gas at equilibrium. There, the properties are expanded in powers of the density by using the method of cluster expansions which keeps the various terms properly ordered. Extension of this idea to the present problem seems a reasonable approach for obtaining not only the low-density limit, but also the extension to higher densities. We shall adopt this method here, although other equivalent approaches provide the same results. In contrast to the equilibrium case, however, use of the cluster expansion method by itself cannot completely solve the problem. We shall still have to give some attention to the difficulties mentioned in the first paragraph.

To begin with, we introduce a set of auxiliary functions D_s, $s = 1, \ldots, N$. Foreach s, D_s is defined as the solution of the Liouville equation associated with the motion of an *isolated* system of s particles in phase space:

$$\partial D_s/\partial t = \{H_s, D_s\}, \quad s = 1, 2, \ldots, N \tag{3.3-1}$$

or, equivalently,

$$(\partial/\partial t)D_s(x_1 \ldots x_s; t) = -\mathfrak{H}_s(x_1 \ldots x_s)D_s(x_1 \ldots x_s; t). \tag{3.3-2}$$

H_s and $\mathfrak{H}_s$ are, respectively, the Hamiltonian function and the Hamiltonian operator for the system of s particles, which were defined in the previous section. In Eq. (3.3-2) we have indicated the relevant variables explicitly as arguments of the operator $\mathfrak{H}_s$. The functions D_s are normalized at all times t in such a way that

$$\int \dots \int D_s(x_1 \dots x_s; t) \mathrm{d}^6 x_1 \dots \mathrm{d}^6 x_s = V^s, \tag{3.3-3}$$

so $V^{-s} D_s(x_1 \dots x_s; t) \mathrm{d}^6 x_1 \dots \mathrm{d}^6 x_s$ gives the probability that the dynamic states of the s molecules of the system under consideration are located, respectively, in the volume elements $\mathrm{d}^6 x_1, \dots, \mathrm{d}^6 x_s$ about $x_1, \dots, x_s$ at time t.

Next, we introduce a set of functions U_s $(s = 1, 2, \dots, N)$, defined through the relations

$$D_1(x_1; t) = U_1(x_1; t), \tag{3.3-4a}$$

$$D_2(x_1 x_2; t) = U_2(x_1 x_2; t) + U_1(x_1; t)\, U_1(x_2; t), \tag{3.3-4b}$$

$$\begin{aligned} D_3(x_1 x_2 x_3; t) = {} & U_3(x_1 x_2 x_3; t) + U_2(x_1 x_2; t)\, U_1(x_3; t) \\ & + U_2(x_2 x_3; t)\, U_1(x_1; t) + U_2(x_3 x_1; t)\, U_1(x_2; t) \\ & + U_1(x_1; t)\, U_1(x_2; t)\, U_1(x_3; t), \end{aligned} \tag{3.3-4c}$$

etc. In general, for $s = 1, 2, \dots, N$,

$$D_s(x_1 \dots x_s; t) = \sum \prod U_l(\quad ; t), \tag{3.3-4}$$

where the sum is carried out over all possible partitions of s molecules into distinct groups and, for each such partition, the product is taken over the U-functions of the molecules in each group.

As each of the equations in (3.3-4) introduces only one new U-function, the functions U_s can be determined in terms of the D-functions by inversion:

$$U_1(x_1; t) = D_1(x_1; t), \tag{3.3-5a}$$

$$U_2(x_1 x_2; t) = D_2(x_1 x_2; t) - D_1(x_1; t) D_1(x_2; t), \tag{3.3-5b}$$

$$\begin{aligned} U_3(x_1 x_2 x_3; t) = {} & D_3(x_1 x_2 x_3; t) - D_2(x_1 x_2; t) D_1(x_3; t) \\ & - D_2(x_2 x_3; t) D_1(x_1; t) - D_2(x_3 x_1; t) D_1(x_2; t) \\ & + 2 D_1(x_1; t) D_1(x_2; t) D_1(x_3; t), \end{aligned} \tag{3.3-5c}$$

etc.; in general, for $s = 1, 2, \dots, N$,

$$U_s(x_1 \dots x_s; t) = \sum (-1)^{k-1} (k-1)! \prod D_l(\quad ; t), \tag{3.3-5}$$

where the sum extends over all possible partitions of the s molecules into k distinct groups, each term being $(-1)^{k-1}(k-1)!$ times the product of the D-functions for the molecules in each group.

The reason for introducing the U-functions is that they exhibit a *cluster property* if the D-functions have a *product property*, i.e. the U-function of given index vanishes whenever the corresponding D-function factorizes into a sum of products of lower order D-functions.

Having defined the D- and U-functions associated with the motion of *isolated* groups of molecules in phase space, we turn to the reduced distribution functions F_s ($s = 1, 2, \ldots, N$), defined in Eq. (3.2-18), which are associated with the motion of groups of molecules *in the system of N molecules*. We introduce a set of G-functions that bear the same relation to the F-functions as do the U-functions to the D-functions:

$$F_1(x_1; t) = G_1(x_1; t), \tag{3.3-6a}$$

$$F_2(x_1 x_2; t) = G_2(x_1 x_2; t) + G_1(x_1; t)\, G_1(x_2; t), \tag{3.3-6b}$$

$$\begin{aligned} F_3(x_1 x_2 x_3; t) = {} & G_3(x_1 x_2 x_3; t) + G_2(x_1 x_2; t)\, G_1(x_3; t) \\ & + G_2(x_2 x_3; t)\, G_1(x_1; t) + G_2(x_3 x_1; t)\, G_1(x_2; t) \\ & + G_1(x_1; t)\, G_1(x_2; t)\, G_1(x_3; t), \end{aligned} \tag{3.3-6c}$$

etc. In general, for $s = 1, 2, \ldots, N$,

$$F_s(x_1 \ldots x_s; t) = \sum \prod G_l(\;\;; t), \tag{3.3-6}$$

where the sum extends over all possible partitions of s molecules into distinct groups and, for each such partition, the product is taken over the G-functions of the molecules in each group. By inversion,

$$G_1(x_1; t) = F_1(x_1; t), \tag{3.3-7a}$$

$$G_2(x_1 x_2; t) = F_2(x_1 x_2; t) - F_1(x_1; t)\, F_1(x_2; t), \tag{3.3-7b}$$

$$\begin{aligned} G_3(x_1 x_2 x_3; t) = {} & F_3(x_1 x_2 x_3; t) - F_2(x_1 x_2; t)\, F_1(x_3; t) \\ & - F_2(x_2 x_3; t)\, F_1(x_1; t) - F_2(x_3 x_1; t)\, F_1(x_2; t) \\ & + 2F_1(x_1; t)\, F_1(x_2; t)\, F_1(x_3; t), \end{aligned} \tag{3.3-7c}$$

etc. In general, for $s = 1, 2, \ldots, N$,

$$G_s(x_1 \ldots x_s; t) = \sum (-1)^k (k-1)! \prod F_l(\;\;; t) \tag{3.3-7}$$

where the sum must be taken over all possible partitions of the s molecules into k distinct groups, each term being $(-1)^k(k-1)!$ times the product of the F-functions for the molecules in each group.

We shall now demonstrate that the method of cluster expansions provides a means of systematically treating the distribution functions in a many-particle system in terms of those for small, isolated groups of particles. We shall do so by deriving expansions for the G-functions in terms of the U-functions. We first treat the special cases $s = 1$ and $s = 2$.

(i) The case $s = 1$. The U-function expansion (3.3-4) of D_N is arranged with respect to the order of the U-function in which x_1 appears:

$$D_N(x_1 \ldots x_N; t) = U_1(x_1; t) D_{N-1}(x_2 \ldots\ldots x_N; t) + \sum_{2 \neq 1} U_2(x_1 x_2; t) D_{N-2}(x_3 \ldots x_N; t) + \sum_{(2,3) \neq 1} U_3(x_1 x_2 x_3; t) D_{N-3}(x_4 \ldots x_N; t) + \ldots \quad (3.3\text{-}8)$$

If this expression is integrated over the phases of all molecules but the first, one obtains, using the normalization (3.3-3) of the functions D_s,

$$F_1(x_1; t) = U_1(x_1; t) + \frac{1}{1!} \frac{N-1}{V} \int U_2(x_1 x_2; t) \, \mathrm{d}^6 x_2 + \frac{1}{2!} \frac{(N-1)(N-2)}{V^2} \iint U_3(x_1 x_2 x_3; t) \, \mathrm{d}^6 x_2 \, \mathrm{d}^6 x_3 + \ldots \quad (3.3\text{-}9)$$

In the thermodynamic limit this yields the following expansion for F_1 or, what is the same, G_1:

$$G_1(x_1; t) = U_1(x_1; t) + \frac{n}{1!} \int U_2(x_1 x_2; t) \, \mathrm{d}^6 x_2 + \frac{n^2}{2!} \iint U_3(x_1 x_2 x_3; t) \, \mathrm{d}^6 x_2 \, \mathrm{d}^6 x_3 + \ldots \quad (3.3\text{-}10)$$

(ii) The case $s = 2$. The U-function expansion (3.3-4) of D_N is arranged with respect to the orders of the U-functions in which x_1 and x_2 appear,

$$
\begin{aligned}
&D_N(x_1 \dots x_N; t) \\
&= U_2(x_1 x_2; t) D_{N-2}(x_3 \dots x_N; t) + \sum_{3 \neq 1,2} U_3(x_1 x_2 x_3; t) D_{N-3}(x_4 \dots x_N; t) \\
&\qquad + \sum_{(3,4) \neq 1,2} U_4(x_1 x_2 x_3 x_4; t) D_{N-4}(x_5 \dots x_N; t) + \dots \\
&\quad + U_1(x_1; t)[U_1(x_2; t) D_{N-2}(x_3 \dots x_N; t) \\
&\qquad + \sum_{3 \neq 1,2} U_2(x_2 x_3; t) D_{N-3}(x_4 \dots x_N; t) \\
&\qquad + \sum_{(3,4) \neq 1,2} U_3(x_2 x_3 x_4; t) D_{N-4}(x_5 \dots x_N; t) + \dots] \\
&\quad + \sum_{3 \neq 1,2} U_2(x_1 x_3; t)[U_1(x_2; t) D_{N-3}(x_4 \dots x_N; t) \\
&\qquad + \sum_{4 \neq 1,2,3} U_2(x_2 x_4; t) D_{N-4}(x_5 \dots x_N; t) \\
&\qquad + \sum_{(4,5) \neq 1,2,3} U_3(x_2 x_4 x_5; t) D_{N-5}(x_6 \dots x_N; t) + \dots] \\
&\quad + \sum_{(3,4) \neq 1,2} U_3(x_1 x_3 x_4; t)[U_1(x_2; t) D_{N-4}(x_5 \dots x_N; t) \\
&\qquad + \sum_{5 \neq 1,2,3,4} U_2(x_2 x_5; t) D_{N-5}(x_6 \dots x_N; t) \\
&\qquad + \sum_{(5,6) \neq 1,2,3,4} U_3(x_2 x_5 x_6; t) D_{N-6}(x_7 \dots x_N; t) + \dots] \dots
\end{aligned}
\tag{3.3-11}
$$

If this expression is integrated over the phases of all molecules but 1 and 2 one obtains, with the normalization (3.3-3) of the functions D_s,

$$
\begin{aligned}
&F_2(x_1 x_2; t) \\
&= U_2(x_1 x_2; t) + \frac{1}{1!} \frac{N-2}{V} \int U_3(x_1 x_2 x_3; t)\, d^6x_3 \\
&\qquad + \frac{1}{2!} \frac{(N-2)(N-3)}{V^2} \iint U_4(x_1 x_2 x_3 x_4; t)\, d^6x_3\, d^6x_4 + \dots \\
&\quad + \left[U_1(x_2; t) + \frac{1}{1!} \frac{N-2}{V} \int U_2(x_2 x_3; t)\, d^6x_3 \right. \\
&\qquad \left. + \frac{1}{2!} \frac{(N-2)(N-3)}{V^2} \iint U_3(x_2 x_3 x_4; t)\, d^6x_3\, d^6x_4 + \dots \right] U_1(x_1; t) \\
&\quad + \frac{1}{1!} \frac{N-2}{V} \left[U_1(x_2; t) + \frac{1}{1!} \frac{N-3}{V} \int U_2(x_2 x_4; t)\, d^6x_4 + \right.
\end{aligned}
$$

$$+\frac{1}{2!}\frac{(N-3)(N-4)}{V^2}\iint U_3(x_2\,x_4\,x_5;t)\,\mathrm{d}^6x_4\,\mathrm{d}^6x_5+\ldots\bigg]$$

$$\times\int U_2(x_1\,x_3;t)\,\mathrm{d}^6x_3$$

$$+\frac{1}{2!}\frac{(N-2)(N-3)}{V^2}\bigg[U_1(x_2;t)+\frac{1}{1!}\frac{N-4}{V}\int U_2(x_2\,x_5;t)\,\mathrm{d}^6x_5$$

$$+\frac{1}{2!}\frac{(N-4)(N-5)}{V^2}\iint U_3(x_2\,x_5\,x_6;t)\,\mathrm{d}^6x_5\,\mathrm{d}^6x_6+\ldots\bigg]$$

$$\times\iint U_3(x_1\,x_3\,x_4;t)\,\mathrm{d}^6x_3\,\mathrm{d}^6x_4+\ldots. \qquad (3.3\text{-}12)$$

In the thermodynamic limit this becomes

$$F_2(x_1\,x_2;t)$$

$$=U_2(x_1\,x_2;t)+\frac{n}{1!}\int U_3(x_1\,x_2\,x_3;t)\,\mathrm{d}^6x_3$$

$$+\frac{n^2}{2!}\iint U_4(x_1\,x_2\,x_3\,x_4;t)\,\mathrm{d}^6x_3\,\mathrm{d}^6x_4+\ldots$$

$$+\bigg[U_1(x_2;t)+\frac{n}{1!}\int U_2(x_2\,x_3;t)\,\mathrm{d}^6x_3$$

$$+\frac{n^2}{2!}\iint U_3(x_2\,x_3\,x_4;t)\,\mathrm{d}^6x_3\,\mathrm{d}^6x_4+\ldots\bigg]\,U_1(x_1;t)$$

$$+\frac{n}{1!}\bigg[U_1(x_2;t)+\frac{n}{1!}\int U_2(x_2\,x_4;t)\,\mathrm{d}^6x_4$$

$$+\frac{n^2}{2!}\iint U_3(x_2\,x_4\,x_5;t)\,\mathrm{d}^6x_4\,\mathrm{d}^6x_5+\ldots\bigg]\int U_2(x_1\,x_3;t)\,\mathrm{d}^6x_3$$

$$+\frac{n^2}{2!}\bigg[U_1(x_2;t)+\frac{n}{1!}\int U_2(x_2\,x_5;t)\,\mathrm{d}^6x_5$$

$$+\frac{n^2}{2!}\iint U_3(x_2\,x_5\,x_6;t)\,\mathrm{d}^6x_5\,\mathrm{d}^6x_6+\ldots\bigg]$$

$$\times\iint U_3(x_1\,x_3\,x_4;t)\,\mathrm{d}^6x_3\,\mathrm{d}^6x_4+\ldots, \qquad (3.3\text{-}13)$$

which can be written in the form

$$F_2(x_1 x_2; t) = G_2(x_1 x_2; t) + G_1(x_1; t) G_1(x_2; t), \qquad (3.3\text{-}14)$$

where G_1 is given by Eq. (3.3-10) and

$$G_2(x_1 x_2; t) = U_2(x_1 x_2; t) + \frac{n}{1!}\int U_3(x_1 x_2 x_3; t)\, d^6 x_3$$
$$+ \frac{n^2}{2!}\iint U_4(x_1 x_2 x_3 x_4; t)\, d^6 x_3\, d^6 x_4 + \ldots . \qquad (3.3\text{-}15)$$

In order to obtain an expansion for a higher order s-particle distribution function, one arranges the U-function expansion (3.3-4) of D_N as follows.

First all products of U-functions in the expansion (3.3-4) that contain the molecules $1 \ldots s$ as an s-particle group are taken together. This results in a term $U_s(x_1 \ldots x_s; t) D_{N-s}(x_{s+1} \ldots x_N; t)$. Then all terms that contain the molecules $1 \ldots s$ in an $(s+1)$-particle group are taken together; this results in a term

$$\sum_{(s+1) \neq 1 \ldots s} U_{s+1}(x_1 \ldots x_{s+1}; t) D_{N-s-1}(x_{s+2} \ldots x_N; t)$$

where the sum extends over all $(N-s)$ molecules $s+1, \ldots, N$. Proceeding in this way one generates the term $G_s(x_1 \ldots x_s; t)$ in the G-expansion of F_s.

Next one takes together all terms containing $s-1$ of the molecules $1 \ldots s$ in an $(s-1)$-particle group and the remaining molecule in a one-particle group, in a two-particle group, etc. This leads to the term $\sum G_{s-1}(x_1 \ldots x_{s-1}; t) G_1(x_s; t)$, where the sum extends over all possible partitions of a group of s molecules into a group of $s-1$ molecules and a group of one molecule.

One continues in this way, dividing the group of s molecules in all possible ways into smaller groups and summing within each of these groups over U-functions of an increasing number of molecules, until at last the division of the s molecules into s one-particle groups is reached. In this way F_s is written as a sum over products of G-functions by partitioning the molecules $1 \ldots s$ in all possible ways into distinct groups and by taking to each partition the corresponding products of the G-functions. (From the way this arrangement is constructed it will be clear that one uses all terms of the U-function expansion (3.3-4) of D_N.) When this result is combined with the G-function expansion (3.3-6) of F_s, the following expansion of G_s in terms of U-functions is obtained in the thermodynamic limit:

$$G_s(x_1 \ldots x_s; t) = U_s(x_1 \ldots x_s; t) + \frac{n}{1!}\int U_{s+1}(x_1 \ldots x_{s+1}; t)\,\mathrm{d}^6x_{s+1}$$

$$+\frac{n^2}{2!}\iint U_{s+2}(x_1 \ldots x_{s+2}; t)\,\mathrm{d}^6x_{s+1}\,\mathrm{d}^6x_{s+2} + \ldots . \quad (3.3\text{-}16)$$

This equation clearly exhibits the relation between the G- and U-functions, both of which have cluster properties. In the next section we shall use these expansions to derive a functional relation between the first and second reduced distribution functions.

3.4. Functional relation between F_1 and F_2

To derive a functional relation between the first and second reduced distribution functions we first solve the Liouville equation (3.3-2). For this purpose we introduce the s-particle streaming operator, $S_t^{(s)}$, defined as

$$S_t^{(s)} \equiv S_t^{(s)}(x_1 \ldots x_s) = \exp\,[t\mathfrak{H}_s(x_1 \ldots x_s)]. \quad (3.4\text{-}1)$$

Thus, $S_t^{(s)}$ is a coordinate transformation which corresponds to the streaming of a system of s particles in $6s$-dimensional phase space under the influence of the Hamiltonian $\mathfrak{H}_s$ of the system during a time interval of length t. For $-\infty < t < \infty$ the operators $S_t^{(s)}$ form an additive Abelian one-parameter group, i.e.,

$$S_{t_1}^{(s)} S_{t_2}^{(s)} = S_{t_1+t_2}^{(s)}, \quad S_0^{(s)} = I, \quad S_t^{(s)} S_{-t}^{(s)} = I, \quad (3.4\text{-}2)$$

where I is the identity operator.

If ψ is an arbitrary function of the phases $x_1 \ldots x_s$ of the particles $1 \ldots s$, $S_t^{(s)}\psi$ is defined as

$$S_t^{(s)}(x_1 \ldots x_s)\psi(x_1 \ldots x_s) = \psi(S_t^{(s)}x_1 \ldots S_t^{(s)}x_s). \quad (3.4\text{-}3)$$

Obviously, for each i, $S_t^{(s)}x_i$ is a function of the phases $x_1 \ldots x_s$ of *all* s particles. Especially, in the absence of external forces, we have

$$S_t^{(1)}(x_i)\psi(x_i) = \psi(S_t^{(1)}x_i) = \psi\left(\boldsymbol{r}_i + \frac{\boldsymbol{p}_i}{m}t, \boldsymbol{p}_i\right), \quad (3.4\text{-}4)$$

because the motion of a single particle is uniform along a straight line.

In terms of the s-particle streaming operator the solution of the Liouville equation (3.3-2) can be written formally:

$$D_s(x_1 \dots x_s; t) = S_{-t}^{(s)}(x_1 \dots x_s) D_s(x_1 \dots x_s; 0)$$
$$= D_s(S_{-t}^{(s)} x_1 \dots S_{-t}^{(s)} x_s; 0). \quad (3.4\text{-}5)$$

It is convenient to characterize the initial correlations in the auxiliary s-particle problem by a separate symbol, a_s, through the definition

$$D_s(x_1 \dots x_s; 0) = a_s(x_1 \dots x_s; 0) \prod_{i=1}^{s} D_1(x_i; 0). \quad (3.4\text{-}6)$$

We note that $a_s(x_1 \dots x_s; 0) = 1$ if and only if there are no correlations between the s particles $1 \dots s$ at $t = 0$. Then, for every value of s, $s = 2, 3, \dots$, the distribution function $D_s(x_1 \dots x_s; t)$ can be related to the distribution functions $D_1(x_1; t), \dots, D_1(x_s; t)$ by means of an operator $\tilde{\mathfrak{S}}_t^{(s)}(x_1 \dots x_s)$:

$$D_s(x_1 \dots x_s; t) = \tilde{\mathfrak{S}}_t^{(s)}(x_1 \dots x_s) \prod_{i=1}^{s} D_1(x_i; t), \quad (3.4\text{-}7)$$

if $\tilde{\mathfrak{S}}_t^{(s)}$ is defined as

$$\tilde{\mathfrak{S}}_t^{(s)}(x_1 \dots x_s) = S_{-t}^{(s)}(x_1 \dots x_s)\, a_s(x_1 \dots x_s; 0) \prod_{i=1}^{s} S_t^{(1)}(x_i). \quad (3.4\text{-}8)$$

Using these results we may rewrite the expansions (3.3-10) and (3.3-15) of G_1 and G_2, respectively, in terms of $D_1(\ \ ; t)$-functions alone. One verifies:

$$G_1(x_1; t) = \tilde{\mathfrak{U}}_t^{(1)}(x_1) D_1(x_1; t) + \frac{n}{1!} \int \tilde{\mathfrak{U}}_t^{(2)}(x_1 x_2) D_1(x_1; t) D_1(x_2; t) \mathrm{d}^6 x_2$$
$$+ \frac{n^2}{2!} \iint \tilde{\mathfrak{U}}_t^{(3)}(x_1 x_2 x_3)\, D_1(x_1; t)\, D_1(x_2; t)\, D_1(x_3; t) \mathrm{d}^6 x_2 \mathrm{d}^6 x_3 + \dots \quad (3.4\text{-}9)$$

$$G_2(x_1 x_2; t) = \tilde{\mathfrak{U}}_t^{(2)}(x_1 x_2) D_1(x_1; t)\, D_1(x_2; t)$$
$$+ \frac{n}{1!} \int \tilde{\mathfrak{U}}_t^{(3)}(x_1 x_2 x_3)\, D_1(x_1; t)\, D_1(x_2; t)\, D_1(x_3; t) \mathrm{d}^6 x_3 + \dots \quad (3.4\text{-}10)$$

where the $\tilde{\mathfrak{U}}$-operators are defined in terms of the $\tilde{\mathfrak{S}}$-operators by

$$\tilde{\mathfrak{U}}_t^{(1)}(x_1) = I, \quad (3.4\text{-}11a)$$
$$\tilde{\mathfrak{U}}_t^{(2)}(x_1 x_2) = \tilde{\mathfrak{S}}_t^{(2)}(x_1 x_2) - I, \quad (3.4\text{-}11b)$$
$$\tilde{\mathfrak{U}}_t^{(3)}(x_1 x_2 x_3) = \tilde{\mathfrak{S}}_t^{(3)}(x_1 x_2 x_3) - \tilde{\mathfrak{S}}_t^{(2)}(x_1 x_2) - \tilde{\mathfrak{S}}_t^{(2)}(x_2 x_3) - \tilde{\mathfrak{S}}_t^{(2)}(x_3 x_1) + 2I. \quad (3.4\text{-}11c)$$

We shall not go into the problem of finding expressions for the general term of the expansions (3.4-9) and (3.4-10) here. They can be found with the aid of the theory of linear graphs as in equilibrium theory. A more interesting question is whether the expansions (3.4-9) and (3.4-10) converge and, if they do, whether they converge rapidly enough to be of practical value. This problem will be discussed in Chapter 13 in connection with the theory of dense gases. For the present, it suffices to note that the successive terms of the expansions are proportional to increasing powers of the density, so that truncating the expansions after the first few terms may be reasonable in systems of sufficiently low density. As such a truncation will lead to Boltzmann's equation, which is the purpose of the present chapter, generalizations of the theory to higher orders in the density and a discussion of the difficulties associated therewith are deferred until dense gases are discussed.

Our aim is to obtain an expression for F_2 in terms of F_1. Therefore, we eliminate D_1 between the expansions (3.4-9) and (3.4-10) to obtain an expression for G_2 in terms of G_1; finding the expression for F_2 is then trivial. For the present purposes it is sufficient to carry this out only for the first two terms of the G_2-expansion. From Eq. (3.4-9) one has

$$D_1(x_1;t) = G_1(x_1;t) - \frac{n}{1!}\int \tilde{\mathfrak{U}}_t^{(2)}(x_1 x_2)\, G_1(x_1;t)\, G_1(x_2;t)\mathrm{d}^6 x_2 - \dots. \tag{3.4-12}$$

Upon substitution of this expression into Eq. (3.4-10) one finds after some trivial algebra,

$$G_2(x_1 x_2;t) = \tilde{\mathfrak{U}}_t^{(2)}(x_1 x_2)\, G_1(x_1;t)\, G_1(x_2;t) + \frac{n}{1!}\int \tilde{\mathfrak{T}}_t^{(3)}(x_1 x_2|x_3)\, G_1(x_1;t)\, G_1(x_2;t)\, G_1(x_3;t)\mathrm{d}^6 x_3 + \dots \tag{3.4-13}$$

with

$$\tilde{\mathfrak{T}}_t^{(3)}(x_1 x_2|x_3) = \tilde{\mathfrak{S}}_t^{(3)}(x_1 x_2 x_3) - \tilde{\mathfrak{S}}_t^{(2)}(x_1 x_2)\,\tilde{\mathfrak{S}}_t^{(2)}(x_1 x_3) - \tilde{\mathfrak{S}}_t^{(2)}(x_1 x_2)\,\tilde{\mathfrak{S}}_t^{(2)}(x_2 x_3) + \tilde{\mathfrak{S}}_t^{(2)}(x_1 x_2). \tag{3.4-14}$$

From Eq. (3.4-13) and the relations (3.3-6) one immediately deduces the following expansion of F_2 in terms of F_1:

$$F_2(x_1 x_2;t) = F_2^{(0)}(x_1 x_2;t) + \frac{n}{1!}F_2^{(1)}(x_1 x_2;t) + \dots, \tag{3.4-15}$$

with

$$F_2^{(0)}(x_1\,x_2;t) = \tilde{\mathfrak{S}}_t^{(2)}(x_1\,x_2)\,F_1(x_1;t)\,F_1(x_2;t), \tag{3.4-16}$$

$$F_2^{(1)}(x_1\,x_2;t) = \int \tilde{\mathfrak{T}}_t^{(3)}(x_1\,x_2|x_3)\,F_1(x_1;t)\,F_1(x_2;t)\,F_1(x_3;t)\,\mathrm{d}^6x_3. \tag{3.4-17}$$

In the next section we shall combine the functional relation (3.4-15) with the first equation of the BBGKY-hierarchy, Eq. (3.2-30), to derive the collision term of Boltzmann's equation.

3.5. The collision operator

The purpose of the previous sections has been to establish a functional relation between the reduced distribution functions F_1 and F_2 in order to write the lowest-order equation of the BBGKY-hierarchy in terms of the function F_1 alone. Eq. (3.4-15) provides such a relation and it should be appreciated that in obtaining this result no approximations have been made. In other words, the relation (3.4-15) holds in *any* infinite system, provided the thermodynamic limits exist.

Consider the right member of Eq. (3.2-30) which is the *exact* expression for the change of F_1 due to collisions. Now, if the relation (3.4-15) is inserted the collision term assumes the form:

$$n\int \Theta_{12}\,F_2^{(0)}(x_1\,x_2;t)\,\mathrm{d}^6x_2 + n^2\int \Theta_{12}\,F_2^{(1)}(x_1\,x_2;t)\,\mathrm{d}^6x_2 + \ldots. \tag{3.5-1}$$

This is to be compared with the collision term in Boltzmann's equation, Eq. (3.1-29), which has been derived on the basis of heuristic arguments. As may be seen from Eq. (3.4-16), $F_2^{(0)}(x_1\,x_2;t)$ is a bilinear form of $F_1(x_1;t)$ and $F_1(x_2;t)$; all higher-order terms in the expansion (3.5-1) involve products of at least three F_1-functions. Thus, we anticipate that the first term in the expansion is related to the collision term of Boltzmann's equation. To investigate this further, we turn to a more detailed discussion of the first term of the expansion (3.5-1).

First we note that the operator $\tilde{\mathfrak{S}}_t^{(2)}(x_1x_2)$ depends on the initial state of the system through the appearance of the function $a_2(x_1x_2;0)$, cf. Eq. (3.4-8). In the language of probability theory, $\tilde{\mathfrak{S}}_t^{(2)}(x_1x_2)$ is a *non-Markovian* operator, i.e., it does not allow prediction of the dynamics of the system through a knowledge of the present state alone. Clearly, the Boltzmann operator *is* Markovian and it is here that the crucial approximation of kinetic theory must be made: In order to make the operator $\tilde{\mathfrak{S}}_t^{(2)}(x_1 x_2)$ Markovian, we must *assume that any correlations present in the initial state may be ignored*. Cohen [1967] has pointed out that it is sufficient to ignore initial correlations between particles initially separated by more than the range of the intermolecular potential, but this does not represent a really appreciable gain. To construct a satisfactory theory it is necessary to show that the initial correlations have no important effect on the behavior of the system for times of interest in kinetic theory–viz., times long compared to the duration of a collision. Thus, if one could show that correlations decay very rapidly in the time required for a collision, then one could use as the initial state required above, the state in which correlations have died out. Unfortunately, this is impossible since one can imagine systems for which correlations *never* decay e.g., a beam of well separated particles of a single velocity. Thus the best one can hope for is a demonstration that systems for which correlations do not decay are exceedingly rare. Then one could say that Boltzmann's equation, which results from the above assumptions, holds for "almost all" systems at "almost all" times. Such a statement would be analogous to the statement of equilibrium statistical mechanics that entropy "almost always" increases. A completely satisfactory discussion of this point has not yet been given. For this reason we shall not go further into it, but shall content ourselves with stating that the situation as depicted is plausible and shall simply adopt the lack of correlation at $t = 0$ as an assumption. The nature of this assumption must be appreciated. In saying that correlations vanish at some time in the *past* we are differentiating between time directions; the future and past are no longer equivalent and, in fact, *irreversibility* has been introduced into the theory. Furthermore, we remark that the assumption of vanishing correlations in the initial state is essentially Boltzmann's *Stosszahlansatz* in a more sophisticated form.

To proceed, we note that when initial correlations are ignored, the operator $\tilde{\mathfrak{S}}_t^{(2)}(x_1 x_2)$ reduces to

$$\mathfrak{S}_t^{(2)}(x_1 x_2) = S_{-t}^{(2)}(x_1 x_2)\, S_t^{(1)}(x_1)\, S_t^{(1)}(x_2). \tag{3.5-2}$$

It remains to investigate this operator in some detail. Its action on the phases of two particles is depicted in Fig. 3.3; $\boldsymbol{r}_1$ and $\boldsymbol{r}_2$ are the positions of particles 1 and 2 at time t, $\boldsymbol{p}_1$ and $\boldsymbol{p}_2$ their respective momenta, and $r_{12} = |\boldsymbol{r}_1 - \boldsymbol{r}_2| < r_0$, where r_0 is the range of the intermolecular potential; $\boldsymbol{r}_1' = S_t^{(1)}(x_1)\boldsymbol{r}_1$

and $\boldsymbol{r}_2' = S_t^{(1)}(x_2)\boldsymbol{r}_2$ are the positions the two particles would have had at $t = 0$ if there were no interaction; $\boldsymbol{r}_1'' = S_{-t}^{(2)}(x_1 x_2)\boldsymbol{r}_1'$ and $\boldsymbol{r}_2'' = S_{-t}^{(2)}(x_1 x_2)\boldsymbol{r}_2'$ are the positions of the particles at time t given that they start at positions $\boldsymbol{r}_1'$ and $\boldsymbol{r}_2'$ at $t = 0$ and interact. Thus $\boldsymbol{r}_1'' = \mathfrak{S}_t^{(2)}(x_1 x_2)\boldsymbol{r}_1$ and $\boldsymbol{r}_2'' = \mathfrak{S}_t^{(2)}(x_1 x_2)\boldsymbol{r}_2$.

For the case depicted in the figure, for t large enough so that $|\boldsymbol{r}_1' - \boldsymbol{r}_2'| > r_0$, the action of the operator $\mathfrak{S}_t^{(2)}(x_1 x_2)$ on the phases of the particles is independent of t. In this case, it is possible to define

$$\mathfrak{S}_\infty^{(2)}(x_1 x_2) = \lim_{t \to \infty} S_{-t}^{(2)}(x_1 x_2)\, S_t^{(1)}(x_1)\, S_t^{(2)}(x_2). \tag{3.5-3}$$

As the ability to define this limit plays an essential role in the following theory it is important to investigate under what conditions its existence is guaranteed.

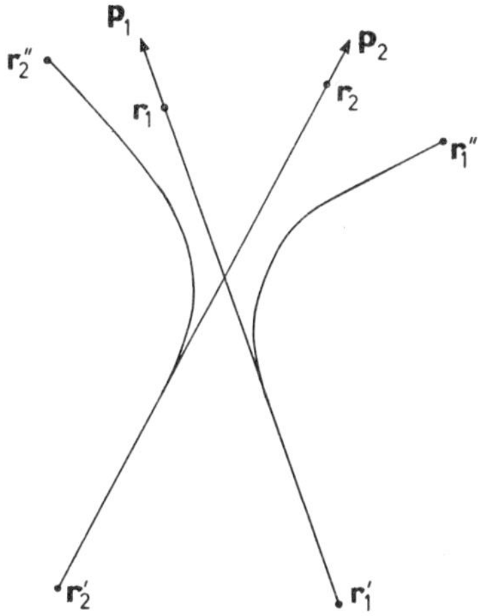

Fig. 3.3. Action of the operator $\mathfrak{S}_t^{(2)}(x_1 x_2) = S_{-t}^{(2)}(x_1 x_2) S_t^{(1)}(x_1) S_t^{(1)}(x_2)$ on the phases of two particles.

Let the *mean collision time*, t_{coll}, be defined as the ratio of the range of the intermolecular force and the mean thermal speed of the molecules. For monatomic molecules and purely repulsive intermolecular potentials the binary interaction pattern is very nearly as depicted in Fig. 3.3. Here, the only possible source of trouble comes from molecules having very small relative velocities which interact during a time interval that is long compared to the mean collision time. However, it may be shown (Frieman and Goldman [1966]) that the perturbations introduced by such molecules are of the order $(t_{\text{coll}}/t)^3$ and, therefore, may be safely neglected. Much more serious is the fact that true intermolecular potentials are never purely repulsive and it is quite possible for situations other than that depicted in Fig. 3.3 to arise. In fact, it is easily seen that permanent orbiting can occur for certain initial phases of the particles (the period of a permanent orbit being of the order of the mean collision time). The volume of phase space occupied by particles in permanent orbits is, of course, constant in time. On the other hand, the number of ordinary collisions (of the type depicted in Fig. 3.3) occurring in the time interval between 0 and t is proportional to the number of particles in the collision cylinder whose length is gt, so the volume of phase space which will produce ordinary collisions grows linearly with time. Thus, the contributions from permanent orbits may be neglected for times large compared to the mean collision time. Finally, anticipating the discussion of Chapter 9, we mention the possibility of collisions which require an indefinitely long time for their completion (so-called "orbiting collisions"). Fortunately, this difficulty is not serious either as may be seen by the following argument. Orbiting collisions occur only at a definite value g^* of the relative velocity of the two colliding molecules, where g^* is a function of the impact parameter. For relative velocities near g^*, the time required for the completion of a collision can be shown to be approximately proportional to $-\log|g-g^*|$. Thus, although the number of orbiting collisions occurring in the time interval between 0 and t grows, as for ordinary collisions, linearly with time, the range of relative velocities which will produce collisions of duration greater than t decays exponentially with time. Then, the volume of phase space which leads to collisions lasting longer than t is proportional to $t \exp(-\alpha t)$, where α is some positive constant, and may also be safely neglected.

Thus, under the assumptions made above, for $t \gg t_{\text{coll}}$, the first term of the expansion (3.5-1) becomes

$$n \int \Theta_{12} F_2^{(0)}(x_1 x_2 | F_1) \, d^6 x_2$$
$$= n \int \Theta_{12} \mathfrak{S}_\infty^{(2)}(x_1 x_2) F_1(x_1; t) F_1(x_2; t) \, d^6 x_2, \quad (3.5\text{-}4)$$

where the notation $F_2^{(0)}(x_1 x_2|F_1)$ is used to indicate that the whole time dependence of the function $F_2^{(0)}$ is through the time dependence of the functions

F_1. To show the connection of the expression (3.5-4) with the collision term in the Boltzmann equation (3.1-29) we proceed as follows.

First we introduce the abbreviations

$$\mathfrak{S}_\infty^{(2)}(x_1\, x_2)x_i = X_i^{(2)}(x_1\, x_2) = (\boldsymbol{R}_i^{(2)}(x_1\, x_2), \boldsymbol{P}_i^{(2)}(x_1\, x_2)), \quad i = 1, 2. \qquad (3.5\text{-}5)$$

Then we note that, in the functions F_1 in the expression (3.5-4), the particles 1 and 2 are taken at *different* positions, viz. $\boldsymbol{R}_1^{(2)}$ and $\boldsymbol{R}_2^{(2)}$ respectively, instead of both at the same position $\boldsymbol{r}_1$ as in Boltzmann's equation. Therefore, we expand $\boldsymbol{R}_1^{(2)}$ and $\boldsymbol{R}_2^{(2)}$ around $\boldsymbol{r}_1$. Introducing center-of-mass and relative coordinates by

$$\bar{\boldsymbol{r}} = \tfrac{1}{2}(\boldsymbol{r}_1+\boldsymbol{r}_2) \quad \text{and} \quad \boldsymbol{r}_{21} = \boldsymbol{r}_2-\boldsymbol{r}_1 \qquad (3.5\text{-}6)$$

and using the straight-forwardly proven identity

$$\mathfrak{S}_\infty^{(2)}(x_1\, x_2)\bar{\boldsymbol{r}} = \bar{\boldsymbol{r}}, \qquad (3.5\text{-}7)$$

one verifies that

$$\boldsymbol{R}_1^{(2)}(x_1\, x_2) = \boldsymbol{r}_1+\tfrac{1}{2}\boldsymbol{r}_{21}-\tfrac{1}{2}\mathfrak{S}_\infty^{(2)}(x_1\, x_2)\boldsymbol{r}_{21}, \qquad (3.5\text{-}8a)$$

$$\boldsymbol{R}_2^{(2)}(x_1\, x_2) = \boldsymbol{r}_1+\tfrac{1}{2}\boldsymbol{r}_{21}+\tfrac{1}{2}\mathfrak{S}_\infty^{(2)}(x_1\, x_2)\boldsymbol{r}_{21}. \qquad (3.5\text{-}8b)$$

Thus we obtain the following expression for the right member of Eq. (3.5-4):

$$n\int \Theta_{12}\,\mathfrak{S}_\infty^{(2)}(x_1\, x_2)F_1(x_1;t)F_1(x_2;t)\,\mathrm{d}^6x_2 = n[J(F_1\,F_1)+J_1(F_1\,F_1)], \qquad (3.5\text{-}9)$$

where $J(F_1\,F_1)$ is given by

$$J(F_1\,F_1) = \int \Theta_{12}F_1(\boldsymbol{r}_1, \boldsymbol{P}_1^{(2)}; t)F_1(\boldsymbol{r}_1, \boldsymbol{P}_2^{(2)}; t)\,\mathrm{d}^6x_2 \qquad (3.5\text{-}10)$$

and $J_1(F_1\,F_1)$ by

$$\begin{aligned} J_1(F_1\,F_1) = \int \Theta_{12}[&\tfrac{1}{2}\boldsymbol{r}_{21}\cdot\nabla_{\boldsymbol{r}_1}F_1(\boldsymbol{r}_1, \boldsymbol{P}_1^{(2)}; t)F_1(\boldsymbol{r}_1, \boldsymbol{P}_2^{(2)}; t) \\ &+\tfrac{1}{2}\mathfrak{S}_\infty^{(2)}(x_1\, x_2)\boldsymbol{r}_{21}\cdot\{F_1(\boldsymbol{r}_1, \boldsymbol{P}_1^{(2)}; t)\nabla_{\boldsymbol{r}_1}F_1(\boldsymbol{r}_1, \boldsymbol{P}_2^{(2)}; t) \\ &-F_1(\boldsymbol{r}_1, \boldsymbol{P}_2^{(2)}; t)\nabla_{\boldsymbol{r}_1}F_1(\boldsymbol{r}_1, \boldsymbol{P}_1^{(2)}; t)\}]\,\mathrm{d}^6x_2. \end{aligned} \qquad (3.5\text{-}11)$$

Here we have restricted ourselves to terms containing no higher than the first-order gradients of F_1.

We now show that the operator J defined by Eq. (3.5-10) is identical with the collision operator occurring in Boltzmann's equation (3.1-29). As the molecular interaction operator Θ_{12}, defined in Eq. (3.2-17), only acts on the momenta $\boldsymbol{p}_1$ and $\boldsymbol{p}_2$, we can treat $\boldsymbol{r}_1$ and t in the following as parameters, which we shall not write down explicitly.

$\boldsymbol{P}_1^{(2)}$ and $\boldsymbol{P}_2^{(2)}$ are the (constant) initial momenta of the particles 1 and 2 that lead to the phase $(x_1 x_2)$ at time t. Being constants of the motion, their Poisson bracket with the Hamiltonian H_2, which governs the motion, must vanish. Then, also the Poisson bracket of H_2 with any function of $\boldsymbol{P}_1^{(2)}$ and $\boldsymbol{P}_2^{(2)}$ must vanish. In particular,

$$\{H_2, F_1(\boldsymbol{P}_1^{(2)}) F_1(\boldsymbol{P}_2^{(2)})\} = 0. \tag{3.5-12}$$

If the effect of external forces which act on the molecules can be neglected during a collision, the relation (3.5-12) yields the identity

$$\begin{aligned}\Theta_{12} F_1(\boldsymbol{P}_1^{(2)}) F_1(\boldsymbol{P}_2^{(2)}) &= \{\varphi_{12}, F_1(\boldsymbol{P}_1^{(2)}) F_1(\boldsymbol{P}_2^{(2)})\} \\ &= \left(\frac{\boldsymbol{p}_1}{m} \cdot \nabla_{\boldsymbol{r}_1} + \frac{\boldsymbol{p}_2}{m} \cdot \nabla_{\boldsymbol{r}_2}\right) F_1(\boldsymbol{P}_1^{(2)}) F_1(\boldsymbol{P}_2^{(2)}). \end{aligned} \tag{3.5-13}$$

However, $\boldsymbol{P}_1^{(2)}$ and $\boldsymbol{P}_2^{(2)}$ depend on $\boldsymbol{r}_1$ and $\boldsymbol{r}_2$ only through the difference $\boldsymbol{r}_{21} = \boldsymbol{r}_2 - \boldsymbol{r}_1$, so instead of (3.5-13) we may write

$$\Theta_{12} F_1(\boldsymbol{P}_1^{(2)}) F_1(\boldsymbol{P}_2^{(2)}) = \frac{\boldsymbol{p}_2 - \boldsymbol{p}_1}{m} \cdot \nabla_{\boldsymbol{r}_{21}} F_1(\boldsymbol{P}_1^{(2)}) F_1(\boldsymbol{P}_2^{(2)}). \tag{3.5-14}$$

Now the integration over $\boldsymbol{r}_2$ in Eq. (3.5-10) can be carried out. By taking the origin for $\boldsymbol{r}_2$ at $\boldsymbol{r}_1$, choosing a cylindrical coordinate system for $\boldsymbol{r}_{21}$ with the axis in the direction of $\boldsymbol{p}_2 - \boldsymbol{p}_1$, denoting the coordinate along the axis by ξ and the polar coordinates in the plane perpendicular to the axis by b and ε (see Fig. 3.4) and carrying out the integration over ξ, the following expression for $J(F_1 F_1)$ is obtained:

$$J(F_1 F_1) = \int \mathrm{d}^3 p_2 \frac{|\boldsymbol{p}_2 - \boldsymbol{p}_1|}{m} \int_0^\infty \int_0^{2\pi} b\, \mathrm{d}b\, \mathrm{d}\varepsilon\, F_1(\boldsymbol{P}_1^{(2)}) F_1(\boldsymbol{P}_2^{(2)}) \Bigg|_{\xi=-\infty}^{\xi=+\infty}. \tag{3.5-15}$$

What are the momenta $\boldsymbol{P}_i^{(2)}$ if, in the two-particle system considered here, $\xi \to \pm\infty$, i.e., if the distance between the two particles is so large that the particles are no longer interacting? We recall the definition of $\boldsymbol{P}_i^{(2)}$,

$$\boldsymbol{P}_i^{(2)} \equiv \boldsymbol{P}_i^{(2)}(\boldsymbol{r}_{21}, \boldsymbol{p}_1, \boldsymbol{p}_2) = \lim_{t\to\infty} S_{-t}^{(2)}(x_1 x_2) S_t^{(1)}(x_1) S_t^{(1)}(x_2) \boldsymbol{p}_i, \quad i = 1, 2. \tag{3.5-16}$$

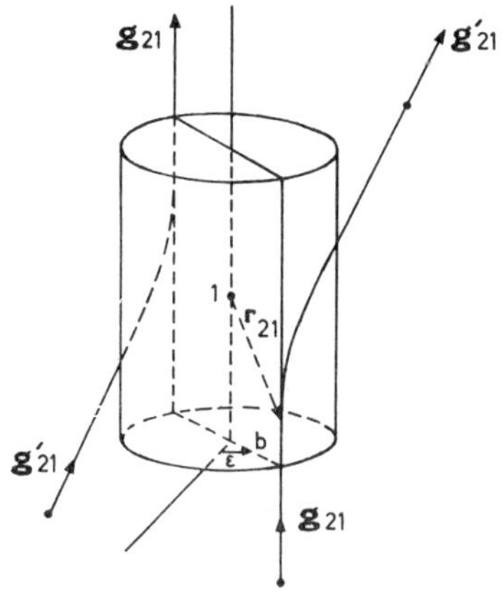

Fig. 3.4. Frame of reference for r_2-integration.

As the effect of external forces is neglected, the operators $S_t^{(1)}$ leave the momenta p_i invariant according to Eq. (3.4-5) and the expression (3.5-16) reduces to

$$P_i^{(2)} = \lim_{t\to\infty} S_{-t}^{(2)}(x_1 x_2) p_i, \quad i = 1, 2. \tag{3.5-17}$$

Then it follows that, for the dynamical state in which r_{21} has coordinates ($\xi = -\infty, b, \varepsilon$), the momenta $P_i^{(2)}$ must be identical with p_i since the particles will never interact and hence $S_{-t}^{(2)}$ has no effect.

On the other hand, after changing the variable in the ε-integration in expression (3.5-15) from ε to $\varepsilon+\pi$, one then has to deal with a dynamical state in which r_{21} has coordinates ($\xi = +\infty, b, \varepsilon+\pi$). In this case, the particles certainly will interact and the momenta after such a collision will be identical with the momenta p_i' of the inverse collision ($p_1', p_2' \to p_1, p_2$) described in Section 3.1. Thus we have

$$\lim_{\xi\to-\infty} P_i^{(2)}(r_{21}, p_1, p_2) = p_i, \quad i = 1, 2, \tag{3.5-18a}$$

$$\lim_{\xi\to+\infty} P_i^{(2)}(r_{21}, p_1, p_2) = p_i', \quad i = 1, 2. \tag{3.5-18b}$$

Substitution of these results into the expression (3.5-15) yields the following result for the operator J,

$$J(F_1 F_1) = \iiint [F_1(p_1')F_1(p_2') - F_1(p_1)F_1(p_2)] \frac{|p_1 - p_2|}{m} b\, \mathrm{d}b\, \mathrm{d}\varepsilon\, \mathrm{d}^3 p_2, \tag{3.5-19}$$

which indeed demonstrates the identity of the J-operator with the collision operator of Boltzmann's equation (3.1-29).

3.6. Concluding remarks

In the previous sections a fundamental, modern derivation of Boltzmann's equation has been given. The length and complexity of the derivation contrast strongly with the attractive and simple intuitive derivation given in the first section of this chapter. One might well ask why we went through all the details of Sections 3.2 to 3.5. The most important reason for this is that an intuitive derivation of a kinetic equation valid at higher densities has not been found except in very special cases. In order to derive such an equation it is first necessary to clarify the assumptions hidden in the heuristic classical arguments of Boltzmann. Once we understand these assumptions to their full extent we should, for example, be able to generalize Boltzmann's equation to cases in which triple and higher-order encounters between the gas molecules are no longer negligible. Another question which may arise after the classical derivation of Boltzmann's equation concerns the phenomenon of *contraction* of the description: Under which circumstances is a *one*-particle distribution function sufficient for the description of a system composed of *many* particles? We have seen in the previous sections how deeply one has to go into the theory in order to answer these and similar questions. By way of summary we review briefly the results.

The equation of motion which describes the behavior in phase space of a system of N particles under the influence of external forces and mutual interactions is Liouville's equation, which is a first-order differential equation for the distribution function, D_N,

$$\mathrm{d}D_N/\mathrm{d}t = 0. \tag{3.6-1}$$

As a first step in the derivation of Boltzmann's equation from Liouville's equation one defines the reduced distribution functions F_s, $s = 1, \ldots, N$,

$$F_s(x_1 \ldots x_s; t) = V^{-(N-s)} \int \ldots \int D_N(x_1 \ldots x_N; t)\,\mathrm{d}^6x_{s+1} \ldots \mathrm{d}^6x_N, \tag{3.6-2}$$

and passes to the thermodynamic limit of an infinite system: $N, V \to \infty$, $n = N/V$ constant. Thus, one derives the so-called BBGKY-hierarchy of equations

$$\mathrm{d}F_s/\mathrm{d}t = n\mathfrak{L}_s F_{s+1}, \quad s = 1, 2, \ldots, \tag{3.6-3}$$

which is an infinite chain of integro-differential equations, each equation being an s-particle Liouville equation supplemented by a term that represents the interaction of the s particles with the remaining particles. The equation for the one-particle distribution function, F_1, is of the form

$$\left(\frac{\partial}{\partial t}+\frac{\boldsymbol{p}_1}{m}\cdot\nabla_{r_1}+\mathfrak{F}_1\cdot\nabla_{p_1}\right)F_1(x_1;t)=n\int\Theta_{12}F_2(x_1\,x_2;t)\,\mathrm{d}^6x_2, \tag{3.6-4}$$

where the inhomogeneous term involves the two-particle distribution function, F_2. The operator Θ_{12} depends upon the mechanism of molecular interactions; $\mathfrak{F}$ is the external force acting on a molecule. Since Boltzmann's equation is an equation in terms of the one-particle distribution function alone, the next step consists of the derivation of an expression for the function F_2 in terms of the function F_1.

The method of cluster expansions provides a systematic approach to the latter problem. Using the density, n, as the ordering parameter throughout the analysis one establishes a relation between F_2 and F_1 of the form

$$F_2(x_1\,x_2;t)=F_2^{(0)}(x_1\,x_2;t)+\frac{n}{1!}F_2^{(1)}(x_1\,x_2;t)+\ldots \tag{3.6-5a}$$

in which the zero-order term, $F_2^{(0)}$, is a quadratic form of F_1,

$$F_2^{(0)}(x_1\,x_2;t)=\tilde{\mathfrak{S}}_t^{(2)}(x_1\,x_2;t)F_1(x_1;t)F_1(x_2;t), \tag{3.6-5b}$$

the first-order term, $F_2^{(1)}$, a cubic form of F_1,

$$F_2^{(1)}(x_1\,x_2;t)=\int\tilde{\mathfrak{T}}_t^{(3)}(x_1\,x_2|x_3)F_1(x_1;t)F_1(x_2;t)F_1(x_3;t)\,\mathrm{d}^6x_3, \tag{3.6-5c}$$

etcetera. In these expressions, $\tilde{\mathfrak{S}}_t^{(2)}$, $\tilde{\mathfrak{T}}_t^{(3)}$ and the higher-order operators are defined in terms of s-particle streaming operators ($s = 2, 3, \ldots$) which, in turn, are associated with the motion of *isolated* systems of s particles in phase space.

To derive the collision term of Boltzmann's equation, which is quadratic in the distribution function, one retains only the term to *lowest* order in the density and approximates the right member of Eq. (3.6-4),

$$n\int\Theta_{12}F_2(x_1\,x_2;t)\,\mathrm{d}^6x_2=n\int\Theta_{12}F_2^{(0)}(x_1\,x_2;t)\,\mathrm{d}^6x_2. \tag{3.6-6}$$

Then, it is observed that the operator $\tilde{\mathfrak{S}}_t$ occurring in $F_2^{(0)}$ is non-Markovian, whereas Boltzmann's collision operator is Markovian, i.e., the latter allows prediction of the dynamics of the system through a knowledge of the present state alone and does not require any knowledge of the initial state of the system. It is not at all obvious that, in a real gas of interacting molecules the non-Markovian character of the collision operator in Eq. (3.6-6) dies out after an initial transient, say. The difficulty that is encountered here, is of a very fundamental nature indeed and there appears only one way out of the dilemma: one *must* assume that any correlations present in the initial state may be ignored (*molecular chaos* in the infinite past) and consider only those systems for which the assumption is satisfied. That this may lead to difficulties will be seen in Chapter 13, when the kinetic theory of dense gases is discussed. An important consequence of the assumption is that, by thus distinguishing between the past and the future, one introduces *irreversibility* into the kinetic theory.

Now, if the time of interest is much larger than the mean collision time, the operator $\mathfrak{S}_t^{(2)}$ – which is the same as $\tilde{\mathfrak{S}}_t^{(2)}$ except for the fact that the initial correlations have been ignored – may be replaced by its asymptotic form, $\mathfrak{S}_\infty^{(2)}$, and, instead of Eq. (3.6-6), one has

$$n\int \Theta_{12}F_2(x_1\,x_2;\,t)\,\mathrm{d}^6x_2 = n\int \Theta_{12}\,\mathfrak{S}_\infty^{(2)}(x_1\,x_2)\,F_1(x_1;\,t)\,F_1(x_2;\,t)\,\mathrm{d}^6x_2\,. \tag{3.6-7}$$

Combining this result with the equation for the one-particle distribution function in the BBGKY-hierarchy, Eq. (3.6-4), one arrives at an integro-differential equation in terms of F_1 alone, which is, in fact, a generalized form of Boltzmann's equation.

A comparison of the collision term (3.6-7) and the collision term of Boltzmann's equation reveals the following important difference. In the expression (3.6-7), the functions F_1 are evaluated at *different* positions whereas, in the collision term of Boltzmann's equation, the distribution functions are always evaluated at the *same* position. This difference disappears if, in Eq. (3.6-7), a Taylor series expansion is performed. Assuming that the variation of F_1 over an intermolecular distance is small one can indeed show that, to zero order, one has

$$\begin{aligned} n\int \Theta_{12}\,\mathfrak{S}_\infty^{(2)}(x_1\,x_2)\,F_1(x_1;\,t)\,F_1(x_2;\,t)\,\mathrm{d}^6x_2 \\ = n\iiint [F_1(\boldsymbol{p}_1')F_1(\boldsymbol{p}_2') - F_1(\boldsymbol{p}_1)F_1(\boldsymbol{p}_2)]\,\frac{|\boldsymbol{p}_1-\boldsymbol{p}_2|}{m}\,b\,\mathrm{d}b\,\mathrm{d}\varepsilon\,\mathrm{d}^3p_2\,, \end{aligned} \tag{3.6-8}$$

where, now, each F_1 is evaluated at the same point $\boldsymbol{r}_1$ of physical space.

Since the one-particle distribution function $F_1(\boldsymbol{r}, \boldsymbol{p}; t)$ is related to the distribution function $f(\boldsymbol{r}, \boldsymbol{c}, t)$ of Section 3.1 through the identity

$$f(\boldsymbol{r}, \boldsymbol{c}, t)\mathrm{d}^3c = nF_1(\boldsymbol{r}, \boldsymbol{p}; t)\mathrm{d}^3p, \tag{3.6-9}$$

we have, from Eqs. (3.6-4), (3.6-7) and (3.6-8),

$$\left(\frac{\partial}{\partial t} + \boldsymbol{c}_1 \cdot \boldsymbol{\nabla}_{r_1} + \boldsymbol{F} \cdot \boldsymbol{\nabla}_{c_1}\right) f(\boldsymbol{r}_1, \boldsymbol{c}_1, t)$$

$$= \iiint [f(\boldsymbol{c}_1')f(\boldsymbol{c}_2') - f(\boldsymbol{c}_1)f(\boldsymbol{c}_2)]\, g_{12}\, b\, \mathrm{d}b\, \mathrm{d}\varepsilon\, \mathrm{d}^3c_2, \tag{3.6-10}$$

which is indeed identical with Boltzmann's equation (3.1-29).

From these observations it is evident how generalizations of Boltzmann's equation may be constructed. By retaining terms of higher order in the density in the functional relation (3.6-5) between F_2 and F_1 one may account for triple, quadruple and higher-order interactions. By retaining terms of higher order in the Taylor series expansion of the function F_1 in Eq. (3.6-7) one may account for higher values of the spatial gradients that may exist in the system. These ideas will be further worked out when we come to treat the kinetic theory of dense gases in Chapters 12 and 13.

Fundamental properties of Boltzmann's equation 4

The Boltzmann equation derived in the previous chapter describes the evolution of the velocity distribution function in phase space. Generally speaking, the equation is made up of two parts: a *streaming* part and a *collision* part. The former describes the motion of molecules along their trajectories in phase space and is represented by a differential operator, the latter describes velocity changes resulting from collisions and is represented by an integral operator. The Boltzmann equation is, therefore, an *integro-differential* equation. The striking feature about it, though, is the *non-linearity* of the collision part. As might be expected, this non-linearity is the major obstacle in the design of methods for solving the Boltzmann equation. The situation is further complicated by the fact that the collision integral is intimately connected with the mechanism of molecular interaction, about which only fragmentary analytical information is available. Thus, it is expedient to start a study of the Boltzmann equation by extracting all possible information from the equation *prior* to attempting a complete solution. This will be done in the present chapter.

From gas dynamics we know that a detailed microscopic description of a gas by means of a distribution function is unnecessary in the majority of problems. It is therefore natural to seek a less detailed description using the macroscopic quantities of fluid dynamics (density, hydrodynamic velocity, temperature) introduced in Chapter 2. Since these quantities are defined in terms of velocity moments of f, we are led to analyze various velocity moments of the Boltzmann equation. Of particular interest are, of course, the moments with respect to the collisional invariants, as the macroscopic quantities of fluid dynamics are directly related to the corresponding moments of f. In fact, we show (Section 4.1) that the equations of transfer for the collisional invariants are identical with the *conservation equations* of fluid dynamics and, thus, establish the formal relationship between kinetic theory and fluid dynamics.

In deriving the Boltzmann equation we made a number of important assumptions. Of these, the *Stosszahlansatz* or assumption of molecular chaos was clearly a non-mechanical (non-deterministic) assumption; it introduced *irreversibility* into kinetic theory. In Section 4.2 we further elaborate upon this observation and prove Boltzmann's *H-theorem*. That is, we prove that there exists a functional H which assigns a real number to every velocity distribution function and which has the property that its value cannot increase in the course of time. Hence, the H-theorem demonstrates the *irreversible, monotonic approach to equilibrium.*

If $\mathrm{d}H/\mathrm{d}t = 0$ and the equilibrium state prevails, the solution of the Boltzmann equation can be found explicitly. It is the well-known *Maxwellian distribution function.* We discuss some of its properties in Section 4.3.

Before generalizing the H-theorem to bounded systems we discuss the nature of boundary conditions in transport problems (Section 4.4). Although no definite statemens can be made about the exact form of the boundary conditions that must be prescribed with the Boltzmann equation in order to get a uniquely determined solution f from given initial data – the problem of the existence theory for Boltzmann's equation is still open – we can still follow intuition (and experience from linear transport theory) and formulate some general principles. In Section 4.5 we show that, if the boundary conditions are chosen in accordance with these principles, then the H-theorem can indeed be generalized to bounded systems.

If the state of the gas is not too far from thermal equilibrium one may assume that a linearized form of the Boltzmann equation provides a reasonably accurate description of the transport phenomena. Such a linearized form is presented in Section 4.6. Its collision operator is, in fact, a linear integral operator with a *symmetric* kernel whose properties depend, of course, on the mechanism of molecular interaction. Certain integral theorems that are related to the linearized collision operator and that will be used later in constructing approximate solutions of the non-linear Boltzmann equation, are derived in Section 4.7. Finally, in Section 4.8, using methods of functional analysis we establish an existence and uniqueness theorem for the linearized Boltzmann equation. This theorem states, in precise mathematical terms, the conditions under which the linearized Boltzmann equation yields a unique solution from given initial data and, thus, provides a rigorous basis for the kinetic theory of transport processes in gases near thermal equilibrium.

4.1. Macroscopic conservation equations

Boltzmann's equation for the ith component of a mixture of gases can be written in the form

$$\mathfrak{D}f_i = \sum_j J(f_i f_j), \tag{4.1-1}$$

where $\mathfrak{D}$ is the streaming operator,

$$\mathfrak{D}f_i = \frac{\partial f_i}{\partial t} + \boldsymbol{c}_i \cdot \nabla_r f_i + \boldsymbol{F}_i \cdot \nabla_{c_i} f_i \tag{4.1-2}$$

and J the collision operator,

$$J(f_i f_j) = \iiint (f_i' f_j' - f_i f_j) g b \, \mathrm{d}b \, \mathrm{d}\varepsilon \, \mathrm{d}^3 c_j . \tag{4.1-3}$$

We recall that, in the expression for $\mathfrak{D}$, $\boldsymbol{F}_i$ is the external force per unit mass acting on a molecule; it is assumed that $\boldsymbol{F}_i$ is independent of the molecular velocity.

Let ϕ be an arbitrary function of $\boldsymbol{r}$, $\boldsymbol{c}$, and t, defined for each component of the mixture. Since $\sum_j J(f_i f_j)$ represents the rate of change of f_i due to collisions – it was denoted earlier by $(\partial f_i/\partial t)_{\mathrm{coll}}$, cf. Eq. (3.1-3) – we define

$$\left(\frac{\partial \phi_i}{\partial t}\right)_{\mathrm{coll}} = \frac{1}{n_i} \sum_j \int \phi_i J(f_i f_j) \, \mathrm{d}^3 c_i \tag{4.1-4}$$

and interpret $(\partial \phi_i/\partial t)_{\mathrm{coll}}$ as the average rate of change of the molecular property ϕ_i due to collisions. Similarly, we define

$$\left(\frac{\partial \phi}{\partial t}\right)_{\mathrm{coll}} = \frac{1}{n} \sum_i n_i \left(\frac{\partial \phi_i}{\partial t}\right)_{\mathrm{coll}} = \frac{1}{n} \sum_{i,j} \int \phi_i J(f_i f_j) \, \mathrm{d}^3 c_i \tag{4.1-5}$$

and interpret $(\partial \phi/\partial t)_{\mathrm{coll}}$ as the average rate of change of the molecular property ϕ for the entire gas due to collisions.

From the definition we have

$$\left(\frac{\partial \phi_i}{\partial t}\right)_{\mathrm{coll}} = \frac{1}{n_i} \sum_j \iiiint \phi_i (f_i' f_j' - f_i f_j) g b \, \mathrm{d}b \, \mathrm{d}\varepsilon \, \mathrm{d}^3 c_j \, \mathrm{d}^3 c_i \tag{4.1-6}$$

which, because of Liouville's law for elastic collisions, Eq. (3.1-20), by interchange of primed and unprimed variables can also be written as

$$\left(\frac{\partial \phi_i}{\partial t}\right)_{\text{coll}} = \frac{1}{n_i} \sum_j \iiiint \phi_i'(f_i f_j - f_i' f_j') g b \, \mathrm{d}b \, \mathrm{d}\varepsilon \, \mathrm{d}^3 c_j \, \mathrm{d}^3 c_i . \quad (4.1\text{-}7)$$

Adding the corresponding members of Eqs. (4.1-6) and (4.1-7) we find

$$\left(\frac{\partial \phi_i}{\partial t}\right)_{\text{coll}} = \frac{1}{2n_i} \sum_j \iiiint (\phi_i - \phi_i')(f_i' f_j' - f_i f_j) g b \, \mathrm{d}b \, \mathrm{d}\varepsilon \, \mathrm{d}^3 c_j \, \mathrm{d}^3 c_i . \quad (4.1\text{-}8)$$

Then, from the definition (4.1-5) and the result (4.1-8),

$$\left(\frac{\partial \phi}{\partial t}\right)_{\text{coll}} = \frac{1}{2n} \sum_{i,j} \iiiint (\phi_i - \phi_i')(f_i' f_j' - f_i f_j) g b \, \mathrm{d}b \, \mathrm{d}\varepsilon \, \mathrm{d}^3 c_j \, \mathrm{d}^3 c_i . \quad (4.1\text{-}9)$$

In the right member, i and j are only dummy indices. Therefore, also

$$\left(\frac{\partial \phi}{\partial t}\right)_{\text{coll}} = \frac{1}{2n} \sum_{i,j} \iiiint (\phi_j - \phi_j')(f_i' f_j' - f_i f_j) g b \, \mathrm{d}b \, \mathrm{d}\varepsilon \, \mathrm{d}^3 c_j \, \mathrm{d}^3 c_i . \quad (4.1\text{-}10)$$

Adding the corresponding members of Eqs. (4.1-9) and (4.1-10) we find

$$\left(\frac{\partial \phi}{\partial t}\right)_{\text{coll}} = \frac{1}{4n} \sum_{i,j} \iiiint (\phi_i + \phi_j - \phi_i' - \phi_j')(f_i' f_j' - f_i f_j) g b \, \mathrm{d}b \, \mathrm{d}\varepsilon \, \mathrm{d}^3 c_j \, \mathrm{d}^3 c_i . \quad (4.1\text{-}11)$$

Now, let us multiply both sides of the Boltzmann equation (4.1-1) by ϕ_i and integrate over $\boldsymbol{c}_i$,

$$\int \phi_i \mathfrak{D} f_i \mathrm{d}^3 c_i = n_i (\partial \phi_i / \partial t)_{\text{coll}} . \quad (4.1\text{-}12)$$

The various terms in the integral are easily transformed by integration by parts. Thus we obtain the following result,

$$\frac{\partial (n_i \overline{\phi}_i)}{\partial t} = -\nabla_r \cdot n_i \overline{\boldsymbol{c}_i \phi_i} + n_i \left\{ \overline{\frac{\partial \phi_i}{\partial t}} + \overline{\boldsymbol{c}_i \cdot \nabla_r \phi_i} + \boldsymbol{F}_i \cdot \overline{\nabla_{c_i} \phi_i} + \left(\frac{\partial \phi_i}{\partial t}\right)_{\text{coll}} \right\}, \quad (4.1\text{-}13)$$

where we have used the definition (2.1-7) for the average value of a function of the molecular velocity.

The interpretation of this equation of change is obvious: In Eq. (4.1-13) the rate of change per unit volume at $\boldsymbol{r}$ of the quantity $n_i \overline{\phi}_i$ associated with the ith component of the mixture is analyzed into various rates of change due to (i) the streaming of molecules of species i, (ii) the dependence of ϕ_i on t, $\boldsymbol{r}$, and $\boldsymbol{c}_i$, and (iii) collisions.

Summation over i gives the equation of change for the property ϕ for the entire gas. If Eq. (2.3-4) is used to define the average value of a function of the molecular velocity for the entire gas, one verifies easily that the equation of change is

$$\frac{\partial(n\bar{\phi})}{\partial t} = -\boldsymbol{\nabla}_r \cdot n\overline{\boldsymbol{c}\phi} + n\left\{\overline{\frac{\partial \phi}{\partial t}} + \overline{\boldsymbol{c}\cdot\boldsymbol{\nabla}_r\phi} + \overline{\boldsymbol{F}\cdot\boldsymbol{\nabla}_c\phi} + \left(\frac{\partial\phi}{\partial t}\right)_{\text{coll}}\right\}. \tag{4.1-14}$$

In particular, if ϕ depends on the velocity only, Eqs. (4.1-13) and (4.1-14) simplify to the equations

$$\frac{\partial(n_i\bar{\phi}_i)}{\partial t} = -\boldsymbol{\nabla}_r \cdot n_i\overline{\boldsymbol{c}_i\phi_i} + n_i\left\{\overline{\boldsymbol{F}_i\cdot\boldsymbol{\nabla}_{c_i}\phi_i} + \left(\frac{\partial\phi_i}{\partial t}\right)_{\text{coll}}\right\} \tag{4.1-15}$$

and

$$\frac{\partial(n\bar{\phi})}{\partial t} = -\boldsymbol{\nabla}_r \cdot n\overline{\boldsymbol{c}\phi} + n\left\{\overline{\boldsymbol{F}\cdot\boldsymbol{\nabla}_c\phi} + \left(\frac{\partial\phi}{\partial t}\right)_{\text{coll}}\right\}, \tag{4.1-16}$$

respectively. These *equations of transfer* for the molecular property ϕ were already known to Maxwell [1867]; the generalizations (4.1-13) and (4.1-14) are due to Enskog [1928].

Of particular importance are the equations of transfer for those functions ϕ whose values summed over the molecules involved in a collision do not change during the collision. Such functions are called *collisional invariants* or *summational invariants*. In the case of binary collisions, ψ is a summational invariant if

$$\psi_i(\boldsymbol{c}_i)+\psi_j(\boldsymbol{c}_j) = \psi_i(\boldsymbol{c}_i')+\psi_j(\boldsymbol{c}_j') \tag{4.1-17}$$

for all possible combinations (i, j), where $(\boldsymbol{c}_i, \boldsymbol{c}_j)$ are the velocities of the molecules before the collision and $(\boldsymbol{c}_i', \boldsymbol{c}_j')$ the velocities of the same molecules after the collision. Hence, from Eq. (4.1-11) it follows that for a summational invariant, ψ,

$$(\partial\psi/\partial t)_{\text{coll}} = 0 \tag{4.1-18}$$

and the equation of transfer (4.1-16) becomes the *general conservation equation* for a mixture,

$$\frac{\partial}{\partial t}\sum_i\int\psi_i f_i \mathrm{d}^3c_i + \boldsymbol{\nabla}_r\cdot\sum_i\int\psi_i\boldsymbol{c}_i f_i\mathrm{d}^3c_i = \sum_i \boldsymbol{F}_i\cdot\int f_i\boldsymbol{\nabla}_{c_i}\psi_i\mathrm{d}^3c_i. \tag{4.1-19}$$

The collision process satisfies the laws of conservation of momentum and energy. Moreover, the mass of a molecule does not change during a collision. Hence, five summational invariants are

$$\psi_1 = m,$$
$$\psi_2 = mc_x, \quad \psi_3 = mc_y, \quad \psi_4 = mc_z, \tag{4.1-20}$$
$$\psi_5 = \tfrac{1}{2}mc^2.$$

Of course, any linear combination of these is again a summational invariant. More independent summational invariants may exist, depending on the type of gas – e.g., in the case of polyatomic molecules, angular momentum. However, for molecules whose energy is purely translational there are no further independent summational invariants. This follows immediately from the fact that, in a binary collision, six relations are needed to fix the molecular velocities after the collision in terms of those before the collision. As the dynamics of the collision yield two such relations, only four more independent relations among the velocities can exist. These are exhausted by the momentum (three) and energy conservation relations. Hence, no additional independent relation valid for all collisions is possible.

Now, let us apply the general conservation equation to each of the summational invariants. Taking into account the definitions of the pressure tensor, $\mathbf{P}$, the heat flow vector, $\boldsymbol{q}$, and the specific internal energy, u, we transform the integrals appearing in Eq. (4.1-19):

$$\sum_i \int m_i \boldsymbol{c}_i \boldsymbol{c}_i f_i \mathrm{d}^3 c_i = \sum_i \int m_i (\boldsymbol{C}_i + \boldsymbol{v})(\boldsymbol{C}_i + \boldsymbol{v}) f_i \mathrm{d}^3 c_i = \mathbf{P} + \rho \boldsymbol{v}\boldsymbol{v}, \tag{4.1-21}$$

$$\sum_i \int \tfrac{1}{2} m_i c_i^2 \boldsymbol{c}_i f_i \mathrm{d}^3 c_i = \sum_i \int \tfrac{1}{2} m_i (C_i^2 + 2\boldsymbol{C}_i \cdot \boldsymbol{v} + v^2)(\boldsymbol{C}_i + \boldsymbol{v}) f_i \mathrm{d}^3 c_i$$
$$= \boldsymbol{q} + \rho(u + \tfrac{1}{2}v^2)\boldsymbol{v} + \mathbf{P} \cdot \boldsymbol{v}. \tag{4.1-22}$$

(Recall that $\overline{\boldsymbol{C}}_i = \boldsymbol{V}_i$, where $\boldsymbol{V}_i$ is the diffusion velocity of species i, and that $\sum_i \rho_i \boldsymbol{V}_i = 0$.)

When $\psi = m$, we obtain the conservation equation for mass,

$$\partial \rho / \partial t + \nabla \cdot \rho \boldsymbol{v} = 0. \tag{4.1-23}$$

When $\psi = m\boldsymbol{c}$, we obtain the conservation equation for momentum,

$$\frac{\partial}{\partial t}(\rho \boldsymbol{v}) + \nabla \cdot (\mathbf{P} + \rho \boldsymbol{v}\boldsymbol{v}) = \sum_i \rho_i \boldsymbol{F}_i \tag{4.1-24}$$

or, using the conservation equation for mass,

$$\rho \frac{\partial \boldsymbol{v}}{\partial t} + \rho \boldsymbol{v} \cdot \nabla \boldsymbol{v} + \nabla \cdot \boldsymbol{P} = \sum_i \rho_i \boldsymbol{F}_i. \tag{4.1-25}$$

Finally, when $\psi = \frac{1}{2}mc^2$ we obtain the conservation equation for kinetic energy,

$$\frac{\partial}{\partial t} \rho(u + \tfrac{1}{2}v^2) + \nabla \cdot \{\boldsymbol{q} + \rho(u + \tfrac{1}{2}v^2)\boldsymbol{v} + \boldsymbol{P} \cdot \boldsymbol{v}\} = \sum_i \rho_i \boldsymbol{F}_i \cdot (\boldsymbol{v} + \boldsymbol{V}_i) \tag{4.1-26}$$

or, using the conservation equations for mass and momentum,

$$\rho \frac{\partial u}{\partial t} + \rho \boldsymbol{v} \cdot \nabla u + \nabla \cdot \boldsymbol{q} + \boldsymbol{P} : \nabla \boldsymbol{v} = \sum_i \rho_i \boldsymbol{F}_i \cdot \boldsymbol{V}_i. \tag{4.1-27}$$

If we introduce the concept of *substantial derivative*,

$$\frac{\mathrm{d}}{\mathrm{d}t} = \frac{\partial}{\partial t} + \boldsymbol{v} \cdot \nabla,$$

so that $(\mathrm{d}/\mathrm{d}t)$ represents the time derivative following the motion of the gas, we can rewrite Eqs. (4.1-23), (4.1-25), and (4.1-27) in the following form:

$$\frac{1}{\rho} \frac{\mathrm{d}\rho}{\mathrm{d}t} = -\nabla \cdot \boldsymbol{v}, \tag{4.1-28}$$

$$\rho \frac{\mathrm{d}\boldsymbol{v}}{\mathrm{d}t} = \sum_i \rho_i \boldsymbol{F}_i - \nabla \cdot \boldsymbol{P}, \tag{4.1-29}$$

$$\rho \frac{\mathrm{d}u}{\mathrm{d}t} = -\nabla \cdot \boldsymbol{q} - \boldsymbol{P} : \nabla \boldsymbol{v} + \sum_i \rho_i \boldsymbol{F}_i \cdot \boldsymbol{V}_i. \tag{4.1-30}$$

In this form, the conservation equations are identical with the *equation of continuity*, the *equation of motion* and the *equation of energy* for a fluid.

The physical interpretation of the conservation equations is well known. Let each of them be multiplied by $\mathrm{d}^3r\,\mathrm{d}t$. Then the equation of continuity states that the rate of change of the mass density in a volume element d^3r during $\mathrm{d}t$ is equal to the net mass of the molecules streaming into the volume element. The equation of motion states that the acceleration of the mass in a volume element d^3r during $\mathrm{d}t$ is the sum of (a) the acceleration resulting from external agents, and (b) the acceleration resulting from the internal stresses. Similarly, the equation of energy analyzes the increase of the energy density in d^3r during dt in terms of (a) the net heat flow into the volume element, (b) the work done by the external forces, and (c) the work done by the internal stresses. The new insight that is brought in by kinetic theory is, that the forces associated with internal stresses can be interpreted in terms of the momentum carried by the molecules and, similarly, that the heat flow can be interpreted in terms of the kinetic energy carried by the molecules.

In addition to the conservation equation for mass for the entire gas we may write a conservation equation for the mass of each individual component of the mixture. From Eq. (4.1-8) it is immediately clear that $(\partial \phi_i/\partial t)_{\text{coll}} = 0$ whenever ϕ is independent of $\boldsymbol{c}$. Thus, in the case $\phi = m$, it follows from Eq. (4.1-13) that, for each component of the mixture,

$$\frac{\partial \rho_i}{\partial t} = -\boldsymbol{\nabla} \cdot \rho_i(\boldsymbol{v}+\boldsymbol{V}_i), \quad i = 1, \ldots, K \tag{4.1-31}$$

where $\boldsymbol{V}_i$ is defined by Eq. (2.3-7), or, in terms of substantial derivatives,

$$\frac{1}{\rho_i}\frac{d\rho_i}{dt} = -\boldsymbol{\nabla} \cdot (\boldsymbol{v}+\boldsymbol{V}_i) - \frac{1}{\rho_i}\boldsymbol{V}_i \cdot \boldsymbol{\nabla}\rho_i, \quad i = 1, \ldots, K. \tag{4.1-32}$$

Furthermore, we note that the last term in the equation of energy, Eq. (4.1-30), vanishes whenever the force per unit mass $\boldsymbol{F}$ is the same for all components, $\boldsymbol{F}_i \equiv \boldsymbol{F}$.

The conservation equations for a simple gas are special cases of the conservation equations for a mixture. Since in a simple gas there is no diffusion, we have, instead of Eqs. (4.1-28), (4.1-29), (4.1-30), and (4.1-32), the following set of equations,

$$\frac{1}{\rho}\frac{d\rho}{dt} = -\boldsymbol{\nabla} \cdot \boldsymbol{v}, \tag{4.1-33}$$

$$\rho\frac{d\boldsymbol{v}}{dt} = \rho\boldsymbol{F} - \boldsymbol{\nabla} \cdot \mathbf{P}, \tag{4.1-34}$$

$$\rho\frac{du}{dt} = -(\boldsymbol{\nabla} \cdot \boldsymbol{q} + \mathbf{P} : \boldsymbol{\nabla}\boldsymbol{v}). \tag{4.1-35}$$

These are, again, identical with the equations of mass, motion, and energy for a simple fluid.

Now, what is the significance of this? The fact that the general conservation equations of kinetic theory are identical with the equations of mass, motion, and energy of fluid dynamics implies, first of all, that the kinetic theory definitions of the pressure tensor, heat flow vector and diffusion velocity are, at least, consistent with the usual definitions in fluid dynamics. There is, though, an important difference. In the equations obtained above, the pressure tensor, heat flow vector, and diffusion velocities have been defined in terms of velocity distribution functions – which, at this stage, are still unknown. Therefore, the conservation equations of kinetic theory have only a *formal* meaning. On the other hand, in fluid dynamics the equations of mass, motion, and energy are supplemented by the so-called *constitutive equations*, which relate the internal stresses, heat flow vector and diffusion velocities to gradients of the macroscopic observables (density, velocity, temperature). For example, *Fourier's law* of heat conduction is invoked to relate the heat flow vector to the temperature gradient through the coefficient of thermal conductivity. Similarly, *Newton's law* states that the pressure tensor is proportional to the rate-of-strain tensor, the proportionality constant being the viscosity coefficient of the fluid, and *Fick's law* states the linear relationship between the diffusion velocity and the density gradient through the diffusion coefficient. Of course, fluid dynamics does not say how the proportionality constants (the so-called *transport coefficients*) in the constitutive equations are computed – in fact, their values are generally determined by experiment –, but the important thing to observe is that, together with the constitutive equations, the equations of mass, motion, and energy form a *closed* system: given the initial data, the system can be solved under the appropriate boundary conditions.

Hence, the following potential application of the kinetic theory of gases emerges. Suppose we can solve Boltzmann's equation, either exactly or after well-defined approximations. Then, it is possible to express the flux vectors – i.e., the pressure tensor, heat flow vector and diffusion velocities – in terms of the molecular properties of the gas, check the validity of the constitutive equations, and close the system of conservation equations obtained above. This will indeed be our main goal throughout the remainder of the book.

4.2. Boltzmann's *H*-theorem

Besides the macroscopic conservation equations we can derive another important relation from the Boltzmann equation. It has the form of an inequality and was first derived by Boltzmann [1872], who referred to it as the *E-theorem* (*E* for entropy); later, it became known as the *H-theorem* (*H*, Greek eta). It is the quantitative formulation of the fact that kinetic theory describes a process which is *irreversible* in time. We illustrate this by considering the effect of time reversal upon Boltzmann's equation.

Let f_0 describe the state of a simple gas in phase space at time t_0 and assume that, for any $t > t_0$, the velocity distribution function f is determined from the Boltzmann equation,

$$\frac{\partial f}{\partial t} + \boldsymbol{c} \cdot \nabla_r f + \boldsymbol{F} \cdot \nabla_c f = \iiint (f' f_1' - f f_1) g b \, \mathrm{d}b \, \mathrm{d}\varepsilon \, \mathrm{d}^3 c_1 . \qquad (4.2\text{-}1)$$

Suppose that, at some time t_1 ($t_1 > t_0$, $t_1 - t_0 \gg t_{\text{coll}}$), we stop the system and try to make it run backward by reversing the velocities of the molecules. We can calculate the result by using the Boltzmann equation with the substitutions $\tilde{t} = -t$, $\tilde{\boldsymbol{c}} = -\boldsymbol{c}$ and $\tilde{f}(\boldsymbol{r}, \tilde{\boldsymbol{c}}, \tilde{t}) \equiv f(\boldsymbol{r}, -\boldsymbol{c}, -t)$. Then, as is easily verified, $\tilde{f}$ is determined from the equation

$$-\left(\frac{\partial \tilde{f}}{\partial t} + \tilde{\boldsymbol{c}} \cdot \nabla_r \tilde{f} + \boldsymbol{F} \cdot \nabla_{\tilde{c}} \tilde{f}\right) = \iiint (\tilde{f}' \tilde{f}_1' - \tilde{f} \tilde{f}_1) g b \, \mathrm{d}b \, \mathrm{d}\varepsilon \, \mathrm{d}^3 \tilde{c}_1 . \qquad (4.2\text{-}2)$$

Comparing Eqs. (4.2-1) and (4.2-2) we see that the Boltzmann equation and its time reversed form are different. Hence, as $\tilde{t}$ increases from the initial value $\tilde{t} = -t_1$ to the value $\tilde{t} = -t_0$ the reversed path in phase space is different from the forward path and, consequently, the final state will be different from the original state f_0.

The situation should be compared with that of a system obeying the deterministic laws of mechanics. There, we have reversibility at any instant, i.e., a time reversed system will retrace its trajectory. Clearly, we have introduced "something" that destroys the reversibility of a purely mechanical system. Since, in the derivation of the Boltzmann equation, we have properly accounted for the free motion of particles and the dynamics of binary collisions is correctly treated, this "something" can only be the one serious assumption of the theory – the *Stosszahlansatz*, i.e., the assumption that the molecules involved in a collision are uncorrelated as they start interacting. Thus was a distinction between the past and the future built into the Boltzmann equation.

Now, if there is irreversibility there are two possibilities: a velocity distribution function which describes the state of a gas in phase space (μ-space) either keeps changing indefinitely as $t \to \infty$, or, in some way, it approaches a limit function. If the velocity distribution function approaches a limit function, the state of the gas must develop into an equilibrium state. The question we are concerned with in the present section is, whether there exists a quantity which, on the kinetic time scale, "measures" the approach to equilibrium.

Consider a gas of N identical molecules. The gas occupies a finite volume, X say, in the six-dimensional phase space. Let the volume X be partitioned into I elements of equal volume Δx and let N_i be the number of molecules in the ith element, the center of which is at $x_i = (\boldsymbol{r}_i, \boldsymbol{c}_i)$, $i = 1, \ldots, I$. Of course, if $f(\boldsymbol{r}, \boldsymbol{c})$ is the velocity distribution of the gas at a fixed time t, $N_i \approx f(\boldsymbol{r}_i, \boldsymbol{c}_i)\Delta x$, so each velocity distribution function corresponds to a set of *occupation numbers* $(N_1, \ldots, N_I)$ in the discretized phase space. Now, an alternative to the description in phase space is provided in the $6N$-dimensional Γ-space, where the state of a gas is represented by a single point. The set of states which defines the particular set of occupation numbers $(N_1, \ldots, N_I)$ corresponds to a volume element in Γ-space and, without going into the technical details, we may say that the probability of finding a certain set of occupation numbers and, thence, a certain distribution function f in phase space is given by the measure of the volume element corresponding to it in Γ-space. We now turn to a computation of this measure.

If the phases of the N molecules are changed so that each molecule still remains in its element of volume Δx, then the point representing the system in Γ-space moves around in an element of volume $(\Delta x)^N$. Furthermore, we can interchange the N identical molecules in $N!$ different permutations, each of which gives a different point in Γ-space with its associated element of volume $(\Delta x)^N$. However, if only molecules belonging to the same group of N_i molecules in the element around x_i are interchanged, then the resulting change in Γ-space is performed within the volume $(\Delta x)^N$ and, therefore, the number of permutations among the N molecules, which give rise to *distinguishable* elements of volume $(\Delta x)^N$ in Γ-space is equal to $N!/(N_1! \ldots N_I!)$. Thus, the total volume in Γ-space corresponding to the set of occupation numbers $(N_1, \ldots, N_I)$ in phase space is given by

$$P_N = \frac{N!}{N_1! \ldots N_I!} (\Delta x)^N, \tag{4.2-3}$$

whence

$$\log P_N = \log N! + N \log \Delta x - \sum_i \log N_i! . \tag{4.2-4}$$

As $N \to \infty$, we have Stirling's formula – see e.g., Whittaker and Watson [1915] –

$$N! = N^{N+\frac{1}{2}} e^{-N} \sqrt{2\pi}[1 + O(N^{-1})], \tag{4.2-5}$$

so for very large N,

$$\log P_N \approx N \log N - N + N \log \Delta x - \sum_i N_i \log N_i + \sum_i N_i \tag{4.2-6}$$

or, since $\sum_i N_i = N$,

$$\log P_N \approx -\sum_i N_i \log (N_i/N\,\Delta x). \tag{4.2-7}$$

Expressing N_i in terms of the distribution function and letting Δx become the infinitesimal element $\mathrm{d}x = \mathrm{d}^3 r \mathrm{d}^3 c$ we obtain for the logarithm of the probability $P(f)$ of the given distribution f,

$$\log P(f) = -\iint f \log f \,\mathrm{d}^3 r\, \mathrm{d}^3 c. \tag{4.2-8}$$

Thus, to each state f we can assign a (real) number

$$H(f) = \iint f \log f \,\mathrm{d}^3 r\, \mathrm{d}^3 c \tag{4.2-9}$$

which is a measure for the negative of the logarithm of the probability of that state. In kinetic theory it is customary to call the quantity H defined in Eq. (4.2-9) the *Boltzmann H-function*; more properly, it should be called the H-functional. In the remainder of this section and in Section 4.4 we will show that the Boltzmann H-function measures the approach to equilibrium and is a generalization of the concept of entropy of equilibrium statistical mechanics.

Consider a gas mixture in which the distribution functions f_i are *independent of* $\mathbf{r}$ and *no external forces* act upon the molecules. The Boltzmann equation (3.1-28) for each f_i reads

$$\frac{\partial f_i}{\partial t} = \sum_j \iiint (f_i' f_j' - f_i f_j) g b \,\mathrm{d}b\, \mathrm{d}\varepsilon\, \mathrm{d}^3 c_j . \tag{4.2-10}$$

Define the integral

$$H = \sum_i \int f_i \log f_i \,\mathrm{d}^3 c_i , \tag{4.2-11}$$

where the sum extends over all constituents of the mixture. Derivation with respect to t yields

$$\frac{\mathrm{d}H}{\mathrm{d}t} = \sum_i \int (\log f_i + 1) \frac{\partial f_i}{\partial t} \mathrm{d}^3 c_i, \tag{4.2-12}$$

which becomes after substitution of $\partial f_i / \partial t$ from Eq. (4.2-10)

$$\frac{\mathrm{d}H}{\mathrm{d}t} = \sum_{i,j} \int (\log f_i + 1) J(f_i f_j) \mathrm{d}^3 c_i. \tag{4.2-13}$$

Now, from Eqs. (4.1-5) and (4.1-11) it follows with $\phi = \log f + 1$, that

$$\sum_{i,j} \int (\log f_i + 1) J(f_i f_j) \mathrm{d}^3 c_i$$
$$= -\tfrac{1}{4} \sum_{i,j} \int\int\int\int (f_i' f_j' - f_i f_j)(\log f_i' f_j' - \log f_i f_j) g b \, \mathrm{d}b \, \mathrm{d}\varepsilon \, \mathrm{d}^3 c_i \, \mathrm{d}^3 c_j. \tag{4.2-14}$$

Since the expression $(x-y)(\log x - \log y)$ is always positive except for $x = y$, when it vanishes ($x > y$ implies $x-y > 0$ and $\log x - \log y > 0$, $x < y$ implies $x-y < 0$ and $\log x - \log y < 0$), it follows that $\mathrm{d}H/\mathrm{d}t$ is always negative except if $f_i f_j = f_i' f_j'$ for all $\mathbf{c}_i$, $\mathbf{c}_j$, when it vanishes. In other words, in the case of a uniform gas, when no external forces act upon the molecules, the function H defined in Eq. (4.2-11) obeys the differential inequality

$$\mathrm{d}H/\mathrm{d}t \leqq 0, \tag{4.2-15}$$

i.e., H is a non-increasing function of time. This is Boltzmann's *H-theorem.* It is equivalent to the second law of thermodynamics which states that entropy cannot decrease. It is, in fact, more general inasmuch as the quantity H is defined for *all* systems, whereas the entropy is defined only at equilibrium; however, the H-theorem has been proven only for dilute gases, whereas the second law applies to *any* system in equilibrium.

If H is a non-increasing function of time, one must have either $H \to -\infty$ as $t \to \infty$ or H is bounded from below. The former case can occur if and only if, for some i, the integral $\int f_i \log f_i \, \mathrm{d}^3 c_i$ fails to converge as $t \to \infty$. Now, as $c_i \to \infty$, $f_i \to 0$ and $\log f_i \to -\infty$. On the other hand, the total energy of translation of the molecules of species i, $\int \frac{1}{2} m_i c_i^2 f_i \mathrm{d}^3 c_i$, is necessarily a finite quantity for all values of t. Hence, the integral for H may fail to converge if and only if, for some i, $\log f_i$ tends to $-\infty$ more rapidly than $-c_i^2$. But then f_i tends to zero more rapidly than $\exp(-c_i^2)$ and when this occurs the integral for H certainly converges. Thus it follows that H is bounded from below.

As $t \to \infty$, H will tend to a finite limit, corresponding to a state of the gas in which $\mathrm{d}H/\mathrm{d}t = 0$. As shown above, one then has, for all $\boldsymbol{c}_i$, $\boldsymbol{c}_j$,

$$f_i f_j = f_i' f_j' \tag{4.2-16}$$

or, equivalently,

$$\log f_i + \log f_j = \log f_i' + \log f_j'. \tag{4.2-17}$$

In other words, if $\mathrm{d}H/\mathrm{d}t = 0$, then the state of the gas is such that $\log f$ is a summational invariant. Conversely, if the state of the gas is such that $\log f$ is a summational invariant, then one has for all i, according to Eq. (4.2-10), $\partial f_i/\partial t = 0$ and, according to Eq. (4.2-12), $\mathrm{d}H/\mathrm{d}t = 0$.

Now, consider the equilibrium form of Eq. (4.2-10),

$$\sum_j \iiint (f_i' f_j' - f_i f_j) g b \,\mathrm{d}b \,\mathrm{d}\varepsilon \,\mathrm{d}^3 c_j = 0 \quad \text{for all } i. \tag{4.2-18}$$

From the arguments presented in the preceding paragraph it follows that Eq. (4.2-18) admits at least one set of solutions, viz., the set for which $\log f$ is a summational invariant. It is easy to show that this is the only set of solutions. For any other set, $\log f$ is, by definition, not a summational invariant and, hence, $\mathrm{d}H/\mathrm{d}t$ does not vanish. But then, from Eq. (4.2-12), at least one of the partial derivatives $\partial f_i/\partial t$ must be non-zero in some finite domain of velocity space, which is in contradiction with Eq. (4.2-18). Therefore, Eq. (4.2-18) admits one and only one set of solutions f, which obey the relations (4.2-17).

If $\log f$ is a summational invariant, it must be a linear combination of the known summational invariants, so that

$$\log f_i = m_i \alpha_i^{(1)} + m_i \boldsymbol{c}_i \cdot \boldsymbol{\alpha}_i^{(2)} + \tfrac{1}{2} m_i c_i^2 \alpha^{(3)}. \tag{4.2-19}$$

The coefficients $\alpha_i^{(1)}$, $\boldsymbol{\alpha}_i^{(2)}$, $\alpha_i^{(3)}$ must be independent of $\boldsymbol{r}$ and t, since the state of the gas is uniform and steady. Also, to satisfy the equations of conservation of momentum and energy in collisions of unlike molecules one has $\boldsymbol{\alpha}_i^{(2)} = \boldsymbol{\alpha}_j^{(2)} = \boldsymbol{\alpha}^{(2)}$ say, and $\alpha_i^{(3)} = \alpha_j^{(3)} = \alpha^{(3)}$ for all i, j. Thus, setting $\boldsymbol{C}_i' = \boldsymbol{c}_i - \boldsymbol{\alpha}^{(2)}/\alpha^{(3)}$, we have

$$f_i = \alpha_i^{(0)} \exp\left(-\tfrac{1}{2} m_i C_i'^2 \alpha^{(3)}\right), \tag{4.2-20}$$

where $\alpha_i^{(0)} = \exp\left(m_i \alpha_i^{(1)} + \tfrac{1}{2} m_i \boldsymbol{\alpha}^{(2)} \cdot \boldsymbol{\alpha}^{(2)}/\alpha^{(3)}\right)$ is a new constant. The constants $\alpha_i^{(0)}$, $\boldsymbol{\alpha}^{(2)}$ and $\alpha^{(3)}$ can be evaluated in terms of the number density n_i, the hydrodynamic velocity, $\boldsymbol{v}$, and the kinetic temperature T. First,

$$n_i = \int f_i \, \mathrm{d}^3 c_i = \int f_i \, \mathrm{d}^3 C_i' = \alpha_i^{(0)} \int \exp\left(-\tfrac{1}{2} m_i C_i'^2 \alpha^{(3)}\right) \mathrm{d}^3 C_i'. \quad (4.2\text{-}21)$$

Using polar coordinates (C_i', θ, φ) one readily obtains

$$n_i = \alpha_i^{(0)} (2\pi / m_i \alpha^{(3)})^{\frac{3}{2}}. \quad (4.2\text{-}22)$$

Second, from the definition of the hydrodynamic velocity,

$$\rho \boldsymbol{v} = \sum_i \int m_i \boldsymbol{c}_i f_i \, \mathrm{d}^3 c_i = \sum_i m_i \left[\int \boldsymbol{C}_i' f_i \, \mathrm{d}^3 C_i' + \frac{\boldsymbol{\alpha}^{(2)}}{\alpha^{(3)}} \int f_i \, \mathrm{d}^3 C_i' \right]. \quad (4.2\text{-}23)$$

The first term between the brackets vanishes because the integrand is an odd function of the components of $\boldsymbol{C}_i'$. The last term between the brackets can be evaluated from Eqs. (4.2-21) and (4.2-22). The result is

$$\boldsymbol{v} = \boldsymbol{\alpha}^{(2)} / \alpha^{(3)}, \quad (4.2\text{-}24)$$

so

$$\boldsymbol{C}_i' = \boldsymbol{c}_i - \boldsymbol{v} = \boldsymbol{C}_i, \quad (4.2\text{-}25)$$

where $\boldsymbol{C}_i$ is the peculiar velocity. Finally, with the definition of the kinetic temperature,

$$\tfrac{3}{2} kT = \frac{1}{n} \sum_i \int \tfrac{1}{2} m_i C_i^2 f_i \, \mathrm{d}^3 c_i = \frac{1}{2n} \sum_i \int m_i C_i^2 f_i \, \mathrm{d}^3 C_i. \quad (4.2\text{-}26)$$

Again, introducing polar coordinates (C_i, θ, φ) and using Eq. (4.2-22) to eliminate $\alpha_i^{(0)}$, one obtains $\frac{3}{2} kT = 3/2\alpha^{(3)}$, hence

$$\alpha^{(3)} = 1/kT. \quad (4.2\text{-}27)$$

Consequently, the expression (4.2-20) is equivalent to

$$f_i(\boldsymbol{c}_i) = n_i (m_i / 2\pi kT)^{\frac{3}{2}} \exp\left(-m_i C_i^2 / 2kT\right) \quad (4.2\text{-}28)$$

with $\boldsymbol{C}_i = \boldsymbol{c}_i - \boldsymbol{v}$, which is the usual form of the *velocity distribution function at thermodynamic equilibrium*; the expression (4.2-28) was first derived by Maxwell [1860]. In the following we shall use the subscript M to denote the Maxwellian velocity distribution function (4.2-28).

Two observations are due. First, the result (4.2-27) that the mean kinetic energies of peculiar motion of molecules of the different constituents are the same and equal to $\frac{3}{2} kT$, is a special case of the statistical-mechanical theorem of *equipartition* of energy. Second, a simple computation shows that, for a gas in the uniform steady state the quantity H can be expressed as

$$H_M = \sum_i n_i[\log n_i + \tfrac{3}{2}\{\log(m_i/2\pi kT) - 1\}], \tag{4.2-29}$$

which becomes, in the case of a simple gas,

$$H_M = n[\log(n/T^{\frac{3}{2}}) + \tfrac{3}{2}\{\log(m/2\pi k) - 1\}]. \tag{4.2-30}$$

Comparing this expression with the thermodynamic expression for the *entropy* S of an ideal gas, which is equal to $-n[k \log(n/T^{\frac{3}{2}}) + \text{const.}]$, we find

$$S = -kH_M + \text{constant}; \tag{4.2-31}$$

the constant merely reflects the arbitrariness of the zero point of entropy. Likewise, Eq. (4.2-29) is identical to $-1/k$ times the expression for the entropy of an ideal gas mixture.

To summarize, we have shown that there exists a functional H which assigns to each velocity distribution function f a real number $H(f)$ measuring (the negative of) the logarithm of its probability. If, in a uniform state, the gas evolves according to Boltzmann's equation, then H is a decreasing function of time. That is, if f_1 and f_2 describe the state of the gas in phase space (μ-space) at times t_1 and t_2 respectively, then the volume of the region in Γ-space corresponding to f_2 is larger than the volume of the region in Γ-space corresponding to f_1 whenever $t_2 > t_1$. As $t \to \infty$, the distribution function will approach a Maxwellian and, when equilibrium prevails, the functional H measures the entropy of the gas.

Recently, the application of the principles of information theory to problems of physics has led to a generalization of the statistical mechanics of Gibbs – e.g., see Katz [1967] and Jaynes [1967] – and to a new interpretation of Boltzmann's H-theorem. From information theory it is inferred that Boltzmann's H-function measures the amount of uncertainty as to the true location of the system in phase space and, thus, refers to our *information* about the *microscopic* state of a physical system. If one would follow the motions of the particles according to the laws of classical mechanics, no information on the microscopic level would ever be lost since the laws of classical mechanics are reversible. However, kinetic theory offers a contracted description of the system and, as $t \to +\infty$, any information about the microscopic state of the system is gradually lost due to the molecular interactions. Maximum uncertainty about the microscopic state prevails when the system reaches equilibrium.

4.3. Maxwell's velocity distribution function

Consider a simple gas in the uniform steady state. Its velocity distribution function is, according to Eq. (4.2-28),

$$f_{\mathrm{M}} = n(m/2\pi kT)^{\frac{3}{2}} \exp(-mC^2/2kT), \tag{4.3-1}$$

where $\boldsymbol{C}$ is the peculiar velocity, $\boldsymbol{C} = \boldsymbol{c} - \boldsymbol{v}$. Since $C^2 = C_x^2 + C_y^2 + C_z^2$, the exponential can be written as a product of three exponentials, each involving only one component of $\boldsymbol{C}$. Thus the distribution function of any component of the peculiar velocity is independent of the values of the remaining two components.

The magnitude of any component of the molecular velocity $\boldsymbol{c}$ is normally distributed about its mean value with variance kT/m.

In a polar coordinate system (C, θ, φ) one has $\mathrm{d}^3c = \mathrm{d}^3C = C^2\mathrm{d}C \times \sin\theta\,\mathrm{d}\theta\,\mathrm{d}\varphi$ and, integrating f with respect to θ and φ one finds the number of molecules per unit volume whose peculiar speeds lie in the range $\mathrm{d}C$ about C, viz., $(2/\pi)^{\frac{1}{2}} n(m/kT)^{\frac{3}{2}} C^2 \exp(-mC^2/2kT)$.

The mean value $\bar{\phi}$ of any molecular property ϕ for a gas in the uniform steady state is given by the expression

$$\bar{\phi} = (m/2\pi kT)^{\frac{3}{2}} \int \phi \exp(-mC^2/2kT)\,\mathrm{d}^3C, \tag{4.3-2}$$

in accordance with the definition (2.1-7). From Eq. (4.3-2) one verifies that the mean value of a function ϕ which is odd in any component of $\boldsymbol{C}$, vanishes. This implies that, in a gas in the uniform steady state the shear stresses are zero; also, since the Maxwellian distribution function is isotropic, the normal pressures are equal, so that the pressure system is indeed hydrostatic (cf. Section 2.2).

From the expression (4.3-2) above it is obvious that integrals of the type

$$I_r = (m/2\pi kT)^{\frac{3}{2}} \int C^r \exp(-mC^2/2kT)\,\mathrm{d}^3C \tag{4.3-3}$$

with $r = 0, 1, 2, \ldots$ play an important role in kinetic theory. They can be evaluated explicitly; the result is

$$I_r = (2/\sqrt{\pi})(2kT/m)^{\frac{1}{2}r}\Gamma(\tfrac{1}{2}r + \tfrac{3}{2}). \tag{4.3-4}$$

Thus one finds for the average values of C and C^2:

$$\bar{C} = (8kT/\pi m)^{\frac{1}{2}}, \tag{4.3-5}$$

$$\overline{C^2} = 3kT/m, \tag{4.3-6}$$

which implies that the root-mean-square of the peculiar speed, $(\overline{C^2})^{\frac{1}{2}}$ is slightly larger than the mean peculiar speed, the ratio being equal to $\sqrt{(\frac{3}{8}\pi)} = 1.0854$. From Eq. (4.3-6) one obtains the pressure at thermal equilibrium

$$p = \tfrac{1}{3}\rho\overline{C^2} = nkT, \tag{4.3-7}$$

i.e., the ideal gas law.

4.4. Boundary conditions

For a complete description of a problem of kinetic theory the Boltzmann equation must be supplemented by the correct initial and/or boundary conditions. A rigorous analysis of the conditions under which existence and uniqueness of a solution can be guaranteed is extremely difficult and has, as yet, not been given for the general case. This question will be discussed later in more detail and from a slightly different point of view after a study of methods for actually solving the Boltzmann equation, see Section 5.10. In the present section, we give a somewhat heuristic discussion of the boundary conditions that one expects to go with the Boltzmann equation in a bounded medium. The reader who desires some insight into the problems related to existence and uniqueness proofs may consult the references cited in Section 5.10.

The collision operator acts only on the velocities of the particles, not on the space or time variables. It is therefore reasonable to expect that the initial and boundary conditions that must be imposed on the distribution function will be governed by the streaming operator, i.e., they will be the same as those which must be used in conjunction with the *collisionless* Boltzmann equation. For a bounded system, one can show without much difficulty that a unique solution to the collisionless Boltzmann equation exists if the distribution function is specified at some particular initial time, say $t = t_0$, for all $\boldsymbol{r}$ in the system and all $\boldsymbol{c}$ and if, for $t > t_0$, the *incoming* distribution at the surface is specified as a function of $\boldsymbol{c}$ and t; the boundary condition may be generalized, for example, to allow the incoming distribution to be specified in terms of the outgoing distribution. An indication that these are indeed the appropriate boundary conditions to apply to the *full* Boltzmann equation is provided by the Boltzmann equation with a *linearized* collision operator (cf. Section 4.8): in this case the conditions described above are indeed sufficient to guarantee the existence of a unique solution in the large. By analogy and for lack of information to the contrary, we shall henceforth assume that they are also appropriate for the full non-linear Boltzmann equation; we emphasize, however, that this has not been proven.

In principle, the boundary conditions can be determined by solving the problem of the interaction of a gas particle with the conglomeration of particles comprising the wall, or they can be found by direct experimental measurement using molecular beams. However, both the calculations and the experiments are extremely difficult and data are available only for a few gas-surface material combinations. For that reason, one often uses a *model* of the boundary condition that is analytically tractable and consistent with what is known about the true boundary condition.

If the gas is contained in a vessel V bounded by a solid surface S, the boundary condition should determine, at any position $\boldsymbol{r} \in S$ and at any instant t, the velocity distribution of the molecules leaving the surface in terms of the velocity distribution of the molecules incident on the surface; it is assumed that the molecules do not adhere to the surface for any appreciable time, so that the variables $\boldsymbol{r}$ and t can be considered as parameters. Since, at any time, relatively few gas molecules adhere to the surface, the interaction is almost entirely with the wall material and it is reasonable to expect that the boundary condition can be described by a (linear) functional relation of the form

$$f^+ = \mathfrak{R} f^- \quad \text{on } S, \tag{4.4-1}$$

where $f^+ = f$ for $\boldsymbol{c} \cdot \boldsymbol{n} > 0$, $f^+ = 0$ for $\boldsymbol{c} \cdot \boldsymbol{n} < 0$, and $f^- = f$ for $\boldsymbol{c} \cdot \boldsymbol{n} < 0$, $f^- = 0$ for $\boldsymbol{c} \cdot \boldsymbol{n} > 0$. Here $\boldsymbol{n}$ is the unit vector which is locally normal to S, oriented toward the interior of V. The relation (4.4-1) is, usually, an integral relation with a kernel which may be singular. It is convenient to write, instead of Eq. (4.4-1)

$$f^+(\boldsymbol{c}) = -\int \frac{\boldsymbol{c}' \cdot \boldsymbol{n}}{\boldsymbol{c} \cdot \boldsymbol{n}} W(\boldsymbol{c}, \boldsymbol{c}') f^-(\boldsymbol{c}') \mathrm{d}^3 c' \quad \text{on } S, \tag{4.4-2}$$

where $W(\boldsymbol{c}, \boldsymbol{c}')\mathrm{d}^3 c$ is the (non-negative) probability that a molecule with velocity $\boldsymbol{c}'$ will be reflected into the range $\mathrm{d}^3 c$ about $\boldsymbol{c}$. The kernel W must be compatible with the following physical observations.

(i) If the boundary surface is not a source or sink for the gas molecules, then, under steady state conditions, it follows that for any $\boldsymbol{c}'$,

$$\int_{\boldsymbol{c} \cdot \boldsymbol{n} > 0} W(\boldsymbol{c}, \boldsymbol{c}') \mathrm{d}^3 c = 1 \tag{4.4-3}$$

i.e., regardless of the element of velocity space from which the incident molecule comes, it must be re-emitted.

(ii) If the distribution of the incident molecules is locally Maxwellian with temperature equal to the local temperature of the surface, then the distribution of the reflected molecules should again be locally Maxwellian with the same temperature. Thus, for any $\boldsymbol{c}$ with $\boldsymbol{c} \cdot \boldsymbol{n} > 0$, W should satisfy the relation

$$\exp(-mc^2/2kT_w) = \int_{\boldsymbol{c}' \cdot \boldsymbol{n} < 0} \frac{-(\boldsymbol{c}' \cdot \boldsymbol{n})}{\boldsymbol{c} \cdot \boldsymbol{n}} W(\boldsymbol{c}, \boldsymbol{c}') \exp(-mc'^2/2kT_w) \mathrm{d}^3c'; \tag{4.4-4}$$

here m is the mass of the gas molecule, T_w the temperature of the wall. The condition (4.4-4) is sometimes referred to as the *thermostat condition.*

Since very little is known about gas-surface interactions, the function W is not in general known. Lacking this information the best one can do is to construct a model of the function W consistent with Eqs. (4.4-3) and (4.4-4) and other information available concerning the interaction. This question will be discussed in greater detail in Chapter 15 in connection with the kinetic theory of rarefied gases.

4.5. The H-theorem for bounded systems

In Section 4.2 we proved the H-theorem for a spatially uniform gas with no external forces acting upon the molecules. Now, we shall drop the restrictive assumptions and show that a similar theorem holds in the case of a gas contained in a stationary vessel V bounded by a surface S, if there is no net energy transfer from the gas to the boundary surface. The proof of this theorem, in the form presented here, was first given by Darrozès and Guiraud [1966].

Consider the integral

$$H(t) = \int_V H(\boldsymbol{r}, t) \mathrm{d}^3r = \int_V \sum_i \int f_i \log f_i \mathrm{d}^3c_i \mathrm{d}^3r, \tag{4.5-1}$$

which is the immediate generalization of the definition (4.2-11) to the case of a gas mixture. Derivation with respect to time yields

$$\frac{\mathrm{d}H(t)}{\mathrm{d}t} = \int_V \frac{\partial H(\boldsymbol{r}, t)}{\partial t} \mathrm{d}^3r \tag{4.5-2}$$

and this expression can be transformed with the aid of Boltzmann's equation, as in Section 4.2. When Green's theorem is used, the result is

$$\frac{\mathrm{d}H}{\mathrm{d}t} = \int_S \mathrm{d}^2S \sum_i \int (\boldsymbol{c}_i \cdot \boldsymbol{n}) f_i \log f_i \mathrm{d}^3 c_i - \int_V \mathrm{d}^3 r \sum_i \int \boldsymbol{F}_i \cdot \boldsymbol{\nabla}_{c_i} f_i \log f_i \mathrm{d}^3 c_i$$
$$+ \int_V \mathrm{d}^3 r \sum_{i,j} \int (\log f_i + 1) J(f_i f_j) \mathrm{d}^3 c_i . \quad (4.5\text{-}3)$$

As external forces are assumed to be independent of the molecular velocity and the velocity distribution function of each species is assumed to vanish sufficiently rapidly as $c \to \infty$, the second term of the right member is identically zero.

With respect to the last term in Eq. (4.5-3), from Eqs. (4.1-5) and (4.1-11) one deduces, taking $\phi = \log f + 1$,

$$\int_V \mathrm{d}^3 r \sum_{i,j} \int (\log f_i + 1) J(f_i f_j) \mathrm{d}^3 c_i$$
$$= -\tfrac{1}{4} \int_V \mathrm{d}^3 r \sum_{i,j} \iiiint (f_i' f_j' - f_i f_j)(\log f_i' f_j' - \log f_i f_j) g b \, \mathrm{d}b \, \mathrm{d}\varepsilon \, \mathrm{d}^3 c_j \mathrm{d}^3 c_i$$
$$= -\tfrac{1}{4} \int_V \sum_{i,j} \Delta(f_i f_j) \mathrm{d}^3 r, \quad (4.5\text{-}4)$$

where $\Delta(f_i f_j)$ is positive except if $f_i f_j = f_i' f_j'$ for all $\boldsymbol{c}_i$, $\boldsymbol{c}_j$, in which case it vanishes.

The first term of the right member of Eq. (4.5-3) expresses the rate of change of H due to interaction with the boundary surface and, thus, depends on the boundary conditions prescribed. As we have seen in the previous section, the boundary conditions on S are, generally, expressed in terms of a linear operator $\mathfrak{K}$, cf. Eq. (4.4-1), or in terms of a transition probability W, cf. Eq. (4.4-2). In considering the boundary term, it is convenient to write, for each constituent of the gas mixture and for $\boldsymbol{r} \in S$,

$$f = \exp(-mc^2/2kT_{\mathrm{w}})(1+\Phi), \quad (4.5\text{-}5)$$

where T_{w} is the local temperature of the wall, and to rewrite the general boundary condition (4.4-1) in terms of Φ and a new linear operator $\mathfrak{H}$,

$$\Phi^+ = \mathfrak{H}\Phi^- \quad \text{on } S. \quad (4.5\text{-}6)$$

One verifies easily that, in terms of the function W, this can be written as

$$\Phi^+(\boldsymbol{c}) = \int \frac{-(\boldsymbol{c}'\cdot\boldsymbol{n})\exp(-mc'^2/2kT_\mathrm{w})}{(\boldsymbol{c}\cdot\boldsymbol{n})\exp(-mc^2/2kT_\mathrm{w})} W(\boldsymbol{c},\boldsymbol{c}')\Phi^-(\boldsymbol{c}')\,\mathrm{d}^3c' \quad \text{on } S. \tag{4.5-7}$$

The properties (4.4-3) and (4.4-4) of W lead to the following two properties of the operator $\mathfrak{H}$:

$$1^+ = \mathfrak{H}1^- \tag{4.5-8}$$

and, for any Φ,

$$\int (\boldsymbol{c}\cdot\boldsymbol{n})\exp(-mc^2/2kT_\mathrm{w})\{\Phi^- - \widetilde{\mathfrak{H}\Phi^-}\}\,\mathrm{d}^3c = 0. \tag{4.5-9}$$

The symbol $\sim$ above a function denotes the function obtained by replacing the argument $\boldsymbol{c}$ by $\boldsymbol{c}-2(\boldsymbol{c}\cdot\boldsymbol{n})\boldsymbol{n}$; it is a transformation which takes a function of type $+$ into one of type $-$, and vice versa.

Now, consider the surface integral in expression (4.5-3). Define, for $\Phi \geqq -1$,

$$I(\Phi) = (1+\Phi)\log(1+\Phi) - \Phi. \tag{4.5-10}$$

For every constituent of the mixture, one has the identity

$$\int (\boldsymbol{c}\cdot\boldsymbol{n}) f \log f\,\mathrm{d}^3c$$
$$= -(1/kT_\mathrm{w})\int \tfrac{1}{2}mc^2(\boldsymbol{c}\cdot\boldsymbol{n})f\,\mathrm{d}^3c + \int (\boldsymbol{c}\cdot\boldsymbol{n})\exp(-mc^2/2kT_\mathrm{w})I(\Phi)\,\mathrm{d}^3c, \tag{4.5-11}$$

where the relation (4.5-9) has been used; the integrals are over all $\boldsymbol{c}$. The integral in the first term of the right member is equal to the normal component of the heat flow vector at the wall, so that

$$-(1/kT_\mathrm{w})\int \tfrac{1}{2}mc^2(\boldsymbol{c}\cdot\boldsymbol{n})f\,\mathrm{d}^3c = -(\boldsymbol{q}\cdot\boldsymbol{n})/kT_\mathrm{w}. \tag{4.5-12}$$

Since $S = -kH$, this is the well-known thermodynamic entropy transfer from the wall. The second term of the right member of Eq. (4.5-11) can be written as

$$\int_{\text{all } \boldsymbol{c}} (\boldsymbol{c}\cdot\boldsymbol{n})\exp(-mc^2/2kT_\mathrm{w})I(\Phi)\,\mathrm{d}^3c$$
$$= \int_{\boldsymbol{c}\cdot\boldsymbol{n}<0} (\boldsymbol{c}\cdot\boldsymbol{n})\exp(-mc^2/2kT_\mathrm{w})\{I(\Phi^-) - I(\widetilde{\mathfrak{H}\Phi^-})\}\,\mathrm{d}^3c. \tag{4.5-13}$$

If the identity (4.5-9) – which holds for *any* function Φ – is applied to $I(\Phi)$

and the result is used in Eq. (4.5-13), the latter yields

$$\int_{\text{all } \boldsymbol{c}} (\boldsymbol{c}\cdot\boldsymbol{n})\exp(-mc^2/2kT_{\mathrm{w}})\,I(\Phi)\,\mathrm{d}^3c$$

$$= -\int_{\boldsymbol{c}\cdot\boldsymbol{n}>0} (\boldsymbol{c}\cdot\boldsymbol{n})\exp(-mc^2/2kT_{\mathrm{w}})\,\Sigma(\Phi^-)\,\mathrm{d}^3c \quad (4.5\text{-}14)$$

with

$$\Sigma(\Phi^-) = \mathfrak{H}I(\Phi^-) - I(\mathfrak{H}\Phi^-). \quad (4.5\text{-}15)$$

From the definition (4.5-10) one may verify that the function $I(x)$, $x \geqq -1$, is *convex*, i.e., for every pair of points P_1, P_2 on the curve $y = I(x)$ the points on the arc P_1P_2 lie below the chord P_1P_2. Then *Jensen's inequality* for Stieltjes integrals can be used, which, for any convex function I defined on a set Ω^* and any function f defined on a measurable set Ω (with measure ω) which maps Ω into Ω^*, takes the form

$$I\left\{\int_\Omega f\,\mathrm{d}\omega \Big/ \int_\Omega \mathrm{d}\omega\right\} \leqq \int_\Omega I(f)\,\mathrm{d}\omega \Big/ \int_\Omega \mathrm{d}\omega, \quad (4.5\text{-}16)$$

if all the integrals in question exist. A proof of this inequality can be found in the literature, e.g., Zygmund [1959]. Taking for Ω the set of all $\boldsymbol{c}'$ with $\boldsymbol{c}'\cdot\boldsymbol{n} < 0$, defining the measure ω by

$$\mathrm{d}\omega = -(\boldsymbol{c}'\cdot\boldsymbol{n})\exp(-mc'^2/2kT_{\mathrm{w}})\,W(\boldsymbol{c},\boldsymbol{c}')\,\mathrm{d}^3c' \quad (4.5\text{-}17)$$

(since W may be singular, the integrals in the inequality (4.5-16) must be treated as Stieltjes integrals) and using the property (4.5-8) of the operator $\mathfrak{H}$, we deduce from Jensen's inequality that, for every value of $\boldsymbol{c}$ with $\boldsymbol{c}\cdot\boldsymbol{n} > 0$,

$$I(\mathfrak{H}\Phi^-)(\boldsymbol{c}) \leqq \int_{\boldsymbol{c}'\cdot\boldsymbol{n}<0} \frac{-(\boldsymbol{c}'\cdot\boldsymbol{n})\exp(-mc'^2/2kT_{\mathrm{w}})}{(\boldsymbol{c}\cdot\boldsymbol{n})\exp(-mc^2/2kT_{\mathrm{w}})}\,W(\boldsymbol{c},\boldsymbol{c}')\,I(\Phi^-(\boldsymbol{c}'))\,\mathrm{d}^3c'. \quad (4.5\text{-}18)$$

The right member of this inequality is exactly equal to $\mathfrak{H}I(\Phi^-)(\boldsymbol{c})$, so that we have proved that, for every value of $\boldsymbol{c}$ with $\boldsymbol{c}\cdot\boldsymbol{n} > 0$,

$$\Sigma(\Phi^-) \geqq 0. \quad (4.5\text{-}19)$$

It is easily verified that the equality sign holds if and only if Φ is constant. Thus, from Eq. (4.5-14),

$$\int (\boldsymbol{c}\cdot\boldsymbol{n})\exp(-mc^2/2kT_{\mathrm{w}})\,I(\Phi)\,\mathrm{d}^3c = -\delta(f), \quad (4.5\text{-}20)$$

where $\delta(f)$ is positive except if f is a constant (possibly depending on $\boldsymbol{r}$ and t) times $\exp(-mc^2/2kT_w)$, in which case it vanishes.

From this discussion one deduces that the first term of the right member of Eq. (4.5-3) can be written as

$$\int_S d^2S \sum_i \int (\boldsymbol{c}_i \cdot \boldsymbol{n}) f_i \log f_i d^3c_i = -\int_S \frac{\boldsymbol{q} \cdot \boldsymbol{n}}{kT_w} d^2S - \int_S \sum_i \delta(f_i) d^2S. \qquad (4.5\text{-}21)$$

Note that this expression vanishes only if the distribution function at the wall is everywhere Maxwellian with the local wall temperature. Furthermore, Eq. (4.5-21) illustrates the statement of the second law of thermodynamics holding that the local entropy transfer to the gas must be at least $(\boldsymbol{q} \cdot \boldsymbol{n})/T_w$.

Thus, for a gas contained in a vessel V bounded by a surface S one has

$$\frac{dH}{dt} + \int_S \frac{\boldsymbol{q} \cdot \boldsymbol{n}}{kT_w} d^2S = -\int_V \sum_{i,j} \Delta(f_i f_j) d^3r - \int_S \sum_i \delta(f_i) d^2S \qquad (4.5\text{-}22)$$

and we see that, if there is no net energy flow from the gas into the solid body that constitutes the boundary of the vessel, we have the differential inequality

$$dH/dt \leqq 0. \qquad (4.5\text{-}23)$$

This is a generalization of Boltzmann's *H-theorem* to spatially non-uniform systems.

From relation (4.5-22) it follows that $dH/dt = 0$ if and only if (i) the gas at the wall has a Maxwellian distribution with the local wall temperature and (ii) $f_i f_j = f_i' f_j'$ for all $\boldsymbol{c}_i$, $\boldsymbol{c}_j$ and $\boldsymbol{r}$. As in Section 4.2, (ii) implies that f_i must be Maxwellian everywhere,

$$f_i(\boldsymbol{c}_i) = n_i(m_i/2\pi kT)^{\frac{3}{2}} \exp(-m_i C_i^2/2kT), \qquad (4.5\text{-}24)$$

where $\boldsymbol{C}_i = \boldsymbol{c}_i - \boldsymbol{v}$. It is important to observe that the quantities n_i, $\boldsymbol{v}$ and T may now depend on $\boldsymbol{r}$. Thus, in contrast to the situation for a uniform gas, the condition $dH/dt = 0$ alone does *not* uniquely determine the distribution function. There may still be macroscopic gradients in the system and, hence, transport of mass, momentum and energy. In other words, in bounded systems the condition $dH/dt = 0$ is necessary but not sufficient for complete thermal equilibrium.

Now, suppose that the distribution functions f_i are such that they all satisfy the stationary Boltzmann equation (i.e., $\partial f_i/\partial t = 0$ for each i) *and* make the functional H stationary. Then each f_i must be of the form (4.5-24) and, in addition, we have, for each i,

$$\boldsymbol{c}_i \cdot \nabla_r f_i + \boldsymbol{F}_i \cdot \nabla_{c_i} f_i = 0 \tag{4.5-25a}$$

or, equivalently,

$$\boldsymbol{c}_i \cdot \nabla_r \log f_i + \boldsymbol{F}_i \cdot \nabla_{c_i} \log f_i = 0. \tag{4.5-25b}$$

This imposes certain restrictions on the possible variation of n_i, $\boldsymbol{v}$ and T with position and, in fact, these parameters are now uniquely determined. From Eq. (4.5-24),

$$\log f_i = \log(n_i/T^{\frac{3}{2}}) - m_i C_i^2/2kT + \text{constant} \tag{4.5-26}$$

and, hence,

$$\nabla_r \log f_i = \nabla_r \log(n_i/T^{\frac{3}{2}}) + (m_i C_i^2/2kT^2)\nabla_r T + (m_i/kT)\boldsymbol{C}_i \cdot \nabla_r \boldsymbol{v} \tag{4.5-27}$$

and

$$\nabla_{c_i} \log f_i = -(m_i/kT)\boldsymbol{C}_i. \tag{4.5-28}$$

Using these expressions and rearranging according to powers of the components of $\boldsymbol{C}_i$, we obtain, instead of Eq. (4.5-25),

$$\begin{aligned}
&[\boldsymbol{v} \cdot \nabla_r \log(n_i/T^{\frac{3}{2}})] \\
&\quad + [\boldsymbol{C}_i \cdot \nabla_r \log(n_i/T^{\frac{3}{2}}) + (m_i/kT)\boldsymbol{C}_i \cdot \{(\boldsymbol{v} \cdot \nabla_r)\boldsymbol{v}\} - (m_i/kT)\boldsymbol{F}_i \cdot \boldsymbol{C}_i] \\
&\quad + [(m_i/kT)\boldsymbol{C}_i\boldsymbol{C}_i : \nabla_r \boldsymbol{v} + (m_i C_i^2/2kT^2)\boldsymbol{v} \cdot \nabla_r T] \\
&\quad + [(m_i C_i^2/2kT^2)\boldsymbol{C}_i \cdot \nabla_r T] = 0.
\end{aligned} \tag{4.5-29}$$

If this is to hold identically in $\boldsymbol{C}_i$, each expression in square brackets must vanish separately. Hence, for all values of $\boldsymbol{C}_i$,

$$(m_i C_i^2/2kT^2)\boldsymbol{C}_i \cdot \nabla_r T = 0, \tag{4.5-30}$$

whence it follows that $\nabla_r T = 0$, that is, the temperature must be uniform throughout the gas. Next, requiring the third term of the right member of Eq. (4.5-29) to vanish, we find

$$(m_i/kT)\boldsymbol{C}_i\boldsymbol{C}_i : \nabla_r \boldsymbol{v} = 0. \tag{4.5-31}$$

Now, $\nabla_r \boldsymbol{v}$ can be uniquely decomposed into a symmetric tensor and antisymmetric tensor. Since $\boldsymbol{C}_i\boldsymbol{C}_i$ is symmetric, the first of these must be zero.

The second, however, need not be zero. The most general form that $\boldsymbol{v}$ can take and which is compatible with these conditions is $\boldsymbol{v} = \boldsymbol{v}'+\boldsymbol{\omega}\times\boldsymbol{r}$, where $\boldsymbol{v}'$ and $\boldsymbol{\omega}$ are arbitrary constant vectors. That is, the gas may move like a rigid body with a screw motion.

Now, choose a Cartesian coordinate system with the z-axis along the axis of rotation. Then $\boldsymbol{v}$ has the components $v_x = v'_x-\omega y$, $v_y = v'_y+\omega x$ and $v_z = v'_z$. Consequently, the vector $(\boldsymbol{v}\cdot\nabla_r)\boldsymbol{v}$ has the components $-\omega^2x+\omega v'_y$, $-\omega^2y+\omega v'_x$ and 0 and, hence, may be derived from a potential,

$$\Psi_0(\boldsymbol{r}) = -\tfrac{1}{2}\omega^2(x^2+y^2), \tag{4.5-32}$$

in the sense that

$$(\boldsymbol{v}\cdot\nabla_r)\boldsymbol{v} = \nabla_r\Psi_0+\boldsymbol{\omega}\times\boldsymbol{v}'. \tag{4.5-33}$$

Thus, setting the second term of the right member of Eq. (4.5-29) equal to zero we arrive at the identity

$$\boldsymbol{C}_i\cdot[\nabla_r\{\log n_i+(m_i/kT)\Psi_0\}+(m_i/kT)\boldsymbol{\omega}\times\boldsymbol{v}'-(m_i/kT)\boldsymbol{F}_i] = 0, \tag{4.5-34}$$

which must hold for all values of $\boldsymbol{C}_i$. Since $\boldsymbol{\omega}\times\boldsymbol{v}'$ is a constant vector this is possible only if

$$\boldsymbol{F}_i = \boldsymbol{\omega}\times\boldsymbol{v}'+\nabla_r\{\Psi_0+(kT/m_i)\log n_i\}. \tag{4.5-35}$$

The first term is just the force needed to balance the Coriolis force that acts on a body simultaneously rotating and translating, cf., Goldstein [1950]; since it is not pertinent to the discussion here, we will assume that $\boldsymbol{\omega}\times\boldsymbol{v}' = 0$. Then, $\boldsymbol{F}_i$ must be the negative of the gradient of a potential, Ψ_i say, and we have

$$n_i = n_{i0}\exp\{-m_i(\Psi_i+\Psi_0)/kT\}. \tag{4.5-36}$$

For example, if the gas is at rest ($\Psi_0 = 0$), but the molecules are subject to a gravitational force $F_z = -g$ per unit mass ($\Psi_i = gz$, where g is the acceleration of gravity and z is the height), Eq. (4.5-36) is identical with the *barometric equation.* In the absence of external forces, Eq. (4.5-36) describes the density non-uniformity due to a centrifugal potential Ψ_0.

Finally, from the first term of the right member of Eq. (4.5-29) and the identity $\boldsymbol{v}\cdot\nabla_r\Psi_0 = 0$, we obtain the condition

$$\boldsymbol{v}\cdot\nabla_r\Psi_i = 0, \tag{4.5-37}$$

expressing that the motion of the ith component of the gas must, at every point, be along an equipotential surface $\Psi_i =$ constant. Thus, if $\boldsymbol{\omega} = 0$

and $\boldsymbol{v}' \neq 0$, Ψ_i must be independent of z; if $\boldsymbol{v}' = 0$ and $\boldsymbol{\omega} \neq 0$, Ψ_i must be invariant under a rotation around the z-axis; if $\boldsymbol{v}' \neq 0$ and $\boldsymbol{\omega} \neq 0$, Ψ_i must be constant along spiral curves around the z-axis. Obviously, if the vessel V which contains the gas, is smooth and stationary, the gas motion must be consistent with the shape of the vessel. Thus, in general the gas must be at rest, but if the vessel possesses symmetry about some axis a rotation about this axis is possible.

To summarize, if a gas is contained in a stationary smooth vessel V bounded by a non-porous surface S, a state of complete thermal equilibrium can exist only if the following two conditions are satisfied: (i) the gas is at rest or, if the vessel possesses symmetry about some axis, the gas may rotate about this axis, (ii) the external force $\boldsymbol{F}_i$ acting upon a particle of species i is derivable from a potential Ψ_i. When the gas is in thermal equilibrium the temperature is constant throughout the gas and equal to the temperature of the wall. In a gas at rest the velocity distribution of each component is of the form

$$f(\boldsymbol{c}) = n_0(m/2\pi kT)^{\frac{3}{2}} \exp\{-(m/kT)(\tfrac{1}{2}c^2 + \Psi)\}. \qquad (4.5\text{-}38)$$

To complete our discussion of the H-theorem in bounded systems we recall that the stationarity of the Boltzmann H-function, i.e., the condition $\mathrm{d}H/\mathrm{d}t = 0$, was not sufficient to uniquely determine the velocity distribution function. In fact, provided that the gas temperature at the wall is equal to the wall temperature, any *local* Maxwellian distribution will result in $\mathrm{d}H/\mathrm{d}t = 0$. Since $\mathrm{d}H/\mathrm{d}t$ cannot be positive, and, as we have seen, the local Maxwellian is not an equilibrium state, the H-function must have a *horizontal inflection point* if the system should pass through a local Maxwellian state. That this is indeed the case can be shown by straight-forward computation of the second and third time-derivatives of H in the vicinity of such a state, see Ferziger [1969].

4.6. The linearized Boltzmann equation

The collision term in the Boltzmann equation is quadratic in the velocity distribution function. This non-linearity is essential if the state of the gas is far from thermal equilibrium. However, if the state of the gas is near thermal equilibrium one may anticipate that a linearized form of the Boltzmann equation – obtained from the Boltzmann equation by assuming that the perturbation of the velocity distribution function from its equilibrium form is small – will provide a reasonably accurate description of the transport phenomena. This situation will be considered in the present section. First we shall restrict ourselves to the case of a simple gas.

Assume that, at every position $\boldsymbol{r}$ in physical space and at every instant t, the velocity distribution function f can be written as

$$f(\boldsymbol{r}, \boldsymbol{c}, t) = f_{\mathrm{M}}[1+\Phi(\boldsymbol{r}, \boldsymbol{c}, t)] \tag{4.6-1}$$

with

$$f_{\mathrm{M}} = n(m/2\pi kT)^{\frac{3}{2}} \exp(-mc^2/2kT), \tag{4.6-2}$$

such that the quantity Φ represents a small correction, i.e., $|\Phi| \ll 1$ for those velocities at which most of the particles are found. Note that, in Eq. (4.6-2), n and T have been taken as constants and $\boldsymbol{v} = 0$, i.e., an *absolute* Maxwellian has been used. Linearization about a *local* Maxwellian, with n, $\boldsymbol{v}$ and T functions of $\boldsymbol{r}$ and t, has also been applied, see Grad [1958]. Terms of second and higher order in Φ can be neglected and, in the absence of external forces, the Boltzmann equation (3.1-29) reduces to

$$f_{\mathrm{M}}\left(\frac{\partial \Phi}{\partial t} + \boldsymbol{c}\cdot\boldsymbol{\nabla}_r \Phi\right) = \iiint f_{\mathrm{M}} f_{\mathrm{M}1}(\Phi' + \Phi_1' - \Phi - \Phi_1) g b \, \mathrm{d}b \, \mathrm{d}\varepsilon \, \mathrm{d}^3 c_1 . \tag{4.6-3}$$

This is the *linearized Boltzmann equation* for a simple gas near thermal equilibrium. It is convenient to denote the collision term – which is now linear in the unknown function Φ – by a separate symbol. If F is any function of the molecular velocity, we define

$$I(F) = \frac{1}{n^2}\iiint f_{\mathrm{M}} f_{\mathrm{M}1}(F + F_1 - F' - F_1') g b \, \mathrm{d}b \, \mathrm{d}\varepsilon \, \mathrm{d}^3 c_1 , \tag{4.6-4}$$

so that Eq. (4.6-3) is equivalent to

$$f_{\mathrm{M}}\left(\frac{\partial \Phi}{\partial t} + \boldsymbol{c}\cdot\boldsymbol{\nabla}_r \Phi\right) = -n^2 I(\Phi). \tag{4.6-5}$$

Similarly, in the case of a gas mixture near thermal equilibrium, the Ansatz

$$f_i(\boldsymbol{r}, \boldsymbol{c}_i, t) = f_{\mathrm{M}_i}[1+\Phi_i(\boldsymbol{r}, \boldsymbol{c}_i, t)] \tag{4.6-6}$$

for every constituent i of the mixture leads, upon neglect of terms of second and higher order in the functions Φ, to the following form of the linearized Boltzmann equation,

$$f_{\mathrm{M}_i}\left(\frac{\partial \Phi_i}{\partial t} + \boldsymbol{c}_i\cdot\boldsymbol{\nabla}_r \Phi_i\right) = -\sum_j n_i n_j I_{ij}(\Phi), \tag{4.6-7}$$

where the operator I_{ij} is defined through the relation

$$I_{ij}(F) = \frac{1}{n_i n_j} \iiint f_{Mi} f_{Mj} (F_i + F_j - F_i' - F_j') g b \, db \, d\varepsilon \, d^3 c_j . \qquad (4.6\text{-}8)$$

In this definition F is any function of the molecular velocity defined in the velocity domain of each component of the mixture.

Next we demonstrate that I is an *integral operator* with a *symmetric kernel.* This fact was first shown by Hilbert [1912] for the case of a gas of rigid elastic spherical molecules and later by Enskog [1917] for the general case.

Consider a simple gas of molecules of mass m. In terms of the collision cross section σ, which has been introduced at the end of Section 3.1, the operator I may be defined by

$$I(F) = \frac{1}{n^2} \iiint f_{M1} f_M (F + F_1 - F' - F_1') |\boldsymbol{c} - \boldsymbol{c}_1| \sigma(|\boldsymbol{c} - \boldsymbol{c}_1|, \chi) \sin \chi \, d\chi \, d\varepsilon \, d^3 c_1 . \qquad (4.6\text{-}9)$$

We recall briefly the kinematics of a binary collision, cf. Fig. 4.1; $\boldsymbol{c}$ and $\boldsymbol{c}_1$ are the initial velocities, $\boldsymbol{c}'$ and $\boldsymbol{c}_1'$ the final velocities. The effect of the collision is a *rotation* of the vector $\boldsymbol{g} = \boldsymbol{c}_1 - \boldsymbol{c}$ into the vector $\boldsymbol{g}' = \boldsymbol{c}_1' - \boldsymbol{c}'$, where the center-of-mass velocity $\boldsymbol{G}$ remains unchanged. Thus, $g = g'$ and $\boldsymbol{G} = \frac{1}{2}(\boldsymbol{c} + \boldsymbol{c}_1) = \frac{1}{2}(\boldsymbol{c}' + \boldsymbol{c}_1')$. The effect of the collision is completely specified in terms of χ and ε, where χ is the angle between $\boldsymbol{g}$ and $\boldsymbol{g}'$ and ε is the angle between the plane through $\boldsymbol{c}$ and $\boldsymbol{c}'$ and the plane through $\boldsymbol{g}$ and $\boldsymbol{g}'$; if $\varepsilon = 0$, all four velocities $\boldsymbol{c}$, $\boldsymbol{c}_1$, $\boldsymbol{c}'$ and $\boldsymbol{c}_1'$ lie in one plane.

In the following demonstration we will consider separately the four terms comprising $n^2 I(F)$, i.e., those involving the functions F, F_1, F' and F_1', respectively.

In the first term, F is independent of the variables of integration. With the abbreviation

$$\begin{aligned} K_0(\boldsymbol{c}) &= n^2 (m/2\pi kT)^3 \iiint \exp\{-(m/2kT)(c^2 + c_1^2)\} \\ &\qquad \times |\boldsymbol{c} - \boldsymbol{c}_1| \sigma(|\boldsymbol{c} - \boldsymbol{c}_1|, \chi) \sin \chi \, d\chi \, d\varepsilon \, d^3 c_1 \\ &= 2\pi n^2 (m/2\pi kT)^3 \iint \exp\{-(m/2kT)(c^2 + c_1^2)\} \\ &\qquad \times |\boldsymbol{c} - \boldsymbol{c}_1| \sigma(|\boldsymbol{c} - \boldsymbol{c}_1|, \chi) \sin \chi \, d\chi \, d^3 c_1 \end{aligned} \qquad (4.6\text{-}10)$$

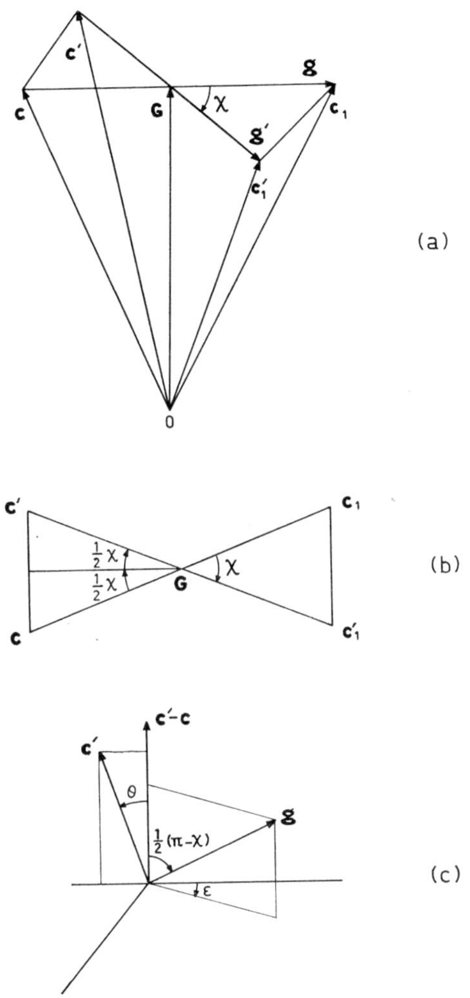

Fig. 4.1. The kinematics of a binary collision; (a) general situation, (b) plane through $\boldsymbol{g}$ and $\boldsymbol{g}'$, (c) cylindrical coordinate system.

this term may be expressed as

$$K_0(\boldsymbol{c})F(\boldsymbol{c}). \tag{4.6-11}$$

The second term may be expressed as

$$\int K_1(\boldsymbol{c}, \boldsymbol{c}_1)F(\boldsymbol{c}_1)\,\mathrm{d}^3c_1, \tag{4.6-12}$$

if K_1 is defined by

$$K_1(\boldsymbol{c}, \boldsymbol{c}_1)$$
$$= n^2(m/2\pi kT)^3 \exp\{-(m/2kT)(c^2+c_1^2)\}\,|\boldsymbol{c}-\boldsymbol{c}_1| \iint \sigma(|\boldsymbol{c}-\boldsymbol{c}_1|, \chi) \sin\chi\, d\chi\, d\varepsilon$$
$$= 2\pi n^2(m/2\pi kT)^3 \exp\{-(m/2kT)(c^2+c_1^2)\}\,|\boldsymbol{c}-\boldsymbol{c}_1| \int \sigma(|\boldsymbol{c}-\boldsymbol{c}_1|, \chi) \sin\chi\, d\chi. \tag{4.6-13}$$

Obviously, $K_1(\boldsymbol{c}, \boldsymbol{c}_1)$ is a symmetric function of its arguments. Notice also the relation between the functions K_0 and K_1,

$$K_0(\boldsymbol{c}) = \int K_1(\boldsymbol{c}, \boldsymbol{c}_1)\, d^3c_1. \tag{4.6-14}$$

Next, we turn to the third term. Since it involves the function $F(\boldsymbol{c}')$, we change the variable of integration from $\boldsymbol{c}_1$ to $\boldsymbol{c}'$. As $\boldsymbol{c}$ is a fixed vector, the element d^3c_1 is equal to $d^3(c_1-c) = g^2\, dg\, d^2e$, where $\boldsymbol{e}$ is a unit vector in the direction of $\boldsymbol{c}_1-\boldsymbol{c}$. Using elementary trigonometry in the plane through $\boldsymbol{g}$ and $\boldsymbol{g}'$, see Fig. 4.1b, one verifies

$$|\boldsymbol{c}'-\boldsymbol{c}| = |\boldsymbol{c}_1-\boldsymbol{c}| \sin\tfrac{1}{2}\chi = g\sin\tfrac{1}{2}\chi, \tag{4.6-15}$$

so $d^3c_1 = g^2\, dg\, d^2e = |\boldsymbol{c}'-\boldsymbol{c}|^2 \sin^{-3}\tfrac{1}{2}\chi\, d|\boldsymbol{c}'-\boldsymbol{c}|\, d^2e$. Now, the change of the variables of integration from $(|\boldsymbol{c}'-\boldsymbol{c}|, \boldsymbol{e})$ to $(|\boldsymbol{c}'-\boldsymbol{c}|, \boldsymbol{k})$, where $\boldsymbol{k}$ is a unit vector in the direction of $\boldsymbol{c}'-\boldsymbol{c}$, is simply a rotation, so $d^2e = d^2k$. Whence it follows that, under the change of the variables of integration from $\boldsymbol{c}_1$ to $\boldsymbol{c}'$, d^3c_1 is replaced by $\sin^{-3}\tfrac{1}{2}\chi\, d^3c'$. Because of the occurrence of the function f_{M1} we must also express c_1^2 in terms of the new variables of integration. Therefore, write $\boldsymbol{c}_1 = \boldsymbol{c}'+(\boldsymbol{c}-\boldsymbol{c}')+(\boldsymbol{c}_1-\boldsymbol{c})$. Then

$$c_1^2 = c'^2+2\boldsymbol{c}'\cdot(\boldsymbol{c}-\boldsymbol{c}')+2\boldsymbol{c}'\cdot(\boldsymbol{c}_1-\boldsymbol{c})+|\boldsymbol{c}_1-\boldsymbol{c}'|^2. \tag{4.6-16}$$

Again using elementary trigonometry in the plane through $\boldsymbol{g}$ and $\boldsymbol{g}'$, see Fig. 4.1b, one verifies

$$|\boldsymbol{c}_1-\boldsymbol{c}'|^2 = |\boldsymbol{c}-\boldsymbol{c}'|^2 \operatorname{cotan}^2\tfrac{1}{2}\chi. \tag{4.6-17}$$

In order to find the cosine of the angle between $\boldsymbol{c}'$ and $\boldsymbol{c}_1-\boldsymbol{c} = \boldsymbol{g}$ – which do not intersect unless $\varepsilon = 0$ or $\varepsilon = \pi$ – we must use spherical trigonometry. Take the endpoint of $\boldsymbol{c}'$ as origin of a cylindrical coordinate system with polar axis in the direction of $\boldsymbol{c}'-\boldsymbol{c}$ and azimuthal angles measured with respect to the plane through $\boldsymbol{c}'-\boldsymbol{c}$ and $\boldsymbol{c}'$, see Fig. 4.1c. Let θ be the angle between $\boldsymbol{c}'$ and $\boldsymbol{c}'-\boldsymbol{c}$. As the vectors $\boldsymbol{c}'$, $\boldsymbol{c}$ and $\boldsymbol{c}'-\boldsymbol{c}$ are coplanar, the plane through the vectors $\boldsymbol{g}$ and $\boldsymbol{g}'$ – which contains the polar axis – has azimuthal angle ε. In the latter plane, the angle between $\boldsymbol{g}$ and $\boldsymbol{c}'-\boldsymbol{c}$ is known, viz. $\frac{1}{2}\pi-\frac{1}{2}\chi$, cf. Fig. 4.1b. Thus, in the coordinate system chosen, the unit vector in the direction of $\boldsymbol{c}'$ has polar coordinates $(\theta, 0)$ and the unit vector in the direction of $\boldsymbol{g}$ has polar coordinates $(\frac{1}{2}\pi-\frac{1}{2}\chi, \varepsilon)$. Then, from spherical geometry it is well known that the cosine of the angle between $\boldsymbol{c}'$ and $\boldsymbol{g}$ is equal to

$$\cos\theta\cos(\tfrac{1}{2}\pi-\tfrac{1}{2}\chi)+\sin\theta\sin(\tfrac{1}{2}\pi-\tfrac{1}{2}\chi)\cos\varepsilon = \cos\theta\sin\tfrac{1}{2}\chi+\sin\theta\cos\tfrac{1}{2}\chi\cos\varepsilon \tag{4.6-18}$$

and, consequently,

$$\begin{aligned}\boldsymbol{c}'\cdot(\boldsymbol{c}_1-\boldsymbol{c}) &= c'|\boldsymbol{c}_1-\boldsymbol{c}|(\sin\theta\cos\tfrac{1}{2}\chi\cos\varepsilon+\cos\theta\sin\tfrac{1}{2}\chi)\\ &= c'|\boldsymbol{c}'-\boldsymbol{c}|(\sin\theta\,\mathrm{cotan}\,\tfrac{1}{2}\chi\cos\varepsilon+\cos\theta)\\ &= |\boldsymbol{c}\times\boldsymbol{c}'|\,\mathrm{cotan}\,\tfrac{1}{2}\chi\cos\varepsilon+\boldsymbol{c}'\cdot(\boldsymbol{c}'-\boldsymbol{c}).\end{aligned} \tag{4.6-19}$$

Therefore,

$$c_1^2 = c'^2+2|\boldsymbol{c}\times\boldsymbol{c}'|\,\mathrm{cotan}\,\tfrac{1}{2}\chi\cos\varepsilon+|\boldsymbol{c}-\boldsymbol{c}'|^2\,\mathrm{cotan}^2\,\tfrac{1}{2}\chi. \tag{4.6-20}$$

Thus, the third term in the expression for $n^2 I(F)$ may be expressed as

$$\int K_2(\boldsymbol{c}, \boldsymbol{c}')F(\boldsymbol{c}')\,\mathrm{d}^3c', \tag{4.6-21}$$

if K_2 is defined as

$$\begin{aligned}K_2(\boldsymbol{c}, \boldsymbol{c}') &= n^2(m/2\pi kT)^3\exp\{-(m/2kT)(c^2+c'^2)\}|\boldsymbol{c}-\boldsymbol{c}'|\\ &\times\iint\exp\{-(m/2kT)(2|\boldsymbol{c}\times\boldsymbol{c}'|\,\mathrm{cotan}\,\tfrac{1}{2}\chi\cos\varepsilon+|\boldsymbol{c}-\boldsymbol{c}'|^2\,\mathrm{cotan}^2\,\tfrac{1}{2}\chi)\}\\ &\times\sin^{-4}\tfrac{1}{2}\chi\,\sigma(|\boldsymbol{c}-\boldsymbol{c}'|/\sin\tfrac{1}{2}\chi, \chi)\sin\chi\,\mathrm{d}\chi\,\mathrm{d}\varepsilon.\end{aligned} \tag{4.6-22}$$

The integral over ε can be expressed in terms of the Bessel function of purely imaginary argument, I_0, see e.g., Whittaker and Watson [1915]. The result is

$$\begin{aligned}K_2(\boldsymbol{c}, \boldsymbol{c}') &= 2\pi n^2(m/2\pi kT)^3\exp\{-(m/2kT)(c^2+c'^2)\}|\boldsymbol{c}-\boldsymbol{c}'|\\ &\times\int_0^\pi\frac{\exp\{-(m/2kT)|\boldsymbol{c}-\boldsymbol{c}'|^2\,\mathrm{cotan}^2\,\tfrac{1}{2}\chi\}}{\sin^4\tfrac{1}{2}\chi}I_0\left(\frac{m}{kT}|\boldsymbol{c}\times\boldsymbol{c}'|\,\mathrm{cotan}\,\tfrac{1}{2}\chi\right)\\ &\times\sigma(|\boldsymbol{c}-\boldsymbol{c}'|/\sin\tfrac{1}{2}\chi, \chi)\sin\chi\,\mathrm{d}\chi.\end{aligned} \tag{4.6-23}$$

Notice that $K_2(\boldsymbol{c}, \boldsymbol{c}')$ is a symmetric function of its arguments.

Finally, the last term in the expression for $n^2 I(F)$ is calculated in a similar manner. It may be expressed as

$$\int K_3(\boldsymbol{c}, \boldsymbol{c}_1') F(\boldsymbol{c}_1') \mathrm{d}^3 c_1', \tag{4.6-24}$$

where the kernel K_3 is defined as

$$K_3(\boldsymbol{c}, \boldsymbol{c}_1') = 2\pi n^2 (m/2\pi kT)^3 \exp\{-(m/2kT)(c^2 + c_1'^2)\} |\boldsymbol{c} - \boldsymbol{c}_1'| \times \int_0^\pi \frac{\exp\{-(m/2kT)|\boldsymbol{c}-\boldsymbol{c}_1'|^2 \tan^2 \frac{1}{2}\chi\}}{\cos^4 \frac{1}{2}\chi} I_0\left(\frac{m}{kT} |\boldsymbol{c} \times \boldsymbol{c}_1'| \tan \tfrac{1}{2}\chi\right) \times \sigma(|\boldsymbol{c}-\boldsymbol{c}_1'|/\cos \tfrac{1}{2}\chi, \chi) \sin \chi \, \mathrm{d}\chi. \tag{4.6-25}$$

By changing the variable of integration from χ to $\pi - \chi$, we can also write the expression for K_3 as

$$K_3(\boldsymbol{c}, \boldsymbol{c}_1') = 2\pi n^2 (m/2\pi kT)^3 \exp\{-(m/2kT)(c^2 + c_1'^2)\} |\boldsymbol{c} - \boldsymbol{c}_1'| \times \int_0^\pi \frac{\exp\{-(m/2kT)|\boldsymbol{c}-\boldsymbol{c}_1'|^2 \operatorname{cotan}^2 \frac{1}{2}\chi\}}{\sin^4 \frac{1}{2}\chi} I_0\left(\frac{m}{kT} |\boldsymbol{c} \times \boldsymbol{c}_1'| \operatorname{cotan} \tfrac{1}{2}\chi\right) \times \sigma(|\boldsymbol{c}-\boldsymbol{c}_1'|/\sin \tfrac{1}{2}\chi, \pi-\chi) \sin \chi \, \mathrm{d}\chi. \tag{4.6-26}$$

Again, $K_3(\boldsymbol{c}, \boldsymbol{c}_1')$ is a symmetric function of its arguments.

Combining the results (4.6-11), (4.6-12), (4.6-21) and (4.6-24) we conclude that

$$n^2 I(F) = K_0(\boldsymbol{c}) F(\boldsymbol{c}) - \int K(\boldsymbol{c}, \boldsymbol{c}_1) F(\boldsymbol{c}_1) \mathrm{d}^3 c_1; \tag{4.6-27}$$

the function K_0 is given by Eq. (4.6-10) and the kernel K is defined by

$$K(\boldsymbol{c}, \boldsymbol{c}_1) = 2\pi n^2 (m/2\pi kT)^3 \exp\{-(m/2kT)(c^2 + c_1^2)\} |\boldsymbol{c} - \boldsymbol{c}_1| \times \int_0^\pi \Bigg[\frac{\exp\{-(m/2kT)|\boldsymbol{c}-\boldsymbol{c}_1|^2 \operatorname{cotan}^2 \frac{1}{2}\chi\}}{\sin^4 \frac{1}{2}\chi} I_0\left(\frac{m}{kT} |\boldsymbol{c} \times \boldsymbol{c}_1| \operatorname{cotan} \tfrac{1}{2}\chi\right) \times \left\{\sigma\left(\frac{|\boldsymbol{c}-\boldsymbol{c}_1|}{\sin \frac{1}{2}\chi}, \chi\right) + \sigma\left(\frac{|\boldsymbol{c}-\boldsymbol{c}_1|}{\sin \frac{1}{2}\chi}, \pi-\chi\right)\right\} - \sigma(|\boldsymbol{c}-\boldsymbol{c}_1|, \chi)\Bigg] \sin \chi \, \mathrm{d}\chi; \tag{4.6-28}$$

$K(\boldsymbol{c}, \boldsymbol{c}_1)$ is a symmetric function of its arguments. The extension of this result to the case of a mixture of molecules of different species offers no specific difficulties and will therefore be omitted.

For an application of the theory of integral equations it is important to know what properties – other than the symmetry property – the kernel K possesses. For example, one may wish to know whether K is square integrable so that the Fredholm alternative is applicable, cf. Courant and Hilbert [1953]. The answer to such questions depends essentially upon the nature of the function σ and, hence, on the molecular interaction. We shall henceforth assume that any realistic model of molecular interaction gives rise to a function σ which allows the Fredholm alternative to be applied to the integral operator I (see Pekeris [1963] and Dorfman [1963] for the case of rigid sphere molecules). In this respect, another remark must be put forward: So far, it has been tacitly supposed that all integrals over the geometrical collision variables (i.e., b and ε, or χ and ε) are convergent. However, for potentials with infinite range integrals like that in Eq. (4.6-10) become infinite if the range of integration with respect to χ extends over the whole interval $0 \leqq \chi \leqq \pi$. In such cases, one may, for example, choose to ignore "grazing" encounters, i.e., all encounters for which the deflection angle χ is less than a small angle δ, which may even vary with the relative speed g of the two molecules (*angular cut-off*). Or, equivalently, one may limit the range of the intermolecular influence by taking an upper limit for b (*radial cut-off*). Although the values of such integrals then depend on the choice of a limit on χ or b, the results turn out to be correct for most actual gases. The reason is that, if the potential falls off fast enough at long distances, the grazing collisions do not change the velocities of the particles much and thus have little effect on the behavior of the gas.

4.7. Integral theorems; $I(F)$, $[F, G]$

The linearized collision operator plays an essential role in the methods of Hilbert, Chapman and Enskog to find approximate solutions to the Boltzmann equation. As we have seen in the previous section, the operator is, in fact, a linear integral operator with a symmetric kernel. In the present section we introduce some other integrals which occur in the Chapman-Enskog theory and which are directly related to the linearized collision operator.

In the case of a simple gas, if F is any function of the molecular velocity, the function $I(F)$ was defined as

$$I(F) = \frac{1}{n^2}\iiint f_{M1}\, f_M(F+F_1-F'-F_1')g b \,\mathrm{d}b\,\mathrm{d}\varepsilon\,\mathrm{d}^3c_1\,. \qquad (4.7\text{-}1)$$

Similarly, in the case of a mixture, if F is any function of the molecular velocity defined in the velocity domain of each constituent of the mixture, the function $I_{ij}(F)$ was defined as

$$I_{ij}(F) = \frac{1}{n_i n_j} \iiint f_{Mi} f_{Mj} (F_i + F_j - F_i' - F_j') g b \, db \, d\varepsilon \, d^3 c_j, \qquad (4.7\text{-}2)$$

where $g = |c_i - c_j|$ and the integrals are evaluated for collisions between molecules of species i and molecules of species j. For later convenience we also define two operators $I_{ij,\,i}$ and $I_{ij,\,j}$ by the relations

$$I_{ij,i}(F) = \frac{1}{n_i n_j} \iiint f_{Mi} f_{Mj} (F_i - F_i') g b \, db \, d\varepsilon \, d^3 c_j, \qquad (4.7\text{-}3a)$$

$$I_{ij,j}(F) = \frac{1}{n_i n_j} \iiint f_{Mi} f_{Mj} (F_j - F_j') g b \, db \, d\varepsilon \, d^3 c_j. \qquad (4.7\text{-}3b)$$

Comparing Eqs. (4.7-2) and (4.7-3) we see that the operator I_{ij} is the sum of the operators $I_{ij,i}$ and $I_{ij,j}$. We notice that all three quantities $I_{ij}(F)$, $I_{ij,\,i}(F)$ and $I_{ij,j}(F)$ are functions of the variable c_i.

Obviously, if I is any of the above operators,

$$I(\lambda F + \mu G) = \lambda I(F) + \mu I(G), \qquad (4.7\text{-}4)$$

λ and μ arbitrary complex or real numbers, i.e., I is a linear operator.

Next we introduce complete integrals that are related to the integral operators I.

First, consider the case of a simple gas. We recall from Eq. (4.7-1) that the operator I maps any function of the molecular velocity into another function of the molecular velocity, so if F and G are both functions of the molecular velocity we may define the *bracket integral* of F and G as

$$[F, G] = \int G I(F) \, d^3 c. \qquad (4.7\text{-}5)$$

Since I is a linear operator, the bracket integral is a *bilinear* form. By symmetry arguments one verifies

$$[F, G] = \frac{1}{4n^2} \iiiint f_M f_{M1} (F + F_1 - F' - F_1')(G + G_1 - G' - G_1') \times g b \, db \, d\varepsilon \, d^3 c \, d^3 c_1, \qquad (4.7\text{-}6)$$

whence it follows that the bracket integral is also *symmetric*,

$$[F, G] = [G, F]. \tag{4.7-7}$$

An important property of the operator I follows from the relation (4.7-6) upon taking $F = G$. In that case, the integrand is always positive except if F is a linear combination of the summational invariants, when it vanishes. This implies that I is a positive semi-definite operator in the sense that, for arbitrary F,

$$[F, F] = \int F\, I(F)\, \mathrm{d}^3c \geqq 0. \tag{4.7-8}$$

The equality sign holds if and only if F is a linear combination of the summational invariants. Consequently, if F is a scalar quantity, the general solution of the homogeneous integral equation $I(F) = 0$ is given by

$$F = \alpha^{(1)} + \boldsymbol{\alpha}^{(2)} \cdot m\boldsymbol{c} + \alpha^{(3)} \tfrac{1}{2} mC^2, \tag{4.7-9}$$

where $\alpha^{(1)}$, $\boldsymbol{\alpha}^{(2)}$ and $\alpha^{(3)}$ are independent of $\boldsymbol{c}$.

Now, we generalize the definition of the bracket integral to the case of a multicomponent gas mixture. Let F and G be two functions of the molecular velocity defined in the velocity domain of each constituent of the mixture. The operators $I_{ij,\,i}$ and $I_{ij,\,j}$ give rise to the definition of the *partial bracket integrals*

$$[F, G]'_{ij} = \int G_i I_{ij,\,i}(F)\, \mathrm{d}^3c_i \tag{4.7-10a}$$

and

$$[F, G]''_{ij} = \int G_i I_{ij,\,j}(F)\, \mathrm{d}^3c_i. \tag{4.7-10b}$$

Again, the partial bracket integrals are *bilinear* forms. By symmetry arguments one verifies

$$[F, G]'_{ij} = \frac{1}{2n_i n_j} \iiiint f_{\mathrm{M}i} f_{\mathrm{M}j} (G_i - G'_i)(F_i - F'_i) g b\, \mathrm{d}b\, \mathrm{d}\varepsilon\, \mathrm{d}^3c_i\, \mathrm{d}^3c_j \tag{4.7-11a}$$

and, similarly,

$$[F, G]''_{ij} = \frac{1}{2n_i n_j} \iiiint f_{\mathrm{M}i} f_{\mathrm{M}j} (G_i - G'_i)(F_j - F'_j) g b\, \mathrm{d}b\, \mathrm{d}\varepsilon\, \mathrm{d}^3c_i\, \mathrm{d}^3c_j, \tag{4.7-11b}$$

whence it follows that the partial bracket integrals have the symmetry properties

$$[F, G]'_{ij} = [G, F]'_{ij}, \quad [F, G]''_{ij} = [G, F]''_{ji}. \tag{4.7-12}$$

It should be noted that the subscript ij on the bracket indicates that the integrals are evaluated for collisions between molecules of species i and j. Finally, we define the complete *bracket integral* of F and G for the mixture as

$$[F, G] = \sum_{i,j} \frac{n_i n_j}{n^2} ([F, G]'_{ij} + [F, G]''_{ij}). \tag{4.7-13}$$

From the earlier results one concludes immediately that this complete bracket integral is a *bilinear* form which is also *symmetric*,

$$[F, G] = [G, F]. \tag{4.7-14}$$

The symmetry relation (4.7-14) also follows from the representation

$$[F, G] = \frac{1}{4n^2} \sum_{i,j} \iiiint f_{Mi} f_{Mj} (F_i + F_j - F'_i - F'_j) \times (G_i + G_j - G'_i - G'_j) g b \, db \, d\varepsilon \, d^3c_i \, d^3c_j, \tag{4.7-15}$$

which is equivalent to Eq. (4.7-13). From the representation (4.7-15) one deduces, as in the case of a simple gas, that the bracket $[F, F]$ is positive semi-definite,

$$[F, F] \geqq 0, \tag{4.7-16}$$

the equality sign holding if and only if F is a linear combination of the summational invariants. Thus, the general solution of the equation $[F, F] = 0$ is given by

$$F_i = \alpha_i^{(1)} + \boldsymbol{\alpha}^{(2)} \cdot m_i \boldsymbol{c}_i + \alpha^{(3)} \tfrac{1}{2} m_i C_i^2, \quad i = 1, \ldots, K, \tag{4.7-17}$$

where $\alpha_i^{(1)}$, $\boldsymbol{\alpha}^{(2)}$ and $\alpha^{(3)}$ are independent of $\boldsymbol{c}_i$.

Finally, with $H = [F, F]G - [F, G]F$ and the inequalities (4.7-8) and (4.7-16) one proves that, in the case of a simple gas as well as in the case of a multicomponent gas mixture, the bracket integral satisfies the inequality

$$[F, G]^2 \leqq [F, F][G, G]. \tag{4.7-18}$$

In conclusion, we remark that in all bracket integrals the arguments may be either scalars or vectors or tensors, as long as they are of like kind. The integrals $I(F)$ are of the same nature as F; the bracket integrals are scalar quantities, so their integrands contain the appropriate scalar products of G and $I(F)$.

4.8. Existence theory for the linearized Boltzmann equation

In Section 4.6 we established the linearized Boltzmann equation and discussed the general structure of its collision operator. In the present section we give some results of the existence theory for the linearized Boltzmann equation and show under which conditions a given initial value problem is well-posed. The methods to be used in this section are considerably more mathematical than any we have utilized thus far. In order that the reader not equipped with the necessary mathematical tools may still acquire some idea of the results of this section, we begin with a heuristic description of the analysis. The ideas presented are precisely those which are made rigorous after the heuristic introduction.

Before proceeding we remark that an existence theory for the non-linear Boltzmann equation itself is extremely difficult and, in fact, only modest progress can be reported in that respect – cf. Section 5.10. For the linearized Boltzmann equation, however, the problems are less formidable and a number of rigorous proofs, all using methods of functional analysis, have recently been presented to establish the existence and uniqueness of solutions to certain initial and boundary value problems – see Cercignani [1967, 1969a, 1969b], Scharf [1967], Guiraud [1968a, 1968b]. By studying these proofs one may, hopefully, gain some insight into the structure of the non-linear operator that is associated with Boltzmann's equation. We note in passing that an existence theory for the linearized Boltzmann equation has its own intrinsic value too, since the equation is used extensively in studies of low velocity flows of rarefied gases, for which the usual equations of hydrodynamics are not valid – see Chapter 15.

We consider a gas contained in a vessel of volume V. To simplify the analysis we assume that the range of the intermolecular potential does not extend to infinity or, if it does, we may artificially cut off the grazing collisions (see the discussion at the end of Section 4.6). Furthermore, we neglect external forces, as they pose no specific problems in the present analysis. At time $t = 0$ a velocity distribution $f_0(\boldsymbol{r}, \boldsymbol{c})$ exists inside the vessel. We ask for the resulting time-dependent velocity distribution function $f(\boldsymbol{r}, \boldsymbol{c}, t)$, assuming that the system is sufficiently close to absolute equilibrium for the time evolution of f to be adequately described – throughout the entire system – by a linearized Boltzmann equation. Consistent with the latter assumption and with the results of Section 4.5, we assume that the vessel is bounded by a surface whose temperature is constant everywhere and that the linearization is performed with respect to an absolute Maxwellian whose temperature is equal to the temperature of the wall. (The latter assumption is introduced to simplify the analysis; the results of the present section can be generalized to the case, in which the linearization is performed with respect to an arbitrary absolute Maxwellian – as may be the case, for example, when the temperature of the wall is a slowly varying function of position.)

Thus, the initial value problem may be formulated as follows. Let

$$f(\boldsymbol{r}, \boldsymbol{c}, t) = f_{\mathrm{M}}[1+\Phi(\boldsymbol{r}, \boldsymbol{c}, t)], \tag{4.8-1a}$$

with

$$f_{\mathrm{M}} \equiv f_{\mathrm{M}}(\boldsymbol{c}) = n(m/2\pi k T_{\mathrm{w}})^{\frac{3}{2}} \exp(-mc^2/2kT_{\mathrm{w}}), \tag{4.8-1b}$$

where n and T_{w} are constants and T_{w} is equal to the temperature of the wall. The perturbation function Φ satisfies the linearized Boltzmann equation

$$\frac{\partial \Phi}{\partial t} = -\boldsymbol{c}\cdot\nabla_r \Phi - \iint f_{\mathrm{M}1}(\Phi+\Phi_1-\Phi'-\Phi_1')g\sigma\, \mathrm{d}^2\Omega\, \mathrm{d}^3 c_1 \tag{4.8-2}$$

for all $\boldsymbol{r} \in V$ and for all $\boldsymbol{c}$. At the initial instant $t = 0$ one has

$$f(\boldsymbol{r}, \boldsymbol{c}, 0) = f_0(\boldsymbol{r}, \boldsymbol{c}) \quad (\boldsymbol{r} \in V, \text{ all } \boldsymbol{c}) \tag{4.8-3}$$

where f_0 is a given function, uniquely defined throughout V and for all $\boldsymbol{c}$. In terms of Φ, the initial condition (4.8-3) is

$$\Phi(\boldsymbol{r}, \boldsymbol{c}, 0) = \Phi_0(\boldsymbol{r}, \boldsymbol{c}), \tag{4.8-4a}$$

where

$$\Phi_0(\boldsymbol{r}, \boldsymbol{c}) = f_{\mathrm{M}}^{-1} f_0(\boldsymbol{r}, \boldsymbol{c}) - 1. \tag{4.8-4b}$$

If the vessel V is finite and bounded by a solid surface S, we have, in addition to the initial condition (4.8-3), local boundary conditions prescribed at every point $\boldsymbol{r}_s \in S$,

$$f^+ = \mathfrak{R} f^-, \quad f^\pm = \begin{cases} f & \text{if } \boldsymbol{c}\cdot\boldsymbol{n} \gtrless 0, \\ 0 & \text{if } \boldsymbol{c}\cdot\boldsymbol{n} \lessgtr 0, \end{cases} \tag{4.8-5}$$

where $\boldsymbol{n}$ is the local normal on S pointing towards the interior of V. We assume that the operator $\mathfrak{R}$, which connects the incoming and emitted distributions at $\boldsymbol{r}_s$, is linear and of the form described in Section 4.4, i.e., instead of Eq. (4.8-5) we may write

$$f^+(\boldsymbol{c}) = -\int \frac{\boldsymbol{c}'\cdot\boldsymbol{n}}{\boldsymbol{c}\cdot\boldsymbol{n}} W(\boldsymbol{c}, \boldsymbol{c}') f^-(\boldsymbol{c}')\, \mathrm{d}^3 c', \tag{4.8-6}$$

where the kernel W is a probability density which satisfies the conditions (4.4-3) and (4.4-4). The boundary condition (4.8-5) must be satisfied at all times $t \geqq 0$. In terms of the perturbation function Φ, the boundary condition (4.8-5) becomes

$$\Phi^+ = \mathfrak{H}\Phi^-, \quad \Phi^\pm = \begin{cases} \Phi & \text{if } \boldsymbol{c}\cdot\boldsymbol{n} \gtrless 0, \\ 0 & \text{if } \boldsymbol{c}\cdot\boldsymbol{n} \lessgtr 0, \end{cases} \tag{4.8-7}$$

where the operator $\mathfrak{H}$ is identical with the operator $\mathfrak{H}$ introduced in Eq. (4.5-6); we assume that $\mathfrak{H}$ has the properties (4.5-8) and (4.5-9).

The problem consists of finding (sufficient) conditions to be imposed on the initial distribution f_0 (or Φ_0) and, possibly, on the operator $\mathfrak{K}$ (or $\mathfrak{H}$), such that the solution to the initial value problem formulated above exists and is uniquely determined.

Our study is based upon Eqs. (4.8-2), (4.8-4a) and (4.8-7). We start by writing Eq. (4.8-2) in the form

$$\frac{\partial \Phi}{\partial t} = A\Phi, \tag{4.8-8a}$$

$$A\Phi = -\boldsymbol{c}\cdot\nabla_r\Phi - \iint f_{M1}(\Phi+\Phi_1-\Phi'-\Phi_1')g\sigma\, \mathrm{d}^2\Omega\, \mathrm{d}^3c_1 . \tag{4.8-8b}$$

The operator A does not depend on t. If A were just a real or complex number, the solution to Eq. (4.8-8a), subject to the initial condition (4.8-4a), would be simply

$$\Phi(\boldsymbol{r}, \boldsymbol{c}, t) = \mathrm{e}^{tA}\,\Phi_0(\boldsymbol{r}, \boldsymbol{c}). \tag{4.8-9}$$

Since A is actually an operator, the first problem is to find out whether the operator $T(t) = \mathrm{e}^{tA}$ exists, to determine its properties, and, particularly, to find out whether Φ as given by Eq. (4.8-9) – with the proper meaning assigned to e^{tA} – actually reduces to Φ_0 as $t \to 0$.

The situation is further complicated because of the boundary condition (4.8-7) which has to be satisfied at all times $t \geqq 0$: Does the operator e^{tA} leave the boundary conditions unaffected – i.e., if the initial function Φ_0 satisfies Eq. (4.8-7), does the function Φ, determined from Eq. (4.8-9), satisfy the same condition (4.8-7) at *all* later times $t > 0$?

In the subsequent analysis it will be shown that the above questions can be answered satisfactorily. By properly defining the domain of the operator A, whose formal expression has already been given in Eq. (4.8-8b), one can prove that A is a so-called *infinitesimal generator of a semigroup of operators*,

$\{T(t),\ 0 \leqq t < \infty\}$, which enjoy certain continuity properties. Without going into the mathematical details, this implies that there exists a one-parameter family of operators $\{T(t), 0 \leqq t < \infty\}$ which satisfy the semi-group property: $T(t_1)T(t_2) = T(t_1+t_2)$ for $t_1, t_2 > 0$, and, furthermore, $T(0) = I$, where I is the identity operator. This result is, of course, in accordance with intuition if one thinks of $T(t)$ as e^{tA}. The continuity properties imply, for example, that $\lim_{t \gtrsim 0} T(t)f = f$ for any f in the domain of the semigroup (again, a precise description of the sense in which the limit is taken, is omitted). In terms of A, $T(t)$ may be represented by

$$T(t)f = \lim_{n\to\infty} \exp\{tA(I-n^{-1}A)^{-1}\}f, \tag{4.8-10}$$

the convergence being uniform in t on every closed subinterval of $0 \leqq t < \infty$. Thus, $T(t)$ has a mathematically well-defined meaning and, instead of Eq. (4.8-9), one obtains the following expression for the solution Φ to the initial value problem,

$$\Phi(\boldsymbol{r}, \boldsymbol{c}, t) = T(t)\Phi_0(\boldsymbol{r}, \boldsymbol{c}). \tag{4.8-11}$$

Moreover, as the domain of the semigroup (i.e., the set of all functions to which the operator $T(t)$, $0 \leqq t < \infty$, can be legitimately applied) is invariant under the application of the semigroup, it follows that, if Φ_0 satisfies the boundary condition (4.8-7) at the initial instant $t = 0$, Φ does so at *all* later times $t > 0$.

We now turn to the rigorous analysis. In order to solve an initial value problem described by Eq. (4.8-8), we must analyze the properties of the operator A. To proceed effectively we must choose a space of functions in which to work. We choose the real Hilbert space $\mathscr{H}$ of all measurable functions $\phi \equiv \phi(\boldsymbol{r}, \boldsymbol{c})$ that are square integrable with weight f_M over the domain $V \times \omega$ of phase space (V is the region of physical space that is occupied by the system and ω is the entire velocity space). The scalar product of two elements ϕ and ψ of $\mathscr{H}$ is defined by

$$(\phi, \psi) = \iint_{V\times\omega} f_M \phi\psi \, \mathrm{d}^3 r \, \mathrm{d}^3 c, \tag{4.8-12}$$

so that the norm of any element $\phi \in \mathscr{H}$ becomes

$$||\phi|| = \left(\iint_{V\times\omega} f_M(\boldsymbol{c})\,(\phi(\boldsymbol{r}, \boldsymbol{c}))^2 \, \mathrm{d}^3 r \, \mathrm{d}^3 c\right)^{\frac{1}{2}}. \tag{4.8-13}$$

Apart from technical reasons this particular choice of $\mathscr{H}$ has a physical justification. Consider a system of particles whose velocity distribution function, f, is such that the velocity moments of f with respect to the summational invariants exist. Suppose, for simplicity, that the number of particles in the system is constant in time. If the macroscopic variables n, $\boldsymbol{v}$, and T of the system are approximately equal to the corresponding parameters of f_{M} – the difference being negligibly small at least to the first order in the perturbation function – then, apart from an additive constant, Boltzmann's H is equal to

$$\begin{aligned} H &= \iint f \log \frac{f}{n(m/2\pi kT)^{\frac{3}{2}}} \, \mathrm{d}^3c\, \mathrm{d}^3r \\ &= \iint f_{\mathrm{M}}(1+\Phi)[-mC^2/2kT + \log(1+\Phi)]\, \mathrm{d}^3c\, \mathrm{d}^3r \\ &= -U/kT + \iint f_{\mathrm{M}}(1+\Phi) \log(1+\Phi)\, \mathrm{d}^3c\, \mathrm{d}^3r \end{aligned} \tag{4.8-14a}$$

which, upon linearization, becomes

$$H = -U/kT + \tfrac{1}{2} \iint f_{\mathrm{M}} \Phi^2\, \mathrm{d}^3c\, \mathrm{d}^3r, \tag{4.8-14b}$$

where U is the total internal energy of the gas. So, if we associate H with $-(1/k)S$, where S is the entropy of the system, then

$$\|\Phi\|^2 = (2/kT)(U - TS). \tag{4.8-15}$$

In other words, $\|\Phi\|^2$ is a measure of the free energy. Thus, studying the linearized Boltzmann equation in the Hilbert space $\mathscr{H}$ introduced above, we restrict our attention to systems with a *finite* amount of free energy.

Because of the assumption on the intermolecular potential, the operator A may be split into three parts,

$$A = -\boldsymbol{c} \cdot \boldsymbol{\nabla}_r - \nu - L_0, \tag{4.8-16a}$$

where ν is a (bounded) multiplicative operator corresponding to multiplication with the collision frequency,

$$\begin{aligned} \nu\phi(\boldsymbol{c}) &= \nu(\boldsymbol{c})\phi(\boldsymbol{c}) \\ &= \iint f_{\mathrm{M}}(\boldsymbol{c}_1) g\sigma\, \mathrm{d}^2\Omega\, \mathrm{d}^3c_1\, \phi(\boldsymbol{c}), \end{aligned} \tag{4.8-16b}$$

and L_0 is a linear integral operator,

$$L_0\,\phi(\boldsymbol{c}) = \iint f_{\mathrm{M}1}\,[\phi(\boldsymbol{c}_1)-\phi(\boldsymbol{c}')-\phi(\boldsymbol{c}_1')]g\sigma\,\mathrm{d}^2\Omega\,\mathrm{d}^3c_1\,. \quad (4.8\text{-}16c)$$

It has been shown by Grad [1963b] that L_0 is a completely continuous operator on $\mathscr{H}$ for rigid spheres and any power-law potential with an angular cut-off. Thus, A is the sum of an unbounded operator $(\boldsymbol{c}\cdot\boldsymbol{\nabla}_r)$, a bounded operator (ν) and a completely continuous operator (L_0).

A basic tool in proving existence and uniqueness theorems in Banach spaces is the Hille-Yosida theorem characterizing the infinitesimal generators of semigroups – see Hille and Phillips [1957] or Yosida [1968]. In a Hilbert space such a characterization is slightly simpler by virtue of the existence of a scalar product. In fact, if A is a linear operator with domain $D(A)$ and range $R(A)$ both in the real Hilbert space $\mathscr{H}$, A is called *dissipative* if $(Af, f) \leqq 0$ whenever $f \in D(A)$, and a theorem of Phillips and Lumer (see Yosida [1968]) states that, if $D(A)$ is dense in $\mathscr{H}$, then A is the infinitesimal generator of a contraction semigroup of class (C_0) in $\mathscr{H}$ if and only if A is dissipative and $R(\lambda I - A) = \mathscr{H}$ for some $\lambda > 0$. As a corollary one proves that, if A is a densely defined closed linear operator, with $D(A)$ and $R(A)$ both in $\mathscr{H}$, and if A and its adjoint A^* are dissipative, then, again, A is the infinitesimal generator of a contraction semigroup of class (C_0).

As we shall see later, for operators which generate a contraction semigroup of class (C_0) the so-called abstract initial value problem can be solved. Hence, our first objective is to define the domain $D(A)$ of the operator A given by Eq. (4.8-8b) in such a way that A is indeed the infinitesimal generator of a contraction semigroup of class (C_0). For the operator A to exist as an operator from $\mathscr{H}$ into itself we require that all elements $\phi \in \mathscr{H}$ which belong to the domain $D(A)$ of A, are absolutely continuous (and, hence, differentiable almost everywhere) with respect to $\boldsymbol{r}$ throughout V, and satisfy the condition: $\boldsymbol{c}\cdot\boldsymbol{\nabla}_r\phi \in \mathscr{H}$. To meet the boundary condition (4.8-7) we further require that every element of $D(A)$ actually satisfies Eq. (4.8-7).

For any $\phi \in \mathscr{H}$ which is absolutely continuous with respect to $\boldsymbol{r}$ throughout V, such that $\boldsymbol{c}\cdot\boldsymbol{\nabla}_r\phi \in \mathscr{H}$, one has

$$(\phi, \boldsymbol{c}\cdot\boldsymbol{\nabla}_r\phi) = -\iint_{S\times\omega} (\boldsymbol{c}\cdot\boldsymbol{n}) f_{\mathrm{M}}|\phi|^2\,\mathrm{d}^2S\,\mathrm{d}^3c - (\boldsymbol{c}\cdot\boldsymbol{\nabla}_r\phi, \phi). \quad (4.8\text{-}17)$$

($\boldsymbol{n}$ is the local normal on S, oriented toward the interior of V.) So

$$(\phi, \boldsymbol{c}\cdot\nabla_r\phi) = -\tfrac{1}{2}\iint\limits_{S\times\omega}(\boldsymbol{c}\cdot\boldsymbol{n})f_{\mathrm{M}}\phi^2\,\mathrm{d}^2S\,\mathrm{d}^3c \tag{4.8-18a}$$

or, equivalently,

$$(\phi, \boldsymbol{c}\cdot\nabla_r\phi) = \tfrac{1}{2}\int_S\left\{\int|\boldsymbol{c}\cdot\boldsymbol{n}|f_{\mathrm{M}}(\phi^-)^2\,\mathrm{d}^3c - \int|\boldsymbol{c}\cdot\boldsymbol{n}|f_{\mathrm{M}}(\phi^+)^2\,\mathrm{d}^3c\right\}\mathrm{d}^2S. \tag{4.8-18b}$$

At every point $\boldsymbol{r}_s \in S$ we introduce the real Hilbert space $\mathscr{H}_S$ of all measurable functions $\phi^+(\boldsymbol{r}_s, \boldsymbol{c})$, defined for all $\boldsymbol{c}$ with $\boldsymbol{c}\cdot\boldsymbol{n} > 0$, which are square integrable with weight $(\boldsymbol{c}\cdot\boldsymbol{n})f_{\mathrm{M}}$. The scalar product of two elements ϕ^+ and ψ^+ of $\mathscr{H}_S$ is

$$(\phi^+, \psi^+)_S = \int_{\boldsymbol{c}\cdot\boldsymbol{n}>0}(\boldsymbol{c}\cdot\boldsymbol{n})f_{\mathrm{M}}(\boldsymbol{c})\,\phi^+(\boldsymbol{c})\psi^+(\boldsymbol{c})\,\mathrm{d}^3c \tag{4.8-19}$$

and the S-norm of any element $\phi^+ \in \mathscr{H}_S$ is

$$\|\phi^+\|_S = (\phi^+, \phi^+)_S^{\frac{1}{2}}. \tag{4.8-20}$$

Thus, for any $\phi \in \mathscr{H}$ which – in addition to being absolutely continuous with respect to $\boldsymbol{r}$ throughout V, with $\boldsymbol{c}\cdot\nabla_r\phi \in \mathscr{H}$ – is such that, at every point $\boldsymbol{r}_s \in S$, its restriction ϕ^+ and its reflected restriction $\widetilde{\phi^-}$ belong to the local Hilbert space $\mathscr{H}_S$, we have the identity

$$(\phi, \boldsymbol{c}\cdot\nabla_r\phi) = \tfrac{1}{2}\int_S\{\|\widetilde{\phi^-}\|_S^2 - \|\phi^+\|_S^2\}\,\mathrm{d}^2S. \tag{4.8-21}$$

(As before, the symbol $\widetilde{\ }$ indicates that the argument $\boldsymbol{c}$ is replaced by $\boldsymbol{c}-2(\boldsymbol{c}\cdot\boldsymbol{n})\boldsymbol{n}$.)

Now, suppose furthermore that ϕ satisfies the boundary condition (4.8-7). Then, instead of Eq. (4.8-21), we have the expression

$$(\phi, \boldsymbol{c}\cdot\nabla_r\phi) = \tfrac{1}{2}\int\{\|\widetilde{\phi^-}\|_S^2 - \|\mathfrak{H}\phi^-\|_S^2\}\,\mathrm{d}^2S. \tag{4.8-22}$$

If the identity (4.5-9) – which holds for *any* function Φ – is applied to the function $(\phi^-)^2$ and the result is used in the definition of $\|\widetilde{\phi^-}\|_S^2$, one obtains

$$\|\widetilde{\phi^-}\|_S^2 = \int(\boldsymbol{c}\cdot\boldsymbol{n})\,f_{\mathrm{M}}\,\mathfrak{H}(\phi^-)^2\,\mathrm{d}^3c. \tag{4.8-23}$$

Moreover, using Jensen's inequality as in Section 4.5, one may verify that the inequality

$$(\mathfrak{H}\phi^-)^2 \leqq \mathfrak{H}(\phi^-)^2 \tag{4.8-24}$$

holds for all $\boldsymbol{c}$ for which $\boldsymbol{c} \cdot \boldsymbol{n} > 0$. Thus, combining Eqs. (4.8-23) and (4.8-24) and using the definition of $||\mathfrak{H}\phi^-||_S^2$ we find that the properties (4.5-8) and (4.5-9) are sufficient to prove the inequality

$$||\widetilde{\phi^-}||_S^2 \geqq ||\mathfrak{H}\phi^-||_S^2 \tag{4.8-25}$$

and, thence, the inequality

$$(\phi, \boldsymbol{c} \cdot \boldsymbol{\nabla}_r \phi) \geqq 0 \tag{4.8-26}$$

for any ϕ which satisfies the assumptions above.

Finally, standard manipulations yield the result

$$\begin{aligned}(\phi, (v+L_0)\phi) &= \iiiint f_{\mathrm{M}} f_{\mathrm{M}1}\, \phi(\phi+\phi_1-\phi'-\phi_1') g\sigma \,\mathrm{d}^2\Omega\, \mathrm{d}^3 c_1\, \mathrm{d}^3 c\, \mathrm{d}^3 r \\ &= \tfrac{1}{4} \iiiint f_{\mathrm{M}} f_{\mathrm{M}1} (\phi+\phi_1-\phi'-\phi_1')^2 g\sigma \,\mathrm{d}^2\Omega\, \mathrm{d}^3 c_1\, \mathrm{d}^3 c\, \mathrm{d}^3 r \geqq 0. \end{aligned} \tag{4.8-27}$$

So, if we define the domain $D(A)$ of A by

$$\begin{aligned} D(A) = \{\phi : \phi \in \mathscr{H},\ &\phi \text{ absolutely continuous with respect to } \boldsymbol{r} \\ &\text{throughout } V,\ \boldsymbol{c} \cdot \boldsymbol{\nabla}_r \phi \in \mathscr{H};\ \phi^+ \in \mathscr{H}_S,\ \widetilde{\phi^-} \in \mathscr{H}_S \text{ and} \\ &\phi^+ = \mathfrak{H}\phi^- \text{ locally for all } \boldsymbol{r}_s \in S\}, \end{aligned}$$

then

$$(\phi, A\phi) = -(\phi, \boldsymbol{c} \cdot \boldsymbol{\nabla}_r \phi) - (\phi, (v+L_0)\phi) \leqq 0 \tag{4.8-28}$$

and, thus, the linear operator A, whose domain $D(A)$ and range $R(A)$ are both in $\mathscr{H}$, is dissipative.

$D(A)$ is certainly dense in $\mathscr{H}$. To find the adjoint A^* of A, we take any $\phi \in D(A)$ and obtain, by elementary manipulations,

$$\begin{aligned}(A\phi, \psi) &= (-\boldsymbol{c} \cdot \boldsymbol{\nabla}_r \phi - v\phi - L_0 \phi, \psi) \\ &= (\phi, \boldsymbol{c} \cdot \boldsymbol{\nabla}_r \psi - v\psi - L_0 \psi) + \iint\limits_{S \times \omega} (\boldsymbol{c} \cdot \boldsymbol{n}) f_{\mathrm{M}} \phi \psi \,\mathrm{d}^3 c\, \mathrm{d}^2 S. \end{aligned} \tag{4.8-29}$$

The integral, in turn, may be transformed into

$$\iint\limits_{S \times \omega} (\boldsymbol{c} \cdot \boldsymbol{n}) f_{\mathrm{M}} \phi \psi \,\mathrm{d}^3 c\, \mathrm{d}^2 S = \int_S \{(\mathfrak{H}\phi^-, \psi^+)_S - (\widetilde{\phi^-}, \widetilde{\psi^-})_S\} \,\mathrm{d}^2 S. \tag{4.8-30}$$

Introducing the adjoint operator $\mathfrak{H}^*$, defined by

$$(\mathfrak{H}\phi^-, \psi^+)_S = (\widetilde{\phi^-}, \widetilde{\mathfrak{H}^*\psi^+})_S \tag{4.8-31}$$

for all ϕ^- such that $\widetilde{\phi^-} \in \mathscr{H}_S$, we finally write Eq. (4.8-30) in the form

$$\iint\limits_{S\times\omega} (\boldsymbol{c}\cdot\boldsymbol{n}) f_M \phi\psi \, d^3c \, d^2S = \int_S (\widetilde{\phi^-}, \widetilde{\mathfrak{H}^*\psi^+} - \widetilde{\psi^-})_S \, d^2S. \tag{4.8-32}$$

Hence, from Eqs. (4.8-29) and (4.8-32) we conclude that A^* is defined by

$$A^* = \boldsymbol{c}\cdot\nabla_r - \nu - L_0 \tag{4.8-33}$$

on the domain $D(A^*)$:

$$D(A^*) = \{\phi : \phi \in \mathscr{H},\ \phi \text{ absolutely continuous with respect to } \boldsymbol{r} \text{ throughout } V,\ \boldsymbol{c}\cdot\nabla_r\phi \in \mathscr{H};\ \phi^+ \in \mathscr{H}_S,\ \widetilde{\phi^-} \in \mathscr{H}_S \text{ and } \phi^- = \mathfrak{H}^*\phi^+ \text{ locally for all } \boldsymbol{r}_s \in S\}.$$

It is easily verified that $A^{**} = A$, hence A is closed. Furthermore, if $\phi \in D(A^*)$,

$$(\phi, \boldsymbol{c}\cdot\nabla_r\phi) = \tfrac{1}{2}\int_S \{\|\widetilde{\mathfrak{H}^*\phi^+}\|_S^2 - \|\phi^+\|_S^2\} \, d^2S, \tag{4.8-34}$$

and since, according to Eq. (4.8-25), $\mathfrak{H}$ is a contractive operator, its adjoint $\mathfrak{H}^*$ is contractive as well, i.e.,

$$\|\widetilde{\mathfrak{H}^*\phi^+}\|_S^2 \leqq \|\phi^+\|_S^2 \tag{4.8-35}$$

and, thus, we have the inequality

$$(\phi, \boldsymbol{c}\cdot\nabla_r\phi) \leqq 0 \tag{4.8-36}$$

for any $\phi \in D(A^*)$. Consequently,

$$(\phi, A^*\phi) = (\phi, \boldsymbol{c}\cdot\nabla_r\phi) - (\phi, (\nu + L_0)\phi) \leqq 0, \tag{4.8-37}$$

i.e., A^* is dissipative.

This completes the proof that the operator A, defined by Eq. (4.8-8b) on the domain $D(A) \subset \mathscr{H}$, is the infinitesimal generator of a contraction semi-group $\{T(t), 0 \leqq t < \infty\}$ of class (C_0) or, in other words, that there exists a one-parameter family of bounded linear operators $T(t)$ on $\mathscr{H}$ which has the following properties,

$$T(t_1)T(t_2) = T(t_1+t_2) \qquad \text{(semigroup property)}$$
$$T(0) = I$$
$$\|T(t)\| \leqq 1 \qquad \text{(contraction property)}$$
$$\operatorname*{s-lim}_{t\to 0} T(t)f = f \text{ for all } f \in \mathscr{H} \qquad \text{(strong continuity)},$$

and which is represented in terms of A by Eq. (4.8-10).

For operators which generate a contraction semigroup of class (C_0) the so-called abstract initial value problem can be solved. Referring the reader to Hille and Phillips [1957] or Yosida [1968] for further details, we formulate the result in the following theorem.

Theorem: The operator A, defined by Eq. (4.8-8b) on the domain $D(A) \subset \mathscr{H}$, is the infinitesimal generator of a contraction semigroup $\{T(t), 0 \leqq t < \infty\}$ of class (C_0), where $T(t)$ is represented in terms of A by Eq. (4.8-10). If $\Phi_0 \in D(A)$, there exists one and only one function $\Phi(t)$ from $[0, \infty)$ to $\mathscr{H}$ such that

(i) $\Phi(t)$ is continuously differentiable, $\Phi(t) \in D(A)$ for each $t \in [0, \infty)$,
(ii) $\Phi(t)$ satisfies the linearized Boltzmann equation (4.8-8),
(iii) $\lim_{t\to 0} \|\Phi(t)-\Phi_0\| = 0$.

The function $\Phi(t)$ is given by $\Phi(t) = T(t)\,\Phi_0$.

Thus we have shown that, under properly chosen initial conditions, the linearized Boltzmann equation yields a unique solution at all times $t \geqq 0$. Boundary conditions of the type discussed in Section 4.4 will remain to be satisfied once they are satisfied by the initial distribution function.

To conclude, we remark that representations other than Eq. (4.8-10) may exist for the semigroup $T(t)$. In particular, if the spectral decomposition of the complex λ-plane by the operator A and the analytical properties of the resolvent $R_\lambda = (\lambda I - A)^{-1}$ are known, one may obtain a representation of $T(t)$ from the expression

$$T(t) = \frac{1}{2\pi \mathrm{i}} \int_{\gamma-\mathrm{i}\infty}^{\gamma+\mathrm{i}\infty} \mathrm{e}^{\lambda t} R_\lambda \, \mathrm{d}\lambda.$$

However, we will not go into a detailed study of the operator A at this point. Some further properties will be discussed in Chapter 15 in the context of rarefied gas dynamics. The main objective of the present section has been to rigorously prove existence and uniqueness of the solution to the linearized Boltzmann equation and this objective has been obtained in the theorem formulated above.

The non-uniform state of a simple gas 5

In this chapter we turn to the solution of the Boltzmann equation following the method developed by Enskog in his doctoral dissertation [1917]. We consider a simple gas, the discussion of a mixture of gases is postponed until Chapter 6.

In the previous chapter it was shown that, in the steady state, the Boltzmann equation is satisfied uniquely by the Maxwellian velocity distribution function. In general, however, the Boltzmann equation does not permit exact solution and, hence, must be solved in some approximate way.

Now, in the large majority of physical situations an adequate and completely satisfactory description of the state of a gas is provided by fluid dynamics. In mathematical terms, this description consists of a set of partial differential equations for the measurable macroscopic properties of the gas – density, hydrodynamic velocity and temperature – with the property that, if the values of these variables are known at *one* instant, their values at *all* later times can be predicted unambiguously. This is so for two reasons : (i) the equations are first-order differential equations in the time variable, (ii) the equations are self-contained, provided only that certain transport coefficients – viscosity and thermal conductivity – are expressed as functions of the macroscopic variables mentioned earlier.

In the previous chapter, Section 4.1, a set of equations similar to the equations of fluid dynamics was derived – viz., the conservation equations of mass, momentum and energy. However, the description provided by these equations is not self-contained, since the pressure tensor and heat flow vector that appear in them have not been expressed in terms of the macroscopic variables. Furthermore, any attempt to find an exact relation of this type directly from the Boltzmann equation will not meet with success. If, for example, one attempts to find such a relation by taking higher-order velocity moments of the Boltzmann equation one finds that still further variables, in addition to the pressure tensor and heat flow vector, are introduced. On the other hand, the success of fluid dynamics suggests that one should look for a procedure to generate from kinetic theory approximate expressions for the pressure tensor and heat flow vector such that, in the regime in which the fluid dynamic description is known to be good, the Newton and Fourier laws of friction and heat conduction, respectively, are obtained as low-order approximations with the transport coefficients explicitly known in terms of the intermolecular potentials. Then, upon substitution, the macroscopic conservation equations would become self-contained and, in fact, identical with the equations of fluid dynamics. These ideas lead to the construction of approximate solutions of the Boltzmann equation presented in this and the following chapters.

The following observations may make the situation clearer. In the Boltzmann equation, the important time scale is the mean free time ($\sim 10^{-9}$ sec. at normal conditions) and the distribution function carries the information. In contrast, the time scale of importance in fluid dynamics is the time required for a sound wave to cross a dimension of macroscopic interest (typically $\sim 10^{-3}$ sec.) and all of the essential information is provided by the small number of macroscopic properties – density, hydrodynamic velocity and temperature. In other words, a transition from kinetic theory to fluid dynamics implies a *contraction* of the formal description. This situation is reminiscent of the situation encountered in Chapter 3. There, the description of the dynamical N-body problem was first given in terms of the N-particle distribution function, which was governed by the Liouville equation, and subsequently contracted to a description in terms of the one-particle distribution function, governed by the (generalized) Boltzmann equation. A satisfactory answer to the present contraction problem was first given by Hilbert [1912], who discussed the existence and uniqueness of solutions of the Boltzmann equation. By suitably restricting the class of functions in which a solution of Boltzmann's equation is sought, Hilbert proved the existence of a one-to-one correspondence between the solution f and the first five velocity moments of f, i.e. the density, hydrodynamic velocity and temperature, at any instant t. It is important to note that this links a unique solution of the Boltzmann equation with each solution of the fluid dynamic equations. Hilbert's theory is dealt with in the first section of the present chapter.

After the structure of the solution to the Boltzmann equation and the unique correspondence between the velocity distribution function and the macroscopic variables n, $\boldsymbol{v}$ and T were established, methods for solving the Boltzmann equation by successive approximation were developed independently by Chapman and Enskog in the 1910's. The principles of these methods are explained in Section 5.2. The zero- and first-order results are given in Sections 5.3 and 5.4, respectively. Section 5.5 deals with certain computational aspects of the first-order approximation.

Since, in the first-order approximation, the Navier-Stokes equations of hydrodynamics turn out to be the appropriate equations for the macroscopic variables, Newton's and Fourier's law are explicitly obtained. Thus, the evaluation of the viscosity and thermal conductivity transport coefficients from first principles is possible; the results are found in Section 5.6. In Section 5.7 an alternative method of evaluating these transport coefficients in the first-order theory, due to Kihara, is given. In order to discuss the qualitative aspects of the second-order approximation, which was fully worked out by Burnett, we have given the main results of this approximation in Section 5.8. Order-of-magnitude estimates of the successive approximations obtained in the Chapman-Enskog theory are to be found in Section 5.9. In a final section to this chapter, Section 5.10, we give an interpretation of the Chapman-Enskog theory and the results obtained with it.

5.1. Hilbert's uniqueness theorem

In a discussion of the equations of kinetic theory, Hilbert [1912] proved a uniqueness theorem, to which we now turn. Before proceeding, we observe that if the motion of molecules from one point to another could be "turned off", collisions would bring the distribution function to the Maxwellian within a few collision times. Furthermore, if macroscopic gradients are small, the molecules drifting into a small region do not differ greatly in velocity distribution from those already there; they will induce variations only on the macroscopic time scale mentioned in the introduction. Thus the dominant effect is that of collisions driving the gas toward thermal equilibrium. To emphasize this we introduce a small parameter ε to give more weight to the collision term. Thus we write, instead of Eq. (4.1-1),

$$\mathfrak{D}f = \varepsilon^{-1} J(ff). \qquad (5.1\text{-}1)$$

Since Eq. (5.1-1) corresponds to the Boltzmann equation for a velocity distribution function $\varepsilon^{-1}f$, ε is merely a scaling factor for the density. During the following analysis we treat ε as a small parameter; of course, eventually ε will be set equal to unity, so that all the results are in formal agreement with the original nomenclature.

Now, if ε is small, Eq. (5.1-1) is a *singular perturbation problem* for f. However, proceeding formally we look for a solution which is *regular* near $\varepsilon = 0$, and adopt the expansion

$$f = f^{(0)} + \varepsilon f^{(1)} + \varepsilon^2 f^{(2)} + \ldots \tag{5.1-2}$$

Thus, we substitute this expansion into Eq. (5.1-1) and equate the coefficients of like powers of ε. We obtain the following equation for $f^{(0)}$,

$$J(f^{(0)}f^{(0)}) = 0, \tag{5.1-3}$$

while, for $r = 1, 2, \ldots$ the function $f^{(r)}$ satisfies the equation

$$J(f^{(0)}f^{(r)}) + J(f^{(r)}f^{(0)}) = \mathfrak{D}f^{(r-1)} - J(f^{(1)}f^{(r-1)}) - \ldots - J(f^{(r-1)}f^{(1)}), \tag{5.1-4}$$

there being r terms on the right (when $r = 1$, the single term on the right is $\mathfrak{D}f^{(0)}$). Eq. (5.1-3) has been discussed in connection with the H-theorem. Its general solution is

$$f^{(0)}(\boldsymbol{c}) = n^{(0)}(m/2\pi kT^{(0)})^{\frac{3}{2}} \exp(-mC^2/2kT^{(0)}), \tag{5.1-5}$$

where $\boldsymbol{C} = \boldsymbol{c} - \boldsymbol{v}^{(0)}$; $n^{(0)}$, $\boldsymbol{v}^{(0)}$, $T^{(0)}$ may be *arbitrary functions of space and time*, related to $f^{(0)}$ in the following manner:

$$n^{(0)} = \int f^{(0)}\, \mathrm{d}^3c, \tag{5.1-6}$$

$$\rho^{(0)}\boldsymbol{v}^{(0)} = \int m\boldsymbol{c} f^{(0)}\, \mathrm{d}^3c, \tag{5.1-7}$$

$$\tfrac{3}{2}n^{(0)}kT^{(0)} = \rho^{(0)}u^{(0)} = \int \tfrac{1}{2}mC^2 f^{(0)}\, \mathrm{d}^3c. \tag{5.1-8}$$

Now consider Eq. (5.1-4). At every value of r, the right-hand side involves functions that are known from the solutions of the previous equations; the unknown function appears only on the left and occurs there linearly. To obtain the general solution, let us put $\phi^{(r)} = f^{(r)}/f^{(0)}$. Because $f^{(0)}$ is Maxwellian,

$$J(f^{(0)}f^{(r)})+J(f^{(r)}f^{(0)}) = -\iiint f^{(0)}f_1^{(0)}(\phi^{(r)}+\phi_1^{(r)}-\phi^{(r)'}-\phi_1^{(r)'})$$
$$\times gb\,\mathrm{d}b\,\mathrm{d}\varepsilon\,\mathrm{d}^3c_1 \quad (5.1\text{-}9)$$

or, taking into account the definition (4.7-1) of the operator I,

$$J(f^{(0)}f^{(r)})+J(f^{(r)}f^{(0)}) = (n^{(0)})^2 I(\phi^{(r)}). \quad (5.1\text{-}10)$$

Then, using expression (4.6-27), it follows that

$$J(f^{(0)}f^{(r)})+J(f^{(r)}f^{(0)}) = -K_0(\boldsymbol{c})\,\phi^{(r)}(\boldsymbol{c})+\int K(\boldsymbol{c},\, \boldsymbol{c}_1)\,\phi^{(r)}(\boldsymbol{c}_1)\,\mathrm{d}^3c_1, \quad (5.1\text{-}11)$$

where $K(\boldsymbol{c}, \boldsymbol{c}_1)$ is a symmetric function of $\boldsymbol{c}$, $\boldsymbol{c}_1$. Thus, Eq. (5.1-4) is a linear, inhomogeneous Fredholm integral equation of the second kind in terms of the function $\phi^{(r)}$. The associated homogeneous equation is

$$I(\phi^{(r)}) = 0. \quad (5.1\text{-}12)$$

The latter has five independent solutions, which may be chosen to be 1, the three components of $m\boldsymbol{c}$ and $\frac{1}{2}mC^2$ and which we shall interpret as the components ψ_k of a vector $\boldsymbol{\psi}$. From the theory of Fredholm integral equations we infer that the inhomogeneous equation (5.1-4) is soluble if and only if the right member is orthogonal to *all* solutions of the homogeneous equation (5.1-12). So, for $r = 1, 2, \ldots$ a solution $f^{(r)}$ will exist if and only if

$$\int \boldsymbol{\psi}\{\mathfrak{D}f^{(r-1)}-J(f^{(1)}f^{(r-1)})-\ \ldots\ -J(f^{(r-1)}f^{(1)})\}\,\mathrm{d}^3c = 0. \quad (5.1\text{-}13)$$

If this condition is satisfied, there is a five-parameter family of solutions of Eq. (5.1-4):

$$\phi^{(r)} = \bar{\phi}^{(r)}+\boldsymbol{\alpha}^{(r)}\cdot\boldsymbol{\psi}, \quad (5.1\text{-}14)$$

where $\bar{\phi}^{(r)}$ is any particular solution of Eq. (5.1-4) and $\boldsymbol{\alpha}^{(r)}$ is a vector whose components – $\alpha^{(1,r)}$, the three components of $\boldsymbol{\alpha}^{(2,r)}$, and $\alpha^{(3,r)}$ – are arbitrary functions of space and time; $\bar{\phi}^{(r)}$ is uniquely determined if we require

$$\int \boldsymbol{\psi}\bar{\phi}^{(r)}f^{(0)}\,\mathrm{d}^3c = 0. \quad (5.1\text{-}15)$$

Thus we have found that, for each value of r ($r = 0$ included) there exists a five-parameter family of solutions $f^{(r)}$. However, to guarantee the existence of a solution for $f^{(1)}$, five conditions must be placed on $\mathfrak{D}f^{(0)}$ and, in turn, to solve for $f^{(r)}$, five conditions have to be imposed on $\mathfrak{D}f^{(r-1)}$. These conditions (5.1-13) take the form of partial differential equations that allow us to express, for each value of r, the *time* derivatives of the parameters in terms of their *space* derivatives.

The case of $f^{(0)}$ is distinct from the others, since the parameters – $n^{(0)}$, $\boldsymbol{v}^{(0)}$, $T^{(0)}$ – enter differently. For $r = 1$, the compatibility condition (5.1-13) is

$$\int \psi \mathfrak{D} f^{(0)} \, d^3c = 0, \tag{5.1-16}$$

which are exactly the five macroscopic conservation equations with the distribution function replaced by the Maxwellian $f^{(0)}$. If $f^{(0)}$ is used to calculate the pressure tensor and heat flow vector one obtains,

$$\mathsf{P}^{(0)} = p^{(0)} \boldsymbol{I} \text{ with } p^{(0)} = n^{(0)} kT^{(0)}, \tag{5.1-17}$$

$$\boldsymbol{q}^{(0)} = 0, \tag{5.1-18}$$

so that the conservation equations become

$$(1/\rho^{(0)})(d\rho^{(0)}/dt) = -\nabla \cdot \boldsymbol{v}^{(0)}, \tag{5.1-19}$$

$$\rho^{(0)}(d\boldsymbol{v}^{(0)}/dt) = \rho^{(0)} \boldsymbol{F} - \nabla p^{(0)}, \tag{5.1-20}$$

$$(d/dt)(\rho^{(0)}/T^{(0)\frac{3}{2}}) = 0, \tag{5.1-21}$$

which are the *ideal*, or *Euler*, *hydrodynamic equations* with an equation of state which is the ideal gas law. In turn, the parameters $n^{(0)}(\boldsymbol{r}, t)$, $\boldsymbol{v}^{(0)}(\boldsymbol{r}, t)$ and $T^{(0)}(\boldsymbol{r}, t)$, and therefore $f^{(0)}(\boldsymbol{r}, \boldsymbol{c}, t)$, are uniquely determined provided that initial values $n^{(0)}(\boldsymbol{r}, t_0)$, $\boldsymbol{v}^{(0)}(\boldsymbol{r}, t_0)$ and $T^{(0)}(\boldsymbol{r}, t_0)$ are prescribed. Thus, if a solution for $f^{(1)}$ is to exist, only the initial values of these parameters need be supplied.

We continue by induction. Assume that the functions $f^{(0)} \ldots f^{(r-1)}$ have been uniquely determined. We have

$$f^{(r)} = f^{(0)} \{ \bar{\phi}^{(r)} + \boldsymbol{\alpha}^{(r)} \cdot \boldsymbol{\psi} \}, \tag{5.1-22}$$

where $\bar{\phi}^{(r)}$ is the particular solution of Eq. (5.1-4) that satisfies the condition (5.1-15) and $\boldsymbol{\alpha}^{(r)}$ is a vector whose components are arbitrary functions of space and time. The inhomogeneous integral equation for $f^{(r+1)}$ has a solution if and only if

$$\int \boldsymbol{\psi} \{ \mathfrak{D} f^{(r)} - J(f^{(1)} f^{(r)}) - \ldots - J(f^{(r)} f^{(1)}) \} \, d^3c = 0 \tag{5.1-23}$$

or, since the components ψ_k of $\boldsymbol{\psi}$ are summational invariants,

$$\int \boldsymbol{\psi} \mathfrak{D} f^{(r)} \, d^3c = 0. \tag{5.1-24}$$

These are exactly the general conservation equations with the distribution function replaced by $f^{(r)}$. They suggest defining a vector $\boldsymbol{\beta}^{(r)}$ whose components are the first few velocity moments of $f^{(r)}$, viz. $n^{(r)}$, the three components of $\rho^{(r)}\boldsymbol{v}^{(r)}$, and $\rho^{(r)}u^{(r)}$,

$$\boldsymbol{\beta}^{(r)} = \int \boldsymbol{\psi} f^{(r)} \,\mathrm{d}^3 c. \tag{5.1-25}$$

A relation between the vectors $\boldsymbol{\alpha}^{(r)}$ and $\boldsymbol{\beta}^{(r)}$ readily follows upon substitution of the expression (5.1-22) for $f^{(r)}$ into the definition (5.1-25). Because of Eq. (5.1-15) one has

$$\boldsymbol{\beta}^{(r)} = \int \boldsymbol{\psi}\boldsymbol{\psi} f^{(0)} \,\mathrm{d}^3 c \cdot \boldsymbol{\alpha}^{(r)}. \tag{5.1-26}$$

When written out, this equation becomes a set of linear algebraic equations for the unknown components of $\boldsymbol{\alpha}^{(r)}$; its determinant does not vanish unless $f^{(0)}$ vanishes, so there exists a one-to-one correspondence between $\boldsymbol{\alpha}^{(r)}$ and $\boldsymbol{\beta}^{(r)}$ and either may be used. Now, if the orthogonality relation (5.1-24) is evaluated using the expression (5.1-22) for $f^{(r)}$, one obtains the following set of partial differential equations,

$$\frac{\partial}{\partial t}\int \boldsymbol{\psi}\boldsymbol{\psi} f^{(0)} \,\mathrm{d}^3 c \cdot \boldsymbol{\alpha}^{(r)} + \nabla_r \cdot \int \boldsymbol{c}\boldsymbol{\psi}\boldsymbol{\psi} f^{(0)} \,\mathrm{d}^3 c \cdot \boldsymbol{\alpha}^{(r)} - \int \boldsymbol{c} \cdot (\nabla_r \boldsymbol{\psi})\boldsymbol{\psi} f^{(0)} \,\mathrm{d}^3 c \cdot \boldsymbol{\alpha}^{(r)}$$
$$- \boldsymbol{F} \cdot \int (\nabla_c \boldsymbol{\psi})\boldsymbol{\psi} f^{(0)} \,\mathrm{d}^3 c \cdot \boldsymbol{\alpha}^{(r)} + \int \boldsymbol{\psi} \mathfrak{D} f^{(0)} \bar{\phi}^{(r)} \,\mathrm{d}^3 c = 0, \tag{5.1-27}$$

which must be satisfied by the vector $\boldsymbol{\alpha}^{(r)}$ if a solution for the function $f^{(r+1)}$ is to exist. The solution $\boldsymbol{\alpha}^{(r)}(\boldsymbol{r}, t)$ is uniquely determined provided that the value of the vector $\boldsymbol{\alpha}^{(r)}$ or, equivalently, of the vector $\boldsymbol{\beta}^{(r)}$ is prescribed at any initial instant $t = t_0$. In other words, the vector $\boldsymbol{\alpha}^{(r)}(\boldsymbol{r}, t)$, and therefore the function $f^{(r)}(\boldsymbol{r}, \boldsymbol{c}, t)$, is uniquely determined once the initial vector of the first few velocity moments of $f^{(r)}$, i.e. $\boldsymbol{\beta}^{(r)}(\boldsymbol{r}, t_0)$, is prescribed. Thus, by induction we conclude that the formal Hilbert expansion technique yields a *unique* solution to the Boltzmann equation provided that initial values are assigned to *all* the vectors $\boldsymbol{\beta}^{(0)}, \boldsymbol{\beta}^{(1)}, \ldots$.

Next, consider the vector $\boldsymbol{\beta}$ whose components are the first few velocity moments of f, viz. n, the three components of $\rho\boldsymbol{v}$, and ρu,

$$\boldsymbol{\beta} = \int \boldsymbol{\psi} f \,\mathrm{d}^3 c. \tag{5.1-28}$$

Substitution of the expansion (5.1-2) yields, because of Eq. (5.1-25),

$$\boldsymbol{\beta} = \boldsymbol{\beta}^{(0)} + \varepsilon \boldsymbol{\beta}^{(1)} + \varepsilon^2 \boldsymbol{\beta}^{(2)} + \ldots \tag{5.1-29}$$

so, by assigning initial values to *all* vectors $\boldsymbol{\beta}^{(0)}, \boldsymbol{\beta}^{(1)}, \ldots$, we assign implicitly an initial value to the vector $\boldsymbol{\beta}$. The question is whether the solution obtained with the formal Hilbert expansion method depends on how the initial value of $\boldsymbol{\beta}$ is allotted to the $\boldsymbol{\beta}^{(r)}$. That is, is the solution generated by the initial vectors, $\boldsymbol{\beta}^{(0)*}, \boldsymbol{\beta}^{(1)*}, \ldots$ say, any different from the solution obtained by choosing for example, the initial value of the vector $\boldsymbol{\beta}^{(0)}$ equal to $\sum \varepsilon^r \boldsymbol{\beta}^{(r)*}$ and the initial values of all remaining vectors $\boldsymbol{\beta}^{(1)}, \boldsymbol{\beta}^{(2)}, \ldots$ equal to zero? (The result can be proven for any other rearrangement of the $\boldsymbol{\beta}^{(r)}$ by the same argument.)

Let the solution in the former case be denoted by $f^*(\boldsymbol{r}, \boldsymbol{c}, t; \varepsilon)$. In the latter case, the initial conditions depend on ε; therefore, we shall denote the expansion parameter by μ and the solution by $f(\boldsymbol{r}, \boldsymbol{c}, t; \varepsilon, \mu)$. Then, we must consider the limit of the latter function as $\mu \to \varepsilon$, where ε has a *fixed* value. Due to its construction, $f(\boldsymbol{r}, \boldsymbol{c}, t; \varepsilon, \varepsilon)$ is an expansion in powers of ε; it has the property that, at $t = t_0$, the vector $\boldsymbol{\beta}$ associated with it, is $\sum \varepsilon^r \boldsymbol{\beta}^{(r)*}$ and coincides with the initial value assigned to the vector $\boldsymbol{\beta}$ associated with $f^*(\boldsymbol{r}, \boldsymbol{c}, t; \varepsilon)$. Thus, $f(\boldsymbol{r}, \boldsymbol{c}, t; \varepsilon, \varepsilon)$ and $f^*(\boldsymbol{r}, \boldsymbol{c}, t; \varepsilon)$ are both solutions of the set of equations (5.1-3), (5.1-4) with the same initial value of $\boldsymbol{\beta}$. Since we have shown that the solution of these equations is uniquely determined by the initial value of $\boldsymbol{\beta}$, these two functions are identical. This proves *Hilbert's uniqueness theorem*: In the class of solutions of the Boltzmann equation which can be represented by power series in ε there is a one-to-one correspondence between solutions $f(\boldsymbol{r}, \boldsymbol{c}, t)$ and the value of the vector $\int \boldsymbol{\psi} f \, \mathrm{d}^3 c$ at any "initial" instant.

However, we must note that at this stage there is no guarantee that the series (5.1-2) converges and thus it is not certain whether solutions of this type do exist at all. We are merely assured that, if they exist, they are uniquely determined by the initial conditions on their first five velocity moments. The convergence of the series is, in fact, still something of an open question. Following Grad [1958], we shall refer to the class of solutions of the Boltzmann equation of the type discussed in the present section as the *Hilbert class* or the *class of normal solutions*.

Now, if the solution $f(\boldsymbol{r}, \boldsymbol{c}, t)$ is uniquely determined by the value of the vector $\int \boldsymbol{\psi} f \, \mathrm{d}^3 c$ at any instant $t = t_0$, then certainly $f(\boldsymbol{r}, \boldsymbol{c}, t_0)$ is determined by the value of the vector $\int \boldsymbol{\psi} f \, \mathrm{d}^3 c$ at $t = t_0$. Since this holds for any t_0, it follows that in the Hilbert class there is a unique correspondence between a distribution function $f(\boldsymbol{r}, \boldsymbol{c}, t)$ and the vector of its first few velocity moments – i.e., between a distribution function $f(\boldsymbol{r}, \boldsymbol{c}, t)$ and the density $n(\boldsymbol{r}, t)$ hydrodynamic velocity $\boldsymbol{v}(\boldsymbol{r}, t)$ and temperature $T(\boldsymbol{r}, t)$. Moreover, the transport operator associated with the Boltzmann equation is a one-to-one mapping of the Hilbert class into itself.

5.2. Chapman-Enskog method

In the previous section we proved that in the Hilbert class of normal solutions to the Boltzmann equation a velocity distribution function is uniquely determined by the values of its first five velocity moments at any initial instant $t = t_0$. (This phenomenon is sometimes referred to as the *Hilbert paradox.*) Consequently, the density, hydrodynamic velocity and temperature at all times t can be determined directly from their values at the initial instant t_0. Moreover, since the one-to-one correspondence between the velocity distribution function and its first five velocity moments is preserved in the course of time, the higher-order velocity moments and, in particular, the pressure tensor and heat flow vector can be determined at any time directly in terms of the density, hydrodynamic velocity and temperature. Thus, upon substitution, the general conservation equations derived in Section 4.1 become a closed system of equations.

Still, this result is not very convincing. Although Hilbert's method for solving the Boltzmann equation yields the Euler equations of fluid dynamics in the lowest order of approximation, it appears to be impossible in this way to generate the classical equations of fluid dynamics – viz., the Navier-Stokes equations. For this reason, Hilbert's method has not been worked out except for very special cases, see Waldmann [1958], and, in fact, there is some doubt as to whether in the general case the method is practical at all, see Struminskii [1964].

Another method for solving the Boltzmann equation, which generates the Euler as well as the Navier-Stokes equations, was worked out independently by Chapman and Enskog in the 1910's. Their original work is to be found in two papers by Chapman [1916, 1917] and the doctoral dissertation of Enskog [1917]. Enskog made the Boltzmann equation the basis of his work, Chapman followed Maxwell in using the equations of transfer (an enlarged set of moments of the Boltzmann equation). Both authors arrived at almost exactly the same results. A detailed comparison between his and Chapman's results was published later by Enskog [1922a]. The identity of the two sets of independently derived results stimulated the general acceptance of kinetic theory.

In the present text we shall confine ourselves to Enskog's method. Alternative, but equivalent, methods have been proposed by Hecke [1922] and Cotter [1952] for the case of a simple gas of hard spherical molecules and by McCune, Morse and Sandri [1963].

As in the previous section, the Boltzman equation is taken in the form (5.1-1); one looks for a solution given by a power series expansion of the form (5.1-2). Again, $\boldsymbol{\beta} = \int \boldsymbol{\psi} f \, d^3c$, where $\boldsymbol{\psi}$ is the vector whose components are 1, the three components of $m\boldsymbol{c}$, and $\frac{1}{2}mC^2$, respectively. Equivalently, one may think of $\boldsymbol{\beta}$ as the vector with components n, the three components of $\boldsymbol{v}$, and T.

In order to evaluate the pressure tensor and heat flow vector in terms of the density, hydrodynamic velocity and temperature, Enskog *postulated* that time does not enter explicitly among the arguments of f and $\partial\boldsymbol{\beta}/\partial t$, but only *implicitly* through $\boldsymbol{\beta}$ and the *spatial* gradients of $\boldsymbol{\beta}$, i.e., f and $\partial\boldsymbol{\beta}/\partial t$ can be written as

$$f(\boldsymbol{r}, \boldsymbol{c}, t) \equiv f(\boldsymbol{r}, \boldsymbol{c}|\boldsymbol{\beta}, \nabla_r \boldsymbol{\beta}, \ldots) \tag{5.2-1}$$

and

$$(\partial/\partial t)\boldsymbol{\beta}(\boldsymbol{r}, t) \equiv \boldsymbol{\Phi}(\boldsymbol{r}|\boldsymbol{\beta}, \nabla_r \boldsymbol{\beta}, \ldots), \tag{5.2-2}$$

respectively; the dots indicate higher-order spatial derivatives of $\boldsymbol{\beta}$.

Mathematically, this postulate can be justified on the basis of Hilbert's results, as we have seen above. Physically, one can argue that a few collision intervals after the initial instant a contracted description of the state of the gas is possible, in which the temporal development is determined by much fewer variables. The reason why these variables are the macroscopic variables n, $\boldsymbol{v}$ and T (or, in other words, the components of the vector $\boldsymbol{\beta}$) is, of course, that they correspond to the velocity moments of the distribution function with respect to the collisional invariants. Therefore, the collisions do not affect these macroscopic variables directly. They will stay constant in a time of the order of the mean free time and in this sense they can be regarded as constants of motion on the kinetic time scale. The macroscopic variables will change with time only secularly through the change of the velocity distribution function and, therefore, on the macroscopic time scale they will completely govern the temporal development of the gas.

Eq. (5.2-2) implies that the time derivatives of the macroscopic observables are determined entirely in terms of the macroscopic observables themselves and their spatial derivatives, i.e., we have a fluid dynamic description. In accord with the considerations presented in the introduction to the present chapter our aim will be to evaluate the expression $\boldsymbol{\Phi}$.

First, we remark that, if a power series expansion of f is adopted as in Eq. (5.1-2), then also the moments of f, i.e., the vector $\boldsymbol{\beta}$, and, *a fortiori*, $\boldsymbol{\Phi}$ must be written as power series. Hence it is consistent to assume that f and $\boldsymbol{\Phi}$ shall be of the form

$$f = f^{(0)} + \varepsilon f^{(1)} + \varepsilon^2 f^{(2)} + \dots \tag{5.2-3}$$

and

$$\boldsymbol{\Phi} = \boldsymbol{\Phi}^{(0)} + \varepsilon \boldsymbol{\Phi}^{(1)} + \varepsilon^2 \boldsymbol{\Phi}^{(2)} + \dots. \tag{5.2-4}$$

Next, consider the time derivative of f. On using the chain rule for differentiation one finds that $\partial f/\partial t$ is a sum of products of derivatives of f with respect to various spatial derivatives of $\boldsymbol{\beta}$ and the time derivatives of these spatial derivatives of $\boldsymbol{\beta}$. When the latter are evaluated via Eq. (5.2-2), one finds

$$\frac{\partial f}{\partial t} = \boldsymbol{\Phi} \cdot \nabla_{\beta} f + (\nabla_r \boldsymbol{\Phi}) : (\nabla_{\nabla_r \beta} f) + \dots. \tag{5.2-5}$$

Now, if the expansions (5.2-3) and (5.2-4) are substituted, this may be written as

$$\frac{\partial f}{\partial t} = \frac{\partial_0 f^{(0)}}{\partial t} + \varepsilon \left\{ \frac{\partial_1 f^{(0)}}{\partial t} + \frac{\partial_0 f^{(1)}}{\partial t} \right\} + \varepsilon^2 \left\{ \frac{\partial_2 f^{(0)}}{\partial t} + \frac{\partial_1 f^{(1)}}{\partial t} + \frac{\partial_0 f^{(2)}}{\partial t} \right\} + \dots, \tag{5.2-6}$$

where, for $i = 0, 1, \dots$, the operator $\partial_i/\partial t$ has been defined as

$$\frac{\partial_i}{\partial t} = \boldsymbol{\Phi}^{(i)} \cdot \nabla_{\beta} + \nabla_r \boldsymbol{\Phi}^{(i)} : \nabla_{\nabla_r \beta} + \dots. \tag{5.2-7}$$

In this way, we arrive at an expansion of the streaming term of the Boltzmann equation, $\mathfrak{D}f$, of the form

$$\mathfrak{D}f = (\mathfrak{D}f)^{(0)} + \varepsilon (\mathfrak{D}f)^{(1)} + \varepsilon^2 (\mathfrak{D}f)^{(2)} + \dots \tag{5.2-8}$$

where, for $i = 0, 1, \dots$,

$$(\mathfrak{D}f)^{(i)} = \frac{\partial_i f^{(0)}}{\partial t} + \dots + \frac{\partial_0 f^{(i)}}{\partial t} + \boldsymbol{c} \cdot \nabla_r f^{(i)} + \boldsymbol{F} \cdot \nabla_c f^{(i)}. \tag{5.2-9}$$

The expansion (5.2-8) is not only unique, but also significantly different from the corresponding result of Hilbert's method inasmuch as, now, the time derivative of f has been divided up in a non-trivial way among the various terms of the expansion. This contrasts with the assignment of $\partial f^{(i)}/\partial t$ to $(\mathfrak{D}f)^{(i)}$ in the Hilbert method and is the key to the present method. As we shall see below, it allows a much greater freedom in choosing $f^{(0)}$.

A formal substitution of the expansion (5.2-3) into the collision term of the Boltzmann equation yields the result

$$J(ff) = J(f^{(0)}f^{(0)})+\varepsilon\{J(f^{(0)}f^{(1)})+J(f^{(1)}f^{(0)})\}$$
$$+\varepsilon^2\{J(f^{(0)}f^{(2)})+J(f^{(1)}f^{(1)})+J(f^{(2)}f^{(0)})\}+\ldots, \qquad (5.2\text{-}10)$$

so that, when we now equate the coefficients of like powers of ε in the Boltzmann equation,

$$\mathfrak{D}f = \frac{1}{\varepsilon} J(ff), \qquad (5.2\text{-}11)$$

we obtain the following equation for $f^{(0)}$,

$$J(f^{(0)}f^{(0)}) = 0 \qquad (5.2\text{-}12)$$

while, for $r = 1, 2, \ldots$

$$J(f^{(0)}f^{(r)})+J(f^{(r)}f^{(0)}) = (\mathfrak{D}f)^{(r-1)}-J(f^{(1)}f^{(r-1)})-\ldots-J(f^{(r-1)}f^{(1)}), \qquad (5.2\text{-}13)$$

there being r terms on the right (when $r = 1$, the single term on the right is $(\mathfrak{D}f)^{(0)}$). Eqs. (5.2-12) and (5.2-13) show a great similarity with Eqs. (5.1-3) and (5.1-4) of the previous section – in fact, they are identical if all time derivatives in Eq. (5.1-4) are eliminated with the aid of the differential equations (5.1-27).

From the discussion of the previous section we infer that, for $r = 1, 2, \ldots$, Eq. (5.2-13) represents an integral equation for $f^{(r)}$, which is soluble if and only if the orthogonality condition

$$\int \psi\{(\mathfrak{D}f)^{(r-1)}-J(f^{(1)}f^{(r-1)})-\ldots-J(f^{(r-1)}f^{(1)})\}\mathrm{d}^3c = 0 \qquad (5.2\text{-}14)$$

is satisfied or, since the components of ψ are the summational invariants, if and only if the condition

$$\int \psi(\mathfrak{D}f)^{(r-1)}\,\mathrm{d}^3c = 0 \tag{5.2-15}$$

is satisfied. Moreover, as we shall see, the solution – if it exists – is such that, at *all* $\boldsymbol{r}$ and t, we are free to set

$$\int \psi f^{(0)}\,\mathrm{d}^3c = \int \psi f\,\mathrm{d}^3c = \boldsymbol{\beta}(\boldsymbol{r}, t), \tag{5.2-16}$$

$$\int \psi f^{(r)}\,\mathrm{d}^3c = 0 \quad \text{for } r = 1, 2, \ldots, \tag{5.2-17}$$

which means that the macroscopic variables are always determined from $f^{(0)}$ alone. Other choices are, of course, possible, but this particular one leads to a more attractive theory.

Now, we consider the orthogonality condition (5.2-15) in greater detail. Upon substitution of the expressions (5.2-9) and (5.2-7) into it, Eq. (5.2-15) becomes

$$\int \psi\{\sum_{i=0}^{r-1}(\boldsymbol{\Phi}^{(r-1-i)}\cdot\nabla_{\beta}f^{(i)}+\nabla_r\boldsymbol{\Phi}^{(r-1-i)}:\nabla_{\nabla_r\beta}f^{(i)}+\ldots)$$
$$+\boldsymbol{c}\cdot\nabla_r f^{(r-1)}+\boldsymbol{F}\cdot\nabla_c f^{(r-1)}\}\,\mathrm{d}^3c = 0. \tag{5.2-18}$$

First, consider the integral

$$\int \psi\boldsymbol{\Phi}^{(r-1-i)}\cdot\nabla_{\beta}f^{(i)}\,\mathrm{d}^3c \tag{5.2-19}$$

for $i = 0, \ldots, r-1$. Here, $\boldsymbol{\Phi}^{(r-1-i)}$ is independent of the variable of integration and from the constraints (5.2-16) and (5.2-17) it readily follows that the integral vanishes identically for all i, except for $i = 0$, when it is equal to $\boldsymbol{\Phi}^{(r-1)}\cdot\nabla_{\beta}\boldsymbol{\beta}$, which, in turn, is equal to $\boldsymbol{\Phi}^{(r-1)}$. So

$$\int \psi\boldsymbol{\Phi}^{(r-1)}\cdot\nabla_{\beta}f^{(0)}\,\mathrm{d}^3c = \boldsymbol{\Phi}^{(r-1)}, \quad r = 1, 2, \ldots \tag{5.2-20a}$$

$$\int \psi\boldsymbol{\Phi}^{(r-1-i)}\cdot\nabla_{\beta}f^{(i)}\,\mathrm{d}^3c = 0, \quad i = 1, \ldots, r-1;\ r = 1, 2, \ldots. \tag{5.2-20b}$$

Next, consider the integral

$$\int \psi(\nabla_r\boldsymbol{\Phi}^{(r-1-i)}):(\nabla_{\nabla_r\beta}f^{(i)})\,\mathrm{d}^3c \tag{5.2-21}$$

for $i = 0, \ldots, r-1$. Again, the tensor $\mathbf{\nabla}_r \boldsymbol{\Phi}^{(r-1-i)}$ is independent of $\boldsymbol{c}$ and, from the constraints (5.2-16) and (5.2-17) it follows that the integral vanishes identically for all i, except for $i = 0$ when it is equal to $\mathbf{\nabla}_r \boldsymbol{\Phi}^{(r-1)} : \mathbf{\nabla}_{\mathbf{\nabla}_r \beta} \boldsymbol{\beta}$. However, since the quantities $\boldsymbol{\beta}$ and $\mathbf{\nabla}_r \boldsymbol{\beta}$ are treated as independent vaiiables, one has $\mathbf{\nabla}_{\mathbf{\nabla}_r \beta} \boldsymbol{\beta} = 0$ and, therefore,

$$\int \psi (\mathbf{\nabla}_r \boldsymbol{\Phi}^{(r-1-i)}) : (\mathbf{\nabla}_{\mathbf{\nabla}_r \beta} f^{(i)}) \mathrm{d}^3 c = 0 \tag{5.2-22}$$

for all $i = 0, \ldots, r-1$ and $r = 1, 2, \ldots$.

Similarly one proves that, in the integral (5.2-18), all terms involving gradients of $f^{(i)}$ with respect to *higher*-order spatial derivatives of $\boldsymbol{\beta}$ vanish identically. Thus, the necessary and sufficient condition for the solubility of the rth order coefficient $f^{(r)}$ in the power series expansion (5.2-4) of f may be put in the form

$$\boldsymbol{\Phi}^{(r-1)} + \int \psi \boldsymbol{c} \cdot \mathbf{\nabla}_r f^{(r-1)} \mathrm{d}^3 c + \int \psi \boldsymbol{F} \cdot \mathbf{\nabla}_c f^{(r-1)} \mathrm{d}^3 c = 0, \tag{5.2-23}$$

which specifies $\boldsymbol{\Phi}^{(r-1)}$ in terms of velocity moments of $f^{(r-1)}$.

Now, if we turn to the actual computation of the distribution function f itself, we may use the following scheme of successive approximations. First, Eq. (5.2-12) is solved for the zero-order coefficient $f^{(0)}$ under the constraint (5.2-16), and the zero-order coefficient $\boldsymbol{\Phi}^{(0)}$ is prescribed according to the condition (5.2-23) for $r = 1$. This clears the way for the first-order approximation, in which one solves $f^{(1)}$ from the integral equation (5.2-13) under the constraint (5.2-17), taking $r = 1$, and prescribes $\boldsymbol{\Phi}^{(1)}$ according to the condition (5.2-23) for $r = 2$. Then the second-order approximation can be worked out, which yields the coefficients $f^{(2)}$ and $\boldsymbol{\Phi}^{(2)}$, etc. As a result, after the rth order approximation has been completed the coefficients $f^{(0)}, \ldots, f^{(r)}$ and $\boldsymbol{\Phi}^{(0)}, \ldots, \boldsymbol{\Phi}^{(r)}$ in the expansions (5.2-3) and (5.2-4), respectively, are known. Then, within an rth order theory, the time-derivatives of the macroscopic observables are completely specified from Eqs. (5.2-2), (5.2-4) and (5.2-23),

$$\frac{\partial \boldsymbol{\beta}}{\partial t} = \sum_{j=0}^{r} \varepsilon^j \boldsymbol{\Phi}^{(j)} = -\sum_{j=0}^{r} \varepsilon^j \int \psi \boldsymbol{c} \cdot \mathbf{\nabla}_r f^{(j)} \mathrm{d}^3 c - \sum_{j=0}^{r} \varepsilon^j \int \psi \boldsymbol{F} \cdot \mathbf{\nabla}_c f^{(j)} \mathrm{d}^3 c, \tag{5.2-24}$$

which may be brought in the form

$$\frac{\partial \boldsymbol{\beta}}{\partial t} + \nabla_r \cdot \sum_{j=0}^{r} \varepsilon^j \int \boldsymbol{c}\boldsymbol{\psi} f^{(j)} \, \mathrm{d}^3 c - \sum_{j=0}^{r} \varepsilon^j \int \boldsymbol{c} \cdot (\nabla_r \boldsymbol{\psi}) f^{(j)} \, \mathrm{d}^3 c$$
$$- \boldsymbol{F} \cdot \sum_{j=0}^{r} \varepsilon^j \int (\nabla_c \boldsymbol{\psi}) f^{(j)} \, \mathrm{d}^3 c = 0. \qquad (5.2\text{-}25)$$

If Eq. (5.2-25) is written out in terms of individual components – recall that $\boldsymbol{\beta} = \int \boldsymbol{\psi} f \, \mathrm{d}^3 c$, where $\boldsymbol{\psi}$ is the vector whose components are 1, the three components of $m\boldsymbol{c}$, and $\frac{1}{2}mC^2$, so that $\boldsymbol{\beta}$ is the vector whose components are n, the three components of $\rho \boldsymbol{v}$, and ρu – one obtains after some trivial manipulation, using the constraints (5.2-16) and (5.2-17), the following set of partial differential equations,

$$\frac{1}{\rho} \frac{\mathrm{d}\rho}{\mathrm{d}t} = -\nabla \cdot \boldsymbol{v}, \qquad (5.2\text{-}26)$$

$$\rho \frac{\mathrm{d}\boldsymbol{v}}{\mathrm{d}t} = \rho \boldsymbol{F} - \sum_{j=0}^{r} \nabla \cdot \mathbf{P}^{(j)}, \qquad (5.2\text{-}27)$$

$$\rho \frac{\mathrm{d}u}{\mathrm{d}t} = -\Big(\sum_{j=0}^{r} \nabla \cdot \boldsymbol{q}^{(j)} + \sum_{j=0}^{r} \mathbf{P}^{(j)} : \nabla \boldsymbol{v}\Big), \qquad (5.2\text{-}28)$$

where, as usual, $\mathrm{d}/\mathrm{d}t$ stands for the substantial derivative, $\mathrm{d}/\mathrm{d}t = \partial/\partial t + \boldsymbol{v} \cdot \nabla_r$ and where $\mathbf{P}^{(j)}$ and $\boldsymbol{q}^{(j)}$ are the jth order contribution to the pressure tensor and heat flow vector, respectively,

$$\mathbf{P}^{(j)} = \varepsilon^j \int m\boldsymbol{C}\boldsymbol{C} f^{(j)} \, \mathrm{d}^3 c, \qquad (5.2\text{-}29)$$

$$\boldsymbol{q}^{(j)} = \varepsilon^j \int \tfrac{1}{2} m C^2 \boldsymbol{C} f^{(j)} \, \mathrm{d}^3 c. \qquad (5.2\text{-}30)$$

Comparing this result with the conservation equations derived in Section 4.1, we conclude that the Chapman-Enskog method not only provides a unique method of determining successive approximations to the solution of the Boltzmann equation but, at the same time, it yields the equations of fluid dynamics in such a form that they can be explicitly evaluated to each order of approximation.

In this and the preceeding section we have described two distinct procedures for obtaining normal solutions of the Boltzmann equation. They are formally equivalent, but there are some subtle differences which we now discuss. In Hilbert's method, successive coefficients $f^{(r)}$ in the power series expansion of the velocity distribution function f are evaluated by an algorithm, involving the solution of (i) a certain integral equation, Eq. (5.1-4), and (ii) a set of higher-order partial differential equations, Eq. (5.4-27). In particular, to specify the formal solution uniquely, initial values of the moments of f are required. In the Chapman-Enskog method, one computes $f^{(r)}$ in terms of $\boldsymbol{\beta}$ (i.e., in terms of the macroscopic observables n, $\boldsymbol{v}$ and T), then evaluates the pressure tensor and heat flow vector from $f^{(r)}$, also in terms of $\boldsymbol{\beta}$ and the spatial gradients of $\boldsymbol{\beta}$, to obtain a sequence of successively higher-order fluid dynamic equations for $\boldsymbol{\beta}$, which contain the expansion parameter ε. The distinction is therefore that Hilbert expands the *solution* f in a power series, whereas Chapman and Enskog expand both the *solution and the equations*, cf. Grad [1958]. If the fluid dynamic equations in the Chapman-Enskog approach are solved as power series in ε, we return to the Hilbert result. However, to make proper use of these higher-order equations we should be more sophisticated than to look for solutions as power series in ε.

On the other hand, the theory of the Euler equations is known, so the Hilbert theory can always be carried out, whereas in general nothing at all is known about the theory of the higher-order fluid dynamic equations of the Chapman-Enskog theory which are *always* of the first-order in the time derivatives of the macroscopic variables, but are successively of *higher-order* in the spatial derivatives. The resolution of this difficulty was pointed out by Grad [1963a] and lies in the fact that any finite result which emerges from either the Hilbert or Chapman-Enskog formalism can be shown to be asymptotic to a solution of the Boltzmann equation. The Hilbert algorithm can *always* be carried out and yields a result which is bounded for any *finite* time; the higher-order fluid dynamic equations and, in particular, the first-order (Navier-Stokes) equations yield a solution which is bounded for *all* time. If a Chapman-Enskog equation fails to have a solution, there is no alternative but to use the Hilbert expansion. We return to this point in Section 5.10.

5.3. Zero-order approximation to f

The general solution of the zero-order equation (5.2-12) is given in Eq. (5.1-5). The arbitrary parameters $n^{(0)}$, $\boldsymbol{v}^{(0)}$ and $T^{(0)}$ are fixed by the constraints (5.2-16) to be the actual *local* number density, hydrodynamic velocity and temperature. Thus

$$f^{(0)} = n(m/2\pi kT)^{\frac{3}{2}} \exp(-mC^2/2kT) \tag{5.3-1}$$

with $\boldsymbol{C} = \boldsymbol{c}-\boldsymbol{v}$. That is, the zero-order approximation to f is a *local* Maxwellian function corresponding to local macroscopic properties. When the pressure tensor and heat flow vector are calculated from the zero-order approximation (5.3-1), one finds

$$\mathsf{P}^{(0)} = p\mathsf{I} \text{ with } p = nkT, \tag{5.3-2}$$

$$\boldsymbol{q}^{(0)} = 0. \tag{5.3-3}$$

Substitution of these zero-order results in the conservation equations (5.2-26), (5.2-27), (5.2-28) leads to the following set of partial differential equations for the mass density, hydrodynamic velocity and temperature:

$$\frac{1}{\rho}\frac{\mathrm{d}\rho}{\mathrm{d}t} = -\nabla \cdot \boldsymbol{v}, \tag{5.3-4}$$

$$\rho\frac{\mathrm{d}\boldsymbol{v}}{\mathrm{d}t} = \rho\boldsymbol{F}-\nabla p, \tag{5.3-5}$$

$$\frac{\mathrm{d}}{\mathrm{d}t}(\rho T^{-\frac{3}{2}}) = 0, \tag{5.3-6}$$

i.e., the parameters n, $\boldsymbol{v}$ and T satisfy the *ideal*, or *Euler*, *hydrodynamic* equations with adiabatic temperature changes and with an equation of state which is the ideal gas law.

5.4. First-order approximation to f

The equation from which $f^{(1)}$ is to be determined, is obtained from Eq. (5.2-13) with $r = 1$,

$$J(f^{(1)} f^{(0)})+J(f^{(0)} f^{(1)}) = (\mathfrak{D}f)^{(0)} \tag{5.4-1}$$

or, if we substitute $f^{(1)} = f^{(0)}\phi^{(1)}$,

$$-n^2 I(\phi^{(1)}) = (\mathfrak{D}f)^{(0)}, \tag{5.4-2}$$

where the integral operator I has been defined in Eq. (4.7-1). This is an inhomogeneous linear integral equation for $\phi^{(1)}$. The right member depends entirely on $f^{(0)}$,

$$(\mathfrak{D}f)^{(0)} = \partial_0 f^{(0)}/\partial t+\boldsymbol{c}\cdot\nabla_r f^{(0)}+\boldsymbol{F}\cdot\nabla_c f^{(0)}. \tag{5.4-3}$$

Since $f^{(0)}$ is given as a function of $\boldsymbol{C}$ rather than $\boldsymbol{c}$, cf. Eq. (5.3-1), it is convenient to express $(\mathfrak{D}f)^{(0)}$ in the following form:

$$(\mathfrak{D}f)^{(0)} = \frac{\mathrm{d}_0 f^{(0)}}{\mathrm{d}t} + \boldsymbol{C}\cdot\nabla_r f^{(0)} + \left(\boldsymbol{F} - \frac{\mathrm{d}_0 \boldsymbol{v}}{\mathrm{d}t}\right)\cdot\nabla_C f^{(0)} - (\nabla_C f^{(0)})\boldsymbol{C} : \nabla_r \boldsymbol{v}, \tag{5.4-4}$$

where $\boldsymbol{C}$ is now regarded as independent of $\boldsymbol{r}$ and $\mathrm{d}_0/\mathrm{d}t = \partial_0/\partial t + \boldsymbol{v}\cdot\nabla_r$. Eq. (5.4-4), in turn, is equivalent to

$$(\mathfrak{D}f)^{(0)} = f^{(0)}\left\{\frac{\mathrm{d}_0 \log f^{(0)}}{\mathrm{d}t} + \boldsymbol{C}\cdot\nabla_r \log f^{(0)} + \left(\boldsymbol{F} - \frac{\mathrm{d}_0 \boldsymbol{v}}{\mathrm{d}t}\right)\cdot\nabla_C \log f^{(0)} - (\nabla_C \log f^{(0)})\boldsymbol{C} : \nabla\boldsymbol{v}\right\}. \tag{5.4-5}$$

From Eq. (5.3-1) we have

$$\log f^{(0)} = \log n - \tfrac{3}{2}\log T - mC^2/2kT + \text{constant}, \tag{5.4-6}$$

and, from the zero-order conservation equations (5.3-4), (5.3-5) and (5.3-6),

$$\frac{\mathrm{d}_0 n}{\mathrm{d}t} = -n\nabla\cdot\boldsymbol{v}, \tag{5.4-7a}$$

$$\frac{\mathrm{d}_0 \boldsymbol{v}}{\mathrm{d}t} = \boldsymbol{F} - \frac{1}{nm}\nabla nkT, \tag{5.4-7b}$$

$$\frac{\mathrm{d}_0 T}{\mathrm{d}t} = -\tfrac{2}{3}T\nabla\cdot\boldsymbol{v}. \tag{5.4-7c}$$

Then, in the zero-order approximation $(\mathrm{d}_0/\mathrm{d}t)\log f$ is given by

$$\begin{aligned}\frac{\mathrm{d}_0 \log f^{(0)}}{\mathrm{d}t} &= \frac{1}{n}\frac{\mathrm{d}_0 n}{\mathrm{d}t} + \left(\frac{mC^2}{2kT} - \frac{3}{2}\right)\frac{1}{T}\frac{\mathrm{d}_0 T}{\mathrm{d}t}\\ &= -\frac{mC^2}{3kT}\nabla\cdot\boldsymbol{v}.\end{aligned} \tag{5.4-8}$$

Further, one verifies easily

$$\nabla_r \log f^{(0)} = \nabla\log n + \left(\frac{mC^2}{2kT} - \frac{3}{2}\right)\nabla\log T \tag{5.4-9}$$

and

$$\nabla_C \log f^{(0)} = -\frac{m}{kT}\boldsymbol{C}, \tag{5.4-10}$$

so the expression (5.4-5) becomes on substitution

$$(\mathfrak{D}f)^{(0)} = f^{(0)} \left\{ \left(\frac{mC^2}{2kT} - \frac{5}{2} \right) \boldsymbol{C} \cdot \nabla \log T + \frac{m}{kT} \boldsymbol{CC} : \nabla \boldsymbol{v} - \frac{mC^2}{3kT} \nabla \cdot \boldsymbol{v} \right\}, \tag{5.4-11}$$

which can also be written as

$$(\mathfrak{D}f)^{(0)} = f^{(0)} \left\{ \left(\frac{mC^2}{2kT} - \frac{5}{2} \right) \boldsymbol{C} \cdot \nabla \log T + \frac{m}{kT} (\boldsymbol{CC} - \tfrac{1}{3}C^2 \mathbf{I}) : \nabla \boldsymbol{v} \right\}. \tag{5.4-12}$$

The external force $\boldsymbol{F}$ does not occur in this expression.

From the relations (5.4-2) and (5.4-12) we conclude that the first-order perturbation function $\phi^{(1)}$ satisfies the inhomogeneous Fredholm integral equation of the second kind,

$$n^2 I(\phi^{(1)}) = -f^{(0)} \left\{ \left(\frac{mC^2}{2kT} - \frac{5}{2} \right) \boldsymbol{C} \cdot \nabla \log T + \frac{m}{kT} (\boldsymbol{CC} - \tfrac{1}{3}C^2 \mathbf{I}) : \nabla \boldsymbol{v} \right\}. \tag{5.4-13}$$

Obviously, $\phi^{(1)}$ is not uniquely determined by Eq. (5.4-13) alone since, as we have seen earlier, to any particular solution one may add a linear combination of the summational invariants which are the solutions of the associated homogeneous equation. Uniqueness, however, is obtained when the constraints (5.2-17) are applied. It is clear from the form of the integral equation (5.4-13) and the constraints (5.2-17) that the scalar quantity $\phi^{(1)}$ is a sum of (i) a linear combination of the components of $\nabla \log T$, (ii) a linear combination of the components of $\nabla \boldsymbol{v}$, (iii) the general (scalar) solution of the associated homogeneous equation $I(\phi^{(1)}) = 0$, i.e.,

$$\phi^{(1)} = -\frac{1}{n} \boldsymbol{A} \cdot \nabla \log T - \frac{1}{n} \mathbf{B} : \nabla \boldsymbol{v} + \boldsymbol{\alpha}^{(1)} \cdot \boldsymbol{\psi}, \tag{5.4-14}$$

where $\boldsymbol{A}$ and $\mathbf{B}$ are vector and tensor functions of $\boldsymbol{C}$, respectively, while $\boldsymbol{\alpha}^{(1)}$ is a vector independent of the velocity variable. $\boldsymbol{A}$, $\mathbf{B}$ and $\boldsymbol{\alpha}^{(1)}$ may depend upon $\boldsymbol{r}$ and t.

The components of $\nabla \log T$ and $\nabla \boldsymbol{v}$ are all linearly independent. Hence, substituting the expression (5.4-14) into Eq. (5.4-13) we may equate the coefficients of each component of $\nabla \log T$ and $\nabla \boldsymbol{v}$ on either side of the equality sign. Thus we obtain the following integral equations for the vector $\boldsymbol{A}$ and the tensor $\mathbf{B}$,

$$n\,I(\boldsymbol{A}) = f^{(0)}\left(\frac{mC^2}{2kT} - \frac{5}{2}\right)\boldsymbol{C}, \tag{5.4-15}$$

$$n\,I(\mathsf{B}) = \frac{m}{kT} f^{(0)}(\boldsymbol{CC} - \tfrac{1}{3}C^2\mathsf{I}). \tag{5.4-16}$$

These equations are soluble if and only if the orthogonality conditions

$$\int \psi f^{(0)}\left(\frac{mC^2}{2kT} - \frac{5}{2}\right)\boldsymbol{C}\,\mathrm{d}^3c = 0, \tag{5.4-17}$$

$$\int \psi f^{(0)}(\boldsymbol{CC} - \tfrac{1}{3}C^2\mathsf{I})\,\mathrm{d}^3c = 0, \tag{5.4-18}$$

are satisfied. By straight-forward computation it is easily verified that this is indeed the case; we leave the details to the reader.

Now we turn to a discussion of the solutions of Eqs. (5.4-15) and (5.4-16). The only variables involved are n, T and $\boldsymbol{C}$, the spatial coordinates or derivatives do not occur explicitly. Consequently, $\boldsymbol{A}$ and B are functions of the spatial coordinates only through the space dependence of n, T and $\boldsymbol{C}$. Since I is a linear, rotationally invariant operator and the right member of Eq. (5.4-15) is a vector in the direction of $\boldsymbol{C}$, it follows that $\boldsymbol{A}$ is also a vector in the direction of $\boldsymbol{C}$. Hence $\boldsymbol{A}$ must be of the form

$$\boldsymbol{A} \equiv A(C)\boldsymbol{C}, \tag{5.4-19}$$

where A is some scalar function of n, T and C. Furthermore, since the right member of Eq. (5.4-16) is a symmetric, traceless tensor it follows that the antisymmetric part of B, i.e., $\frac{1}{2}(\mathsf{B}-\mathsf{B}^T)$, and the trace of B, i.e., $\mathsf{B} : \mathsf{I}$, both satisfy the homogeneous equation $I(\phi) = 0$ and, hence, can be absorbed into the term $\boldsymbol{\alpha}^{(1)} \cdot \boldsymbol{\psi}$ in Eq. (5.4-14). The only symmetric traceless tensors that can be formed from n, T and $\boldsymbol{C}$ are scalar multiples of $\boldsymbol{CC} - \frac{1}{3}C^2\mathsf{I}$, so B must be of the form

$$\mathsf{B} \equiv B(C)(\boldsymbol{CC} - \tfrac{1}{3}C^2\mathsf{I}), \tag{5.4-20}$$

where B is some scalar function of n, T and C.

The solution (5.4-14) is made unique by the constraint (5.2-17). In fact, with the results (5.4-19) and (5.4-20), Eq. (5.2-17) becomes

$$-\frac{1}{n}\int \boldsymbol{\psi} f^{(0)} A(C)\boldsymbol{C}\cdot\nabla \log T \,\mathrm{d}^3c - \frac{1}{n}\int \boldsymbol{\psi} f^{(0)} B(C)(\boldsymbol{CC}-\tfrac{1}{3}C^2\boldsymbol{I}) : \nabla\boldsymbol{v}\,\mathrm{d}^3c$$

$$+\int \boldsymbol{\psi} f^{(0)}(\boldsymbol{\alpha}^{(1)}\cdot\boldsymbol{\psi})\,\mathrm{d}^3c = 0. \quad (5.4\text{-}21)$$

Recall that $\boldsymbol{\psi}$ is the vector whose components are the summational invariants 1, the three components of $m\boldsymbol{c}$, and $\frac{1}{2}mC^2$, or, more conveniently, 1, $m\boldsymbol{C}$ and $\frac{1}{2}mC^2$; correspondingly, the vector $\boldsymbol{\alpha}^{(1)}$ has components $\alpha^{(1,1)}$, $\boldsymbol{\alpha}^{(2,1)}$ and $\alpha^{(3,1)}$. Thus, Eq. (5.4-21) decomposes into two scalar equations and a vector equation, which may be put in the following form,

$$\frac{1}{n}\int f^{(0)} A(C)\boldsymbol{C}\cdot\nabla \log T\,\mathrm{d}^3c + \frac{1}{n}\int f^{(0)} B(C)(\boldsymbol{CC}-\tfrac{1}{3}C^2\boldsymbol{I}) : \nabla\boldsymbol{v}\,\mathrm{d}^3c$$

$$= \int f^{(0)}\{\alpha^{(1,1)} + m\boldsymbol{\alpha}^{(2,1)}\cdot\boldsymbol{v} + m\boldsymbol{\alpha}^{(2,1)}\cdot\boldsymbol{C} + \tfrac{1}{2}m\alpha^{(3,1)}C^2\}\,\mathrm{d}^3c, \quad (5.4\text{-}22)$$

$$\frac{1}{n}\int f^{(0)} A(C)C^2\boldsymbol{C}\cdot\nabla \log T\,\mathrm{d}^3c + \frac{1}{n}\int f^{(0)} B(C)C^2(\boldsymbol{CC}-\tfrac{1}{3}C^2\boldsymbol{I}) : \nabla\boldsymbol{v}\,\mathrm{d}^3c$$

$$= \int f^{(0)}C^2\{\alpha^{(1,1)} + m\boldsymbol{\alpha}^{(2,1)}\cdot\boldsymbol{v} + m\boldsymbol{\alpha}^{(2,1)}\cdot\boldsymbol{C} + \tfrac{1}{2}m\alpha^{(3,1)}C^2\}\,\mathrm{d}^3c, \quad (5.4\text{-}23)$$

$$\frac{1}{n}\int f^{(0)} A(C)\boldsymbol{CC}\cdot\nabla \log T\,\mathrm{d}^3c + \frac{1}{n}\int f^{(0)} B(C)\boldsymbol{C}(\boldsymbol{CC}-\tfrac{1}{3}C^2\boldsymbol{I}) : \nabla\boldsymbol{v}\,\mathrm{d}^3c$$

$$= \int f^{(0)}\boldsymbol{C}\{\alpha^{(1,1)} + m\boldsymbol{\alpha}^{(2,1)}\cdot\boldsymbol{v} + m\boldsymbol{\alpha}^{(2,1)}\cdot\boldsymbol{C} + \tfrac{1}{2}m\alpha^{(3,1)}C^2\}\,\mathrm{d}^3c. \quad (5.4\text{-}24)$$

The left members of both Eqs. (5.4-22) and (5.4-23) are identically zero: In the integrals involving A the integrands are odd functions of the components of $\boldsymbol{C}$; in the integrals involving B one applies the identity (A.14-1). Also, in the right members, the integrals involving the product $m\boldsymbol{\alpha}^{(2,1)}\cdot\boldsymbol{C}$ vanish because their integrands are again odd functions of the components of $\boldsymbol{C}$, so Eqs. (5.4-22) and (5.4-23) form a homogeneous system of two linear equations for the unknown scalars $\alpha^{(1,1)}+m\boldsymbol{\alpha}^{(2,1)}\cdot\boldsymbol{v}$ and $\alpha^{(3,1)}$:

$$\left(\int f^{(0)}\,\mathrm{d}^3c\right)(\alpha^{(1,1)}+m\boldsymbol{\alpha}^{(2,1)}\cdot\boldsymbol{v})+\left(\int \tfrac{1}{2}mC^2 f^{(0)}\,\mathrm{d}^3c\right)\alpha^{(3,1)} = 0, \qquad (5.4\text{-}25)$$

$$\left(\int \tfrac{1}{2}mC^2 f^{(0)}\,\mathrm{d}^3c\right)(\alpha^{(1,1)}+m\boldsymbol{\alpha}^{(2,1)}\cdot\boldsymbol{v})+\left(\int (\tfrac{1}{2}mC^2)^2 f^{(0)}\,\mathrm{d}^3c\right)\alpha^{(3,1)} = 0. \qquad (5.4\text{-}26)$$

Knowing $f^{(0)}$ we can calculate the determinant of the coefficients. It has the value $\frac{3}{2}(nkT)^2$, so the only solution of Eqs. (5.4-25) and (5.4-26) is the trivial solution,

$$\alpha^{(1,1)}+m\boldsymbol{\alpha}^{(2,1)}\cdot\boldsymbol{v} = 0 \text{ and } \alpha^{(3,1)} = 0. \qquad (5.4\text{-}27)$$

Turning to Eq. (5.4-24) we find that the integral involving A can be transformed with the aid of the identity (A.14-2); the integral involving B vanishes identically because the integrand is an odd function of the components of $\boldsymbol{C}$; for the same reason, all the integrals of the right member vanish except the one involving the product $m\boldsymbol{\alpha}^{(2,1)}\cdot\boldsymbol{C}$; the latter integral can, in turn, be transformed with the aid of the identity (A.14-2). Thus, Eq. (5.4-24) yields

$$\boldsymbol{\alpha}^{(2,1)} = \frac{(1/n)\displaystyle\int f^{(0)}A(C)C^2\,\mathrm{d}^3c}{m\displaystyle\int f^{(0)}C^2\,\mathrm{d}^3c}\,\boldsymbol{\nabla}\log T; \qquad (5.4\text{-}28)$$

in other words, $\boldsymbol{\alpha}^{(2,1)}$ is proportional to $\boldsymbol{\nabla}\log T$.

The conclusion is, therefore, that the constraint (5.2-17) implies that, in the expression (5.4-14) for $\phi^{(1)}$ the scalar product $\boldsymbol{\alpha}^{(1)}\cdot\boldsymbol{\psi}$ reduces to $m\boldsymbol{\alpha}^{(2,1)}\cdot\boldsymbol{C}$, while $\boldsymbol{\alpha}^{(2,1)}$ is proportional to $\boldsymbol{\nabla}\log T$. But then, in Eq. (5.4-14), the term $\boldsymbol{\alpha}^{(1)}\cdot\boldsymbol{\psi}$ may be absorbed into the term $\boldsymbol{A}\cdot\boldsymbol{\nabla}\log T$ and, without loss of generality, we may set $\boldsymbol{\alpha}^{(2,1)}$ equal to zero.

Thus we find that the integral equation (5.4-13) has a unique solution $\phi^{(1)}$,

$$\phi^{(1)} = -\frac{1}{n}A(C)\boldsymbol{C}\cdot\boldsymbol{\nabla}\log T-\frac{1}{n}B(C)(\boldsymbol{CC}-\tfrac{1}{3}C^2\boldsymbol{I}):\boldsymbol{\nabla v}, \qquad (5.4\text{-}29)$$

where the vector $\boldsymbol{A} = A(C)\boldsymbol{C}$ and the tensor $\mathsf{B} = B(C)(\boldsymbol{CC}-\frac{1}{3}C^2\boldsymbol{I})$ are the solutions of the integral equations (5.4-15) and (5.4-16), respectively, and

A is subject to the constraint

$$\int f^{(0)} A(C)\, C^2\, \mathrm{d}^3c = 0. \tag{5.4-30}$$

Before giving the details of the method by which A and $\mathbf{B}$ are determined, we consider the conservation equations (5.2-26), (5.2-27), (5.2-28) in the first-order approximation. For this, we must calculate the corrections to the pressure tensor and heat flow vector, i.e., $\mathbf{P}^{(1)}$ and $\boldsymbol{q}^{(1)}$, to this order of approximation. With the solution (5.4-29) for $\phi^{(1)}$, $\mathbf{P}^{(1)}$ becomes

$$\mathbf{P}^{(1)} = \int m\boldsymbol{CC} f^{(1)}\, \mathrm{d}^3c$$
$$= -\frac{m}{n}\int f^{(0)}\boldsymbol{CC}\{A(C)\boldsymbol{C}\cdot\nabla \log T + B(C)(\boldsymbol{CC}-\tfrac{1}{3}C^2\mathbf{I}) : \nabla\boldsymbol{v}\}\, \mathrm{d}^3c. \tag{5.4-31}$$

(The parameter ε has been set equal to one.) The term containing A does not contribute, since the integrand is odd in the components of $\boldsymbol{C}$. In the term containing B, the tensor $\boldsymbol{CC}$ may be replaced by the traceless tensor $\boldsymbol{CC}-\frac{1}{3}C^2\boldsymbol{I}$, since the subtracted integral is zero as a consequence of the identity (A.14-1). Then, using the identity (A.14-6) we obtain

$$\mathbf{P}^{(1)} = -\frac{m}{5n}\int f^{(0)} B(C)(\boldsymbol{CC}-\tfrac{1}{3}C^2\mathbf{I}) : (\boldsymbol{CC}-\tfrac{1}{3}C^2\mathbf{I})\, \mathrm{d}^3c\ \boldsymbol{S} \tag{5.4-32}$$

where $\boldsymbol{S}$ is the symmetric traceless component of $\nabla\boldsymbol{v}$,

$$S_{\alpha\beta} = \frac{1}{2}\left(\frac{\partial v_\beta}{\partial x_\alpha} + \frac{\partial v_\alpha}{\partial x_\beta}\right) - \tfrac{1}{3}\nabla\cdot\boldsymbol{v}\, \delta_{\alpha\beta}\,. \tag{5.4-33}$$

The tensor $\boldsymbol{S}$ is called the *rate-of-shear tensor.* (The symmetric tensor $\boldsymbol{S}+\frac{1}{3}\nabla\cdot\boldsymbol{v}\boldsymbol{I}$ is sometimes referred to as the *rate-of-strain tensor.*) Hence the first-order correction to the pressure tensor is proportional to the rate-of-shear tensor. The proportionality constant may be transformed with the aid of the integral equation (5.4-16), which is satisfied by the tensor $\mathbf{B} \equiv B(C)\times(\boldsymbol{CC}-\frac{1}{3}C^2\boldsymbol{I})$. In fact, instead of Eq. (5.4-32) we may write

$$\mathbf{P}^{(1)} = -\tfrac{1}{5}kT\int \mathbf{B} : I(\mathbf{B})\, \mathrm{d}^3c \tag{5.4-34}$$

or, recalling the definition (4.7-5) of the bracket integral,

$$\mathbf{P}^{(1)} = -\tfrac{1}{5}kT[\mathbf{B}, \mathbf{B}]. \tag{5.4-35}$$

Thus, if we define

$$\eta = \tfrac{1}{10}kT[\mathbf{B}, \mathbf{B}], \tag{5.4-36}$$

the expression for $\mathbf{P}^{(1)}$ becomes

$$\mathbf{P}^{(1)} = -2\eta\mathbf{S} \tag{5.4-37}$$

and, combining Eqs. (5.2-29), (5.3-2) and (5.4-37) we have the following result for the pressure tensor in the first-order Chapman-Enskog theory,

$$\mathbf{P} = p\mathbf{I} - 2\eta\mathbf{S}. \tag{5.4-38}$$

Equation (5.4-38) is well-known as Newton's law; η is the *coefficient of viscosity*.

Similarly, with the expression (5.4-29) for $\phi^{(1)}$, we obtain for the first-order correction to the heat flow vector (again, taking $\varepsilon = 1$),

$$\begin{aligned} \boldsymbol{q}^{(1)} &= \int \tfrac{1}{2}mC^2\boldsymbol{C}f^{(1)}\mathrm{d}^3c \\ &= -\frac{m}{2n}\int f^{(0)}C^2\boldsymbol{C}\{A(C)\boldsymbol{C}\cdot\nabla\log T + B(C)(\boldsymbol{CC}-\tfrac{1}{3}C^2\mathbf{I}):\nabla\boldsymbol{v}\}\,\mathrm{d}^3c. \end{aligned} \tag{5.4-39}$$

Now, the term involving B does not contribute and we have, using the identity (A.14-2)

$$\boldsymbol{q}^{(1)} = -\frac{m}{6n}\int f^{(0)}A(C)\,C^4\mathrm{d}^3c\,\nabla\log T. \tag{5.4-40}$$

Because A is subject to the constraint (5.4-30) we may replace C^4 under the integral sign by $C^4-(5kT/m)C^2 = (2kT/m)C^2(mC^2/2kT-\tfrac{5}{2})$. Then, using the integral equation (5.4-15), which is satisfied by the vector $\boldsymbol{A} \equiv A(C)\boldsymbol{C}$, and the definition (4.7-5) of the bracket integral we obtain

$$\boldsymbol{q}^{(1)} = -\tfrac{1}{3}kT[\boldsymbol{A}, \boldsymbol{A}]\,\nabla\log T. \tag{5.4-41}$$

Thus, if we define

$$\lambda = \tfrac{1}{3}k[\boldsymbol{A}, \boldsymbol{A}], \tag{5.4-42}$$

and combine Eqs. (5.2-30), (5.3-3) and (5.4-41), we find the following result for the heat flow vector in the first-order Chapman-Enskog theory,

$$\boldsymbol{q} = -\lambda \nabla T. \tag{5.4-43}$$

Equation (5.4-43) is well known as *Fourier's law*; λ is the *coefficient of thermal conductivity*.

When the first-order expressions for the pressure tensor and heat flow vector are substituted into the general conservation equations one obtains the following set of partial differential equations for the mass density, hydrodynamic velocity and temperature:

$$\frac{1}{\rho}\frac{\mathrm{d}\rho}{\mathrm{d}t} = -\nabla \cdot \boldsymbol{v}, \tag{5.4-44}$$

$$\rho \frac{\mathrm{d}\boldsymbol{v}}{\mathrm{d}t} = \rho \boldsymbol{F} - \nabla p + 2\eta \nabla \cdot \mathsf{S}, \tag{5.4-45}$$

$$\rho \frac{\mathrm{d}T}{\mathrm{d}t} = -\frac{2m}{3k}(-\nabla \cdot \lambda \nabla T + p\nabla \cdot \boldsymbol{v} - 2\eta\, \mathsf{S} : \nabla \boldsymbol{v}), \tag{5.4-46}$$

where S is given by Eq. (5.4-33). That is, the parameters n, $\boldsymbol{v}$ and T satisfy the *Navier-Stokes equations of hydrodynamics*.

The remainder of the task now consists of solving the integral equations (5.4-15) and (5.4-16) for the unknown functions $\boldsymbol{A}$ and B and finding explicit expressions for the transport coefficients η and λ. We conclude this section by noting that, since the quantities $\boldsymbol{A}$ and B are independent of the number density of the gas, the same is true of η and λ. This rather surprising result was, in fact, one of the early triumphs of kinetic theory. It was predicted by Maxwell on the basis of mean free path arguments and later experimentally verified by him.

5.5. Sonine polynomials; a variational principle

As the kernel of the integral operator I depends on the potential of intermolecular forces, Eqs. (5.4-15) and (5.4-16) cannot be solved in the general case; there is, in fact, only one special case (Maxwell molecules) for which an exact solution has been obtained. Therefore, an approximation method must be used – the method of series expansions being the most obvious choice. In his original work, Enskog [1917] expanded in powers of C^2. The advantages of the use of associated Legendre polynomials or *Sonine polynomials* were pointed out by Burnett [1935a, b]. Interest in other problems of kinetic theory has led to the use of other types of polynomials. In considering several schemes for solving the Maxwell transfer equations, Ikenberry and Truesdell [1956] used a system of polynomials developed specially for this purpose by Ikenberry [1955]. In a different type of treatment of the Boltzmann equation, Grad [1949] used a series expansion in Hermite polynomials, again specially devised for this purpose. All these three systems of polynomials are, of course, equivalent, see Jancel and Kahan [1962a, b]. A unified discussion has been given recently by Kumar [1966] who, using the theory of irreducible tensorial sets and certain transformations taken from nuclear physics, also showed that the system suggested by Burnett and based on the Sonine polynomials is the most economical from the point of view of the algebraic manipulations involved.

The Sonine polynomial $S_\nu^{(n)}(x)$ of order n (integer) and index ν is defined as

$$S_\nu^{(n)}(x) = \sum_{p=0}^{n} \frac{\Gamma(\nu+n+1)}{(n-p)!\,p!\,\Gamma(\nu+p)}(-x)^p. \tag{5.5-1}$$

In particular, for any value of the index ν,

$$S_\nu^{(0)}(x) = 1, \quad S_\nu^{(1)}(x) = \nu+1-x. \tag{5.5-2}$$

These polynomials are the coefficients in the expansion

$$(1-s)^{-\nu-1} \exp\left(-\frac{xs}{1-s}\right) = \sum_{n=0}^{\infty} S_\nu^{(n)}(x)\, s^n. \tag{5.5-3}$$

It is easily verified that the $S_\nu^{(n)}(x)$ satisfy the orthogonality relation

$$\int_0^\infty \mathrm{e}^{-x} S_\nu^{(p)}(x)\, S_\nu^{(q)}(x)\, x^\nu \,\mathrm{d}x = 0 \quad \text{if } p \neq q, \tag{5.5-4}$$

while the normalization is such that

$$\int_0^\infty \mathrm{e}^{-x} \{S_\nu^{(n)}(x)\}^2\, x^\nu \,\mathrm{d}x = \frac{\Gamma(\nu+n+1)}{n!}. \tag{5.5-5}$$

Now, we observe that, for the computation of the transport coefficients we are not so much interested in the *complete solutions* of Eqs. (5.4-15) and (5.4-16), but, rather, in particular *functionals* of the solutions – viz., the bracket integrals that occur in the definitions of η and λ. Therefore, the methods of variational calculus may be especially convenient. In the remainder of the present section we shall establish a variational principle which, in the next section, will be used for the evaluation of successive approximations to the transport coefficients.

Consider the quantity

$$s = -k \int f \log f \, d^3c \tag{5.5-6}$$

which is the local entropy density of the gas. The rate of change of s due to collisions is given by

$$(\partial s/\partial t)_{\text{coll}} = \tfrac{1}{4}k \iiiint (f'f'_1 - ff_1) \log (f'f'_1/ff_1) g b \, db \, d\varepsilon \, d^3c \, d^3c_1 . \tag{5.5-7}$$

In a first-order theory, i.e., when the velocity distribution function is of the form $f = f^{(0)}(1+\phi^{(1)})$ and terms of second and higher order in $\phi^{(1)}$ are neglected throughout, this reduces to

$$(\partial s/\partial t)_{\text{coll}} = n^2 k \int \phi^{(1)} I(\phi^{(1)}) d^3c = n^2 k[\phi^{(1)}, \phi^{(1)}]. \tag{5.5-8}$$

Now, substituting $\phi^{(1)}$ from Eq. (5.4-29), taking into account the forms (5.4-19) and (5.4-20) of $\boldsymbol{A}$ and $\mathbf{B}$, and neglecting vanishing integrals we find, after some algebra,

$$\begin{aligned}(\partial s/\partial t)_{\text{coll}} &= k\{\tfrac{1}{3}[\boldsymbol{A}, \boldsymbol{A}]\,|\nabla \log T|^2 + \tfrac{1}{5}[\mathbf{B}, \mathbf{B}]\mathbf{S} : \mathbf{S}\} \\ &= \lambda|\nabla \log T|^2 + (2\eta/T)\mathbf{S} : \mathbf{S},\end{aligned} \tag{5.5-9}$$

which shows that λ and η are directly related to the rate of change of the entropy density due to collisions.

Suppose that the vector $\boldsymbol{a} \equiv a(C)\boldsymbol{C}$ satisfies the condition

$$[\boldsymbol{a}, \boldsymbol{a}] = [\boldsymbol{a}, \boldsymbol{A}]. \tag{5.5-10a}$$

Then, from the inequality

$$[\boldsymbol{a}-\boldsymbol{A}, \boldsymbol{a}-\boldsymbol{A}] \geqq 0 \tag{5.5-11}$$

it follows that

$$[\boldsymbol{a}, \boldsymbol{a}] \leqq [\boldsymbol{A}, \boldsymbol{A}], \tag{5.5-12}$$

with equality holding if and only if $a(C) \equiv A(C)$.

Similarly, if the tensor $\mathbf{b} \equiv b(C)(\boldsymbol{CC} - \tfrac{1}{3}C^2\mathbf{I})$ satisfies the condition

$$[\boldsymbol{b}, \boldsymbol{b}] = [\boldsymbol{b}, \boldsymbol{B}], \tag{5.5-13}$$

then

$$[\boldsymbol{b}, \boldsymbol{b}] \leqq [\boldsymbol{B}, \boldsymbol{B}], \tag{5.5-14}$$

with equality holding if and only if $b(C) \equiv B(C)$.

Thus we can formulate the following *maximum principle*: In non-equilibrium systems the distribution of the molecular velocities is such that, for given temperature and velocity gradients, the rate of change of the entropy density due to collisions is as large as possible. This maximum principle, together with a similar minimum principle, was first given by Kohler [1948]. A discussion of these and other variational principles can be found in a paper by Ziman [1956] or in a paper by Snider [1964a].

In the next section we use the maximum principle and Ritz's method of variational calculus to solve Eqs. (5.4-15) and (5.4-16) for $\boldsymbol{A}$ and $\boldsymbol{B}$, respectively, and compute the transport coefficients η and λ. First, we choose two sets of trial functions $\boldsymbol{a} \equiv a(C)\boldsymbol{C}$ and $\boldsymbol{b} \equiv b(C)(\boldsymbol{CC}-\frac{1}{3}C^2\boldsymbol{I})$, which contain a number of arbitrary parameters. Then, we consider only those trial functions $\boldsymbol{a}$ which satisfy the constraint (5.5-10a) and, in analogy with the constraint (5.4-30) on $\boldsymbol{A}$, the constraint

$$\int f^{(0)}\, a(C)\, C^2\, \mathrm{d}^3c = 0. \tag{5.5-10b}$$

From the inequality (5.5-12) we conclude that the best approximation to the exact solution $\boldsymbol{A}$ of Eq. (5.4-15) is given by the vector $\boldsymbol{a}$ which maximizes the bracket $[\boldsymbol{a}, \boldsymbol{a}]$ with respect to all the available parameters in the trial function $\boldsymbol{a}$. Similarly, by considering only those trial functions $\boldsymbol{b}$ which satisfy the constraint (5.5-13), we find from the inequality (5.5-14) that the best approximation to the exact solution $\boldsymbol{B}$ of Eq. (5.4-16) is given by the tensor $\boldsymbol{b}$ which maximizes the bracket $[\boldsymbol{b}, \boldsymbol{b}]$ with respect to all the available parameters in the trial function $\boldsymbol{b}$. To determine these best approximations to $\boldsymbol{A}$ and $\boldsymbol{B}$ is, of course, a simple problem of differential calculus. Having found these best approximations, we then calculate the transport coefficients by substituting $\boldsymbol{a}$ and $\boldsymbol{b}$ for $\boldsymbol{A}$ and $\boldsymbol{B}$, respectively, in the definitions (5.4-36) and (5.4-42) of η and λ.

5.6. Evaluation of η and λ

As the trial functions $\boldsymbol{a}$ in the computation of $\boldsymbol{A} \equiv A(C)\boldsymbol{C}$ we take *finite* linear combinations of Sonine polynomials,

$$\boldsymbol{a} \equiv a(C)\boldsymbol{C} = -\left(\frac{m}{2kT}\right)^{\frac{1}{2}} \sum_{p=0}^{n} a_p^{(n)} S_{\frac{3}{2}}^{(p)}(\mathscr{C}^2)\boldsymbol{\mathscr{C}}, \tag{5.6-1}$$

where $\boldsymbol{\mathscr{C}}$ is a dimensionless velocity variable,

$$\boldsymbol{\mathscr{C}} = (m/2kT)^{\frac{1}{2}}\boldsymbol{C} \tag{5.6-2}$$

and $\mathscr{C}$ is the magnitude of the vector $\boldsymbol{\mathscr{C}}$. The index $\frac{3}{2}$ has been chosen for later convenience. The dependence of the expansion coefficients on the number of terms used in the finite series has been indicated explicitly by the superscript (n).

The trial functions have to satisfy the condition (5.5-10b). With the choice of index made the latter condition yields

$$\begin{aligned} 0 = \frac{1}{n}\int f^{(0)} a(C) C^2 \,\mathrm{d}^3c &= -\frac{1}{n}\sum_{p=0}^{n} a_p^{(n)} \int f^{(0)} S_{\frac{3}{2}}^{(p)}(\mathscr{C}^2)\mathscr{C}^2 \,\mathrm{d}^3c \\ &= -\frac{1}{\pi^{\frac{3}{2}}}\sum_{p=0}^{n} a_p^{(n)} \int \mathrm{e}^{-\mathscr{C}^2} S_{\frac{3}{2}}^{(p)}(\mathscr{C}^2)\mathscr{C}^2 \,\mathrm{d}^3\mathscr{C} \\ &= -\frac{4}{\pi^{\frac{1}{2}}}\sum_{p=0}^{n} a_p^{(n)} \int_0^\infty \mathrm{e}^{-\mathscr{C}^2} S_{\frac{3}{2}}^{(p)}(\mathscr{C}^2)\mathscr{C}^4 \,\mathrm{d}\mathscr{C} \\ &= -\frac{2}{\pi^{\frac{1}{2}}}\sum_{p=0}^{n} a_p^{(n)} \Gamma(\tfrac{5}{2})\delta_{p0} \\ &= -\tfrac{3}{2}a_0^{(n)}. \end{aligned} \tag{5.6-3}$$

(Here we have used the orthogonality relation (5.5-4) and the normalization (5.5-5) for Sonine polynomials.) Hence, it is only necessary to require that $a_0^{(n)} = 0$ and, instead of Eq. (5.6-1), we take

$$\boldsymbol{a} \equiv a(C)\boldsymbol{C} = -\left(\frac{m}{2kT}\right)^{\frac{1}{2}} \sum_{p=1}^{n} a_p^{(n)} S_{\frac{3}{2}}^{(p)}(\mathscr{C}^2)\boldsymbol{\mathscr{C}}. \tag{5.6-4}$$

For a given n, the statement of the variational criterion is

$$\delta\{g\} = 0 \tag{5.6-5a}$$

where g is the bracket integral formed from the trial function (5.6-4),

$$g \equiv [\boldsymbol{a}, \boldsymbol{a}] = \frac{75k}{16}\sum_{q=1}^{n}\sum_{r=1}^{n} \Lambda^{qr} a_q^{(n)} a_r^{(n)}. \tag{5.6-5b}$$

The coefficient Λ^{qr} is proportional to a bracket integral involving the Sonine polynomials of order q and r,*

$$\Lambda^{qr} = \frac{8m}{75k^2T}[S_{\frac{3}{2}}^{(q)}(\mathscr{C}^2)\mathscr{C}, S_{\frac{3}{2}}^{(r)}(\mathscr{C}^2)\mathscr{C}], \tag{5.6-6}$$

the proportionality constant being chosen for later convenience. The Λ^{qr} are symmetric,

$$\Lambda^{qr} = \Lambda^{rq}. \tag{5.6-7}$$

In addition, the trial function (5.6-4) must satisfy the constraint (5.5-10a). The bracket integral of the right member of Eq. (5.5-10a) can be calculated in terms of the unknown coefficients $a_p^{(n)}$. Since the vector $\boldsymbol{A}$ obeys the integral equation (5.4-15) we have

$$\begin{aligned}[\boldsymbol{a}, \boldsymbol{A}] &= \int \boldsymbol{a}\cdot I(\boldsymbol{A})\,\mathrm{d}^3c \\ &= -\frac{1}{n}\sum_{p=1}^{n} a_p^{(n)} \int f^{(0)} S_{\frac{3}{2}}^{(p)}(\mathscr{C}^2)(\mathscr{C}^2-\tfrac{5}{2})\mathscr{C}^2\,\mathrm{d}^3c \\ &= \tfrac{15}{4}a_1^{(n)}. \end{aligned} \tag{5.6-8}$$

(Here we have used the relation $\mathscr{C}^2-\frac{5}{2} = -S_{\frac{3}{2}}^{(1)}(\mathscr{C}^2)$, which follows immediately from Eq. (5.5-2), and the orthogonality and normalization properties of Sonine polynomials, Eqs. (5.5-4) and (5.5-5). The reason for the choice of index $\frac{3}{2}$ is now obvious.) Hence, the variational criterion (5.6-5) is supplemented by the constraint

$$w = 0 \tag{5.6-9a}$$

where

$$w \equiv \frac{75k}{16}\sum_{q=1}^{n}\sum_{r=1}^{n}\Lambda^{qr}a_q^{(n)}a_r^{(n)} - \tfrac{15}{4}a_1^{(n)}. \tag{5.6-9b}$$

The extremum of g subject to the constraint (5.6-9) can be determined by the method of Lagrangian multipliers.

Let λ be a Lagrangian multiplier. Then Eq. (5.6-9) and the equations

$$(\partial/\partial a_p^{(n)})[g+\lambda w] = 0, \quad p = 1, \ldots, n \tag{5.6-10a}$$

or, equivalently,

$$2(1+\lambda)\frac{75k}{16}\sum_{q=1}^{n}\Lambda^{pq}a_q^{(n)} = \tfrac{15}{4}\lambda\delta_{p1}, \quad p = 1, \ldots, n, \tag{5.6-10b}$$

* The definitions of the constants Λ and H (see Eq. (5.6-15)) are similar to (but not identical with) the definitions of the constants Q of Hirschfelder, Curtiss and Bird [1954].

are sufficient to determine λ and the expansion coefficients $a_p^{(n)}$. Multiplying Eq. (5.6-10b) by $a_p^{(n)}$, summing over p, and comparing with Eq. (5.6-9) we obtain the result

$$\lambda = -2, \tag{5.6-11}$$

so the coefficients $a_p^{(n)}$ can be solved from the following set of linear algebraic equations,

$$\sum_{q=1}^{n} \Lambda^{pq} a_q^{(n)} = \frac{4}{5k} \delta_{p1}, \quad p = 1, \ldots, n. \tag{5.6-12}$$

Thus, the function $\boldsymbol{a}$ introduced in Eq. (5.6-4) is uniquely determined.

Next, we use the extremum principle (5.5-14) to find the best solution to the integral equation (5.4-16) for the tensor $\mathbf{B} \equiv B(C)(\boldsymbol{CC} - \frac{1}{3}C^2\mathbf{I})$. As the trial functions $\mathbf{b}$ we take again *finite* linear combinations of Sonine polynomials,

$$\mathbf{b} \equiv b(C)(\boldsymbol{CC} - \tfrac{1}{3}C^2\,\mathbf{I}) = \sum_{p=0}^{n-1} b_p^{(n)} S_{\frac{5}{2}}^{(p)}(\mathscr{C}^2)(\mathscr{C}\mathscr{C} - \tfrac{1}{3}\mathscr{C}^2\mathbf{I}), \tag{5.6-13}$$

where the dimensionless velocity variable $\mathscr{C}$ has been defined in Eq. (5.6-2). The index $\frac{5}{2}$ has been chosen for later convenience. Notice that the number of terms (n) in the trial function $\mathbf{b}$, Eq. (5.6-13), is equal to the number of terms (n) in the trial function $\boldsymbol{a}$, Eq. (5.6-4).

For a given n, the statement of the variational criterion is

$$\delta\{g\} = 0 \tag{5.6-14a}$$

where g is the bracket integral formed from the trial function (5.6-13),

$$g \equiv [\mathbf{b}, \mathbf{b}] = \tfrac{5}{2}kT \sum_{q=0}^{n-1} \sum_{r=0}^{n-1} H^{qr} b_q^{(n)} b_r^{(n)}. \tag{5.6-14b}$$

The coefficient H^{qr}(Gr. eta) is proportional to a bracket integral involving the Sonine polynomials of order q and r,*

$$H^{qr} = \frac{2}{5kT} [S_{\frac{5}{2}}^{(q)}(\mathscr{C}^2)(\mathscr{C}\mathscr{C} - \tfrac{1}{3}\mathscr{C}^2\mathbf{I}),\ S_{\frac{5}{2}}^{(r)}(\mathscr{C}^2)(\mathscr{C}\mathscr{C} - \tfrac{1}{3}\mathscr{C}^2\mathbf{I})], \tag{5.6-15}$$

the proportionality constant being chosen for later convenience. The H^{qr} are symmetric,

$$H^{qr} = H^{rq}. \tag{5.6-16}$$

* See footnote on p. 136.

The trial function (5.6-13) must satisfy the constraint (5.5-13). The bracket integral of the right member of Eq. (5.5-13) can be calculated in terms of the unknown coefficients $b_p^{(n)}$. Since the tensor $\boldsymbol{B}$ obeys the integral equation (5.4-16) we have,

$$\begin{aligned}[\boldsymbol{b}, \boldsymbol{B}] &= \int \boldsymbol{b} : I(\boldsymbol{B})\, \mathrm{d}^3 c \\ &= \frac{2}{n} \sum_{p=0}^{n-1} b_p^{(n)} \int f^{(0)} S_{\frac{5}{2}}^{(p)}(\mathscr{C}^2)(\mathscr{C}\mathscr{C} - \tfrac{1}{3}\mathscr{C}^2 \mathsf{I}) : (\mathscr{C}\mathscr{C} - \tfrac{1}{3}\mathscr{C}^2 \mathsf{I})\, \mathrm{d}^3 c \\ &= 5 b_0^{(n)}, \end{aligned} \tag{5.6-17}$$

so the variational criterion (5.6-14) is supplemented by the constraint

$$w = 0, \tag{5.6-18a}$$

where

$$w \equiv \tfrac{5}{2} kT \sum_{q=0}^{n-1} \sum_{r=0}^{n-1} H^{qr} b_q^{(n)} - 5 b_0^{(n)}. \tag{5.6-18b}$$

Again, using the method of Lagrangian multipliers one verifies that the coefficients $b_p^{(n)}$ can be solved from the following set of linear algebraic equations,

$$\sum_{q=0}^{n-1} H^{pq} b_q^{(n)} = \frac{2}{kT} \delta_{p0}, \quad p = 0, \ldots, n-1. \tag{5.6-19}$$

Thus, the function $\boldsymbol{b}$ introduced in Eq. (5.6-13) is uniquely determined.

Turning finally to the transport coefficients η and λ, we deduce from the foregoing analysis that the coefficient of thermal conductivity λ, which is proportional to the bracket integral $[A, A]$, cf. Eq. (5.4-42), can, in the nth approximation, be written as

$$[\lambda]_n = \tfrac{1}{3} k [\boldsymbol{a}, \boldsymbol{a}] = \tfrac{5}{4} k a_1^{(n)} \tag{5.6-20}$$

and the viscosity coefficient η, which is proportional to the bracket integral $[\boldsymbol{B}, \boldsymbol{B}]$, cf. Eq. (5.4-36), as

$$[\eta]_n = \tfrac{1}{10} kT [\boldsymbol{b}, \boldsymbol{b}] = \tfrac{1}{2} kT b_0^{(n)}. \tag{5.6-21}$$

In particular, for $n = 1$,

$$[\lambda]_1 = \tfrac{5}{4} k a_1^{(1)} = (\Lambda^{11})^{-1} \tag{5.6-22}$$

and

$$[\eta]_1 = \tfrac{1}{2} kT b_0^{(1)} = (H^{00})^{-1}, \tag{5.6-23}$$

and, for $n = 2$,

$$[\lambda]_2 = \tfrac{5}{4}ka_1^{(2)} = \left(1+\frac{\Lambda^{12}\Lambda^{21}}{\Lambda^{11}\Lambda^{22}-\Lambda^{12}\Lambda^{21}}\right)\frac{1}{\Lambda^{11}} \tag{5.6-24}$$

and

$$[\eta]_2 = \tfrac{1}{2}kTb_0^{(2)} = \left(1+\frac{H^{01}H^{10}}{H^{00}H^{11}-H^{01}H^{10}}\right)\frac{1}{H^{00}}. \tag{5.6-25}$$

Thus, taking $n = 1, 2, \ldots$ one generates sequences of numerical approximations to the transport coefficients λ and η. Because the integral equations (5.4-15) and (5.4-16) for $\boldsymbol{A}$ and $\mathbf{B}$ are self-adjoint, these sequences are monotonically increasing and, because the Sonine polynomials form a complete set of functions, they converge to the value of the corresponding transport coefficient predicted by the first-order Chapman-Enskog theory. In each approximation, λ and η are obtained in the form of algebraic combinations of bracket integrals containing Sonine polynomials. These bracket integrals are eight-fold integrals which must be evaluated for the method to be of practical value. This question will be taken up in Chapter 7.

5.7. Kihara approximation

In the previous section we have seen that the variational method, when used in conjunction with the Sonine polynomial trial functions (5.6-4) and (5.6-13), leads to the sets of linear algebraic equations (5.6-12) and (5.6-19). Solution of the latter sets of equations then provides the quantities needed to evaluate the transport coefficients. This approach was developed in 1948 by Kohler. Prior to Kohler's work Chapman and Cowling [1939] and Burnett [1935a,b] had used a somewhat more heuristic approach which leads to precisely the same equations. In their method one simply expands the unknown functions $\boldsymbol{A}$ and $\mathbf{B}$ in *infinite* series of Sonine polynomials similar to the expansions (5.6-1) and (5.6-13), i.e.,

$$\boldsymbol{A} \equiv A(C)\boldsymbol{C} = -\left(\frac{m}{2kT}\right)^{\frac{1}{2}}\sum_{p=1}^{\infty} a_p S_{\frac{3}{2}}^{(p)}(\mathscr{C}^2)\boldsymbol{\mathscr{C}} \tag{5.7-1}$$

$$\mathbf{B} \equiv B(C)(\boldsymbol{C}\boldsymbol{C}-\tfrac{1}{3}C^2\,\mathbf{I}) = \sum_{p=0}^{\infty} b_p S_{\frac{5}{2}}^{(p)}(\mathscr{C}^2)(\boldsymbol{\mathscr{C}}\boldsymbol{\mathscr{C}}-\tfrac{1}{3}\mathscr{C}^2\mathbf{I}). \tag{5.7-2}$$

Here, we have already used the result that the constraint (5.5-10b) requires that $a_0 = 0$. The expansions (5.7-1) and (5.7-2) are then substituted into the equations obeyed by $\boldsymbol{A}$ and $\mathbf{B}$, viz., (5.4-15) and (5.4-16), respectively. When the scalar products of the resulting equations are taken with $S_{\frac{3}{2}}^{(q)}(\mathscr{C}^2)\mathscr{C}$ and $S_{\frac{5}{2}}^{(q)}(\mathscr{C}^2)(\mathscr{C}\mathscr{C}-\frac{1}{3}\mathscr{C}^2\mathbf{I})$ respectively, and integrated over velocity, it is found that the coefficients a_p and b_p obey infinite sets of equations similar to Eqs. (5.6-12) and (5.6-19),

$$\sum_{q=1}^{\infty} \Lambda^{pq} a_q = \frac{4}{5k} \delta_{p1}, \quad p = 1, 2, \ldots \tag{5.7-3}$$

$$\sum_{q=0}^{\infty} H^{pq} b_q = \frac{2}{kT} \delta_{p0}, \quad p = 0, 1, \ldots. \tag{5.7-4}$$

The desired coefficients a_1 and b_0 may now be expressed as ratios of infinite determinants. Chapman and Cowling [1939] proceed by expanding the determinants in infinite series and obtain an nth approximation by keeping n terms of the series. This method is completely equivalent to truncating the series (5.7-1) and (5.7-2) at the nth term which is, in turn, equivalent to replacing Eqs. (5.7-3) and (5.7-4) by Eqs. (5.6-12) and (5.6-19), respectively. Thus the Chapman-Cowling method is equivalent to the variational method and we shall refer to the nth approximation of the previous section as the nth *Chapman-Cowling approximation.*

Kihara [1949] developed a somewhat different and, in fact, simpler approximation which is based on the use of Eqs. (5.7-3) and (5.7-4). He noted that, independently of the molecular model used,

$$\Lambda^{pp} \gg \Lambda^{p,p\pm 1} \gg \Lambda^{p,p\pm 2} \ldots \tag{5.7-5}$$

and

$$H^{pp} \gg H^{p,p\pm 1} \gg H^{p,p\pm 2} \ldots. \tag{5.7-6}$$

(In fact, for the so-called Maxwell molecule model, Λ^{pq} and H^{pq} for $q \neq p$ are identically zero.) The inequalities (5.7-5) and (5.7-6) result from numerical experimentation and have not been proven; Mason [1957a], however, has given an intuitive argument that these inequalities should hold. Kihara's method is a perturbation scheme. A first approximation is obtained by setting all off-diagonal elements equal to zero. Thus, one obtains

$$a_1 = a_1^{(1)K} = \frac{4}{5k}(\Lambda^{11})^{-1}, \tag{5.7-7}$$

$$b_0 = b_0^{(1)K} = \frac{2}{kT}(H^{00})^{-1}, \tag{5.7-8}$$

whence

$$[\lambda]_1^K = (\Lambda^{11})^{-1}, \tag{5.7-9}$$

$$[\eta]_1^K = (H^{00})^{-1}. \tag{5.7-10}$$

These are identical to the results (5.6-22) and (5.6-23) obtained by the Chapman-Cowling method so, to the lowest order, there is no difference between the two methods.

In the second approximation only the elements of Λ^{pq} and H^{pq} with $q = p$ and $q = p \pm 1$ are retained. The set of equations so obtained is not readily soluble, so one introduces a perturbation-theoretic approach at this point. In Eq. (5.7-3) for $p = 2$ one replaces a_1 by $a_1^{(1)K}$ and, similarly, in Eq. (5.7-4) for $p = 1$ one replaces b_0 by $b_0^{(1)K}$. Thus one has

$$\Lambda^{11} a_1 + \Lambda^{12} a_2 = \frac{4}{5k}, \tag{5.7-11a}$$

$$\Lambda^{21} a_1^{(1)K} + \Lambda^{22} a_2 = 0, \tag{5.7-11b}$$

and

$$H^{00} b_0 + H^{01} b_1 = \frac{2}{kT}, \tag{5.7-12a}$$

$$H^{10} b_0^{(1)K} + H^{11} b_1 = 0. \tag{5.7-12b}$$

Solving for the coefficients a_1 and b_0 one has

$$a_1 = a_1^{(2)K} = \left(1 + \frac{\Lambda^{12}\Lambda^{21}}{\Lambda^{11}\Lambda^{22}}\right) a_1^{(1)K}, \tag{5.7-13}$$

$$b_0 = b_0^{(2)K} = \left(1 + \frac{H^{01}H^{10}}{H^{00}H^{11}}\right) b_0^{(1)K}, \tag{5.7-14}$$

whence

$$[\lambda]_2^K = \left(1 + \frac{\Lambda^{12}\Lambda^{21}}{\Lambda^{11}\Lambda^{22}}\right) \frac{1}{\Lambda^{11}}, \tag{5.7-15}$$

$$[\eta]_2^K = \left(1 + \frac{H^{01}H^{10}}{H^{00}H^{11}}\right) \frac{1}{H^{00}}. \tag{5.7-16}$$

Comparing Eqs. (5.7-15) and (5.7-16) with Eqs. (5.6-24) and (5.6-25), respectively, we see that the Kihara approximations are indeed simpler than the corresponding Chapman-Cowling approximations. The results (5.7-15) and (5.7-16) are called the second Kihara approximations by some authors, while other authors introduce a further approximation for Λ^{22} and H^{11} and call the results thus obtained the second Kihara approximations. As these latter approximations depend on the explicit form of the Λ^{pq} and H^{pq} we shall defer discussion of them until Chapter 7.

In the higher-order *Kihara approximations* one proceeds in a similar manner. For example, in the nth approximation for the thermal conductivity only those matrix elements Λ^{pq} with $|p-q| < n$ are retained and, in Eq. (5.7-3) with index p, the unknown a_q is replaced by the quantity $a_q^{(n-p+q)K}$ for each $q : q = 1, 2, \ldots, p-1$. (The latter quantities are known from the lower-order Kihara approximations.) The resulting equations are then readily solved for the unknowns $a_1 = a_1^{(n)K}, \ldots, a_n = a_n^{(n)K}$. The reason for the substitution is that the Kihara approach regards Λ^{pq} as a quantity of order $\varepsilon^{|p-q|}$, where ε is a small parameter (in accord with inequality (5.7-5)) and this procedure assures that in the nth approximation terms of order $\varepsilon^{2(n-1)}$ and no higher are retained. In the Kihara method, the nth order approximation $a_1^{(n)K}$ contains only terms of even order in ε up to order $\varepsilon^{2(n-1)}$; this is readily checked for $n = 1, 2$ from the results given above.

The Kihara approximation can be derived in another way which displays its nature a little more clearly. Equation (5.7-3) may be written in matrix from

$$\boldsymbol{\Lambda a} = \boldsymbol{\alpha}, \tag{5.7-17}$$

where $\boldsymbol{\alpha}$ is the vector $(4/5k, 0, 0, \ldots)$. Then, we define

$$\Lambda_n^{pq} = \begin{cases} \Lambda^{pq} & \text{if} \quad p = q \pm n \\ 0 & \text{otherwise} \end{cases} \tag{5.7-18}$$

and, following the above discussion, $\boldsymbol{\Lambda}_n$ may be regarded as being of order ε^n. Then, we may write

$$\boldsymbol{\Lambda} = \boldsymbol{\Lambda}_0[1 + \varepsilon \boldsymbol{\Lambda}_0^{-1} \boldsymbol{\Lambda}_1 + \varepsilon^2 \boldsymbol{\Lambda}_0^{-1} \boldsymbol{\Lambda}_2 + \ldots]. \tag{5.7-19}$$

The solution to Eq. (5.7-17) is

$$\boldsymbol{a} = \boldsymbol{\Lambda}^{-1} \boldsymbol{\alpha}. \tag{5.7-20}$$

If, now, $\boldsymbol{\Lambda}^{-1}$ is expanded as a power series in ε and terms up to order $\varepsilon^{2(n-1)}$ are retained, we obtain the nth order Kihara approximation. (It turns out that the coefficients of the odd powers of ε make no contribution to $a_1^{(n)K}$ due to the peculiar structure of the $\boldsymbol{\Lambda}_n$.) In fact, the Kihara approximation results may also be obtained by expanding the Chapman-Cowling results in powers of ε.

All of the arguments given above carry over to the treatment of Eq. (5.7-4). The case of the coefficient of viscosity offers, therefore, no specific difficulties and will not be discussed here.

5.8. Second-order approximation to f

In Section 5.4 we have shown that, in the first-order approximation, the Navier-Stokes equations of hydrodynamics provide the appropriate equations for the description of the state of a non-uniform gas. In this section, we shall discuss the second-order approximation. As a result, we shall obtain the so-called Burnett equations, in which the expressions of the heat flow vector and pressure tensor contain second-order derivatives of T and $\boldsymbol{v}$ as well as squares and products of the first derivatives. It may be noted that in contrast to the case for the Navier-Stokes equations, the Burnett equations have never been derived on a heuristic basis; their validity is discussed later. The calculations, although algebraically complicated, follow the same lines as in the first-order approximation. Therefore, we shall not give all details, but merely indicate the essential points. The reader who wants more details may consult the original paper by Burnett [1935a,b] or the monograph of Chapman and Cowling [1952]. Some vector and tensor relations which will be used in the calculations, as well as integrals involving vectors and tensors are given in Appendix A.

The equations from which the second-order coefficient $f^{(2)}$ in the expansion (5.1-2) of the velocity distribution function is to be determined, is obtained from Eq. (5.2-13) with $r = 2$,

$$J(f^{(2)}f^{(0)})+J(f^{(0)}f^{(2)}) = (\mathfrak{D}f)^{(1)}-J(f^{(1)}f^{(1)}) \qquad (5.8\text{-}1)$$

or, if we substitute $f^{(2)} = f^{(0)}\phi^{(2)}$,

$$-n^2 I(\phi^{(2)}) = (\mathfrak{D}f)^{(1)}-J(f^{(1)}f^{(1)}). \qquad (5.8\text{-}2)$$

This is an inhomogeneous integral equation for $\phi^{(2)}$. The right member depends on $f^{(0)}$ and $f^{(1)}$,

$$(\mathfrak{D}f)^{(1)}-J(f^{(1)}f^{(1)}) = \frac{\partial_1 f^{(0)}}{\partial t}+\frac{\partial_0 f^{(1)}}{\partial t}+\boldsymbol{c}\cdot\nabla_r f^{(1)}+\boldsymbol{F}\cdot\nabla_c f^{(1)} -J(f^{(1)}f^{(1)}). \qquad (5.8\text{-}3)$$

From Section 5.3 we recall the expression of $f^{(0)}$, which is the local Maxwellian,

$$f^{(0)} = n(m/2\pi kT)^{\frac{3}{2}} \exp(-mC^2/2kT) = n(m/2\pi kT)^{\frac{3}{2}} \exp(-\mathscr{C}^2) \qquad (5.8\text{-}4)$$

and, from Section 5.4, $f^{(1)}$ is given by

$$f^{(1)} = -\frac{1}{n} f^{(0)} \{\boldsymbol{A} \cdot \nabla \log T + \mathbf{B} : \nabla \boldsymbol{v}\}. \qquad (5.8\text{-}5)$$

In the notation of Section 5.6, first approximations to $\boldsymbol{A}$ and $\mathbf{B}$ are

$$\boldsymbol{A} = -(m/2kT)^{\frac{1}{2}} a_1^{(1)} S_{\frac{3}{2}}^{(1)}(\mathscr{C}^2)\boldsymbol{\mathscr{C}} = (m/2kT) a_1^{(1)} (\mathscr{C}^2 - \tfrac{5}{2})\boldsymbol{C}, \qquad (5.8\text{-}6a)$$

$$\mathbf{B} = b_0^{(1)} S_{\frac{5}{2}}^{(0)}(\mathscr{C}^2)(\boldsymbol{\mathscr{C}\mathscr{C}} - \tfrac{1}{3}\mathscr{C}^2\mathbf{I}) = (m/2kT) b_0^{(1)} (\boldsymbol{CC} - \tfrac{1}{3}C^2\mathbf{I}), \qquad (5.8\text{-}6b)$$

where $a_1^{(1)}$ and $b_0^{(1)}$ are determined from the equations

$$[S_{\frac{3}{2}}^{(1)}(\mathscr{C}^2)\boldsymbol{\mathscr{C}}, S_{\frac{3}{2}}^{(1)}(\mathscr{C}^2)\boldsymbol{\mathscr{C}}] a_1^{(1)} = \frac{15kT}{2m} \qquad (5.8\text{-}7a)$$

$$[(\boldsymbol{\mathscr{C}\mathscr{C}} - \tfrac{1}{3}\mathscr{C}^2\mathbf{I}), (\boldsymbol{\mathscr{C}\mathscr{C}} - \tfrac{1}{3}\mathscr{C}^2\mathbf{I})] b_0^{(1)} = 5, \qquad (5.8\text{-}7b)$$

respectively. In Chapter 7 it will be shown that both bracket integrals in Eq. (5.8-7) have the same value. Therefore, from these equations we deduce that the ratio $a_1^{(1)}/b_0^{(1)}$ is equal to $(3kT/2m)$. Moreover, from Eq. (5.6-23) we deduce that $b_0^{(1)}$ can be expressed in terms of the first approximation to the viscosity coefficient, since $[\eta]_1 = \tfrac{1}{2}kTb_0^{(1)}$. Hence, to a first approximation, $a_1^{(1)} = 3\eta/m$ and $b_0^{(1)} = 2\eta/kT$, and

$$\boldsymbol{A} = \frac{3\eta}{2kT}(\mathscr{C}^2 - \tfrac{5}{2})\boldsymbol{C}, \qquad (5.8\text{-}8a)$$

$$\mathbf{B} = \frac{m\eta}{k^2T^2}(\boldsymbol{CC} - \tfrac{1}{3}C^2\mathbf{I}). \qquad (5.8\text{-}8b)$$

Since we are interested primilarly in the *qualitative* aspects of the second-order approximation to the velocity distribution function, we shall assume that the approximate expression (5.8-8) for $\boldsymbol{A}$ and $\mathbf{B}$ are sufficiently accurate for this purpose (the error is only a few percent in any case). Then, the expression (5.8-5) becomes

$$f^{(1)} = -\left(\frac{m}{2\pi kT}\right)^{\frac{3}{2}} e^{-\mathscr{C}^2} \left\{\frac{3\eta}{2kT}(\mathscr{C}^2 - \tfrac{5}{2})\boldsymbol{C} \cdot \nabla \log T + \frac{m\eta}{k^2T^2}(\boldsymbol{CC} - \tfrac{1}{3}C^2\mathbf{I}) : \nabla \boldsymbol{v}\right\}, \qquad (5.8\text{-}9)$$

which can be written as

$$f^{(1)} = A'(C)\boldsymbol{C}\cdot\nabla T + B'(C)(\boldsymbol{CC}-\tfrac{1}{3}C^2\boldsymbol{I}) : \nabla\boldsymbol{v}, \tag{5.8-10}$$

where A' and B' are defined by

$$A'(C) = -\left(\frac{m}{2\pi kT}\right)^{\frac{3}{2}} \frac{3\eta}{2kT^2}(\mathscr{C}^2-\tfrac{5}{2})\mathrm{e}^{-\mathscr{C}^2}, \tag{5.8-11a}$$

$$B'(C) = -\left(\frac{m}{2\pi kT}\right)^{\frac{3}{2}} \frac{m\eta}{k^2T^2}\mathrm{e}^{-\mathscr{C}^2}. \tag{5.8-11b}$$

Since $f^{(1)}$ is given as a function of $\boldsymbol{C}$ rather than $\boldsymbol{c}$, it is convenient to rewrite the expression (5.8-3) as

$$(\mathfrak{D}f)^{(1)} - J(f^{(1)}f^{(1)}) = \frac{\partial_1 f^{(0)}}{\partial t} + \frac{\mathrm{d}_0 f^{(1)}}{\mathrm{d}t} + \boldsymbol{C}\cdot\nabla_r f^{(1)}$$
$$+\left(\boldsymbol{F}-\frac{\mathrm{d}_0\boldsymbol{v}}{\mathrm{d}t}\right)\cdot\nabla_C f^{(1)} - (\nabla_C f^{(1)})\boldsymbol{C} : \nabla\boldsymbol{v} - J(f^{(1)}f^{(1)}), \tag{5.8-12}$$

where $\boldsymbol{C}$ is now regarded as independent of $\boldsymbol{r}$. From the first-order conservation equations (5.4-44), (5.4-45), (5.4-46) we extract the following expressions for the time-derivatives of n, $\boldsymbol{v}$ and T:

$$\frac{\partial n}{\partial t} = -\nabla\cdot n\boldsymbol{v}, \tag{5.8-13a}$$

$$\frac{\partial \boldsymbol{v}}{\partial t} = \boldsymbol{F} - \frac{1}{\rho}\nabla p - \boldsymbol{v}\cdot\nabla\boldsymbol{v} + \frac{2}{\rho}\nabla\cdot\eta\boldsymbol{S}, \tag{5.8-13b}$$

$$\frac{\partial T}{\partial t} = -\tfrac{2}{3}T\nabla\cdot\boldsymbol{v} - \boldsymbol{v}\cdot\nabla T + \frac{2}{3nk}\nabla\cdot\lambda\nabla T + \frac{4\eta}{3nk}\boldsymbol{S}:\boldsymbol{S}. \tag{5.8-13c}$$

Here we have used the identity $\boldsymbol{S}:\nabla\boldsymbol{v} = \boldsymbol{S}:\boldsymbol{S}$, which follows immediately from the definition (5.4-33) of the tensor $\boldsymbol{S}$.

Hence, since $\partial/\partial t = \partial_0/\partial t + (\varepsilon)\partial_1/\partial t$, comparing Eqs. (5.8-13) with Eqs. (5.4-7), we find

$$\frac{\partial_1 n}{\partial t} = 0, \tag{5.8-14a}$$

$$\frac{\partial_1 \boldsymbol{v}}{\partial t} = \frac{2}{\rho}\nabla\cdot\eta\boldsymbol{S}, \tag{5.8-14b}$$

$$\frac{\partial_1 T}{\partial t} = \frac{2}{3nk} \nabla \cdot \lambda \nabla T + \frac{4\eta}{3nk} \mathbf{S} : \mathbf{S}. \tag{5.8-14c}$$

Thus we obtain for the partial derivative $\partial_1 f^{(0)}/\partial t$:

$$\frac{\partial_1 f^{(0)}}{\partial t} = \frac{1}{p} f^{(0)} \{(\tfrac{2}{3}\mathscr{C}^2 - 1)(\nabla \cdot \lambda \nabla T + 2\eta \mathbf{S} : \mathbf{S}) + 2\boldsymbol{C}\nabla : \eta \mathbf{S}\}. \tag{5.8-15}$$

Furthermore, using the expression (5.4-7c) for $\mathrm{d}_0 T/\mathrm{d}t$,

$$\frac{\mathrm{d}_0 f^{(1)}}{\mathrm{d}t} = -\frac{2}{3}\left(\frac{\partial A'}{\partial T} \boldsymbol{C} \cdot \nabla T + \frac{\partial B'}{\partial T} \boldsymbol{CC} : \mathbf{S}\right) T \nabla \cdot \boldsymbol{v}$$
$$+ A' \boldsymbol{C} \cdot \frac{\mathrm{d}_0 \nabla T}{\mathrm{d}t} + B' \boldsymbol{CC} : \frac{\mathrm{d}_0 \mathbf{S}}{\mathrm{d}t}, \tag{5.8-16}$$

where we have used the identity $(\boldsymbol{CC} - \frac{1}{3}C^2 \mathbf{I}) : \nabla \boldsymbol{v} = \boldsymbol{CC} : \mathbf{S}$, which is trivially proven. Of course, the time-derivatives on the right-hand side of Eq. (5.8-16) can be determined in terms of space-derivatives with the aid of the expressions (5.4-7); however, it is more convenient to retain them in their present form.

The derivatives of $f^{(1)}$ with respect to $\boldsymbol{r}$ and $\boldsymbol{C}$ are easily found,

$$\nabla_r f^{(1)} = \left(\frac{\partial A'}{\partial T} \boldsymbol{C} \cdot \nabla T + \frac{\partial B'}{\partial T} \boldsymbol{CC} : \mathbf{S} + A' \boldsymbol{C} \cdot \nabla\right) \nabla T + B' \nabla \boldsymbol{CC} : \mathbf{S}, \tag{5.8-17}$$

$$\nabla_C f^{(1)} = 2\left(\frac{\partial A'}{\partial C^2} \boldsymbol{C} \cdot \nabla T + \frac{\partial B'}{\partial C^2} \boldsymbol{CC} : \mathbf{S}\right) \boldsymbol{C} + A' \nabla T + 2B' \boldsymbol{C} \cdot \mathbf{S}. \tag{5.8-18}$$

Finally, since J is a bilinear operator,

$$J(f^{(1)} f^{(1)}) = J(A' \boldsymbol{C} \cdot \nabla T, A' \boldsymbol{C} \cdot \nabla T) + J(B' \boldsymbol{CC} : \mathbf{S}, B' \boldsymbol{CC} : \mathbf{S})$$
$$+ J(A' \boldsymbol{C} \cdot \nabla T, B' \boldsymbol{CC} : \mathbf{S}) + J(B' \boldsymbol{CC} : \mathbf{S}, A' \boldsymbol{C} \cdot \nabla T). \tag{5.8-19}$$

Combining the above results, we obtain an expression for $-n^2 I(\phi^{(2)})$ as the sum of three groups of terms

$$-n^2 I(\phi^{(2)}) = \Delta_1 + \Delta_2 + \Delta_3, \tag{5.8-20}$$

where Δ_1 involves only the scalar C (or $\mathscr{C}$),

$$\Delta_1 = \frac{1}{p} f^{(0)} (\tfrac{2}{3}\mathscr{C}^2 - 1)(\nabla \cdot \lambda \nabla T + 2\eta \mathbf{S} : \mathbf{S}) + \frac{1}{\rho} A' (\nabla p) \cdot (\nabla T), \tag{5.8-21}$$

Δ_2 is of odd degree in $\boldsymbol{C}$ (or $\mathscr{C}$),

$$\Delta_2 = \frac{2}{p} f^{(0)} \boldsymbol{C}\nabla : \eta \mathbf{S} - \tfrac{2}{3}(\nabla \cdot \boldsymbol{v}) \left(T \frac{\partial A'}{\partial T} + C^2 \frac{\partial A'}{\partial C^2} \right) \boldsymbol{C} \cdot \nabla T$$
$$+ A' \boldsymbol{C} \cdot \left(\frac{\mathrm{d}_0 \nabla T}{\mathrm{d}t} - (\nabla \boldsymbol{v}) \cdot \nabla T \right) + \frac{2}{\rho} \frac{\partial B'}{\partial C^2} (\boldsymbol{CC} : \mathbf{S})(\boldsymbol{C} \cdot \nabla p)$$
$$+ \frac{2}{\rho} B' \boldsymbol{C}(\nabla p) : \mathbf{S} + B' \boldsymbol{C} \cdot \nabla(\boldsymbol{CC} : \mathbf{S}) + \left(\frac{\partial B'}{\partial T} - 2 \frac{\partial A'}{\partial C^2} \right) (\boldsymbol{CC} : \mathbf{S})(\boldsymbol{C} \cdot \nabla T)$$
$$- J(A' \boldsymbol{C} \cdot \nabla T, B' \boldsymbol{CC} : \mathbf{S}) - J(B' \boldsymbol{CC} : \mathbf{S}, A' \boldsymbol{C} \cdot \nabla T), \qquad (5.8\text{-}22)$$

and where Δ_3 contains only terms which are of even degree in $\boldsymbol{C}$ (or $\mathscr{C}$),

$$\Delta_3 = -\tfrac{2}{3}(\nabla \cdot \boldsymbol{v}) \left(T \frac{\partial B'}{\partial T} + C^2 \frac{\partial B'}{\partial C^2} \right) (\boldsymbol{CC} : \mathbf{S}) + B' \boldsymbol{CC} : \left(\frac{\mathrm{d}_0 \mathbf{S}}{\mathrm{d}t} - 2(\nabla \boldsymbol{v}) \cdot \mathbf{S} \right)$$
$$+ A' \boldsymbol{CC} : \nabla\nabla T + \frac{2}{\rho} \frac{\partial A'}{\partial C^2} \boldsymbol{CC} : (\nabla p)(\nabla T) + \frac{\partial A'}{\partial T} \boldsymbol{CC} : (\nabla T)(\nabla T)$$
$$-2 \frac{\partial B'}{\partial C^2} (\boldsymbol{CC} : \mathbf{S})(\boldsymbol{CC} : \mathbf{S}) - J(A' \boldsymbol{C} \cdot \nabla T, A' \boldsymbol{C} \cdot \nabla T)$$
$$- J(B' \boldsymbol{CC} : \mathbf{S}, B' \boldsymbol{CC} : \mathbf{S}). \qquad (5.8\text{-}23)$$

The integral equation (5.8-20) for the unknown function $\phi^{(2)}$ can be solved by methods similar to those applied to Eq. (5.4-13). The solution is obtained as a sum of several terms, each term having a different tensor character. Again, successive approximations to the unknown functions of C may be determined with the aid of infinite series expansions or by variational methods, using Sonine polynomials, see Burnett [1935a, b].

For the computation of the second-order corrections to the pressure tensor and heat flow vector it is, however, not necessary to know the complete solution of the integral equation (5.8-20). Recall the definition of $\mathbf{P}^{(2)}$,

$$\mathbf{P}^{(2)} = \int m\boldsymbol{CC} f^{(2)} \, \mathrm{d}^3 c. \qquad (5.8\text{-}24)$$

Because of the constraint (5.2-17) on $f^{(2)}$, i.e.,

$$\int \boldsymbol{\psi} f^{(2)} \, \mathrm{d}^3 c = 0, \qquad (5.8\text{-}25)$$

where $\boldsymbol{\psi}$ is the vector whose components are the summational invariants 1,

mc and $\frac{1}{2}mC^2$, the definition (5.8-24) may also be written as

$$\mathbf{P}^{(2)} = \int m(\boldsymbol{CC} - \tfrac{1}{3}C^2\,\mathbf{I}) f^{(2)}\,\mathrm{d}^3c, \tag{5.8-26}$$

which, in turn, can be transformed in the following way:

$$\begin{aligned}\mathbf{P}^{(2)} &= m \int f^{(0)} \phi^{(2)} (\boldsymbol{CC} - \tfrac{1}{3}C^2\mathbf{I})\,\mathrm{d}^3c \\ &= p \int \phi^{(2)} I(\mathbf{B})\,\mathrm{d}^3c \\ &= p \int \mathbf{B}\, I(\phi^{(2)})\,\mathrm{d}^3c \\ &= -\frac{p}{n^2} \int \mathbf{B}(\Delta_1 + \Delta_2 + \Delta_3)\,\mathrm{d}^3c,\end{aligned} \tag{5.8-27}$$

where the integral equation (5.4-16) for the tensor $\mathbf{B}$ and the symmetry property of the bracket integral, Eq. (4.7-7), have been used.

Recalling the identity (A.14-1) we conclude that the contribution to the integral in the right member of Eq. (5.8-27) from the term Δ_1 is zero. The contribution involving Δ_2 likewise vanishes, since the integrand is of odd degree in the components of $\boldsymbol{C}$, so

$$\mathbf{P}^{(2)} = -\frac{p}{n^2} \int \mathbf{B}\Delta_3\,\mathrm{d}^3c. \tag{5.8-28}$$

Thus, since $\mathbf{P}^{(2)}$ is symmetric and traceless, we must have

$$\begin{aligned}\mathbf{P}^{(2)} &= \omega_1 \frac{\eta^2}{p} (\nabla \cdot \boldsymbol{v})\mathbf{S} \\ &+ \omega_2 \frac{\eta^2}{p} \left\{ \frac{\mathrm{d}_0 \mathbf{S}}{\mathrm{d}t} - (\nabla \boldsymbol{v}) \cdot \mathbf{S} - ((\nabla \boldsymbol{v}) \cdot \mathbf{S})^T + \tfrac{2}{3}(((\nabla \boldsymbol{v}) \cdot \mathbf{S}) : \mathbf{I})\mathbf{I} \right\} \\ &+ \omega_3 \frac{\eta^2}{\rho T} (\nabla\nabla T - \tfrac{1}{3}\Delta T\,\mathbf{I}) \\ &+ \omega_4 \frac{\eta^2}{\rho p T} \{ \tfrac{1}{2}(\nabla p)(\nabla T) + \tfrac{1}{2}(\nabla T)(\nabla p) - \tfrac{1}{3}(\nabla p) \cdot (\nabla T)\mathbf{I} \} \\ &+ \omega_5 \frac{\eta^2}{\rho T^2} ((\nabla T)(\nabla T) - \tfrac{1}{3}|\nabla T|^2\,\mathbf{I}) + \omega_6 \frac{\eta^2}{p} (\mathbf{S} \cdot \mathbf{S} - \tfrac{1}{3}((\mathbf{S} \cdot \mathbf{S}) : \mathbf{I})\mathbf{I},\end{aligned} \tag{5.8-29}$$

the tensors in this expression being the only symmetric and traceless tensors that can be formed from the elements involved in Δ_3. The coefficients of the tensors are so chosen that the ω_i are pure numbers. The coefficients ω_5 and ω_6, and no others, depend on the integrals J in Δ_3. With the consistent use of the approximation (5.8-8) for A and B it can be shown that

$$\omega_1 = \frac{4}{3}\left(\frac{7}{2} - \frac{T}{\eta}\frac{\mathrm{d}\eta}{\mathrm{d}T}\right), \quad \omega_2 = 2, \quad \omega_3 = 3, \quad \omega_4 = 0,$$

$$\omega_5 = 3\frac{T}{\eta}\frac{\mathrm{d}\eta}{\mathrm{d}T}, \quad \omega_6 = 8. \tag{5.8-30}$$

Thus, since $\boldsymbol{S}$ depends on the space derivatives of $\boldsymbol{v}$, the second-order approximation to the pressure tensor depends on products of first-order derivatives of the hydrodynamic velocity $\boldsymbol{v}$, on first-order derivatives of $\mathrm{d}_0\boldsymbol{v}/\mathrm{d}t = \boldsymbol{F}-(1/\rho)\nabla p$, on products of first-order derivatives of the temperature with each other and with the first-order derivatives of the pressure, and on second-order derivatives of the temperature.

Similarly, for the second-order correction to the heat flow vector,

$$\begin{aligned}
\boldsymbol{q}^{(2)} &= \int \tfrac{1}{2}mC^2\,\boldsymbol{C}f^{(2)}\,\mathrm{d}^3c \\
&= kT\int f^{(0)}\phi^{(2)}\left(\frac{mC^2}{2kT}-\frac{5}{2}\right)\boldsymbol{C}\,\mathrm{d}^3c \\
&= p\int \phi^{(2)}I(\boldsymbol{A})\,\mathrm{d}^3c \\
&= p\int \boldsymbol{A}\,I(\phi^{(2)})\,\mathrm{d}^3c \\
&= -\frac{p}{n^2}\int \boldsymbol{A}(\Delta_1+\Delta_2+\Delta_3)\,\mathrm{d}^3c.
\end{aligned} \tag{5.8-31}$$

Here, only the term with Δ_2 contributes to the integral, since in both other terms the integrand is an odd function of the components of $\boldsymbol{C}$. Furthermore, since $\boldsymbol{A}$ is such that the integral $\int f^{(0)}\boldsymbol{A}(C)C^2\,\mathrm{d}^3c$ vanishes, cf. Eq. (5.4-30), it is easily seen that the contribution from the first term of Δ_2 in Eq. (5.8-22) is identically zero. Thus we must have

$$\boldsymbol{q}^{(2)} = \theta_1\frac{\eta^2}{\rho T}(\nabla\cdot\boldsymbol{v})\nabla T + \theta_2\frac{\eta^2}{\rho T}\left(\frac{\mathrm{d}_0\nabla T}{\mathrm{d}t} - (\nabla\boldsymbol{v})\cdot\nabla T\right) + \theta_3\frac{\eta^2}{p\rho}(\nabla p)\cdot\boldsymbol{S}$$
$$+\theta_4\frac{\eta^2}{\rho}\nabla\cdot\boldsymbol{S} + \theta_5\frac{\eta^2}{\rho T}\boldsymbol{S}\cdot\nabla T, \tag{5.8-32}$$

the vectors in this expression being the only vectors that can be formed from the elements involved in Δ_2. The coefficients of the vectors are so chosen that the θ_i are pure numbers. The coefficient θ_5, and no others, depends on the integrals J in Δ_2. With the consistent use of the approximations (5.8-8) to A and B it can be shown that

$$\theta_1 = \frac{15}{4}\left(\frac{7}{2} - \frac{T}{\eta}\frac{\mathrm{d}\eta}{\mathrm{d}T}\right), \quad \theta_2 = \tfrac{45}{8}, \quad \theta_3 = -3, \quad \theta_4 = 3,$$

$$\theta_5 = 3\left(\frac{35}{4} + \frac{T}{\eta}\frac{\mathrm{d}\eta}{\mathrm{d}T}\right). \qquad (5.8\text{-}33)$$

Notice that the second-order correction to the heat flow vector, $\boldsymbol{q}^{(2)}$, vanishes whenever the gas is at rest or in uniform motion in space.

The values of the coefficients ω and θ given above are exact for Maxwell molecules (for which the expressions (5.8-8) are exact), and it may be expected that they will be not far from the true values for other models of molecular interaction.

When the second-order approximations to the pressure tensor and heat flow vector,

$$\boldsymbol{P} = p\boldsymbol{I} - 2\eta\boldsymbol{S} + \boldsymbol{P}^{(2)}, \qquad (5.8\text{-}34)$$

$$\boldsymbol{q} = -\lambda\nabla T + \boldsymbol{q}^{(2)}, \qquad (5.8\text{-}35)$$

are inserted in the general conservation equations, one obtains a set of partial differential equations for the mass density, hydrodynamic velocity and temperature, which are commonly designated as the *Burnett equations*.

As is well known from fluid dynamics, there is a considerable, largely indirect, evidence to support the use of the Navier-Stokes equations to describe the motion of a gas as viscous compressible fluid. Ever since the set of equations was first proposed by Navier [1822] and later improved and clarified by Stokes [1845], considerable attention has been given to its mathematical theory, see e.g., Ladyzhenskaia [1969]. In comparison, the use of the Burnett equations has never led to any noticeable success. A general mathematical theory of this type of partial differential equations is unknown. The occurrence of higher-order space derivatives of macroscopic observables requires the specification of more boundary conditions; however, as no heuristic derivation of the Burnett equations has ever been found, it is not at all clear what form these additional boundary conditions must have. This reason is, in fact, strong enough to explain why, in fluid dynamics, the Burnett equations are viewed with grave suspicion.

In this connection it may also be worth while to point out that Grad [1958] has shown that the Chapman-Enskog procedure generates a series which is asymptotic in the parameter ε, i.e., in the ratio of the mean free path and a characteristic macroscopic dimension of the system – cf. Section 5.10. It is likely, although unproven, that this series does not converge for any ε. Nevertheless, as with other asymptotic series, it yields satisfactory approximations provided the number of terms retained is not too large (normally, the exact number depends on and decreases with the value of ε). Thus one can speculate that the Burnett equations are an improvement on the Navier-Stokes equations only in the range of ε for which the latter are already satisfactory. If this is indeed the case, the Burnett equations are not likely to be of any practical value.

5.9. Order-of-magnitude estimates

Before turning to a discussion of the nature of the Chapman-Enskog method we give some order-of-magnitude estimates of successive terms in the approximate solution of the Boltzmann equation obtained with it.

In Section 5.3 it was found that the zero-order approximation, $f^{(0)}$, is identical with a local Maxwellian,

$$f^{(0)} = n(m/2\pi kT)^{\frac{3}{2}} \exp(-\mathscr{C}^2), \tag{5.9-1}$$

cf. Eq. (5.3-1). In Section 5.4 it was found that, if the first-order approximation is written as $f^{(0)}(1+\phi^{(1)})$, the perturbation function $\phi^{(1)}$ has the form

$$\phi^{(1)} = -\frac{1}{n}\{\boldsymbol{A}\cdot\nabla\log T+\mathbf{B}:\nabla\boldsymbol{v}\}, \tag{5.9-2}$$

cf. Eq. (5.4-29). It is our purpose to estimate $\phi^{(1)}$.

To this end we use the first approximations to the functions $\boldsymbol{A}$ and $\mathbf{B}$,

$$\boldsymbol{A} = (3\eta/2kT)(\mathscr{C}^2-\tfrac{5}{2})\boldsymbol{C}, \tag{5.9-3a}$$

$$\mathbf{B} = (m\eta/k^2T^2)(\boldsymbol{C}\boldsymbol{C}-\tfrac{1}{3}C^2\mathbf{I}), \tag{5.9-3b}$$

cf. Eq. (5.8-8), so

$$\phi^{(1)} \approx -\frac{3\eta}{2p}\left(\frac{2k}{mT}\right)^{\frac{1}{2}}(\mathscr{C}^2-\tfrac{5}{2})\boldsymbol{\mathscr{C}}\cdot\nabla T-\frac{2\eta}{p}(\boldsymbol{\mathscr{C}}\boldsymbol{\mathscr{C}}-\tfrac{1}{3}\mathscr{C}^2\mathbf{I}):\nabla\boldsymbol{v}. \tag{5.9-4}$$

Taking helium at normal temperature and pressure as representative of the elemental gases, we have $p = 1.0 \times 10^6$, $\eta = 1.9 \times 10^{-4}$ and $(2k/mT)^{\frac{1}{2}} = 3.9 \times 10^2$ (c.g.s. units) and, therefore, if $\mathscr{C}$ is of unit order of magnitude, the numerical values of the two terms in $\phi^{(1)}$ are $1.1 \times 10^{-7}|\nabla T|$ and $3.8 \times 10^{-10}(\mathbf{S} : \mathbf{S})^{\frac{1}{2}}$. Clearly, in this case $\phi^{(1)}$ is small if the gradients of the temperature and the hydrodynamic velocity are of order one. However, if the density is, say 10^{-6} of the normal density or if the gradients of the temperature and hydrodynamic velocity become very large, then $\phi^{(1)}$ becomes comparable with unity. Hence, the first-order correction $f^{(1)}$ becomes relatively more important as the density decreases. This might have been anticipated since the fluid dynamic description arising from the zero-order approximation is known to be accurate at normal densities.

Although the second-order correction $f^{(2)}$ has not been worked out in complete detail, it is not difficult to see from the results of Section 5.8, particularly Eq. (5.8-20), that $\phi^{(2)} = f^{(2)}/f^{(0)}$ is proportional to n^{-2} and, in fact, the successive approximation scheme is such that, for any r, $r = 1, 2, \ldots$, the ratio $f^{(r)}/f^{(0)}$ is proportional to n^{-r}. Therefore, in rarefied gases the method of Chapman and Enskog cannot be an effective means of obtaining solutions to the Boltzmann equation. Even if convergence of the method is assumed, an inconveniently large number of terms would have to be retained.

Finally, we consider the magnitude of the various terms in the expressions for $\mathbf{P}^{(2)}$ and $\boldsymbol{q}^{(2)}$. Since the quantity $(T/\eta)(\mathrm{d}\eta/\mathrm{d}T)$ is of order unity for ordinary gases, the coefficients ω and θ are all of the same order of magnitude. In the expression (5.8-29) for $\mathbf{P}^{(2)}$, the first, second and sixth terms involve products of velocity-gradients; as their coefficients are all of the form $\omega(\eta^2/p)$, these terms are, therefore, comparable in order of magnitude. If $(\eta/p)(\nabla \cdot \boldsymbol{v})$ is small, the ratio of the first of them to $\mathbf{P}^{(1)}$, cf. Eq. (5.4-37), is small; thus, the first, second and sixth terms are comparable with $\mathbf{P}^{(1)}$ only if $(\eta/p)(\nabla \cdot \boldsymbol{v})$ is comparable with unity, which is true only if $\nabla \cdot \boldsymbol{v}$ is so large that the mass-velocity of the gas changes by an appreciable fraction of the velocity of sound over a distance of order of the mean free path. Such extreme cases, however, do not ordinarily occur in gas dynamics (exceptions are shock waves and highly rarefied gases); therefore, the terms in question can normally be disregarded. The temperature stresses given by the third and fifth terms and, also, under special conditions, by the fourth term in the expression (5.8-29) for $\mathbf{P}^{(2)}$ are more important. For example, if the components of the tensors $\mathbf{S}$ and $(\nabla\nabla T - \frac{1}{3}\Delta T \mathbf{I})$ are of order unity, the ratio of the third term to $\mathbf{P}^{(1)}$ is determined by a factor $\omega_3(\eta/\rho T)$, which is of the order 10^{-3} for gases at normal temperature and pressure. Hence, the temperature stresses are not completely negligible compared with ordinary viscous stresses even at ordinary pressures.

A similar analysis can be given with respect to the different terms in the expression (5.8-32) for $\boldsymbol{q}^{(2)}$. If the scales of the space-variations of temperature and velocity are of the same order, all terms are of the same order of magnitude except the third, which is, in general, smaller than the first. The ratio of the first term of $\boldsymbol{q}^{(2)}$ to $\boldsymbol{q}^{(1)}$, cf. Eq. (5.4-41), is determined by the factor $\theta_1(\eta^2/\rho T\lambda)(\nabla \cdot \boldsymbol{v})$, which is of the order of $(\eta/p)(\nabla \cdot \boldsymbol{v})$. As η is of order 10^{-4} (c.g.s. units) for ordinary gases, the thermal flux given by the second-order approximation is, therefore, completely negligible at normal pressure when the velocity gradients are of order one. When pressure becomes of order 10^{-6} atmosphere, $\boldsymbol{q}^{(2)}$ is no longer negligible; however, under such circumstances the mean free path is comparable to the macroscopic dimension of the system and the usual theory must be modified.

5.10. The relation between the Boltzmann equation and the equations of fluid dynamics

We conclude the present chapter with a discussion of the relation between the Boltzmann equation and the equations of fluid dynamics. This relationship is extremely singular, firstly, because the variables used to describe the state of the gas in the two approaches are so dissimilar and, secondly, because the relevant time scales in the two descriptions are, normally, vastly different. A rigorous mathematical exposition of this aspect requires the proof of strong existence theorems in the large and this is exactly what is absent from kinetic theory. For the spatially homogeneous problem, there is such an existence theorem for hard spheres due to Carleman [1933] and one by Wild [1951], modified by Morgenstern [1954], for a cut-off Maxwellian potential; also, there is an existence theorem by Povzner [1962] for an artificial equation which, in the space-independent case, reduces to the Boltzmann equation with a general cut-off potential. However, the spatially homogeneous case is not very interesting physically, as there is no change at all in any of the macroscopic variables and there is, of course, no gas dynamics without spatial variation. On the other hand, the extreme difficulties encountered in existence and uniqueness problems in non-linear inhomogeneous kinetic theory, are not surprising when one notes that this theory should be at least as difficult as the corresponding non-linear theory of the Navier-Stokes equations (which should arise as a special limiting case) – for which a complete theory is also lacking. Of course, it must always be remembered that there is no *a priori* reason to expect the existence of solutions in the large to any *non-linear* equation.

Existence of solutions to the linearized Boltzmann equation for hard spheres was shown by Carleman [1957]. Recently, the theory has been generalized to include a wide class of intermolecular potentials, see Grad [1963b; 1965a, b], Cercignani [1969a], and Guiraud [1968b]. It has been shown that, in these cases, solutions exist which are bounded for all time and which decay to equilibrium. Even more important, the bounds have been shown to be independent of the mean free path and, therefore, carry over to the macroscopic regime. These strong estimates have been used by Grad [1963a] to show the relation between the Boltzmann equation and the Navier-Stokes equations and, again by Grad [1965c], to derive existence results for the non-linear Boltzmann equation which depend on the amplitude of the initial data. For sufficiently small initial amplitude, a unique solution exists for interesting macroscopic times, but it will be approximated by the solution to the linearized equation for only a short (microscopic) part of this time. If in addition the mean free path is small, then the linear approximation is valid for much longer macroscopic times. Furthermore, both solutions are approximated by fluid dynamics. These results will be further discussed below.

Turning now to the connection between the Boltzmann equation and the equations of fluid dynamics we first mention a plausibility argument given by Grad [1958] which shows that the expansion found with Hilbert's method is *asymptotic* in ε except for a short *initial* boundary layer, a narrow *spatial* boundary layer and, possibly, internal *shock* layers, the thicknesses of which are of the order ε, i.e., of the order of the mean free time or mean free path. To show this, the Boltzmann equation is conveniently written in the form of an integral equation for the velocity distribution function, f. For this to be possible we must put certain restrictions on the long-range behavior of the intermolecular forces, e.g., we must require that the latter have a finite range. Then the integrals

$$\nu = \iiint f_1 g b \, \mathrm{d}b \, \mathrm{d}\varepsilon \, \mathrm{d}^3 c_1 , \tag{5.10-1}$$

$$Lf = \iiint f' f_1' g b \, \mathrm{d}b \, \mathrm{d}\varepsilon \, \mathrm{d}^3 c_1 , \tag{5.10-2}$$

are well-defined and, consequently, the collision operator, J, of the Boltzmann equation may be formally written as the difference of a multiplicative operator, ν, and an integral operator, L,

$$J(ff) = -\nu f + Lf. \tag{5.10-3}$$

Thus, in the absence of external forces, the Boltzmann equation (5.1-1) becomes

$$\frac{\partial f}{\partial t} + \boldsymbol{c} \cdot \nabla_r f + \frac{1}{\varepsilon} \nu f = \frac{1}{\varepsilon} Lf. \tag{5.10-4}$$

To get an integral equation we solve Eq. (5.10-4), considering ν and L as given functions of $\boldsymbol{r}$, $\boldsymbol{c}$ and t. We obtain

$$\begin{aligned} f(\boldsymbol{r}, \boldsymbol{c}, t) &= f(\boldsymbol{r}-\boldsymbol{c}(t-t_0), \boldsymbol{c}, t_0) \exp\left\{-\frac{1}{\varepsilon}\int_{t_0}^{t} \nu(\boldsymbol{r}-\boldsymbol{c}(t-\tau), \boldsymbol{c}, \tau)\,\mathrm{d}\tau\right\} \\ &+ \frac{1}{\varepsilon}\int_{t_0}^{t} Lf(\boldsymbol{r}-\boldsymbol{c}(t-s), \boldsymbol{c}, s) \exp\left\{-\frac{1}{\varepsilon}\int_{s}^{t} \nu(\boldsymbol{r}-\boldsymbol{c}(t-\tau), \boldsymbol{c}, \tau)\,\mathrm{d}\tau\right\} \mathrm{d}s. \end{aligned} \tag{5.10-5}$$

As the effects of the boundaries are very similar to the effects of the initial conditions, we will not consider the former explicitly and, in the present discussion, assume that the boundaries may be neglected. Then, the physical interpretation of Eq. (5.10-5) is clear. If we trace the trajectory of a particle having velocity $\boldsymbol{c}$ back from its position $\boldsymbol{r}$ at time t, we find that the particle either came from the initial distribution (and, therefore, contributes to the first term in the right-hand side of Eq. (5.10-5)) or obtained the velocity $\boldsymbol{c}$ due to a collision at some time $t > t_0$ (and, therefore, contributes to the second term in the right-hand side of Eq. (5.10-5)). An alternative form of Eq. (5.10-5) is obtained if we use the functions

$$\begin{aligned} \mu(\boldsymbol{r}-\boldsymbol{c}(t-s), \boldsymbol{c}, s) &= \frac{1}{\varepsilon} \nu(\boldsymbol{r}-\boldsymbol{c}(t-s), \boldsymbol{c}, s) \exp\left\{-\frac{1}{\varepsilon}\int_{s}^{t} \nu(\boldsymbol{r}-\boldsymbol{c}(t-\tau), \boldsymbol{c}, \tau)\mathrm{d}\tau\right\} \end{aligned} \tag{5.10-6}$$

and

$$g(\boldsymbol{r}, \boldsymbol{c}, t) = Lf(\boldsymbol{r}, \boldsymbol{c}, t)/\nu(\boldsymbol{r}, \boldsymbol{c}, t). \tag{5.10-7}$$

Then, instead of Eq. (5.10-5), we have

$$f(\boldsymbol{r}, \boldsymbol{c}, t) = f(\boldsymbol{r}-\boldsymbol{c}(t-t_0), \boldsymbol{c}, t_0) \exp\left\{-\frac{1}{\varepsilon}\int_{t_0}^{t} \nu(\boldsymbol{r}-\boldsymbol{c}(t-\tau), \boldsymbol{c}, \tau)\,\mathrm{d}\tau\right\}$$
$$+\int_{t_0}^{t} g(\boldsymbol{r}-\boldsymbol{c}(t-s), \boldsymbol{c}, s)\, \mu(\boldsymbol{r}-\boldsymbol{c}(t-s), \boldsymbol{c}, s)\mathrm{d}s. \tag{5.10-8}$$

This equation has an interesting physical interpretation. Since, from Eq. (5.10-6),

$$\int_{t_0}^{t} \mu(\boldsymbol{r}-\boldsymbol{c}(t-s), \boldsymbol{c}, s)\,\mathrm{d}s = 1-\exp\left\{-\frac{1}{\varepsilon}\int_{t_0}^{t} \nu(\boldsymbol{r}-\boldsymbol{c}(t-s), \boldsymbol{c}, s)\,\mathrm{d}s\right\}, \tag{5.10-9}$$

the identity (5.10-8) expresses the fact that $f(\boldsymbol{r}, \boldsymbol{c}, t)$ is a mean value (with combined weight equal to 1) of two functions, viz. a distributed mean of the function g with weight μ taken over the time interval $[t_0, t]$, and of the particle distribution function at the initial instant taken with the remaining weight.

Now, for small ε, the weight in Eq. (5.10-8) is concentrated mostly on g and, on g, it is concentrated near $s = t$. In the limit of vanishing ε we obtain $f(\boldsymbol{r}, \boldsymbol{c}, t) = g(\boldsymbol{r}, \boldsymbol{c}, t)$ which is the same as

$$J(ff) = 0, \tag{5.10-10}$$

so in the limit, as $\varepsilon \to 0$, f is locally Maxwellian, in accordance with previous results. Further approximations to f may be obtained from Eq. (5.10-8) by integration by parts. To this end, we introduce some abbreviations,

$$f(t_1, t_2) \equiv f(\boldsymbol{r}-\boldsymbol{c}(t_1-t_2), \boldsymbol{c}, t_2), \tag{5.10-11a}$$

$$g(t_1, t_2) \equiv g(\boldsymbol{r}-\boldsymbol{c}(t_1-t_2), \boldsymbol{c}, t_2), \tag{5.10-11b}$$

$$\nu(t_1, t_2) \equiv \nu(\boldsymbol{r}-\boldsymbol{c}(t_1-t_2), \boldsymbol{c}, t_2), \tag{5.10-11c}$$

$$\bar{\nu}(t_1, t_2) = \int_{t_2}^{t_1} \nu(t_1, \tau)\,\mathrm{d}\tau, \tag{5.10-11d}$$

for any two values t_1, t_2 ($\geqq t_0$). Then, from Eq. (5.10-8),

$$f(\boldsymbol{r}, \boldsymbol{c}, t) = f(t, t_0) \exp\{-\bar{\nu}(t, t_0)/\varepsilon\} + \int_{t_0}^{t} g(t, s)\frac{\partial}{\partial s} \exp\{-\bar{\nu}(t, s)/\varepsilon\}\,\mathrm{d}s. \tag{5.10-12a}$$

By partial integration we obtain

$$f(\boldsymbol{r}, \boldsymbol{c}, t) = f(t, t_0) \exp\{-\bar{\nu}(t, t_0)/\varepsilon\} + [g(t, s) \exp\{-\bar{\nu}(t, s)/\varepsilon\}]_{s=t_0}^{t}$$
$$-\varepsilon \left[\frac{1}{\nu(t, s)} \frac{\partial g(t, s)}{\partial s} \exp\{-\bar{\nu}(t, s)/\varepsilon\}\right]_{s=t_0}^{t}$$
$$+\varepsilon^2 \left[\frac{1}{\nu(t, s)} \frac{\partial}{\partial s}\left\{\frac{1}{\nu(t, s)} \frac{\partial g(t, s)}{\partial s}\right\} \exp\{-\bar{\nu}(t, s)/\varepsilon\}\right]_{s=t_0}^{t} - \ldots . \quad (5.10\text{-}12b)$$

Thus, f can be written as the sum of two series, one of which is proportional to $\exp\{-\bar{\nu}(t, t_0)/\varepsilon\}$. It should be remembered that the derivation given here is purely formal and, consequently, it cannot claim any rigor. Still, it gives rise to some interesting qualitative results. If, in the right member of Eq. (5.10-12b) we neglect the terms proportional to $\exp\{-\bar{\nu}(t, t_0)/\varepsilon\}$ – which presupposes $\bar{\nu}$ to be bounded near $\varepsilon = 0$ – and keep terms up to first order in ε only, we have

$$f(\boldsymbol{r}, \boldsymbol{c}, t) = g(\boldsymbol{r}, \boldsymbol{c}, t) - \frac{\varepsilon}{\nu(\boldsymbol{r}, \boldsymbol{c}, t)} \left[\frac{\partial g(t, s)}{\partial s}\right]_{s=t} \quad (5.10\text{-}13)$$

which is equivalent to

$$J(ff) = \varepsilon \left(\frac{\partial g}{\partial t} + \boldsymbol{c} \cdot \boldsymbol{\nabla}_r g\right). \quad (5.10\text{-}14)$$

Next we write

$$f = f^{(0)} + \varepsilon f^{(1)}, \quad (5.10\text{-}15)$$

where $f^{(0)}$ is the solution to Eq. (5.10-10). Then, $g = f^{(0)} + O(\varepsilon)$ and, to first order in ε, Eq. (5.10-14) yields the following equation for $f^{(1)}$,

$$J(f^{(0)} f^{(1)}) + J(f^{(1)} f^{(0)}) = \frac{\partial f^{(0)}}{\partial t} + \boldsymbol{c} \cdot \boldsymbol{\nabla}_r f^{(0)} = \mathfrak{D} f^{(0)}, \quad (5.10\text{-}16)$$

which is exactly Eq. (5.1-4) for $r = 1$. Thus, to first order in ε we obtain the results that were obtained in Section 5.1 and, as might be expected, as long as we neglect the terms proportional to $\exp\{-\bar{\nu}(t, t_0)/\varepsilon\}$, a similar conclusion holds to *any* order in ε. This result suggests the possibility of studying the validity of the power series expansion method by estimating the error term that is obtained after integrating by parts a certain number of times. Of course, since the error term contains the unknown velocity distribution function, such an estimate will require *ad hoc* assumptions about the properties that solutions are likely to have, but even the qualitative information obtained in this fashion may prove valuable. Therefore, suppose that we are given initial values and boundary values for the velocity distribution function which are independent of ε, and that, in the neighborhood of a given point $\boldsymbol{r}$ at time t, the solution $f(\boldsymbol{r}, \boldsymbol{c}, t; \varepsilon)$ found from the Boltzmann equation by the power series expansion technique, converges as $\varepsilon \to 0$. In this neighborhood, f will become locally Maxwellian and Hilbert's fluid dynamical equations become a good approximation. However, the power series expansion technique requires the neglect of terms proportional to $\exp\{-\bar{\nu}(t, t_0)/\varepsilon\}$ – which is justified only as long as $t-t_0$ is large compared to ε. In other words, there is an *initial* boundary layer of thickness ε (on the order of the mean free time) and, similarly, in the case of a bounded volume, also a *spatial* boundary layer of thickness ε (on the order of the mean free path) in which the description of the state of a gas by Hilbert's fluid dynamical equations ceases to be possible. Another possibility for such a breakdown occurs if the power series expansion obtained for $f(\boldsymbol{r}, \boldsymbol{c}, t; \varepsilon)$ fails to converge at some point $(\boldsymbol{r}, t)$ of space–time. The locus of such points would be a natural boundary at which solutions to Hilbert's equations obtained on either side can be matched only by some discontinuity condition. Indeed, such situations are well known from compressible gas dynamics, where shock layers, whose thickness is on the order of the mean free path, appear and across which solutions to the equations of fluid dynamics must be pieced together by appropriately prescribed jump conditions. We conclude that, in addition to the initial and spatial boundary layers considered before, there may exist internal *shock* layers of thickness ε (on the order of the mean free path) in which, again, the description of the state of a gas by the equations of fluid dynamics breaks down.

These qualitative arguments lead to the following observation. Given a fixed point in space–time, $(\boldsymbol{r}, t)$, which is not in a spatial or initial boundary layer and not in a shock layer, the power series expansion $\sum_r \varepsilon^r f^{(r)}$ obtained by the techniques presented in this chapter, if truncated after n terms, satisfies the Boltzmann equation with a remainder that is $O(\varepsilon^n)$ in an appropriate norm for any finite time interval. In other words, the series $\sum_r \varepsilon^r f^{(r)}$ is *asymptotic* in ε and the equations of fluid dynamics which are satisfied by each partial sum ($r = 0, \ldots, n$) are *asymptotic approximations* to the Boltzmann equation, except for the short initial boundary layer, the narrow spatial boundary layer and, possibly, internal shock layers.

Then, the following difficulty arises. Solving the equations of fluid dynamics requires the specification of initial and/or boundary conditions. The initial values of f certainly determine initial values of the macroscopic variables n, $\boldsymbol{v}$ and T. However, a boundary layer intervenes between the initial instant and the time domain in which the equations of fluid dynamics are applicable. The initial values to be used in conjunction with the hydrodynamic equations must, therefore, be the "asymptotic initial values" at a time of order ε after the initial instant; the difference between the true initial values and the asymptotic initial values is termed "initial slip". A similar argument applies to the boundary conditions ("boundary slip") and to the conditions across a shock layer ("shock slip"). The problem of matching boundary conditions is a familiar one in the theory of differential equations in which a small parameter multiplies the highest order derivative. Techniques have been designed to solve this problem, the principal one being the technique of coordinate stretching (see e.g. Davies and James [1966]), but the rigorous justification of these techniques is still lacking. This similarity between the problem of kinetic theory and the problem of the theory of differential equations could have been expected from the particular form (5.1-1) of the Boltzmann equation, which may be interpreted formally as a differential equation in terms of f in which the part involving the derivatives of f with respect to time and space, i.e., $\mathfrak{D}f$, is multiplied by a small parameter, ε.

The problem of matching the boundary conditions of kinetic theory with the asymptotic regime of fluid dynamics has barely been touched upon in the literature. Grad [1963b] has studied the initial slip problem, i.e., the problem of relating a given initial distribution function to the Hilbert or Chapman-Enskog solution which takes over after an initial transient. A study of the boundary slip problem, i.e., the problem of relating the solution in the inviscid region, the boundary layer and the Knudsen layer for the steady state flow past bodies of moderate curvature, has been presented recently by Darrozès [1968a, b]. Finally, we mention a recent paper by Scharf [1969], in which the problems of initial and boundary conditions in hydrodynamics are discussed on the basis of the linearized Boltzmann equation. We now turn to a brief discussion of the initial slip flow problem, referring to Grad's paper quoted above for further details.

The result (5.10-12) obtained above by non-rigorous arguments, suggests that Hilbert's theory may be generalized if we look for solutions to the Boltzmann equation that are composed of two parts: one representing the Hilbert (or Chapman-Enskog) series solution, the other being a "remainder" which is concentrated in the initial boundary layer and whose variation is, therefore, significant on a time scale of the order ε only. Thus, we look for a solution to the Boltzmann equation (5.1-1) of the form

$$f(\boldsymbol{r}, \boldsymbol{c}, t; \varepsilon) = f_0(\boldsymbol{r}, \boldsymbol{c}, t; \varepsilon) + \varepsilon f_\lambda(\boldsymbol{r}, \boldsymbol{c}, \tau; \varepsilon), \tag{5.10-17}$$

where $\tau = t/\varepsilon$ (i.e., the coordinate t has been stretched by a factor $1/\varepsilon$). Here the leading term, f_0, is the Hilbert (or Chapman-Enskog) series solution, εf_λ represents the "remainder". For simplicity, we shall assume that the initial distribution is locally Maxwellian. If the leading term in the expansion of f_0 is associated with this initial distribution, then we may assume f_λ to be of the form

$$\varepsilon f_\lambda(\boldsymbol{r}, \boldsymbol{c}, \tau; \varepsilon) = \varepsilon f_\lambda^{(1)}(\boldsymbol{r}, \boldsymbol{c}, \tau) + \varepsilon^2 f_\lambda^{(2)}(\boldsymbol{r}, \boldsymbol{c}, \tau) + \ldots . \tag{5.10-18}$$

Substituting Eq. (5.10-17) into Eq. (5.1-1) and making use of the fact that f_0 is a solution, we obtain an equation for f_λ,

$$\frac{\partial f_\lambda}{\partial \tau} + \varepsilon \boldsymbol{c} \cdot \boldsymbol{\nabla}_r f_\lambda = J(f_0 f_\lambda) + J(f_\lambda f_0) + \varepsilon J(f_\lambda f_\lambda). \tag{5.10-19}$$

To convert this into an equation which involves τ alone, we write $f_0(\boldsymbol{r}, \boldsymbol{c}, t; \varepsilon) = f_0(\boldsymbol{r}, \boldsymbol{c}, \varepsilon\tau; \varepsilon)$ and expand in powers of ε,

$$f(\boldsymbol{r}, \boldsymbol{c}, t; \varepsilon) = f_0^{(0)}(\boldsymbol{r}, \boldsymbol{c}) + \varepsilon f_0^{(1)}(\boldsymbol{r}, \boldsymbol{c}, \tau) + \ldots, \tag{5.10-20}$$

where the leading term is time-independent and locally Maxwellian. Now, substituting Eqs. (5.10-18) and (5.10-20) into Eq. (5.10-19) and equating coefficients of like powers of ε in both members of the equations, we obtain, for $n = 1, 2, \ldots$

$$-\frac{\partial f_\lambda^{(n)}}{\partial \tau} + J(f_0^{(0)} f_\lambda^{(n)}) + J(f_\lambda^{(n)} f_0^{(0)}) = g^{(n)}, \tag{5.10-21}$$

where

$$g^{(1)} = 0, \tag{5.10-22a}$$

$$g^{(n)} = \boldsymbol{c} \cdot \boldsymbol{\nabla}_r f_\lambda^{(n-1)} - \sum_{p=1}^{n-1} \{J(f_0^{(p)} f_\lambda^{(n-p)}) + J(f_\lambda^{(n-p)} f_0^{(p)})\}$$
$$- \sum_{p=1}^{n-1} \{J(f_\lambda^{(p)} f_\lambda^{(n-p)}) + J(f_\lambda^{(n-p)} f_\lambda^{(p)})\}, \quad n = 2, 3, \ldots . \tag{5.10-22b}$$

The equation (5.10-21) is a generalization of Eq. (5.1-4). At every value of n, the unknown function appears in the left member only; the right member is known from the solutions of the previous equations. The formal procedure for solving the functions $f_\lambda^{(n)}$ is, indeed, virtually identical with that given in Section 5.1. Since $f_0^{(0)}$ is locally Maxwellian we are led to put $f_\lambda^{(n)} = f_0^{(0)}\phi_\lambda^{(n)}$ and to introduce the linear operator I, as in Section 5.1. In terms of $\phi_\lambda^{(n)}$, Eq. (5.10-21) becomes

$$f_0^{(0)} \frac{\partial \phi_\lambda^{(n)}}{\partial \tau} + n^2 I(\phi_\lambda^{(n)}) = -g^{(n)}. \tag{5.10-23}$$

(Recall that $f_0^{(0)}$ is time-independent.) So the eigenvalue spectrum of I determines the evolution of each $\phi_\lambda^{(n)}$ and, therefore, of f_λ on the τ-scale. Of course $f_0^{(0)}$ is space-dependent (but not τ-dependent!) so that I contains $\boldsymbol{r}$ as a parameter. As a consequence, the eigenvalue spectrum will vary with position. The eigenvalues may be determined from the equation

$$n^2 I(\phi) = \lambda f_0^{(0)}\phi. \tag{5.10-24}$$

$\lambda = 0$ is a multiple eigenvalue in consequence of the conservation laws. The corresponding eigenfunctions contribute to the Hilbert expansion, f_0, in Eq. (5.10-17).

The present knowledge of the eigenvalue spectrum of the operator I is too incomplete to allow for any general conclusion. In the case of a Maxwellian intermolecular force law the operator is known to have a complete point spectrum and the corresponding eigenfunctions are Sonine polynomials, as discussed in Chapter 15. For other potentials, the spectrum is entirely confined to the positive half of the real axis and is partially a point spectrum between 0 and λ^*, where λ^* is the minimum value of the collision frequency, $\nu(c)$, taken over all values of c, partially a continuous spectrum extending to the right of λ^*. Little is known about the details, however, see Grad [1963b], Ferziger [1965], Kuščer and Williams [1967], Su and Young-Ping Pao [1969]. If, for simplicity, it is assumed that I possesses a complete point spectrum, the expansion of f_λ in eigenfunctions can be accomplished by inspection of Eq. (5.10-23). The τ-dependence of $f_\lambda^{(1)}$ is determined by factors $\exp(-\lambda_s\tau)$, the τ-dependence of $f_\lambda^{(n)}$ for $n = 2, 3, \ldots$ by nth order polynomials in τ times factors $\exp[-(\lambda_{s_1} + \ldots + \lambda_{s_n})\tau]$. Since the eigenvalues λ_s of I are all real and non-negative, and the eigenvalue $\lambda = 0$ contributes to f_0 only, it follows that the remainder εf_λ in Eq. (5.10-17) decays exponentially on the τ-scale.

The formal procedure for solving Eq. (5.10-21) is identical with the procedure of Hilbert's theory. Each of the $\phi_\lambda^{(n)}$ is obtained in turn by solving the inhomogeneous integral equation (5.10-23). Each $\phi_\lambda^{(n)}$ contains an adjustable parameter, ψ_λ (or a vector of adjustable parameters if the eigenvalue

λ is degenerate, as was the case for $\lambda = 0$ in Hilbert's theory), which enters linearly as in Eq. (5.1-14). Compatibility conditions for the solution of successive $\phi_\lambda^{(n)}$ result in partial differential equations in terms of certain macroscopic *eigenstates*, σ_λ, which are the Fourier coefficients

$$\sigma_\lambda(\boldsymbol{r}, t) = \int \psi_\lambda f \, \mathrm{d}^3 c. \tag{5.10-25}$$

These partial differential equations are generalizations of Hilbert's fluid dynamical equations, which were given in Section 5.1,cf. Eq. (5.1-27).

The above arguments lead to the following observation. If the spectrum and eigenfunctions of the linear collision operator are known and completeness of the set of eigenfunctions can be proven, we may expect to be able to match an arbitrary initial f by a linear combination of the eigensolutions f_λ. Thus, there would be a finite contribution to the fluid dynamical state carried by the higher-order, exponentially decaying, eigensolutions, i.e., by those f_λ for which $\lambda > 0$ and, by suitably distributing the initial state among the various eigensolutions, we could find an appropriate f_0 that could be matched asymptotically – i.e., after a short transient – to general smooth solutions of the Boltzmann equation. This would resolve the Hilbert paradox: What is truly *causal* (i.e., self-determined by its own initial values) is not the set of macroscopic observables of fluid dynamics, but the set of somewhat intuitive macroscopic variables that appear in the Hilbert theory.

In the foregoing paragraphs we have restricted ourselves to a discussion of the relation between the Boltzmann equation as treated in Hilbert's theory and the equations of fluid dynamics. We shall not go into a detailed examination of the same relation if, instead, the Chapman-Enskog procedure is used to treat the Boltzmann equation. Here, additional difficulties arise since there is no general theory for the Burnett and higher-order equations of fluid dynamics, so that it is impossible to estimate the solutions to these equations. A discussion of the Navier-Stokes approximation can be found in the paper by Grad [1963b], cited before.

The non-uniform state of a gas mixture

6

In Section 2.3 it was shown that, in a gas mixture, the average velocity of the molecules of a particular species may be different from the hydrodynamic velocity. The difference is the diffusion velocity of the particular component of the mixture. Due to the phenomenon of diffusion the situation in a mixture is obviously more complicated than in a simple gas. However, the general problem of evaluating the velocity distribution function in a gas mixture can still be solved in a manner analogous to that employed for a simple gas. It will turn out that composition effects become manifest in two different ways, each giving rise to a new transport phenomenon (in addition to viscosity and thermal conductivity), viz. *diffusion* and *thermal diffusion*. The transport coefficients associated with these phenomena are the coefficients of diffusion and thermal diffusion; they relate the diffusion velocity to gradients of the number densities and temperature respectively. In the following sections, the Chapman-Enskog theory will be used to derive first-order approximations to the flux vectors of a gas mixture of K components; as before, it is assumed that the molecules have no internal degrees of freedom.

6.1. Chapman-Enskog method

The function f_i, which specifies the distribution of velocities of molecules of the ith constituent of the mixture ($i = 1, \ldots, K$), satisfies the Boltzmann equation

$$\mathfrak{D} f_i = \sum_{j=1}^{K} J(f_i f_j), \tag{6.1-1}$$

with the streaming term $\mathfrak{D}f_i$ defined as

$$\mathfrak{D}f_i = \frac{\partial f_i}{\partial t} + \boldsymbol{c}_i \cdot \boldsymbol{\nabla}_r f_i + \boldsymbol{F}_i \cdot \boldsymbol{\nabla}_{c_i} f_i \tag{6.1-2}$$

and the collision term $J(f_i f_j)$ as

$$J(f_i f_j) = \iiint (f_i' f_j' - f_i f_j) g b \, \mathrm{d}b \, \mathrm{d}\varepsilon \, \mathrm{d}^3 c_j . \tag{6.1-3}$$

As in the previous chapter, a parameter ε is introduced to give more weight to the collision term. Instead of Eq. (6.1-1) we write,

$$\mathfrak{D}f_i = \varepsilon^{-1} \sum_{j=1}^{K} J(f_i f_j). \tag{6.1-4}$$

Again, since Eq. (6.1-4) corresponds to the Boltzmann equation for a velocity distribution function $\varepsilon^{-1} f_i$, ε is merely a scaling factor for the density. Eventually, the value of ε will be taken equal to unity, so that all the results are in formal agreement with the original nomenclature.

The equations (6.1-4) for $i = 1, \ldots, K$ are solved, as in the previous chapter, by Enskog's method. As has been shown in Section 4.1, in a K-component mixture, the summational invariants are related to a set of $K+4$ basic macroscopic observables, viz. $n_1, \ldots, n_K$ (n_i being the number density of the ith component), v_x, v_y, v_z (the three components of the hydrodynamic velocity, $\boldsymbol{v}$) and T (temperature). These quantities, or rather the quantities $n_1, \ldots, n_K, \rho v_x, \rho v_y, \rho v_z$ and ρu will be taken as the components of a $(K+4)$-dimensional vector $\boldsymbol{\beta}$, which will be called the vector of macroscopic observables. Following Enskog, we postulate that none of the f_i's, nor $\partial \boldsymbol{\beta}/\partial t$, depend explicitly on time and that time enters only implicitly through $\boldsymbol{\beta}$ and the spatial gradients of $\boldsymbol{\beta}$. That is, we assume the existence of a solution to the Boltzmann equation (6.1-4) of the form

$$f_i(\boldsymbol{r}, \boldsymbol{c}_i, t) \equiv f_i(\boldsymbol{r}, \boldsymbol{c}_i | \boldsymbol{\beta}, \boldsymbol{\nabla}_r \boldsymbol{\beta}, \ldots) \tag{6.1-5}$$

and, also, the existence of a vector $\boldsymbol{\Phi}$ such that

$$(\partial/\partial t)\boldsymbol{\beta}(\boldsymbol{r}, t) = \boldsymbol{\Phi}(\boldsymbol{r} | \boldsymbol{\beta}, \boldsymbol{\nabla}_r \boldsymbol{\beta}, \ldots), \tag{6.1-6}$$

where dots indicate higher-order spatial derivatives of $\boldsymbol{\beta}$.

Now, if, for $i = 1, \ldots, K$, the solution f_i to Eq. (6.1-4) is represented by the power series

$$f_i = f_i^{(0)} + \varepsilon f_i^{(1)} + \varepsilon^2 f_i^{(2)} + \dots \tag{6.1-7}$$

then, also,

$$\Phi = \Phi^{(0)} + \varepsilon \Phi^{(1)} + \varepsilon^2 \Phi^{(2)} + \dots. \tag{6.1-8}$$

Thus, each time derivative $\partial f_i/\partial t$ may be split up as in the case of a simple gas, cf. Eq. (5.2-6), and the streaming term $\mathfrak{D} f_i$ becomes

$$\mathfrak{D} f_i = (\mathfrak{D} f_i)^{(0)} + \varepsilon (\mathfrak{D} f_i)^{(1)} + \varepsilon^2 (\mathfrak{D} f_i)^{(2)} + \dots \tag{6.1-9}$$

where each term in the expansion is defined as in Eq. (5.2-9). Next, the collision term $J(f_i f_j)$ is expanded formally by substituting the expansions of both f_i and f_j,

$$J(f_i f_j) = J(f_i^{(0)} f_j^{(0)}) + \varepsilon \{J(f_i^{(0)} f_j^{(1)}) + J(f_i^{(1)} f_j^{(0)})\} + \varepsilon^2 \{J(f_i^{(0)} f_j^{(2)}) + J(f_i^{(1)} f_j^{(1)}) + J(f_i^{(2)} f_j^{(0)})\} + \dots, \tag{6.1-10}$$

so that, by equating coefficients of like powers of ε in both members of the Boltzmann equation (6.1-4), the following set of equations is obtained:

$$\sum_{j=1}^{K} J(f_i^{(0)} f_j^{(0)}) = 0, \quad i = 1, \dots, K, \tag{6.1-11}$$

$$\sum_{j=1}^{K} J(f_i^{(0)} f_j^{(r)}) + \sum_{j=1}^{K} J(f_i^{(r)} f_j^{(0)}) = (\mathfrak{D} f_i)^{(r-1)} - \sum_{j=1}^{K} J(f_i^{(1)} f_j^{(r-1)}) - \dots - \sum_{j=1}^{K} J(f_i^{(r-1)} f_j^{(1)}), \quad i = 1, \dots, K, \tag{6.1-12}$$

the latter set of equations holding for $r = 1, 2, \dots$.

The system of equations (6.1-11) has been discussed in connection with the H-theorem, cf. Section 4.5. Its general solution is

$$f_i^{(0)}(\boldsymbol{c}_i) = n_i (m_i/2\pi kT)^{\frac{3}{2}} \exp(-m_i C_i^2/2kT), \quad i = 1, \dots, K \tag{6.1-13}$$

with $\boldsymbol{C}_i = \boldsymbol{c}_i - \boldsymbol{v}$, where n_i, $\boldsymbol{v}$ and T are arbitrary functions of $\boldsymbol{r}$ and t. As in Section 5.2, these quantities are chosen such that, at all $\boldsymbol{r}$ and t, n_i is the number density of the ith component of the mixture, $\boldsymbol{v}$ the hydrodynamic velocity and T the temperature of the gas mixture. This choice implies that

$$\int f_i^{(0)} \mathrm{d}^3 c_i = \int f_i \mathrm{d}^3 c_i = n_i, \quad i = 1, \dots, K, \tag{6.1-14a}$$

$$\sum_{i=1}^{K} \int m_i \boldsymbol{c}_i f_i^{(0)} \mathrm{d}^3 c_i = \sum_{i=1}^{K} \int m_i \boldsymbol{c}_i f_i \mathrm{d}^3 c_i = \rho \boldsymbol{v}, \tag{6.1-14b}$$

$$\sum_{i=1}^{K} \int \tfrac{1}{2} m_i C_i^2 f_i^{(0)} \mathrm{d}^3 c_i = \sum_{i=1}^{K} \int \tfrac{1}{2} m_i C_i^2 f_i \mathrm{d}^3 c_i = \rho u = \tfrac{3}{2} nkT, \tag{6.1-14c}$$

so the vector $\boldsymbol{\beta}$ of macroscopic observables is determined from the functions $f_i^{(0)}$ $(i = 1, \ldots, K)$ alone.

To obtain the general solution to the set of equations (6.1-12) we put $\phi_i^{(r)} = f_i^{(r)}/f_i^{(0)}$. Because the $f^{(0)}$ are Maxwellian,

$$\sum_{j=1}^{K} J(f_i^{(0)} f_j^{(r)}) + \sum_{j=1}^{K} J(f_i^{(r)} f_j^{(0)})$$

$$= -\sum_{j=1}^{K} \iiint f_i^{(0)} f_j^{(0)} (\phi_i^{(r)} + \phi_j^{(r)} - \phi_i^{(r)\prime} - \phi_j^{(r)\prime}) g b \,\mathrm{d}b \,\mathrm{d}\varepsilon \,\mathrm{d}^3 c_j \tag{6.1-15}$$

or, taking into account the definition (4.7-2) of the integral operator I in the case of a mixture,

$$\sum_{j=1}^{K} J(f_i^{(0)} f_j^{(r)}) + \sum_{j=1}^{K} J(f_i^{(r)} f_j^{(0)}) = -\sum_{j=1}^{K} n_i n_j I_{ij}(\phi^{(r)}). \tag{6.1-16}$$

Thus, Eq. (6.1-12) represents a set of K linear inhomogeneous integral equations of the second kind for the unknown functions $\phi_1^{(r)}, \ldots, \phi_K^{(r)}$,

$$-\sum_{j=1}^{K} n_i n_j I_{ij}(\phi^{(r)}) = (\mathfrak{D} f_i)^{(r-1)} - \sum_{j=1}^{K} J(f_i^{(1)} f_j^{(r-1)}) - \ldots - \sum_{j=1}^{K} J(f_i^{(r-1)} f_j^{(1)}),$$
$$i = 1, \ldots, K. \tag{6.1-17}$$

This set is soluble if and only if the vector whose components are the right members of these equations is orthogonal to the vector whose components are the general solutions of the associated homogeneous equations. Now, the homogeneous form of the set of equations (6.1-17) is

$$\sum_{j=1}^{K} n_i n_j I_{ij}(\phi^{(r)}) = 0, \quad i = 1, \ldots, K. \tag{6.1-18}$$

If, for each i, this equation is multiplied by $\phi_i^{(r)}$ and integrated over $\boldsymbol{c}_i$, and the result is summed over all values of i, one obtains the equation

$$[\phi^{(r)}, \phi^{(r)}] = 0, \tag{6.1-19}$$

where the bracket is defined by Eq. (4.7-13). The solution of this equation was shown in Section 4.7 to be

$$\bar{\phi}_i^{(r)} = \alpha_i^{(r,1)} + \boldsymbol{\alpha}^{(r,2)} \cdot m_i \boldsymbol{c}_i + \tfrac{1}{2}\alpha^{(r,3)} m_i C_i^2, \quad i = 1, \ldots, K. \qquad (6.1\text{-}20)$$

In this expression, $\alpha_i^{(r,1)}$ $(i = 1, \ldots, K)$, $\boldsymbol{\alpha}^{(r,2)}$ and $\alpha^{(r,3)}$ are arbitrary functions of $\boldsymbol{r}$ and t, which are independent of $\boldsymbol{c}_i$. Therefore, the set of equations (6.1-17) is soluble if and only if the orthogonality condition

$$\sum_{i=1}^{K} \int \{\alpha_i^{(r,1)} + \boldsymbol{\alpha}^{(r,2)} \cdot m_i \boldsymbol{c}_i + \alpha^{(r,3)} \tfrac{1}{2} m_i C_i^2\}$$
$$\times \{(\mathfrak{D}f_i)^{(r-1)} - \sum_{j=1}^{K} J(f_i^{(1)} f_j^{(r-1)}) - \ldots - \sum_{j=1}^{K} J(f_i^{(r-1)} f_j^{(1)})\} \, \mathrm{d}^3 c_i = 0 \qquad (6.1\text{-}21)$$

is satisfied, or, since the α's are arbitrary functions of $\boldsymbol{r}$ and t, if and only if the conditions

$$\int (\mathfrak{D}f_i)^{(r-1)} \, \mathrm{d}^3 c_i = 0, \quad i = 1, \ldots, K, \qquad (6.1\text{-}22a)$$

$$\sum_{i=1}^{K} \int m_i \boldsymbol{c}_i (\mathfrak{D}f_i)^{(r-1)} \, \mathrm{d}^3 c_i = 0, \qquad (6.1\text{-}22b)$$

$$\sum_{i=1}^{K} \int \tfrac{1}{2} m_i C_i^2 (\mathfrak{D}f_i)^{(r-1)} \mathrm{d}^3 c_i = 0, \qquad (6.1\text{-}22c)$$

are satisfied. We have used the fact that the number of molecules of each component, the total momentum and total kinetic energy are summational invariants to eliminate the terms arising from the collision operators in Eq. (6.1-21).

Uniqueness of the solution to the set of equations (6.1-17) is obtained if, at all $\boldsymbol{r}$ and t, it is required that

$$\int f_i^{(r)} \, \mathrm{d}^3 c_i = 0, \quad i = 1, \ldots, K, \qquad (6.1\text{-}23a)$$

$$\sum_{i=1}^{K} \int m_i \boldsymbol{c}_i f_i^{(r)} \, \mathrm{d}^3 c_i = 0, \qquad (6.1\text{-}23b)$$

$$\sum_{i=1}^{K} \int \tfrac{1}{2} m_i C_i^2 f_i^{(r)} \, \mathrm{d}^3 c_i = 0, \qquad (6.1\text{-}23c)$$

for all $r = 1, 2, \ldots$. This is consistent with Eq. (6.1-14). Then, by a method which is entirely analogous to the method used in Section 5.2 for the evaluation of the conditions (5.2-15), it may be verified that the solubility conditions (6.1-22) for the case of a mixture may be written as

$$\Phi_i^{(r-1)} + \int \boldsymbol{c}_i \cdot \nabla_r f_i^{(r-1)} \mathrm{d}^3 c_i + \int \boldsymbol{F}_i \cdot \nabla_{c_i} f_i^{(r-1)} \mathrm{d}^3 c_i = 0, \quad i = 1, \ldots, K, \tag{6.1-24a}$$

$$\Phi_{K+1,2,3}^{(r-1)} + \sum_{i=1}^{K} \int m_i c_{i\alpha} \boldsymbol{c}_i \cdot \nabla_r f_i^{(r-1)} \mathrm{d}^3 c_i + \sum_{i=1}^{K} \int m_i c_{i\alpha} \boldsymbol{F}_i \cdot \nabla_{c_i} f_i^{(r-1)} \mathrm{d}^3 c_i = 0,$$
$$\alpha = x, y, z, \tag{6.1-24b}$$

$$\Phi_{K+4}^{(r-1)} + \sum_{i=1}^{K} \int \tfrac{1}{2} m_i C_i^2 \boldsymbol{c}_i \cdot \nabla_r f_i^{(r-1)} \mathrm{d}^3 c_i + \sum_{i=1}^{K} \int \tfrac{1}{2} m_i C_i^2 \boldsymbol{F}_i \cdot \nabla_{c_i} f_i^{(r-1)} \mathrm{d}^3 c_i = 0, \tag{6.1-24c}$$

where $\Phi_1^{(r-1)}, \ldots, \Phi_{K+4}^{(r-1)}$ are the $K+4$ components of the vector $\boldsymbol{\Phi}^{(r-1)}$ defined earlier, cf. Eq. (6.1-6). Hence, for each value of r, $r = 1, 2, \ldots$, $\boldsymbol{\Phi}^{(r+1)}$ is completely specified in terms of velocity moments of $f_i^{(r-1)}$, $i = 1, \ldots, K$.

Then, for the actual computation of the distribution functions f_i a scheme of successive approximations may be set up as in the case of a simple gas. As we have seen in Section 5.2, the scheme is such that, after completion of the rth order approximation, all coefficients $f_i^{(0)}, \ldots, f_i^{(r)}$ ($i = 1, \ldots, K$), as well as $\boldsymbol{\Phi}^{(0)}, \ldots, \boldsymbol{\Phi}^{(r)}$ are known.

Then, within an rth order theory, the time-derivatives of the macroscopic observables are completely specified from Eqs. (6.1-6), (6.1-8), and (6.1-24). An analysis which is entirely similar to the analysis of Section 5.2 leading from Eq. (5.2-24) to Eqs. (5.2-26), (5.2-27), and (5.2-28), leads to the following set of partial differential equations for the macroscopic observables,

$$\frac{1}{\rho_i} \frac{\mathrm{d}\rho_i}{\mathrm{d}t} = -\nabla \cdot \boldsymbol{v} - \sum_{j=0}^{r} \nabla \cdot \boldsymbol{V}_i^{(j)} - \sum_{j=0}^{r} \boldsymbol{V}_i^{(j)} \cdot \frac{1}{\rho_i} \nabla \rho_i, \quad i = 1, \ldots, K, \tag{6.1-25}$$

$$\rho \frac{\mathrm{d}\boldsymbol{v}}{\mathrm{d}t} = \sum_{i=1}^{K} \rho_i \boldsymbol{F}_i - \sum_{j=0}^{r} \nabla \cdot \mathsf{P}^{(j)}, \tag{6.1-26}$$

$$\rho \frac{\mathrm{d}u}{\mathrm{d}t} = -\sum_{j=0}^{r} \nabla \cdot \boldsymbol{q}^{(j)} - \sum_{j=0}^{r} \mathsf{P}^{(j)} : \nabla \boldsymbol{v} + \sum_{i=1}^{K} \sum_{j=0}^{r} \rho_i \boldsymbol{F}_i \cdot \boldsymbol{V}_i^{(j)}, \tag{6.1-27}$$

where, as usual, $\mathrm{d}/\mathrm{d}t$ stands for the substantial derivative, $\mathrm{d}/\mathrm{d}t = \partial/\partial t + \boldsymbol{v} \cdot \nabla_r$ and where $\boldsymbol{V}_i^{(j)}$, $\mathsf{P}^{(j)}$ and $\boldsymbol{q}^{(j)}$ are the jth order contributions to the diffusion velocity vector of species i, the pressure tensor and the heat flow vector of the mixture, respectively,

$$n_i \boldsymbol{V}_i^{(j)} = \varepsilon^j \int \boldsymbol{C}_i f_i^{(j)} \mathrm{d}^3 c_i, \tag{6.1-28}$$

$$\mathsf{P}^{(j)} = \sum_{i=1}^{K} \mathsf{P}_i^{(j)} = \varepsilon^j \sum_{i=1}^{K} \int m_i \boldsymbol{C}_i \boldsymbol{C}_i f_i^{(j)} \mathrm{d}^3 c_i, \tag{6.1-29}$$

$$\boldsymbol{q}^{(j)} = \sum_{i=1}^{K} \boldsymbol{q}_i^{(j)} = \varepsilon^j \sum_{i=1}^{K} \int \tfrac{1}{2} m_i C_i^2 \boldsymbol{C}_i f_i^{(j)} \mathrm{d}^3 c_i. \tag{6.1-30}$$

Finally, if Eq. (6.1-25) is multiplied by ρ_i and summed over all i, one obtains, since $\sum_{i=1}^{K} \rho_i \boldsymbol{V}_i^{(j)} = 0$ for all j,

$$\frac{1}{\rho} \frac{\mathrm{d}\rho}{\mathrm{d}t} = -\nabla \cdot \boldsymbol{v}. \tag{6.1-31}$$

Thus, as in the case of a simple gas, the Chapman-Enskog method provides a unique method of obtaining successive approximations to the velocity distribution function of each component of a mixture and, at the same time, yields the equations of fluid dynamics in such a form that the coefficients can be explicitly evaluated to each order of approximation. In the next section the method will be further worked out for the zero and first-order approximations.

6.2. Zero-order approximation

The general solution of the zero-order equation (6.1-11) is given by

$$f_i^{(0)} = n_i (m_i/2\pi kT)^{\frac{3}{2}} \exp(-m_i C_i^2/2kT), \; i = 1, \ldots, K, \tag{6.2-1}$$

with $\boldsymbol{C}_i = \boldsymbol{c}_i - \boldsymbol{v}$. That is, to zero order, each f_i is a *local* Maxwellian corresponding to local macroscopic properties. When the diffusion velocities, pressure tensor and heat flow vector are calculated from the zero-order approximation (6.2-1), one finds

$$\boldsymbol{V}_i^{(0)} = 0, \tag{6.2-2}$$

$$\mathsf{P}^{(0)} = p\mathsf{I} \quad \text{with} \quad p = nkT, \tag{6.2-3}$$

$$\boldsymbol{q}^{(0)} = 0. \tag{6.2-4}$$

Substitution of these zero-order results in the general conservation equations leads to the following set of partial differential equations for the mass densities, hydrodynamic velocity and temperature,

$$\frac{1}{\rho_i}\frac{\mathrm{d}\rho_i}{\mathrm{d}t} = -\nabla \cdot \boldsymbol{v}, \quad i = 1, \ldots, K, \tag{6.2-5}$$

$$\frac{1}{\rho}\frac{\mathrm{d}\rho}{\mathrm{d}t} = -\nabla \cdot \boldsymbol{v}, \tag{6.2-6}$$

$$\rho \frac{\mathrm{d}\boldsymbol{v}}{\mathrm{d}t} = \sum_{i=1}^{K} \rho_i \boldsymbol{F}_i - \nabla p, \tag{6.2-7}$$

$$\frac{\mathrm{d}}{\mathrm{d}t}(\rho T^{-\frac{3}{2}}) = 0, \tag{6.2-8}$$

i.e., the macroscopic observables satisfy the *Euler hydrodynamic equations.*

6.3. First-order approximation

The equations from which the coefficients $f_i^{(1)}$ $(i = 1, \ldots, K)$ are to be found, have been derived earlier, cf. Eq. (6.1-17); the $\phi_i^{(1)}$ are found from the set of integral equations

$$-\sum_{j=1}^{K} n_i n_j I_{ij}(\phi^{(1)}) = (\mathfrak{D}f_i)^{(0)}, \quad i = 1, \ldots, K. \tag{6.3-1}$$

For each i, the right member depends entirely on $f_i^{(0)}$,

$$(\mathfrak{D}f_i)^{(0)} = \partial_0 f_i^{(0)}/\partial t + \boldsymbol{c}_i \cdot \nabla_r f_i^{(0)} + \boldsymbol{F}_i \cdot \nabla_{c_i} f_i^{(0)} \tag{6.3-2}$$

or, if $f_i^{(0)}$ is regarded as a function of $\boldsymbol{r}$, $\boldsymbol{C}_i$ and t, instead of $\boldsymbol{r}$, $\boldsymbol{c}_i$ and t,

$$(\mathfrak{D}f_i)^{(0)} = f_i^{(0)} \left\{ \frac{\mathrm{d}_0 \log f_i^{(0)}}{\mathrm{d}t} + \boldsymbol{C}_i \cdot \nabla_r \log f_i^{(0)} + \left(\boldsymbol{F}_i - \frac{\mathrm{d}_0 \boldsymbol{v}}{\mathrm{d}t} \right) \cdot \nabla_{C_i} \log f_i^{(0)} \right.$$
$$\left. - (\nabla_{C_i} \log f_i^{(0)}) \boldsymbol{C}_i : \nabla \boldsymbol{v} \right\}, \tag{6.3-3}$$

as in Section 5.4; also, as in that section,

$$\frac{\mathrm{d}_0 \log f_i^{(0)}}{\mathrm{d}t} = -\frac{m_i C_i^2}{3kT} \nabla \cdot \boldsymbol{v}. \tag{6.3-4}$$

Further,

$$\nabla_r \log {}_J{}_i^{(0)} = \nabla \log n_i + \left(\frac{m_i C_i^2}{2kT} - \frac{3}{2} \right) \nabla \log T, \tag{6.3-5}$$

and

$$\nabla_{C_i} \log f_i^{(0)} = -\frac{m_i}{kT} C_i, \tag{6.3-6}$$

so

$$(\mathfrak{D}f_i)^{(0)} = f_i^{(0)} \Big\{ C_i \cdot \Big[\nabla \log n_i + \frac{m_i}{\rho kT}(-\nabla p + \sum_k \rho_k F_k) - \frac{m_i}{kT} F_i + \left(\frac{m_i C_i^2}{2kT} - \frac{3}{2}\right) \nabla \log T \Big] + \frac{m_i}{kT}(C_i C_i - \tfrac{1}{3} C_i^2 I) : \nabla v \Big\}. \tag{6.3-7}$$

Now, it is convenient to work in terms of the mole fractions, n_i/n, and the hydrostatic pressure, p. Since $p = nkT$, one has

$$\nabla \log n_i = \nabla \log (n_i/n) + \nabla \log p - \nabla \log T \tag{6.3-8}$$

and one may verify that the expression (6.3-7) is equivalent to

$$(\mathfrak{D}f_i)^{(0)} = f_i^{(0)} \Big\{ \frac{n}{n_i} C_i \cdot d_i + \left(\frac{m_i C_i^2}{2kT} - \frac{5}{2}\right) C_i \cdot \nabla \log T + \frac{m_i}{kT}(C_i C_i - \tfrac{1}{3} C_i^2 I) : \nabla v \Big\}, \tag{6.3-9}$$

where d_i stands for the vector

$$d_i = \nabla \left(\frac{n_i}{n}\right) + \left(\frac{n_i}{n} - \frac{\rho_i}{\rho}\right) \nabla \log p - \frac{\rho_i}{p}\left(F_i - \sum_j \frac{\rho_j}{\rho} F_j\right). \tag{6.3-10}$$

For reasons that will become obvious later, d_i is called the *diffusion driving force* of species i. Since $\sum_i (n_i/n) = 1$ and $\sum_i (\rho_i/\rho) = 1$, one has the identity

$$\sum_i d_i = 0. \tag{6.3-11}$$

Equation (6.3-9) is the generalization of Eq. (5.4-12) to the case of a multicomponent gas mixture. Thus, we obtain the following set of K inhomogeneous integral equations of the second kind for the first-order perturbation functions $\phi_1^{(1)}, \ldots, \phi_K^{(1)}$:

$$\sum_{j=1}^{K} n_i n_j I_{ij}(\phi^{(1)}) = -f_i^{(0)} \Big\{ \frac{n}{n_i} C_i \cdot d_i + \left(\frac{m_i C_i^2}{2kT} - \frac{5}{2}\right) C_i \cdot \nabla \log T + \frac{m_i}{kT}(C_i C_i - \tfrac{1}{3} C_i^2 I) : \nabla v \Big\}. \tag{6.3-12}$$

Since the vectors $\boldsymbol{d}_1, \ldots, \boldsymbol{d}_K$ are linearly dependent, cf. Eq. (6.3-11), it is preferable first to introduce a new, linearly independent, set of vector variables, $\boldsymbol{d}_1^*, \ldots, \boldsymbol{d}_K^*$, say, in place of $\boldsymbol{d}_1, \ldots, \boldsymbol{d}_K$. This may be done by writing

$$\boldsymbol{d}_i = \boldsymbol{d}_i^* - \gamma_i \sum_j \boldsymbol{d}_j^*, \quad i = 1, \ldots, K, \tag{6.3-13}$$

where $\gamma_1, \ldots, \gamma_K$ are arbitrary constants such that

$$\sum_i \gamma_i = 1. \tag{6.3-14}$$

A convenient choice is given by

$$\gamma_i = \rho_i/\rho. \tag{6.3-15}$$

Then, since I is a linear, rotationally invariant, operator it follows, much as in Section 5.4, that the solution $\phi_i^{(1)}$ is of the form

$$\phi_i^{(1)} = -\frac{1}{n}\sum_j \boldsymbol{D}_i^j \cdot \boldsymbol{d}_j^* - \frac{1}{n}\boldsymbol{A}_i \cdot \boldsymbol{\nabla} \log T - \frac{1}{n}\mathbf{B}_i : \boldsymbol{\nabla}\boldsymbol{v}, \tag{6.3-16}$$

the quantities $\boldsymbol{D}_i^j$ and $\boldsymbol{A}_i$ being vector functions of $\boldsymbol{C}_i$, and $\mathbf{B}_i$ being a traceless tensor function of $\boldsymbol{C}_i$; they must take the form

$$\boldsymbol{D}^j = D^j(C)\boldsymbol{C}, \quad \boldsymbol{A} = A(C)\boldsymbol{C},$$
$$\mathbf{B} = B(C)(\boldsymbol{C}\boldsymbol{C} - \tfrac{1}{3}C^2\,\mathbf{I}). \tag{6.3-17}$$

(The subscript i on the velocity variable will be suppressed wherever its presence is obvious.)

The integral equations for the functions D^j, A, and B are found in the usual way by substituting the expression (6.3-16) into Eq. (6.3-12) and comparing the coefficients of corresponding terms,

$$\sum_j \frac{n_i n_j}{n^2} I_{ij}(\boldsymbol{D}^k) = \frac{1}{n_i} f_i^{(0)} \left(\delta_{ik} - \frac{\rho_i}{\rho}\right) \boldsymbol{C}_i, \quad i, k = 1, \ldots, K, \tag{6.3-18}$$

$$\sum_j \frac{n_i n_j}{n^2} I_{ij}(\boldsymbol{A}) = \frac{1}{n} f_i^{(0)} \left(\frac{m_i C_i^2}{2kT} - \frac{5}{2}\right) \boldsymbol{C}_i, \quad i = 1, \ldots, K, \tag{6.3-19}$$

$$\sum_j \frac{n_i n_j}{n^2} I_{ij}(\mathbf{B}) = \frac{m_i}{nkT} f_i^{(0)} (\boldsymbol{C}_i\boldsymbol{C}_i - \tfrac{1}{3}C_i^2\,\mathbf{I}), \quad i = 1, \ldots, K. \tag{6.3-20}$$

It is not difficult to verify that, for these equations, the solubility conditions are satisfied. Uniqueness of the solutions is obtained if Eq. (6.1-23) is required to be satisfied for $r = 1$. This leads to the following conditions on the functions D^k and A,

$$\sum_i m_i \int f_i^{(0)} C_i^2 D_i^k \, \mathrm{d}^3 c_i = 0 \qquad (6.3\text{-}21)$$

and

$$\sum_i m_i \int f_i^{(0)} C_i^2 A_i \mathrm{d}^3 c_i = 0. \qquad (6.3\text{-}22)$$

We observe that the vectors $\boldsymbol{D}^k$ are not uniquely determined by Eqs. (6.3-18) and (6.3-21) alone. In fact, if we multiply both members of Eq. (6.3-18) by (ρ_k/ρ) and sum over k we find that, for each value of i,

$$\sum_j \frac{n_i n_j}{n^2} I_{ij} \left(\sum_k \frac{\rho_k}{\rho} \boldsymbol{D}^k \right) = 0. \qquad (6.3\text{-}23)$$

In other words, the set of vectors $\boldsymbol{D}^k$ is such that the linear combination $\sum_k (\rho_k/\rho) \boldsymbol{D}^k$ is a summational invariant. Without restricting the validity of our arguments we may set

$$\sum_k (\rho_k/\rho) \boldsymbol{D}^k = 0. \qquad (6.3\text{-}24)$$

As a consequence of this choice, the vector $\boldsymbol{d}_j^*$ in the expression (6.3-16) may be replaced by the vector $\boldsymbol{d}_j^* - (\rho_j/\rho) \sum_k \boldsymbol{d}_k^*$, which, in turn, is identical with the vector $\boldsymbol{d}_j$. Thus we find that the solution $\phi_i^{(1)}$ of Eq. (6.3-12) is given by

$$\phi_i^{(1)} = -\frac{1}{n} \sum_j \boldsymbol{D}_i^j \cdot \boldsymbol{d}_j - \frac{1}{n} \boldsymbol{A}_i \cdot \nabla \log T - \frac{1}{n} \mathbf{B}_i : \nabla \boldsymbol{v}, \qquad (6.3\text{-}25)$$

where the vectors $\boldsymbol{D}^j (j = 1, \ldots, K)$ are uniquely determined by Eqs. (6.3-18), (6.3-21) and (6.3-24), the vector $\boldsymbol{A}$ is uniquely determined by Eqs. (6.3-19) and (6.3-22), and the tensor $\mathbf{B}$ is uniquely determined by Eq. (6.3-20).

By virtue of Eqs. (6.3-18), (6.3-19), and (6.3-20), if $\boldsymbol{d}^l$ and $\boldsymbol{a}$ are any vector functions and $\boldsymbol{b}$ is any tensor function defined for each constituent of the mixture, then

$$[\boldsymbol{D}^k, \boldsymbol{d}^l] = \frac{1}{n_k} \int f_k^{(0)} \boldsymbol{d}_k^l \cdot \boldsymbol{C}_k \mathrm{d}^3 c_k - \frac{1}{\rho} \sum_i m_i \int f_i^{(0)} \boldsymbol{d}_i^l \cdot \boldsymbol{C}_i \mathrm{d}^3 c_i, \qquad (6.3\text{-}26)$$

$$[\boldsymbol{A}, \boldsymbol{a}] = \frac{1}{n}\sum_i \int f_i^{(0)}\left(\frac{m_i C_i^2}{2kT} - \frac{5}{2}\right) \boldsymbol{a}_i \cdot \boldsymbol{C}_i \, \mathrm{d}^3 c_i, \tag{6.3-27}$$

$$[\boldsymbol{B}, \boldsymbol{b}] = \frac{1}{nkT}\sum_i m_i \int f_i^{(0)} (\boldsymbol{C}_i \boldsymbol{C}_i - \tfrac{1}{3} C_i^2 \boldsymbol{I}) : \boldsymbol{b}_i \, \mathrm{d}^3 c_i. \tag{6.3-28}$$

From the results of Section 6.2 it follows that the zero-order approximations to the diffusion velocity $\boldsymbol{V}_i$ and the heat flow vector $\boldsymbol{q}$ are both zero. As we shall not go into a discussion of the second-order Chapman-Enskog approximation for multicomponent gas mixtures, there will be no confusion if, henceforth, we write $\boldsymbol{V}$ and $\boldsymbol{q}$ instead of $\boldsymbol{V}^{(1)}$ and $\boldsymbol{q}^{(1)}$ to denote the first-order approximation to the diffusion velocity and heat flow vector, respectively.

Neglecting vanishing integrals and using the identity (A.14-2) we find the following expression for the diffusion velocity $\boldsymbol{V}_i$,

$$\begin{aligned}\boldsymbol{V}_i &= \frac{1}{n_i}\int \boldsymbol{C}_i f_i^{(1)} \mathrm{d}^3 c_i \\ &= -\frac{1}{3nn_i}\sum_j \int f_i^{(0)} C_i^2 D_i^j \mathrm{d}^3 c_i \, \boldsymbol{d}_j - \frac{1}{3nn_i}\int f_i^{(0)} C_i^2 A_i \mathrm{d}^3 c_i \nabla \log T. \end{aligned} \tag{6.3-29}$$

Now, comparing this result with Eqs. (6.3-26) and (6.3-27) and using the constraints (6.3-21) and (6.3-22) we find that the coefficient of $\boldsymbol{d}_j$ may be expressed in terms of the bracket integral $[\boldsymbol{D}^i, \boldsymbol{D}^j]$, and, similarly, the coefficient of $\nabla \log T$ in terms of the bracket integral $[\boldsymbol{D}^i, \boldsymbol{A}]$. In fact, if we define

$$D_{ij} = \frac{1}{3n}[\boldsymbol{D}^i, \boldsymbol{D}^j], \tag{6.3-30}$$

$$D_{Ti} = \frac{1}{3n}[\boldsymbol{D}^i, \boldsymbol{A}], \tag{6.3-31}$$

then, in the first-order Chapman-Enskog theory, the diffusion velocity for the molecules of species i may be written as

$$\boldsymbol{V}_i = -\sum_j D_{ij} \boldsymbol{d}_j - D_{Ti} \nabla \log T. \tag{6.3-32}$$

The coefficients D_{ij} and D_{Ti} are called the *multicomponent diffusion coef-*

ficients and the *multicomponent thermal diffusion coefficients*, respectively.

From the result (6.3-32) we see that the diffusion velocity contains terms proportional to the concentration gradients, the pressure gradient, the difference between the external forces acting on the various molecular species, and the temperature gradient. The occurrence of the first three components of the diffusion velocity is not surprising; the first component corresponds to ordinary diffusion which tends to reduce the inhomogeneity of a gas whose composition is not uniform; the second component shows that, when the pressure is not uniform, the heavier molecules tend to diffuse towards the regions of greater pressure; the third component indicates that diffusion also occurs when the accelerative effects of the forces acting on the molecules of different species are unequal. Thermal diffusion is, however, an unexpected phenomenon. Before the work of Chapman and Enskog, thermal diffusion in the gas phase had been unknown theoretically and unobserved experimentally. Enskog [1911] was the first to predict the phenomenon of thermal diffusion on purely theoretical grounds, although his discovery remained unrecognized. Later, Chapman [1917] made the same prediction. The existence of thermal diffusion was subsequently confirmed experimentally on a binary mixture by Chapman and Dootson [1917]. An accurate physical explanation of the phenomenon was not given until the work of Monchick and Mason [1967]; a description of their picture will be given in Section 6.7.

From the definition (6.3-30) of the diffusion coefficient it follows that D_{ij} is symmetric in its indices,

$$D_{ij} = D_{ji} \tag{6.3-33}$$

and that $D_{ii} > 0$. From Eq. (6.3-24) it follows that the diffusion coefficients and thermal diffusion coefficients are not linearly independent:

$$\sum_i (\rho_i/\rho) D_{ij} = 0, \tag{6.3-34}$$

$$\sum_i (\rho_i/\rho) D_{Ti} = 0, \tag{6.3-35}$$

so in a K-component mixture the number of independent diffusion coefficients is $\frac{1}{2}K(K-1)$ and the number of independent thermal diffusion coefficients is $K-1$.

We remark that there is a considerable variation among authors in the nomenclature and definition of the multicomponent diffusion and thermal diffusion coefficients. The definition (6.3-30) of the multicomponent diffusion coefficients is consistent with Onsager's reciprocity relations of irreversible thermodynamics, cf. Waldmann [1958], De Groot and Mazur [1962]. This consistency is of particular importance if one desires to generalize the kinetic theory of gas mixtures to systems in which the number density of each species is not a conserved quantity – e.g., in gas mixtures in which chemical reactions occur between the different constituents or in polyatomic gases in which transitions occur between states with different internal degrees of freedom (cf. Chapter 11). It has been shown by Van de Ree [1967] that, in these cases, it is of importance to choose the definition of the transport coefficients in such a way that they are consistent with Onsager's reciprocity relations and that the definition of the multicomponent diffusion coefficients given above is therefore to be preferred to, e.g., the definition given by Hirschfelder, Curtiss and Bird [1954]. Recently, J. Curtiss [1968] has supported this point of view. The definition (6.3-31) of the multicomponent thermal diffusion coefficients was introduced by Waldmann [1947].

It is convenient to introduce the *thermal diffusion ratios* k_{Ti},

$$\sum_j D_{ij} k_{Tj} = D_{Ti}, \quad i = 1, \ldots, K, \tag{6.3-36a}$$

$$\sum_i k_{Ti} = 0. \tag{6.3-36b}$$

Because of the relations (6.3-34), the determinant of the coefficients of Eq. (6.3-36a) vanishes. The only solution to the system of homogeneous equations associated with Eq. (6.3-36a) is the K-component vector of the fractional mass densities (ρ_i/ρ), $i = 1, \ldots, K$, but from Eq. (6.3-35) it is seen that the latter vector is, in turn, orthogonal to the K-component vector of the thermal diffusion coefficients – viz., $\sum_i(\rho_i/\rho)D_{Ti} = 0$. So Eq. (6.3-36a) admits a solution vector whose components k_{Ti} are determined up to a constant multiple of the ratio ρ_i/ρ, and the constraint (6.3-36b) serves to guarantee the uniqueness of the k_{Ti}. In terms of the thermal diffusion ratios, the expression (6.3-32) for $\boldsymbol{V}_i$ may also be written as

$$\boldsymbol{V}_i = -\sum_j D_{ij}(\boldsymbol{d}_j + k_{Tj}\nabla \log T). \tag{6.3-37}$$

Next, we turn to the evaluation of the first-order correction to the pressure tensor. Formally, no fundamental changes are involved compared with the case of a simple gas. If the expression (6.3-25) for $\phi_i^{(1)}$ is used and vanishing integrals are neglected, the equation giving $\mathbf{P}^{(1)}$, cf. Eq. (6.1-29), becomes

$$\mathbf{P}^{(1)} = -\frac{1}{n}\sum_i \int m_i f_i^{(0)} \boldsymbol{C}_i \boldsymbol{C}_i B_i(C_i)(\boldsymbol{C}_i\boldsymbol{C}_i - \tfrac{1}{3}C_i^2\,\mathbf{I})\,\mathrm{d}^3c_i : \nabla\boldsymbol{v}. \qquad (6.3\text{-}38)$$

This may also be written as

$$\mathbf{P}^{(1)} = -\frac{1}{5n}\sum_i \int m_i f_i^{(0)} B_i(C_i)(\boldsymbol{C}_i\boldsymbol{C}_i - \tfrac{1}{3}C_i^2\,\mathbf{I}) : (\boldsymbol{C}_i\boldsymbol{C}_i - \tfrac{1}{3}C_i^2\,\mathbf{I})\,\mathrm{d}^3c_i\,\mathbf{S}, \qquad (6.3\text{-}39)$$

cf. Eq. (5.4-32), where the rate-of-shear tensor $\mathbf{S}$ has been defined in Eq. (5.4-33). Thus, using the identity (6.3-28) we find

$$\mathbf{P}^{(1)} = -\tfrac{1}{5}kT[\mathbf{B}, \mathbf{B}]\mathbf{S}. \qquad (6.3\text{-}40)$$

Hence, if we define the *coefficient of viscosity*, η, as

$$\eta = \tfrac{1}{10}kT[\mathbf{B}, \mathbf{B}], \qquad (6.3\text{-}41)$$

the first-order approximation to the pressure tensor may be written as

$$\mathbf{P} = p\mathbf{I} - 2\eta\mathbf{S}. \qquad (6.3\text{-}42)$$

As was the case for a simple gas, the multicomponent viscosity coefficient is essentially positive.

Finally, we evaluate the first-order approximation to the heat flow vector, $\boldsymbol{q}$. With the solution (6.3-25) for $\phi_i^{(1)}$ we have

$$\boldsymbol{q} = -\frac{1}{3n}\sum_{i,j}\int \tfrac{1}{2}m_i C_i^4 f_i^{(0)} D_i^j\,\mathrm{d}^3c_i\,\boldsymbol{d}_j - \frac{1}{3n}\sum_i \int \tfrac{1}{2}m_i C_i^4 f_i^{(0)} A_i\,\mathrm{d}^3c_i\,\nabla\log T \qquad (6.3\text{-}43)$$

or, equivalently,

$$\begin{aligned}\boldsymbol{q} = &-\frac{kT}{3n}\sum_{i,j}\int f_i^{(0)}\left(\frac{m_iC_i^2}{2kT} - \frac{5}{2}\right) D_i^j C_i^2\,\mathrm{d}^3c_i\,\boldsymbol{d}_j\\ &-\frac{kT}{3n}\sum_i\int f_i^{(0)}\left(\frac{m_iC_i^2}{2kT} - \frac{5}{2}\right) A_i C_i^2\,\mathrm{d}^3c_i\,\nabla\log T\\ &-\frac{5kT}{6n}\sum_{i,j}\int f_i^{(0)} D_i^j C_i^2\,\mathrm{d}^3c_i\,\boldsymbol{d}_j - \frac{5kT}{6n}\sum_i\int f_i^{(0)} A_i C_i^2\,\mathrm{d}^3c_i\,\nabla\log T. \end{aligned} \qquad (6.3\text{-}44)$$

Using the identity (6.3-27) we can express the coefficient of $\boldsymbol{d}_j$ in the first term and the coefficient of $\nabla \log T$ in the second term in terms of the bracket integrals $[\boldsymbol{A}, \boldsymbol{D}^j]$ and $[\boldsymbol{A}, \boldsymbol{A}]$, respectively. Comparing with the expression (6.3-29) for the diffusion velocity $\boldsymbol{V}_i$ we see that the third and fourth terms together contribute a component $\frac{5}{2}kT\sum_i n_i \boldsymbol{V}_i$. Thus, defining the *partial coefficient of thermal conductivity*, λ', of a multicomponent gas mixture as

$$\lambda' = \tfrac{1}{3}k[\boldsymbol{A}, \boldsymbol{A}], \tag{6.3-45}$$

we obtain the following result for the first-order approximation to the heat flow vector,

$$\boldsymbol{q} = -\lambda' \nabla T - p\sum_i D_{Ti}\boldsymbol{d}_i + \tfrac{5}{2}kT\sum_i n_i \boldsymbol{V}_i. \tag{6.3-46}$$

With respect to this result two remarks are in order.

(i) The component $\frac{5}{2}kT\sum_i n_i \boldsymbol{V}_i$ corresponds to the flow of enthalpy due to the diffusion of molecules in the coordinate system which moves with the hydrodynamic velocity ($\boldsymbol{v}$). Since the net flow of molecules through any surface element which moves with the average molecular velocity ($\bar{\boldsymbol{c}}$) is zero, this component accounts for the difference between the heat flow vector in the $\boldsymbol{v}$-coordinate system and the heat flow vector in the $\bar{\boldsymbol{c}}$-coordinate system.

(ii) The coefficient λ' is not accessible to direct experimental measurement since, in a mixture of gases, a temperature gradient induces thermal diffusion and thus concentration gradients. So, even if the process is stationary, $\boldsymbol{d}_i$ will be different from zero and the heat flow due to a temperature gradient will always be accompanied by a heat flow due to the concentration gradient.

It is convenient to eliminate $\boldsymbol{d}_i$ from Eq. (6.3-46) with the aid of Eqs. (6.3-36a), (6.3-33), and (6.3-32):

$$\begin{aligned}\sum_i D_{Ti}\boldsymbol{d}_i &= \sum_{i,j} D_{ij}k_{Tj}\boldsymbol{d}_i = \sum_{i,j} D_{ji}k_{Tj}\boldsymbol{d}_i \\ &= -\sum_i k_{Ti}(\boldsymbol{V}_i + D_{Ti}\nabla \log T). \end{aligned}\tag{6.3-47}$$

Thus, instead of Eq. (6.3-46) we obtain the following expression for the heat flow vector,

$$\boldsymbol{q} = -\lambda \nabla T + p\sum_i \left(k_{Ti} + \frac{5}{2}\frac{n_i}{n}\right)\boldsymbol{V}_i, \tag{6.3-48}$$

where λ is the *coefficient of thermal conductivity* of a multicomponent mixture,

$$\lambda = \lambda' - nk \sum_i k_{Ti} D_{Ti} \tag{6.3-49}$$

or, in terms of a bracket integral,

$$\lambda = \tfrac{1}{3}k[\boldsymbol{A} - \sum_i k_{Ti}\boldsymbol{D}^i, \boldsymbol{A} - \sum_i k_{Ti}\boldsymbol{D}^i]. \tag{6.3-50}$$

From Eqs. (6.3-45) and (6.3-50) and the positivity of the bracket integrals it follows that both λ' and λ are positive quantities. The coefficient λ may be measured directly in a heat conduction experiment because, in the stationary case, the vectors $\boldsymbol{V}_i$ all vanish if the gas is at rest. The difference between λ' and λ is seen to be of the order of D_T^2. Since the thermal diffusion coefficients are small, the difference is negligible under normal circumstances. The component $p \sum_i k_{Ti} \boldsymbol{V}_i$ of the heat flow vector in Eq. (6.3-48) represents the *diffusion heat flow*.

With the results (6.3-32), (6.3-42), and (6.3-48) for the diffusion velocity, the pressure tensor and the heat flow vector, respectively, substituted into the general conservation equations for a multicomponent gas mixture, one arrives at the Navier-Stokes equations of fluid dynamics.

6.4. Evaluation of the multicomponent transport coefficients

To compute the transport coefficients in a multicomponent gas mixture one must solve the integral equations (6.3-18), (6.3-19), and (6.3-20) for the unknown functions $\boldsymbol{D}^k$, $\boldsymbol{A}$, and $\mathbf{B}$, respectively. Since we are interested primarily in particular functionals of $\boldsymbol{D}^k$, $\boldsymbol{A}$ and $\mathbf{B}$, – viz., the bracket integrals which occur in the definitions of the transport coefficients – we shall use the methods of variational calculus to solve Eqs. (6.3-18), (6.3-19), and (6.3-20).

As in the case of a simple gas, we begin by considering the entropy production due to collisions. To first order, the rate of change of the local entropy density, s, due to collisions is proportional to the bracket integral $[\phi^{(1)}, \phi^{(1)}]$,

$$(\partial s/\partial t)_{\text{coll}} = n^2 k[\phi^{(1)}, \phi^{(1)}]. \tag{6.4-1}$$

If one uses Eq. (6.3-12) to eliminate the functions $I_{ij}(\phi^{(1)})$ and expresses the remaining integrals in terms of the diffusion velocity $\boldsymbol{V}_i$, heat flow vector

$\boldsymbol{q}$ and pressure tensor $\mathsf{P}^{(1)}$ one finds

$$\left(\frac{\partial s}{\partial t}\right)_{\text{coll}} = -nk\sum_i \boldsymbol{V}_i \cdot \boldsymbol{d}_i - \frac{1}{T}\left(\boldsymbol{q}-\tfrac{5}{2}p\sum_i \frac{n_i}{n}\boldsymbol{V}_i\right)\cdot \nabla \log T - \frac{1}{T}\mathsf{P}^{(1)} : \mathsf{S}. \quad (6.4\text{-}2)$$

Then, using the results of the previous section it follows that

$$\left(\frac{\partial s}{\partial t}\right)_{\text{coll}} = nk\sum_{i,j} D_{ij}(\boldsymbol{d}_i + k_{Ti}\nabla \log T)\cdot(\boldsymbol{d}_j + k_{Tj}\nabla \log T)$$
$$+\lambda|\nabla \log T|^2 + (2\eta/T)\mathsf{S} : \mathsf{S}. \quad (6.4\text{-}3)$$

Since the rate of change of s due to collisions must be positive under all circumstances it follows that the coefficients D_{ij} are such that the inequality

$$\sum_{i,j} D_{ij}\boldsymbol{x}_i \cdot \boldsymbol{x}_j \geqq 0 \quad (6.4\text{-}4)$$

is satisfied for all $\boldsymbol{x}_i$ for which $\sum_i \boldsymbol{x}_i = 0$. Furthermore, by the same argument it follows from Eq. (6.4-3) that

$$\lambda > 0 \text{ and } \eta > 0. \quad (6.4\text{-}5)$$

Suppose that the vectors $\boldsymbol{d}^k \equiv d^k(C)\boldsymbol{C}$ satisfy the conditions

$$\sum_j \frac{n_i n_j}{n^2}\int \boldsymbol{d}_i^k \cdot I_{ij}(\boldsymbol{d}^l)\,\mathrm{d}^3c_i = \sum_j \frac{n_i n_j}{n^2}\int \boldsymbol{d}_i^k \cdot I_{ij}(\boldsymbol{D}^l)\,\mathrm{d}^3c_i, \quad i = 1, \ldots, K \quad (6.4\text{-}6a)$$

for all pairs of indices k, l ($k, l = 1, \ldots, K$), and that, in addition,

$$\sum_i m_i \int f_i^{(0)} C_i^2\, d_i^k\,\mathrm{d}^3c_i = 0 \quad (6.4\text{-}6b)$$

for each k. Then it is easily verified that, for each pair of indices k, l,

$$[\boldsymbol{d}^k, \boldsymbol{d}^l] \leqq [\boldsymbol{D}^k, \boldsymbol{D}^l], \quad (6.4\text{-}7)$$

with equality holding if and only if, for each k, $d^k(C) \equiv D^k(C)$ – apart from a linear combination of the summational invariants, which is irrelevant in the present discussion.

Similarly, if the vectors $\boldsymbol{a} \equiv a(C)\boldsymbol{C}$ satisfy the conditions

$$\sum_j \frac{n_i n_j}{n^2}\int \boldsymbol{a}_i \cdot I_{ij}(\boldsymbol{a})\,\mathrm{d}^3c_i = \sum_j \frac{n_i n_j}{n^2}\int \boldsymbol{a}_i \cdot I_{ij}(\boldsymbol{A})\,\mathrm{d}^3c_i, \quad i = 1, \ldots, K \quad (6.4\text{-}8a)$$

and if, in addition,

$$\sum_i m_i \int f_i^{(0)} C_i^2 a_i \mathrm{d}^3 c_i = 0 \tag{6.4-8b}$$

then

$$[\boldsymbol{a}, \boldsymbol{a}] \leqq [\boldsymbol{A}, \boldsymbol{A}], \tag{6.4-9}$$

with equality holding if and only if $a(C) \equiv A(C)$, and if the tensors $\mathbf{b} \equiv b(C)(\boldsymbol{CC}-\frac{1}{3}C^2\mathbf{I})$ satisfy the conditions

$$\sum_j \frac{n_i n_j}{n^2} \int \mathbf{b}_i : I_{ij}(\mathbf{b}) \mathrm{d}^3 c_i = \sum_j \frac{n_i n_j}{n^2} \int \mathbf{b}_i : I_{ij}(\mathbf{B}) \mathrm{d}^3 c_i, \quad i = 1, \ldots, K, \tag{6.4-10}$$

then

$$[\mathbf{b}, \mathbf{b}] \leqq [\mathbf{B}, \mathbf{B}], \tag{6.4-11}$$

with equality holding if and only if $b(C) \equiv B(C)$.

The problem is thus one of finding the trial functions for $\boldsymbol{d}_i^k$, $\boldsymbol{a}_i$, and $\mathbf{b}_i$, which maximize the bracket integrals $[\boldsymbol{d}^k, \boldsymbol{d}^l]$, $[\boldsymbol{a}, \boldsymbol{a}]$, and $[\mathbf{b}, \mathbf{b}]$, respectively, subject to the corresponding constraints (6.4-6), (6.4-8), and (6.4-10). As in the simple gas case, this implies that the distribution of the molecular velocities is such that, for given gradients of concentration, pressure, temperature and velocity, the rate of change of the entropy density due to collisions is as large as possible. This *maximum principle*, which is a generalization of the maximum principle of Section 5.5, provides a means of obtaining successive approximations to the multicomponent transport coefficients.

As the trial functions $\boldsymbol{d}_i^1, \ldots, \boldsymbol{d}_i^K$ in the computation of $\boldsymbol{D}_i^1 \equiv D_i^1(C_i)\boldsymbol{C}_i$, $\ldots, \boldsymbol{D}_i^K \equiv D_i^K(C_i)\boldsymbol{C}_i$ we take finite linear combinations of Sonine polynomials,

$$\boldsymbol{d}_i^k \equiv d_i^k(C_i)\boldsymbol{C}_i = \left(\frac{m_i}{2kT}\right)^{\frac{1}{2}} \sum_{p=0}^{n-1} d_{i,p}^{k(n)} S_{\frac{3}{2}}^{(p)}(\mathscr{C}_i^2)\boldsymbol{\mathscr{C}}_i, \tag{6.4-12}$$

where $\boldsymbol{\mathscr{C}}_i$ is the dimensionless velocity variable,

$$\boldsymbol{\mathscr{C}}_i = (m_i/2kT)^{\frac{1}{2}} \boldsymbol{C}_i \tag{6.4-13}$$

and $\mathscr{C}_i$ is the magnitude of the vector $\boldsymbol{\mathscr{C}}_i$. The dependence of the expansion coefficients on the number of terms used in the finite series has been indicated explicitly by the superscript (n).

For a given n the statement of the variational criterion is

$$\delta\{g^{kl}\} = 0 \tag{6.4-14a}$$

where g^{kl} is the bracket formed from the trial function (6.4-12),

$$g^{kl} \equiv [\boldsymbol{d}^k, \boldsymbol{d}^l] = \frac{75k}{16} \sum_{i,j=1}^{K} \sum_{q,r=0}^{n-1} \Lambda_{ij}^{qr} d_{i,q}^{k(n)} d_{j,r}^{l(n)}. \tag{6.4-14b}$$

Here the Λ_{ij}^{qr} are proportional to bracket integrals involving Sonine polynomials of order q and r,*

$$\Lambda_{ij}^{qr} = \frac{8m_i^{\frac{1}{2}} m_j^{\frac{1}{2}}}{75k^2 T} \left\{ \delta_{ij} \sum_h \frac{n_i n_h}{n^2} [S_{\frac{3}{2}}^{(q)}(\mathscr{C}^2)\mathscr{C}, S_{\frac{3}{2}}^{(r)}(\mathscr{C}^2)\mathscr{C}]'_{ih} + \frac{n_i n_j}{n^2} [S_{\frac{3}{2}}^{(q)}(\mathscr{C}^2)\mathscr{C}, S_{\frac{3}{2}}^{(r)}(\mathscr{C}^2)\mathscr{C}]''_{ij} \right\}. \tag{6.4-15}$$

From the symmetry properties (4.7-12) of the partial bracket integrals it is readily verified that the constants Λ satisfy the symmetry relation

$$\Lambda_{ij}^{qr} = \Lambda_{ji}^{rq}. \tag{6.4-16}$$

Moreover, from the principle of conservation of momentum one finds that, for each value of q and r,

$$\sum_i \Lambda_{ij}^{q0} = 0, \quad \sum_i \Lambda_{ij}^{0r} = 0. \tag{6.4-17}$$

The trial functions (6.4-12) must satisfy the constraints (6.4-6). The sum in the right member of Eq. (6.4-6a) can be calculated in terms of the unknown coefficients $d_{i,p}^{k(n)}$. Since the vectors $\boldsymbol{D}^k$ obey the integral equations (6.3-18) we have

$$\sum_j \frac{n_i n_j}{n^2} \int \boldsymbol{d}_i^k \cdot I_{ij}(\boldsymbol{D}^l) \mathrm{d}^3 c_i = \frac{1}{n_i} \left(\delta_{il} - \frac{\rho_i}{\rho} \right) \sum_{p=0}^{n-1} d_{i,p}^{k(n)} \int f_i^{(0)} S_{\frac{3}{2}}^{(p)}(\mathscr{C}_i^2) \mathscr{C}_i^2 \mathrm{d}^3 c_i$$
$$= \frac{3}{2} \left(\delta_{il} - \frac{\rho_i}{\rho} \right) d_{i,0}^{k(n)}. \tag{6.4-18}$$

* It should be noted that the subscripts (i, j) on Λ refer to the molecular species under consideration, whereas the superscripts (q, r) refer to the order of the Sonine polynomials in the bracket integrals. As a general rule, the indices h, i, j, k, and l have been reserved to indicate molecular species, the indices p, q, and r to indicate orders of Sonine polynomials. The definitions of the constants Λ and H (to be defined later, cf. Eq. (6.4-36)) are similar to (but not identical with) the definition of the constants Q by Hirschfelder, Curtiss and Bird [1954].

Here we have used the orthogonality and normalization properties of the Sonine polynomials, Eqs. (5.5-4) and (5.5-5). Hence, the variational criterion (6.4-14) is supplemented by the constraints

$$w_i^{kl} = 0, \quad i = 1, \ldots, K, \tag{6.4-19a}$$

where

$$w_i^{kl} = \frac{75k}{16} \sum_{j=1}^{K} \sum_{q,r=0}^{n-1} \Lambda_{ij}^{qr} d_{i,q}^{k(n)} d_{j,r}^{l(n)} - \frac{3}{4}\left\{\left(\delta_{il} - \frac{\rho_i}{\rho}\right) d_{i,0}^{k(n)} + \left(\delta_{ik} - \frac{\rho_i}{\rho}\right) d_{i,0}^{l(n)}\right\}. \tag{6.4-19b}$$

To obtain a symmetric expression for w^{kl} we have used the fact that the constraint (6.4-6a) must hold for the pair (k, l) as well as for the pair (l, k).

The extremum of g^{kl} subject to the constraint (6.4-19) is determined in the usual way. Let $\lambda_1^{kl}, \ldots, \lambda_K^{kl}$ be a set of K Lagrangian multipliers. Then, Eq. (6.4-19) and the equations

$$(\partial/\partial d_{h,p}^{k(n)})[g^{kl} + \sum_i \lambda_i^{kl} w_i^{kl}] = 0, \tag{6.4-20a}$$

$$(\partial/\partial d_{h,p}^{l(n)})[g^{kl} + \sum_i \lambda_i^{kl} w_i^{kl}] = 0, \tag{6.4-20b}$$

$h = 1, \ldots, K; p = 0, \ldots, n-1$, are $(2n+1)K$ linear equations for the $(2nK)$ unknown coefficients $d_{h,p}^{k(n)}$, $d_{h,p}^{l(n)}$ and the (K) multipliers λ_i^{kl}. Eq. (6.4-20) yields

$$(1+\lambda_h^{kl}) \frac{75k}{16} \sum_{j,q} \Lambda_{hj}^{pq} d_{j,q}^{l(n)} = \tfrac{3}{4}\lambda_h^{kl} \left(\delta_{hl} - \frac{\rho_h}{\rho}\right) \delta_{p0}, \tag{6.4-21a}$$

$$\frac{75k}{16} \sum_{j,q} (1+\lambda_j^{kl}) \Lambda_{hj}^{pq} d_{j,q}^{k(n)} = \tfrac{3}{4}\lambda_h^{kl} \left(\delta_{hk} - \frac{\rho_h}{\rho}\right) \delta_{p0}. \tag{6.4-21b}$$

Multiplying Eq. (6.4-21a) by $d_{h,p}^{k(n)}$ and Eq. (6.4-21b) by $d_{h,p}^{l(n)}$, adding the two resulting equations and subsequently performing a summation over the indices p and h, we obtain the equation

$$\frac{75k}{16} \sum_{hj,pq} \Lambda_{hj}^{pq} \{(1+\lambda_h^{kl}) d_{h,p}^{k(n)} d_{j,q}^{l(n)} + (1+\lambda_j^{kl}) d_{h,p}^{l(n)} d_{j,q}^{k(n)}\}$$
$$= \tfrac{3}{4} \sum_h \lambda_h^{kl} \left\{\left(\delta_{hl} - \frac{\rho_h}{\rho}\right) d_{h,0}^{k(n)} + \left(\delta_{hk} - \frac{\rho_h}{\rho}\right) d_{h,0}^{l(n)}\right\}. \tag{6.4-22}$$

With the aid of the constraints $w_h^{kl} = 0$, $h = 1, \ldots, K$, and the symmetry relations (6.4-16) this equation can be reduced to

$$\sum_{hj,\,pq} (1+\lambda_j^{kl})\, \Lambda_{hj}^{pq}\, d_{h,\,p}^{l(n)}\, d_{j,\,q}^{k(n)} = -\sum_{hj,\,pq} \Lambda_{hj}^{pq}\, d_{h,\,p}^{l(n)}\, d_{j,\,q}^{k(n)}, \tag{6.4-23}$$

from which it is seen that

$$\lambda_1^{kl} = \ldots = \lambda_K^{kl} = -2. \tag{6.4-24}$$

Thus, Eqs. (6.4-21a) and (6.4-21b) become identical and it follows that the coefficients $d_{i,\,p}^{k(n)}$ in the expansion (6.4-12) can be solved from the following set of linear algebraic equations,

$$\sum_{j=1}^{K} \sum_{q=0}^{n-1} \Lambda_{ij}^{pq}\, d_{j,\,q}^{k(n)} = \frac{8}{25k}\left(\delta_{ik} - \frac{\rho_i}{\rho}\right) \delta_{p0}, \quad i = 1, \ldots, K,$$
$$p = 0, \ldots, n-1. \tag{6.4-25}$$

For $p = 0$ the system of equations (6.4-25) is linearly dependent, as may be readily verified by a summation over the index i and a subsequent comparison with the identity (6.4-17). So, for $p = 0$ the system must be supplemented by another independent relation among the unknowns $d_{i,\,p}^{k(n)}$. Such a relation is, of course, provided by the constraint (6.4-6b), which reads

$$\sum_i (\rho_i/\rho)\, d_{i,\,0}^{k(n)} = 0. \tag{6.4-26}$$

Thus, the vectors $\boldsymbol{d}_i^1, \ldots, \boldsymbol{d}_i^K$ are uniquely determined.

Next, we use the extremum principle (6.4-9) to find the best solutions to the integral equations (6.3-19) for the vectors $\boldsymbol{A} \equiv A(C)\boldsymbol{C}$. As the trial functions $\boldsymbol{a}_i$ we take finite linear combinations of Sonine polynomials,

$$\boldsymbol{a}_i \equiv a_i(C_i)\boldsymbol{C}_i = -\left(\frac{m_i}{2kT}\right)^{\frac{1}{2}} \sum_{p=0}^{n} a_{i,\,p}^{(n)}\, S_{\frac{3}{2}}^{(p)}(\mathscr{C}_i^2)\, \boldsymbol{\mathscr{C}}_i. \tag{6.4-27}$$

From here on, the analysis is entirely analogous to the preceding one for finding the trial functions $\boldsymbol{d}_i^1, \ldots, \boldsymbol{d}_i^K$. Because now we deal with only one set of functions, the analysis is, in fact, much simpler.

For a given n, the variational statement reads

$$\delta\{g\} = 0, \tag{6.4-28a}$$

where

$$g \equiv [\boldsymbol{a}, \boldsymbol{a}] = \frac{75k}{16} \sum_{i,\,j=1}^{K} \sum_{q,\,r=0}^{n} \Lambda_{ij}^{qr}\, a_{i,\,q}^{(n)}\, a_{j,\,r}^{(n)}. \tag{6.4-28b}$$

The constraints (6.4-8a) are evaluated with the aid of the integral equations (6.3-19). One has

$$\sum_j \frac{n_i n_j}{n^2} \int \boldsymbol{a}_i \cdot I_{ij}(A) \, \mathrm{d}^3 c_i = -\frac{1}{n} \sum_{p=0}^{n} a_{i,p}^{(n)} \int f_i^{(0)} S_{\frac{3}{2}}^{(p)}(\mathscr{C}_i^2)(\mathscr{C}_i^2 - \tfrac{5}{2}) \mathscr{C}_i^2 \, \mathrm{d}^3 c_i$$

$$= \tfrac{15}{4}(n_i/n) \, a_{i,1}^{(n)}, \qquad (6.4\text{-}29)$$

so that the variational criterion (6.4-28) is supplemented by the constraints

$$w_i = 0, \quad i = 1, \ldots, K, \qquad (6.4\text{-}30a)$$

where

$$w_i = \frac{75k}{16} \sum_{j=1}^{K} \sum_{q,r=0}^{n} \Lambda_{ij}^{qr} \, a_{i,q}^{(n)} a_{j,r}^{(n)} - \frac{15}{4} \frac{n_i}{n} a_{i,1}^{(n)}. \qquad (6.4\text{-}30b)$$

Let $\lambda_1, \ldots, \lambda_K$ be a set of K Lagrangian multipliers. Then Eq. (6.4-30) and the equations

$$(\partial/\partial a_{h,p}^{(n)})[g + \sum_i \lambda_i w_i] = 0, \quad h = 1, \ldots, K; \, p = 0, \ldots, n, \qquad (6.4\text{-}31)$$

are sufficient to determine the unknown coefficients $a_{i,p}^{(n)}$ and the multipliers λ_i. One finds, in the usual way, that $\lambda_i = -2$ for all $i = 1, \ldots, K$, while the $a_{i,p}^{(n)}$ can be solved from the following set of linear algebraic equations,

$$\sum_{j=1}^{K} \sum_{q=0}^{n} \Lambda_{ij}^{pq} a_{j,q}^{(n)} = \frac{4}{5k} \frac{n_i}{n} \delta_{p1}, \quad i = 1, \ldots, K; \, p = 0, \ldots, n. \qquad (6.4\text{-}32)$$

For $p = 0$ this set of equations must be supplemented by the constraint (6.4-8b), which reads

$$\sum_i (\rho_i/\rho) \, a_{i,0}^{(n)} = 0. \qquad (6.4\text{-}33)$$

Thus, the vectors $\boldsymbol{a}_i$ are uniquely determined.

Finally, the solution of the integral equations (6.3-20) for the tensors $\mathbf{B} \equiv B(C)(\boldsymbol{CC} - \frac{1}{3}C^2 \boldsymbol{I})$ proceeds along similar lines. As the trial functions $\mathbf{b}_i$ we take again finite linear combinations of Sonine polynomials,

$$\mathbf{b}_i \equiv b_i(C_i)(C_i C_i - \tfrac{1}{3} C_i^2 \, \mathbf{I})$$

$$= \sum_{p=0}^{n-1} b_{i,p}^{(n)} S_{\frac{5}{2}}^{(p)}(\mathscr{C}_i^2)(\mathscr{C}_i \mathscr{C}_i - \tfrac{1}{3} \mathscr{C}_i^2 \, \mathbf{I}). \qquad (6.4\text{-}34)$$

For a given n, the variational criterion is

$$\delta\{g\} = 0, \qquad (6.4\text{-}35a)$$

where

$$g \equiv [\boldsymbol{b}, \boldsymbol{b}] = \tfrac{5}{2}kT \sum_{i,j=1}^{K} \sum_{q,r=0}^{n-1} H_{ij}^{qr} b_{i,q}^{(n)} b_{j,r}^{(n)}. \tag{6.4-35b}$$

Here, the H_{ij}^{qr} (Gr. eta) are proportional to bracket integrals involving Sonine polynomials of order q and r,*

$$H_{ij}^{qr} = \frac{2}{5kT}\left\{\delta_{ij} \sum_h \frac{n_i n_h}{n^2} [S_{\frac{5}{2}}^{(q)}(\mathscr{C}^2)(\mathscr{C}\mathscr{C} - \tfrac{1}{3}\mathscr{C}^2 \mathsf{I}), S_{\frac{5}{2}}^{(r)}(\mathscr{C}^2)(\mathscr{C}\mathscr{C} - \tfrac{1}{3}\mathscr{C}^2 \mathsf{I})]'_{ih}\right.$$
$$\left. + \frac{n_i n_j}{n^2} [S_{\frac{5}{2}}^{(q)}(\mathscr{C}^2)(\mathscr{C}\mathscr{C} - \tfrac{1}{3}\mathscr{C}^2 \mathsf{I}), S_{\frac{5}{2}}^{(r)}(\mathscr{C}^2)(\mathscr{C}\mathscr{C} - \tfrac{1}{3}\mathscr{C}^2 \mathsf{I})]''_{ij}\right\}. \tag{6.4-36}$$

They satisfy the symmetry relation

$$H_{ij}^{qr} = H_{ji}^{rq}. \tag{6.4-37}$$

The variational criterion (6.4-35) is supplemented by the constraints (6.4-10), which read

$$w_i = 0, \quad i = 1, \ldots, K, \tag{6.4-38a}$$

where

$$w_i = \tfrac{5}{2}kT \sum_{j=1}^{K} \sum_{q,r=0}^{n-1} H_{ij}^{qr} b_{i,q}^{(n)} b_{j,r}^{(n)} - 5\frac{n_i}{n} b_{i,0}^{(n)}. \tag{6.4-38b}$$

Applying the variational procedure in the usual way we find that the coefficients $b_{i,p}^{(n)}$ can be solved from the following set of linear algebraic equations,

$$\sum_{j=1}^{K} \sum_{q=0}^{n-1} H_{ij}^{pq} b_{j,q}^{(n)} = \frac{2}{kT} \frac{n_i}{n} \delta_{p0}, \quad i = 1, \ldots, K; \quad p = 0, \ldots, n-1. \tag{6.4-39}$$

Finally, we turn to the calculation of the bracket integrals that occur in the definitions of the multicomponent transport coefficients. First, we calculate the bracket $[\boldsymbol{d}^k, \boldsymbol{d}^l]$ which will replace the bracket $[\boldsymbol{D}^k, \boldsymbol{D}^l]$ in the definition (6.3-30) of the diffusion coefficient D_{kl}. With Eqs. (6.4-12), using the system of equations (6.4-25) as well as the constraint (6.4-26), we find

$$[\boldsymbol{d}^k, \boldsymbol{d}^l] = \tfrac{3}{2} d_{l,0}^{k(n)} \tag{6.4-40a}$$

or, equivalently,

$$[\boldsymbol{d}^k, \boldsymbol{d}^l] = \tfrac{3}{2} d_{k,0}^{l(n)}, \tag{6.4-40b}$$

* See footnote on p. 182.

so, in the nth approximation, D_{kl} can be calculated from the identity *

$$[D_{kl}]_n = \frac{1}{2n} d_{l,0}^{k(n)} \tag{6.4-41a}$$

or, equivalently,

$$[D_{kl}]_n = \frac{1}{2n} d_{k,0}^{l(n)}. \tag{6.4-41b}$$

Next, we calculate the bracket $[\boldsymbol{d}^k, \boldsymbol{a}]$ which will replace the bracket $[\boldsymbol{D}^k, \boldsymbol{A}]$ in the definition (6.3-31) of the thermal diffusion coefficient D_{Tk}. With Eqs. (6.4-12) and (6.4-27), using the system of equations (6.4-25) and the constraint (6.4-33), we find

$$[\boldsymbol{d}^k, \boldsymbol{a}] = -\tfrac{3}{2} a_{k,0}^{(n)}, \tag{6.4-42a}$$

so, in the nth approximation, D_{Tk} can be calculated from the identity *

$$[D_{Tk}]_n = -\frac{1}{2n} a_{k,0}^{(n)}. \tag{6.4-43a}$$

On the other hand, using the system of equations (6.4-32) we find

$$[\boldsymbol{d}^k, \boldsymbol{a}] = -\tfrac{15}{4} \sum_{i=1}^{K} (n_i/n) d_{i,1}^{k(n)}, \tag{6.4-42b}$$

so, alternately, in the nth approximation D_{Tk} may be calculated from the identity *

$$[D_{Tk}]_n = -\frac{5}{4n} \sum_{i=1}^{K} (n_i/n)\, d_{i,1}^{k(n)}. \tag{6.4-43b}$$

Notice that, to determine the first approximation $[D_{Tk}]_1$, two terms are required in either the trial function (6.4-12) or the trial function (6.4-27). In the case of the trial function (6.4-12) this is obvious from the expression (6.4-43b). In the case of the trial function (6.4-27) this is because the system of equations (6.4-32), together with the constraint (6.4-33), admits only the zero solution if the sum in Eq. (6.4-27) consists of one single term. For this reason, thermal diffusion is often referred to as a *second-order* transport effect. The nomenclature we have adopted, though, is such that the *first* approximation to any transport coefficient corresponds to the *first non-zero* approximation. For each transport coefficient the first approximation is obtained by taking $n = 1$ in the relevant trial function.

* Observe the distinction between n (regular size type) which stands for the number density, and the super- and subscripts $_n$ (small type) which stand for the order of the approximation.

It is also interesting to note that D_{Tk} is the only transport coefficient of a neutral gas that is not essentially positive. It may, in fact, change sign as the temperature changes, see Section 7.3.

For the bracket $[\boldsymbol{a}, \boldsymbol{a}]$, which will replace the bracket $[\boldsymbol{A}, \boldsymbol{A}]$ in the definition (6.3-45) of the partial coefficient of thermal conductivity, we find with Eq. (6.4-27), using the system of equations (6.4-32),

$$[\boldsymbol{a}, \boldsymbol{a}] = \tfrac{15}{4} \sum_{i=1}^{K} (n_i/n)\, a_{i,1}^{(n)}, \tag{6.4-44}$$

so, in the nth approximation, λ' can be calculated from the identity *

$$[\lambda']_n = \tfrac{5}{4} k \sum_{i=1}^{K} (n_i/n)\, a_{i,1}^{(n)}. \tag{6.4-45}$$

Finally, taking Eq. (6.4-34) and using the system of equations (6.4-39) we find that the bracket $[\boldsymbol{b}, \boldsymbol{b}]$, which will replace the bracket $[\boldsymbol{B}, \boldsymbol{B}]$ in the definition (6.3-41) of the coefficient of viscosity, is given by

$$[\boldsymbol{b}, \boldsymbol{b}] = 5 \sum_{i=1}^{K} (n_i/n)\, b_{i,0}^{(n)}, \tag{6.4-46}$$

so, in the nth approximation, η can be calculated from the identity *

$$[\eta]_n = \tfrac{1}{2} kT \sum_{i=1}^{K} (n_i/n)\, b_{i,0}^{(n)}. \tag{6.4-47}$$

Thus, letting n assume the values 1, 2, 3, . . . we may generate sequences of numerical approximations to each of the various transport coefficients of a multicomponent gas mixture. Except for the thermal diffusion coefficients, these sequences are monotonically increasing and, in all cases, they converge to the value of the corresponding transport coefficient obtained in the first-order Chapman-Enskog theory. In any approximation, the transport coefficients can be written as algebraic combinations of partial bracket integrals containing Sonine polynomials. These partial bracket integrals are eightfold integrals which must be evaluated for the method to be of practical value. This question will be further taken up in the next chapter.

6.5. The case of a binary mixture. Self-diffusion

In a mixture of two gases there is only one independent diffusion coefficient and one independent thermal diffusion coefficient. If we define the *binary diffusion coefficient*, $\mathscr{D}_{12}$, as

* See footnote on p. 187.

$$\mathscr{D}_{12} = -\frac{(\rho/n)^2}{m_1 m_2} D_{12} = -\frac{(\rho/n)^2}{m_1 m_2} D_{21}, \tag{6.5-1}$$

cf. Eq. (6.3-33), then it follows from Eq. (6.3-34) that

$$D_{11} = \frac{\rho_2}{\rho_1}\frac{m_1 m_2}{(\rho/n)^2}\mathscr{D}_{12} \quad \text{and} \quad D_{22} = \frac{\rho_1}{\rho_2}\frac{m_1 m_2}{(\rho/n)^2}\mathscr{D}_{12}. \tag{6.5-2}$$

We observe that the binary diffusion coefficient thus defined is a positive quantity. Similarly, we define the *binary thermal diffusion coefficient*, D_T, as

$$D_T = \frac{n_1}{n}\frac{\rho/n}{m_2} D_{T1} = -\frac{n_2}{n}\frac{\rho/n}{m_1} D_{T2}, \tag{6.5-3}$$

cf. Eq. (6.3-35), and the *binary thermal diffusion ratio*, k_T, as

$$k_T = k_{T1} = -k_{T2}, \tag{6.5-4}$$

cf. Eq. (6.3-36b). From these definitions and the relation (6.3-36a) it follows that the thermal diffusion coefficient and the thermal diffusion ratio are connected through the identity

$$k_T = D_T/\mathscr{D}_{12}. \tag{6.5-5}$$

In terms of $\mathscr{D}_{12}$, D_T and k_T one derives the following expression for the diffusion velocities $\boldsymbol{V}_1$ and $\boldsymbol{V}_2$ from Eq. (6.3-32),

$$\rho_1 \boldsymbol{V}_1 = -\rho_2 \boldsymbol{V}_2 = -\rho\frac{m_1 m_2}{(\rho/n)^2}(\mathscr{D}_{12}\,\boldsymbol{d}_1 + D_T \nabla \log T) \tag{6.5-6a}$$

or, equivalently, from Eq. (6.3-37)

$$\rho_1 \boldsymbol{V}_1 = -\rho_2 \boldsymbol{V}_2 = -\rho\frac{m_1 m_2}{(\rho/n)^2}\mathscr{D}_{12}(\boldsymbol{d}_1 + k_T \nabla \log T). \tag{6.5-6b}$$

Thus we find that, in a binary mixture, the two components diffuse relative to one another with a diffusion velocity $\boldsymbol{V}_1 - \boldsymbol{V}_2$ which is given by

$$\boldsymbol{V}_1 - \boldsymbol{V}_2 = -\frac{n^2}{n_1 n_2}(\mathscr{D}_{12}\,\boldsymbol{d}_1 + D_T \nabla \log T) \tag{6.5-7a}$$

or, equivalently,

$$\boldsymbol{V}_1 - \boldsymbol{V}_2 = -\frac{n^2}{n_1 n_2}\mathscr{D}_{12}(\boldsymbol{d}_1 + k_T \nabla \log T). \tag{6.5-7b}$$

As will be shown later (Section 7.3), the coefficient $\mathscr{D}_{12}$ is, at least to first order, independent of the concentration ratio of the two components in the mixture.

The heat flow vector in a binary mixture is found from Eq. (6.3-48),

$$\boldsymbol{q} = -\lambda \nabla T + pk_T(\boldsymbol{V}_1 - \boldsymbol{V}_2) + \tfrac{5}{2}kT\,(n_1\boldsymbol{V}_1 + n_2\boldsymbol{V}_2). \tag{6.5-8}$$

Instead of the thermal diffusion ratio, k_T, one sometimes uses the *(binary) thermal diffusion factor*, α_T, which is defined as

$$\alpha_T = (n^2/n_1 n_2)\,k_T. \tag{6.5-9}$$

The factor α_T can be given a well-defined physical meaning. Consider the expression for the heat flow vector in the coordinate system which moves with the average molecular velocity,

$$\boldsymbol{q}^{(W)} = -\lambda \nabla T + pk_T(\boldsymbol{W}_1 - \boldsymbol{W}_2). \tag{6.5-10}$$

($\boldsymbol{W}$ is the diffusion velocity in the $\bar{\boldsymbol{c}}$-coordinate system, cf. Eq. (2.3-8).) Since, in the $\bar{\boldsymbol{c}}$-coordinate system the net flow of molecules through any surface element at rest is zero, one has $n_1\boldsymbol{W}_1 + n_2\boldsymbol{W}_2 = 0$ so, instead of Eq. (6.5-10) one may write

$$\boldsymbol{q}^{(W)} = -\lambda \nabla T + \alpha_T kTn_1\boldsymbol{W}_1. \tag{6.5-11}$$

Thus, if a molecule of type 1 is interchanged with a molecule of type 2 by diffusion, then the amount of energy transported is equal to $\alpha_T kT$.

Finally we mention that the phenomenon of diffusion of a gas through itself is usually called *self-diffusion*. The appropriate transport coefficient, $\mathscr{D}$, is called the *coefficient of self-diffusion*; it is obtained from the binary diffusion coefficient by considering the case in which the two gases are identical. Hence, from Eq. (6.5-2),

$$\mathscr{D} = (n_i/n)\,D_{ii}. \tag{6.5-12}$$

Self-diffusion corresponds, for example, to the situation in which different isotopes of the same gas diffuse through each other; in this case, the mass difference between the isotopes may usually be neglected. The coefficient of self-diffusion can then be measured by using a radioactive isotope whose motion through the gas is readily followed. Another interesting and important application of self-diffusion occurs in the case of polyatomic molecules. If the potential between the molecules is independent of their internal states, the relative motion of molecules in different internal states (which has an important effect on thermal conduction) is determined by the self-diffusion coefficient. In either case, the diffusion velocity of the molecules of species i, i.e., isotope i or internal state i, is given by

$$V_i = -(n/n_i)\mathscr{D}\nabla(n_i/n), \tag{6.5-13}$$

where $\mathscr{D}$ is the coefficient of self-diffusion of the gas.

6.6. The Lorentz gas

A particularly interesting, albeit very specialized, case of a binary gas mixture that permits complete solution of the Chapman-Enskog integral equations was first investigated by Lorentz [1905]. In this model it is presumed that the molecules of one of the components are much lighter than the molecules of the other. Furthermore, it is presumed that the number density of the lighter component is much smaller than that of the heavier component, so that collisions of the lighter molecules with each other are unimportant. Lorentz intended this model to describe electrons in a solid, but, because the Coulomb potential has a very long range and because quantum effects are important, it is not well suited to this purpose. It does, however, provide an exactly soluble case which may be used as a standard, or benchmark, to test approximate methods. Also, a perturbation procedure based on the Lorentz gas as the first approximation has application to some physically interesting cases.

For the sake of definiteness, we assume that component 1 consists of the lighter molecules, so $m_1 \ll m_2$ and $n_1 \ll n_2$. These strong inequalities permit substantial simplification of the problem of finding the velocity distributions of the two components of the mixture. In particular, the low-density assumption i.e., $n_1 \ll n_2$, implies that, in the Boltzmann equation for species 1, only 1–2 collisions need be considered. Furthermore, since the ratio of thermal speeds is given by $\bar{C}_1/\bar{C}_2 = (m_2/m_1)^{\frac{1}{2}}$, the assumption on the mass ratio implies that a particle of species 2 appears almost stationary to an approaching collision partner of species 1. Finally, the kinematics is such that, although the lighter particle may suffer a large directional deflection in a collision, its speed is practically unchanged. The situation is similar to that of a billiard ball hitting a side rail or a bumper; the heavier particle is essentially unaffected by the collision, because the energy exchange is negligible.

In treating the lighter gas one must exercize a little care. Allowing the mass ratio (m_1/m_2) to vanish in the Boltzmann equation itself is tantamount to allowing no energy exchange at all between the two species and, consequently, the average thermal energies (temperatures) of the two species may differ. This is an important and realizable possibility, but it is an unnecessary complication at this juncture; it will be taken up in Section 14.4. Here, we assume that there is just enough energy transfer between the two species to bring the lighter component to a Maxwellian velocity distribution at the temperature of the heavier component. The theory then follows that of Sections 6.2 and 6.3 down to the point at which the solution of Eqs. (6.3-18), (6.3-19) and (6.3-20) is taken up. For species 1 these equations read,

$$I_{12}(\boldsymbol{D}^1) = \frac{n}{n_1^2} f_1^{(0)} \boldsymbol{C}_1, \tag{6.6-1}$$

$$I_{12}(\boldsymbol{A}) = \frac{1}{n_1} f_1^{(0)} \left(\frac{m_1 C_1^2}{2kT} - \tfrac{5}{2} \right) \boldsymbol{C}_1, \tag{6.6-2}$$

$$I_{12}(\mathsf{B}) = \frac{m_1}{n_1 kT} f_1^{(0)} (\boldsymbol{C}_1 \boldsymbol{C}_1 - \tfrac{1}{3} C_1^2 \mathsf{I}). \tag{6.6-3}$$

Once $\boldsymbol{D}^1$ is known, the function $\boldsymbol{D}^2$ can be determined from the relation

$$\boldsymbol{D}^2 = -(\rho_1/\rho_2)\boldsymbol{D}^1. \tag{6.6-4}$$

Notice that, in Eqs. (6.6-1), (6.6-2) and (6.6-3) we have neglected all those terms which are O (n_1/n_2) with respect to those which are O(1) as $n_1/n_2 \to 0$.

Now, consider the integral operator I_{12}; for example,

$$I_{12}(\boldsymbol{A}) = \frac{1}{n_1 n_2} \iiint f_{M1} f_{M2} (\boldsymbol{A}_1 + \boldsymbol{A}_2 - \boldsymbol{A}_1' - \boldsymbol{A}_2') g b \, \mathrm{d}b \, \mathrm{d}\varepsilon \, \mathrm{d}^3 c_2 . \tag{6.6-5}$$

From the discussion above we may conclude that, in a 1–2 collision, $\boldsymbol{C}_2' \approx \boldsymbol{C}_2$. Hence, since $\boldsymbol{A} = A(C)\boldsymbol{C}$ we have

$$I_{12}(\boldsymbol{A}) \approx \frac{1}{n_1 n_2} \iiint f_{M1} f_{M2} (\boldsymbol{A}_1 - \boldsymbol{A}_1') g b \, \mathrm{d}b \, \mathrm{d}\varepsilon \, \mathrm{d}^3 c_2$$

$$= \frac{1}{n_1} \iint f_{M1} (\boldsymbol{A}_1 - \boldsymbol{A}_1') g b \, \mathrm{d}b \, \mathrm{d}\varepsilon. \tag{6.6-6}$$

Furthermore, $C_1' \approx C_1$, so $A_1(C_1) \approx A_1(C_1')$ and, also, $g \approx C_1$. Thus, the expression (6.6-6) is further simplified,

$$I_{12}(A) \approx \frac{1}{n_1}\iint f_{M1}\, A_1(C_1) C_1 (\boldsymbol{C}_1 - \boldsymbol{C}_1') b\, \mathrm{d}b\, \mathrm{d}\varepsilon, \tag{6.6-7}$$

and the function A_1 may be taken outside the integral sign. This reduces the integral equation for $\boldsymbol{A}$ to an algebraic equation and permits trivial analytical solution. All that remains is to evaluate the integral in Eq. (6.6-7).

The portion of the integral coming from $\boldsymbol{C}_1'$ in the integrand is a vector which must be proportional to $\boldsymbol{C}_1$, as this is the only remaining vector in the problem. (More rigorously, one can decompose $\boldsymbol{C}_1'$ into a vector parallel to $\boldsymbol{C}_1$ and one normal to $\boldsymbol{C}_1$. Since the integration over ε is equivalent to integration over the component of $\boldsymbol{C}_1'$ normal to $\boldsymbol{C}_1$ and the integrand is odd in this component, only the component parallel to $\boldsymbol{C}_1$ contributes to the integral.) As the component of $\boldsymbol{C}_1'$ parallel to $\boldsymbol{C}_1$ is equal to $\boldsymbol{C}_1 \cos\chi$, the integral of Eq. (6.6-7) becomes

$$I_{12}(\boldsymbol{A}) \approx \frac{1}{n_1} f_{M1}\, \boldsymbol{A}_1\, C_1\, Q_{12}^{(1)}(C_1), \tag{6.6-8}$$

where

$$Q_{12}^{(1)}(C) = 2\pi \int \{1 - \cos\chi(b, C)\} b\, \mathrm{d}b. \tag{6.6-9}$$

Similarly,

$$I_{12}(\boldsymbol{D}^1) \approx \frac{1}{n_1} f_{M1}\, \boldsymbol{D}_1^1\, C_1\, Q_{12}^{(1)}(C_1). \tag{6.6-10}$$

The simplification of $I_{12}(\mathbf{B})$ proceeds in an analogous manner. We have

$$I_{12}(\mathbf{B}) \approx \frac{1}{n_1}\iint f_{M1}\, B_1(C_1) C_1 [(\boldsymbol{C}_1 \boldsymbol{C}_1 - \tfrac{1}{3}C_1^2\, \mathbf{I}) - (\boldsymbol{C}_1' \boldsymbol{C}_1' - \tfrac{1}{3}C_1'^2 \mathbf{I})] b\, \mathrm{d}b\, \mathrm{d}\varepsilon. \tag{6.6-11}$$

The portion of the integral arising from $\boldsymbol{C}_1' \boldsymbol{C}_1' - \frac{1}{3}C_1'^2 \mathbf{I}$ must be a symmetric traceless tensor and the only such tensor available is $\boldsymbol{C}_1 \boldsymbol{C}_1 - \frac{1}{3}C_1^2 \mathbf{I}$. Therefore,

$$I_{12}(\mathbf{B}) \approx \frac{3}{2n_1} f_{M1}\, \mathbf{B}_1\, C_1\, Q_{12}^{(2)}(C_1), \tag{6.6-12}$$

where

$$Q_{12}^{(2)}(C) = 2\pi \int \{1 - \cos^2\chi(b, C)\} b\, \mathrm{d}b. \tag{6.6-13}$$

The quantities $Q_{12}^{(1)}$ and $Q_{12}^{(2)}$ may be interpreted as transport cross sections, cf. Eq. (7.1-22).

With these approximations for the collision integrals the solutions to Eqs. (6.6-1) through (6.6-4) are found to be,

$$\boldsymbol{D}_1^1 \equiv D_1^1(C_1)\boldsymbol{C}_1 \approx \frac{n}{n_1}\frac{1}{C_1 Q_{12}^{(1)}(C_1)}\boldsymbol{C}_1, \tag{6.6-14}$$

$$\boldsymbol{D}_1^2 \equiv D_1^2(C_1)\boldsymbol{C}_1 \approx -\frac{m_1}{m_2}\frac{1}{C_1 Q_{12}^{(1)}(C_1)}\boldsymbol{C}_1, \tag{6.6-15}$$

$$\boldsymbol{A}_1 \equiv A_1(C_1)\boldsymbol{C}_1 \approx \left(\frac{m_1 C_1^2}{2kT}-\tfrac{5}{2}\right)\frac{1}{C_1 Q_{12}^{(1)}(C_1)}\boldsymbol{C}_1, \tag{6.6-16}$$

$$\mathsf{B}_1 \equiv B_1(C_1)(\boldsymbol{C}_1\boldsymbol{C}_1-\tfrac{1}{3}C_1^2\boldsymbol{I}) \approx \frac{2m_1}{3kT}\frac{1}{C_1 Q_{12}^{(2)}(C_1)}(\boldsymbol{C}_1\boldsymbol{C}_1-\tfrac{1}{3}C_1^2\boldsymbol{I}). \tag{6.6-17}$$

Finally, we may calculate the transport properties either directly or by using the results of Sections 6.3 and 6.5. In the present approximation, the diffusion velocity of the heavier particles is negligible compared to that of the lighter particles. The latter is given by

$$\boldsymbol{V}_1 = -\frac{n}{n_1}(\mathscr{D}_{12}\boldsymbol{d}_1 + D_T \nabla \log T), \tag{6.6-18}$$

where the binary diffusion coefficient is given by

$$\mathscr{D}_{12} = \frac{1}{3n^2}\int f_{M1} C_1^2 D_1^1 \, d^3c_1$$

$$= \frac{1}{3nn_1}\int f_{M1}\frac{C_1}{Q_{12}^{(1)}(C_1)}\, d^3c_1, \tag{6.6-19}$$

and the thermal diffusion coefficient by

$$D_T = \frac{1}{3n^2}\int f_{M1} C_1^2 A_1 \, d^3c_1$$

$$= \frac{1}{3nn_1}\int f_{M1}\left(\frac{m_1 C_1^2}{2kT}-\tfrac{5}{2}\right)\frac{C_1}{Q_{12}^{(1)}(C_1)}\, d^3c_1. \tag{6.6-20}$$

The first-order correction to the pressure tensor due to the lighter gas is usually of small importance, because most of the momentum is carried by the heavier particles. If, however, the pressure system due to the light molecules alone is considered, then the viscosity coefficient is given by

$$\eta = \frac{m_1}{15n}\int f_{\mathrm{M}1}\, C_1^4 B_1\, \mathrm{d}^3 c_1$$

$$= \tfrac{2}{45}\,\frac{m_1^2}{nkT}\int f_{\mathrm{M}1}\,\frac{C_1^3}{Q_{12}^{(2)}(C_1)}\,\mathrm{d}^3 c_1 . \qquad (6.6\text{-}21)$$

Finally, in the conduction of heat, the lighter molecules are most effective because of their large thermal velocities. If the conduction is assumed to be due predominantly to the lighter molecules, then the heat flow vector is given by

$$\boldsymbol{q} = \tfrac{5}{2} n_1 kT\boldsymbol{V}_1 - \left(\lambda' - \frac{n_1 k D_T^2}{\mathscr{D}_{12}}\right)\nabla T, \qquad (6.6\text{-}22)$$

where

$$\lambda' = \frac{k}{3n}\int f_{\mathrm{M}1}\, C_1^2\left(\frac{m_1 C_1^2}{2kT} - \tfrac{5}{2}\right) A_1\, \mathrm{d}^3 c_1$$

$$= \frac{k}{3n}\int f_{\mathrm{M}1}\left(\frac{m_1 C_1^2}{2kT} - \tfrac{5}{2}\right)\frac{C_1}{Q_{12}^{(1)}(C_1)}\,\mathrm{d}^3 c_1 . \qquad (6.6\text{-}23)$$

It is interesting to note that each of the functions $\boldsymbol{D}$, $\boldsymbol{A}$, and $\mathbf{B}$ is inversely proportional to a transport cross section Q. Since all the $Q^{(l)}$ have approximately the same functional behavior, cf. Fig. 9.4, this means that the perturbed part of the distribution function, i.e., ϕ, is also inversely proportional to a cross section or collision frequency. In the case of a Lorentz gas the reason for this is fairly easy to see. At equilibrium, the rate at which particles are scattered into a given velocity interval is equal to the rate at which particles are scattered out (detailed balance), and both are proportional to the collision frequency. The perturbation created by a temperature gradient is $f_{\mathrm{M}}(mC^2/2kT - \frac{5}{2})\boldsymbol{C}\cdot\nabla \log T$, and is *not* proportional to the collision frequency, but the rate at which the perturbation is eliminated in the vicinity of velocity $\boldsymbol{C}$ does remain proportional to the collision frequency. The perturbation function itself must therefore vary inversely to a cross section. Although not quite so obvious, the same statement is true in the general case. This result helps explain the rate of convergence of the Chapman-Cowling procedure, which we will discuss later in more detail (Section 10.2). The Sonine polynomial expansion on which this procedure is based, is just an attempt to fit functions similar to those given by Eqs. (6.6-14) through (6.6-17) by a low-order polynomial. Clearly, the smoother the function, particularly at high velocities, the easier it is to fit and the better the convergence. Since molecules with hard cores produce cross sections which are slowly varying at high velocity, the rate of convergence of the Chapman-Cowling method is good for these models. It is also no surprise that the Coulomb potential, which possesses the softest core of all models, yields the slowest convergence.

In conclusion, we mention that Sandler and Mason [1969] have discussed a perturbation method based on the Lorentz gas. For a binary gas mixture, in which $m_1/m_2 \ll 1$, they wrote the collision operators as the sum of the Lorentz gas approximations to these operators and the difference between the exact collision operators and their Lorentz approximations. The latter were treated as small and the solution was developed as a perturbation series. For sufficiently small mass ratio ($\sim 10^{-3}$) the rate of convergence of this method proved to be better than that of the Chapman-Cowling method. However, due to the relatively limited applicability of this method we shall not discuss it further.

6.7. Phenomenological theory of transport coefficients

It is possible to arrive at expressions for the transport coefficients via a phenomenological theory – one which is based on a physical model of the phenomena involved. Historically, the phenomenological, or mean free path, theory predates the exact theory presented in the last two chapters by about half a century.

Usually, the arguments are made in terms of the hard sphere model of molecules because the model allows easy visualization, but they can be extended to more general models (Monchick [1962], Monchick and Mason [1967]) and, of course, when the extension is properly carried out, the results agree with those of the exact theory based on the Boltzmann equation. We shall treat only the simplest version of this theory and indicate how the extensions may be made. The viscosity, thermal conductivity and binary diffusion coefficients are readily calculated. Since the calculations of the viscosity and thermal conductivity are very similar, we shall treat only the viscosity in detail.

Imagine a simple gas of hard spheres (of diameter σ), uniform in temperature and density, which is flowing parallel to the z-axis of a rectangular coordinate system, with a hydrodynamic velocity v_z which is a function of x alone. Thus, the gas is in laminar motion parallel to the plane $x = 0$. Further, imagine a plane of unit area whose normal is in the positive x-direction, embedded in the flow. We want to calculate the shear stress at the plane; this is just the net flow of momentum across the plane per unit time. In the simplest possible model, we assume that each molecule crossing the plane carries with it precisely the momentum of an average molecule in the layer in which it made its last collision and, further, that all the molecules crossing the plane made their last collision precisely one mean free path (l) from the plane. Then, a molecule crossing in the direction of increasing x carries with it the z-momentum

$$mv_z(x-l) \approx m(v_z - l\,\mathrm{d}v_z/\mathrm{d}x), \tag{6.7-1a}$$

and one crossing in the direction of decreasing x, the z-momentum

$$mv_z(x+l) \approx m(v_z + l\,\mathrm{d}v_z/\mathrm{d}x); \tag{6.7-1b}$$

the values of v_z and $\mathrm{d}v_z/\mathrm{d}x$ in the right members of these expressions are to be taken at the position x of the plane.

Now, the number of molecules crossing the plane per unit time in the direction of increasing x is

$$\int_+ c_x f\,\mathrm{d}^3c = \int_+ C_x f\,\mathrm{d}^3C, \tag{6.7-2a}$$

where the integral extends over all values of $\boldsymbol{C}$ for which C_x is positive, and, likewise, the number of molecules crossing the plane per unit time in the direction of decreasing x is

$$\int_- (-c_x) f\,\mathrm{d}^3c = \int_- (-C_x) f\,\mathrm{d}^3C, \tag{6.7-2b}$$

where the integral extends over all values of $\boldsymbol{C}$ for which C_x is negative. Since $v_x = 0$, these two numbers are equal. To a first approximation, their value is the same as for a gas in a uniform steady state, cf. Eq. (4.3-2), and equal to $\frac{1}{4}n\bar{C}$. Thus, the shear stress at the plane is $\frac{1}{4}n\bar{C}$ times the difference of the momenta (6.7-1a) and (6.7-1b),

$$P_{xz} = -\tfrac{1}{2}n\bar{C}ml\frac{\mathrm{d}v_z}{\mathrm{d}x} = -\tfrac{1}{2}\rho l\bar{C}\frac{\mathrm{d}v_z}{\mathrm{d}x}, \tag{6.7-3}$$

so that

$$\eta = \tfrac{1}{2}\rho l\bar{C} = \frac{1}{\pi\sigma^2}\left(\frac{mkT}{\pi}\right)^{\frac{1}{2}}, \tag{6.7-4}$$

where we have used the expressions (2.4-7) and (4.3-5) for the mean free path and the mean peculiar velocity, respectively.

The above derivation is, of course, very crude. A more elaborate analysis based on the concepts embodied in this derivation can be carried out and one finds that the expression (6.7-4) for the viscosity is multiplied by a numerical factor of order unity.

In a similar way the thermal conductivity is found to be

$$\lambda = \tfrac{1}{2}\rho l\bar{C}c_v = \eta c_v, \tag{6.7-5}$$

where c_v is the specific heat (i.e., the heat capacity per unit mass) of the gas.

To compute the binary diffusion coefficient, we do a similar calculation for a mixture. Again, we consider only the simplest case, in which the gas is at rest and the total number density and temperature are constant. Then, the net flow of molecules of type 1 across a plane at x is, by arguments similar to those presented previously,

$$n_1 V_{1x} = \tfrac{1}{4}\bar{C}_1 n_1(x-l_1) - \tfrac{1}{4}\bar{C}_1 n_1(x+l_1) \approx -\tfrac{1}{2}\bar{C}_1 l_1 \frac{dn_1}{dx}, \tag{6.7-6}$$

and, likewise, the net flow of molecules of type 2 is

$$n_2 V_{2x} = -\tfrac{1}{2}\bar{C}_2 l_2 \frac{dn_2}{dx} = \tfrac{1}{2}\bar{C}_2 l_2 \frac{dn_1}{dx}. \tag{6.7-7}$$

Then, defining the binary diffusion coefficient as in Eq. (6.5-7a), we have

$$\mathscr{D}_{12} = \frac{1}{2n}(n_2 l_1 \bar{C}_1 + n_1 l_2 \bar{C}_2). \tag{6.7-8}$$

The inadequacy of the theory is reflected here in the fact that the flow rates predicted are not consistent with the restriction of zero hydrodynamic velocity. Note, too, that for self-diffusion we have

$$\mathscr{D} = \tfrac{1}{2}l\bar{C} = \eta/\rho. \tag{6.7-9}$$

These relations are useful in giving the density-independence of the properties and in suggesting relationships among the properties. However, the numerical factors are incorrect as we shall see when the exact results for the hard sphere model are given. Also, the temperature dependence of the properties is incorrectly predicted, but this is actually a failing of the hard sphere model and not of the approach used.

It has not been found possible to predict even the existence of thermal diffusion using this model. This should not be too surprising as the thermal diffusion coefficient can be positive, negative or zero depending upon the molecular model chosen and upon the temperature. The reason for this has been pointed out by Monchick and Mason [1967] and is that, while the properties previously calculated arise because of the *existence* of collisions, thermal diffusion arises only because the *collision frequency* is *dependent on speed.* In fact, as we shall see in Chapter 9, the thermal diffusion coefficient vanishes identically when the collision frequency is speed-independent.

A simple formula (and accompanying explanation) for thermal diffusion may be derived in the following way. Consider a binary mixture of gases at zero hydrodynamic velocity, in which the composition and pressure are uniform but in which there is a small temperature gradient. Since $p = nkT$ and the ratios n_i/n are constant we have, for each component $i(i = 1, 2)$,

$$\nabla \log n_i = \nabla \log n = -\nabla \log T \tag{6.7-10}$$

and, hence, to a first approximation,

$$\nabla \log f_i = \nabla \log n_i + (m_i C_i^2/2kT - \tfrac{3}{2}) \nabla \log T = (m_i C_i^2/2kT - \tfrac{5}{2}) \nabla \log T. \tag{6.7-11}$$

We thus see that there is a *decrease* in the density of slow molecules (i.e., those with $m_i C_i^2/2kT = \mathscr{C}_i^2 < \frac{5}{2}$) in the direction of the temperature gradient and an *increase* in the density of the fast molecules (i.e., those with $\mathscr{C}_i^2 > \frac{5}{2}$) in the same direction. Now, suppose the collision frequency ν_i is a function of the speed C_i. Consider those molecules of type i which have their speed in the interval $\mathrm{d}C_i$ around C_i. If we use the argument presented in the derivation of the binary diffusion coefficient, but restrict our attention to those molecules within the velocity range of interest, we find that the net flow of molecules of type 1 is

$$-\tfrac{1}{2} C_1 l_1(C_1) \frac{\partial f_1}{\partial x} = -\frac{1}{2} \frac{C_1^2}{\nu_1(C_1)} \frac{\partial f_1}{\partial x}. \tag{6.7-12}$$

Since the model used in deriving Eq. (6.7-8) for the diffusion coefficient involves following the flight of the molecule for only a single flight and since a molecule does not change its speed during a free flight, we are justified in using the expression (6.7-12) for each speed C_1. The net flow of all molecules of type 1 is thus

$$n_1 V_{1x} = -\frac{1}{2}\int \frac{C_1^2}{v_1(C_1)}(\mathscr{C}_1^2 - \tfrac{5}{2}) f_1(C_1) \mathrm{d}^3 C_1 \, \frac{\partial \log T}{\partial x}$$

$$= -\frac{4n_1}{\sqrt{\pi}} \frac{kT}{m_1}\left\{\int_0^\infty \frac{\mathscr{C}_1^4}{v_1(\mathscr{C}_1)}(\mathscr{C}_1^2 - \tfrac{5}{2}) \exp(-\mathscr{C}_1^2) \mathrm{d}\mathscr{C}_1\right\} \frac{\partial \log T}{\partial x}. \quad (6.7\text{-}13)$$

In the same way one obtains the analogous expression for the net flow of molecules of type 2. Then, using Eq. (6.5-7) to define the thermal diffusion coefficient, and defining the integral in braces in Eq. (6.7-13) to be $\bar{\lambda}_1$, we obtain the following expression for the thermal diffusion coefficient of a binary mixture,

$$D_T = \frac{n_1 n_2}{n^2} \frac{4}{\sqrt{\pi}} kT \left\{\frac{\bar{\lambda}_1}{m_1} - \frac{\bar{\lambda}_2}{m_2}\right\}. \quad (6.7\text{-}14)$$

By a more careful calculation, Monchick and Mason [1967] obtained as a first approximation, Eq. (6.7-14) with the factor 4 replaced by $\frac{8}{3}$; this difference can be traced to the fact that, in Eq. (6.7-8), a more careful calculation would give $\frac{1}{3}$ in place of $\frac{1}{2}$. Note that, if v_i is independent of $\mathscr{C}_i$, the integral in Eq. (6.7-13) is zero and there is no net flux. On the other hand, in the case in which v_i is an increasing function of $\mathscr{C}_i$, such as obtains for rigid spheres, the integral is negative and we have a net flow of the light molecules in the direction of increasing temperature.

Monchick and Mason have shown that, for the case of an isotopic mixture (all interactions indentical, masses only slightly different), Eq. (6.7-13) is qualitatively reasonable, but numerically incorrect by a factor of approximately two.

From the above simple calculation it follows that the thermal diffusion factor α_T, defined by Eq. (6.5-9), is nearly independent of composition; also, the variation of α_T with mass is predicted correctly, at least qualitatively.

The transport coefficients 7

In the previous two chapters it was shown that the transport coefficients of a gas may be obtained to any required degree of accuracy by solving sets of linear algebraic equations, whose coefficients are proportional to bracket integrals of Sonine polynomials. Since the integrands contain $\boldsymbol{c}'$ and $\boldsymbol{c}_1'$ as arguments as well as $\boldsymbol{c}$ and $\boldsymbol{c}_1$, and the former are related to the latter via the dynamics of binary collisions as discussed in Chapter 3, the bracket integrals cannot be computed without specification of the intermolecular force law. Clearly, it is this dependence of the bracket integrals on the intermolecular force law that yields the interesting (and essential) relationship between the force law and the transport properties.

Further, the bracket integrals defined by Eq. (4.7-5) are eight-fold integrals; they contain triple integrals over each of the velocity variables $\boldsymbol{c}$ and $\boldsymbol{c}_1$, as well as integrals over the collision parameters b and χ. Thus, even when the intermolecular force law is given, these expressions are not particularly useful. For simple molecules this situation is mitigated by the fact that the scattering angle, χ, depends only on the magnitude of the relative velocity, g, of the two colliding particles and on the impact parameter, b. As $\boldsymbol{c}'$ and $\boldsymbol{c}_1'$ can be expressed in terms of $\boldsymbol{c}$ and $\boldsymbol{c}_1$ once the scattering angle χ is given, there is the possibility that the unwieldy bracket integrals can be reduced to double integrals over g and b, whatever the force law. This reduction can, in fact, be carried out and renders the calculation of transport properties feasible. The calculation is further simplified by the fact that the Sonine polynomial expansions converge rapidly for almost any force law, as we shall see in Chapter 10.

The first part of the present chapter is devoted to the reduction of the bracket integrals. While the reduction can be carried out in such manner that many integrals are reduced simultaneously, the algebraic manipulations become very complicated and the nature of the underlying calculation is obscured. For this reason, the reduction of only two of the simpler, and more important, integrals is given in greater detail (Section 7.1), while the general method of reduction is given in Appendix B.

The result of the reduction procedure is that the bracket integrals for simple molecules are written as linear combinations of *Ω-integrals*, – i.e., double integrals over the dimensionless relative velocity g and the impact parameter b, which can be evaluated once the intermolecular force law is specified. In Section 7.2 we list the expressions for some of the most important bracket integrals.

Thus, by combining these expressions with the results of Chapters 5 and 6 we find that, in any order of approximation, the transport coefficients may be written as algebraic functions of the Ω-integrals. This is further worked out in Section 7.3.

7.1. Reduction of the integrals $[S_{\frac{3}{2}}^{(1)}(\mathscr{C}^2)\boldsymbol{\mathscr{C}}, S_{\frac{3}{2}}^{(1)}(\mathscr{C}^2)\boldsymbol{\mathscr{C}}]$ and $[\boldsymbol{\mathscr{C}}\boldsymbol{\mathscr{C}}-\frac{1}{3}\mathscr{C}^2\boldsymbol{I}, \boldsymbol{\mathscr{C}}\boldsymbol{\mathscr{C}}-\frac{1}{3}\mathscr{C}^2\boldsymbol{I}]$

For purposes of illustrating how the reduction of the bracket integrals is carried out, we shall perform the reduction for the integrals which occur in the lowest-order approximation to the thermal conductivity and viscosity of a simple gas, viz.,

$$I_1 = [S_{\frac{3}{2}}^{(1)}(\mathscr{C}^2)\boldsymbol{\mathscr{C}}, S_{\frac{3}{2}}^{(1)}(\mathscr{C}^2)\boldsymbol{\mathscr{C}}] \tag{7.1-1}$$

and

$$I_2 = [\boldsymbol{\mathscr{C}}\boldsymbol{\mathscr{C}}-\tfrac{1}{3}\mathscr{C}^2\boldsymbol{I}, \boldsymbol{\mathscr{C}}\boldsymbol{\mathscr{C}}-\tfrac{1}{3}\mathscr{C}^2\boldsymbol{I}]. \tag{7.1-2}$$

As mentioned earlier, the general case is considered in Appendix B.

To begin, we recall the definition of the first of these integrals, cf. Eq. (4.7-5),

$$\begin{aligned} I_1 &= \int S_{\frac{3}{2}}^{(1)}(\mathscr{C}^2)\boldsymbol{\mathscr{C}}\cdot I(S_{\frac{3}{2}}^{(1)}(\mathscr{C}^2)\boldsymbol{\mathscr{C}})\mathrm{d}^3c \\ &= \frac{1}{n^2}\iiiint f_{\mathrm{M}}f_{\mathrm{M}1}\, S_{\frac{3}{2}}^{(1)}(\mathscr{C}^2)\boldsymbol{\mathscr{C}}\cdot[S_{\frac{3}{2}}^{(1)}(\mathscr{C}'^2)\boldsymbol{\mathscr{C}}'+S_{\frac{3}{2}}^{(1)}(\mathscr{C}_1'^2)\boldsymbol{\mathscr{C}}_1' \\ &\qquad -S_{\frac{3}{2}}^{(1)}(\mathscr{C}^2)\boldsymbol{\mathscr{C}}-S_{\frac{3}{2}}^{(1)}(\mathscr{C}_1^2)\boldsymbol{\mathscr{C}}_1]g b\,\mathrm{d}b\,\mathrm{d}\varepsilon\,\mathrm{d}^3c\,\mathrm{d}^3c_1\,. \end{aligned} \tag{7.1-3}$$

Now, as the transformation from $\boldsymbol{c}$ to $\boldsymbol{C}$ is just a translation ($\boldsymbol{C} = \boldsymbol{c}-\boldsymbol{v}$), we have $d^3c = d^3C$ and, likewise, $d^3c_1 = d^3C_1$. Thus, recalling that $f_M f_{M1} = n^2(m/2\pi kT)^3 \exp -(\mathscr{C}^2+\mathscr{C}_1^2)$ where $\mathscr{C} = (m/2kT)^{\frac{1}{2}}\boldsymbol{C}$, and using symmetry arguments as in Section 4.1, we obtain the following expression for the bracket integral,

$$I_1 = \frac{1}{4\pi^3}\iiiint e^{-(\mathscr{C}^2+\mathscr{C}_1{}^2)}[S_{\frac{3}{2}}^{(1)}(\mathscr{C}'^2)\boldsymbol{\mathscr{C}}'+S_{\frac{3}{2}}^{(1)}(\mathscr{C}_1'^2)\boldsymbol{\mathscr{C}}_1'$$
$$-S_{\frac{3}{2}}^{(1)}(\mathscr{C}^2)\boldsymbol{\mathscr{C}}-S_{\frac{3}{2}}^{(1)}(\mathscr{C}_1^2)\boldsymbol{\mathscr{C}}_1]^2 gb\,db\,d\varepsilon\,d^3\mathscr{C}\,d^3\mathscr{C}_1. \qquad (7.1\text{-}4)$$

The square on the vector between the square brackets indicates that the scalar product of the vector with itself is to be taken. In the integral (7.1-4) g is the only remaining dimensional variable.

Recalling that $S_{\frac{3}{2}}^{(1)}(\mathscr{C}^2) = \frac{5}{2}-\mathscr{C}^2$ we see that the Sonine polynomials under the integral sign may be replaced by their arguments. This makes it natural to transform the pair of integration variables $\mathscr{C}$, $\mathscr{C}_1$ to center-of-mass and relative velocities. Defining a dimensionless center-of-mass velocity, $\boldsymbol{\mathscr{G}}_0$,

$$\boldsymbol{\mathscr{G}}_0 = \tfrac{1}{2}(\boldsymbol{\mathscr{C}}+\boldsymbol{\mathscr{C}}_1) = (m/2kT)^{\frac{1}{2}}(\boldsymbol{G}-\boldsymbol{v}), \qquad (7.1\text{-}5)$$

and a dimensionless relative velocity, $\boldsymbol{\mathscr{g}}$,

$$\boldsymbol{\mathscr{g}} = \boldsymbol{\mathscr{C}}-\boldsymbol{\mathscr{C}}_1 = (m/2kT)^{\frac{1}{2}}\boldsymbol{g}, \qquad (7.1\text{-}6)$$

one may verify that the following expression for I_1 is equivalent to the one given in Eq. (7.1-4),

$$I_1 = \frac{1}{4\pi^3}\iiiint e^{-2\mathscr{G}_0{}^2}\exp(-\tfrac{1}{2}\mathscr{g}^2)[\mathscr{g}^2(\boldsymbol{\mathscr{g}}\cdot\boldsymbol{\mathscr{G}}_0)^2+\mathscr{g}'^2(\boldsymbol{\mathscr{g}}'\cdot\boldsymbol{\mathscr{G}}_0)^2$$
$$-2(\boldsymbol{\mathscr{g}}\cdot\boldsymbol{\mathscr{g}}')(\boldsymbol{\mathscr{g}}\cdot\boldsymbol{\mathscr{G}}_0)(\boldsymbol{\mathscr{g}}'\cdot\boldsymbol{\mathscr{G}}_0)]gb\,db\,d\varepsilon\,d^3\mathscr{G}_0\,d^3\mathscr{g}. \qquad (7.1\text{-}7)$$

Now, the integration over $\boldsymbol{\mathscr{G}}_0$ can be carried out. For example, one has

$$\int e^{-2\mathscr{G}_0{}^2}(\boldsymbol{\mathscr{g}}\cdot\boldsymbol{\mathscr{G}}_0)(\boldsymbol{\mathscr{g}}'\cdot\boldsymbol{\mathscr{G}}_0)d^3\mathscr{G}_0 = \int e^{-2\mathscr{G}_0{}^2}\,\boldsymbol{\mathscr{g}}\boldsymbol{\mathscr{g}}' : \boldsymbol{\mathscr{G}}_0\boldsymbol{\mathscr{G}}_0\,d^3\mathscr{G}_0$$
$$= \tfrac{1}{3}(\boldsymbol{\mathscr{g}}\boldsymbol{\mathscr{g}}' : \boldsymbol{I})\int e^{-2\mathscr{G}_0{}^2}\mathscr{G}_0^2\,d^3\mathscr{G}_0 = \frac{\pi^{\frac{3}{2}}}{8\sqrt{2}}(\boldsymbol{\mathscr{g}}\cdot\boldsymbol{\mathscr{g}}'), \qquad (7.1\text{-}8)$$

where the integral relation (A.14-1) has been applied. Thus we obtain

$$I_1 = \frac{\sqrt{2}}{64\pi^{\frac{3}{2}}}\iiint \exp(-\tfrac{1}{2}\mathscr{g}^2)[\mathscr{g}^4+\mathscr{g}'^4-2(\boldsymbol{\mathscr{g}}\cdot\boldsymbol{\mathscr{g}}')^2]gb\,db\,d\varepsilon\,d^3\mathscr{g} \quad (7.1\text{-}9)$$

or, since $g^2 = g'^2$ and $\boldsymbol{g} \cdot \boldsymbol{g}' = g^2 \cos \chi$, where $\chi \equiv \chi(b, g)$ is the scattering angle defined in Chapter 3,

$$I_1 = \frac{\sqrt{2}}{32\pi^{\frac{3}{2}}} \left(\frac{2kT}{m}\right)^{\frac{1}{2}} \iiint \exp(-\tfrac{1}{2}g^2) g^5 (1-\cos^2 \chi) b \, \mathrm{d}b \, \mathrm{d}\varepsilon \, \mathrm{d}^3 g. \quad (7.1\text{-}10)$$

Finally, using spherical coordinates for $\boldsymbol{g}$ we can easily perform the integrations over the angular variables. The integration over ε is likewise trivial and we have the result

$$I_1 = \frac{1}{4}\left(\frac{kT}{\pi m}\right)^{\frac{1}{2}} \int_0^\infty \exp(-\tfrac{1}{2}g^2) g^7 \left[2\pi \int (1-\cos^2 \chi) \, b \, \mathrm{d}b\right] \mathrm{d}g, \quad (7.1\text{-}11)$$

which, in turn, may be transformed into

$$I_1 = 4\left(\frac{kT}{\pi m}\right)^{\frac{1}{2}} \int_0^\infty \exp(-g^2) g^7 \left[2\pi \int (1-\cos^2 \chi) \, b \, \mathrm{d}b\right] \mathrm{d}g. \quad (7.1\text{-}12)$$

The reduction of the second bracket integral, I_2, proceeds in a similar manner. From the definition, using the devices that led from Eq. (7.1-1) to Eq. (7.1-4), we have

$$I_2 = \frac{1}{4\pi^3} \iiiint \mathrm{e}^{-(\mathscr{C}^2+\mathscr{C}_1^2)} [\mathscr{C}'\mathscr{C}' + \mathscr{C}_1'\mathscr{C}_1' - \mathscr{C}\mathscr{C} - \mathscr{C}_1\mathscr{C}_1]^2 g b \, \mathrm{d}b \, \mathrm{d}\varepsilon \, \mathrm{d}^3\mathscr{C} \, \mathrm{d}^3\mathscr{C}_1, \quad (7.1\text{-}13)$$

where, now, the square on the tensor between the square bracket indicates that the scalar product of the tensor with itself is to be taken. Transforming the integration variables from $(\mathscr{C}, \mathscr{C}_1)$ to $(\mathscr{G}_0, \boldsymbol{g})$ by use of Eqs. (7.1-5) and (7.1-6) we obtain

$$I_2 = \frac{1}{16\pi^3} \iiiint \mathrm{e}^{-2\mathscr{G}_0^2} \exp(-\tfrac{1}{2}g^2) [\boldsymbol{g}'\boldsymbol{g}' - \boldsymbol{g}\boldsymbol{g}]^2 g b \, \mathrm{d}b \, \mathrm{d}\varepsilon \, \mathrm{d}^3\mathscr{G}_0 \, \mathrm{d}^3 g. \quad (7.1\text{-}14)$$

The integral over $\mathscr{G}_0$ can be carried out. The result is

$$I_2 = \frac{\sqrt{2}}{64\pi^{\frac{3}{2}}} \iiint \exp(-\tfrac{1}{2}g^2) [g'^4 + g^4 - 2(\boldsymbol{g}' \cdot \boldsymbol{g})^2] g b \, \mathrm{d}b \, \mathrm{d}\varepsilon \, \mathrm{d}^3 g. \quad (7.1\text{-}15)$$

Comparing this result with Eq. (7.1-9) we conclude that the bracket integrals I_1 and I_2 are identical, so

$$I_2 = 4\left(\frac{kT}{\pi m}\right)^{\frac{1}{2}} \int_0^\infty \exp(-g^2) g^7 \left[2\pi \int (1-\cos^2 \chi) \, b \, \mathrm{d}b\right] \mathrm{d}g. \quad (7.1\text{-}16)$$

The reduction of bracket integrals other than the two selected above follows similar lines. Clearly, the computation is more difficult when higher-order polynomials are involved or when the brackets are not symmetric in their arguments, but the difficulty is computational rather than conceptual.

Since all bracket integrals involved in transport property calculations for simple gases can be reduced to linear combinations of integrals of the type occurring in Eq. (7.1-12) – with only the powers of $\mathscr{g}$ and cos χ varying – it is convenient to define *Ω-integrals*,

$$\Omega^{(l,r)} = \left(\frac{kT}{\pi m}\right)^{\frac{1}{2}} \int_0^\infty \exp(-\mathscr{g}^2)\mathscr{g}^{2r+3} Q^{(l)} \mathrm{d}\mathscr{g}, \tag{7.1-17}$$

where

$$Q^{(l)} \equiv Q^{(l)}(g) = 2\pi \int \{1 - \cos^l \chi(b, g)\}\, b \,\mathrm{d}b. \tag{7.1-18}$$

Thus, from Eqs. (7.1-1) and (7.1-12) and from Eqs. (7.1-2) and (7.1-16) we find that, in terms of Ω-integrals, we have

$$[S_{\frac{3}{2}}^{(1)}(\mathscr{C}^2)\mathscr{C}, S_{\frac{3}{2}}^{(1)}(\mathscr{C}^2)\mathscr{C}] = 4\Omega^{(2,2)}, \tag{7.1-19}$$

$$[\mathscr{C}\mathscr{C} - \tfrac{1}{3}\mathscr{C}^2\mathbf{I}, \mathscr{C}\mathscr{C} - \tfrac{1}{3}\mathscr{C}^2\mathbf{I}] = 4\Omega^{(2,2)}. \tag{7.1-20}$$

For gas mixtures, the bracket integrals resulting from collisions between unlike or like molecules can be reduced to slightly generalized *Ω-integrals.* For interactions between molecules of species i and j these Ω-integrals are defined as

$$\Omega_{ij}^{(l,r)} = \left(\frac{kT}{2\pi m_{ij}}\right)^{\frac{1}{2}} \int_0^\infty \exp(-\mathscr{g}^2)\mathscr{g}^{2r+3} Q_{ij}^{(l)} \mathrm{d}\mathscr{g}, \tag{7.1-21}$$

where $Q_{ij}^{(l)}$ is defined as

$$Q_{ij}^{(l)} \equiv Q_{ij}^{(l)}(g) = 2\pi \int \{1 - \cos^l \chi_{ij}(b, g)\}\, b \,\mathrm{d}b \tag{7.1-22}$$

and m_{ij} is the *reduced mass* of the *i-j* system,

$$m_{ij}^{-1} = m_i^{-1} + m_j^{-1}. \tag{7.1-23}$$

In the integral (7.1-21) the dimensionless variable $\mathscr{g}$ is related to the magnitude of the relative velocity $\boldsymbol{g} = \boldsymbol{c}_j - \boldsymbol{c}_i$ by

$$\mathscr{g} = (m_{ij}/2kT)^{\frac{1}{2}} g. \tag{7.1-24}$$

Notice that for interactions between molecules of the same species i one has $m_{ii} = \frac{1}{2}m_i$ and the definition (7.1-21) is identical with the definition (7.1-17). Again, for the details of the reduction of bracket integrals for mixtures to the Ω-integrals the reader is referred to Appendix B.

From Eqs. (7.1-18) and (7.1-22) it may be seen that $Q^{(l)}$ and $Q_{ij}^{(l)}$ represent averaged cross sections. They are sometimes called *transport cross sections.* The quantities $(kT/\pi m)^{-\frac{1}{2}}\Omega^{(l,r)}$ and $(kT/2\pi m_{ij})^{-\frac{1}{2}}\Omega_{ij}^{(l,r)}$ – which are obtained by integrating $Q^{(l)}$ and $Q_{ij}^{(l)}$, respectively, over all relative velocities with a certain weight function – may therefore be interpreted as *temperature dependent effective cross sections* for the particular type of interaction.

In Section 9.1 it will be shown that the scattering angle χ is a function of $m_{ij}g^2$, i.e., of $\mathscr{g}^2$, rather than of g. Hence, $Q^{(l)}$ is independent of the mass of the molecules and so is the integral in Eq. (7.1-17). Therefore, the Ω-integral is inversely proportional to the square root of the molecular mass. Similarly, in a mixture, for interactions between molecules of species i and j the Ω_{ij}-integral is inversely proportional to the square root of the reduced mass of the i-j system.

From the definitions (7.1-17) and (7.1-21) it is easily proven that the Ω-integrals satisfy the recursion relation

$$T\frac{\partial\Omega_{ij}^{(l,r)}}{\partial T} + (r+\tfrac{3}{2})\Omega_{ij}^{(l,r)} = \Omega_{ij}^{(l,r+1)}. \qquad (7.1\text{-}25)$$

From this relation one could obtain all the $\Omega_{ij}^{(l,r)}$ for a given l once one of them has been computed. However, as this would require differentiation – a process which tends to amplify errors – this method has not been commonly used. The relation (7.1-25) has proven useful, however, in obtaining relations among transport properties.

In general, the Ω-integrals defined by Eqs. (7.1-17) and (7.1-21) cannot be evaluated analytically due to the complicated dependence of χ on b and g; in fact, for realistic potentials χ is itself defined by an integral which cannot be evaluated in closed form. There is one simple model which does, however, permit evaluation of these integrals in a simple manner – viz., the *rigid sphere model.* The details of the calculation are given in Section 9.2, but we may simply note here that, for this model,

$$\cos\tfrac{1}{2}\chi = b/\sigma_{ij}, \qquad (7.1\text{-}26)$$

where $\sigma_{ij} = \frac{1}{2}(\sigma_i+\sigma_j)$ is the separation of the centers of the two molecules – whose diameters are σ_i and σ_j, respectively – at the instant of collision.

Using this, one readily finds

$$Q_{ij}^{(l)} = \left[1 - \frac{1+(-1)^l}{2(l+1)}\right] \pi\sigma_{ij}^2 \tag{7.1-27}$$

and, therefore,

$$[\Omega_{ij}^{(l,r)}]_{\text{r.s.}} = \left(\frac{kT}{2\pi m_{ij}}\right)^{\frac{1}{2}} \frac{(r+1)!}{2} \left[1 - \frac{1+(-1)^l}{2(l+1)}\right] \pi\sigma_{ij}^2 . \tag{7.1-28}$$

The Ω-integral for a simple gas is simply obtained by setting $m_{ii} = \frac{1}{2}m$ in the expression (7.1-28),

$$[\Omega^{(l,r)}]_{\text{r.s.}} = \left(\frac{kT}{\pi m}\right)^{\frac{1}{2}} \frac{(r+1)!}{2} \left[1 - \frac{1+(-1)^l}{2(l+1)}\right] \pi\sigma^2 . \tag{7.1-29}$$

Although it is far from realistic, the rigid sphere model frequently permits rapid approximate evaluation of various properties and thus serves as a useful testing-ground for approximation methods (including the Sonine polynomial expansion method). It also allows one to define *reduced Ω-integrals* by

$$\Omega_{ij}^{(l,r)\star} = \Omega_{ij}^{(l,r)}/[\Omega_{ij}^{(l,r)}]_{\text{r.s.}}, \tag{7.1-30}$$

where the denominator represents the Ω-integral for the rigid sphere model, Eq. (7.1-28). These reduced integrals have the advantages of being more smoothly varying functions of temperature than the unnormalized integrals and of being closer to unity, thus allowing simpler presentation.

Finally, for later use, we give special symbols to certain combinations of the reduced Ω-integrals which occur frequently in transport property calculations, viz.,

$$\mathrm{A}_{ij}^{\star} = \Omega_{ij}^{(2,2)\star}/\Omega_{ij}^{(1,1)\star}, \tag{7.1-31}$$

$$\mathrm{B}_{ij}^{\star} = [5\Omega_{ij}^{(1,2)\star} - 4\Omega_{ij}^{(1,3)\star}]/\Omega_{ij}^{(1,1)\star}, \tag{7.1-32}$$

$$\mathrm{C}_{ij}^{\star} = \Omega_{ij}^{(1,2)\star}/\Omega_{ij}^{(1,1)\star}, \tag{7.1-33}$$

$$\mathrm{E}_{ij}^{\star} = \Omega_{ij}^{(2,3)\star}/\Omega_{ij}^{(2,2)\star}, \tag{7.1-34}$$

$$\mathrm{F}_{ij}^{\star} = \Omega_{ij}^{(3,3)\star}/\Omega_{ij}^{(1,1)\star}. \tag{7.1-35}$$

These ratios may be defined for a simple gas in exactly the same manner. Note that the definitions of $\mathrm{A}_{ij}^{\star}$, $\mathrm{B}_{ij}^{\star}$ and $\mathrm{C}_{ij}^{\star}$ differ from those given by Chapman and Cowling [1952] by numerical factors; however, they are consistent with those of Hirschfelder, Curtiss and Bird [1957].

7.2. Expressions for some of the bracket integrals

In this section we list expressions for some of the most important bracket integrals in terms of Ω-integrals. These formulae may be obtained by the methods of Appendix B.

(a) *Bracket integrals for simple gases*

We recall the definition (4.7-5) of the bracket integral of two functions F and G of the molecular velocity,

$$[F, G] = \int GI(F)\,\mathrm{d}^3c \tag{7.2-1}$$

or, with the definition (4.7-1) of the integral operator I taken into account,

$$[F, G] = \frac{1}{n^2}\int\!\!\int\!\!\int\!\!\int f_{\mathrm{M}} f_{\mathrm{M}1}\, G[F+F_1-F'-F_1']\,gb\,\mathrm{d}b\,\mathrm{d}\varepsilon\,\mathrm{d}^3c\,\mathrm{d}^3c_1 . \tag{7.2-2}$$

Here, f_{M} stands for the Maxwellian distribution function. It is to be noted that the bracket integral (7.2-1) is symmetric in its arguments, $[F, G] = [G, F]$, so a number of other integrals can be generated from the formulae below.

The Ω-integrals for simple gases have been defined in the previous section, Eq. (7.1-17).

TABLE 7.1

The bracket integral $[S^{(p)}_{\frac{3}{2}}(\mathscr{C}^2)\mathscr{C}, S^{(q)}_{\frac{3}{2}}(\mathscr{C}^2)\mathscr{C}]$ for $p, q = 0, 1, 2, 3$

	$[S^{(p)}_{\frac{3}{2}}(\mathscr{C}^2)\mathscr{C}, S^{(q)}_{\frac{3}{2}}(\mathscr{C}^2)\mathscr{C}]$
$p = 0, q = 0, 1, \ldots$	$0,$
$p = 1, q = 1$	$4\Omega^{(2,2)},$
$p = 1, q = 2$	$7\Omega^{(2,2)} - 2\Omega^{(2,3)},$
$p = 2, q = 2$	$\frac{77}{4}\Omega^{(2,2)} - 7\Omega^{(2,3)} + \Omega^{(2,4)},$
$p = 1, q = 3$	$\frac{63}{8}\Omega^{(2,2)} - \frac{9}{2}\Omega^{(2,3)} + \frac{1}{2}\Omega^{(2,4)},$
$p = 2, q = 3$	$\frac{945}{32}\Omega^{(2,2)} - \frac{261}{16}\Omega^{(2,3)} + \frac{25}{8}\Omega^{(2,4)} - \frac{1}{4}\Omega^{(2,5)},$
$p = 3, q = 3$	$\frac{14553}{256}\Omega^{(2,2)} - \frac{1215}{32}\Omega^{(2,3)} + \frac{313}{32}\Omega^{(2,4)} - \frac{9}{8}\Omega^{(2,5)} + \frac{1}{16}\Omega^{(2,6)} + \frac{1}{6}\Omega$

TABLE 7.2

The bracket integral $[S_{\frac{5}{2}}^{(p)}(\mathscr{C}^2)(\mathscr{C}\mathscr{C}-\frac{1}{3}\mathscr{C}^2\mathbf{I}),\ S_{\frac{5}{2}}^{(q)}(\mathscr{C}^2)(\mathscr{C}\mathscr{C}-\frac{1}{3}\mathscr{C}^2\mathbf{I})]$ for $p, q = 0, 1, 2$

	$[S_{\frac{5}{2}}^{(p)}(\mathscr{C}^2)(\mathscr{C}\mathscr{C}-\frac{1}{3}\mathscr{C}^2\mathbf{I}),\ S_{\frac{5}{2}}^{(q)}(\mathscr{C}^2)(\mathscr{C}\mathscr{C}-\frac{1}{3}\mathscr{C}^2\mathbf{I})]$
$p = 0,\ q = 0$	$4\Omega^{(2,2)}$,
$p = 0,\ q = 1$	$7\Omega^{(2,2)}-2\Omega^{(2,3)}$,
$p = 1,\ q = 1$	$\frac{301}{12}\Omega^{(2,2)}-7\Omega^{(2,3)}+\Omega^{(2,4)}$,
$p = 0,\ q = 2$	$\frac{63}{8}\Omega^{(2,2)}-\frac{9}{2}\Omega^{(2,3)}+\frac{1}{2}\Omega^{(2,4)}$,
$p = 1,\ q = 2$	$\frac{1365}{32}\Omega^{(2,2)}-\frac{321}{16}\Omega^{(2,3)}+\frac{25}{8}\Omega^{(2,4)}-\frac{1}{4}\Omega^{(2,5)}$,
$p = 2,\ q = 2$	$\frac{25137}{256}\Omega^{(2,2)}-\frac{1755}{32}\Omega^{(2,3)}+\frac{381}{32}\Omega^{(2,4)}-\frac{9}{8}\Omega^{(2,5)}+\frac{1}{16}\Omega^{(2,6)}+\frac{1}{2}\Omega^{(4,4)}$.

(b) *Bracket integrals for gas mixtures*

We recall the definitions (4.7-5) of the partial bracket integrals of two functions F and G defined in the velocity domain of species i and j,

$$[F, G]'_{ij} = \int G_i I_{ij,i}(F)\mathrm{d}^3c_i, \tag{7.2-16a}$$

$$[F, G]''_{ij} = \int G_i I_{ij,j}(F)\mathrm{d}^3c_i, \tag{7.2-16b}$$

or, with the definitions (4.7-3) of the integral operators $I_{ij,i}$ and $I_{ij,j}$ taken into account,

$$[F, G]'_{ij} = \frac{1}{n_i n_j}\iiiint f_{\mathrm{M}i}f_{\mathrm{M}j}G_i[F_i-F'_i]gb\,\mathrm{d}b\,\mathrm{d}\varepsilon\,\mathrm{d}^3c_i\,\mathrm{d}^3c_j, \tag{7.2-17a}$$

$$[F, G]''_{ij} = \frac{1}{n_i n_j}\iiiint f_{\mathrm{M}i}f_{\mathrm{M}j}G_i[F_j-F'_j]gb\,\mathrm{d}b\,\mathrm{d}\varepsilon\,\mathrm{d}^3c_i\,\mathrm{d}^3c_j. \tag{7.2-17b}$$

It is to be remembered that one has the symmetry relations

$$[F, G]'_{ij} = [G, F]'_{ij}, \tag{7.2-18a}$$

$$[F, G]''_{ij} = [G, F]''_{ji}, \tag{7.2-18b}$$

so a number of other integrals can be generated from the formulae given below.

The Ω-integrals for gas mixtures have been defined in the previous section, Eq. (7.1-21).

In the formulae below we have introduced the dimensionless mass ratios,

$$\mu_i = \frac{m_i}{m_i+m_j}, \quad \mu_j = \frac{m_j}{m_i+m_j}. \tag{7.2-19}$$

TABLE 7.3

The partial bracket integral $[S_{\frac{3}{2}}^{(p)}(\mathscr{C}^2)\mathscr{C},\ S_{\frac{3}{2}}^{(q)}(\mathscr{C}^2)\mathscr{C}]'_{ij}$ for $p, q = 0, 1, 2$

	$[S_{\frac{3}{2}}^{(p)}(\mathscr{C}^2)\mathscr{C}, S_{\frac{3}{2}}^{(q)}(\mathscr{C}^2)\mathscr{C}]'_{ij}$
$p = 0, q = 0$	$8\mu_j \Omega_{ij}^{(1,1)}$,
$p = 0, q = 1$	$8\mu_j^2(\frac{5}{2}\Omega_{ij}^{(1,1)} - \Omega_{ij}^{(1,2)})$,
$p = 1, q = 1$	$8\mu_j[\frac{5}{4}(6\mu_i^2+5\mu_j^2)\Omega_{ij}^{(1,1)} - 5\mu_j^2\Omega_{ij}^{(1,2)} + \mu_j^2\Omega_{ij}^{(1,3)} + 2\mu_i\mu_j\Omega_{ij}^{(2,2)}]$,
$p = 0, q = 2$	$4\mu_j^3(\frac{35}{4}\Omega_{ij}^{(1,1)} - 7\Omega_{ij}^{(1,2)} + \Omega_{ij}^{(1,3)})$,
$p = 1, q = 2$	$8\mu_j^2[\frac{35}{16}(12\mu_i^2+5\mu_j^2)\Omega_{ij}^{(1,1)} - \frac{21}{8}(4\mu_i^2+5\mu_j^2)\Omega_{ij}^{(1,2)} + \frac{19}{4}\mu_j^2\Omega_{ij}^{(1,3)}$ $- \frac{1}{2}\mu_j^2\Omega_{ij}^{(1,4)} + 7\mu_i\mu_j\Omega_{ij}^{(2,2)} - 2\mu_i\mu_j\Omega_{ij}^{(2,3)}]$,
$p = 2, q = 2$	$8\mu_j[\frac{35}{64}(40\mu_i^4 + 168\mu_i^2\mu_j^2 + 35\mu_j^4)\Omega_{ij}^{(1,1)} - \frac{7}{8}\mu_j^2(84\mu_i^2+35\mu_j^2)\Omega_{ij}^{(1,2)}$ $+ \frac{1}{8}\mu_j^2(108\mu_i^2+133\mu_j^2)\Omega_{ij}^{(1,3)} - \frac{7}{2}\mu_j^4\Omega_{ij}^{(1,4)} + \frac{1}{4}\mu_j^4\Omega_{ij}^{(1,5)}$ $+ \frac{7}{2}\mu_i\mu_j(4\mu_i^2+7\mu_j^2)\Omega_{ij}^{(2,2)} - 14\mu_i\mu_j^3\Omega_{ij}^{(2,3)} + 2\mu_i\mu_j^3\Omega_{ij}^{(2,4)}$ $+ 2\mu_i^2\mu_j^2\Omega_{ij}^{(3,3)}]$.

TABLE 7.4

The partial bracket integral $[S_{\frac{3}{2}}^{(p)}(\mathscr{C}^2)\mathscr{C},\ S_{\frac{3}{2}}^{(q)}(\mathscr{C}^2)\mathscr{C}]''_{ij}$ for $p, q = 0, 1, 2$

	$[S_{\frac{3}{2}}^{(p)}(\mathscr{C}^2)\mathscr{C}, S_{\frac{3}{2}}^{(q)}(\mathscr{C}^2)\mathscr{C}]''_{ij}$
$p = 0, q = 0$	$-8\mu_i^{\frac{1}{2}}\mu_j^{\frac{1}{2}}\Omega_{ij}^{(1,1)}$,
$p = 0, q = 1$	$-8\mu_i^{\frac{3}{2}}\mu_j^{\frac{1}{2}}(\frac{5}{2}\Omega_{ij}^{(1,1)} - \Omega_{ij}^{(1,2)})$,
$p = 1, q = 1$	$-8\mu_i^{\frac{3}{2}}\mu_j^{\frac{3}{2}}(\frac{55}{4}\Omega_{ij}^{(1,1)} - 5\Omega_{ij}^{(1,2)} + \Omega_{ij}^{(1,3)} - 2\Omega_{ij}^{(2,2)})$,
$p = 0, q = 2$	$-4\mu_i^{\frac{5}{2}}\mu_j^{\frac{1}{2}}(\frac{35}{4}\Omega_{ij}^{(1,1)} - 7\Omega_{ij}^{(1,2)} + \Omega_{ij}^{(1,3)})$,
$p = 1, q = 2$	$-8\mu_i^{\frac{5}{2}}\mu_j^{\frac{3}{2}}(\frac{595}{16}\Omega_{ij}^{(1,1)} - \frac{189}{8}\Omega_{ij}^{(1,2)} + \frac{19}{4}\Omega_{ij}^{(1,3)} - \frac{1}{2}\Omega_{ij}^{(1,4)} - 7\Omega_{ij}^{(2,2)} + 2\Omega_{ij}^{(2,3)}$
$p = 2, q = 2$	$-8\mu_i^{\frac{5}{2}}\mu_j^{\frac{5}{2}}(\frac{8505}{64}\Omega_{ij}^{(1,1)} - \frac{833}{8}\Omega_{ij}^{(1,2)} + \frac{241}{8}\Omega_{ij}^{(1,3)} - \frac{7}{2}\Omega_{ij}^{(1,4)}$ $+ \frac{1}{4}\Omega_{ij}^{(1,5)} - \frac{77}{2}\Omega_{ij}^{(2,2)} + 14\Omega_{ij}^{(2,3)} - 2\Omega_{ij}^{(2,4)} + 2\Omega_{ij}^{(3,3)})$.

TABLE 7.5

The partial bracket integral $[S_{\frac{5}{2}}^{(p)}(\mathscr{C}^2)(\mathscr{C}\mathscr{C}-\frac{1}{3}\mathscr{C}^2\mathbf{I}),\ S_{\frac{5}{2}}^{(q)}(\mathscr{C}^2)(\mathscr{C}\mathscr{C}-\frac{1}{3}\mathscr{C}^2\mathbf{I})]'_{ij}$ for $p,\ q = 0,\ 1,\ 2$

$$[S_{\frac{5}{2}}^{(p)}(\mathscr{C}^2)(\mathscr{C}\mathscr{C}-\tfrac{1}{3}\mathscr{C}^2\mathbf{I}),\ S_{\frac{5}{2}}^{(q)}(\mathscr{C}^2)(\mathscr{C}\mathscr{C}-\tfrac{1}{3}\mathscr{C}^2\mathbf{I})]'_{ij}$$

$= 0,\ q = 0$
$$\tfrac{16}{3}\mu_j(5\mu_i\Omega_{ij}^{(1,1)}+\tfrac{3}{2}\mu_j\Omega_{ij}^{(2,2)}),$$

$= 0,\ q = 1$
$$\tfrac{16}{3}\mu_j^2(\tfrac{35}{2}\mu_i\Omega_{ij}^{(1,1)}-7\mu_i\Omega_{ij}^{(1,2)}+\tfrac{21}{4}\mu_j\Omega_{ij}^{(2,2)}-\tfrac{3}{2}\mu_j\Omega_{ij}^{(2,3)}),$$

$= 1,\ q = 1$
$$\tfrac{16}{3}\mu_j[\tfrac{1}{4}\mu_i(140\mu_i^2+245\mu_j^2)\Omega_{ij}^{(1,1)}-49\mu_i\mu_j^2\Omega_{ij}^{(1,2)}+8\mu_i\mu_j^2\Omega_{ij}^{(1,3)}$$
$$+\tfrac{1}{8}\mu_j(154\mu_i^2+147\mu_j^2)\Omega_{ij}^{(2,2)}-\tfrac{21}{2}\mu_j^3\Omega_{ij}^{(2,3)}+\tfrac{3}{2}\mu_j^3\Omega_{ij}^{(2,4)}+3\mu_i\mu_j^2\Omega_{ij}^{(3,3)}],$$

$= 0,\ q = 2$
$$\tfrac{16}{3}\mu_j^3(\tfrac{315}{8}\mu_i\Omega_{ij}^{(1,1)}-\tfrac{63}{2}\mu_i\Omega_{ij}^{(1,2)}+\tfrac{9}{2}\mu_i\Omega_{ij}^{(1,3)}+\tfrac{189}{16}\mu_j\Omega_{ij}^{(2,2)}$$
$$-\tfrac{27}{4}\mu_j\Omega_{ij}^{(2,3)}+\tfrac{3}{4}\mu_j\Omega_{ij}^{(2,4)}),$$

$= 1,\ q = 2$
$$\tfrac{16}{3}\mu_j^2[\tfrac{1}{16}\mu_i(2520\mu_i^2+2205\mu_j^2)\Omega_{ij}^{(1,1)}-\tfrac{1}{8}\mu_i(504\mu_i^2+1323\mu_j^2)\Omega_{ij}^{(1,2)}$$
$$+\tfrac{207}{4}\mu_i\mu_j^2\Omega_{ij}^{(1,3)}-\tfrac{9}{2}\mu_i\mu_j^2\Omega_{ij}^{(1,4)}+\tfrac{1}{32}\mu_j(2772\mu_i^2+1323\mu_j^2)\Omega_{ij}^{(2,2)}$$
$$-\tfrac{1}{16}\mu_j(396\mu_i^2+567\mu_j^2)\Omega_{ij}^{(2,3)}+\tfrac{75}{8}\mu_j^3\Omega_{ij}^{(2,4)}-\tfrac{3}{4}\mu_j^3\Omega_{ij}^{(2,5)}$$
$$+\tfrac{27}{2}\mu_i\mu_j^2\Omega_{ij}^{(3,3)}-3\mu_i\mu_j^2\Omega_{ij}^{(3,4)}],$$

$= 2,\ q = 2$
$$4\mu_j[\tfrac{1}{16}\mu_i(2520\mu_i^4+15120\mu_i^2\mu_j^2+6615\mu_j^4)\Omega_{ij}^{(1,1)}$$
$$-\tfrac{1}{2}\mu_i\mu_j^2(1512\mu_i^2+1323\mu_j^2)\Omega_{ij}^{(1,2)}+\tfrac{1}{2}\mu_i\mu_j(252\mu_i^2+621\mu_j^2)\Omega_{ij}^{(1,3)}$$
$$-54\mu_i\mu_j^4\Omega_{ij}^{(1,4)}+3\mu_i\mu_j^4\Omega_{ij}^{(1,5)}+\tfrac{1}{32}\mu_j(4536\mu_i^4+16632\mu_i^2\mu_j^2$$
$$+3969\mu_j^4)\Omega_{ij}^{(2,2)}-\tfrac{1}{4}\mu_j^3(1188\mu_i^2+567\mu_j^2)\Omega_{ij}^{(2,3)}$$
$$+\tfrac{1}{4}\mu_j^3(156\mu_i^2+225\mu_j^2)\Omega_{ij}^{(2,4)}-9\mu_j^5\Omega_{ij}^{(2,5)}+\tfrac{1}{2}\mu_j^5\Omega_{ij}^{(2,6)}$$
$$+\mu_i\mu_j^2(42\mu_i^2+81\mu_j^2)\Omega_{ij}^{(3,3)}-36\mu_i\mu_j^4\Omega_{ij}^{(3,4)}+4\mu_i\mu_j^4\Omega_{ij}^{(3,5)}$$
$$+4\mu_i\mu_j^3\Omega_{ij}^{(4,4)}].$$

7.3. Explicit expressions for the transport coefficients

By combining the results of Sections 5.5 and 6.4 with the results of the previous section we see that, in any approximation, the transport coefficients of a gas can be obtained by solving sets of linear algebraic equations in which the coefficients are linear combinations of Ω-integrals. In the present section we discuss the solution of these equations and write the transport coefficients as algebraic functions of Ω-integrals. To some extent, the present section summarizes the results of the foregoing chapters.

TABLE 7.6

The partial bracket integral $[S_{\frac{5}{2}}^{(p)}(\mathscr{C}^2)(\mathscr{C}\mathscr{C}-\frac{1}{3}\mathscr{C}^2\mathbf{I}),\ S_{\frac{5}{2}}^{(q)}(\mathscr{C}^2)(\mathscr{C}\mathscr{C}-\frac{1}{3}\mathscr{C}^2\mathbf{I})]_{ij}''$ for $p, q = 0, 1, 2$

	$[S_{\frac{5}{2}}^{(p)}(\mathscr{C}^2)(\mathscr{C}\mathscr{C}-\frac{1}{3}\mathscr{C}^2\mathbf{I}),\ S_{\frac{5}{2}}^{(q)}(\mathscr{C}^2)(\mathscr{C}\mathscr{C}-\frac{1}{3}\mathscr{C}^2\mathbf{I})]_{ij}''$
$p = 0, q = 0$	$-\frac{16}{3}\mu_i\mu_j(5\Omega_{ij}^{(1,1)}-\frac{3}{2}\Omega_{ij}^{(2,2)})$,
$p = 0, q = 1$	$\frac{16}{3}\mu_i^2\mu_j(-\frac{35}{2}\Omega_{ij}^{(1,1)}+7\Omega_{ij}^{(1,2)}+\frac{21}{4}\Omega_{ij}^{(2,2)}-\frac{3}{2}\Omega_{ij}^{(2,3)})$,
$p = 1, q = 1$	$-\frac{16}{3}\mu_i^2\mu_j^2(\frac{385}{4}\Omega_{ij}^{(1,1)}-49\Omega_{ij}^{(1,2)}+8\Omega_{ij}^{(1,3)}-\frac{301}{8}\Omega_{ij}^{(2,2)}+\frac{21}{2}\Omega_{ij}^{(2,3)}$ $-\frac{3}{2}\Omega_{ij}^{(2,4)}+3\Omega_{ij}^{(3,3)})$,
$p = 0, q = 2$	$\frac{16}{3}\mu_i^3\mu_j(-\frac{315}{8}\Omega_{ij}^{(1,1)}+\frac{63}{2}\Omega_{ij}^{(1,2)}-\frac{9}{2}\Omega_{ij}^{(1,3)}+\frac{189}{16}\Omega_{ij}^{(2,2)}$ $-\frac{27}{4}\Omega_{ij}^{(2,3)}+\frac{3}{4}\Omega_{ij}^{(2,4)})$,
$p = 1, q = 2$	$\frac{16}{3}\mu_i^3\mu_j^2(-\frac{4725}{16}\Omega_{ij}^{(1,1)}+\frac{1827}{8}\Omega_{ij}^{(1,2)}-\frac{207}{4}\Omega_{ij}^{(1,3)}+\frac{9}{2}\Omega_{ij}^{(1,4)}$ $+\frac{4095}{32}\Omega_{ij}^{(2,2)}-\frac{963}{16}\Omega_{ij}^{(2,3)}+\frac{75}{8}\Omega_{ij}^{(2,4)}-\frac{3}{4}\Omega_{ij}^{(2,5)}-\frac{27}{2}\Omega_{ij}^{(3,3)}+3\Omega_{ij}^{(3,4)})$
$p = 2, q = 2$	$4\mu_i^3\mu_j^3(-\frac{24255}{16}\Omega_{ij}^{(1,1)}+\frac{2835}{2}\Omega_{ij}^{(1,2)}-\frac{873}{2}\Omega_{ij}^{(1,3)}+54\Omega_{ij}^{(1,4)}-3\Omega_{ij}^{(1,5)}$ $+\frac{25137}{32}\Omega_{ij}^{(2,2)}-\frac{1755}{4}\Omega_{ij}^{(2,3)}+\frac{381}{4}\Omega_{ij}^{(2,4)}-9\Omega_{ij}^{(2,5)}+\frac{1}{2}\Omega_{ij}^{(2,6)}$ $-123\Omega_{ij}^{(3,3)}+36\Omega_{ij}^{(3,4)}-4\Omega_{ij}^{(3,5)}+4\Omega_{ij}^{(4,4)})$.

(a) *Viscosity of a simple gas*

Equation (5.6-21) expresses the coefficient of viscosity of a simple gas in terms of a single Sonine expansion coefficient. In the nth approximation,

$$[\eta]_n = \tfrac{1}{2}kTb_0^{(n)}, \tag{7.3-1}$$

where $b_0^{(n)}$ is obtained from the solution of the following set of n linear algebraic equations for the unknowns $b_0^{(n)}, \ldots, b_{n-1}^{(n)}$,

$$\sum_{q=0}^{n-1} H^{pq}b_q^{(n)} = \frac{2}{kT}\,\delta_{p0}\,,\ p = 0, \ldots, n-1. \tag{7.3-2}$$

The coefficient H^{pq} is defined in terms of a bracket integral,

$$H^{pq} = \frac{2}{5kT}\,[S_{\frac{5}{2}}^{(p)}(\mathscr{C}^2)(\mathscr{C}\mathscr{C}-\tfrac{1}{3}\mathscr{C}^2\mathbf{I}),\ S_{\frac{5}{2}}^{(q)}(\mathscr{C}^2)(\mathscr{C}\mathscr{C}-\tfrac{1}{3}\mathscr{C}^2\mathbf{I})] \tag{7.3-3}$$

and can be writen as a linear combination of Ω-integrals with the aid of the formulae given in Table 7.2. The H^{pq} satisfy the symmetry relation

$$H^{pq} = H^{qp}. \tag{7.3-4}$$

The first approximation ($n = 1$) is particularly simple,

$$[\eta]_1 = \frac{1}{H^{00}}, \tag{7.3-5}$$

so, in terms of Ω-integrals,

$$[\eta]_1 = \frac{5kT}{8\Omega^{(2,2)}} \tag{7.3-6a}$$

or, in terms of reduced Ω-integrals,

$$[\eta]_1 = \frac{5}{16}\frac{(\pi mkT)^{\frac{1}{2}}}{\pi\sigma^2\Omega^{(2,2)\star}}. \tag{7.3-6b}$$

One may verify that, in the nth order Chapman-Cowling approximation the coefficient of viscosity can be written in the form

$$[\eta]_n = [\eta]_1 f_\eta^{(n)}, \tag{7.3-7a}$$

where

$$f_\eta^{(1)} = 1, \tag{7.3-7b}$$

$$f_\eta^{(2)} = f_\eta^{(1)} + \frac{(H^{01})^2}{H^{00}H^{11} - (H^{01})^2}, \tag{7.3-7c}$$

$$f_\eta^{(3)} = f_\eta^{(2)} + H^{00}(H^{01}H^{12} - H^{11}H^{02})^2 / \{[H^{00}H^{11} - (H^{01})^2] \times [H^{00}H^{11}H^{22} + 2H^{01}H^{02}H^{12} - H^{00}(H^{12})^2 - H^{11}(H^{02})^2 - H^{22}(H^{01})^2]\}. \tag{7.3-7d}$$

The constants H^{pq} required by these expressions can be obtained from Eq. (7.3-3) and Table 7.2. In fact, the factors $(2/5kT)$ in H^{pq} cancel and the bracket integrals can be used in place of H^{pq}.

Similarly, from Kihara's method,

$$[\eta]_n^K = [\eta]_1 f_\eta^{(n)K}, \tag{7.3-8a}$$

where

$$f_\eta^{(1)K} = 1, \tag{7.3-8b}$$

$$f_\eta^{(2)K} = f_\eta^{(1)K} + \frac{(H^{01})^2}{H^{00}H^{11}}, \tag{7.3-8c}$$

$$f_\eta^{(3)K} = f_\eta^{(2)K} + \left\{ \frac{(H^{01})^4}{(H^{00}H^{11})^2} + \frac{(H^{01}H^{12} - H^{11}H^{02})^2}{H^{00}(H^{11})^2 H^{22}} \right\}. \tag{7.3-8d}$$

As was mentioned in Section 5.7, these expressions are regarded as representing the nth order Kihara approximations by some authors, but, more commonly, a further approximation is made. As a consequence of Eq. (7.1-25), the $\Omega^{(l,r)}$ with $l \neq r$ can be expressed in terms of $\Omega^{(l,l)}$ and its derivatives with respect to temperature. When these expressions are substituted into the formulae for H^{pq}, it is found that, for any pair (p, q), H^{pq} is a finite sum of the form

$$H^{pq} = \frac{2}{5kT} \sum_{j \geq |p-q|} T^j \frac{\partial^j}{\partial T^j} \Omega_\eta^{pq}, \tag{7.3-9}$$

where Ω_η^{pq} is a finite linear combination of the integrals $\Omega^{(2,2)}$, $\Omega^{(3,3)}$, ... Since the Ω-integrals vary relatively slowly with temperature (the variation is quite model-dependent, however; see Chapter 10), we may conclude from Eq. (7.3-9) that the order of magnitude of H^{pq} is determined primarily by the temperature derivative of order $|p-q|$ of the Ω-integrals. In fact, this allows us to identify the expansion parameter of the Kihara method with the temperature differential operator. Then one obtains, instead of Eq. (7.3-8),

$$[\eta]_n^K = [\eta]_1 \tilde{f}_\eta^{(n)K}, \tag{7.3-10a}$$

where

$$\tilde{f}_\eta^{(1)K} = 1, \tag{7.3-10b}$$

$$\tilde{f}_\eta^{(2)K} = \tilde{f}_\eta^{(1)K} + \frac{3}{49}\left(\frac{\Omega^{(2,3)}}{\Omega^{(2,2)}} - \frac{7}{2}\right)^2. \tag{7.3-10c}$$

The expression for $\tilde{f}_\eta^{(3)K}$ is very complicated and will not be given here.

(b) *Thermal conductivity of a simple gas*

Equation (5.6-20) expresses the coefficient of thermal conductivity of a simple gas in terms of a single Sonine expansion coefficient. In the nth approximation,

$$[\lambda]_n = \tfrac{5}{4}k\, a_1^{(n)}, \tag{7.3-11}$$

where $a_1^{(n)}$ is obtained from the solution of the following set of n linear algebraic equations for the unknowns $a_1^{(n)}, \ldots, a_n^{(n)}$,

$$\sum_{q=1}^{n} \Lambda^{rq} a_q^{(n)} = \frac{4}{5k} \delta_{p1}, \qquad p = 1, \ldots, n. \tag{7.3-12}$$

The coefficient Λ^{pq} is defined in terms of a bracket integral,

$$\Lambda^{pq} = \frac{8m}{75k^2 T} [S_{\frac{3}{2}}^{(p)}(\mathscr{C}^2)\mathscr{C}, S_{\frac{3}{2}}^{(q)}(\mathscr{C}^2)\mathscr{C}] \tag{7.3-13}$$

and can be written as a linear combination of Ω-integrals with the aid of the formulae given in Table 7.1. The Λ^{pq} satisfy the symmetry relation

$$\Lambda^{pq} = \Lambda^{qp}. \tag{7.3-14}$$

The first approximation ($n = 1$) is particularly simple,

$$[\lambda]_1 = \frac{1}{\Lambda^{11}}, \tag{7.3-15}$$

so, in terms of Ω-integrals,

$$[\lambda]_1 = \frac{25 c_v kT}{16 \Omega^{(2,2)}} \tag{7.3-16a}$$

or, in terms of reduced Ω-integrals,

$$[\lambda]_1 = \frac{25}{32} \frac{(\pi mkT)^{\frac{1}{2}}}{\pi \sigma^2 \Omega^{(2,2)\star}} c_v . \tag{7.3-16b}$$

Here, we have introduced the *specific heat* (i.e., the heat capacity per unit mass) of the gas at constant volume, c_v, which is defined by

$$c_v = 3k/2m. \tag{7.3-17}$$

One may verify that, in the nth order Chapman-Cowling approximation the coefficient of thermal conductivity can be written in the form

$$[\lambda]_n = [\lambda]_1 f_\lambda^{(n)}, \tag{7.3-18a}$$

where

$$f_\lambda^{(1)} = 1, \tag{7.3-18b}$$

$$f_\eta^{(2)} = f_\lambda^{(1)} + \frac{(\Lambda^{12})^2}{\Lambda^{11}\Lambda^{22} - (\Lambda^{12})^2}, \tag{7.3-18c}$$

$$f_\lambda^{(3)} = f_\lambda^{(2)} + \Lambda^{11}(\Lambda^{12}\Lambda^{23} - \Lambda^{22}\Lambda^{13})^2 / \{[\Lambda^{11}\Lambda^{22} - (\Lambda^{12})^2] \times [\Lambda^{11}\Lambda^{22}\Lambda^{33} + 2\Lambda^{12}\Lambda^{13}\Lambda^{23} - \Lambda^{11}(\Lambda^{23})^2 - \Lambda^{22}(\Lambda^{13})^2 - \Lambda^{33}(\Lambda^{12})^2]\}. \tag{7.3-18d}$$

The constants Λ^{pq} required by these expressions can be obtained from Eq. (7.3-13) and Table 7.1. Again, the bracket integrals can be used in place of Λ^{pq}.

Similarly, from Kihara's method,

$$[\lambda]_n^K = [\lambda]_1 f_\lambda^{(n)K}, \tag{7.3-19a}$$

where

$$f_\lambda^{(1)K} = 1, \tag{7.3-19b}$$

$$f_\lambda^{(2)K} = f_\lambda^{(1)K} + \frac{(\Lambda^{12})^2}{\Lambda^{11}\Lambda^{22}}, \tag{7.3-19c}$$

$$f_\lambda^{(3)K} = f_\lambda^{(2)K} + \left\{\frac{(\Lambda^{12})^4}{(\Lambda^{11}\Lambda^{22})^2} + \frac{(\Lambda^{12}\Lambda^{23} - \Lambda^{22}\Lambda^{13})^2}{\Lambda^{11}(\Lambda^{22})^2\Lambda^{33}}\right\}. \tag{7.3-19d}$$

Approximating the Λ^{pq} still further, in the same manner as the H^{pq}, one finds,

$$[\lambda]_n^K = [\lambda]_1 \tilde{f}_\lambda^{(n)K}, \tag{7.3-20a}$$

where

$$\tilde{f}_\lambda^{(1)K} = 1, \tag{7.3-20b}$$

$$\tilde{f}_\lambda^{(2)K} = \tilde{f}_\lambda^{(1)K} + \frac{2}{21}\left(\frac{\Omega^{(2,3)}}{\Omega^{(2,2)}} - \frac{7}{2}\right)^2. \tag{7.3-20c}$$

Again, the expression for $\tilde{f}_\lambda^{(3)K}$ is very complicated and will not be given here.

It is interesting to note that the first approximations for both the viscosity and the thermal conductivity are inversely proportional to the same Ω-integral. As an immediate consequence of this we see that, in the lowest-order approximation, the *Eucken ratio*, f, which is defined by

$$\mathrm{f} = \lambda/\eta c_v, \tag{7.3-21}$$

has the constant value $\frac{5}{2}$, independent of the nature of the gas and independent of the temperature,

$$[\mathrm{f}]_1 = [\lambda]_1/[\eta]_1 c_v = \tfrac{5}{2}. \tag{7.3-22}$$

This remarkable property is only very slightly modified by higher order approximations, when the Eucken ratio becomes

$$[\mathrm{f}]_n = [\lambda]_n/[\eta]_n c_v = \tfrac{5}{2}(f_\lambda^{(n)}/f_\eta^{(n)}). \tag{7.3-23}$$

Before turning to the transport coefficients of a multicomponent gas mixture we introduce the following notation. Let $[\eta_i]_1$ be the first approximation to the viscosity coefficient of the simple gas i,

$$[\eta_i]_1 = \frac{5}{16}\frac{(\pi m_i kT)^{\frac{1}{2}}}{\pi\sigma_i^2\Omega_i^{(2,2)\star}} \tag{7.3-24}$$

and define, by analogy, a quantity $[\eta_{ij}]_1$ thus:

$$[\eta_{ij}]_1 = \frac{5}{16}\frac{(2\pi m_{ij} kT)^{\frac{1}{2}}}{\pi\sigma_{ij}^2\Omega_{ij}^{(2,2)\star}}, \tag{7.3-25}$$

where m_{ij} is the reduced mass of the i-j system defined in Eq. (7.1-23). Similarly, let $[\lambda_i]_1$ be the first approximation to the coefficient of thermal conductivity of the simple gas i,

$$[\lambda_i]_1 = \frac{25}{32}\frac{(\pi m_i kT)^{\frac{1}{2}}}{\pi\sigma_i^2\Omega_i^{(2,2)\star}}\frac{3k}{2m_i} \tag{7.3-26}$$

and define, by analogy, a quantity $[\lambda_{ij}]_1$ thus:

$$[\lambda_{ij}]_1 = \frac{25}{32}\frac{(2\pi m_{ij} kT)^{\frac{1}{2}}}{\pi\sigma_{ij}^2\Omega_{ij}^{(2,2)\star}}\frac{3k}{4m_{ij}}. \tag{7.3-27}$$

The quantities $[\eta_{ij}]_1$ and $[\lambda_{ij}]_1$ are related to the binary diffusion coefficient of a mixture of molecules of species i and j, cf. Eqs. (7.3-47) and (7.3-48), but have no direct physical meaning otherwise. One has the following analog of the Eucken formula (7.3-22),

$$[\lambda_{ij}/\eta_{ij}]_1 = \tfrac{5}{2}(3k/4m_{ij}). \tag{7.3-28}$$

(c1) *Multicomponent diffusion coefficients*

The diffusion coefficients of a K-component gas mixture are given in terms of Sonine expansion coefficients by Eq. (6.4-41),

$$[D_{kl}]_n = [D_{lk}]_n = \frac{1}{2n}\, d_{l,0}^{k(n)}, \tag{7.3-29}$$

where the $d_{l,0}^{k(n)}$ are obtained from the solution of the following set of linear algebraic equations for the unknowns $d_{1,0}^{k(n)}, \ldots, d_{1,n-1}^{k(n)}, d_{2,0}^{k(n)}, \ldots, d_{2,n-1}^{k(n)}, \ldots d_{K,n-1}^{k(n)}$,

$$\sum_{j=1}^{K} \sum_{q=0}^{n-1} \Lambda_{ij}^{pq} d_{j,q}^{k(n)} = \frac{8}{25k} \left(\delta_{ik} - \frac{\rho_i}{\rho} \right) \delta_{p0}, \qquad i = 1, \ldots, K; \quad p = 0, \ldots, n-1. \tag{7.3-30a}$$

For $p = 0$ the set of Eqs. (7.3-30a) is linearly dependent and must be supplemented by the equation

$$\sum_{i=1}^{K} (\rho_i/\rho) d_{i,0}^{k(n)} = 0. \tag{7.3-30b}$$

The coefficients Λ_{ij}^{pq} are defined in terms of partial bracket integrals,

$$\Lambda_{ij}^{pq} = \frac{8 m_i^{\frac{1}{2}} m_j^{\frac{1}{2}}}{75 k^2 T} \left\{ \delta_{ij} \sum_{h=1}^{K} \frac{n_i n_h}{n^2} [S_{\frac{3}{2}}^{(p)}(\mathscr{C}^2)\mathscr{C}, S_{\frac{3}{2}}^{(q)}(\mathscr{C}^2)\mathscr{C}]'_{ih} + \frac{n_i n_j}{n^2} [S_{\frac{3}{2}}^{(p)}(\mathscr{C}^2)\mathscr{C}, S_{\frac{3}{2}}^{(q)}(\mathscr{C}^2)\mathscr{C}]''_{ij} \right\} \tag{7.3-31}$$

and can be written as linear combinations of Ω-integrals with the aid of the formulae given in Tables 7.3 and 7.4. The Λ_{ij}^{pq} satisfy the symmetry relation

$$\Lambda_{ij}^{pq} = \Lambda_{ji}^{qp} \tag{7.3-32}$$

and, moreover, because of the conservation of momentum,

$$\sum_{i=1}^{K} \Lambda_{ij}^{0q} = 0 \quad \text{and} \quad \sum_{j=1}^{K} \Lambda_{ij}^{p0} = 0. \tag{7.3-33}$$

The coefficients Λ_{ij}^{pq} for $p = 0, 1$ and $q = 0, 1$, which are required in the computation of the first and second approximations to the diffusion coefficients as well as, later, in the computation of the first approximations to the thermal diffusion coefficients and coefficient of thermal conductivity, can be written in terms of the quantities $[\lambda_i]_1$ and $[\lambda_{ij}]_1$ introduced above – see Eqs. (7.3-26) and (7.3-27). One may verify the following expressions:

$$\Lambda_{ii}^{00} = \sum_{\substack{l=1 \\ (l \neq i)}}^{K} \frac{x_i x_l}{2 A_{il}^{\star} [\lambda_{il}]_1}, \tag{7.3-34a}$$

$$\Lambda_{ij}^{00} = - \frac{x_i x_j}{2 A_{ij}^{\star} [\lambda_{ij}]_1} \quad (i \neq j), \tag{7.3-34b}$$

$$\Lambda_{ii}^{01} = \Lambda_{ii}^{10} = -\sum_{\substack{l=1\\(l\neq i)}}^{K} \frac{x_i x_l}{4\mathrm{A}_{il}^{\star}[\lambda_{il}]_1} \frac{m_l}{m_i+m_l}(6\mathrm{C}_{il}^{\star}-5), \tag{7.3-34c}$$

$$\Lambda_{ij}^{01} = \Lambda_{ji}^{10} = \frac{x_i x_j}{4\mathrm{A}_{ij}^{\star}[\lambda_{ij}]_1} \frac{m_i}{m_i+m_j}(6\mathrm{C}_{ij}^{\star}-5) \quad (i \neq j), \tag{7.3-34d}$$

$$\Lambda_{ii}^{11} = \frac{x_i^2}{[\lambda_i]_1} + \sum_{\substack{l=1\\(l\neq i)}}^{K} \frac{x_i x_l}{2\mathrm{A}_{il}^{\star}[\lambda_{il}]_1} \frac{\frac{15}{2}m_i^2 + \frac{25}{4}m_l^2 - 3m_l^2\mathrm{B}_{il}^{\star} + 4m_i m_l \mathrm{A}_{il}^{\star}}{(m_i+m_l)^2}, \tag{7.3-34e}$$

$$\Lambda_{ij}^{11} = -\frac{x_i x_j}{2\mathrm{A}_{ij}^{\star}[\lambda_{ij}]_1} \frac{m_i m_j}{(m_i+m_j)^2}(\tfrac{55}{4} - 3\mathrm{B}_{ij}^{\star} - 4\mathrm{A}_{ij}^{\star}) \quad (i \neq j). \tag{7.3-34f}$$

Here we have introduced the *molecular fractions*, x_i, defined by

$$x_i = n_i/n, \tag{7.3-35}$$

for each component of the mixture.

In a K-component mixture there are $\frac{1}{2}K(K-1)$ independent diffusion coefficients. They satisfy the relations

$$D_{kl} = D_{lk} \tag{7.3-36a}$$

and

$$\sum_{k=1}^{K} (\rho_k/\rho) D_{kl} = 0. \tag{7.3-36b}$$

(c2) *Binary diffusion coefficient*

In the case of a mixture of two gases there is only one independent diffusion coefficient. From Eqs. (6.5-1) and (6.5-2) and the result (7.3-29) one derives the following expression for the binary diffusion coefficient in the nth approximation,

$$[\mathscr{D}_{12}]_n = -\frac{1}{2n} \frac{(\rho/n)^2}{m_1 m_2} d_{1,0}^{2(n)}. \tag{7.3-37}$$

In particular, for $n = 1$, one finds after some trivial algebra,

$$[\mathscr{D}_{12}]_1 = \frac{4n_1 n_2}{25kn^3 \Lambda_{11}^{00}} = \frac{3kT}{16nm_{12}\Omega_{12}^{(1,1)}} \tag{7.3-38a}$$

or, in terms of the reduced Ω-integral,

$$[\mathscr{D}_{12}]_1 = \frac{3}{16nm_{12}} \frac{(2\pi m_{12}kT)^{\frac{1}{2}}}{\pi\sigma_{12}^2 \Omega_{12}^{(1,1)\star}}. \tag{7.3-38b}$$

In the nth order Chapman-Cowling approximation the binary diffusion coefficient can be written in the form

$$[\mathscr{D}_{12}]_n = [\mathscr{D}_{12}]_1 f_D^{(n)}. \tag{7.3-39}$$

One may verify the following expression for $f_D^{(2)}$,

$$f_D^{(2)} = (1-\Delta)^{-1}, \tag{7.3-40}$$

where

$$\Delta = \frac{\Lambda_{11}^{10}(\Lambda_{11}^{01}\Lambda_{22}^{11}-\Lambda_{12}^{01}\Lambda_{21}^{11})-\Lambda_{22}^{10}(\Lambda_{12}^{01}\Lambda_{11}^{11}-\Lambda_{11}^{01}\Lambda_{12}^{11})}{\Lambda_{11}^{00}(\Lambda_{11}^{11}\Lambda_{22}^{11}-\Lambda_{12}^{11}\Lambda_{21}^{11})}. \tag{7.3-41}$$

Using the expressions (7.3-34) for the coefficients Λ_{ij}^{pq} we find

$$\Delta = \tfrac{1}{10}(6\mathrm{C}_{12}^{\star}-5)^2 \frac{P_1 x_1^2+P_2 x_2^2+P_{12} x_1 x_2}{Q_1 x_1^2+Q_2 x_2^2+Q_{12} x_1 x_2}, \tag{7.3-42}$$

with

$$P_1 = \frac{m_1}{m_2}\frac{[\lambda_{12}]_1}{[\lambda_1]_1}, \tag{7.3-43a}$$

$$P_2 = \frac{m_2}{m_1}\frac{[\lambda_{12}]_1}{[\lambda_2]_1}, \tag{7.3-43b}$$

$$P_{12} = \frac{15}{4\mathrm{A}_{12}^{\star}}\frac{(m_1-m_2)^2}{m_1 m_2}+2, \tag{7.3-43c}$$

and

$$Q_1 = \frac{[\lambda_{12}]_1}{[\lambda_1]_1}\left[3\frac{m_2}{m_1}+(\tfrac{5}{2}-\tfrac{6}{5}\mathrm{B}_{12}^{\star})\frac{m_1}{m_2}+\tfrac{8}{5}\mathrm{A}_{12}^{\star}\right], \tag{7.3-44a}$$

$$Q_2 = \frac{[\lambda_{12}]_1}{[\lambda_2]_1}\left[3\frac{m_1}{m_2}+(\tfrac{5}{2}-\tfrac{6}{5}\mathrm{B}_{12}^{\star})\frac{m_2}{m_1}+\tfrac{8}{5}\mathrm{A}_{12}^{\star}\right], \tag{7.3-44b}$$

$$Q_{12} = \tfrac{16}{5}\mathrm{A}_{12}^{\star}\frac{(m_1+m_2)^2}{4m_1 m_2}\frac{[\lambda_{12}]_1^2}{[\lambda_1]_1[\lambda_2]_1}+(11-\tfrac{12}{5}\mathrm{B}_{12}^{\star}) + \frac{15}{8\mathrm{A}_{12}^{\star}}\frac{(m_1-m_2)^2}{m_1 m_2}(5-\tfrac{12}{5}\mathrm{B}_{12}^{\star}). \tag{7.3-44c}$$

In Eq. (7.3-42), x_1 and x_2 are the molecular fractions of species 1 and 2, respectively.

In the nth order Kihara approximation one has, similarly,

$$[\mathscr{D}_{12}]_n^K = [\mathscr{D}_{12}]_1 f_D^{(n)K}. \tag{7.3-45}$$

For $n = 2$,

$$f_D^{(2)K} = 1 + \Delta^K, \tag{7.3-46}$$

where Δ^K is identical to Δ except that $B_{12}^\star$, wherever it appears, is to be replaced by $\frac{5}{4}$, which happens to be its value for Maxwell molecules. The results (7.3-40) and (7.3-46) were first given by Mason [1957a].

From Eq. (7.3-38) we see that, in the first approximation, the binary diffusion coefficient is independent of the composition of the mixture. Also, only the interaction between unlike molecules enters into the expression (7.3-38), so the binary diffusion coefficient affords an excellent means of obtaining the force law beween unlike molecules.

We note that the quantities $[\eta_{ij}]_1$ and $[\lambda_{ij}]_1$, introduced in Eqs. (7.3-25) and (7.3-27), are related to the binary diffusion coefficient $\mathscr{D}_{ij}$ in the following manner:

$$[\eta_{ij}]_1 = \frac{5nm_{ij}}{3A_{ij}^\star}[\mathscr{D}_{ij}]_1, \tag{7.3-47}$$

$$[\lambda_{ij}]_1 = \frac{25nk}{8A_{ij}^\star}[\mathscr{D}_{ij}]_1. \tag{7.3-48}$$

(c3) *Coefficient of self-diffusion*

The coefficient of self-diffusion – i.e., the coefficient of diffusion for a gas through itself – is obtained from Eq. (7.3-27) by considering the case in which the two gases are identical, so that $m_{12} = \frac{1}{2}m$, and all the force laws are identical. In this limit Eq. (7.3-38b) reduces to

$$[\mathscr{D}]_1 = \frac{3}{8nm}\frac{(\pi mkT)^{\frac{1}{2}}}{\pi\sigma^2\Omega^{(1,1)}}, \tag{7.3-49}$$

while, from Eqs. (7.3-39), (7.3-40) and (7.3-42),

$$[\mathscr{D}]_2 = [\mathscr{D}]_1\left[1 - \frac{(6C^\star-5)^2}{55-12B^\star+32A^\star}\right]^{-1}. \tag{7.3-50}$$

The second Kihara approximation is given by

$$[\mathscr{D}]_2^K = [\mathscr{D}]_1\left[1 + \frac{(6C^\star-5)^2}{40+32A^\star}\right]. \tag{7.3-51}$$

We remark that the ratio

$$\rho[\mathscr{D}]_1/[\eta]_1 = \tfrac{6}{5}A^\star \tag{7.3-52}$$

is almost independent of temperature (see Chapter 10); the ratio does depend, however, on the molecular force law.

(d) *Thermal diffusion coefficients*

The thermal diffusion coefficients in a K-component mixture are given by Eq. (6.4-43a),

$$[D_{Tk}]_n = -\frac{1}{2n} a_{k,0}^{(n)}, \tag{7.3-53a}$$

or, alternatively, by Eq. (6.4-43b),

$$[D_{Tk}]_n = -\frac{5}{4n} \sum_{i=1}^{K} \frac{n_i}{n} d_{i,1}^{k(n)}. \tag{7.3-53b}$$

In the latter expression, the coefficients $d_{i,1}^{k(n)}$ are obtained from the solution of the set of equations (7.3-30). In the former, the coefficients $a_{k,0}^{(n)}$ are obtained from the solution of the set of equations (6.4-32),

$$\sum_{j=1}^{K} \sum_{q=0}^{n} \Lambda_{ij}^{pq} a_{j,q}^{(n)} = \frac{4}{5k} \frac{n_i}{n} \delta_{p1}, \quad i = 1, \ldots, K; \; p = 0, \ldots, n. \tag{7.3-54a}$$

For $p = 0$, the set of Eqs. (7.3-54a) is linearly dependent and must be supplemented by the equation

$$\sum_{i=1}^{K} (\rho_i/\rho) a_{i,0}^{(n)} = 0. \tag{7.3-54b}$$

In a K-component mixture there are $K-1$ independent thermal diffusion coefficients. They satisfy the relation

$$\sum_{k=1}^{K} (\rho_k/\rho) D_{Tk} = 0. \tag{7.3-55}$$

In the case of a binary mixture there is only one independent thermal diffusion coefficient, see Eq. (6.5-3). However, since no significant simplifications occur, we shall not discuss this particular case in greater detail. The relevant formulae can be easily derived from the results above.

(e1) *Thermal diffusion ratios – multicomponent mixtures*

The thermal diffusion ratios for a K-component mixture are defined by Eq. (6.3-36). Some care must be taken in defining their nth Chapman-Cowling approximations, because the thermal diffusion ratios relate the diffusion coefficients, which express first-order transport effects, to the thermal diffusion coefficients, which express second-order transport effects. From the results (7.3-29) and (7.3-53a) we conclude that a consistent definition of the nth approximation to the thermal diffusion ratios $[k_{Ti}]_n$ is given by

$$\sum_{j=1}^{K} d_{j,0}^{i(n+1)}[k_{Tj}]_n = -a_{i,0}^{(n)}, \quad i = 1, \ldots, K. \tag{7.3-56a}$$

Alternatively, using the results (7.3-29) and (7.3-53b), we have

$$\sum_{j=1}^{K} d_{j,0}^{i(n+1)}[k_{Tj}]_n = -\tfrac{5}{2}\sum_{j=1}^{K} \frac{n_j}{n} d_{j,1}^{i(n+1)}, \quad i = 1, \ldots, K. \tag{7.3-56b}$$

Both sets of Eqs. (7.3-56) must be supplemented by the constraint

$$\sum_{i=1}^{K} k_{Ti} = 0. \tag{7.3-57}$$

Hence, in general, the thermal diffusion ratios can be computed only after the diffusion coefficients and the thermal diffusion coefficients have been found. However, in the first approximation ($n = 1$), they can be computed directly, as we will now show.

For $n = 1$, Eq. (7.3-56b) reads

$$\sum_{j=1}^{K} d_{j,0}^{i(2)}[k_{Tj}]_1 = -\tfrac{5}{2}\sum_{j=1}^{K} (n_j/n) d_{j,1}^{i(2)}, \quad i = 1, \ldots, K. \tag{7.3-58}$$

The coefficients $d_{j,0}^{i(2)}$ and $d_{j,1}^{i(2)}$ are to be found from Eq. (7.3-30) for $n = 2$,

$$\sum_j \Lambda_{lj}^{00} d_{j,0}^{i(2)} + \sum_j \Lambda_{lj}^{01} d_{j,1}^{i(2)} = \frac{8}{25k}\left(\delta_{li} - \frac{\rho_l}{\rho}\right), \quad l = 1, \ldots, K, \tag{7.3-59a}$$

$$\sum_j \Lambda_{lj}^{10} d_{j,0}^{i(2)} + \sum_j \Lambda_{lj}^{11} d_{j,1}^{i(2)} = 0, \qquad l = 1, \ldots, K. \tag{7.3-59b}$$

Now, the matrix (Λ_{lj}^{11}) is non-singular, as may be verified from Eq. (7.3-34e, f). Hence, from Eq. (7.3-59b) we derive

$$d_{j,1}^{i(2)} = -\sum_{h,l} (\Lambda^{11})_{jh}^{-1} \Lambda_{hl}^{10} d_{l,0}^{i(2)}, \tag{7.3-60}$$

where $(\Lambda^{11})_{jh}^{-1}$ is the (j, h)-element of the inverse of the matrix Λ^{11},

$$\sum_{h=1}^{K} (\Lambda^{11})_{jh}^{-1} \Lambda_{hi}^{11} = \delta_{ij}, \quad i, j = 1, \ldots, K. \tag{7.3-61}$$

Substituting the expression (7.3-60) into Eq. (7.3-58) we obtain the following set of homogeneous equations,

$$\sum_{j=1}^{K} \left\{ [k_{Tj}]_1 - \tfrac{5}{2} \sum_{h,l=1}^{K} \frac{n_l}{n} (\Lambda^{11})^{-1}_{lh} \Lambda^{10}_{hj} \right\} d^{i(2)}_{j,0} = 0. \tag{7.3-62}$$

Since the coefficients $d^{i(2)}_{j,0}$ satisfy the constraint (7.3-30b), but are linearly independent otherwise, it follows that Eq. (7.3-62) can have a non-trivial solution if and only if the expression between the braces is some multiple of the mass ratio (ρ_j/ρ), for each j:

$$[k_{Tj}]_1 = \tfrac{5}{2} \sum_{h,l=1}^{K} \frac{n_l}{n} (\Lambda^{11})^{-1}_{lh} \Lambda^{10}_{hj} + \alpha \frac{\rho_j}{\rho}, \tag{7.3-63}$$

where α is an arbitrary constant. However, with Eq. (7.3-33) one verifies that the only value of α compatible with the constraint (7.3-57), is $\alpha = 0$; hence, instead of Eq. (7.3-63) we have the result

$$[k_{Ti}]_1 = \tfrac{5}{2} \sum_{h,j=1}^{K} \frac{n_j}{n} (\Lambda^{11})^{-1}_{jh} \Lambda^{10}_{hi} \tag{7.3-64a}$$

or, equivalently,

$$[k_{Ti}]_1 = \tfrac{5}{2} \sum_{h,j=1}^{K} \Lambda^{01}_{ih} (\Lambda^{11})^{-1}_{hj} (n_j/n), \quad i = 1, \ldots, K, \tag{7.3-64b}$$

which shows that, in the first approximation, the thermal diffusion ratios in a multicomponent mixture can be calculated directly, i.e., without first calculating the diffusion coefficients and thermal diffusion coefficients and then solving the system of equations (6.3-36).

Instead of the thermal diffusion ratios k_{Ti} one often works in terms of the *(polynary) thermal diffusion factors*, α_{ij}, which are defined by

$$k_{Ti} = \sum_{j=1}^{K} (n_i n_j / n^2) \alpha_{ij}, \quad i = 1, \ldots, K. \tag{7.3-65}$$

In a K-component mixture there are $\frac{1}{2}K(K-1)$ independent thermal diffusion factors, since

$$\alpha_{ij} + \alpha_{ji} = 0. \tag{7.3-66}$$

The thermal diffusion factors are purely theoretical quantities, except in the case of a binary mixture, see below. In the first approximation one has, from Eq. (7.3-64),

$$[\alpha_{ij}]_1 = \tfrac{5}{2}(n/n_i)\sum_{h=1}^{K} \Lambda_{ih}^{01}(\Lambda^{11})_{hj}^{-1}. \tag{7.3-67}$$

(e2) *Thermal diffusion ratio – binary mixture*

In a binary mixture there is only one independent thermal diffusion ratio, cf. Eq. (6.5-4). We shall not consider its general form in the nth approximation. However, the first approximation involves a reasonable amount of algebra and is worth being studied from both the theoretical and the practical point of view. From Eq. (7.3-64b) we have

$$[k_T]_1 = \tfrac{5}{2}\{(\Lambda_{11}^{01}(\Lambda^{11})_{11}^{-1}+\Lambda_{12}^{01}(\Lambda^{11})_{21}^{-1})x_1+(\Lambda_{11}^{01}(\Lambda^{11})_{12}^{-1}+\Lambda_{12}^{01}(\Lambda^{11})_{22}^{-1})x_2\} \tag{7.3-68a}$$

or

$$[k_T]_1 = \frac{5}{2}\frac{(\Lambda_{11}^{01}\Lambda_{22}^{11}-\Lambda_{12}^{01}\Lambda_{21}^{11})x_1-(\Lambda_{11}^{01}\Lambda_{12}^{11}-\Lambda_{12}^{01}\Lambda_{11}^{11})x_2}{\Lambda_{11}^{11}\Lambda_{22}^{11}-\Lambda_{12}^{11}\Lambda_{21}^{11}}, \tag{7.3-68b}$$

and thence, using the expressions (7.3-34e, f),

$$[k_T]_1 = x_1 x_2(6\mathrm{C}_{12}^{*}-5)\frac{S_1 x_1 - S_2 x_2}{Q_1 x_1^2+Q_2 x_2^2+Q_{12}x_1 x_2}, \tag{7.3-69}$$

where

$$S_1 = \frac{m_1+m_2}{2m_2}\frac{[\lambda_{12}]_1}{[\lambda_1]_1} - \frac{15}{4\mathrm{A}_{12}^{*}}\frac{m_2-m_1}{2m_1} - 1, \tag{7.3-70a}$$

$$S_2 = \frac{m_1+m_2}{2m_1}\frac{[\lambda_{12}]_1}{[\lambda_2]_1} - \frac{15}{4\mathrm{A}_{12}^{*}}\frac{m_1-m_2}{2m_2} - 1 \tag{7.3-70b}$$

and the Q's have been defined before, see Eq. (7.3-44); x_1 and x_2 are the molecular fractions of species 1 and 2, respectively.

The thermal diffusion ratio, α_T, of a binary mixture is defined by

$$k_T = (n_1 n_2/n^2)\alpha_T. \tag{7.3-71}$$

From Eq. (7.3-69) we conclude that, to a first approximation, α_T is given by

$$[\alpha_T]_1 = (6\mathrm{C}_{12}^{*}-5)\frac{S_1 x_1 - S_2 x_2}{Q_1 x_1^2+Q_2 x_2^2+Q_{12}x_1 x_2}. \tag{7.3-72}$$

From this result it is obvious that the use of the thermal diffusion factor may be preferable to the use of the thermal diffusion ratio: the latter goes to zero when the molecular fraction of either species in the mixture goes to zero, whereas the former remains finite in that case.

The thermal diffusion ratio is thus a very complex function of the temperature, the concentration, and the molecular masses, and depends parametrically on the force law of the molecules. The primary concentration dependence is given by $x_1 x_2$ and to a lesser extent by $S_1 x_1 - S_2 x_2$. The main dependence on the masses of the molecules is given by S_1 and S_2, the principal temperature dependence by the factor $(6\,\text{C}^\star_{12} - 5)$. The thermal diffusion ratio can be positive or negative. If k_T is positive, then component 1 tends to move into the cooler region and component 2 into the warmer region. The temperature at which k_T changes sign is called the *inversion temperature.*

(e3) *Thermal diffusion ratio – binary mixture of isotopes*

The result (7.3-69) can be considerably simplified for the special case of a binary mixture of isotopes. Then, the difference between the molecular masses of the two components is small ($m_1 \approx m_2$) so that an expansion in powers of $(m_1 - m_2)/(m_1 + m_2)$ can be made, and, moreover, all intermolecular forces are identical. Since the primary concentration dependence of the thermal diffusion ratio k_T is given by the factor $x_1 x_2$, one may take $x_1 = x_2 = \frac{1}{2}$ in the last factor of the right member of Eq. (7.3-69). Thus, the following approximate formula for k_T results:

$$k_T \approx \frac{15(6\text{C}^\star - 5)(2\text{A}^\star + 5)}{2\text{A}^\star(55 + 16\text{A}^\star - 12\text{B}^\star)} \frac{m_1 - m_2}{m_1 + m_2} x_1 x_2 . \tag{7.3-73}$$

For rigid spherical molecules,

$$[k_T]_{\text{r.s.}} \approx \frac{105}{118} \frac{m_1 - m_2}{m_1 + m_2} x_1 x_2 , \tag{7.3-74}$$

a formula which was first found by Furry, Jones and Onsager [1939]. If k_T is reduced with respect to its rigid sphere value one has

$$k_T^\star = k_T/[k_T]_{\text{r.s.}} \approx \frac{59(6\text{C}^\star - 5)(2\text{A}^\star + 5)}{7\text{A}^\star(55 + 16\text{A}^\star - 12\text{B}^\star)} . \tag{7.3-75}$$

Thus, $k_T^\star$ (which is usually called R_T in the literature) is a universal function of reduced temperature.

(f) *Viscosity of a multicomponent gas mixture*

The coefficient of viscosity of a K-component gas mixture is given by

Eq. (6.4-47),

$$[\eta]_n = \tfrac{1}{2}kT\sum_{i=1}^{K}(n_i/n)b_{i,0}^{(n)}, \tag{7.3-76}$$

where the coefficients $b_{i,0}^{(n)}$ are obtained from the solution of the following set of linear algebraic equations for the unknowns $b_{i,0}^{(n)}, \ldots, b_{i,n-1}^{(n)}$:

$$\sum_{j=1}^{K}\sum_{q=0}^{n-1} H_{ij}^{pq} b_{j,q}^{(n)} = \frac{2}{kT}(n_i/n)\delta_{p0}, \quad i = 1, \ldots, K;\ p = 0, \ldots, n-1. \tag{7.3-77}$$

The coefficients H_{ij}^{pq} (Gr.eta) are defined in terms of partial bracket integrals,

$$H_{ij}^{pq} = \frac{2}{5kT}\left\{\delta_{ij}\sum_{h=1}^{K}\frac{n_i n_h}{n^2}[S_{\frac{5}{2}}^{(p)}(\mathscr{C}^2)(\mathscr{C}\mathscr{C}-\tfrac{1}{3}\mathscr{C}^2\mathbf{I}), S_{\frac{5}{2}}^{(q)}(\mathscr{C}^2)(\mathscr{C}\mathscr{C}-\tfrac{1}{3}\mathscr{C}^2\mathbf{I})]'_{ih}\right.$$

$$\left.+\frac{n_i n_j}{n^2}[S_{\frac{5}{2}}^{(p)}(\mathscr{C}^2)(\mathscr{C}\mathscr{C}-\tfrac{1}{3}\mathscr{C}^2\mathbf{I}), S_{\frac{5}{2}}^{(q)}(\mathscr{C}^2)(\mathscr{C}\mathscr{C}-\tfrac{1}{3}\mathscr{C}^2\mathbf{I})]''_{ih}\right\} \tag{7.3-78}$$

and can be written as linear combinations of Ω-integrals with the aid of the formulae given in Tables 7.5 and 7.6. The H_{ij}^{pq} satisfy the symmetry relation

$$H_{ij}^{pq} = H_{ji}^{qp}. \tag{7.3-79}$$

The coefficients H_{ij}^{pq} for $p = 0$ and $q = 0$, which are required in the computation of the first approximation to the viscosity coefficient can be written in terms of the quantitities $[\eta_i]_1$and $[\eta_{ij}]_1$ introduced in Eqs. (7.3-24) and (7.3-25). One may verify the following expressions

$$H_{ii}^{00} = \frac{x_i^2}{[\eta_i]_1} + \sum_{\substack{l=1\\(l\neq i)}}^{K}\frac{2x_i x_l}{[\eta_{il}]_1}\frac{m_i m_l}{(m_i+m_l)^2}\left(\frac{5}{3A_{il}^*}+\frac{m_l}{m_i}\right), \tag{7.3-80a}$$

$$H_{ij}^{00} = -\frac{2x_i x_j}{[\eta_{ij}]_1}\frac{m_i m_j}{(m_i+m_j)^2}\left(\frac{5}{3A_{ij}^*}-1\right) \quad (i \neq j), \tag{7.3-80b}$$

where x_i is the molecular fraction of species i, see Eq. (7.3-35).

In the first approximation ($n = 1$) the coefficients $b_{i,0}^{(1)}$ needed in Eq. (7.3-76), are calculated from the set of equations

$$\sum_{j=1}^{K} H_{ij}^{00} b_{j,0}^{(1)} = \frac{2}{kT}(n_i/n), \quad i = 1, \ldots, K. \tag{7.3-81}$$

Solving by Cramer's rule and substituting $b_{i,0}^{(1)}$ into the expression (7.3-76) for $n = 1$, we find that the coefficient of viscosity of a K-component mixture may be written as the ratio of two determinants, the one in the numerator being of the order $K+1$ and the one in the demoninator of the order K,

$$[\eta]_1 = - \begin{vmatrix} H_{11}^{00} \dots H_{1K}^{00}\, x_1 \\ \dots\dots\dots \\ H_{K1}^{00} \dots H_{KK}^{00}\, x_K \\ x_1 \dots x_K \quad 0 \end{vmatrix} \Bigg/ \begin{vmatrix} H_{11}^{00} \dots H_{1K}^{00} \\ \dots\dots\dots \\ \\ H_{K1}^{00} \dots H_{KK}^{00} \end{vmatrix} . \tag{7.3-82}$$

For $K = 1$, the expression (7.3-82) reduces to the simple gas result, Eq. (7.3-5).

(g) *Thermal conductivity of a multicomponent gas mixture*

The computation of the thermal conductivity of a gas mixture requires somewhat more care than the computation of other properties. As was mentioned in Section 6.3, the expression (6.3-46) for the heat flow vector contains not only the ordinary heat conduction term, $-\lambda'\nabla T$, but also terms involving the diffusion driving forces $\boldsymbol{d}_i$. As the latter contain terms proportional to the temperature gradient and the thermal conductivity is normally measured under conditions in which the diffusion velocities $\boldsymbol{V}_i$ are all zero, the thermal conductivity, as normally defined, is not λ', but λ, defined by Eq. (6.3-49). Thus,

$$\lambda = \lambda' - nk \sum_{i=1}^{K} k_{Ti} D_{Ti}, \tag{7.3-83}$$

where k_{Ti} and D_{Ti} are the thermal diffusion ratio and thermal diffusion coefficient of species i, respectively. The Chapman-Cowling approximations for the latter transport coefficients have already been discussed.

The partial coefficient of thermal conductivity, λ', is, in the nth approximation, given by Eq. (6.4-45),

$$[\lambda']_n = \tfrac{5}{4} k \sum_{i=1}^{K} (n_i/n) a_{i,1}^{(n)} \tag{7.3-84}$$

and the coefficients $a_{i,1}^{(n)}$ are obtained by solving the set of equations (7.3-54) above. Notice that, in the first approximation, it is necessary to retain both $a_{i,0}^{(1)}$ and $a_{i,1}^{(1)}$. In a simple gas, $a_0^{(1)}$ vanishes due to the constraint imposed on the solution of the integral equation for $\boldsymbol{A}$ and, thus, the thermal conductivity is determined by $a_1^{(1)}$ alone. In a gas mixture, however, we have only the constraint (6.4-33) and, thus, only a single relation between the coefficients

$a_{i,0}^{(1)}$; the latter are in fact non-zero and are related to the thermal diffusion coefficients.

In the first approximation ($n = 1$), λ can be evaluated explicitly. Using the expression (7.3-53a) for $[D_{Ti}]_1$ and the expression (7.3-64b) for $[k_{Ti}]_1$ we obtain from Eq. (7.3-83),

$$[\lambda]_1 = \tfrac{5}{4}k \sum_i \frac{n_i}{n} a_{i,1}^{(1)} + \tfrac{5}{4}k \sum_{h,i,j} \Lambda_{ih}^{01} (\Lambda^{11})_{hj}^{-1} \frac{n_j}{n} a_{i,0}^{(1)}. \tag{7.3-85}$$

This expression can be simplified if, under the second summation sign, Λ_{ih}^{01} is replaced by Λ_{hi}^{10} and Eq. (7.3-54a) for $n = 1, p = 0$, is used to eliminate the sum $\sum_i \Lambda_{hi}^{10} a_{i,0}^{(1)}$. The result is simply

$$[\lambda]_1 = \sum_{h,j=1}^{K} \frac{n_h}{n} (\Lambda^{11})_{hj}^{-1} \frac{n_j}{n}, \tag{7.3-86}$$

which may also be written as the ratio of two determinants, the one in the numerator being of the order $K+1$ and the one in the denominator of the order K,

$$\lambda = - \begin{vmatrix} \Lambda_{11}^{11} & \dots & \Lambda_{1K}^{11} & x_1 \\ \dots & \dots & \dots & \dots \\ \Lambda_{K1}^{11} & \cdots & \Lambda_{KK}^{11} & x_K \\ x_1 & \cdots & x_K & 0 \end{vmatrix} \Bigg/ \begin{vmatrix} \Lambda_{11}^{11} & \dots & \Lambda_{1K}^{11} \\ \dots & \dots & \dots \\ \Lambda_{K1}^{11} & \cdots & \Lambda_{KK}^{11} \end{vmatrix}. \tag{7.3-87}$$

For $K = 1$, the expression (7.3-87) reduces to the simple gas result, Eq. (7.3-15). The result (7.3-87) was first found by Muckenfuss and Curtiss [1958b].

Intermolecular forces and atomic collisions

8

Although the computation of transport properties has now been reduced to the computation of Ω-integrals, the latter cannot be explicitly evaluated until the scattering angle, χ, is specified as a function of the impact parameter, b, and the magnitude of the relative velocity, g. Equivalently, one may ask for the cross section for turning the relative velocity through an angle between χ and $\chi+\mathrm{d}\chi$, as a function of relative speed (see Chapter 3). In classical mechanics these two means of expressing the physics of the scattering process merely differ in language; in quantum mechanics the concept of a cross section must be used. We now turn to a discussion of molecular scattering.

If the range of the force between the particles is finite, a two particle collision can be envisioned as a process in which two particles first approach each other with their pre-collision velocities, then interact for a time t_{coll} (exchanging momentum and energy) and finally separate with their post-collision velocities. The time t_{coll} is of the order of the range of the intermolecular force, r_0, divided by the mean thermal speed of the molecules, which is of the order $(kT/m)^{\frac{1}{2}}$. In such a collision, if the particles may be envisioned as interacting point masses, the equations of motion yield the post-collision velocities once the interaction potential is given; the details of this calculation are given in the next chapter, Section 9.1.

Of course, the particles of interest in the present context, being atoms or molecules, possess internal structure and this internal structure (particularly, the electron distribution in the atoms or molecules) is severely distorted when the particles are in close proximity. In fact, it is even possible for the particles to be left in internal energy states which are different from those which they had before the collision. In that case we speak of inelastic collisions. Also, chemical change (ionization or chemical reaction) may occur during a collision. However, for the present we shall ignore the possibility of inelastic and/or reactive collisions; they will be discussed more fully in Chapter 11. On the other hand it must be understood that the atoms or molecules may be distorted *during* a collision. At low temperatures, the relative velocity of two colliding particles – which is of the order $(kT/m)^{\frac{1}{2}}$, where m is the mass of the particles – is very much smaller than the speed of an orbital electron – which is of the order $(E_i/m_e)^{\frac{1}{2}}$, where E_i is the energy needed to ionize an atom and m_e the electron mass. The ratio is typically of the order 10^{-2}. This means that, at every instant during a collision, the electron distributions in the colliding particles are essentially what they would be if the two particles were held statically at the instantaneous separation. Then the instantaneous force between the particles can be computed from a static calculation and the collision may be treated as one between two point masses interacting according to the force law obtained from the static calculation, see Mott and Massey [1965]. Thus, by means of this semi-classical argument one greatly reduces the difficulty of the calculation.

There still remains the question of whether the dynamics of the collision itself must be computed by means of the classical or quantum equations of motion. As is shown, for example, in the monograph of Landau and Lifshitz [1958], the two calculations give nearly the same result for the collision cross section except at very small scattering angles (at small angles the classical cross section becomes infinite for any infinite-range potential, whereas the quantum mechanical cross section remains finite if the potential goes to zero fast enough). However, the integrand of the integral defining $\Omega^{(l,r)}$ is proportional to $1-\cos^l\chi$, which behaves like χ^2 for small χ. Thus, small angle collisions are unimportant as far as transport property calculations are concerned and the classical results may be used to very good approximation (see also the discussion of Section 9.3).

The plan of this chapter is as follows. In Sections 8.1 and 8.2 we discuss the nature of the forces between monatomic and polyatomic molecules, respectively, in terms of intermolecular potentials. In Section 8.3 the case of charged particles is briefly discussed. Next, in Section 8.4 we introduce several *model potentials* that are commonly used in transport property calculations. These model potentials are more or less consistent with general features of intermolecular potentials and contain a number of adjustable parameters that can be used for fitting the experimental data. Combination rules which allow one to obtain transport properties of a gas mixture from known properties of the individual gases of which the mixture is composed, are discussed in the final section, Section 8.5.

8.1. Forces between monatomic molecules

The simplest molecules are those which contain only a single atom per molecule. Only the noble gases – helium, neon, argon, krypton and xenon – exist in monatomic form ordinarily, but dissociated gases, which can be studied in the laboratory, are also of this type. Monatomic gases are clearly the easiest to study theoretically and many of the results obtained can be applied, at least qualitatively, to other gases. Thus, we shall consider forces between monatomic molecules first.

Because there exists no preferred direction in space, a monatomic gas atom in its free state, i.e., with no external disturbance acting on it, will have its electrons distributed isotropically about the nucleus. However, when an electric field is applied to the atom, the forces on the nucleus and the electrons are in opposite directions; the nucleus is displaced in the direction of the field while the electrons are displaced in the direction opposed to it. Consequently, there is an effective charge separation along the field direction and the atom possesses an *induced electric dipole moment*. We say that the atom has been polarized; for weak electric fields, this *polarization* is proportional to the electric field and one can write

$$\boldsymbol{\mu}^{(\mathrm{ind})} = \alpha \boldsymbol{E}, \tag{8.1-1}$$

where $\boldsymbol{\mu}^{(\mathrm{ind})}$ is the induced dipole moment (polarization), $\boldsymbol{E}$ is the applied electric field and α a proportionality constant called the *polarizability*. In general, α is a tensor. However, for simplicity we shall assume that α is isotropic and, hence, may be considered a scalar.

In turn, the dipole moment of the polarized atom generates an electric field $\boldsymbol{E}^{(\mathrm{ind})}$ of its own. Using polar coordinates with the nucleus at the origin and the polar axis in the direction of the induced dipole we have, as a first approximation,

$$\boldsymbol{E}^{(\mathrm{ind})} = \frac{\mu^{(\mathrm{ind})}}{r^3}(2\boldsymbol{e}_r \cos\theta + \boldsymbol{e}_\theta \sin\theta), \tag{8.1-2}$$

where $\boldsymbol{e}_r$ and $\boldsymbol{e}_\theta$ are the unit vectors in the directions of increasing r and θ, respectively.

Finally, the potential energy φ of interaction of any dipole $\boldsymbol{\mu}$ (induced or permanent) with an electric field $\boldsymbol{E}$ is given by

$$\varphi = -(\boldsymbol{\mu} \cdot \boldsymbol{E}). \tag{8.1-3}$$

These simple facts suffice for an understanding of the forces between atoms at large distances.

Consider two atoms separated by several atomic diameters. If they are both in their normal states there will be no fields and hence no force between them. Suppose, for the sake of argument, that one of the atoms suddenly develops a small dipole moment perpendicular to the line between the centers. This will create an electric field at the second atom which will induce a polarization in the latter opposite in direction to that of the first atom. In turn this polarization causes a field that further polarizes the first atom, and so on. The net result is that the two atoms become polarized in opposite directions and exert an attractive force on each other; the extent of polarization and, hence, the strength of the attraction is governed by the interplay of the mutual induction described above and the attraction between the electrons and the nucleus in each atom. An exact calculation of the strength of this interaction is beyond the scope of this book, but its most important features are readily uncovered. From Eq. (8.1-2) we see that the electric field at each atom is proportional to r^{-3} and from Eq. (8.1-1) that the polarization is proportional to this field. Then Eq. (8.1-3) immediately shows that the potential energy of interaction behaves like

$$\varphi(r) \sim -cr^{-6} \qquad (r \to \infty), \tag{8.1-4}$$

where the constant c is necessarily proportional to the product of the polarizabilities of the two atoms, and the minus sign indicates that the potential is attractive. The force described here is called the *dispersion force*, see London [1930, 1937], and for neutral atoms it is the dominant force at large separation; it is also known as the *Van der Waals force*, cf. Pauling and Wilson [1935]. Of course, this picture of the interaction is oversimplified and cannot be used to compute the proportionality constant in Eq. (8.1-4). A correct calculation must be made on a quantum mechanical basis and may be found, for example, in the monograph of Hirschfelder, Curtiss and Bird [1954]. Such a calculation shows that Eq. (8.1-4) gives the first term in an asymptotic expansion of φ for large separations; the succeeding terms may be ascribed to the creation of quadrupole and higher-order electric moments.

At shorter distances between the atoms the description given above is no longer valid, as the distortion of the electron distribution in the atoms cannot be effectively described in terms of a superposition of electric multipoles. The computation of the interaction energy at short separations becomes a very difficult quantum mechanical many body problem. For simple atoms like hydrogen and helium an approximate calculation is possible. The results of such calculations are quantitatively incorrect, but do yield important qualitative information on the nature of the interaction at short separation. In particular, it is found that the potential becomes strongly repulsive; in fact, it must become infinite at zero separation. At separations from one tenth of an atomic diameter to one atomic diameter the potential is found to be attractive and approximately exponentially decreasing. It is reasonable to expect similar qualitative behavior for other monatomic molecules and many of the features described here hold for other types of molecules as well. A qualitative picture of an intermolecular potential is displayed in Fig. 8.1.

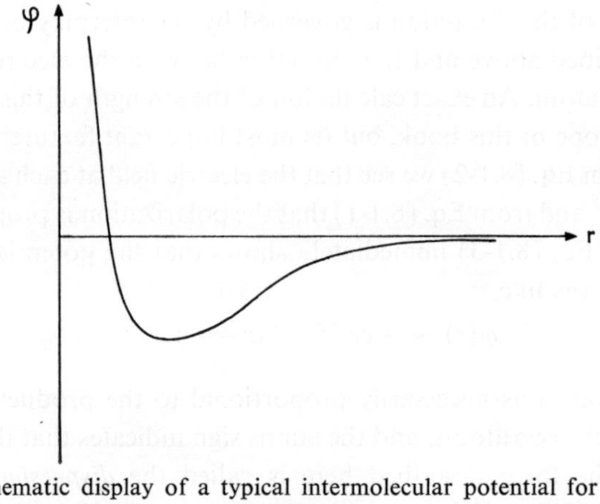

Fig. 8.1. Schematic display of a typical intermolecular potential for neutral non-polar molecules.

8.2. Forces between polyatomic molecules

Although the kinetic theory which has been presented so far is applicable only to gases without internal degrees of freedom, i.e., monatomic gases, many results for polyatomic gases can be obtained by the application of heuristic arguments or approximations to the results already obtained – see Chapter 11, Sections 11.1 and 11.2. For this reason and for purposes of unifying the discussion of intermolecular forces, interactions involving polyatomic molecules will be discussed here.

The simplest non-monatomic gas molecules are the diatomic elemental gases such as O_2, N_2 and Cl_2. Being symmetric, they have symmetric electronic distributions and thus have no permanent electric multipole moments (as opposed to induced multipole moments). This means that at long distances their interactions with each other and with monatomic gases should be of the induced dipole–induced dipole type already discussed and, thus, the potential should again vary as r^{-6} at large separations. Information about the potential at short range is difficult to obtain, but strong repulsive forces must be expected and it is reasonable to assume that the repulsive potential is of roughly the same shape as for monatomic molecules. Thus the entire potential can be expected to be similar to that for monatomic molecules. On the other hand, during collisions involving diatomic molecules, rotation of the molecules and, possibly, their internal vibrations, should be taken into consideration. A discussion of these points must wait until a discussion of the kinetic theory of polyatomic gases is presented in Chapter 11.

Asymmetric molecules such as HCl and H_2O possess *permanent* dipole moments. The electron density tends to be highest near the most electronegative atoms (Cl in HCl, O in H_2O, for example) and this asymmetry produces a dipole moment. The interaction between two such molecules at large distance is again of the dipole–dipole type but, because the dipoles are permanent, the dipole moment of each atom does not depend on the field caused by the second particle. Consequently, the interaction potential behaves as r^{-3} at large separations, instead of r^{-6}. Furthermore, the interaction is dependent on the orientation of the two dipoles. In fact, it is easily shown from Coulomb's law that a first approximation to the potential energy of interaction is given by

$$\varphi(r) \sim -\frac{\mu_1 \mu_2}{r^3}[2\cos\theta_1\cos\theta_2 - \sin\theta_1\sin\theta_2\cos(\phi_1-\phi_2)], \qquad (8.2\text{-}1)$$

where θ_1 and θ_2 are the angles that the dipole axes make with the line of centers, and ϕ_1 and ϕ_2 are azimuthal angles measured about the line of centers; μ_1 and μ_2 are the two dipole moments. Of course, the molecules will also induce further dipole moments in each other, so that there will be an r^{-6} contribution to the potential in addition to the permanent dipole contribution given by Eq. (8.2-1). Finally, the short-range potential may be expected to be similar to that for the types of molecules already discussed.

Interactions of polar molecules with those not possessing a permanent dipole moment are dependent on the induction of a dipole moment in the non-polar molecule. The induced dipole moment will be parallel to the electric field produced by the permanent dipole. Thus the potential will vary as r^{-6} and depend only on the orientation of the permanent dipole. Carrying out the calculation one finds

$$\varphi(r) \sim -\frac{\mu_1^2 \alpha_2}{2r^6}(3\cos^2\theta + 1), \tag{8.2-2}$$

where μ_1 is the dipole moment of the polar molecule, α_2 the polarizability of the non-polar molecule and θ the angle the dipole makes with the line of centers. Again, there is another r^{-6} component due to mutual induction, but this is normally small, and there is, of course, a strong short-range repulsive potential.

The above formulae are based on idealized dipoles, i.e., the charge separation distance in the dipole was assumed small compared to the distance between the particles. In real atoms, the charge separation distance is of the order of the separation between the nuclei in the molecule. This distance can be estimated by the atomic diameter so that the long-range potentials given above can be expected to be valid only when the separation is at least several atomic diameters. Thus, there is an intermediate range of separations for which little information is available.

In addition to dipole moments, molecules may possess quadrupole moments. Normally, these are not very important and, therefore, we merely mention that intermolecular potentials due to quadrupole interactions are highly orientation dependent and fall off with distance somewhat faster than the dipole interaction terms. As example, (permanent) dipole–quadrupole interactions behave like r^{-4}, quadrupole–quadrupole interactions like r^{-5}, and induced quadrupole–induced quadrupole interactions fall of as r^{-8}.

8.3. Forces in systems containing charged particles

In ionized gases, even those which are only slightly ionized, the charged particles may affect the transport properties strongly. The interactions of charged particles with each other or with neutral atoms is not difficult to discuss; the principles are the same as before.

Between pairs of electrons or protons there exists only the Coulomb force, and the mutual potential energy is

$$\varphi(r) = \frac{e_1 e_2}{r} \tag{8.3-1}$$

where e_1 and e_2 are the charges of the two particles. Although this potential is very simple, it falls off much more slowly at large distances than any of those so far discussed. This has important consequences in the computation of transport properties of ionized gases which will be discussed in Chapter 14.

When electrons collide with ions, or ions with each other, there will be short-range repulsive forces in addition to the Coulomb contribution (8.3-1); these are similar in nature to those discussed earlier.

In collisions between electrons and non-polar neutral molecules the long-range part of the interaction is due to polarization of the atom by the field of the electron. In this case the polarization is along the line of centers; the polarization is now proportional to r^{-2} and the interaction potential is, by the previous argument,

$$\varphi(r) \sim -\frac{e^2\alpha}{r^4}. \tag{8.3-2}$$

Finally, the interaction between an electron and a molecule possessing a permanent dipole moment is easily computed and is

$$\varphi(r) \sim -\frac{e\mu}{r^2}\cos\theta, \tag{8.3-3}$$

where θ is the angle the dipole makes with the line of centers.

8.4. Model potentials

It is clear that, except for the simplest molecules, an accurate calculation of the intermolecular potential is beyond present computational capability. In fact, not much of even a qualitative nature is known about the intermolecular potential at intermediate separations (one atomic diameter, say). As a result, the relations between transport properties (and thermodynamic properties) and intermolecular potentials are more frequently used to obtain information on the potentials than *vice versa*. The usual procedure is to construct *model potentials* which are consistent with general features of intermolecular potentials and contain a small number of adjustable parameters which may be used for fitting data. Of course, with a sufficient number of such parameters one cannot but help fitting experimental data. The procedure becomes reasonable only when the number of parameters is restricted, while, at the same time, it is required that the same set of parameters predicts all of the various properties depending on the intermolecular potential (other than the transport properties, examples are virial coefficients, Joule-Thomson coefficients, compressibilities, and optical and microwave line broadening, see Hirschfelder, Curtiss and Bird [1954]).

In the present section we give a list of some of the model potentials which have been proposed and a short discussion of their advantages and disadvantages. The models in this list are those which have proved most useful and which are most frequently used. Schematic diagrams of them are given in Fig. 8.2.

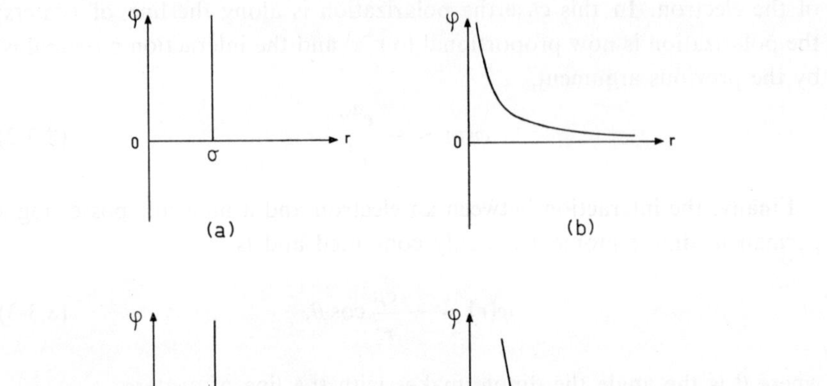

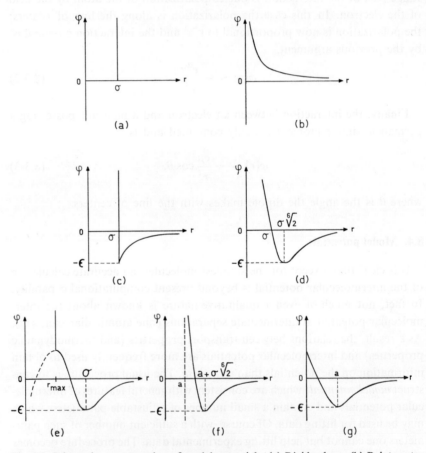

Fig. 8.2. Schematic representation of model potentials. (a) Rigid sphere. (b) Point center of repulsion. (c) Sutherland's model. (d) Lennard-Jones potential. (e) Modified Buckingham (6-exp) potential (the dashed curve represents the unmodified Buckingham potential). (f) Kihara potential. (g) Morse potential.

(a) *Rigid sphere*

In this model molecules are pictured as impenetrable billiard balls of diameter σ so that

$$\varphi(r) = \begin{cases} \infty & r < \sigma, \\ 0 & r > \sigma. \end{cases} \tag{8.4-1}$$

This is the simplest of all the models. It possesses only a single parameter, σ. It has the distinct advantages of allowing easy visualization and of allowing the acquisition of exact analytic results. As a consequence of the first feature this model was frequently used in early, heuristic, kinetic theory calculations. The latter feature (exact calculation) renders this model useful for exploratory calculation and also as a comparison standard. The Ω-integrals can be computed exactly and, in fact, in Section 7.1 they have already been used in the definitions of the reduced Ω-integrals.

The rigid sphere model is, of course, very unrealistic; it represents the repulsive part of the potential by a function which is much too steep and completely ignores the long range part of the potential. Consequently, it predicts the temperature dependence of transport properties (and other properties as well) very poorly.

(b) *Point center of repulsion*

This potential is also purely repulsive but corrects for the exaggerated steepness of the rigid sphere model. It is given by

$$\varphi(r) = (\sigma/r)^{\nu}, \tag{8.4-2}$$

where ν is called the index of repulsion. This model provides the diffuseness lacking in the rigid sphere model and, by its possession of two parameters, permits a better fit to the transport properties. The temperature dependence of transport properties is governed entirely by the index ν; for best fit to real gas properties the value of ν should be between $\nu = 9$ (*soft* molecules) and $\nu = 15$ (*hard* molecules). This model again lacks the important long range attractive portion of the real intermolecular potential and, thus, contains no "well".

An important special case of the model (8.4-2) is obtained if one takes $\nu = 4$ (*Maxwell molecules*) because, then, many calculations can be done explicitly. This is a consequence of the fact that the Sonine polynomials are exact eigenfunctions of the scattering operator for Maxwell molecules, a fact that was essentially discovered by Burnett [1935a] and first pointed out explicitly by Wang-Chang and Uhlenbeck [1952]. These properties have also made this model very popular for kinetic theory calculations involving rarefied gases.

It should be noted that in the limit $\nu \to \infty$ the model (8.4-2) becomes identical to the rigid sphere model (8.4-1).

(c) *Sutherland's model*

As we have seen, the disadvantage of the previous models is that they cannot be made to allow for both the attractive and repulsive portions of the actual potentials. The next two models to be discussed are essentially arrived at by superposing attractive and repulsive potentials of the types already discussed. Sutherland [1893, 1909] proposed a model in which the repulsive potential is that of a hard sphere and the attractive potential obeys a power law:

$$\varphi(r) = \begin{cases} \infty & r < \sigma, \\ -\epsilon(\sigma/r)^{\nu} & r > \sigma. \end{cases} \tag{8.4-3}$$

This potential, which contains three parameters, ϵ, σ and ν, is fairly realistic. It can be made to yield a good fit to the temperature dependence of the transport properties and a single set of parameters can fit several types of data reasonably well; it is also not too difficult to handle computationally. Its principal drawback is that the repulsive potential is too steep and, since the repulsive potential determines the transport properties at high temperatures, the Sutherland model does not yield good results at high temperatures. In order to obtain the correct long range behavior, one usually sets $\nu = 6$.

(d) *Lennard-Jones potential*

To correct for excessive steepness of the repulsive potential in the Sutherland model, Lennard-Jones [1924] proposed using a power law potential for the repulsive as well as the attractive portion,

$$\varphi(r) = dr^{-\delta} - cr^{-\gamma}, \tag{8.4-4}$$

where $dr^{-\delta}$ is the repulsive part and $-cr^{-\gamma}$ the attractive part. In order for the attractive part to dominate at large r and for the repulsive part to dominate at small r, one must have $\delta > \gamma$; again, γ is generally taken to be six. The most commonly used form of this potential is that with $\delta = 12$. Then Eq. (8.4-4) may be written

$$\varphi(r) = 4\epsilon[(\sigma/r)^{12} - (\sigma/r)^{6}], \tag{8.4-5}$$

where σ now has the significance of being the point at which the potential changes from repulsive to attractive, and $-\epsilon$ is the potential at the minimum which occurs at $r = 2^{\frac{1}{6}}\sigma$.

The Lennard-Jones potential is the potential which has found the widest application in the literature. Undoubtedly, this is due to its ability to fit a sizeable amount of experimental data with only two parameters. The potential has a fairly realistic shape except for its repulsive portion, which is not steep enough. To correct this, some authors have recently used larger values of δ; it was found by Dymond, Rigby and Smith [1964] that $\delta = 18$ gives a better fit to second virial coefficients of the noble gases than does $\delta = 12$, and it is expected that the same will be true for transport properties.

As we have seen, the physical potential is approximately exponential at short range and no power law can adequately substitute for an exponential. The potentials discussed below attempt to overcome this shortcoming.

(e) *Modified Buckingham (6-exp) potential*

We have noted that none of the above potentials provides the expected exponential behavior of the attractive potential at moderately short ranges. However, the simple introduction of an exponential does not suffice to overcome this problem, as the exponential does not become infinite as $r \to 0$. The modified Buckingham (6-exp) potential (sometimes called the 6-exp potential) avoids this difficulty by including a hard core center:

$$\varphi(r) = \begin{cases} \varphi_1(r) = \dfrac{\epsilon}{1-(6/\alpha)}\left[\dfrac{6}{\alpha}\exp\left\{\alpha\left(1-\dfrac{r}{\sigma}\right)\right\} - \left(\dfrac{\sigma}{r}\right)^6\right], & r > r_{\max}, \\ \infty, & r < r_{\max}. \end{cases} \tag{8.4-6}$$

The function φ_1, considered on the interval $0 \leqq r < \infty$, possesses both a maximum and a minimum which are located, respectively, at the smaller and larger roots of the equation

$$(r/\sigma)^7 \exp\{-\alpha(r/\sigma)\} = \exp(-\alpha) \tag{8.4-7}$$

(assuming that α is large enough for any root to exist); also, φ_1 becomes negatively infinite as $r \to 0$. The hard core center is added to the potential to remove this unphysical behavior. Normally, $r_{\max}$ is taken as the value of r at which φ_1 has its maximum, i.e., at the smaller root of Eq. (8.4-7). Thus, there are three parameters in the model, viz., ϵ, α and σ, and this allows variation in the shape of the potential well along with its depth and width. In this form the potential is capable of fitting experimental data quite well, but is computationally somewhat difficult to handle. To avoid excessive difficulty, α is frequently treated as a fixed parameter (it always lies in the range 12–16) and the potential is treated as a two-parameter model.

(f) *Kihara potential*

Another means of overcoming the incorrect behavior of the repulsive part of the Lennard-Jones potential was suggested by Kihara and consists of simply using a displaced distance in place of the intermolecular separation itself. Thus,

$$\varphi(r) = 4\epsilon[(\sigma/\rho)^{12} - (\sigma/\rho)^{6}] \quad \text{with } \rho = r - a, \tag{8.4-8}$$

where a is a minimum intermolecular separation at which the potential becomes infinite. Thus, this potential possesses three parameters – the diameter a, in addition to the usual Lennard-Jones potential parameters. It has been found to give a better fit to the temperature dependence of the transport properties and the virial coefficients than the normal Lennard-Jones potential. This may be attributed to the greater effective steepness of the potential at short range and to the fact that its asymptotic expansion contains a term proportional to r^{-8} that does not occur in the Lennard-Jones 6–12 potential.

(g) *Morse potential and its modifications*

As we shall see in Chapter 10, the above potentials yield good agreement between calculated and measured values of transport coefficients. An exception is helium which, due to its low mass and stable electronic configuration, has a very small potential well and a weak long-range attractive potential, and which, coincidentally, must be treated quantum mechanically. Furthermore, the relative simplicity of the helium atom has made it a favorite object of theoretical study and its potential is therefore of greater than normal interest. Because of the properties mentioned above, it is important to describe the well of the helium potential accurately and, for this reason, the Morse potential, which is based on theoretical calculations of the short-range part of the potential, is frequently used. This potential is given by

$$\varphi(r) = \epsilon[e^{-2\alpha\rho} - 2e^{-\alpha\rho}] \quad \text{with } \rho = r - a. \tag{8.4-9}$$

Although the Morse potential has been used with moderate success, it has obvious shortcomings. The principal of these is that it does not give the long-range r^{-6} dependence. Also, the potential does not become infinite as $r \to 0$ and, therefore, it cannot be used if $kT > \varphi(0) = \epsilon[\exp(2\alpha a) - 2\exp(\alpha a)]$.

To improve on the poor behavior at infinite separation, Bruch and McGee [1967] have suggested modifying the Morse potential by joining it to a long-range potential which is given by the first two non-zero terms in the multipole expansion (i.e., the dipole–dipole and dipole–quadrupole interactions),

$$\varphi(r) = \begin{cases} \epsilon[e^{-2\alpha\rho} - 2e^{-\alpha\rho}], & r \leqq r_m, \\ -c_6 r^{-6} - c_8 r^{-8}, & r \geqq r_m, \end{cases} \tag{8.4-10}$$

where, of course, the constants are chosen to produce a smooth joint at $r = r_m$, i.e., the functions and their first derivatives are matched. Thus, there are four adjustable parameters: six parameters (ϵ, α, a, c_6, c_8, r_m) are linked by two matching constraints.

(h) *Model potentials for polar and charged particles*

To construct potentials for polar molecules or charged particles one usually adopts one of the potentials already discussed in order to account for the short-range repulsion and the long-range induced dipole–induced dipole interaction, and adds to it an additional potential to account for the interaction potential energy due to the permanent dipole or charges. These additional terms have been discussed in Section 2 of this chapter. One of the most commonly used potentials of this type is the *Stockmayer potential* which is simply a superposition of the Lennard-Jones potential (8.4-5) and the permanent dipole–permanent dipole interaction (8.2-1),

$$\varphi(r) = 4\epsilon\left[\left(\frac{\sigma}{r}\right)^{12} - \left(\frac{\sigma}{r}\right)^{6}\right] - \frac{\mu_1\mu_2}{r^3}[2\cos\theta_1\cos\theta_2 - \sin\theta_1\sin\theta_2\cos(\phi_1 - \phi_2)]. \tag{8.4-11}$$

This potential is frequently referred to in the literature as the 12–6–3 potential. Many other similar superpositions have been employed but, as they involve no new principles, we shall postpone discussion of them until the discussion of transport properties of polar gases in Chapter 11.

Since all polar gases are necessarily polyatomic, they also have internal degrees of freedom (vibrational, rotational etc.). The energy of internal motion is, of course, transported with the molecule and leads to an important contribution to the thermal conductivity. While a fuller discussion of transport phenomena in polyatomic gases is reserved for Chapter 11, we should mention here that models have also found frequent use in treating the transport theory of polyatomic gases. For example, molecules have been visualized as rotating, perfectly rough, spheres, as spherocylinders (cylinders capped by hemispheres at either end), as ellipsoids, and so on. Again, these are discussed in Chapter 11.

8.5. Combination rules

All of the above discussion applies to collisions between pairs of unlike molecules as well as to collisions between pairs of like molecules. However, it must be recalled that in practice the parameters in a given model are obtained by fitting the calculated transport properties to experimental data.

In this manner, one obtains information about the potential rather than the transport properties. From the transport properties of simple gases one obtains potentials describing interactions between like pairs. Information about potentials between unlike pairs must be obtained from mixture transport properties; diffusion and thermal diffusion coefficients are especially suited to this task.

Now suppose that one wishes to calculate the transport properties of a gas mixture for which not all the potentials are known. The question arises as to whether these properties can be obtained from known properties of the individual gases of which the mixture is composed. Procedures of this sort are known as *combination* or *mixing rules* and, while they cannot be expected to be exact, they are frequently useful in obtaining reasonably accurate estimates of the desired properties. Mixing rules are best applied to the potential (rather than the transport properties), as physical arguments are more readily available in this case.

Mixing rules naturally take the form of relations for the parameters for unlike-molecule potentials in terms of the parameters for like-molecule potentials. This means that the rules will depend on the model chosen and are sensible only if the same model is used for all potentials involved. We shall restrict our attention to (i) the rigid sphere model, because of its simplicity, and (ii) the Lennard-Jones and modified Buckingham (6-exp) models, because they are the most frequently used models. (In the latter model, the parameter α will be kept fixed for the present discussion.) A discussion of a simple set of mixing rules for the potential parameters in mixtures containing polar gases may be found in a paper by Mason and Monchick [1962b].

For rigid spheres it is easily seen that, at contact, the distance σ_{ij} between the centers of two molecules of diameters σ_i and σ_j is simply

$$\sigma_{ij} = \tfrac{1}{2}(\sigma_i + \sigma_j) \tag{8.5-1}$$

and this is the logical combination rule for this model.

Both the Lennard-Jones and modified Buckingham (6-exp) models behave asymptotically as

$$\varphi_{ij}(r) \sim c_{ij} r^{-6} \quad (r \to \infty) \tag{8.5-2}$$

and, of course, this relation holds equally well for like – or unlike – molecule pairs. Now, from general theoretical arguments (cf. below Eq. (8.1-4)) it was found that the proportionality constant c_{ij} is proportional to the product of the polarizabilities of the two atoms, i.e., $c_{ij} \propto \alpha_i \alpha_j$. Thus, it is expected that

$$c_{ij}^2 = c_{ii} c_{jj} \tag{8.5-3}$$

and, from this relation, we can obtain rules for the two models under consideration. For either of the two models Eq. (8.5-3) becomes

$$\epsilon_{ij} \sigma_{ij}^6 = (\epsilon_i \sigma_i^6 \epsilon_j \sigma_j^6)^{\frac{1}{2}}. \tag{8.5-4}$$

This mixing rule is the best founded and, not surprisingly, has been found to give the best agreement with experiment, see Van Heijningen [1967].

Of course, a single rule is not sufficient to determine both parameters ϵ_{ij} and σ_{ij}. Generally, one adopts as a second rule an analog of the mixing rule for rigid spheres,

$$\sigma_{ij} = \tfrac{1}{2}(\sigma_i + \sigma_j). \tag{8.5-5}$$

This rule has less theoretical justification and is not as accurately verified by experimental data. Finally, we note that, if $\sigma_i \approx \sigma_j$, Eq. (8.5-4) becomes

$$\epsilon_{ij} = (\epsilon_i \epsilon_j)^{\frac{1}{2}} \tag{8.5-6}$$

and this rule has occasionally been used instead of Eq. (8.5-4).

Calculation of transport properties for specific models 9

The task now remaining is essentially computational. In Chapter 7, the expressions for the transport properties were reduced to the computation of Ω-integrals, while in Chapter 8 the intermolecular potentials which occur implicitly in the Ω-integrals were discussed. In this chapter we carry out the task of deriving explicit expressions for the Ω-integrals for those simple models which permit such derivation, and indicate the computational procedure for the more realistic potentials that do not admit explicit expressions for the Ω-integrals. Only Ω-integrals for collisions between molecules of different gases will be discussed, as the Ω-integrals for a simple gas are trivially obtained from the former.

9.1. Dynamics of binary collisions

As has been mentioned in Section 7.1, the Ω-integrals represent effective cross sections for the scattering processes in a gas. As a valid first approximation two colliding molecules may be treated as classical point centers of force whose interaction is described by an intermolecular potential function which depends solely upon the distance between the centers of the two molecules. Then, it is a simple problem of dynamics to derive the scattering angle, χ, in terms of the impact parameter, b, and the magnitude of the relative velocity of the two particles, g – assuming that the intermolecular potential, φ, is given. (Note that the following discussion applies to simple molecules only, i.e., internal degrees of freedom are neglected altogether; discussion of the more general case is deferred until Chapter 11.)

Consider two molecules of species i and j, respectively. Let $\boldsymbol{F}$ be the intermolecular force derived from the potential φ. The equations of motion for the two particles are

$$m_i \frac{\mathrm{d}^2 \boldsymbol{r}_i}{\mathrm{d}t^2} = \boldsymbol{F}(|\boldsymbol{r}_i - \boldsymbol{r}_j|), \quad m_j \frac{\mathrm{d}^2 \boldsymbol{r}_j}{\mathrm{d}t^2} = -\boldsymbol{F}(|\boldsymbol{r}_i - \boldsymbol{r}_j|). \tag{9.1-1}$$

It is natural to make a transition to center-of-mass and relative coordinates,

$$\boldsymbol{R} = \frac{m_i \boldsymbol{r}_i + m_j \boldsymbol{r}_j}{m_i + m_j}, \qquad \boldsymbol{r} = \boldsymbol{r}_i - \boldsymbol{r}_j, \tag{9.1-2}$$

respectively. In these coordinates the equations of motion (9.1-1) are

$$\frac{\mathrm{d}^2 \boldsymbol{R}}{\mathrm{d}t^2} = 0, \quad m_{ij} \frac{\mathrm{d}^2 \boldsymbol{r}}{\mathrm{d}t^2} = \boldsymbol{F}(r); \tag{9.1-3}$$

m_{ij} is the reduced mass of the i–j system, $m_{ij} = m_i m_j/(m_i + m_j)$. The first of the equations (9.1-3) simply gives the uniform motion of the center of mass and is of no interest. The second equation gives the equivalent one-body problem of a mass m_{ij} moving under the influence of a force $\boldsymbol{F}$.

It is well known that the trajectory $\boldsymbol{r}(t)$ lies in a plane. Introducing polar coorinates (r, θ), recalling that $\boldsymbol{F} = -\nabla\varphi$, and taking the scalar product of the second expression of Eq. (9.1-3) with $\mathrm{d}\boldsymbol{r}/\mathrm{d}t$, we find on integrating,

$$\tfrac{1}{2} m_{ij} \frac{\mathrm{d}\boldsymbol{r}}{\mathrm{d}t} \cdot \frac{\mathrm{d}\boldsymbol{r}}{\mathrm{d}t} = -\varphi(r) + \text{constant} \tag{9.1-4a}$$

or, equivalently,

$$\tfrac{1}{2} m_{ij} \left[\left(\frac{\mathrm{d}r}{\mathrm{d}t} \right)^2 + r^2 \left(\frac{\mathrm{d}\theta}{\mathrm{d}t} \right)^2 \right] + \varphi(r) = \tfrac{1}{2} m_{ij} g^2, \tag{9.1-4b}$$

which is easily recognized as the conservation law for total energy. The constant has been evaluated by noting that the pre-collision total energy is equal to $\frac{1}{2} m_{ij} g^2$.

Another constant of the motion is the angular momentum. The conservation law for this quantity is obtained from the second expression of Eq. (9.1-3) by taking the vector product with $\boldsymbol{r}$. Since $\boldsymbol{F}$ is radial, $\boldsymbol{r} \times \boldsymbol{F} = 0$ and we have on integrating,

$$m_{ij} \left| \boldsymbol{r} \times \frac{\mathrm{d}\boldsymbol{r}}{\mathrm{d}t} \right| = \text{constant} \tag{9.1-5a}$$

or, equivalently,

$$m_{ij} r^2 \frac{d\theta}{dt} = m_{ij} g b, \tag{9.1-5b}$$

where the constant has been evaluated by noting that the pre-collision angular momentum is $m_{ij}gb$.

An interesting result is obtained by using Eq. (9.1-5b) to eliminate $d\theta/dt$ from Eq. (9.1-4b):

$$\tfrac{1}{2}m_{ij}\left(\frac{dr}{dt}\right)^2 + \left[\varphi(r) + \frac{1}{2}\frac{m_{ij}g^2b^2}{r^2}\right] = \tfrac{1}{2}m_{ij}g^2, \tag{9.1-6}$$

which is an equation for one-dimensional motion with a potential $\varphi(r)+\frac{1}{2}m_{ij}g^2b^2/r^2$. The term $\frac{1}{2}m_{ij}g^2b^2/r^2$ may be interpreted as an effective potential due to the centrifugal force and plays an important role in the computational aspects to be discussed later.

Now, we are not interested in the time dependence of the motion, only in the deflection angle. To obtain an equation involving only r and θ, we first transpose the effective potential term in Eq. (9.1-6) to the right-hand side and solve for dr/dt. Then, dividing the resulting expression by $d\theta/dt$ as found from Eq. (9.1-5b), we have

$$\frac{dr}{d\theta} = \left[\frac{r^4}{b^2} - r^2 - \frac{2r^4}{m_{ij}g^2b^2}\varphi(r)\right]^{\frac{1}{2}}. \tag{9.1-7}$$

Finally, we note that the trajectory is symmetric about the point of closest approach so that the deflection angle, χ, defined earlier is π minus twice the change in θ that occurs when r goes from infinity to the distance of closest approach, r_0. The latter is obtained by setting the right-hand side of Eq. (9.1-7) equal to zero:

$$r_0^2 - b^2 - \frac{2r_0^2}{m_{ij}g^2}\varphi(r_0) = 0, \tag{9.1-8}$$

and thus

$$\chi = \pi - 2b\int_{r_0}^{\infty} \frac{dr/r^2}{[1-(b^2/r^2)-2\varphi(r)/m_{ij}g^2]^{\frac{1}{2}}}. \tag{9.1-9}$$

This is the formula desired, and all that remains to be done in the computation of transport properties is a straight-forward application of this result to the calculation of the Ω-integrals and, thence, the transport properties.

9.2. The Ω-integrals for model potentials

(a) *Rigid spheres*

The rigid sphere model is sufficiently simple that χ may be obtained by geometric arguments without resort to Eq. (9.1-9), although the latter may, of course, be used. From Fig. 9.1, we see that

$$\cos \tfrac{1}{2}\chi = \frac{b}{\tfrac{1}{2}(\sigma_i+\sigma_j)} = \frac{b}{\sigma_{ij}}, \tag{9.2-1}$$

with $\sigma_{ij} = \tfrac{1}{2}(\sigma_i+\sigma_j)$.

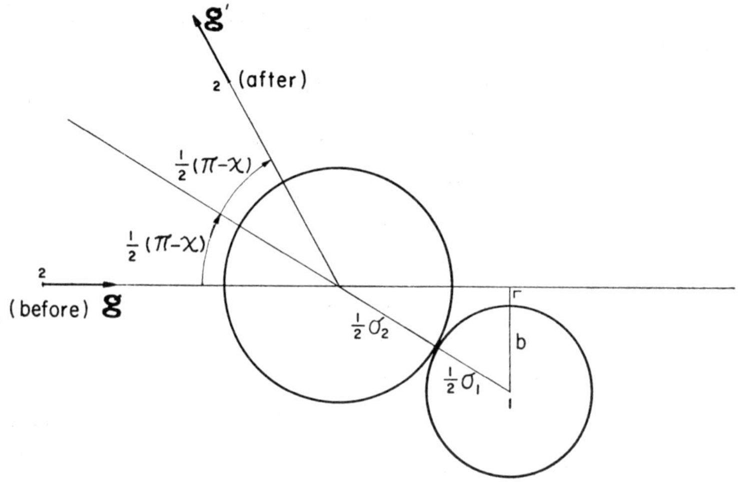

Fig. 9.1. Geometry of a collision of two rigid spheres.

Note that for the rigid sphere model the impact parameter b must be less than σ_{ij} for a collision to occur. Using Eq. (9.2-1) one immediately finds:

$$Q_{ij}^{(l)} = 2\pi \int_0^{\sigma_{ij}} (1-\cos^l \chi) b\, \mathrm{d}b = \left[1 - \frac{1}{2}\,\frac{1+(-1)^l}{l+1}\right] \pi\sigma_{ij}^2 \tag{9.2-2}$$

and, hence,

$$\begin{aligned} \Omega_{ij}^{(l,r)} &= \left(\frac{kT}{2\pi m_{ij}}\right)^{\frac{1}{2}} \int_0^\infty Q_{ij}^{(l)} g^{2r+3} \exp(-g^2) \mathrm{d}g \\ &= \left(\frac{kT}{2\pi m_{ij}}\right)^{\frac{1}{2}} \tfrac{1}{2}(r+1)! \left[1 - \frac{1}{2}\,\frac{1+(-1)^l}{l+1}\right] \pi\sigma_{ij}^2 . \end{aligned} \tag{9.2-3}$$

For a simple gas, we have, setting $m_{ij} = \frac{1}{2}m$,

$$\Omega^{(l,r)} = \left(\frac{kT}{\pi m}\right)^{\frac{1}{2}} \tfrac{1}{2}(r+1)! \left[1 - \frac{1}{2}\,\frac{1+(-1)^l}{l+1}\right] \pi\sigma^2. \tag{9.2-4}$$

Thus, all of the Ω-integrals can be expressed in a very simple analytic form and, hence the transport properties can be found to any degree of accuracy. In particular, since $\Omega^{(l,r)} \sim T^{\frac{1}{2}}$ for all l and r, we see that the rigid sphere model predicts that the viscosity and thermal conductivity vary as $T^{\frac{1}{2}}$ with temperature.

(b) *Point centers of repulsion*

For this model, for which the potential is given by Eq. (8.4-2), the deflection angle (9.1-9) becomes

$$\chi = \pi - 2b \int_{r_0}^{\infty} \frac{dr/r^2}{[1-(b^2/r^2)-2(\sigma_{ij}/r)^\nu/m_{ij}g^2]^{\frac{1}{2}}}. \tag{9.2-5}$$

Now, introducing the new variables

$$y = b/r, \quad y_0 = b/r_0, \quad z = (b/\sigma_{ij})(m_{ij}g^2/2\nu)^{1/\nu}, \tag{9.2-6}$$

we find that Eq. (9.2-5) becomes

$$\chi = \pi - 2\int_0^{y_0} [1-y^2-\nu^{-1}(y/z)^\nu]^{-\frac{1}{2}} dy. \tag{9.2-7}$$

Notice that y_0 is a function of z, defined by

$$1-y_0^2-(y_0/z)^\nu \nu^{-1} = 0, \tag{9.2-8}$$

which is Eq. (9.1-8) written in terms of the variables defined by Eqs. (9.2-6). Thus, for a given ν, χ is a function of the single varible z and not of b or g individually.

It is now a simple matter to obtain the form of the Ω-integrals. First, we note that

$$\begin{aligned} Q_{ij}^{(l)} &= 2\pi \int_0^\infty (1-\cos^l \chi) b\, db \\ &= 2\pi\sigma_{ij}^2 \left(\frac{m_{ij}g^2}{2\nu}\right)^{-2/\nu} \int_0^\infty (1-\cos^l \chi(z,\nu)) z\, dz \\ &= 2\pi\sigma_{ij}^2 \left(\frac{m_{ij}g^2}{2\nu}\right)^{-2/\nu} A_l(\nu), \end{aligned} \tag{9.2-9}$$

where the integral over z, being a number dependent only on the parameter ν, has been defined as $A_l(\nu)$. Some values of $A_l(\nu)$ are given in Table 9.1.

TABLE 9.1

Values of $A_1(\nu)$, $A_2(\nu)$ for point centers of repulsion*

ν	$A_1(\nu)$	$A_2(\nu)$
4	0.298	0.308
6	0.306	0.283
8	0.321	0.279
10	0.333	0.278
12	0.346	0.279
14	0.356	0.280
20	—	0.286
24	—	0.289
∞	0.500	0.333

* These values of $A_l(\nu)$ were taken from Chapman and Cowling [1952]. However, our notation conforms to that of Hirschfelder, Curtiss and Bird [1954] so that the $A_l(\nu)$ correspond to the $2^{-2/\nu}A_l(\nu+1)$ of Chapman and Cowling.

Then one finds for the Ω-integrals

$$\Omega_{ij}^{(l,r)} = \left(\frac{kT}{2\pi m_{ij}}\right)^{\frac{1}{2}} \left(\frac{\nu}{kT}\right)^{2/\nu} A_l(\nu)\,\Gamma\left(r+2-\frac{2}{\nu}\right)\pi\sigma_{ij}^2. \qquad (9.2\text{-}10)$$

Thus, in this model $\Omega^{(l,r)} \sim T^{(\nu-4)/2\nu}$. Since, for neutral gases, the physically interesting values of ν vary from 4 (Maxwell molecules) to infinity we see that the temperature dependence of the Ω-integrals varies from T^0 to $T^{\frac{1}{2}}$. Furthermore, since the temperature dependence of the viscosity and thermal conductivity is given by $T/\Omega^{(2,2)}$, we see that these transport properties vary as T^n with n varying from unity for $\nu = 4$ to one half for $\nu \to \infty$. The latter again corresponds to the fact that in the limit $\nu \to \infty$, the point center of repulsion model degenerates to the hard sphere model.

The physical origin of these results is eaily seen. Reducing ν corresponds to making the potential softer, i.e., to making the molecular potential more diffuse. Since at higher temperatures the molecules are in more rapid motion, they tend to approach each other more closely and the diameter of the molecules is effectively reduced. This reduction in the effective hard sphere diameter in turn reflects itself in an increased temperature dependence of the transport properties.

We note in passing that, using Eq. (9.2-10) it is possible to show that the bracket integrals for Maxwell molecules ($\nu = 4$) vanish unless $p = q$.* This means that the first approximation to the transport properties is in fact the exact result for this particular case. A direct proof of these properties is given by Chapman and Cowling [1952], but as Maxwell molecules are of little physical interest we shall not reproduce it here.

Finally, we note that the case $\nu = 1$, corresponding to Coulomb forces is also of great importance. However, this case requires special treatment as the integral defining $A_l(1)$ does not converge. It will be further discussed in Chapter 14, where ionized gases are considered.

(c) *Attractive-repulsive models*

For models other than those already treated, exact analytical results have not been obtained. Some results can be had by making approximations (for example, assuming that the repulsive part of the potential is weak), but with the availability of fast digital computers the value of such approximate results is considerably diminished. Thus, we shall concentrate on a description of some of the more important aspects of the computations of the Ω-integrals and a discussion of methods of fitting the curves so obtained to experimental data. To simplify the discussion, the Lennard-Jones (6–12) potential will be used as an example, but the methods and qualitative features apply equally well to the modified Buckingham (6-exp) and Sutherland potentials. All these potential functions may be characterized by two parameters, σ and ϵ, and may be written in the form $\varphi(r) = \epsilon f(r/\sigma_{ij})$, in which f is assumed to be the same function for all substances. In order to lend the discussion a more universal character we introduce a set of non-dimensional reduced quantities:

$r^* = r/\sigma_{ij}$,	reduced intermolecular distance,
$b^* = b/\sigma_{ij}$,	reduced impact parameter,
$\varphi^* = \varphi/\epsilon$,	reduced intermolecular potential energy,
$T^* = kT/\epsilon$,	reduced temperature,
$g^{*2} = \frac{1}{2}m_{ij}g^2/\epsilon$,	reduced relative kinetic energy.

* In fact, one can prove that $[S^{(r)}_{l+\frac{1}{2}}(\mathscr{C}^2)\mathscr{C}^{(l)}, S^{(r')}_{l'+\frac{1}{2}}(\mathscr{C}^2)\mathscr{C}^{(l')}]$, where $\mathscr{C}^{(l)}$ is the irreducible homogeneous tensor of degree l, is zero unless $l = l'$ and $r = r'$ simultaneously; the bracket integrals occurring in the transport property calculations are special cases of these for $l = 1, 2$. Since $[A, B] = \int A I(B) \mathrm{d}^3 c$, this is equivalent to the statement that $S^{(r)}_{l+\frac{1}{2}}(\mathscr{C}^2)\mathscr{C}^{(l)}$ is an eigenfunction of the I operator for Maxwell molecules. For further details, see Chapter 15 or Wang-Chang and Uhlenbeck [1952], Uhlenbeck and Ford [1962].

(Notice that the variable g^* introduced above is different from the dimensionless relative velocity variable $\mathscr{g}$ introduced in Eq. (7.1-24); in fact, the two are related through the identity $g^{*2}/T^* = \mathscr{g}^2$.)

Now, after Eq. (9.1-6) it was pointed out that the dynamics of a binary collision may be studied in terms of the *effective potential energy*, φ_{eff}, which is the sum of the intermolecular potential energy and the centrifugal potential energy,

$$\varphi_{\text{eff}}(r) = \varphi(r)+\tfrac{1}{2}m_{ij}g^2b^2/r^2. \tag{9.2-11}$$

In terms of the dimensionless reduced quantities the effective potential, φ^*_{eff}, is defined as

$$\varphi^*_{\text{eff}}(r^*) = \varphi^*(r^*)+g^{*2}b^{*2}/r^{*2}. \tag{9.2-12}$$

In particular, for the Lennard-Jones (6–12) potential we have

$$\varphi^*_{\text{eff}}(r^*) = 4r^{*-12}-4r^{*-6}+(g^*b^*)^2r^{*-2}. \tag{9.2-13}$$

Thus, the reduced effective potential is a function of r^* which depends parametrically on the quantity g^*b^*. In Fig. 9.2 we have plotted φ^*_{eff} for a few values of $(g^*b^*)^2$.

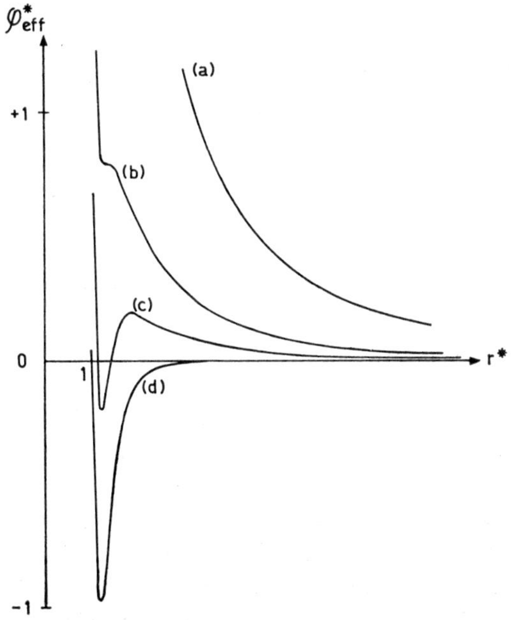

Fig. 9.2. The reduced effective potential for the Lennard-Jones (6–12) potential. (a) $(g^*b^*)^2 = 10$; (b) $(g^*b^*)^2 = 2.4624$; (c) $(g^*b^*)^2 = 1$; (d) $(g^*b^*)^2 = 0.008$. (From Hirschfelder, Curtiss and Bird [1954].)

Recall that the last term of the effective potential represents the effect of the centrifugal force. This term also falls off much more slowly at large r^* than either of the other terms in Eq. (9.2-13), and so provides the dominant contribution to the effective potential at large distance. Thus, the effective potential is positive at large r^* and, for g^*b^* not too large, it has both a maximum and a minimum. Of course, for very small g^*b^* the centrifugal potential is negligible and the effective potential is essentially identical to the Lennard-Jones potential. The position of the minimum changes only slightly as g^*b^* is increased but the position of the maximum moves rapidly toward smaller r^*. For $(g^*b^*)^2 = 2.4624$ the maximum and minimum coincide, i.e., there is an inflection point. For larger values of g^*b^* the effective potential is a smooth monotonically decreasing function of r^*.

The changing character of the effective potential with changing g^*b^* results in a change in the nature of the collision. Note, first of all, that the turning point defined by Eq. (9.1-8) is, in terms of the reduced parameters, the point at which

$$\varphi^*_{\text{eff}}(r^*_0) = g^{*2}. \tag{9.2-14}$$

In the event that there is more than one solution of this equation (as may be the case for $(g^*b^*)^2 < 2.4624$), the largest solution is to be taken. Also, in terms of reduced parameters, the deflection angle (9.1-9) becomes

$$\chi(b^*, g^*) = \pi - 2b^* \int_{r_0^*}^{\infty} \frac{dr^*/r^{*2}}{[1-\varphi^*_{\text{eff}}/g^{*2}]^{\frac{1}{2}}}. \tag{9.2-15}$$

The collisions which occur may be classified as follows:

(a) For very small values of $(g^*b^*)^2$, the condition (9.2-14) is normally met by a value of r_0^* which falls on the steeply rising portion of the curve. Since small values of g^* are of little importance in the Ω-integrals due to the factor g^{*2r+3}, these are essentially those collisions with small values of b^*, i.e., head-on collisions. Now, note that the integrand in Eq. (9.2-15) is inversely proportional to the square root of $g^{*2} - \varphi^*_{\text{eff}}$. Even though the turning point is determined by the repulsive potential, the major contribution to the deflection angle comes therefore from those values of r^* for which $g^{*2} - \varphi^*_{\text{eff}}$ is small. Thus, in this case the deflection angle is determined mainly by the attractive portion of the potential. However, this type of collision is relatively unimportant because it is weighted by the factor b in the Ω-integrals; to state it otherwise, the cross section for this type of collision is small.

(b) When $(g^*b^*)^2$ is large enough for φ^*_{eff} to have a pronounced maximum, but still below 2.4624, several types of collision are possible.

(i) For small g^{*2}, condition (9.2-14) will be satisfied by a value of r^*_0 to the right of the maximum. In such a collision the repulsive part of the potential plays essentially no role.

(ii) When g^{*2} approaches the value of φ^*_{eff} at its maximum, the relative kinetic energy $g^{*2}-\varphi^*_{\text{eff}}$ and, hence, the relative velocity is very small near the turning point. In fact, the particles may spin about their common center of gravity many times before separating so that *orbiting* occurs. The principle of time reversibility does not allow the formation of a stable bound system, but the deflection angle actually does blow up as $-\log|g^{*2}-(\varphi^*_{\text{eff}})_{\max}|$ as g^{*2} approaches $(\varphi^*_{\text{eff}})_{\max}$. Although very interesting, orbiting collisions are not very important because they occur for a limited range of values of g^* and b^* and the oscillations in $\cos\chi$ that occur tend to cancel out on integration.

(iii) For g^{*2} much larger than the maximum of φ^*_{eff}, the situation is very similar to that described under case (a). This type of collision is not very important for reasons similar to those given under case (a).

(c) When $(g^*b^*)^2$ is large, the centrifugal term in φ^*_{eff}, i.e., the last term of Eq. (9.2-13) completely dominates the attractive potential, i.e., the second term of Eq. (9.2-13), and the latter plays almost no role. The repulsive term is very important in this case, as a large part of the integral (9.2-15) comes from values of r^* near r^*_0.

The *temperature dependence* of the transport properties is intimately connected with the properties of the various types of collisions. Of particular importance is the fact that the Ω-integrals contain the factor $(g^{*2}/T^*)^{r+\frac{3}{2}}\times \exp(-g^{*2}/T^*)$, which is very sharply peaked near $g^{*2} = (r+\frac{3}{2})T^*$. Thus, at *low temperatures*, the major contribution to the Ω-integrals comes from small values of g^*. From the above discussion we see that collisions of type b-(i) are most important, and Ω-integrals and transport properties are therefore determined mainly by the *attractive* portion of the potential in this case. In fact, for the Lennard-Jones potential, the low-temperature transport properties very nearly match those obtained from a point center of attraction model with index 6. At *high temperatures*, on the other hand, collisions of types b-(iii) and c are the most important, and the *repulsive* portion of the potential dominates. Thus, the high-temperature transport properties obtained from the Lennard-Jones potential are very similar to those for a point center of repulsion model with index 12. This discussion is, of course, just an elaboration of the idea that at high temperatures the higher relative velocities which occur allow the particles to penetrate each other's potentials more deeply and thus the repulsive core potential becomes more important. A closer look at the above discussion shows, however, that this simple argument does not tell the entire story – relatively deep penetrations do occur at low temperatures and it is the low relative velocity of the particles in such collisions that causes the attractive potential to have a much greater influence on the deflection angle than the repulsive potential.

For any of the attractive-repulsive potentials the integral (9.2-15) must be evaluated numerically. Various numerical methods have been proposed – see De Boer and Van Kranendonk [1948] and Hirschfelder, Curtiss and Bird [1954] – and, in fact, any standard numerical quadrature procedure may be used. Typical results are shown in Fig. 9.3. For small values of g^*b^* the re

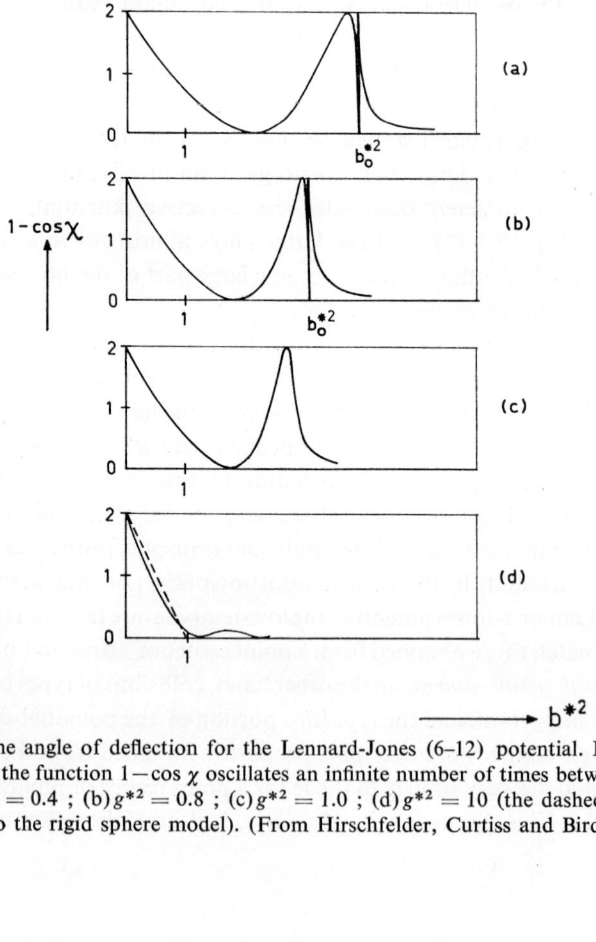

Fig. 9.3. The angle of deflection for the Lennard-Jones (6–12) potential. In the region around b_0^{*2} the function $1-\cos\chi$ oscillates an infinite number of times between zero and two. (a) $g^{*2} = 0.4$; (b) $g^{*2} = 0.8$; (c) $g^{*2} = 1.0$; (d) $g^{*2} = 10$ (the dashed line corresponds to the rigid sphere model). (From Hirschfelder, Curtiss and Bird [1954].)

sults are nearly those for an attractive inverse sixth power potential. At large values of g^{*2} the results are practically those for rigid spheres. The phenomenon of orbiting also clearly shows up in the first two graphs of Fig. 9.3. For a fixed value of g^{*2}, in the region of the critical value of b^{*2}, $1-\cos\chi$ oscillates an infinite number of times between zero and two.

A further numerical integration is necessary to produce the transport cross sections,

$$Q_{ij}^{(l)} = 2\pi\int_0^\infty (1-\cos^l\chi)b\,\mathrm{d}b = 2\pi\sigma_{ij}^2\int_0^\infty (1-\cos^l\chi)b^*\mathrm{d}b^*. \qquad (9.2\text{-}16)$$

They may be normalized with respect to the rigid sphere value. In fact, let $[Q_{ij}^{(l)}]_{\text{r.s.}}$ be the value of the rigid sphere cross section, Eq. (9.2-2), with σ_{ij} taken to be that of the Lennard-Jones potential. Then we define the redu-

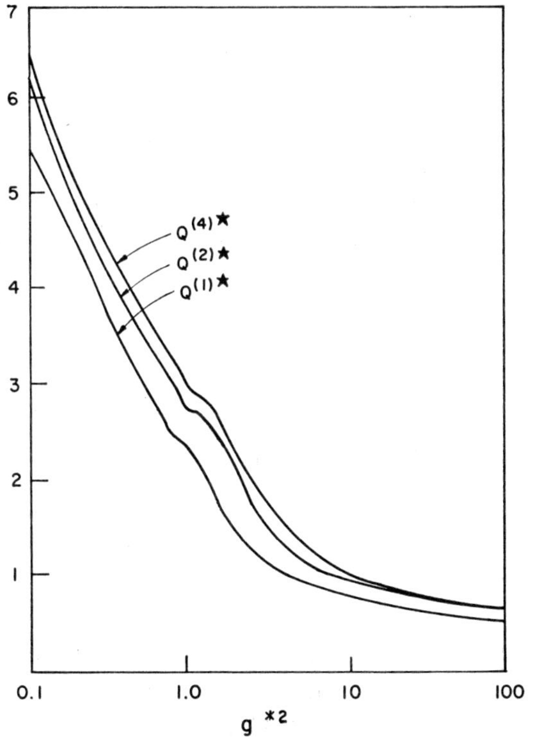

Fig. 9.4. The transport cross sections $Q^{(1)*}$, $Q^{(2)*}$ and $Q^{(4)*}$ for the Lennard-Jones (6–12) potential. The rigid sphere model corresponds to $Q^* = 1$. (From Hirschfelder, Curtiss and Bird [1954].)

ced transport cross section $Q^{(l)\star}$,

$$Q^{(l)\star} = Q_{ij}^{(l)}/[Q_{ij}^{(l)}]_{\text{r.s}}\,, \tag{9.2-17}$$

with $Q_{ij}^{(l)}$ taken from Eq. (9.2-16). It is easily verified that $Q^{(l)\star}$ is a universal function of the reduced kinetic energy of relative motion, g^{*2}.

Graphs of a few cross sections are given in Fig. 9.4. It is seen that at low velocities the Lennard-Jones potential gives a much larger cross section than does the rigid sphere model. Furthermore, the cross section falls off rapidly with increasing relative velocity. This is the expected effect of the long range attractive tail of the potential. At high velocities the cross sections become more nearly those of the rigid sphere model but still display a decrease with increasing velocity which is a consequence of the diffuseness of the repulsive core of this potential model.

One final numerical integration produces the $\Omega^{(l,r)\star}$:

$$\Omega^{(l,r)\star} = \frac{2}{(r+1)!}\int_0^\infty Q^{(l)\star}(g^*)\, \mathscr{g}^{2r+3}\exp(-\mathscr{g}^2)\mathrm{d}\mathscr{g}, \tag{9.2-18}$$

where the numerical factor in front of the integral again effects a normalization to the rigid sphere value of $\Omega^{(l,r)}$. Note that g^{*2} is non-dimensionalized with respect to ϵ/m_{ij}, while $\mathscr{g}^2$ is non-dimensionalized with respect to kT/m_{ij}. A simple transformation then shows that $\Omega^{(l,r)\star}$ is a function only of $T^* =$

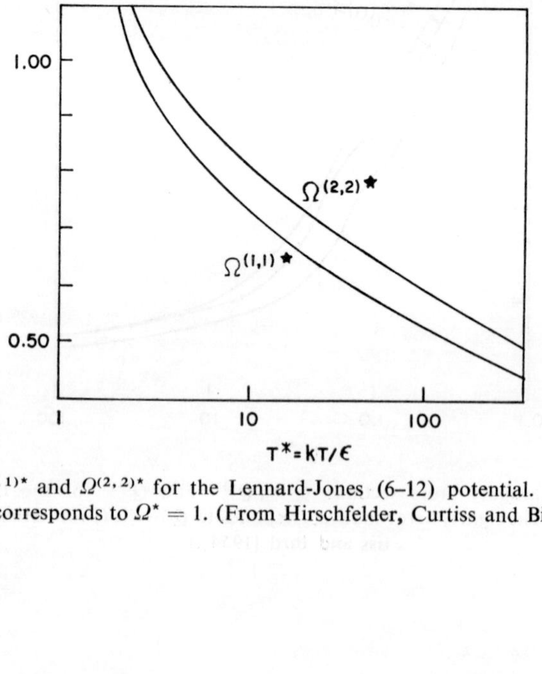

Fig. 9.5. $\Omega^{(1,1)\star}$ and $\Omega^{(2,2)\star}$ for the Lennard-Jones (6–12) potential. The rigid sphere model corresponds to $\Omega^* = 1$. (From Hirschfelder, Curtiss and Bird [1954].)

kT/ϵ. This is essentially a law of corresponding states for the transport properties and allows the properties of all gases to be plotted on a single chart. Also, since the critical temperature may be approximately related to ϵ by $kT/\epsilon \approx 1.3$, T^* can be considered a reduced temperature defined with respect to the critical temperature. The two most important of the integrals $\Omega^{(l,r)*}$ are displayed in Fig. 9.5. As a consequence of the decrease of $Q^{(l)*}$ with increasing g^{*2}, the $\Omega^{(l,r)*}$ fall off with increasing T^*. Tabulations of the Ω-integrals for the Lennard-Jones (6–12) and modified Buckingham (6-exp) potentials (as well as some others) have been given by Hirschfelder, Curtiss and Bird [1954]. A more recent tabulation of the Ω-integrals for the Lennard-Jones (6–12) potential including quantum effects was given by Munn, Smith and Mason [1965], while Mason [1954] has given a tabulation for the modified Buckingham (6-exp) potential. These are reproduced in Appendix C.

9.3. Quantum mechanical calculations

Although classical mechanics suffices for the treatment of the large majority of gases whose transport properties we wish to calculate, quantum mechanical corrections are non-negligible for a few gases of interest. To see this, we recall from Chapter 8 that classical mechanics is correct when the scattering angle χ is large compared to $(h/\sigma_{ij})(m_{ij}kT)^{\frac{1}{2}}$. For quantum corrections to be small, this must be the case for all but the smallest scattering angles, i.e.,

$$\frac{h}{\sigma_{ij}(m_{ij}kT)^{\frac{1}{2}}} \ll 1. \tag{9.3-1}$$

On the whole, σ_{ij} does not vary much from molecule to molecule, so quantum effects will be important only if $(m_{ij}T)^{\frac{1}{2}}$ is small, i.e., at low temperatures or for systems containing an appreciable quantity of a light gas. In practice this means that quantum effects are important only for systems containing hydrogen or helium.

Here we are interested only in quantum mechanical effects on scattering (diffraction effects) and their effect in turn on the transport integrals $\Omega^{(l,r)}$. Another effect – that at high densities or low temperatures the Maxwell-Boltzmann distribution must be replaced by the Fermi-Dirac or Bose-Einstein distributions – has not proved of importance in computing the transport properties of gases* and will not be discussed here. The discussion given here will be brief; for a detailed development of the quantum mechanical theory of scattering, the reader is referred to well-known texts on that subject as e.g., Schiff [1949] and Landau and Lifshitz [1958].

Recall that in quantum mechanics the relation between impact parameter and deflection angle cannot be defined. Instead, we must ask for the cross section for producing a deflection angle in a range $d\chi$ about χ. Since the deflection angle χ is defined in the center-of-mass coordinate system, we confine our attention to that coordinate system (the separation into center-of-mass and relative coordinates used for classical mechanics in Section 9.1 produces the analogous equivalent one-body problem in quantum mechanics). Thus we are faced with solving the Schroedinger equation:

$$-\frac{\hbar^2}{2m_{ij}}\nabla^2\psi+\varphi(r)\psi = E\psi, \tag{9.3-2}$$

where $\hbar$ is the reduced Planck constant $(h/2\pi)$, ψ is the wave function of the relative motion, m_{ij} and φ are the reduced mass and potential defined earlier and E is the total energy of the system $(=\frac{1}{2}m_{ij}g^2)$. We can represent the incoming particle as a plane wave travelling in the z-direction e^{ikz}, where $k(=m_{ij}g/\hbar)$ is the wave number, and may expect that the total wave function will be the sum of this incoming wave and a spherical, outgoing wave representing the scattered particles. Thus:

$$\psi(\boldsymbol{r}) \approx e^{ikz}+f(\chi)\frac{e^{ikr}}{r}, \tag{9.3-3}$$

where $r^{-1}e^{ikr}$ represents a spherically symmetric wave travelling outward from the scattering center and f contains the information about the angular distribution of the scattered particles.

The cross section for scattering in solid angle $d^2\Omega(=\sin\chi\, d\chi\, d\varepsilon)$ is defined as

$$\sigma_{ij}(\chi)\, d^2\Omega = \frac{\text{Number of particles scattered into } d^2\Omega \text{ per second}}{\text{Number of particles incident per unit area per second}} \tag{9.3-4}$$

i.e., the ratio of two particle fluxes. Now, as the particle flux per unit area

* These effects are important, however, when the electrons in a metal are treated as a gas and in the highly dense white dwarf stars.

is given in quantum mechanics by $(i\hbar/2m_{ij})(\psi\nabla\psi^* - \psi^*\nabla\psi)$, where the asterisk indicates complex conjugation, we see that the incoming wave function e^{ikz} is normalized to g particles per unit area per second; this is the denominator of (9.3-4). The outgoing flux into $d^2\Omega$ is also easily evaluated using the wave function (9.3-3) and yields $g|f(\chi)|^2 d^2\Omega$. Thus the cross section is simply

$$\sigma_{ij}(\chi) = |f(\chi)|^2. \tag{9.3-5}$$

Now the problem remaining is the calculation of $f(\chi)$. We follow the development of Landau and Lifshitz [1958].

The wave function ψ can be expanded in Legendre polynomials, cf. Whittaker and Watson [1915],

$$\psi(\boldsymbol{r}) = \sum_l A_l R_l(r) P_l(\cos\chi), \tag{9.3-6}$$

where we have taken advantage of the symmetry about the z axis. Substituting this into the wave equation (9.3-2) we find that the radial functions satisfy the ordinary differential equations

$$\frac{1}{r^2}\frac{d}{dr}\left(r^2\frac{dR_l}{dr}\right) + \left[k^2 - \frac{l(l+1)}{r^2} - \frac{2m_{ij}}{\hbar^2}\varphi(r)\right]R_l(r) = 0. \tag{9.3-7}$$

We note in passing that this is a one-dimensional wave equation with an effective potential $\varphi(r) + (\hbar^2/2m_{ij})l(l+1)/r^2$. Comparing with the effective potential of the classical case, Eq. (9.1-6), we see that the second term in this expression is just the centrifugal potential of the classical case with the angular momentum $m_{ij}gb$ replaced by $\sqrt{l(l+1)}\hbar$, a well known result in quantum mechanics. A complete solution of the problem requires integration of Eq. (9.3-7). This can be done analytically only for a few particular potentials just as Eq. (9.1-7) can be completely integrated only in special cases. In other cases numerical and various approximate methods have been used. More will be said about this later.

A useful form of the scattering cross section can be obtained by noting that the potential and centrifugal terms in Eq. (9.3-7) vanish at large r and the solution asymptotically takes the form

$$R_l(r) \sim A_l(kr)^{-1}\sin(kr - \tfrac{1}{2}l\pi + \delta_l) \qquad (r \to \infty). \tag{9.3-8}$$

Here δ_l is an unknown phase shift and the factor $\frac{1}{2}l\pi$ is inserted for convenience. On the other hand, taking $z = r \cos \chi$ one finds in the asymptotic region

$$e^{ikz} \sim \sum_l i^l(2l+1) P_l(\cos \chi)(kr)^{-1} \sin (kr - \tfrac{1}{2}l\pi) \quad (r \to \infty), \tag{9.3-9}$$

so that

$$\psi - e^{ikz} \sim \sum_l P_l(\cos \chi)(kr)^{-1}\{A_l \sin (kr - \tfrac{1}{2}l\pi + \delta_l) - i^l(2l+1) \sin (kr - \tfrac{1}{2}l\pi)\}. \tag{9.3-10}$$

According to Eq. (9.3-3) this function must represent an outgoing spherical wave, i.e., the coefficient of e^{-ikr} must vanish for every l. This leads to an expression for A_l:

$$A_l = i^l(2l+1) \exp (i\delta_l). \tag{9.3-11}$$

When this is inserted into Eq. (9.3-10) and the result compared with Eq. (9.3-3), we have

$$f(\chi) = (2ik)^{-1} \sum_{l=0}^{\infty} (2l+1)(\exp (2i\delta_l) - 1) P_l(\cos \chi), \tag{9.3-12}$$

which is the sought for expression. From this result and Eq. (9.3-5) we see that the cross section is determined once the phase shifts δ_l have been calculated.

As mentioned above, the calculation of the δ_l may be carried out in a number of ways. The availability of fast computers has made direct integration of Eq. (9.3-7) feasible and this procedure has been adopted in most recent calculations, e.g., by Munn, Smith and Mason [1965]. Another technique which has been used, is approximate integration using the WKB method, see Hirschfelder, Curtiss and Bird [1954]; this method has the advantage that it gives the classical result as the first approximation and the succeeding terms as systematic quantum corrections. Although instructive, this method does not possess any computational advantage over the former method. Finally, variational methods which are frequently used in quantum mechanics, could be applied for computing δ_l, although this method appears not to have been used in this particular application.

In the above calculation it was assumed that the two colliding particles are distinguishable. In collisions between like molecules, this calculation is no longer valid. In such collisions in which the total spin of the system, i.e., the vector sum of the individual particle spins, is even, the wave function describing the relative motion must be symmetric with respect to an interchange of the coordinates of the two particles. Likewise, if the total spin is odd, the wave function must be antisymmetric with respect to the interchange, see Landau and Lifshitz [1958]. In the former case we speak of Bose-Einstein statistics, in the latter, of Fermi-Dirac statistics. As the Legendre polynomials are symmetric or antisymmetric in χ according to whether their order, l, is even or odd, this means that only even l contribute to the cross section in the Bose-Einstein case, and only odd l contribute in the Fermi-Dirac case. By retracing the calculation one can show that this again leads to Eq. (9.3-5) for the cross section, but $f(\chi)$ is given by

$$f_{\mathrm{BE}}(\chi) = (\mathrm{i}k\sqrt{2})^{-1} \sum_{l\ \mathrm{even}} (2l+1)(\exp(2\mathrm{i}\delta_l)-1)\, P_l(\cos\chi) \tag{9.3-13}$$

and

$$f_{\mathrm{FD}}(\chi) = (\mathrm{i}k\sqrt{2})^{-1} \sum_{l\ \mathrm{odd}} (2l+1)(\exp(2\mathrm{i}\delta_l)-1)\, P_l(\cos\chi), \tag{9.3-14}$$

respectively; the phase shifts δ_l are precisely those defined above.

In terms of the scattering cross section, the expressions for the transport cross sections become,

$$Q_{ij}^{(l)}(g) = 2\pi \int_0^{\pi} (1-\cos^l\chi)\sigma_{ij}(g,\chi)\sin\chi\, \mathrm{d}\chi. \tag{9.3-15}$$

Using the recurrence relation for Legendre polynomials, these expressions can be integrated to give, for the case of distinguishable particles:

$$Q_{ij}^{(1)}(g) = \frac{4\pi}{k^2} \sum_{l=0}^{\infty} (l+1)\sin^2(\delta_{l+1}-\delta_l), \tag{9.3-16}$$

$$Q_{ij}^{(2)}(g) = \frac{2\pi}{k^2} \sum_{l=0}^{\infty} \frac{(l+1)(l+2)}{(l+\frac{3}{2})} \sin^2(\delta_{l+2}-\delta_l). \tag{9.3-17}$$

For the Bose-Einstein (Fermi-Dirac) case, the odd-l (even-l) terms must be suppressed, the formulae multiplied by two, and $(l+1)\sin^2(\delta_{l+1}-\delta_l)$ in Eq. (9.3-16) must be replaced by $(2l+1)\sin^2\delta_l$.

Finally, we note that in a collision of two particles each of which has spin s, the total spin S may have any integral value between zero and $2s$. Thus both odd and even total spins may occur. Noting that there are $2S+1$ independent quantum states of spin S, we can easily calculate the probability of odd or even spin. The cross section is then the sum of the Bose-Einstein and Fermi-Dirac cross sections described above weighted by these probabilities. For even s one has

$$(2s+1)Q^{(n)} = (s+1)Q^{(n)}_{\mathrm{BE}} + sQ^{(n)}_{\mathrm{FD}}, \tag{9.3-18}$$

while for odd s

$$(2s+1)Q^{(n)} = sQ^{(n)}_{\mathrm{BE}} + (s+1)Q^{(n)}_{\mathrm{FD}}. \tag{9.3-19}$$

The Ω-integrals can then be computed using Eq. (9.2-13). One thus sees that the quantum mechanical calculation is of no greater difficulty than the classical calculation. At each g, one needs to compute the phase shifts for various angular momenta l, rather than the scattering angle for each impact parameter b, and then sum over l, rather than integrate over b.

Comparison of theory and experiment - Noble gases and their mixtures

10

In comparing theoretical predictions with experimental results we shall, first of all, limit the discussion to the noble gases – helium, neon, argon, krypton and xenon – and their mixtures. Although these gases are of relatively minor practical importance, they are the only ones to which the theory presented thus far can be applied without the introduction of further approximations. Thus, they provide an unambiguous test of the theory.

As was pointed out in Chapter 8, the most desirable procedure – viz., computing an intermolecular potential from first principles and thence the transport properties – is beyond the capability of molecular physics in its present state and is likely to remain so for some time. Thus, the invariable procedure has been to use one of the model potentials with two or three adjustable parameters and thence compute the transport properties. When used in this manner, the theory can degenerate into a sophisticated form of curve fitting insofar as transport properties are concerned; however, it does provide one of the best means of obtaining the intermolecular potential. Also, this procedure enables one to use the results obtained from the measurement of one property to calculate other, as yet unknown, properties. It is well to bear in mind that *any* property which depends functionally on the potential may be used for this purpose; the list of such properties is quite large and includes, in addition to the transport properties, all thermodynamic state functions such as the equation of state (generally in the form of virial coefficients) and the thermodynamic derivative functions (of which the Joule-Thompson coefficient $(\partial T/\partial p)_H$ is a good example). Since the statistical mechanical theory of thermodynamic properties is as well developed as that of transport coefficients, the ability of a single model potential to predict all properties or a comparison of the potentials obtained by fitting the various properties provide tests of the theory and of the model potentials employed.

In this chapter we shall first briefly describe some of the curve fitting procedures used. Then the experimental values of some relevant properties of the pure noble gases will be compared with the theoretical values and the potentials obtained will be compared with those obtained from other properties. A discussion of diffusion and thermal diffusion will be given in an effort to shed some light on the unlike-particle potential; methods of obtaining properties of mixtures from simple gas properties and of obtaining various properties directly from measurements of other properties will also be discussed and evaluated.

10.1. Curve fitting procedures and sensitivity of parameters to data

A number of procedures have been used to obtain model potential parameters from experimental data. For single-parameter models, such as hard spheres, the easiest method is to choose the parameter so as to obtain agreement between the theoretical and experimental curves at some particular temperature, preferably a temperature near the center of the range for which data are available. A better and less arbitrary procedure is to choose the parameter so that some measure of the deviation between the theoretical and experimental curves is minimized; the mean square deviation is the conventional choice.

For two-parameter potentials analogous methods can be used. The simplest method is to fit the theoretical and experimental curves at two temperatures, the preferred choice being two temperatures near the ends of the experimentally accessible range. For example, for the viscosity of a simple gas, η, this is done as follows. In the first approximation, η is given by Eq. (7.3-6),

$$\eta(T) = \frac{5(\pi m k T)^{\frac{1}{2}}}{16\pi\sigma^2 \Omega^{(2,2)*}(T^*)}. \tag{10.1-1}$$

As we shall see later, this approximation is correct to a few per cent and can be used with considerable accuracy to obtain values of model potential parameters. Agreeing to do so, we then see that the ratio

$$\frac{\eta(T_2)}{\eta(T_1)} = \left(\frac{T_2}{T_1}\right)^{\frac{1}{2}} \frac{\Omega^{(2,2)*}(T_1^*)}{\Omega^{(2,2)*}(T_2^*)} \tag{10.1-2}$$

does not involve the parameter σ. With the experimental value of $\eta(T_2)/\eta(T_1)$ Eq. (10.1-2) is solved for ϵ by trial and error using tabulated values of the Ω-integrals. Having a value of ϵ, one trivially finds σ from Eq. (10.1-1) with either $T = T_1$ or $T = T_2$.

The following simple, but non-rigorous, method of finding ϵ and σ by trial and error has been attributed to Keesom. From Eq. (10.1-1), one has

$$\log\left[\eta(T)/\sqrt{T}\right] = \log\left[1/\Omega^{(2,2)*}(T^*)\right] - \log\sigma^2 + \text{const.} \qquad (10.1\text{-}3)$$

Then, if some value of ϵ is assumed and $T^* = kT/\epsilon$, it is possible to plot $\log(\eta(T^*)/\sqrt{T^*})$ and $\log[1/\Omega^{(2,2)*}(T^*)]$ vs. $\log T^*$ on the same graph. The two curves should have the same shape and can be brought into agreement by shifting one of the curves both horizontally and vertically. The horizontal shift required then determines ϵ, while the vertical shift is simply $\log \sigma^2$.

A number of other methods of curve fitting have also been used, including the straight-forward least squares method.

If the purpose of the analysis is to obtain information about the intermolecular potential, one must inquire as to how accurately the potential can be determined from transport property measurements. Two separate points must be considered in such a discussion: (i) the accuracy of the experimental data, and (ii) the manner in which experimental errors are propagated in the computation of the potential parameters, i.e., the sensitivity of the parameters to the transport properties or *vice versa.*

A detailed discussion of experimental technique will not be given here, but some mention of the accuracy must be made. Viscosity is generally more easily measured than is thermal conductivity. With the most recent viscosimeters an accuracy of one per cent is easily obtained and accuracies of 0.1 % are obtainable if some precaution is taken. Far higher accuracies (as high as 10^{-5} of one per cent) have been reported, but such accuracy is not necessary for the present purposes. For the thermal conductivity the best recent measurements report accuracies of approximately one per cent and this seems to be the limit with current technique. For the diffusion and thermal diffusion coefficients, typical experimental errors are of the order of five per cent, and some data with errors as low as one per cent have been reported.

Next, the sensitivity of the potential parameters and transport properties to each other will be discussed.

First, consider the coefficients of viscosity and thermal conductivity of a pure gas. Since only rough estimates are desired, the first approximations to these coefficients are sufficient. Then, the Eucken ratio $(\lambda/\eta c_v)$ may be treated as a constant. Hence, the coefficients of viscosity and thermal conductivity are directly proportional and we can restrict ourselves to just one of them – the coefficient of viscosity, say. Now, from Eq. (10.1-1) it follows that

$$\frac{\Delta\eta}{\eta} = \frac{1}{\eta}\frac{\partial\eta}{\partial\sigma}\Delta\sigma + \frac{1}{\eta}\frac{\partial\eta}{\partial\epsilon}\Delta\epsilon = -2\frac{\Delta\sigma}{\sigma} + \omega(T^*)\frac{\Delta\epsilon}{\epsilon}, \tag{10.1-4}$$

where $\omega(T^*) = \mathrm{d}\log\Omega^{(2,2)*}(T^*)/\mathrm{d}\log T^*$.

Assuming that the two point method described above is used, we can write Eq. (10.1-4) for each of the two temperatures and solve the two resulting equations for $(\Delta\sigma/\sigma)$ and $(\Delta\epsilon/\epsilon)$. We find

$$\frac{\Delta\epsilon}{\epsilon} = (\omega_2-\omega_1)^{-1}[(\Delta\eta/\eta)_2-(\Delta\eta/\eta)_1] \approx \sqrt{2}(\omega_2-\omega_1)^{-1}(\Delta\eta/\eta), \tag{10.1-5}$$

where the subscripts refer to temperature; we have assumed that the expected values of $(\Delta\eta/\eta)$ are the same at both temperatures and of random sign. Similarly, we find

$$\frac{\Delta\sigma}{\sigma} = \tfrac{1}{2}(\omega_1-\omega_2)^{-1}[\omega_2(\Delta\eta/\eta)_1-\omega_1(\Delta\eta/\eta)_2] \approx \frac{1}{\sqrt{2}}\frac{\omega_1+\omega_2}{\omega_1-\omega_2}\left(\frac{\Delta\eta}{\eta}\right). \tag{10.1-6}$$

Values of $(\Delta\epsilon/\epsilon)$ and $(\Delta\sigma/\sigma)$ for a particular value of T_1^* are given in Fig. 10.1; they were taken from the work of Kestin [1964]. As may be seen from the figures, ϵ is approximately ten times as sensitive to variations in the viscosity as $\sigma : \Delta\sigma/\sigma \approx 2\Delta\eta/\eta$, while $\Delta\epsilon/\epsilon \approx 20\,\Delta\eta/\eta$ in a reasonable temperature range. The lack of sensitivity of the viscosity to changes in ϵ is due to the fact that the well of the potential does not contribute much to the transport integrals. In turn, this is due to the rapid relative motion of the colliding particles when they are in the vicinity of the potential minimum and to the relative narrowness of the well. As a consequence, the values of ϵ obtained by fitting viscosity and thermal conductivity data are not very accurate and this explains why the values of ϵ reported in the literature show much wider deviations and uncertainties than do those of σ. Conversely, we may infer that good calculations of the viscosity and thermal conductivity can be made with model potentials with relatively large uncertainties in ϵ. Finally we remark that Fig. 10.1 provides convincing evidence that data must be fit over a wide temperature range if accurate parameters are to be obtained.

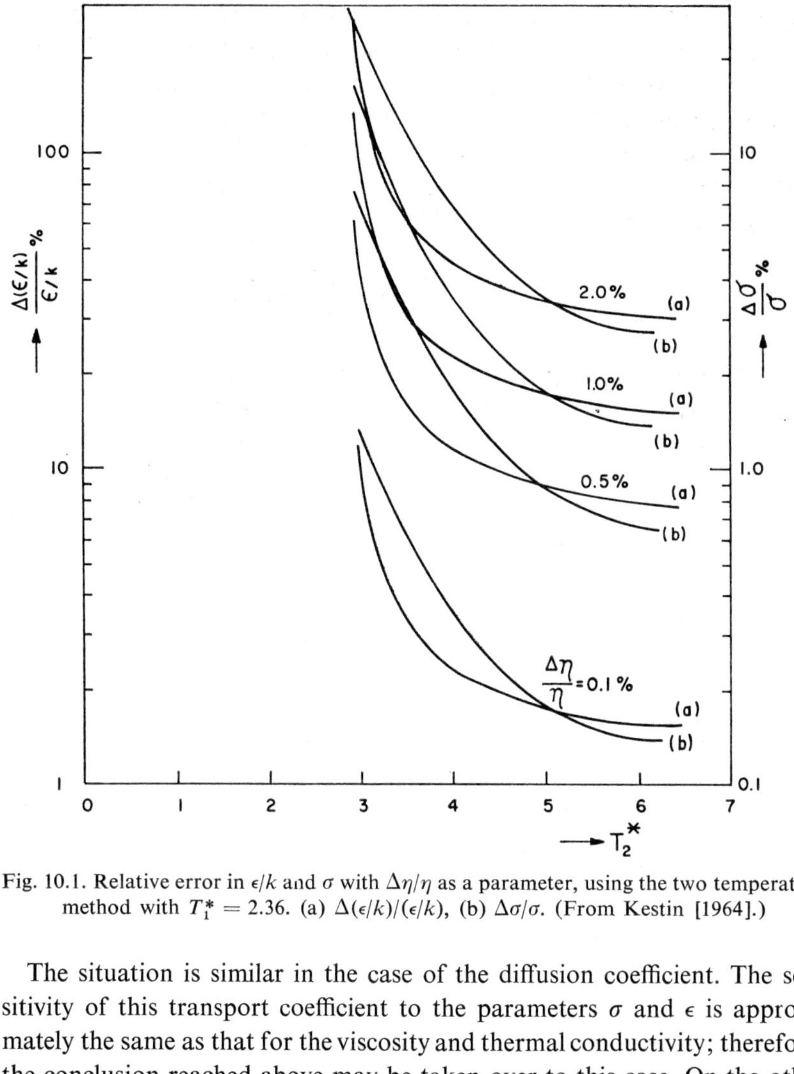

Fig. 10.1. Relative error in ϵ/k and σ with $\Delta\eta/\eta$ as a parameter, using the two temperature method with $T_1^* = 2.36$. (a) $\Delta(\epsilon/k)/(\epsilon/k)$, (b) $\Delta\sigma/\sigma$. (From Kestin [1964].)

The situation is similar in the case of the diffusion coefficient. The sensitivity of this transport coefficient to the parameters σ and ϵ is approximately the same as that for the viscosity and thermal conductivity; therefore, the conclusion reached above may be taken over to this case. On the other hand, the situation is reversed in the case of the thermal diffusion coefficient. This transport coefficient is more sensitive to ϵ than to σ. The reason for this lies in the factor $6C^* - 5$ to which the thermal diffusion coefficient is proportional, cf. Eq. (7.3-69), and which is responsible for most of the temperature variation of this property. As this quantity is essentially a small difference of two Ω-integrals (recall Eq. (7.1-33), $C^* = \Omega^{(1,2)*}/\Omega^{(1,1)*}$), the above argument as to why transport properties are more sensitive to σ than to ϵ is no longer valid and the reverse is, in fact, the case. Unfortunately, curves similar to those of Fig. 10.1 are not available for this property.

Finally, comparison will later be made with potential parameters obtained from measurements of thermodynamic properties – in particular, the virial coefficients. The latter are defined as the coefficients in the series expansion of pV/RT in decreasing powers of V,

$$\frac{pV}{RT} = 1 + \frac{B(T)}{V} + \frac{C(T)}{V^2} + \dots. \tag{10.1-7}$$

(R is the gas constant per gram-molecule; it is directly related to Boltzmann's constant k through the identity $R = Nk$, where N is Avogadro's number, $N = 6.023 \dots \times 10^{23}$.) The first virial coefficient, $B(T)$, is explicitly given in terms of the intermolecular potential, φ, by the integral

$$B(T) = 2\pi \int_0^\infty [\mathrm{e}^{-\varphi(r)/kT} - 1] r^2 \mathrm{d}r. \tag{10.1-8}$$

We note that B is equally sensitive to changes in σ and ϵ; the sensitivity is of the same magnitude as the sensitivity of the viscosity to σ.

10.2. Convergence of the Sonine polynomial expansions

Because of the variational nature of the approximation method one expects that the calculational scheme used in Chapters 5 and 6 to generate the expressions for the transport coefficients will be rapidly converging. Variational techniques are designed to produce good results for average quantities (in this case, the transport coefficients) even when the trial functions used only roughly approximate the true solutions of the equations. Furthermore, since the integral equations of the Chapman-Enskog method are self-adjoint, the variational method produces monotonically decreasing or monotonically increasing sequences of values for the transport coefficients (except, possibly, for the thermal diffusion coefficient). Thus, each approximation must produce a better value than the previous one and no oscillations occur. The latter situation does not necessarily occur if the Kihara approximation method is used. On the other hand, as we shall see later, Kihara's method frequently produces more rapid convergence than the variational method – particularly for the thermal diffusion coefficient, for which the expressions obtained with the variational method converge relatively slowly.

We recall that the results for the transport coefficients obtained with the variational methods of Chapters 5 and 6 are identical with those obtained by Chapman and Cowling with a series expansion technique. Hence, in studying the convergence of the Sonine polynomial expansions we can use the results presented by these authors in their monograph, Chapman and Cowling [1952]. Also, we shall use the results of higher-order calculations for binary mixtures that were obtained by Mason [1957a, 1957b].

First we consider the convergence for the thermal conductivity, viscosity and the self-diffusion coefficient of a simple gas. As can be seen in Table 10.1, for hard spheres, the first approximation is incorrect by only two per cent, while the second approximation is exact within the limits of present experimental precision. For other molecular models the convergence has been found to be slightly more rapid and the values in Table 10.1 may be taken as representing an unfavorable case.

TABLE 10.1

Convergence of the Chapman-Cowling expansion for a simple gas of hard spheres

Approximation	$[\lambda]_n/[\lambda]_4$	$[\eta]_n/[\eta]_4$	$[\mathscr{D}]_n/[\mathscr{D}]_3$
1	0.976	0.984	0.982
2	0.998	0.9989	0.9985
3	0.9997	0.9999	

Clearly, for a mixture in which the masses of the various species are approximately equal and the force laws between the various types of molecules are similar (for example, a mixture of isotopes of a single gas), the convergence of the approximation method should be approximately as rapid as that for a similar simple gas; an exception may be the thermal diffusion coefficient. To emphasize the role of mass disparity on the rate of convergence, we shall take the extreme case of a binary mixture in which the ratio of masses is vanishingly small. Further, in order to eliminate the effects of concentration dependence, we shall study the limit in which the concentration of the light component is also vanishingly small and study only the properties of the light gas. This is the Lorentz gas studied in Section 6.6. Results for a Lorentz gas of hard spheres are given in Table 10.2. It is seen that the convergence is considerably slower than that for a pure gas and, as expected, the thermal diffusion ratio k_T is the most slowly converging property. Although this represents an extremely unfavorable case, one sees that for a mixture containing gases of greatly disparate mass several terms in the series may be needed.

TABLE 10.2

Convergence of the Chapman-Cowling and Kihara approximations for a Lorentz gas of hard spheres

Approximation		$[\lambda]_n/\lambda$	$[\eta]_n/\eta$	$[\mathscr{D}_{12}]_n/\mathscr{D}_{12}$	$[k_T]_n/k_T$
C-C	1	0.85	0.92	0.883	0.769
	2	0.93	0.98	0.957	0.894
	3			0.978	0.939
Kihara	1			0.883	1.000
	2			0.972	1.062

TABLE 10.3

Convergence of approximations to the binary diffusion coefficient of a Lorentz gas

Approximation		$[\mathscr{D}_{12}]_n/\mathscr{D}_{12}$					
		$\nu = \infty$	12	10	8	6	4
C-C	1	0.883	0.947	0.957	0.970	0.986	1.000
	2	0.957	0.984	0.988	0.992	0.996	1.000
	3	0.978	0.992	0.994	0.996	0.998	1.000
Kihara	1	0.883	0.947	0.957	0.970	0.986	1.000
	2	0.972	0.989	0.991	0.994	0.997	1.000

TABLE 10.4

Convergence of approximations to the thermal diffusion ratio of a Lorentz gas

Approximation		$[k_T]_n/k_T$					
		$\nu = \infty$	12	10	8	6	4
C-C	1	0.769	0.849	0.865	0.889	0.928	1.000
	2	0.894	0.939	0.947	0.959	0.975	1.000
	3	0.939	0.968	0.972	0.979	0.998	1.000
Kihara	1	1.000	1.000	1.000	1.000	1.000	1.000
	2	1.062	1.030	1.025	1.018	1.008	1.000

Results for a Lorentz gas of point repulsive centers were obtained by Mason and are given in Tables 10.3 and 10.4. It is clearly seen that the convergence improves as the repulsion index ν varies from infinity (hard spheres) to four (Maxwell molecules), the appropriate range for neutral gases.

Also given in the tables are some results for the Kihara approximation for the diffusion coefficient and thermal diffusion ratio. For the diffusion coefficient the first Chapman-Cowling and Kihara approximations are identical but, beyond the first approximation, results obtained by Kihara's method are seen to be more rapidly convergent. For the thermal diffusion ratio, the first Kihara approximation accidentally gives the correct result but it is to be noted that the second Kihara approximation is as accurate as the third Chapman-Cowling approximation.

Mason [1957b] also investigated the transport properties of a Lorentz gas using the modified Buckingham (6-exp) potential and found that the convergence is much more rapid for this potential than for the inverse power potential, and that for a given order, the Kihara method is again superior to the Chapman-Cowling method. Over a wide range of temperatures the first Chapman-Cowling approximation was found to be accurate to within one per cent for the diffusion coefficient and about ten per cent for the thermal diffusion factor. Although the Kihara approximation yields better re-

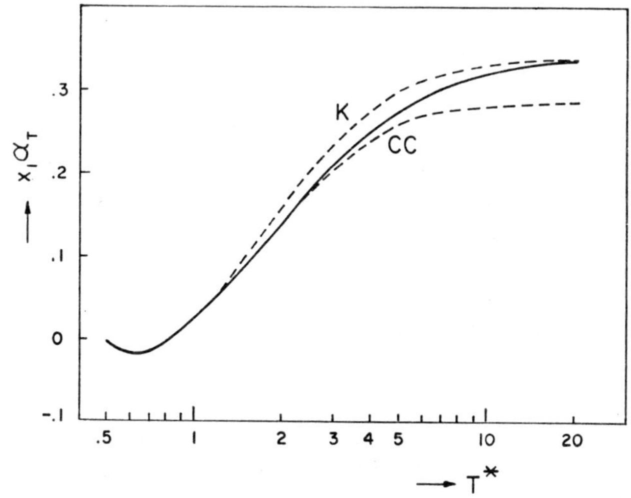

Fig. 10.2. The thermal diffusion factor of a Lorentz gas obeying a modified Buckingham (6-exp) potential ($\alpha = 14$). The solid line is the exact result and the dashed lines are the first approximations of Kihara (K) and Chapman and Cowling (CC). (From Mason [1957b].)

sults than the Chapman-Cowling approximation in the cases cited above, the results are temperature dependent and one should not conclude that the Kihara method is always better. This is clearly shown in Fig. 10.2. However, the greater simplicity of the formulae produced by Kihara's method is a considerable advantage.

10.3. Transport properties of the noble gases

Because the noble gases provide the most severe test of theoretical calculations for thermodynamic properties as well as transport properties, they have been studied in considerable detail and a great quantity of experimental measurements is available. Although the best and most numerous data exist for the viscosity, thermal conductivity and virial coefficients, a strenuous test of the theory must also include a study of other properties which depend only on the intermolecular potential of a single molecular species, in particular, the self-diffusion and isotopic thermal diffusion coefficients.

One of the most startling results of the theory is the prediction that, to first order, the Eucken ratio $\lambda/\eta c_v$ has the constant value 2.5 for all monatomic gases independent of the temperature. Since, as we have seen in the previous section, the first approximation to the transport properties of simple gases is quite accurate, this result is only very slightly modified by higher approximations.

A careful study of the Eucken ratio was made by O'Neal and Brokaw [1962]. They found, on the basis of 73 measurements for argon at varying temperatures, that the experimental and theoretical values of the Eucken ratio differed by 0.19 % compared to an experimental standard deviation of 0.17 %. For helium, on the basis of 45 measurements, they found a difference of 0.20 % compared to an experimental standard deviation of 0.29 %. In both cases the calculated values were corrected for higher polynomial approximations using the modified Buckingham (6-exp) potential. Thus it may be concluded that the prediction of constant Eucken ratio is confirmed by experiment.

Since this prediction is confirmed by experiment, we can, for purposes of further comparison, consider only one of the two transport properties appearing in the Eucken ratio. The viscosity is chosen because, being the more readily measured property, there exist more and better data for it. Before making the comparison, however, it must be pointed out that the limits of experimental accuracy given earlier can be misleading. Although measurements with better than one per cent precision are possible, the properties of all gases have not been measured to this precision at all temperatures. Furthermore, the measurements of different experimenters often differ by more than the quoted experimental errors.

In making a comparison between theory and experiment for the two-parameter potentials, we may take advantage of the law of corresponding states that results from the use of such potentials. Thus, the viscosities of all monatomic gases shoud fall on a single curve when plotted against the reduced temperature. The results are shown in Fig. 10.3, in which the solid line represents calculations based on the Lennard-Jones (6–12) potential. As can be seen, the results are quite good, although discrepancies of as much as ten per cent occur. Some of this discrepancy is due to the experimental inconsistency mentioned above, but even when only the most accurate data are taken into account, a discrepancy outside the experimental error remains. For the modified Buckingham (6-exp) potential a slightly better fit may be obtained, but the agreement is still not within the experimental uncertainties.

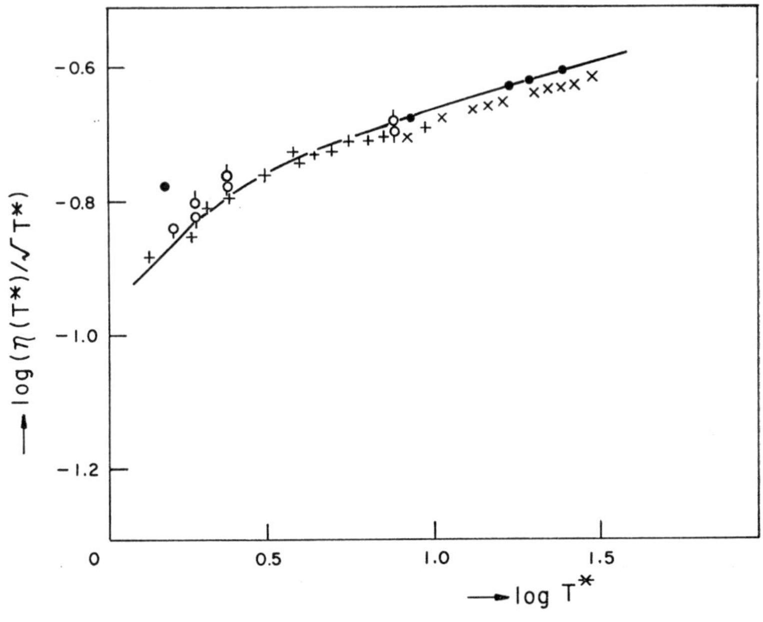

Fig. 10.3. Theoretical curves for $\log_{10}(\eta(T^*)/\sqrt{T^*})$ as a function of $\log_{10} T^*$. Experimental points: × Neon, + Argon, ♁ Hydrogen, ♀ Deuterium, ● Helium. (From Cohen [1966b].)

Thus, it may be concluded that, although the two-parameter Lennard-Jones and 6-exp potentials are sufficiently accurate for *estimating* transport properties, they do not give an accurate representation of the true intermolecular potential and the "law" of corresponding states resulting from them is useful only as a means of estimating properties. A recent set of Lennard-Jones parameters which were obtained by Van Heijningen [1967] from fitting both viscosity and virial coefficient data is given in Table 10.5; they fit both types of data equally well, that is, to within about five per cent on the average.

TABLE 10.5

Lennard-Jones parameters for the noble gases

Gas	ϵ/k(°K)	σ(Å)
Helium	10.5	2.56
Neon	35.9	2.77
Argon	121	3.42
Krypton	173	3.63
Xenon	218	4.05

To further test the theory one can now apply the potentials obtained in the manner described above to the calculation of the self-diffusion and isotopic thermal diffusion coefficients. These properties are special cases of the diffusion and thermal diffusion coefficients that will be studied in the following section. They are especially useful for testing the theory, as they depend on different sets of Ω-integrals than do the viscosity and thermal conductivity. In particular, the isotopic thermal diffusion factor, in contrast to all of the other transport properties, depends only slightly on the range parameter σ of the potential and very strongly on the strength parameter ϵ. It is found that, if self-diffusion coefficients are computed using model potentials obtained by fitting viscosity and thermal conductivity data, the discrepancy with experimental data is the same as in the case of the transport coefficients from which they are derived. For the isotopic thermal diffusion factor the agreement is not quite so good, but still within reasonable bounds considering the greater experimental uncertainty in the measurement of this property. The values of the virial coefficients obtained by using potentials which fit transport properties are in agreement with experiment, but not within the experimental error. The single exception is krypton, for which a two-parameter potential can be made to fit either all the transport properties but not the virial coefficient, or the virial coefficient and the transport properties except the thermal diffusion coefficient; it appears impossible to fit all of the properties simultaneously with a two-parameter potential. Why this should be so is somewhat puzzling, since krypton follows the trends of the other noble gases in all other respects and the potential parameters obtained by fitting partial sets of properties fall in the sequence formed by other gases in the noble gas series. The difficulty can probably be ascribed to the inability of a two-parameter potential to fit the true potential accurately in the well. (Note that, of the properties we are dealing with, the isotopic thermal diffusion factor and the second virial coefficient – the two properties that can not be reconciled – are the two which are most sensitive to the well shape.)

Taking a different point of view, Munn [1964], and Munn and Smith [1965] used a potential model with a large number of adjustable constants and were able to fit the data for neon, argon, krypton and xenon almost within the scatter of experimental data (helium is discussed in the following paragraph); their potential also fits the virial coefficient data and the small amount of molecular beam data that is available. The potential, which is shown in Fig. 10.4, is somewhat deeper than the Lennard-Jones potential and has a steeper repulsive part. These variations from the Lennard-Jones model are in line with expectations.

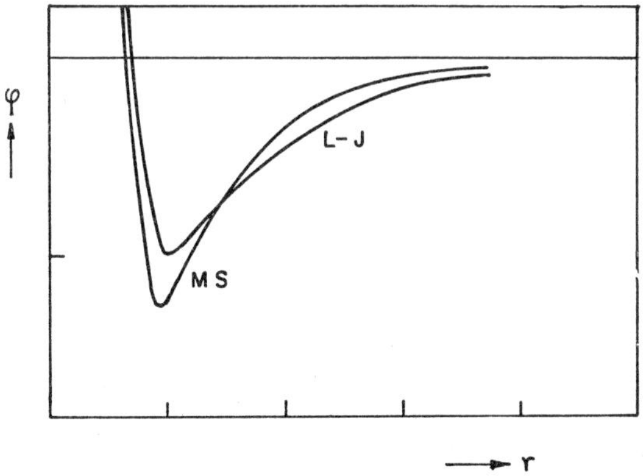

Fig. 10.4. A comparison of the Munn-Smith and Lennard-Jones (6–12) potentials. (From Munn and Smith [1965].)

Among the noble gases, helium must be singled out for special attention. As was mentioned in Chapter 8, the helium potential is much weaker than those of the other noble gases (this is borne out by Table 10.5) and must be

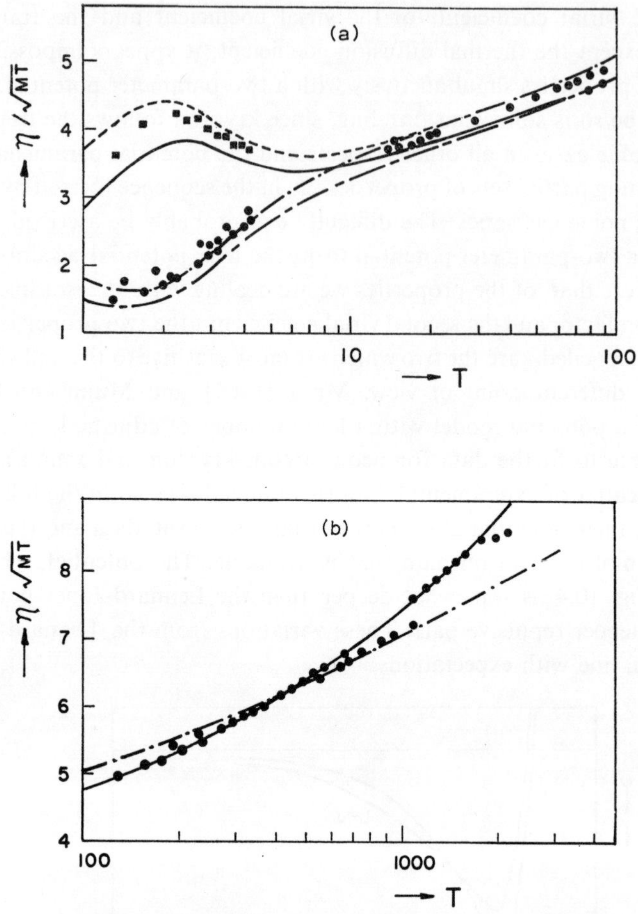

Fig. 10.5(a) Viscosity of ^{3}He and ^{4}He below 100° K. Units: T(°K), η (10^{-6} g cm^{-1} sec^{-1}); M is the molecular weight.

——— calculated viscosity for ^{3}He, using the modified Morse potential,

– – — calculated viscosity for ^{4}He, using the modified Morse potential,

– – – – calculated viscosity for ^{3}He, using the Lennard-Jones (6–12) potential,

—·—· calculated viscosity for ^{4}He, using the Lennard-Jones (6–12) potential,

■ experimental points for ^{3}He,

● experimental points for ^{4}He.

(b) Viscosity of ^{4}He above 100° K. Units: T(°K), η (10^{-6} g cm^{-1} sec^{-1}); M is the molecular weight.

——— calculated viscosity, using the modified Morse potential,

—·—· calculated viscosity, using the Lennard-Jones (6–12) potential,

● experimental points.

(From Keller and Taylor [1969].)

treated quantum mechanically, particularly at low temperatures. Also, it is of greater than normal interest, because it is the object of considerable theoretical investigation. The low-temperature viscosity data for helium have been compared with the predictions of the Lennard-Jones and modified Morse potential (8.4-10) by Keller and Taylor [1969]. Their results are given in Fig. 10.5. One sees that the modified Morse potential is indeed an improvement over the Lennard-Jones potential (as might be expected because of the greater number of adjustable parameters), but is still in disagreement with the data for ^{3}He by about five per cent. We also note in passing that Fig. 10.5 dramatically displays the quantum effects; ^{4}He is a Bose-Einstein gas, while ^{3}He is a Fermi-Dirac gas and the difference is quite obvious. (The difference is even more dramatic in the liquid state: ^{4}He becomes a superfluid, while ^{3}He does not.) The potentials used by Keller and Taylor are shown in Fig. 10.6.

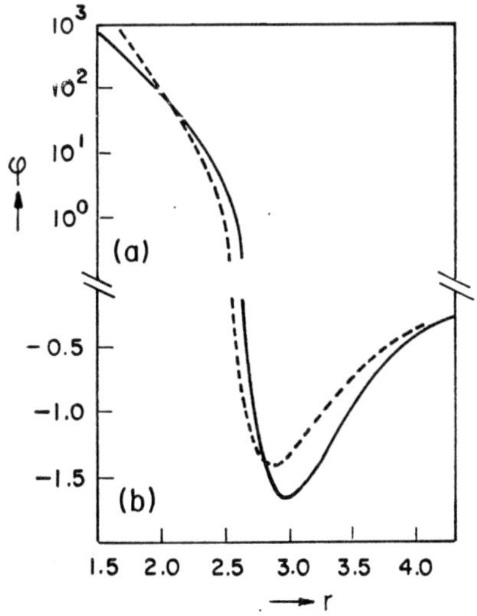

Fig. 10.6. A comparison of the modified Morse potential and Lennard-Jones (6–12) potential. (a) The repulsive portion, (b) the potential well. Units: r(Å), $\varphi(10^{-16}$ erg). ——— modified Morse potential, - - - - Lennard–Jones potential. (From Keller and Taylor [1969].)

10.4. Noble gas mixtures – Diffusion and thermal diffusion

Now we turn to a study of the properties of mixtures of the noble gases. Although it seems natural to discuss viscosity and thermal conductivity first as the more important and more easily measured properties, we shall treat the diffusion and thermal diffusion coefficients first. The reason for so doing is that the diffusion properties are almost entirely due to unlike-particle collisions and thus depend almost entirely on the potential associated with such collisions. Consequently, although they are not measured to as high accuracy as the viscosity and thermal conductivity, they provide more accurate information with respect to the unlike-particle potential. Furthermore, the composition dependence of the diffusive properties is generally somewhat simpler than that of the viscosity and thermal conductivity.

The first Chapman-Cowling or Kihara approximation for the binary diffusion coefficient is given by Eq. (7.3-38),

$$[\mathscr{D}_{12}]_1 = \frac{3}{16nm_{12}} \frac{(2\pi m_{12} kT)^{\frac{1}{2}}}{\pi\sigma_{12}^2 \Omega_{12}^{(1,1)*}(T^*)}. \tag{10.4-1}$$

In the nth Chapman-Cowling approximation, $[\mathscr{D}_{12}]_1$ is multiplied by the factor $f_D^{(n)}$,

$$[\mathscr{D}_{12}]_n = [\mathscr{D}_{12}]_1 f_D^{(n)}, \tag{10.4-2}$$

and, similarly, in the Kihara approximation, by the factor $f_D^{(n)K}$,

$$[\mathscr{D}_{12}]_n^K = [\mathscr{D}_{12}]_1 f_D^{(n)K}. \tag{10.4-3}$$

The expressions for $n = 2$ were given in Section 7.3, Eqs. (7.3-40) and (7.3-46). Since the approximation techniques converge very rapidly, both $f_D^{(n)}$ and $f_D^{(n)K}$ differ by only a few per cent from unity. However, they are strongly dependent on composition and temperature and must be included in the analysis if the proper potentials are to be obtained. For example, for the noble gases the factor $f_D^{(2)}$ varies between 1.00 and 1.10. It is found to be very nearly unity (within 10^{-4}) for mixtures containing only a trace of the heavy component. (It is exactly unity in the limit $m_1/m_2 \to \infty$.) Generally, $f_D^{(2)}$ is an increasing function of the mole fraction of the heavy component; thus it takes its largest values when the light component is present in trace amounts. Also, it is an increasing function of temperature. Thus, in performing diffusion coefficient measurements it is best from a theoretical view point to use mixtures containing only a trace of the heavy component. When the experimental method prohibits measurements with trace components, the results must be extrapolated to the limit $x_1 = 1$. It is then possible to analyze the data using Eq. (10.4-2) with $f_D^{(n)} = 1$.

Then, restricting our attention to the first approximation, we note that, since $\Omega_{12}^{(1,1)*}$ is a universal function of $T^* = kT/\epsilon_{12}$, the diffusion coefficients of various gas mixtures obey a *law of corresponding states*, viz.,

$$n\sigma_{12}^2(2m_{12}/kT)^{\frac{1}{2}}[\mathscr{D}_{12}]_1 = \frac{3}{8\sqrt{\pi}}[\Omega_{12}^{(1,1)*}]^{-1}. \tag{10.4-4}$$

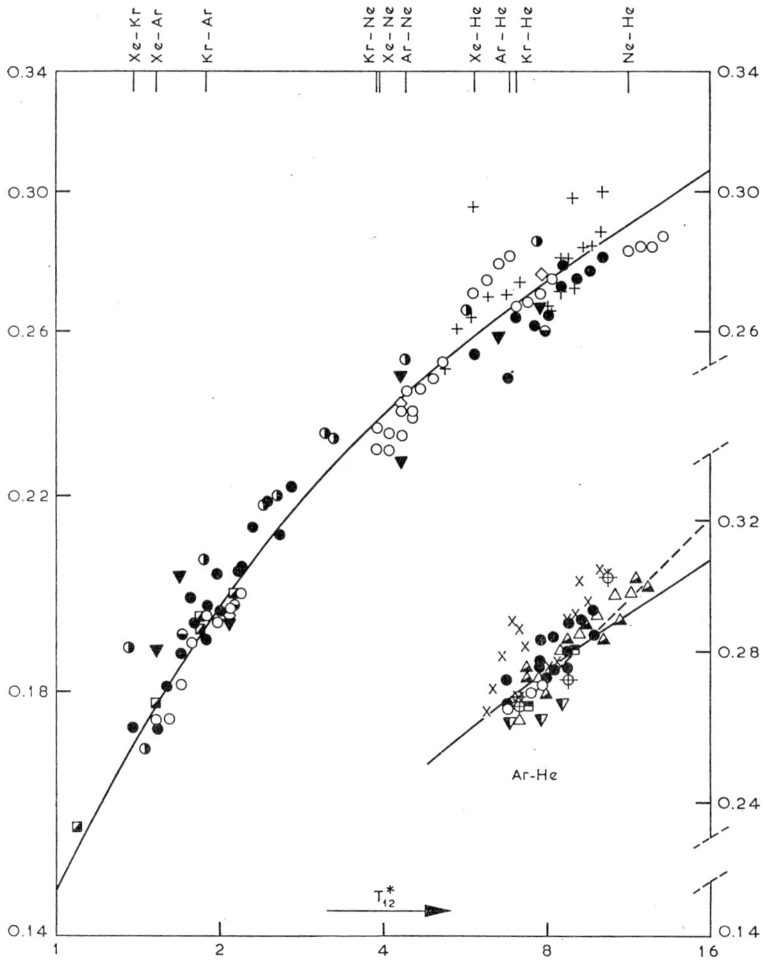

Fig. 10.7. The diffusion coefficient of the noble gaseous mixtures. The solid line represents the theory for the Lennard-Jones (6–12) potential. The marks denoting different mixtures refer to $T = 273$ °K. (The data for the Ar–He mixture are plotted separately by shifting the ordinate.) (From Van Heijningen [1967].)

Thus, in a plot of $n\sigma_{12}^2[\mathscr{D}_{12}]_1/\sqrt{T}$ or $\sigma_{12}^2[\mathscr{D}_{12}]_1 p/T^{\frac{3}{2}}$ vs. T^*, the data for all gas mixtures should fall on a single curve described by $[\Omega_{12}^{(1,1)*}(T^*)]^{-1}$. Such a plot is given in Fig. 10.7 and it is seen that, again within the scatter of the data, the theoretical curve is in agreement with experiment. Since the data are not as accurate as those for the viscosity and thermal conductivity, a serious evaluation of the appropriateness of the potentials cannot be given.

The most recent and most careful measurements of the diffusion coefficients of binary noble gas mixtures were made by Van Heijningen [1967], who measured the diffusion coefficients of all ten mixtures of the five noble gases and from them obtained potential parameters for the Lennard-Jones (6–12) potential and the modified Buckingham (6-exp) potential with four different values of α (12, 13, 14, 15). Van Heijningen was able to obtain good fits to his data with all of these potentials and was therefore unable to choose a "best" potential. A very useful consequence of these measurements is, that they provide a significant test of the combination rules for the potential parameters. For this purpose, recall Eqs. (8.5-4), (8.5-5), (8.5-6),

$$\epsilon_{12}\sigma_{12}^6 = [\epsilon_1\sigma_1^6\epsilon_2\sigma_2^6]^{\frac{1}{2}}, \tag{10.4-5}$$

$$\sigma_{12} = \tfrac{1}{2}(\sigma_1+\sigma_2), \tag{10.4-6}$$

$$\epsilon_{12} = (\epsilon_1\epsilon_2)^{\frac{1}{2}}. \tag{10.4-7}$$

Further recall that Eq. (10.4-5) was derived by means of semi-quantitative theoretical arguments, whereas the other two relations have less firm bases.

Now, by selecting a set of three noble gases and using the measured parameters for all three pairs that can be formed from the triplet, one obtains from the above equations a set of three equations for the parameters $\epsilon\sigma^6$, σ, and ϵ of the individual pure gases, respectively. In this way Van Heijningen computed the Lennard-Jones (6–12) parameters for the noble gases from each of the ten possible triplets; the results are shown in Table 10.6.

From the last two rows it appears that the agreement in all cases is fairly good, except for $\epsilon\sigma^6$ for helium. However, it is seen that the individual determinations of ϵ for each gas differ by so much that the comparison of averages is not meaningful. In part, the variation in the values of ϵ obtained is simply a reflection of the sensitive dependence of ϵ on the transport properties that was discussed earlier, so that no definite conclusion, other than that one should exercise caution, can be reached with respect to the ϵ-mixing rule. The other two rules are rather closely obeyed except for helium. This is not surprising in view of the earlier discussion of pure helium; the Lennard-

Jones potential is simply not appropriate for this gas. Use of the modified Morse potential, which improves the results considerably for the transport properties of pure helium, should improve the agreement between theory and experiment. Although such calculations have not been carried out, Fig. 10.8 displays the remarkable improvement obtained when this potential is applied to the calculation of the isotopic thermal diffusion factor in ^3He–

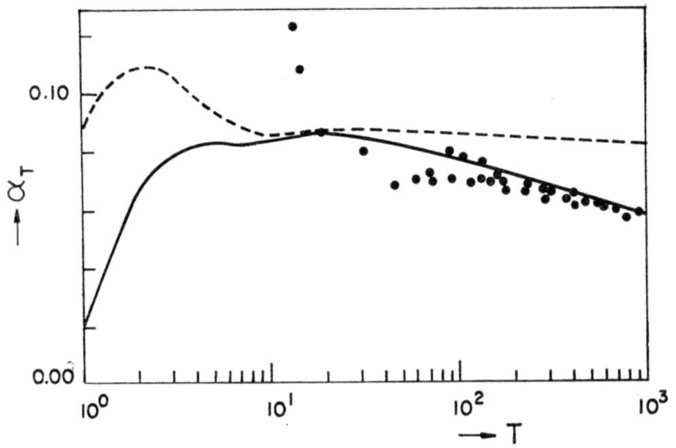

Fig. 10.8. Thermal diffusion factor of an equimolar ^3He–^4He mixture.
——— $[\alpha_T]_3$, using the modified Morse potential,
– – – – $[\alpha_T]_3$, using the Lennard-Jones (6–12) potential,
● experimental points. (From Keller and Taylor [1969].)

^4He mixtures. For the other gases, the agreement between the mixing rule for σ and experiment may be ascribed to the fact that σ-values for the various gases do not differ by very much and almost any averaging procedure will yield accurate results. On examination, one sees that the results are worst in gas mixtures with the largest differences in σ (note the last entries for krypton and xenon) and one should thus conclude that the σ-mixing rule, although not valid, may prove useful in mixtures for which the radii of the individual species are approximately equal. The comparison of $\epsilon\sigma^6$ is the most strenuous, as the variation in the values from element to element eliminates the possibility of accidental good agreement such as was found for the σ-rule. It is seen that, again with the exception of helium, the rule passes the test rather successfully. Of course, the rule is not perfect, but the agreement is good in both the comparison of absolute values and in the smallness of deviations in the individual determinations.

TABLE 10.6

Test of the combination rules for Lennard-Jones (6–12) parameters

Triplet	Parameter		Derived value for simple gas				
			He	Ne	Ar	Kr	Xe
He–Ne–Ar	ϵ/k	°K	15.4	36.4	105		
	σ	Å	2.51	2.77	3.45		
	$(\epsilon/k)\sigma^6$	°KÅ^6	4.12	1.56	1.94		
He–Ne–Kr	ϵ/k	°K	13.2	42.4		115	
	σ	Å	2.57	2.71		3.77	
	$(\epsilon/k)\sigma^6$	°KÅ^6	4.00	1.60		4.05	
He–Ne–Xe	ϵ/k	°K	15.9	35.2			136
	σ	Å	2.53	2.75			4.21
	$(\epsilon/k)\sigma^6$	°KÅ^6	4.48	1.43			1.04
He–Ar–Kr	ϵ/k	°K	10.8		149	141	
	σ	Å	2.62		3.34	3.72	
	$(\epsilon/k)\sigma^6$	°KÅ^6	4.05		1.97	3.99	
He–Ar–Xe	ϵ/k	°K	10.5		154		206
	σ	Å	2.70		3.26		4.04
	$(\epsilon/k)\sigma^6$	°KÅ^6	4.66		1.75		1.03
He–Kr–Xe	ϵ/k	°K	9.2			165	235
	σ	Å	2.69			3.65	4.05
	$(\epsilon/k)\sigma^6$	°KÅ^6	4.31			3.75	1.08

Ne–Ar–Kr	ϵ/k	°K		29.7	128	164	
	σ	Å		2.82	3.40	3.66	
	$(\epsilon/k)\sigma^6$	°KÅ^6		1.58	1.91	4.11	
Ne–Ar–Xe	ϵ/k	°K		24.0	159		199
	σ	Å		2.94	3.28		4.02
	$(\epsilon/k)\sigma^6$	°KÅ^6		1.58	1.91		0.94
Ne–Kr–Xe	ϵ/k	°K		24.5		199	195
	σ	Å		2.87		3.61	4.09
	$(\epsilon/k)\sigma^6$	°KÅ^6		1.54		4.20	0.97
Ar–Kr–Xe	ϵ/k	°K			131	161	242
	σ	Å			3.33	3.73	3.97
	$(\epsilon/k)\sigma^6$	°KÅ^6			1.87	4.22	0.96
Average for simple gas	ϵ/k	°K	12.5 ± 1.0	32.3 ± 2.3	138 ± 8	158 ± 10	202 ± 16
	σ	Å	2.60 ± 0.03	2.81 ± 0.03	3.34 ± 0.03	3.69 ± 0.03	4.06 ± 0.04
	$(\epsilon/k)\sigma^6$	°KÅ^6	4.24 ± 0.17	1.56 ± 0.03	1.90 ± 0.04	4.10 ± 0.08	0.98 ± 0.03
Average from simple gas properties	ϵ/k	°K	10.5	35.9	121	173	218
	σ	Å	2.56	2.77	3.42	3.63	4.05
	$(\epsilon/k)\sigma^6$	°KÅ^6	2.95	1.62	1.93	3.95	0.96

Note: Results for $\epsilon\sigma^6$ have been divided by the following factors: 10^3 for He, 10^4 for Ne, 10^5 for Ar and Kr, and 10^6 for Xe.

A special case of the binary diffusion coefficient, the self-diffusion coefficient, was discussed in the previous section. Results for this property are obtained from the results given here by allowing the masses of the two gases to become equal and the force laws between all pairs of molecules to be the same. In practice, one measures the diffusion of isotopes of the same gas through each other, but the difference in properties between the isotopes can generally be neglected.

Finally, some mention should be made of multicomponent diffusion. Relatively few measurements of multicomponent diffusion coefficients have been made and their complicated dependence on composition makes them rather difficult to analyze. However, once the necessary potential parameters are available, there is no difficulty in principle in calculating these properties and the availability of computers makes the task rather straightforward. As far as is known, the theory is as accurate in this case as in those discussed previously.

The thermal diffusion coefficient which we now turn to is, perhaps, the most interesting of all the transport properties. Its behavior is strongly dependent on the nature of the intermolecular potential and on the masses of the particles, and it is the most difficult property to calculate accurately. Thermal diffusion was so little understood that an accurate physical picture of the phenomenon was given only recently by Monchick and Mason [1967]. An excellent review devoted solely to this property was given by Mason, Munn and Smith [1966].

It is more convenient, for a binary mixture, to consider the thermal diffusion factor α_T defined by

$$\alpha_T = (D_T/\mathscr{D}_{12})(x_1 x_2)^{-1}, \tag{10.4-8}$$

cf. Eqs. (6.5-5) and (7.3-71). The thermal diffusion factor is a less rapidly varying function of composition than the thermal diffusion coefficient itself. Its first Chapman-Cowling approximation $[\alpha_T]_1$ is given by Eq. (7.3-72). Higher-order approximations can be written in the form

$$\alpha_T = [\alpha_T]_1(1+K_{12}) = (6\mathrm{C}^\star_{12}-5)\frac{S_1 x_1 - S_2 x_2}{Q_1 x_1^2 + Q_2 x_2^2 + Q_{12} x_1 x_2}(1+K_{12}). \tag{10.4-9}$$

To obtain the Kihara approximations one merely replaces $\mathrm{B}^\star_{12}$ by $\frac{5}{4}$ in the expressions (7.3-44) for the coefficients Q. Of course, the expressions for K_{12} in the Chapman-Cowling and Kihara approximations are different.

It was shown earlier that calculations of the thermal diffusion factor converge relatively slowly so that K_{12} is not small. This makes analysis of thermal diffusion in terms of the intermolecular potential rather difficult. A means of avoiding this difficulty was indicated by Laranjeira [1960] who found, on the basis of mean free path arguments, that the reciprocal of α_T is a linear function of x_1 and x_2. Although there is no reason to expect this result to be valid in general, it was found to be in rather close agreement with experiment. This led Mason, Weissman and Wendt [1964] to manipulate Eq. (10.4-9) into the form:

$$\alpha_T^{-1} = (6c_{12}^{\star}-5)^{-1}[(Q_1/S_1)(1+K_1)^{-1}x_1-(Q_2/S_2)(1+K_2)^{-1}x_2]+R, \qquad (10.4\text{-}10)$$

where K_1 and K_2 are the values of K_{12} when $x_1 = 1$ and $x_2 = 1$, respectively. The factor R is a rather complicated function, but Mason et al. showed by numerical experimentation that it never contributes more than about one per cent to α_T^{-1}. Thus, K_{12} need be computed only for the two extreme compositions in order to obtain α_T to within the experimental error; these results are well borne out by experiment. We note in passing that Eq. (10.4-2) for the binary diffusion coefficient can be put into a similar form, but it is not as important to do so as the composition depencence of this quantity is considerably weaker than that of K_{12}.

Unfortunately, a systematic set of measurements of the thermal diffusion factors of the noble gas mixtures has not yet been made. There exist partial results, but the data required for a systematic evaluation of the usefulness and validity of the various potential models and the combination rules do not yet exist. Furthermore, with the important exception of isotopic thermal diffusion, there exists no law of corresponding states for thermal diffusion of the type found for binary diffusion; while this makes a systematic discusion difficult, it also provides a much more severe test of the theory.

Results obtained by Grew and Mundy [1961] for argon–krypton and krypton–xenon mixtures with the heavier element present in low concentration follow the trends predicted by the Lennard-Jones (6–12) and modified Buckingham (6-exp) potentials, but cannot be properly fit by either. It is reasonable to expect that the discrepancy can be removed by the use of more flexible potentials.

Munn and Smith [1965] tested their potential against isotopic thermal diffusion factor data for the noble gases – except helium – and found good agreement. However, the agreement is not as good as that obtained for the

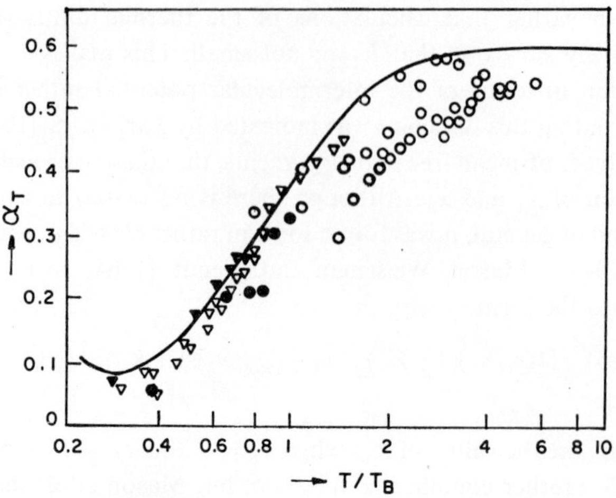

Fig. 10.9. Comparison of isotopic thermal diffusion factors of Ne (○), Ar (▽), Kr (●) and Xe (▼) with predictions of the Munn-Smith potential. The abscissa is T/T_B, where T_B is the Boyle temperature, i.e., the temperature at which the second virial coefficient vanishes; if $T_B^* = kT_B/\epsilon$, then $T_B^* = 3.42$ for the Lennard-Jones potential, $T_B^* = 2.61$ for the Munn-Smith potential. (From Munn and Smith [1965].)

viscosity and thermal conductivity, see Fig. 10.9. On the other hand, the Lennard-Jones potential produces a curve which lies about fifty per cent above the experimental data. The use of the modified Morse potential for helium has already been discussed.

It is clear then, that thermal diffusion provides an excellent tool for studying intermolecular potentials. However, much further work must be done before its usefulness is fully exploited.

10.5. Noble gas mixtures – Viscosity and thermal conductivity

In contrast to the properties just discussed, the viscosity and thermal conductivity of gas mixtures depend as strongly on the like-particle potential as they do on the unlike-particle potential. Consequently, they are of lesser utility in determining potentials and, therefore, in testing the theory – although they can be and have been used for that purpose. From the previous discussion, we can be reasonably well assured that if all of the necessary potentials were available, the viscosity and thermal conductivity of a gas mixture could be computed with satisfactory accuracy. In fact, for the noble gases at least, this condition is met and there is no serious problem in computing transport properties of mixtures of them; use of the mixing rules produces results which are sufficiently accurate for most purposes.

Comparisons of theory and experiment illustrating the above remarks are important and will be given later. However, we must also remember that in many applications the transport properties of a complex gas mixture for which not all the required potentials are known, must be computed. For such mixtures, further approximations must be introduced. These approximations should, of course, be based as nearly as possible on the rigorous theory, should take advantage of whatever data are available and should be as simple as possible. As both excellent experimental results and accurate calculations are available for the noble gases, they provide an excellent testing ground for these approximation methods.

Numerous approximation methods have been suggested in the literature. These may be roughly classified in the following manner:

1. Completely empirical methods in which an assumed form of the composition and/or temperature dependence of the property is fitted to experimental data.

2. Heuristic methods in which a relation among various transport properties is assumed. In this category fall methods which assume the transport properties of a mixture to be averages (of various sorts) of the transport properties of the pure components and also, for example, the method in which the Eucken ratio is assumed to retain its pure monatomic gas value.

3. Methods based on simplified theories such as the mean free path theory. These methods can usually be relied on to produce a formula giving the correct functional dependence on composition, but the constants are invariably incorrect. However, they can be used as a rational basis for formulating semi-empirical formulae.

4. Methods based on derived relations between the property in question and other transport properties.

5. Methods based on systematic approximation of the rigorous formulae.

Of course, not all of the formulae which have been suggested fall into just one of these categories and, indeed, some of the formulae can be derived by more than one method. As might be expected, the empirical and heuristic approaches are of the least accuracy and generally require fitting adjustable constants to the data for each special case. Thus they should be regarded merely as *ad hoc* formulae and, as they have been only moderately successful, we shall not discuss them in detail.

One of the most successful formulae was derived on the basis of the mean free path theory by Wassiljewa [1904] who found that the thermal conductivity and viscosity of a K-component mixture can be represented in the form:

$$\lambda = \sum_{i=1}^{K} \lambda_i \left[1 + \sum_{\substack{j=1 \\ j \neq i}}^{K} G_{ij}^{\lambda} \frac{x_j}{x_i}\right]^{-1}, \tag{10.5-1a}$$

$$\eta = \sum_{i=1}^{K} \eta_i \left[1 + \sum_{\substack{j=1 \\ j \neq i}}^{K} G_{ij}^{\eta} \frac{x_j}{x_i}\right]^{-1}, \tag{10.5-1b}$$

where the G_{ij} are universal constants for each *gas pair* which may, however, be functions of temperature; they were interpreted by Cowling, Gray and Wright [1963] as the relative effectiveness of molecules of type j in impeding the transport of energy or momentum by molecules of type i. The formulae (10.5-1) provide an enormous simplification in that the properties of complex gas mixtures may be computed using constants obtained by fitting experimental data for simpler mixtures. If it is accepted that Eqs. (10.5-1) are a reasonable and accurate means of fitting the data, then the problem is one of computing the coefficients G_{ij}. Many authors have given means of calculating these coefficients, but they almost invariably rely on questionable arguments. Rather than go into these arguments, we shall turn our attention to the rigorous Chapman-Cowling formula, show that Eq. (10.5-1) is in fact an approximation to it and then compare the results with experiment and the exact results.

It was found in Chapter 7 that the first approximation for the viscosity and thermal conductivity of gas mixtures could be written as the ratio of two determinants. As an example, the first approximation for the viscosity of a K-component mixture is given by (cf. Eq. (7.3-82)):

$$[\eta]_1 = \frac{-1}{\det |H_{ij}|} \begin{vmatrix} H_{11} & H_{12} & \ldots & H_{1K} & x_1 \\ \vdots & \vdots & & \vdots & \vdots \\ H_{K1} & H_{K2} & \ldots & H_{KK} & x_K \\ x_1 & x_2 & \ldots & x_K & 0 \end{vmatrix}. \tag{10.5-2}$$

(Note that we have omitted the upper indices on the elements H_{ij}, $H_{ij} \equiv H_{ij}^{00}$.) The elements H_{ij} appearing in this determinant are different according to whether one uses the Chapman-Cowling or Kihara approximation method. In fact, by suitably redefining the matrix elements, higher-order approximations for the viscosity may be put into the form (10.5-2); naturally, the matrix elements become more complicated as the order of approximation increases. The first approximation for the thermal conductivity can be represented by an expression similar to the right member of Eq. (10.5-2) with appropriate definitions of the matrix elements, cf. Eq. (7.3-87).

Thus it can be seen that the viscosity and the thermal conductivity are bilinear functions of the molecular fractions x_i. However, the coefficient of $x_i x_j$ is not readily determined, especially if there are many components in the gas. On the other hand, it has been found by numerical computation that the diagonal elements of the matrix $\boldsymbol{H}$ are much larger than the off-diagonal elements and this can be used as the basis of a perturbation expansion. The expansion is readily worked out if one notes that Eq. (10.5-2) is equivalent to

$$[\eta]_1 = \sum_{ij} H_{ij}^{-1} x_i x_j , \tag{10.5-3}$$

where H_{ij}^{-1} is the i–j element of the matrix inverse to $\boldsymbol{H}$.

To expand this, we write

$$\boldsymbol{H} = \boldsymbol{H}_\mathrm{d} + \boldsymbol{H}' , \tag{10.5-4}$$

where $\boldsymbol{H}_\mathrm{d}$ is a diagonal matrix whose elements are H_{ii}, and $\boldsymbol{H}'$ is the matrix obtained from $\boldsymbol{H}$ by setting all the diagonal elements equal to zero. Then, by elementary matrix manipulation,

$$\begin{aligned} \boldsymbol{H}^{-1} &= [\boldsymbol{H}_\mathrm{d}(1+\boldsymbol{H}_\mathrm{d}^{-1}\boldsymbol{H}')]^{-1} = (1+\boldsymbol{H}_\mathrm{d}^{-1}\boldsymbol{H}')^{-1}\boldsymbol{H}_\mathrm{d}^{-1} \\ &= (1-\boldsymbol{H}_\mathrm{d}^{-1}\boldsymbol{H}'+\boldsymbol{H}_\mathrm{d}^{-1}\boldsymbol{H}'\boldsymbol{H}_\mathrm{d}^{-1}\boldsymbol{H}'- \ldots)\boldsymbol{H}_\mathrm{d}^{-1}. \end{aligned} \tag{10.5-5}$$

Substitution of this expansion into Eq. (10.5-2) yields the following expansion for the viscosity,

$$[\eta]_1 = \sum_i \frac{x_i^2}{H_{ii}} - \sum_i \sum_{\substack{j \\ (j \neq i)}} \frac{x_i H_{ij} x_j}{H_{ii} H_{jj}} + \sum_i \sum_{\substack{j \\ (j\neq i)}} \sum_{\substack{k \\ k(\neq j)}} \frac{x_i H_{ij} H_{jk} x_k}{H_{ii} H_{jj} H_{kk}} - \ldots . \tag{10.5-6}$$

As a first approximation, one retains only the first term of this expression and obtains

$$[\eta]_1 = \sum_i \frac{x_i^2}{H_{ii}} . \tag{10.5-7}$$

This expression can, of course, be obtained more simply by neglecting the off-diagonal elements of $\boldsymbol{H}$ in Eq. (10.5-2). However, the derivation given allows for systematic extension. Brokaw [1958] has shown that by re-arranging the series (10.5-6) and neglecting different terms one may also arrive at the approximate result

$$[\eta]_1 = \sum_i x_i^2 (\sum_j H_{ij})^{-1}. \tag{10.5-8}$$

This formula is of about the same accuracy as the previous one and, because of the form of the H_{ij}, is actually simpler than the formula (10.5-7). It is particularly important to note that in the first Chapman-Cowling or Kihara approximations for the viscosity, Eqs. (10.5-7) and (10.5-8) can be brought to the form (10.5-1b) with the coefficients G_{ij}^{η} determined. Thus, using the first Chapman-Cowling or Kihara approximations in conjunction with Eq. (10.5-7) one obtains, using Eq. (7.3-80a) for $H_{ii}(\equiv H_{ii}^{00})$,

$$G_{ij}^{\eta} = \frac{2m_j^2}{(m_i+m_j)^2} \frac{[\eta_i]_1}{[\eta_{ij}]_1} \left[1 + \frac{5}{3\mathrm{A}_{ij}^{\star}} \frac{m_i}{m_j}\right]. \tag{10.5-9}$$

Eq. (10.5-8) yields, on the other hand,

$$G_{ij}^{\eta} = \frac{2m_j}{m_i+m_j} \frac{[\eta_i]_1}{[\eta_{ij}]_1}. \tag{10.5-10}$$

Since setting $\mathrm{A}_{ij}^{\star} = \frac{5}{3}$ makes the off-diagonal element H_{ij}, $i \neq j$, vanish, some authors have suggested that this approximation be used consistently throughout. Then all terms but the first in Eq. (10.5-6) are removed and one obtains Eq. (10.5-7) as an immediate consequence. If the substitution of $\frac{5}{3}$ for $\mathrm{A}_{ij}^{\star}$ is made in Eq. (10.5-9), one obtains for G_{ij}^{η} a formula identical to Eq. (10.5-10). However, this approach has the important disadvantage that it is not readily generalized. Furthermore, Brokaw [1958] has pointed out that for most molecular pairs $\frac{6}{5}\mathrm{A}_{ij}^{\star} \approx 1.32$ and that an empirical fit to data by Buddenberg and Wilke [1949] using a G_{ij}^{η} of the form (10.5-10), treating $\frac{6}{5}\mathrm{A}_{ij}^{\star}$ as an adjustable constant, yielded a value of 1.38 for this parameter. Thus it is apparent that the effectiveness of the approximation method given above derives from the smallness of the off-diagonal elements of $\boldsymbol{H}$, and that it is better to leave the diagonal elements unaltered. The small difference between the empirical constant 1.38 and the theoretical value of 1.32 is evidently a compensation for eliminating the higher-order terms in Eq. (10.5-6) and, probably, for higher-order terms in the Sonine polynomial approximation. That the empirical constant is greater than 1.32 is in consonance with the above remarks; the non-diagonal H_{ij} are negative and thus the second term in Eq. (10.5-6) is also negative.

The formulae given above have proven useful but they still have the disadvantage that, for each pair (i, j), the binary diffusion coefficient $[\mathscr{D}_{ij}]_1$ is required for the computation of $[\eta_{ij}]_1$, cf. Eq. (7.3-47). These are not likely to be available for all gas pairs in a mixture and it is even less likely that the intermolecular potentials needed to compute them will be available. One must then fall back on the mixing rules for the potentials which were discussed earlier. For the rigid sphere model, the result of using the mixing rule (8.5-1) can be put into a particularly simple form first given by Wilke [1950]:

$$\frac{[\eta_i]_1}{[\eta_{ij}]_1} = \frac{1}{4}\left(\frac{m_i+m_j}{2m_j}\right)^{\frac{1}{2}}\left[1+\left(\frac{m_j}{m_i}\right)^{\frac{1}{4}}\left(\frac{[\eta_i]_1}{[\eta_j]_1}\right)^{\frac{1}{2}}\right]^2, \quad (10.5\text{-}11)$$

which has the important advantage of relating the ratio required by Eqs. (10.5-9) and (10.5-10) to the properties of the pure gases. The relation (10.5-11) is useful as a first approximation for other force laws and yields surprisingly good results.

Most of the foregoing applies equally well to the thermal conductivity. In Chapter 7, Eq. (7.3-87), we have seen that the expression obtained in the first approximation can be put precisely in the form (10.5-2) or (10.5-3) with the elements $H_{ij}(\equiv H_{ij}^{00})$ replaced by $\Lambda_{ij} \equiv \Lambda_{ij}^{11}$. Thus, by following the procedures that led to Eq. (10.5-7) one obtains

$$[\lambda]_1 = \sum_i \frac{x_i^2}{\Lambda_{ii}} \quad (10.5\text{-}12)$$

and, in analogy to Eq. (10.5-8) one has

$$[\lambda]_1 = \sum_i x_i^2 (\sum_j \Lambda_{ij})^{-1}. \quad (10.5\text{-}13)$$

The expressions for $\Lambda_{ij}\ (\equiv \Lambda_{ij}^{11})$ were given by Eq. (7.3-34) and, using them one obtains an expression of the form (10.5-1a) with G_{ij}^{λ} given by:

$$G_{ij}^{\lambda} = \frac{1}{2\mathrm{A}_{ij}^{\star}} \frac{\frac{15}{2}m_i^2 + \frac{25}{4}m_j^2 - 3\mathrm{B}_{ij}^{\star} m_j^2 + 4\mathrm{A}_{ij}^{\star} m_i m_j}{(m_i+m_j)^2} \frac{[\lambda_i]_1}{[\lambda_{ij}]_1} \quad (10.5\text{-}14)$$

from Eq. (10.5-12), and

$$G_{ij}^{\lambda} = \frac{1}{2\mathrm{A}_{ij}^{\star}} \frac{(\frac{5}{4}(6m_i - 5m_j) + 3m_j \mathrm{B}_{ij}^{\star})(m_i - m_j) + 8m_i m_j \mathrm{A}_{ij}^{\star}}{(m_i+m_j)^2} \frac{[\lambda_i]_1}{[\lambda_{ij}]_1} \quad (10.5\text{-}15)$$

from Eq. (10.5-13). Furthermore, by using the relation (7.3-28) between $[\lambda_{ij}]_1$ and $[\eta_{ij}]_1$ and the Eucken relation (7.3-22) between $[\lambda_i]_1$ and $[\eta_i]_1$, one can easily convert Eq. (10.5-11) into a relation between $[\lambda_i]_1$ and $[\lambda_{ij}]_1$,

$$\frac{[\lambda_i]_1}{[\lambda_{ij}]_1} = \frac{1}{4}\left(\frac{2m_j}{m_i+m_j}\right)^{\frac{1}{2}}\left[1+\left(\frac{m_i}{m_j}\right)^{\frac{1}{4}}\left(\frac{[\lambda_i]_1}{[\lambda_j]_1}\right)^{\frac{1}{2}}\right]^2. \qquad (10.5\text{-}16)$$

This result was first given by Mason and Saxena [1958] who also showed, on an empirical basis, that its accuracy is improved somewhat by inserting a numerical factor 1.065 on the right hand side. Clearly, this is essentially equivalent to the empirical correction that was applied earlier to Eq. (10.5-9), i.e., changing the factor 1.32 to 1.38.

Finally, we note that the results for the second Kihara approximation are obtained by replacing B^*_{ij} wherever it occurs in the formulae by $\frac{5}{4}$.

Further approximations of the type discussed above were given by Brokaw [1965]. However, the complexity of the formulae is such that the computation involved is as great as that for the exact formulae (10.5-2) or (10.5-3). They are therefore of very restricted utility and will not be discussed further here.

TABLE 10.7
Percent errors in approximate calculations of mixture viscosities*

Gas mixture	Method 1	Method 2	Method 3
He–Ne	2.9	3.5	3.2
He–Ar	2.1	0.4	2.7
Ne–Ar	0.5	2.4	0.1
He–Ne–Ar	2.2	1.2	2.4

* All calculations are based on Eq. (10.5-1b). Method 1 uses Eq. (10.5-10) with $\frac{6}{5}A^*_{ij} = 1.385$; method 2 uses Eq. (10.5-11) and method 3 is based on Eq. (10.5-10) without the replacement used in method 1. (From Brokaw [1965].)

TABLE 10.8
Percent errors in approximate calculations of mixture thermal conductivities*

Gas mixture	Method 1	Method 2
He–Kr	1.7	3.6
He–Xe	10.1	2.8
Kr–Xe	1.0	1.4
He–Kr–Xe	6.5	3.2

* Both methods use Eq. (10.5-1a). In method 1, G^λ_{ij} is obtained from Eqs. (10.5-14) and (10.5-16), with a numerical correction factor 1.065 inserted (see text). Method 2 uses G^λ_{ij} obtained from Eqs. (10.5-15). (From Brokaw [1965].)

TABLE 10.9

Comparison of diffusion coefficients obtained from mixture viscosity data with theoretical values*

Mixture	T(°K)	$p\mathscr{D}_{12}$ (at cm²/sec) From η_{mix}	Theoretical
He–Ne	20.4	0.0102±0.0001	0.0086
	90.1	0.144 ±0.002	0.136
	194.0	0.532 ±0.001	0.506
	293.2	1.06 ±0.01	1.02
	473.2	2.37 ±0.02	2.31
He–Ar	72.0	0.0681±0.0010	0.0604
	192.5	0.349 ±0.004	0.347
	291.2	0.698 ±0.007	0.706
	373.2	1.07 ±0.01	1.08
	473.2	1.59 ±0.02	1.62
He–Kr	291.2	0.616 ±0.007	0.633
He–Xe	291.3	0.512 ±0.008	0.581
	400.0	0.880 ±0.004	1.01
	550.0	1.49 ±0.01	1.74
Ne–Ar	72.3	0.0259±0.0001	0.0232
	193.4	0.147 ±0.001	0.149
	291.2	0.292 ±0.007	0.308
	373.2	0.466 ±0.002	0.474
	473.2	0.695 ±0.005	0.711
	523.2	0.822 ±0.003	0.845
Ne–Kr	291.2	0.252 ±0.004	0.243
Ne–Xe	291.2	0.208 ±0.001	0.220
Ar–Kr	291.2	0.132 ±0.001	0.132
Ar–Xe	291.2	0.107 ±0.000	0.110
Kr–Xe	291.2	0.0725±0.0006	0.728

* The errors quoted in the experimental results are simply those derived by statistical analysis applied to the stated errors in viscosity measurements. Thus, systematic errors are *not* included. This table was taken from Weissman and Mason [1962]; see the original paper for the sources of experimental data.

Since the G_{ij} can be expressed in terms of the diffusion coefficients of binary mixtures, it is clear that the formulae can be used to obtain diffusion coefficients if mixture viscosity or thermal conductivity data are available.

TABLE 10.10

Comparison of calculated and experimental viscosities of binary mixtures at 20 °C *

Mole fraction of light gas	$\eta(10^{-6}$ g/cm sec) Calculated	Experimental
He–Ne		
0	307.0	309.2
0.2041	296.0	300.4
0.5624	265.9	270.2
0.7621	239.5	242.9
1	193.1	194.1
He–Ar		
0	222.0	221.1
0.3405	229.3	227.8
0.4906	231.3	229.6
0.5966	231.4	230.4
0.7565	227.4	227.0
1	193.1	197.3
Ne–Ar		
0	222.0	221.3
0.2580	241.2	240.1
0.3909	252.0	250.4
0.5382	264.6	263.5
0.7230	281.2	281.1
1	307.0	309.2

* From Brokaw [1965]; see Brokaw's paper for sources of experimental data.

This has been done by Weissman and Mason [1962] and Storvick and Mason [1966] with good success, as we shall demonstrate later. For this purpose, viscosity data are preferable to thermal conductivity data as they are generally more accurate and the procedure for obtaining diffusion coefficients tends to magnify errors in the data.

As was stated earlier, numerous other approximations for the transport properties of mixtures have been proposed. Those which have not been treated here have generally proven less successful than those which have been given, and are no longer extensively used. A listing of a number of these formulae was given by Gandhi and Saxena [1966]; the reader is referred to their article, in which also comparisons with experiment and exact calculations can be found.

We now give some numerical results to illustrate that the theory predicts

TABLE 10.11

Comparison of calculated and experimental thermal conductivities of binary mixtures at 29 °C *

Mole fraction of light gas	$\lambda(10^{-6}$ cal/cm sec °K) Calculated	Experimental
He–Kr		
0	22.61	23.2
0.240	54.70	54.1
0.490	104.7	103
0.750	192.8	193
0.880	265.4	261
1	367.1	367
He–Xe		
0	13.98	14.27
0.202	38.81	35.7
0.418	76.38	71.7
0.787	199.31	188.2
1	367.1	367
Kr–Xe		
0	13.98	14.27
0.158	14.98	14.9
0.276	15.80	15.8
0.510	17.63	18.6
0.785	20.20	20.6
1	22.61	23.2

* From Brokaw [1965]; see Brokaw's paper for sources of experimental data.

the transport properties of mixtures as accurately as those of simple gases and to indicate the accuracy of the approximation formulae just given. Some care is needed here, as the uncertainties in the currently available data are generally of the same order of magnitude as the difference between exact and approximate calculations. Thus, to show the error in the approximations more clearly, we give, in Tables 10.7 and 10.8, the error in the approximate methods as compared to the exact first Chapman-Cowling approximation for the viscosity and the thermal conductivity. The errors quoted are root-mean-square errors obtained by averaging over a range of compositions; the Lennard-Jones (6–12) potential was used throughout.

A still better test is to compare the mixing coefficients G_{ij} obtained in various ways. Some precaution is necessary here as large errors in empirically derived mixing coefficients can result from relatively small errors in the viscosity. For this reason it is better to bypass the computation of mixing coefficients and directly compute binary diffusion coefficients from mixture property data. This approach was taken by Weissman and Mason [1962] and Storvick and Mason [1966]. The former authors used Eq. (10.5-9) in conjunction with Eq. (10.5-1) for binary mixtures and thus obtained a quadratic equation for the binary diffusion coefficient. Then, noting that A^*_{12} is not strongly dependent on the force law and adopting a reasonable value for it, they obtained the binary diffusion coefficient. A comparison of the values obtained with theoretical values of the diffusion coefficient (which we have already seen are in good agreement with experiment) is given in Table 10.9. As can be seen the results are in very good agreement. Storvick and Mason [1966] state that, from viscosity data of 0.1 % accuracy, binary diffusion coefficients of 1 % accuracy can be derived. Thus, if this accuracy is experimentally feasible, mixture viscosity data may be as useful for analyzing unlike-molecule potentials as diffusion coefficient data.

Finally, in Tables 10.10 and 10.11, we give some direct comparisons of the calculations with experiment. The calculated values are the first Chapman-Cowling approximations for the viscosity and the thermal conductivity. As is to be expected from the foregoing, the agreement is good and similar agreement is found for multicomponent mixtures.

Polyatomic gases 11

The Chapman-Enskog theory presented in the preceeding chapters is remarkably successful in providing a derivation of the equations of fluid dynamics and in relating the transport properties of gases to the forces between molecules. Although some work remains to be done, it can be said that the theory is in a rather mature state. Remarkable though it is, the theory possesses important limitations that greatly reduce its usefulness. Most of these restrictions are actually inherent in the Boltzmann equation itself; foremost among them are the limitations to *monatomic molecules* and *low densities*. The theory for polyatomic and dense gases is currently far from complete and is undergoing extensive development. Considerable progress has been made already and we will attempt to describe the present state of knowledge and indicate what further progress may reasonably be expected. Although neither problem is simple, the treatment of polyatomic gases is perhaps a little simpler than that of dense gases, because it allows somewhat greater use of intuition (which may, however, be misleading). We therefore discuss the problem of polyatomic gases first and turn our attention to dense gases in the following chapter. Still other extensions of the theory – to ionized and rarefied gases – are reserved until later chapters.

Polyatomic gases possess two features that do not appear in the treatment of monatomic gases: (i) they have *internal degrees of freedom* (or modes of motion) with which energy may be associated, and (ii) they may have various shapes and, consequently, the molecular interaction is, in general, *not spherically symmetric*. These two factors affect the transport properties in different ways. The existence of internal molecular energy is likely to have its greatest effects on those properties which are related to the energy that a molecule carries with it – i.e., the thermal conductivity and the thermal diffusion coefficient –, and less important effects on the other transport properties. The lack of symmetry of the molecular potential affects all transport properties to approximately the same degree; however, it is difficult to treat in a unified manner, as the nature of the intermolecular potential varies considerably from one molecular species to another.

Of course, the effects of internal molecular energy and asymmetry cannot be separated completely. An important phenomenon which prevents such a clear separation is that of *inelastic collisions*, i.e., collisions producing changes in the internal state of either or both molecules. However, for purposes of developing a simple theory it is convenient to treat the two features separately. We begin this chapter by giving some of the simpler methods of treating polyatomic gases. In Section 11.1 we discuss the effects of asymmetric potentials on the viscosity. If molecules possessed no internal energy, the Eucken formula (7.3-22) between the coefficients of viscosity and thermal conductivity would still apply and the latter transport coefficient could be trivially computed once the former was available. In the case of a polyatomic gas this condition is not met and the effects of transport of internal energy must be included in a discussion of thermal conductivity. In Section 11.2 we give a simple method for treating the internal degrees of freedom.

As the methods of Sections 11.1 and 11.2 do not rely on a modified Boltzmann equation, they allow full use of the theory already at hand. However, they cannot pretend to be rigorous and they rely heavily on intuitive arguments. One should not be surprised, therefore, that the results are not accurate. To the contrary, their moderate success can be considered a tribute to the ingenuity of the authors of these methods. They are presented here for a twofold reason: firstly, they are of some value in predicting transport properties, especially since the complete theory is still lacking, and, secondly, the ideas introduced help in understanding aspects of the problems which arise in constructing the rigorous theory.

To improve on the simple theories, one must essentially begin all over again, by developing a modified Boltzmann equation appropriate to polyatomic molecules. The difficulty here is, that the dynamics of the collisions between polyatomic molecules is at present rather poorly understood and very little information on the collision cross sections is available. Two approaches have been tried. In the first, one accepts *a priori* a model to describe the molecules – e.g., rough spheres or ellipsoids, for which the dynamics of a collision can be worked out – and then proceeds via a modification of the Chapman–Enskog method to obtain the transport properties; there is usually a new parameter or two available for fitting the results to data. In the other approach, which we shall take in Sections 11.3 and 11.4, one uses the modified Boltzmann equation with assumed, but unknown, cross sections and formally obtains the transport properties in terms of averages of the cross sections. By means of further approximations one then obtains expressions that may be used to interpolate or extrapolate data. Since a great many cross sections are involved, it is impossible to use the data to obtain more than a very limited amount of information about the cross sections. This difficulty is mitigated, however, by the existence of a greater number of transport coefficients for polyatomic gases than for monatomic gases. For example, in Section 11.5 we demonstrate that, in a magnetic field, there are several coefficients of viscosity, whose measurement can provide considerable information on the interactions of polyatomic molecules.

11.1. Viscosity – The effect of asymmetric potentials

As polyatomic molecules exist in practically every variety of size and shape, it is impossible to give a simple treatment of their properties and, especially, of their interactions with other molecules. However, by restricting ourselves to relatively simple molecules (a broad enough category to include most of the common gases), we are able to make some general statements. The potentials of these gases have already been discussed in Chapter 8.

It will be recalled that the homopolar molecules, i.e., the diatomic molecules of the elemental gases, possess no permanent dipole moment and, consequently, their interaction potential at long range is proportional to r^{-6}. The only difference between these molecules and those of the noble gases is that, since the polarizability of the molecules is not a scalar but a tensor, the potential energy of interaction of two molecules will depend on their orientation. Also, since the polarizability is generally larger than in the case of the noble gases, the r^{-6} portion of the potential can be expected to be somewhat larger; the polarizabilities of some common gases are given in Table 11.1. The difficulty in treating such molecules lies, of course, in the orientation dependence of the potential. Since the time required for a molecule to rotate once is approximately equal to the duration of the collision,*

* The duration of a collision can be estimated by $t_{coll} \approx (2kT/m)^{-\frac{1}{2}} r_0$, where r_0 is the range of the potential. On the other hand, the rotational angular frequency is given by $\omega^2 = 2E_r/I$ where E_r is the energy associated with the rotational motion (which can be estimated by kT for a diatomic molecule) and I is the moment of inertia ($= \frac{1}{2}mR^2$, where R is the separation of two atoms). Thus the rotation time $\tau = 2\pi/\omega \approx \pi(kT/m)^{-\frac{1}{2}}R$. Since r_0 is a few times R, $t_{coll} \approx \tau$.

TABLE 11.1
Molecular polarizability*

Molecule	α_1	α_2	α_3	$\bar{\alpha} = \frac{1}{3}(\alpha_1+\alpha_2+\alpha_3)$	Remarks
		(10^{-25} cm^3)			
Diatomic molecules					
H_2	9.3	7.1	7.1	7.9	α_1 symmetry axis
N_2	23.8	14.5	14.5	17.6	α_1 symmetry axis
O_2	23.5	12.1	12.1	16.0	α_1 symmetry axis
Cl_2	66.0	36.2	36.2	46.1	α_1 symmetry axis
HF				24.6	α_1 symmetry axis
HCl	31.3	23.9	23.9	26.3	α_1 symmetry axis
HBr	42.2	33.1	33.1	36.1	α_1 symmetry axis
HI	65.8	48.9	48.9	54.4	α_1 symmetry axis
CO	26.0	16.3	16.3	19.5	α_1 symmetry axis
Linear molecules					
N_2O	48.6	20.7	20.7	30.0	α_1 symmetry axis; molecules linear and unsymmetric N≡N=O
CO_2	40.1–41.0	19.7–19.3	19.7–19.3	26.5	α_1 symmetry axis; linear
CS_2	151.4	55.4	55.4	87.4	α_1 symmetry axis; linear
HCN	39.2	19.2	19.2	25.9	α_1 symmetry axis; linear
C_2H_2	51.2	24.3	24.3	33.3	α_1 symmetry axis; linear
Nearly spherical molecules					
CH_4	26.0	26.0	26.0	26.0	regular tetrahedron
CCl_4	105	105	105	105	regular tetrahedron
Other molecules					
H_2S	40.4	34.4	40.1	37.8	$\mu = \mu_3$; $\alpha_2 \perp$ HSH plane
SO_2	54.9	27.2	34.9	37.2	$\mu = \mu_3$; $\alpha_2 \perp$ OSO plane
C_2H_6	54.8	39.7	39.7	44.7	α_1 symmetry axis
C_2H_4				42.6	α_1 symmetry axis
C_3H_8	50.1	69.3	69.3	62.9	$\alpha_1 \perp$ CCC plane
CH_3Cl	54.2	41.4	41.4	45.6	$\mu = \mu_1$; α_1 symmetry axis
CH_2Cl_2	50.2	84.7	59.6	64.8	$\mu = \mu_3$; $\alpha_1 \perp$ ClCCl plane; α_3 through C and bisects ClCl bond
$CHCl_3$	66.8	90.1	90.1	82.3	$\mu = \mu_1$; α_1 symmetry axis

* From Landolt-Börnstein [1951], Vol. I, Part 3, pp. 510, 511. μ is the permanent dipole moment.

the dynamical problem is rather difficult and has, in fact, not been worked out except for models in which the molecules are assumed to be rigid bodies.

The argument generally adopted for avoiding this difficulty is that, since the particles have random orientations before the collision and since they are rotating during the collision, the dynamics of a collision are affected only by the potential averaged over orientations. Since averaging the potential between two non-polar diatomic molecules over orientation simply produces a Lennard-Jones (6–12) potential, one has, at least crudely, justified the use of the Lennard-Jones (6–12) potential for these gases; the same argument may be applied to other non-polar polyatomic molecules such as the symmetric linear molecules of carbon dioxide and nitrous oxide and the nearly spherical molecules of methane. Although this procedure is certainly not the best one which can be devised (see the treatment of polar molecules below) it has achieved considerable success and has therefore been rather widely adopted. A typical result of fitting the viscosities of non-polar gases with the Lennard-Jones potential is given in Fig. 11.1. The curve with $\delta^* = 0$ (corresponding to a Lennard-Jones potential) gives a good fit to the viscosity

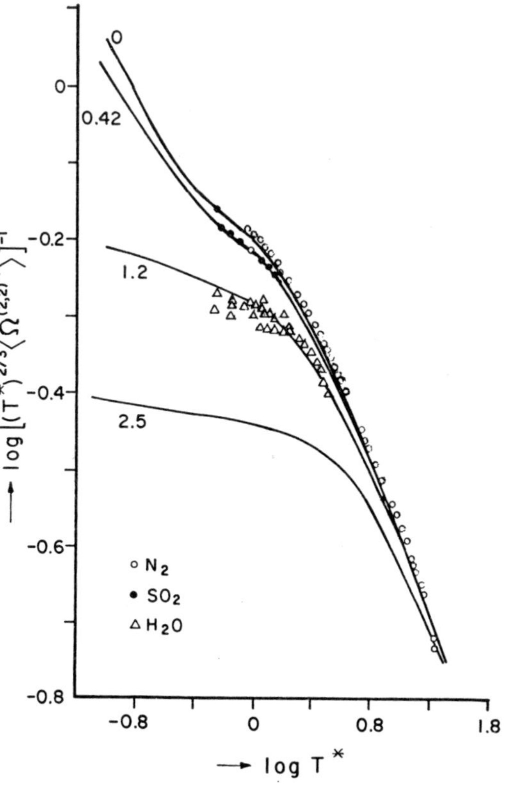

Fig. 11.1. Reduced viscosity vs. reduced temperature for polar gases. The curves are computed using the Stockmayer potential with fixed relative orientation during a collision (see text). The numbers 0, 0.42, 1.2, 2.5 are the values of δ^* used. (From Monchick and Mason [1961].)

of nitrogen and a reasonably good fit to the viscosity of sulphur dioxide.

For those gases which have a permanent dipole moment, we have seen in Chapter 8 that the potential should consist of a short range exponential or r^{-12} term, an induced dipole–permanent dipole term varying as r^{-6}, and a permanent dipole–permanent dipole term varying as r^{-3}. Both of the latter are orientation dependent but, since the permanent dipole–permanent dipole term is more strongly so and is more important at large distances, the orientation dependence of the induction term is usually ignored. One then has the Stockmayer potential (8.4-11), which is a combination of the Lennard-Jones (6–12) potential and the dipole–dipole interaction given by Eq. (8.2-1),

$$\varphi(r) = 4\epsilon \left[\left(\frac{\sigma}{r}\right)^{12} - \left(\frac{\sigma}{r}\right)^{6} - \delta \left(\frac{\sigma}{r}\right)^{3}\right], \tag{11.1-1}$$

where

$$\delta = \frac{\mu^2}{4\epsilon\sigma^3} [2 \cos \theta_1 \cos \theta_2 - \sin \theta_1 \sin \theta_2 \cos(\phi_1 - \phi_2)], \tag{11.1-2}$$

where, again, μ is the permanent dipole moment of the molecule and θ_1, ϕ_1 and θ_2, ϕ_2 are angles describing the orientation of the two molecules, respectively. Note that σ and ϵ are no longer the zero point and depth of the potential.

Transport properties have not been computed with this potential due to the difficulties mentioned above. A number of methods have been proposed, however, which incorporate some of the features of this potential. In one of the earliest proposals, the molecules were assumed to be aligned head to tail ($\theta_1 = \theta_2 = 0$), so that $\delta = \delta^* = \mu^2/2\epsilon\sigma^3$. The potential becomes

$$\varphi(r) = 4\epsilon \left[\left(\frac{\sigma}{r}\right)^{12} - \left(\frac{\sigma}{r}\right)^{6} - \delta^* \left(\frac{\sigma}{r}\right)^{3}\right]. \tag{11.1-3}$$

However, the molecules will line up as prescribed only if the relative velocity of the two molecules is very low, i.e., at low temperatures, so this argument cannot be correct. To avoid this difficulty, Hirschfelder, Curtiss and Bird [1954] suggested treating δ^* as an adjustable parameter and, with a third adjustable parameter available, a rather good fit to the data can be obtained. The objection to this procedure is that the dipole moment of a molecule can be measured independently (for example, it is simply obtained by measuring the dielectric constant), and δ^* is therefore known once σ and ϵ are determined. The value of δ^* obtained by these two procedures will, in general, be quite different.

Another proposal is based on the idea used above that, since the molecules are randomly oriented prior to the collision and since they rotate during the collision, an average of the dipole–dipole potential over orientations should be used. However, the simple average of this potential is zero and so one must use the concept of an "effective" orientation. The function δ of Eq. (11.1-2) is then replaced by a constant – its value at the "effective" orientation. This leads again to Eq. (11.1-3) and is thus simply another way of viewing the adjustable δ^* method. One can argue that the "effective" orientation should be temperature dependent (Barua, Saran and Singh [1967]) but in so doing one gives the model so much flexibility that its ability to fit the data loses much of its significance.

The most accurate procedure that still manages to avoid the need for a full calculation of the dynamics of an encounter was suggested by Monchick and Mason [1961]. They argue that most of the deflection in an encounter occurs when the particles are near the distance of closest approach and, hence, in a rather small portion of the total interaction time. Since the molecules, on the average, rotate once during the entire collision, they are in a fixed relative orientation during that part of the collision which is responsible for most of the deflection. Monchick and Mason have computed trajectories assuming that the particles remain in a fixed relative orientation during a collision (which means that in any given collision the molecules interact with a potential of the type (11.1-1)). Then an average of the resulting cross sections or Ω-integrals over molecular orientation is performed; in the absence of better information, Monchick and Mason chose the average based on equal probability for all orientations:

$$\langle \Omega^{(l,r)} \rangle = \frac{1}{8\pi} \iiint \Omega^{(l,r)}(\theta_1, \theta_2, \phi) \sin\theta_1 \, d\theta_1 \sin\theta_2 \, d\theta_2 \, d\phi \quad (11.1\text{-}4)$$

and these averaged Ω-integrals replace the ordinary Ω-integrals in all of the formulae already developed.

With this method, the viscosity data of polar gases can be fit rather well, as is shown in Fig. 11.1. Potential parameters for various gases are given in Table 11.2. In fitting the data, δ^* was treated as an adjustable parameter but it was found that the dipole moment obtained by this method differed by only a few percent from independent measurements of the dipole moment. Lest too much significance be attached to this success, we note that fitting the virial coefficient data with the Stockmayer potential also produces good results and a reasonable value of the dipole moment but, as seen in Table

TABLE 11.2
Stockmayer potential parameters‡

Gas	μ (debye)†	δ^*	σ(Å)	ϵ/k(°K)
Inorganic gases				
H_2O	1.85	1.0	2.52	775
NH_3	1.47	0.7	3.15	358
HCl	1.08	0.34	3.36	328
HBr	0.80	0.14	3.41	417
HI	0.42	0.029	4.13	313
SO_2	1.63	0.42	4.04	347
H_2S	0.92	0.21	3.49	343
Halogenated hydrocarbons				
$CHCl_3$	1.013	0.07	5.31	355
CH_2Cl_2	1.57	0.2	4.52	483
CH_3Cl	1.87	0.5	3.94	414
CH_3Br	1.80	0.4	4.25	382
C_2H_5Cl	2.03	0.4	4.45	423
Other organic compounds				
CH_3OH	1.70	0.5	3.69	417
C_2H_5OH	1.69	0.3	4.31	431
n-C_3H_7OH	1.69	0.2	4.71	495
i-C_3H_7OH	1.69	0.2	4.64	518
$(CH_3)_2O$	1.30	0.19	4.21	432
$(C_2H_5)_2O$	1.15	0.08	5.49	362
$(CH_3)_2CO$	2.88	0.06	4.50	549
CH_3COOCH_3	1.72	0.2	5.04	418
$CH_3COOC_2H_5$	1.78	0.16	5.24	499

‡ From Monchick and Mason [1961].
† 1 debye = 10^{-18} e.s.u.

TABLE 11.3
Comparison of Stockmayer potential parameters obtained from viscosity and virial coefficient data‡

Gas	μ_v	μ_{vc}	δ^*_v	δ^*_{vc}	σ_v	σ_{vc}	$(\epsilon/k)_v$	$(\epsilon/k)_{vc}$
	(debye)				(Å)		(°K)	
H_2O	1.85	1.83	1.0	1.70	2.52	2.65	775	380
NH_3	1.47	1.47	0.7	1.41	3.15	2.60	358	320
$CHCl_3$	1.013	1.05	0.07	0.14	5.31	2.98	355	1060
CH_3Cl	1.87	1.89	0.5	0.85	3.94	3.43	414	380
C_2H_5Cl	2.03	2.02	0.4	0.28	4.45	5.41	423	320
CH_3OH	1.70	1.66	0.5	1.13	3.69	2.40	417	630
$(CH_3)_2CO$	2.88	2.74	0.06	0.99	4.50	3.76	549	520

‡ From Monchick and Mason [1961].

11.3, the parameters σ and ϵ obtained are considerably different. Likewise, if the parameters obtained by fitting the viscosity are used to compute the virial coefficients the results are far from satisfactory. Monchick and Mason ascribe this to the strong dependence of the virial coefficients on the higher multipole moments in the molecular potential of real molecules. Smith, Munn and Mason [1967] have also discussed the effects of quadrupole moments on transport properties but, in view of the approximations made, it is not clear that these effects can be separated from the effects of the approximations made above.

The only property discussed in this section is the viscosity. This is a result of the poorer accuracy of the thermal conductivity data and the need for a careful study of the effects of transport of molecular internal energy before the thermal conductivity can be computed; we shall look at this property more carefully in the following section. Roughly similar arguments hold for the thermal diffusion coefficient. Diffusion coefficients have been omitted from the discussion simply because the data available are, except in a few cases, not accurate enough to allow one to obtain accurate values of the parameters in the potentials.

11.2. Thermal conductivity – The Eucken correction

Much of what was said about the viscosity in the previous section applies to the thermal conductivity *mutatis mutandis*. In fact, if molecules possessed no internal energy the relation (7.3-22) between the coefficients of viscosity and thermal conductivity,

$$\lambda = \tfrac{5}{2}\eta c_v, \tag{11.2-1}$$

would still apply and the latter transport coefficient could be trivially computed once the former was available. Indeed, one of the earliest suggestions was that this relationship be applied to polyatomic as well as monatomic gases. There is no justification for this idea and, not surprisingly, it doesn't work very well.

Even before the Chapman-Enskog theory was completely worked out, Eucken [1913] pointed out that, while the relationship (11.2-1) is obeyed by all monatomic gases, the mean free path approach predicts a numerical factor of unity, instead of $\frac{5}{2}$. The discrepancy is due to the correlation between the energy of a molecule and its velocity, which is entirely neglected in the mean free path approach. Since there is no reason to expect any correlation between the *internal* energy of a molecule and its velocity, it is clear that the part of the thermal conductivity which is due to the internal molecular energy indeed should not be enhanced by the factor $\frac{5}{2}$. Thus, Eucken was led to divide λ into two parts, λ_{tr} and λ_{int}, which are respectively the thermal conductivity due to the transport of translational and internal energy, and to write

$$\lambda = \lambda_{tr} + \lambda_{int} = (c_{v,tr}\,\mathrm{f}_{tr} + c_{v,int}\,\mathrm{f}_{int})\eta, \tag{11.2-2}$$

where $c_{v,tr}$ is the portion of the specific heat due to the translational energy of the molecules,

$$c_{v,tr} = 3k/2m, \tag{11.2-3}$$

and $c_{v,int}$ is the portion due to the internal energy of the molecules,

$$c_{v,int} = c_v - c_{v,tr}. \tag{11.2-4}$$

According to the above discussion we should have

$$\mathrm{f}_{tr} = \tfrac{5}{2}, \quad \mathrm{f}_{int} = 1. \tag{11.2-5}$$

Hence, instead of Eq. (11.2-1), we have

$$\lambda = \tfrac{5}{2}(1 - \tfrac{3}{5}c_{v,int}/c_v)\eta c_v. \tag{11.2-6}$$

Under normal circumstances of temperature and pressure the specific heat at constant volume, c_v, may be expressed as

$$c_v = \frac{k/m}{\gamma - 1}, \tag{11.2-7}$$

where γ is the ratio of the specific heats, $\gamma = c_p/c_v$. Thus, from Eqs. (11.2-3), (11.2-4) and (11.2-7) we obtain

$$c_{v,int}/c_v = \tfrac{5}{2} - \tfrac{3}{2}\gamma. \tag{11.2-8}$$

Hence, Eq. (11.2-6) is usually written in the form

$$\lambda = \tfrac{1}{4}(9\gamma - 5)\eta c_v. \tag{11.2-9}$$

The factor $\frac{1}{4}(9\gamma - 5)$ is known as the *Eucken correction* for the thermal conductivity of a polyatomic gas.

The Chapman-Enskog theory provides a simple method of improving the result (11.2-6). One notes that a polyatomic gas molecule may have only definite, fixed, internal energy states. For purposes of computation we may consider the molecules in each state as different species and, in the absence of information to the contrary, we assume that the interaction between any pair of molecules is independent of their internal energy state. Considered in this way, a polyatomic gas becomes a very simple type of mixture. The heat flow vector for such a mixture has the form

$$\boldsymbol{q} = \tfrac{1}{2} \sum_i \rho_i \overline{C_i^2 \boldsymbol{C}_i} + \sum_i \rho_i \overline{u_{\text{int},i} \boldsymbol{C}_i}. \tag{11.2-10}$$

The first term is just the energy transport that would exist if the molecules possessed no internal energy; $u_{\text{int},i}$ is the specific internal energy (i.e., the internal energy per unit mass) for the molecules in the state i. In a gas of molecules of equal mass with the same force law between every pair of molecules, thermal diffusion does not arise and use has been made of this in writing Eq. (11.2-10).

The first term may be substituted from Eq. (6.3-48). In fact, because all molecules have the same mass, $\sum_i n_i \boldsymbol{V}_i = 0$ and we have

$$\tfrac{1}{2} \sum_i \rho_i \overline{C_i^2 \boldsymbol{C}_i} = -\lambda_0 \nabla T, \tag{11.2-11}$$

where λ_0 is the thermal conductivity of the gas obtained if internal degrees of freedom of the molecules are ignored,

$$\lambda_0 = \tfrac{5}{2} \eta c_{v,\text{tr}} = \tfrac{15}{4} k\eta/m. \tag{11.2-12}$$

Furthermore, neglecting the correlation between internal energy and thermal velocity, we have

$$\sum_i \rho_i \overline{u_{\text{int},i} \boldsymbol{C}_i} = \sum_i \rho_i u_{\text{int},i} \boldsymbol{V}_i, \tag{11.2-13}$$

where $\boldsymbol{V}_i$ is the diffusion velocity for molecules in the state i. Now, $\boldsymbol{V}_i$ can be computed from Eq. (6.5-13):

$$\boldsymbol{V}_i = -\frac{n}{n_i} \mathscr{D} \nabla \left(\frac{n_i}{n} \right) = -\frac{\rho}{\rho_i} \mathscr{D} \nabla \left(\frac{\rho_i}{\rho} \right), \tag{11.2-14}$$

where $\mathscr{D}$ is the coefficient of self-diffusion. However, an expression for $\nabla(\rho_i/\rho)$ cannot be obtained without further theoretical development. In the absence of a theory for computing this quantity we shall assume that the composition is everywhere that which would exist at equilibrium at the local temperature. With this assumption, ρ_i becomes a function of local temperature only and we have

$$\nabla\left(\frac{\rho_i}{\rho}\right) = \frac{\mathrm{d}(\rho_i/\rho)}{\mathrm{d}T}\nabla T. \tag{11.2-15}$$

Now, the specific internal energy of the gas, u_{int}, is

$$u_{\text{int}} = \sum_i (\rho_i/\rho) u_{\text{int},\,i}. \tag{11.2-16}$$

Hence, the specific heat at constant volume, associated with the internal energy of the molecules, is just

$$c_{v,\,\text{int}} = \frac{\mathrm{d}u_{\text{int}}}{\mathrm{d}T} = \sum_i u_{\text{int},\,i}\frac{\mathrm{d}(\rho_i/\rho)}{\mathrm{d}T}, \tag{11.2-17}$$

so, from Eqs. (11.2-13), (11.2-14) and (11.2-17), we obtain

$$\sum_i \rho_i \overline{u_{\text{int},\,i}\boldsymbol{C}_i} = \rho\mathscr{D}c_{v,\,\text{int}}\nabla T. \tag{11.2-18}$$

Thus, substituting Eqs. (11.2-11) and (11.2-18) into Eq. (11.2-10) we find that the heat flow vector may be written in the form

$$\boldsymbol{q} = -(\lambda_0 + \rho\mathscr{D}c_{v,\,\text{int}})\nabla T, \tag{11.2-19}$$

from which one sees that the thermal conductivity is given by

$$\lambda = \lambda_0 + \rho\mathscr{D}c_{v,\,\text{int}}. \tag{11.2-20}$$

Using Eq. (11.2-12) one may verify that this expression, in turn, is equivalent to

$$\lambda = \frac{5}{2}\left\{\frac{c_{v,\,\text{tr}}}{c_v} + \frac{2}{5}\frac{\rho\mathscr{D}}{\eta}\frac{c_{v,\,\text{int}}}{c_v}\right\}\eta c_v \tag{11.2-21}$$

or, using Eq. (11.2-4), to

$$\lambda = \frac{5}{2}\left\{1 - \left(1 - \frac{2}{5}\frac{\rho\mathscr{D}}{\eta}\right)\frac{c_{v,\,\text{int}}}{c_v}\right\}\eta c_v. \tag{11.2-22}$$

Comparing this result with Eq. (11.2-6) we find that the Eucken correction is obtained by setting the dimensionless ratio $(\rho\mathscr{D}/\eta)$ equal to unity.

Furthermore, from Eq. (11.2-20) it follows that the ratio λ/λ_0 is equal to

$$\frac{\lambda}{\lambda_0} = 1 + \frac{\rho\mathscr{D}}{\lambda_0}c_{v,\,\text{int}}. \tag{11.2-23}$$

Now, the dimensionless ratio $(\rho\mathscr{D}c_{v,\,\mathrm{int}}/\lambda_0)$ is temperature dependent, but Hirschfelder [1957] has shown that, for the Lennard-Jones (6–12) and the modified Buckingham (6-exp) potentials it is well approximated by a constant. Adopting an average value for this quantity, which corresponds to a value 1.328 for the ratio $\rho\mathscr{D}/\eta$, one has

$$\lambda/\lambda_0 = 0.469+0.354\ C_v/R, \tag{11.2-24}$$

where C_v is the heat capacity per mole at constant volume, and R is the gas constant. In terms of these variables, Eucken's result can be written as

$$\lambda/\lambda_0 = \tfrac{3}{5}+\tfrac{4}{15}C_v/R. \tag{11.2-25}$$

The expression (11.2-22) is more accurate than Eucken's expression (11.2-6), as might be expected from a comparison of the methods of derivation. As is easily seen, the improvement essentially comes from using the Chapman-Enskog expression for the diffusion coefficient in place of the mean free path value. However, the formula (11.2-22) does not fit experimental data to within the experimental uncertainty and is certainly not a correct result. Generally, the fit is better at high temperatures than at low temperatures, but the formula is useful mainly as a means of obtaining approximate values quickly. Hirschfelder [1957] used a somewhat different approach to arrive at the above result. In his calculation the molecules in the various energy states were treated as different chemical species and inelastic collisions as chemical reactions. This is very similar to the theory given in Section 11.3 and it will become clear there why Hirschfelder obtains this result.

There are two major sources of error in the derivation. The first is the assumption that the populations of the various internal energy states are everywhere those which would obtain if equilibrium were reached at the local temperature. This will be the case only if the internal energy states come to thermal equilibrium rapidly, i.e., after a few collisions. In real gases, the rotational states do indeed reach equilibrium rather rapidly, but the vibrational states generally require many collisions to reach equilibrium. This is the major reason why the above expressions are not very accurate. Some improvement can perhaps be obtained by including only the contribution of rotational states in computing $c_{v,\,\mathrm{int}}$, the specific heat due to internal degrees of freedom, but it is sounder to begin by constructing a theory in which the internal states of the molecules are accounted for *ab initio*. The other major source of error is one which we can do nothing about at this point. It is simply that inelastic collisions are dynamically different from elastic collisions; a change in internal energy state has an important effect on the kinematics of a collision. Clearly this can be corrected only by constructing a more exact theory and we now turn to this.

11.3. Classical and quasi-classical treatment of polyatomic gases

The treatment given in the previous sections provides effective means of estimating the transport properties of polyatomic gases – at least, of the simpler of these gases. In contrast to the methods developed for monatomic gases, however, it does not provide effective means of obtaining molecular properties from transport coefficient measurements. For the accurate computation of the transport properties of polyatomic gases a more rigorous theory is required. It is clear that such a theory must have as its starting point a Boltzmann equation generalized to account for the internal states of the molecule. The basic idea is one that was introduced in the previous section – molecules in different internal energy states are treated as separate classes and a separate equation is written for each state.

The next question one must face is whether the internal states may be treated by classical mechanics or a rigorous quantum mechanical treatment is required. The first attempt by Wang Chang and Uhlenbeck [1951] actually utilized a semi-classical approach. A completely classical treatment was given later by Taxman [1958]. There have also been a number of works dealing with specific classical models. Since the models can be regarded as a means of computing collision frequencies, it is possible to develop the theory in general terms and then use the models to compute the required properties. All of the models used to date idealize molecules as rigid bodies whose internal energy is purely rotational. Some of the models used are:

1. Rough spheres (spheres which grip without slipping in a collision) – Pidduck [1922], Chapman and Cowling [1952].
2. Loaded spheres (spheres whose center of gravity does not coincide with the geometric center) – Jeans [1901], Dahler [1959].
3. Spherocylinders (cylinders with hemispherical caps at either end) – Curtiss [1956], Curtiss and Muckenfuss [1957], Muckenfuss and Curtiss [1958].
4. Ellipsoids (prolate and oblate) – Dahler [1959].

Somewhat surprisingly, the quasi-classical theory is a little simpler than either the fully classical or the quantum mechanical theory. This is a consequence of the fact that it is possible to define inverse collisions in the quasi-classical case but not in the other cases. On the other hand, the quasi-classical approach is not correct in some very important situations and the quantum mechanical approach is then required. The quasi-classical approach and its relation to the fully classical case will be treated in this section, the treatment of the quantum mechanical case will be deferred to the following section.

In the quasi-classical approach one assumes that the molecules may have only certain *discrete* internal energy states, which are labelled with an index. Then, in accord with the above arguments, we define $f_i(\boldsymbol{r}, \boldsymbol{c}, t)\mathrm{d}^3r\mathrm{d}^3c$ to be the number of molecules in internal state i in d^3r about $\boldsymbol{r}$ and d^3c about $\boldsymbol{c}$ at time t. The derivation of the Boltzmann equation for these functions follows the derivation of the Boltzmann equation for a mixture very closely. The essential difference arises from the existence of inelastic collisions i.e., collisions in which the internal state of one or both molecules is changed. For such collisions it is convenient to define a differential collision cross section which is the straight-forward generalization of the cross section introduced in Eqs. (3.1-30) and (3.1-31). We let $\sigma_{ij}^{kl}(g, \chi, \varepsilon)$ be the cross section for scattering molecules in internal states i and j with relative speed g, such that, after the encounter, the molecules have internal states k and l, respectively, and such that the relative velocity (whose magnitude is no longer a constant) is rotated through a polar angle χ and azimuthal angle ε; the cross section may depend on ε as the molecules are no longer spherically symmetric. Thus,

$$g\sigma_{ij}^{kl}(g, \chi, \varepsilon)\, f_i(\boldsymbol{r}, \boldsymbol{c}, t)\, f_j(\boldsymbol{r}, \boldsymbol{c}_1, t)\,\mathrm{d}^3c\,\mathrm{d}^3c_1 \sin\chi\,\mathrm{d}\chi\,\mathrm{d}\varepsilon, \qquad (11.3\text{-}1)$$

where $g = |\boldsymbol{c}-\boldsymbol{c}_1|$, is the number of collisions per unit volume and unit time involving molecules in state i with velocities in d^3c about $\boldsymbol{c}$ and molecules in state j with velocities in d^3c_1 about $\boldsymbol{c}_1$, such that, after the collision, the molecules are in the states k and l, respectively, and have their relative velocity oriented in a direction in $\sin\chi\,\mathrm{d}\chi$ about χ and $\mathrm{d}\varepsilon$ about ε with respect to their initial relative velocity. Although the relative velocity $\boldsymbol{g}'$ after the collision does in general not have the same magnitude as $\boldsymbol{g}$, it is fixed by giving its orientation with respect to $\boldsymbol{g}$ if the energies of the various states are given.

The symmetry properties of the Schrödinger equation lead to the following reciprocity relation for the cross-sections,

$$g\sigma_{ij}^{kl}(g, \chi, \varepsilon)\mathrm{d}^3c\,\mathrm{d}^3c_1 = g'\sigma_{kl}^{ij}(g', \chi, \varepsilon)\mathrm{d}^3c'\,\mathrm{d}^3c_1', \qquad (11.3\text{-}2)$$

which is essentially a relation between the rates of direct and inverse collisions. We note that relation (11.3-2) strictly holds only if the internal energy states are not degenerate, that is, if no two states have the same energy. This is not so for the important case of rotating molecules, but the implications of this are put off until the next section.

With the use of Eqs. (11.3-1) and (11.3-2) it is now a relatively easy matter to derive the Boltzmann equation(s) for a polyatomic gas. As the derivation follows closely that given in Section 3.1 for a monatomic gas we shall simply state the equation:

$$\frac{\partial f_i}{\partial t} + \boldsymbol{c} \cdot \boldsymbol{\nabla}_r f_i + \boldsymbol{F}_i \cdot \boldsymbol{\nabla}_c f_i = \sum_{jkl} \iint [f'_k f'_l - f_i f_j] g \sigma_{ij}^{kl}(g, \chi, \varepsilon) \mathrm{d}^2\Omega \, \mathrm{d}^3 c_j, \quad (11.3\text{-}3)$$

which may be abbreviated in the usual way,

$$\mathfrak{D} f_i = \sum_{jkl} J_{ij}^{kl}(ff). \quad (11.3\text{-}4)$$

Thus, the collision term $J_{ij}^{kl}(ff)$ is defined by

$$J_{ij}^{kl}(ff) = \iint (f'_k f'_l - f_i f_j) g \sigma_{ij}^{kl}(g, \chi, \varepsilon) \mathrm{d}^2\Omega \, \mathrm{d}^3 c_j. \quad (11.3\text{-}5)$$

We have suppressed any functional dependence of f_i that is clear from the context; as in Chapter 3, $\mathrm{d}^2\Omega$ denotes $\sin\chi \, \mathrm{d}\chi \, \mathrm{d}\varepsilon$. Eq. (11.3-3) is known as the *Wang Chang-Uhlenbeck equation*; it was established independently by Wang Chang and Uhlenbeck, and De Boer in the 1940's, see Wang Chang, Uhlenbeck and De Boer [1964].

The line of attack now closely follows the one for the monatomic gas. Therefore, we shall emphasize the new points only. First, consider the macroscopic conservation equations. These are obtained in the usual way by multiplying the Boltzmann equation by a summational invariant and integrating throughout velocity space. In the present case we must also sum over the internal states of the molecules.

Although the number of molecules in each state is not conserved in a collision, the *total* number of molecules is and the equation for conservation of mass for the entire gas takes the same form as before:

$$\frac{1}{\rho} \frac{\mathrm{d}\rho}{\mathrm{d}t} = -\boldsymbol{\nabla} \cdot \boldsymbol{v}, \quad (11.3\text{-}6)$$

where ρ is the total density. Fortunately, the equations for individual species densities are not required in the calculation.

The situation for momentum conservation is very similar; again, one obtains the equation

$$\rho \frac{\mathrm{d}\boldsymbol{v}}{\mathrm{d}t} = \rho \boldsymbol{F} - \nabla \cdot \mathbf{P}, \tag{11.3-7}$$

where $\mathbf{P}$ is the pressure tensor,

$$\mathbf{P} = m \sum_i \int f_i \boldsymbol{CC} \mathrm{d}^3 c. \tag{11.3-8}$$

For the energy the situation is slightly different from that encountered earlier, due to the presence of energy associated with internal states. Let E_i be the energy associated with the internal state i. We define the *average internal energy* per unit mass, u, by

$$\rho u = \sum_i \int (\tfrac{1}{2}mC^2 + E_i) f_i \mathrm{d}^3 c. \tag{11.3-9}$$

At this stage we can state only that u is a function of the temperature, but we cannot define the temperature as easily as in the monatomic gas case. We further define the heat flow vector in terms of the total energy flow across a surface element moving with the fluid:

$$\boldsymbol{q} = \sum_i \int (\tfrac{1}{2}mC^2 + E_i) \boldsymbol{C} f_i \mathrm{d}^3 c. \tag{11.3-10}$$

Then, on multiplying the Boltzmann equation by $\frac{1}{2}mC^2 + E_i$, integrating over velocity and summing over states, one has

$$\rho \frac{\mathrm{d}u}{\mathrm{d}t} = -\nabla \cdot \boldsymbol{q} - \mathbf{P} : \nabla \boldsymbol{v}. \tag{11.3-11}$$

To express this equation in terms of the temperature one notes that $\mathrm{d}u/\mathrm{d}t = c_v \mathrm{d}T/\mathrm{d}t$, where c_v is the specific heat at constant volume.

It is also possible to derive other conservation equations from the Wang Chang-Uhlenbeck equation. For instance, one could find an equation governing the flow of internal energy or, in the case of rotating molecules, an equation for the angular momentum density.

An H-theorem can be proven for Eq. (11.3-3) but, as the proof follows that for monatomic gases, we shall not give it here and proceed directly to the Chapman-Enskog theory. Again, this follows the theory of Chapter 6 and we will take the equations over directly. In zeroth order one finds that the distribution functions must be such that the equation

$$\sum_{jkl} J_{ij}^{kl}(f^{(0)}f^{(0)}) = 0 \tag{11.3-12}$$

is satisfied for all i. The H-theorem shows that this is so if and only if

$$f_k^{(0)\prime} f_l^{(0)\prime} = f_i^{(0)} f_j^{(0)} \tag{11.3-13}$$

for all i, j, k, l. Thus, $\log f_i$ must be a summational invariant and, since the only summational invariants are m, $m\boldsymbol{C}$ and $\frac{1}{2}mC^2+E_i$ we have

$$\log f_i^{(0)} = m\alpha_i^{(1)} + m\boldsymbol{\alpha}_i^{(2)} \cdot \boldsymbol{C} - \alpha_i^{(3)}(\tfrac{1}{2}mC^2+E_i). \tag{11.3-14}$$

Eqs. (11.3-13) require that the α's be the same for all i, so we may drop the subscripts. Furthermore, imposing the usual requirements of the Chapman-Enskog method,

$$n = \sum_i \int f_i \mathrm{d}^3c = \sum_i \int f_i^{(0)} \mathrm{d}^3c, \tag{11.3-15a}$$

$$\rho \boldsymbol{v} = \sum_i \int m\boldsymbol{c} f_i \mathrm{d}^3c = \sum_i \int m\boldsymbol{c} f_i^{(0)} \mathrm{d}^3c, \tag{11.3-15b}$$

$$\rho u = \sum_i \int (\tfrac{1}{2}mC^2+E_i) f_i \mathrm{d}^3c = \sum_i \int (\tfrac{1}{2}mC^2+E_i) f_i^{(0)} \mathrm{d}^3c, \tag{11.3-15c}$$

we obtain

$$f_i^{(0)} = n\left(\frac{m\alpha^{(3)}}{2\pi}\right)^{\frac{3}{2}} \frac{\exp\left[-\alpha^{(3)}(\frac{1}{2}mC^2+E_i)\right]}{\sum_i \exp(-\alpha^{(3)}E_i)}, \tag{11.3-16}$$

where $\alpha^{(3)}$ is determined by the relation

$$\rho u = n\left(\frac{3}{2\alpha^{(3)}} + \frac{\sum_i E_i \exp(-\alpha^{(3)}E_i)}{\sum_i \exp(-\alpha^{(3)}E_i)}\right). \tag{11.3-17}$$

For consistency with the definition of temperature for the case in which internal degrees of freedom are absent (or do not change in collisions) we must have $\alpha^{(3)} = 1/kT$ and, instead of Eq. (11.3-16), we have

$$f_i^{(0)} = n\left(\frac{m}{2\pi kT}\right)^{\frac{3}{2}} \frac{\exp\left[-(mC^2/2kT)-(E_i/kT)\right]}{\sum_i \exp(-E_i/kT)}. \tag{11.3-18}$$

The function

$$Z_{\mathrm{int}} = \sum_i \exp(-E_i/kT) \tag{11.3-19}$$

is recognized as the statistical mechanical partition function for the internal degrees of freedom. With the choice of $\alpha^{(3)}$ we find from Eq. (11.3-17), that the internal energy density ρu is the sum of a contribution from the translational degrees of freedom and a contribution from the internal degrees of freedom,

$$\rho u = \tfrac{3}{2}nkT + \rho u_{\text{int}}, \tag{11.3-20a}$$

with

$$\rho u_{\text{int}} = n\bar{E} = n\,\frac{\sum_i E_i \exp(-E_i/kT)}{\sum_i \exp(-E_i/kT)}. \tag{11.3-20b}$$

If the pressure tensor and heat flow vector are calculated from the expression (11.3-18) one finds $\mathbf{P} = nkT\mathbf{I} = p\mathbf{I}$ and $\boldsymbol{q} = 0$ so that, in the zero-order approximation, the macroscopic conservation equations again become the Euler equations.

The first-order Chapman-Enskog approximation is somewhat modified by the presence of the factor $\exp(-E_i/kT)$ in $f_i^{(0)}$. One finds the following integral equation for the first-order perturbation function, ϕ,

$$\begin{aligned}\sum_{jkl}\iint f_i^{(0)} f_j^{(0)}(\phi_i+\phi_j-\phi_k'-\phi_l')g\sigma_{ij}^{kl}(g,\chi,\varepsilon)\mathrm{d}^2\Omega\,\mathrm{d}^3c_j \\ = -f_i^{(0)}\Big\{[(\mathscr{C}^2-\tfrac{5}{2})+(\mathscr{E}_i-\bar{\mathscr{E}})]\boldsymbol{C}\cdot\nabla\log T+2(\mathscr{C}\mathscr{C}-\tfrac{1}{3}\mathscr{C}^2\mathbf{I}):\nabla\boldsymbol{v} \\ +\left[\frac{2}{3}\frac{c_{v,\text{int}}}{c_v}(\mathscr{C}^2-\tfrac{3}{2})-\frac{k/m}{c_v}(\mathscr{E}_i-\bar{\mathscr{E}})\right]\nabla\cdot\boldsymbol{v}\Big\},\end{aligned} \tag{11.3-21}$$

where $\mathscr{E}_i$ and $\bar{\mathscr{E}}$ are dimensionless energy variables,

$$\mathscr{E}_i = \frac{E_i}{kT}, \quad \bar{\mathscr{E}} = \frac{\bar{E}}{kT} = \frac{\sum_i \mathscr{E}_i \exp(-\mathscr{E}_i)}{\sum_i \exp(-\mathscr{E}_i)}; \tag{11.3-22}$$

c_v and $c_{v,\text{int}}$ are the coefficients of specific heat defined by

$$c_v = (\partial u/\partial T) = (3k/2m)+c_{v,\text{int}}, \tag{11.3-23a}$$

and

$$c_{v,\text{int}} = \frac{\partial u_{\text{int}}}{\partial T} = \frac{k}{m}(\overline{\mathscr{E}^2}-\bar{\mathscr{E}}^2) = \frac{k}{m}\frac{\sum_i(\mathscr{E}_i-\bar{\mathscr{E}})^2\exp(-\mathscr{E}_i)}{\sum_i\exp(-\mathscr{E}_i)}, \tag{11.3-23b}$$

respectively. By analogy with the monatomic gas case the expression in the left member of Eq. (11.3-21) will be denoted by $n^2 I_i(\phi)$. Thus, in the present case the integral operator I_i is defined by

$$I_i(F) = \frac{1}{n^2}\sum_{jkl}\int\int f_i^{(0)} f_j^{(0)}(F_i + F_j - F'_k - F'_l) g\sigma_{ij}^{kl}(g, \chi, \varepsilon) \mathrm{d}^2\Omega \, \mathrm{d}^3 c_j, \tag{11.3-24}$$

where the functions $f_i^{(0)}$ and $f_j^{(0)}$ are defined by Eq. (11.3-20). We note that there appear on the right-hand side of Eq. (11.3-21) some terms that were not encountered in the monatomic gas case. The most interesting and important of these is the $\nabla \cdot \boldsymbol{v}$ term which, as we shall see, leads to new physical effects.

The solution of Eq. (11.3-21) proceeds in a manner similar to that of the monatomic gas case. We first note that, for a solution to exist, the right-hand side must be orthogonal to the solutions of the homogeneous equation which is obtained by setting the right-hand side equal to zero. These solutions are again the summational invariants, in this case, m, $m\boldsymbol{C}$ and $\frac{1}{2}mC^2 + E_i$, and it is easily checked that the orthogonality conditions are indeed satisfied. Then the solution of Eq. (11.3-21) must be of the form:

$$\phi_i = -\frac{1}{n}\boldsymbol{A}_i \cdot \nabla \log T - \frac{1}{n}\mathbf{B}_i : \nabla\boldsymbol{v} - \frac{1}{n}\Gamma_i \nabla \cdot \boldsymbol{v}. \tag{11.3-25}$$

To this should be added a solution of the homogeneous equation i.e., a sum of summational invariants but, as in the monatomic gas case, these play no role in the subsequent development and will be ignored.

Substitution of this Ansatz into Eq. (11.3-21) decomposes it into three equations for the vector $\boldsymbol{A}$, the tensor $\mathbf{B}$ and the scalar Γ. Arguing as in Section 5.4 we note that $\boldsymbol{A}$ must be proportional to $\boldsymbol{C}$ while $\mathbf{B}$, which is symmetric and traceless, must be proportional to $\boldsymbol{\mathscr{C}\mathscr{C}} - \frac{1}{3}\mathscr{C}^2\mathbf{I}$,

$$\boldsymbol{A} = A(\mathscr{C}, \mathscr{E})\boldsymbol{C}, \tag{11.3-26a}$$

$$\mathbf{B} = B(\mathscr{C}, \mathscr{E})(\boldsymbol{\mathscr{C}\mathscr{C}} - \tfrac{1}{3}\mathscr{C}^2\mathbf{I}), \tag{11.3-26b}$$

$$\Gamma = \Gamma(\mathscr{C}, \mathscr{E}). \tag{11.3-26c}$$

It is now rather straight-forward to evaluate the pressure tensor and the heat flow vector in the first Chapman-Enskog approximation. Using the notation of the bracket integrals,

$$[F, G] = \sum_i \int G_i I_i(F) \, \mathrm{d}^3 c, \tag{11.3-27}$$

one easily finds

$$\mathbf{P} = p\mathbf{I} - \tfrac{1}{5}kT[\mathbf{B}, \mathbf{B}]\mathbf{S} - kT[\Gamma, \Gamma]\nabla \cdot \boldsymbol{v}\mathbf{I}, \tag{11.3-28}$$

$$\boldsymbol{q} = -\tfrac{2}{3}(k^2T/m)[\boldsymbol{A}, \boldsymbol{A}]\nabla T. \tag{11.3-29}$$

From the latter of these one immediately deduces the coefficient of thermal conductivity,

$$\lambda = \frac{2k^2T}{3m}[\boldsymbol{A}, \boldsymbol{A}], \tag{11.3-30}$$

in strict analogy with the monatomic gas case. In the pressure tensor we note that there are now two viscosities: the *shear viscosity*, with coefficient

$$\eta = \tfrac{1}{10}kT[\mathbf{B}, \mathbf{B}], \tag{11.3-31}$$

and a new property, the *bulk viscosity* or *dilatation viscosity*, with coefficient

$$\kappa = kT[\Gamma, \Gamma]. \tag{11.3-32}$$

Note that for the monatomic gas the latter coefficient is zero. In the older literature one finds a coefficient $\eta_2 = \kappa - \frac{2}{3}\eta$, called the second coefficient of viscosity. For monatomic gases, $3\eta_2 + 2\eta = 0$ and this is often called the *Stokes relation*. Thus, with the definitions (11.3-31) and (11.3-32) the pressure tensor may be written as

$$\mathbf{P} = p\mathbf{I} - 2\eta\mathbf{S} - \kappa\nabla \cdot \boldsymbol{v}\mathbf{I}. \tag{11.3-33}$$

The physical origin of the bulk viscosity coefficient is interesting and we shall devote a brief aside to it. In a gas with internal degrees of freedom the energy partition between the translation and internal degrees of freedom in non-equilibrium need not be the same as in equilibrium. Thus, if a small element of a gas is suddenly compressed the energy deposited in the gas initially becomes energy associated with the translational motion and only later do the internal degrees of freedom receive any of it. Since the pressure arises entirely from the translational motion of the molecules, it will be somewhat higher initially than it would be if equilibrium between the translational and internal degrees of freedom were instantaneously established. Thus the effect is equivalent to having an enhanced hydrostatic pressure in a compressed gas (for which $\nabla \cdot \boldsymbol{v} < 0$) and a reduced hydrostatic pressure in an expanded gas (for which $\nabla \cdot \boldsymbol{v} > 0$). The bulk viscosity also arises in dense gases and liquids so that its absence in dilute monatomic gases is an exception rather than a rule. Further justification of these remarks will be given after the expression for the bulk viscosity has been developed.

Now, the equations satisfied by $\boldsymbol{A}$, $\mathbf{B}$ and Γ are,

$$nI_i(\boldsymbol{A}) = f_i^{(0)}[(\mathscr{C}^2-\tfrac{5}{2})+(\mathscr{E}_i-\bar{\mathscr{E}})]\mathbf{C}, \tag{11.3-34a}$$

$$nI_i(\mathbf{B}) = 2f_i^{(0)}[\mathscr{C}\mathscr{C}-\tfrac{1}{3}\mathscr{C}^2\mathbf{I}], \tag{11.3-34b}$$

$$nI_i(\Gamma) = f_i^{(0)}\left[\frac{2}{3}\,\frac{c_{v,\,\mathrm{int}}}{c_v}(\mathscr{C}^2-\tfrac{3}{2})-\frac{k/m}{c_v}(\mathscr{E}_i-\bar{\mathscr{E}})\right], \tag{11.3-34c}$$

and these must be solved subject to the auxiliary Chapman-Enskog conditions which, for a solution of the form (11.3-25) become

$$\sum_i \int f_i^{(0)}A_i(\mathscr{C},\mathscr{E}_i)\mathscr{C}^2\,\mathrm{d}^3c = 0, \tag{11.3-35a}$$

$$\sum_i \int f_i^{(0)}\Gamma_i(\mathscr{C},\mathscr{E}_i)\mathrm{d}^3c = 0, \tag{11.3-35b}$$

$$\sum_i \int f_i^{(0)}\Gamma_i(\mathscr{C},\mathscr{E}_i)(\mathscr{C}^2+\mathscr{E}_i)\mathrm{d}^3c = 0. \tag{11.3-35c}$$

Now one can formally proceed to solve these equations. The variational method of Chapter 5 may be applied but, since we shall not proceed further than the lowest-order approximation, the following method – which is simpler and which produces the same results – will be applied.

The viscosity equation i.e., the equation for $\mathbf{B}$, is the simplest to handle. To lowest order, one assumes that $\mathbf{B}$ has the form

$$\mathbf{B} = b(\mathscr{C}\mathscr{C}-\tfrac{1}{3}\mathscr{C}^2\mathbf{I}), \tag{11.3-36}$$

where b is a constant, cf. Eq. (5.6-13). If one substitutes this Ansatz into Eq. (11.3-34b), takes the scalar product with $\mathscr{C}\mathscr{C}-\frac{1}{3}\mathscr{C}^2\mathbf{I}$, integrates over $\boldsymbol{c}$ and sums over i, one finds

$$b = 5[\mathscr{C}\mathscr{C}-\tfrac{1}{3}\mathscr{C}^2\mathbf{I},\,\mathscr{C}\mathscr{C}-\tfrac{1}{3}\mathscr{C}^2\mathbf{I}]^{-1} \tag{11.3-37}$$

and, thus,

$$\eta = \tfrac{5}{2}kT[\mathscr{C}\mathscr{C}-\tfrac{1}{3}\mathscr{C}^2\mathbf{I},\,\mathscr{C}\mathscr{C}-\tfrac{1}{3}\mathscr{C}^2\mathbf{I}]^{-1}. \tag{11.3-38}$$

The bracket represents an eightfold integral as well as a sum over the internal states. As in the monatomic gas case most of the integrals can be carried out without knowledge of the cross section. In the present case, the integrals over all variables but χ, ε and g may be carried out; again, this is due to the dependence of the cross section on these quantities. One finds

$$[\mathscr{C}\mathscr{C}-\tfrac{1}{3}\mathscr{C}^2\mathbf{I}, \mathscr{C}\mathscr{C}-\tfrac{1}{3}\mathscr{C}^2\mathbf{I}] = 4(kT/\pi m)^{\frac{1}{2}} Z_{\text{int}}^{-2} \sum_{ijkl} \exp\{-(\mathscr{E}_i+\mathscr{E}_j)\}$$
$$\times \iint \exp(-g^2) g^3 [g^4 \sin^2\chi - \tfrac{1}{2}(\Delta\mathscr{E})^2 \sin^2\chi + \tfrac{1}{3}(\Delta\mathscr{E})^2] \sigma_{ij}^{kl}(g, \chi, \varepsilon)\, \mathrm{d}^2\Omega\, \mathrm{d}g, \tag{11.3-39}$$

where, as before, $g^2 = mg^2/4kT$ and $\Delta\mathscr{E} = \mathscr{E}_k+\mathscr{E}_l-\mathscr{E}_i-\mathscr{E}_j = g^2-g'^2$. One cannot proceed further without making assumptions about the cross section.

Next, we consider the equation related to the bulk viscosity (11.3-34c). In view of the right-hand side of this equation, the simplest approximation that one can make is

$$\Gamma = \gamma_1(\mathscr{C}^2-\tfrac{3}{2})+\gamma_2(\bar{\mathscr{E}}-\mathscr{E}), \tag{11.3-40}$$

where γ_1 and γ_2 are constants. When this is substituted into Eq. (11.3-34c) and moments are taken with respect to $\mathscr{C}^2-\frac{3}{2}$ and $\bar{\mathscr{E}}-\mathscr{E}$, there result two algebraic equations for γ_1 and γ_2:

$$[\mathscr{C}^2-\tfrac{3}{2}, \mathscr{C}^2-\tfrac{3}{2}]\gamma_1+[\mathscr{C}^2-\tfrac{3}{2}, \bar{\mathscr{E}}-\mathscr{E}]\gamma_2 = \frac{c_{v,\,\text{int}}}{c_v}, \tag{11.3-41a}$$

$$[\mathscr{C}^2-\tfrac{3}{2}, \bar{\mathscr{E}}-\mathscr{E}]\gamma_1+[\bar{\mathscr{E}}-\mathscr{E}, \bar{\mathscr{E}}-\mathscr{E}]\gamma_2 = \frac{c_{v,\,\text{int}}}{c_v}. \tag{11.3-41b}$$

In deriving the last relation use has been made of Eq. (11.3-23b). Now, since $\mathscr{C}^2-\frac{3}{2}+\mathscr{E}-\bar{\mathscr{E}}$ is a summational invariant, $I_i(\mathscr{C}^2-\frac{3}{2}+\mathscr{E}-\bar{\mathscr{E}}) = 0$ and one sees that the bracket integrals in Eqs. (11.3-41) are all equal. Since the right-hand sides are also equal, the equations are, in fact, identical. However, the auxiliary conditions (11.3-35b, c) must still be satisfied. For the trial function (11.3-40) the first of these is an identity, while the second yields

$$\frac{3k}{2m}\gamma_1 = c_{v,\,\text{int}}\gamma_2, \tag{11.3-42}$$

which, when solved simultaneously with either of Eqs. (11.3-41) yields

$$\gamma_1 = \left(\frac{c_{v,\,\text{int}}}{c_v}\right)^2 [\mathscr{C}^2-\tfrac{3}{2}, \mathscr{C}^2-\tfrac{3}{2}]^{-1}, \tag{11.3-43a}$$

$$\gamma_2 = \frac{c_{v,\,\text{int}}}{c_v}\,\frac{3k/2m}{c_v}\,[\mathscr{C}^2-\tfrac{3}{2}, \mathscr{C}^2-\tfrac{3}{2}]^{-1}. \tag{11.3-43b}$$

Again, use has been made of the equality of the bracket integrals in Eqs. (11.3-41). Using these results one finds that the bulk viscosity in the first approximation is given by

$$\kappa = \left(\frac{c_{v,\,\mathrm{int}}}{c_v}\right)^2 kT[\mathscr{C}^2-\tfrac{3}{2}, \mathscr{C}^2-\tfrac{3}{2}]^{-1}. \tag{11.3-44}$$

The bracket integral which occurs in this expression may be reduced,

$$[\mathscr{C}^2-\tfrac{3}{2}, \mathscr{C}^2-\tfrac{3}{2}] = 2(kT/\pi m)^{\frac{1}{2}} \mathrm{Z}_{\mathrm{int}}^{-2} \sum_{ijkl} \exp\{-(\mathscr{E}_i+\mathscr{E}_j)\} \times \iint \exp(-g^2) g^3 (\Delta\mathscr{E})^2 \sigma_{ij}^{kl}(g, \chi, \varepsilon) \mathrm{d}^2\Omega\, \mathrm{d}g, \tag{11.3-45}$$

where all of the symbols are as defined earlier.

At this point we again digress briefly to point out the relation between the bulk viscosity and the time required for the rotational and translational degrees of freedom to come to equilibrium with each other. To do this we consider a gas which is slightly removed from equilibrium but, since we are interested only in dilatation effects, we assume that $\boldsymbol{A}$ and $\boldsymbol{B}$ vanish identically. Thus, the distribution function is

$$f_i = f_i^{(0)}(1+\phi_i) = f_i^{(0)}\{1-n^{-1}[\gamma_1(\mathscr{C}^2-\tfrac{3}{2})+\gamma_2(\bar{\mathscr{E}}-\mathscr{E}_i)]\nabla\cdot\boldsymbol{v}\}. \tag{11.3-46}$$

This may be recognized as the linearization of a Maxwellian with different translational and internal temperatures about the Maxwellian with equal temperatures i.e.,

$$f_i \approx [\mathrm{Z}_{\mathrm{int}}(T_{\mathrm{int}})]^{-1} n(m/2\pi kT_{\mathrm{tr}})^{\frac{3}{2}} \exp[-(mC^2/2kT_{\mathrm{tr}})-E_i/kT_{\mathrm{int}}], \tag{11.3-47}$$

provided

$$\frac{T_{\mathrm{tr}}-T}{T} = -\frac{\gamma_1}{n}\nabla\cdot\boldsymbol{v}, \quad \frac{T_{\mathrm{int}}-T}{T} = -\frac{\gamma_2}{n}\nabla\cdot\boldsymbol{v}. \tag{11.3-48}$$

Now, from the Boltzmann equation we may derive equations for the rate of change of T_{tr} and T_{int}. Multiplying the Boltzmann equation by $\frac{1}{2}mC^2$ and E_i, respectively, integrating over velocity and summing over all internal states one finds (ignoring spatial effects):

$$\tfrac{3}{2}nk\frac{\partial T_{\mathrm{tr}}}{\partial t} = -\sum_i \int \tfrac{1}{2}mC^2 I_i(\phi)\,\mathrm{d}^3c, \tag{11.3-49a}$$

$$\rho c_{v,\,\mathrm{int}}\frac{\partial T_{\mathrm{int}}}{\partial t} = -\sum_i \int E_i I_i(\phi)\,\mathrm{d}^3c, \tag{11.3-49b}$$

where gradient terms have been ignored. With ϕ_i as in Eq. (11.3-46), the integrals on the right-hand sides of these equations may be written in terms of the bracket integrals of Eq. (11.3-41). Since the latter are all equal, Eqs. (11.3-49) reduce to

$$\frac{\partial T_{\mathrm{tr}}}{\partial t} = -\frac{T_{\mathrm{tr}}-T}{\tau_\kappa}, \tag{11.3-50a}$$

$$\frac{\partial T_{\mathrm{int}}}{\partial t} = -\frac{T_{\mathrm{int}}-T}{\tau_\kappa}, \tag{11.3-50b}$$

where

$$\tau_\kappa = \frac{3}{2n}\frac{c_{v,\,\mathrm{int}}}{c_v}[\mathscr{C}^2-\tfrac{3}{2}, \mathscr{C}^2-\tfrac{3}{2}]^{-1}, \tag{11.3-51a}$$

so that

$$\kappa = \tfrac{2}{3}p\frac{c_{v,\,\mathrm{int}}}{c_v}\tau_\kappa. \tag{11.3-51b}$$

In the derivation of Eq. (11.3-50b) the relation (11.3-42) between γ_1 and γ_2 has been used.

Thus it is seen that τ_κ is the *relaxation time* required for the translational and internal degrees of freedom to come to equilibrium, and that the coefficient of bulk viscosity is directly proportional to this relaxation time.

We now turn to the treatment of the thermal conduction equation (11.3-34a). Here $\boldsymbol{A}$ will be approximated by

$$\boldsymbol{A} = (m/2kT)\{a_1(\mathscr{C}^2-\tfrac{5}{2})+a_2(\mathscr{E}-\bar{\mathscr{E}})\}\boldsymbol{C}, \tag{11.3-52}$$

where a_1 and a_2 are constants, cf. Eq. (5.6-4). Substitution of this expression into Eq. (11.3-34a) and taking moments with respect to $(\mathscr{C}^2-\frac{5}{2})\boldsymbol{C}$ and $(\mathscr{E}-\bar{\mathscr{E}})\boldsymbol{C}$ gives

$$\alpha_{11}a_1+\alpha_{12}a_2 = 15kT/2m \tag{11.3-53a}$$

and

$$\alpha_{21}a_1+\alpha_{22}a_2 = 3c_{v,\,\mathrm{int}}T, \tag{11.3-53b}$$

respectively, where we have introduced the following symbols for the bracket integrals:

$$\alpha_{11} = [(\mathscr{C}^2-\tfrac{5}{2})\boldsymbol{\mathscr{C}}, (\mathscr{C}^2-\tfrac{5}{2})\boldsymbol{\mathscr{C}}], \tag{11.3-54a}$$

$$\alpha_{12} = \alpha_{21} = [(\mathscr{C}^2-\tfrac{5}{2})\boldsymbol{\mathscr{C}}, (\mathscr{E}-\bar{\mathscr{E}})\boldsymbol{\mathscr{C}}], \tag{11.3-54b}$$

$$\alpha_{22} = [(\mathscr{E}-\bar{\mathscr{E}})\boldsymbol{\mathscr{C}}, (\mathscr{E}-\bar{\mathscr{E}})\boldsymbol{\mathscr{C}}]. \tag{11.3-54c}$$

In this case the auxiliary condition (11.3-35a) is satisfied identically by the trial function and we are left with the problem of solving the two algebraic equations (11.3-53). One has

$$a_1 = \left[\frac{15kT}{2m}\alpha_{22} - 3c_{v,\,\mathrm{int}}T\alpha_{12}\right][\alpha_{11}\alpha_{22} - \alpha_{12}^2]^{-1}, \qquad (11.3\text{-}55a)$$

$$a_2 = \left[3c_{v,\,\mathrm{int}}T\alpha_{11} - \frac{15kT}{2m}\alpha_{21}\right][\alpha_{11}\alpha_{22} - \alpha_{12}^2]^{-1}. \qquad (11.3\text{-}55b)$$

The thermal conductivity can be broken up into two parts, one arising from the first term on the right-hand side of Eq. (11.3-52) and the other from the second term,

$$\lambda = \lambda_1 + \lambda_2 \qquad (11.3\text{-}56a)$$

and for λ_1 and λ_2 one finds:

$$\lambda_1 = \frac{75k^2T}{8m\alpha_{11}} \frac{1 - \frac{3}{5}(\alpha_{12}/\alpha_{22})(c_{v,\,\mathrm{int}}/(3k/2m))}{1 - \alpha_{12}^2/\alpha_{11}\alpha_{22}}, \qquad (11.3\text{-}56b)$$

$$\lambda_2 = \frac{9kT}{4\alpha_{22}} \frac{c_{v,\,\mathrm{int}}^2}{3k/2m} \frac{1 - \frac{5}{3}(\alpha_{12}/\alpha_{11})((3k/2m)/c_{v,\,\mathrm{int}})}{1 - \alpha_{12}^2/\alpha_{11}\alpha_{22}}; \qquad (11.3\text{-}56c)$$

λ_1 and λ_2 may be interpreted as the coefficients of thermal conductivity due to translational energy and internal energy, respectively, for reasons that will become apparent shortly.

The bracket integrals needed here may be reduced,

$$\alpha_{11} = 4\left(\frac{kT}{\pi m}\right)^{\frac{1}{2}} Z_{\mathrm{int}}^{-2} \sum_{ijkl} \mathrm{e}^{-(\mathscr{E}_i + \mathscr{E}_j)} \iint \exp(-\ ^2)g^3[g^4 \sin^2\chi - \tfrac{1}{2}(\Delta\mathscr{E})^2 \sin^2\chi + \tfrac{11}{8}(\Delta\mathscr{E})^2]\sigma_{ij}^{kl}(g, \chi, \varepsilon)\mathrm{d}^2\Omega\,\mathrm{d}g, \qquad (11.3\text{-}57a)$$

$$\alpha_{12} = -\frac{5}{2}\left(\frac{kT}{\pi m}\right)^{\frac{1}{2}} Z_{\mathrm{int}}^{-2} \sum_{ijkl} \mathrm{e}^{-(\mathscr{E}_i + \mathscr{E}_j)} \iint \exp(-g^2)g^3(\Delta\mathscr{E})^2\sigma_{ij}^{kl}(g, \chi, \varepsilon)\mathrm{d}^2\Omega\,\mathrm{d}g, \qquad (11.3\text{-}57b)$$

$$\alpha_{22} = \left(\frac{kT}{\pi m}\right)^{\frac{1}{2}} Z_{\mathrm{int}}^{-2} \sum_{ijkl} \mathrm{e}^{-(\mathscr{E}_i + \mathscr{E}_j)} \iint \exp(-g^2)g^3[|\boldsymbol{g}(\mathscr{E}_i - \mathscr{E}_j) - \boldsymbol{g}'(\mathscr{E}_k - \mathscr{E}_l)|^2 + \tfrac{3}{2}(\Delta\mathscr{E})^2]\sigma_{ij}^{kl}(g, \chi, \varepsilon)\mathrm{d}^2\Omega\,\mathrm{d}g. \qquad (11.3\text{-}57c)$$

This completes the exposition of the basic theory and we now turn to some applications.

Firstly, we note that in molecular collisions, the inelastic energy exchange is generally small. Available evidence indicates (Mason and Monchick [1962]) that in collisions between rotating molecules generally only one or a few quanta of rotational energy are exchanged and since the rotational quantum for gases at ordinary temperatures is much smaller than the relative kinetic energy of a colliding pair (roughly, kT), the assumption that $\Delta\mathscr{E}$ is small is justified for rotational internal energy. Furthermore, for the vibrational internal energy, we note that at normal temperatures only the lowest vibrational energy states are populated and evidence indicates that approximately one collision in a hundred involves vibrational energy transfer. Thus it is reasonable to expect $\Delta\mathscr{E}$ to be small and we shall treat it as a perturbation.

To lowest order we assume $\Delta\mathscr{E} = 0$. If this is done, inelastic scattering is completely ignored and, if one further assumes that the cross section is independent of the internal states of the molecules involved in the collision one finds that the bracket integrals reduce to those for the monatomic gas. One finds

$$\eta = 5kT/8\Omega^{(2,2)}, \tag{11.3-58a}$$

$$\kappa = 0, \tag{11.3-58b}$$

$$\lambda_1 = 75k^2T/32m\Omega^{(2,2)}, \tag{11.3-58c}$$

$$\lambda_2 = (3kT/8\Omega^{(1,1)})c_{v,\,\mathrm{int}}. \tag{11.3-58d}$$

The first three of these coefficients are precisely the shear viscosity, the bulk viscosity and the thermal conductivity coefficients of the monatomic gas. The last coefficient is

$$\lambda_2 = \rho\mathscr{D}c_{v,\,\mathrm{int}}. \tag{11.3-58e}$$

This result is equivalent to Eq. (11.2-20) and provides further justification for that formula. It is interesting to note the approximations made in deriving the expressions (11.3-58). Inelastic cross sections have been completely ignored for the computation of the transport properties; however, the zeroth order distribution function (11.3-20) corresponds to *local* equilibrium and the establishment of that equilibrium obviously requires the existence of inelastic collisions. Thus, for the formulae (11.3-58) to hold we see that inelastic collisions must be rare but not so rare as to make local equilibrium impossible to attain.

A better approximation has been introduced by Mason and Monchick [1962a]; they retain some of the effects of inelastic collisions. For the viscosity, their approach is particularly simple. One must evaluate the integral in Eq. (11.3-39); the difficult term is $(\Delta\mathscr{E})^2 \sin^2\chi$ which couples the internal and translational degrees of freedom. To treat this term, Mason and Monchick assume that $\sin^2 \chi$ may be approximated by its average value over the unit sphere, $\frac{2}{3}$. This is equivalent to using the rigid spere model (which predicts a cross section independent of χ and ε). When this approximation is used the terms in $(\Delta\mathscr{E})^2$ cancel and one again finds Eq. (11.3-58a) i.e., the viscosity remains unaffected by the presence of inelastic collisions to this order of approximation.

With respect to the bulk viscosity little can be done aside from rewriting it in terms of the relaxation time as in Eq. (11.3-51b). Thus we proceed to the thermal conductivity which requires estimation of the three bracket integrals (11.3-57). In the first of these we again replace $\sin^2 \chi$ by its average value over the unit sphere, $\frac{2}{3}$, in the terms involving $(\Delta\mathscr{E})^2$, to obtain

$$\alpha_{11} = 4\left(\frac{kT}{\pi m}\right)^{\frac{1}{2}} Z_{\text{int}}^{-2} \sum_{ijkl} e^{-(\mathscr{E}_i+\mathscr{E}_j)} \iint \exp(-g^2) g^3 [g^4 \sin^2\chi + \tfrac{25}{24}(\Delta\mathscr{E})^2] \sigma_{ij}^{kl} d^2\Omega\, dg$$

$$= 4\Omega^{(2,2)} + \frac{25}{12}\frac{mc_{v,\,\text{int}}}{nk\tau_\kappa}. \qquad (11.3\text{-}59a)$$

Similarly

$$\alpha_{12} = -\frac{5}{4}\frac{mc_{v,\,\text{int}}}{nk\tau_\kappa}. \qquad (11.3\text{-}59b)$$

The approximation of α_{22} is a little more difficult. The term involving $(\Delta\mathscr{E})^2$ in Eq. (11.3-57c) causes no difficulty and is readily converted to $\frac{3}{4}(mc_{v,\,\text{int}}/nk\tau_\kappa)$. With the further approximation that $(\mathscr{E}_k - \mathscr{E}_l)g' = (\mathscr{E}_i - \mathscr{E}_j)g$ inside the integrals the other two terms reduce to

$$2\left(\frac{kT}{\pi m}\right)^{\frac{1}{2}} Z_{\text{int}}^{-2} \sum_{ijkl} e^{-(\mathscr{E}_i+\mathscr{E}_j)} \iint (\mathscr{E}_i - \mathscr{E}_j)^2 g^5 (1-\cos\chi) \exp(-g^2) \sigma_{ij}^{kl} d^2\Omega\, dg$$

and one thus finds

$$\alpha_{22} = 4\left(\frac{kT}{\pi m}\right)^{\frac{1}{2}} \frac{mc_{v,\,\text{int}}}{k} \Omega^{(1,1)} + \frac{3}{4}\frac{mc_{v,\,\text{int}}}{nk\tau_\kappa}. \qquad (11.3\text{-}59c)$$

Using the relation (11.3-58a) between the viscosity and $\Omega^{(2,2)}$ and the relation (7.3-49) between the self-diffusion coefficient and $\Omega^{(1,1)}$ we can now write the two components of the thermal conductivity coefficient as:

$$\lambda_1 = \tfrac{5}{2}\eta \frac{3k}{2m}\left[1-\frac{c_{v,\,\mathrm{int}}}{3k/2m}\,\frac{\tfrac{5}{2}\eta-\rho\mathscr{D}}{2p\tau_\kappa}\right], \tag{11.3-60a}$$

$$\lambda_2 = \rho\mathscr{D}\,c_{v,\,\mathrm{int}}\left[1+\frac{\tfrac{5}{2}\eta-\rho\mathscr{D}}{2p\tau_\kappa}\right]. \tag{11.3-60b}$$

Finally, the ratio of the coefficients of thermal conductivity and viscosity can be written in the form

$$\frac{\lambda}{\eta} = \mathrm{f}_{\mathrm{tr}}\frac{3k}{2m} + \mathrm{f}_{\mathrm{int}}\,c_{v,\,\mathrm{int}} \tag{11.3-61a}$$

as before, cf. Eq. (11.2-2), but now

$$\mathrm{f}_{\mathrm{tr}} = \frac{5}{2}\left\{1-\frac{5}{6}\left[1-\frac{2}{5}\frac{\rho\mathscr{D}}{\eta}\right]\frac{c_{v,\,\mathrm{int}}}{k/m}\left(\frac{\eta}{p\tau_\kappa}\right)\right\}, \tag{11.3-61b}$$

$$\mathrm{f}_{\mathrm{int}} = \frac{\rho\mathscr{D}}{\eta}\left\{1+\frac{5}{4}\left[1-\frac{2}{5}\frac{\rho\mathscr{D}}{\eta}\right]\frac{\eta}{p\tau_\kappa}\right\}. \tag{11.3-61c}$$

In these formulae one sees that the first terms in the curly brackets represent the result of the zeroth order or heuristic approximation and the second terms, which are not always small, represent corrections. Furthermore, since τ_κ can be related to the bulk viscosity, Eqs. (11.3-61) may be regarded as a theoretical relation between measurable properties. However, in many cases accurate data for this property are not available. One must then rely on theoretical calculations of τ_κ or infer it from other measurements, such as light scattering experiments. Mason and Monchick [1962] have compared the approximations derived in the previous section and the approximation just developed with each other and with experiment for a number of gases. A typical case, that of nitrogen, is shown in Fig. 11.2. Even though the data show considerable scatter, it is clear that the approximation given by Eqs. (11.3-61) is superior to the other two and is, in fact, as good a fit as one could hope for with the given data. Similar accuracy was obtained for several other gases. Only hydrogen gave a poor fit to the formula. This could be a consequence of quantum effects which are important in hydrogen; however, the fit becomes poorer with increasing temperature, which is contrary to what one would expect if the discrepancy were in fact quantum mechanical.

As has been pointed out by Mason and Monchick [1962], the above re-

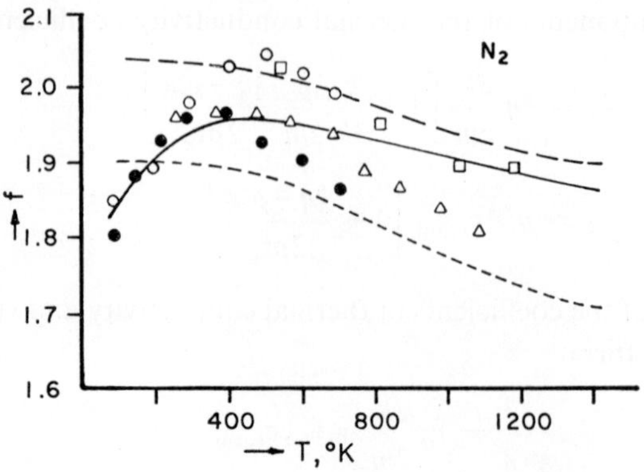

Fig. 11.2. The Eucken factor f $= \lambda/\eta c_v$ for nitrogen. The points are experimental data while the curves represent three theories. The light dashed curve is the Eucken approximation, Eq. (11.2-9); the heavy dashed curve is the modified Eucken approximation, Eq. (11.2-22); the heavy curve is obtained from Eqs. (11.3-61). (From Mason and Monchick [1962a].)

sult (11.3-61) cannot be expected to be as accurate for polar gases as it is for non-polar gases. The reason is that, in polar gases, collisions involving an exchange of a quantum of rotational energy between the molecules become rather common. In such a collision the molecule initially in state i is in state j after the collision and *vice versa*. Consequently, such a collision looks exactly like an elastic collision with deflection angle $\pi - \chi$. The result of this is that the diffusion of rotational energy is no longer governed by the self-diffusion coefficient. One then obtains Eqs. (11.3-61) with $\mathscr{D}$ replaced by a smaller effective diffusion coefficient. However, as the reduction factors depend on the nature of the molecule we shall not discuss the procedure here; we only mention that again good results can be obtained.

The discussion so far has been restricted to the semi-classical case. To obtain the classical results, one needs only to replace the discrete energy levels of the quasi-quantum mechanical case by a continuous set of levels and make the corresponding appropriate change in the cross sections. There remains, however, the difficulty that inverse collisions do not always exist in classical systems, e.g., they do not exist for rough spheres (Chapman and Cowling [1952]). However, Taxman [1958] has shown that in such cases, there exists a finite set of collisions by means of which the molecules can return from their post-collision states to their pre-collision states and, when this fact is used, one obtains substantially the same results as when the existence of inverse collisions is improperly assumed.

As was mentioned earlier, calculations have been carried out for a number of classical models in which the molecules are conceived as being rigid bodies. The simplest of these is the rough sphere model in which the molecules are conceived as spheres that grip each other without slipping during a collision. In this model only one parameter other than the sphere diameter, σ, enters, viz. the moment of inertia, I. The results obtained from this model are most conveniently written in terms of a non-dimensional moment of inertia parameter

$$K = 4I/m\sigma^2. \tag{11.3-62}$$

This parameter varies between zero, when all the mass is concentrated at the center of the sphere, to $\frac{2}{3}$, when all the mass is at the surface. For this model one finds (Chapman and Cowling [1952]) for the viscosity:

$$\eta = \frac{15}{8\sigma^2}\left(\frac{mkT}{\pi}\right)^{\frac{1}{2}}\frac{(1+K)^2}{6+13K}. \tag{11.3-63}$$

For the thermal conductivity one obtains

$$\lambda_1 = \frac{225}{16\sigma^2}\left(\frac{k^3T}{\pi m}\right)^{\frac{1}{2}}\frac{(1+2K)(1+K)^3}{(12+75K+101K^2+102K^3)}, \tag{11.3-64a}$$

$$\lambda_2 = \frac{9}{4\sigma^2}\left(\frac{k^3T}{\pi m}\right)^{\frac{1}{2}}\frac{(3+19K)(1+K)^2}{(12+75K+101K^2+102K^3)}. \tag{11.3-64b}$$

One finds that the viscosity varies between 0.994 and 1.136 of its rigid sphere value, while the thermal conductivity varies between 1.480 and 1.555 of its rigid sphere value, for values of K between 0 and $\frac{2}{3}$.

This model has not met with success in fitting experimental data but, since it is a model for which the transport properties can be found exactly, it provides an interesting test of the previous approximations. This is conveniently done by studying both the results of this model and the approximate methods in the limit of small K. Expanding Eqs. (11.3-63) and (11.3-64) in powers of K one has

$$\eta = \frac{5}{16\sigma^2}\left(\frac{mkT}{\pi}\right)^{\frac{1}{2}}(1-\tfrac{1}{6}K), \tag{11.3-65a}$$

$$\lambda_1 = \frac{75}{64\sigma^2}\left(\frac{k^3T}{\pi m}\right)^{\frac{1}{2}}(1-\tfrac{5}{4}K), \tag{11.3-65b}$$

$$\lambda_2 = \frac{9}{16\sigma^2}\left(\frac{k^3T}{\pi m}\right)^{\frac{1}{2}}(1+\tfrac{25}{12}K). \tag{11.3-65c}$$

On the other hand, from Eqs. (11.3-60) and (11.3-61) one finds for this model

$$\eta = \frac{5}{16\sigma^2}\left(\frac{mkT}{\pi}\right)^{\frac{1}{2}}, \tag{11.3-66a}$$

$$\lambda_1 = \frac{75}{64\sigma^2}\left(\frac{k^3T}{\pi m}\right)^{\frac{1}{2}}(1-\tfrac{13}{12}K), \tag{11.3-66b}$$

$$\lambda_2 = \frac{9}{16\sigma^2}\left(\frac{k^3T}{\pi m}\right)^{\frac{1}{2}}(1+\tfrac{13}{12}K). \tag{11.3-66c}$$

One sees from these results that the approximate expressions are reasonably accurate for the viscosity and the translational portion of the thermal conductivity; in fact, when the Eucken ratio is computed, the factors multiplying the translational specific heat $(3k/2m)$ are identical. For the internal contribution to the thermal conductivity, however, the approximate expression predicts only half of the correction. The discrepancy must be due to the approximations made in computing α_{22}, but improved expressions have not been developed. Fortunately, it is apparent from the comparison with experiment given earlier that the error made in calculating α_{22} is not as serious for real molecules as it is for rough spheres.

11.4. Quantum mechanical effects

The theory of the previous section is deficient in at least two respects. Firstly, it employs, without justification, quantum mechanical cross sections in an essentially classical equation. Secondly, in solving the Boltzmann equation, the possibility that the zero-order approximation to the distribution function might depend on a new summational invariant – the angular momentum – was overlooked.

For most purposes, it appears that these deficiencies are not very serious. An indication of this may be found in the good agreement with experiment that was obtained in the latter part of the previous section. This may, however, have been fortuitous, and a fundamental examination of the problem should certainly be worthwhile. Furthermore, there are important phenomena which cannot be properly explained by a classical theory. By far the most important of these is the effect of an applied magnetic field on the transport properties of a gas – the Senftleben-Beenakker effect. This effect provides an important tool for understanding the interactions between polyatomic molecules, but its potential cannot be fully realized until a correct theory has been developed. Such a theory cannot be based on classical ideas, because the key role in the Senftleben-Beenakker effect is played by the interaction between the magnetic field and the magnetic moment of the molecule. As the latter is intimately connected with the rotation of the molecule, it is of crucial importance that the rotation of the molecule be treated correctly. This is impossible in the Wang Chang-Uhlenbeck theory because no allowance is provided for the degeneracy of rotational states in the modified Boltzmann equation. We shall delay consideration of the Senftleben-Beenakker effect until the following section and consider the quantum mechanical case with no magnetic field in this section.

A qualitative reason why the Wang Chang-Uhlenbeck equation fails for rotating molecules was pointed out by Waldmann [1965], and the argument runs as follows. For the Wang Chang-Uhlenbeck equation to be correct the particles must be in well-defined quantum states before the collision. In quantum mechanical terms this means that the waves representing scattered particles from a collision must become uncorrelated before the following collision. From the uncertainty principle this can occur only if

$$\Delta E \, t_{\text{coll}} \gg \hbar, \tag{11.4-1}$$

where ΔE is the energy difference between two states of the molecule. The condition (11.4-1) is obeyed for all molecular states except those which happen to have identical energies, that is, for all but degenerate states. Now it is well known that rotational levels are highly degenerate, i.e., the lth rotational level is composed of $2l+1$ independent states correponding to different magnetic quantum numbers and, thus, the Wang Chang-Uhlenbeck equation cannot be applied to the treatment of such states.

We must begin again and derive a quantum mechanically correct equation analogous to the Boltzmann equation. Unfortunately, this is not simple, because, in quantum mechanics, one cannot define the position and momentum of a particle simultaneously and, therefore, a single particle distribution function $f(\boldsymbol{r}, \boldsymbol{c}, t)$ cannot be defined. Instead, we must choose a function whose classical limit is the desired distribution function. Such a function will be introduced below, cf. Eq. (11.4-6). It is related to the single particle density matrix $\rho(\boldsymbol{r}, \sigma; \boldsymbol{r}', \sigma')$, which is defined as

$$\rho(\boldsymbol{r}, \sigma; \boldsymbol{r}', \sigma') = \sum_k \psi_k(\boldsymbol{r}, \sigma)\psi_k^*(\boldsymbol{r}', \sigma'). \tag{11.4-2}$$

Here, $\boldsymbol{r}$ is a position coordinate, σ represents all of the coordinates of the internal degrees of freedom of the molecule, k is an index representing the quantum state of the molecule (more correctly, k is a set of indices, one index for each degree of freedom), ψ_k is the wave function for that state, and the sum is taken over all states. Beginning with the quantum mechanical analog of the Liouville equation for the N-particle density matrix and proceeding very much along the lines presented in Chapter 3, Waldmann [1957] and Snider [1960] have derived a Boltzmann-like equation for the single particle density matrix $\rho(\boldsymbol{r}, \sigma; \boldsymbol{r}', \sigma')$. However, as the derivation is lengthy, it will not be given here.

Since we are treating *free* particles, each wave function can be factored into a wave function for the motion of the molecule as a whole (a plane wave) and an internal wave function,

$$\psi_k(\boldsymbol{r}, \sigma) = \psi_\alpha(\boldsymbol{r})\,\varphi_l(\sigma), \tag{11.4-3}$$

where, again, α and l may represent sets of indices. Now, for non-degenerate states there will only be one function φ_l for each energy level. However, as stated earlier, rotational states are highly degenerate and, since all polyatomic molecules possess rotational degrees of freedom, there will always be several wave functions for each energy level. With the usual notation for rotational states, we have

$$\varphi_l(\sigma) = \sum_m c_{lm}\,\varphi_{lm}(\sigma), \tag{11.4-4}$$

where φ_{lm} represents the wave function for the molecule in the lth rotational level with magnetic quantum number (projection of angular momentum on some arbitrary axis), m. Thus, the density matrix takes the form

$$\rho(\boldsymbol{r}, \sigma; \boldsymbol{r}', \sigma') = \sum_{\alpha lmm'} \psi_\alpha(\boldsymbol{r})\psi_\alpha^*(\boldsymbol{r}')\,c_{lm}\,c_{lm'}^*\,\varphi_{lm}(\sigma)\,\varphi_{lm'}^*(\sigma'), \tag{11.4-5a}$$

or, alternatively,

$$\rho(\boldsymbol{r}, \sigma; \boldsymbol{r}', \sigma') = \sum_{lmm'} \rho\begin{pmatrix} lm & lm' \\ \boldsymbol{r} & \boldsymbol{r}' \end{pmatrix} \varphi_{lm}(\sigma)\,\varphi_{lm'}^*(\sigma'), \tag{11.4-5b}$$

where the definition of the density matrix $\rho\left(\begin{smallmatrix} lm \\ r \end{smallmatrix}\middle|\begin{smallmatrix} lm' \\ r' \end{smallmatrix}\right)$ is obvious. Note that, since the molecule cannot be said to be in a definite m-state,* the density matrix $\rho\left(\begin{smallmatrix} lm \\ r \end{smallmatrix}\middle|\begin{smallmatrix} lm' \\ r' \end{smallmatrix}\right)$ depends on a pair of magnetic quantum numbers, m and m', although it depends on only one rotational quantum number, l.

An equation for the density matrix $\rho\left(\begin{smallmatrix} lm \\ r \end{smallmatrix}\middle|\begin{smallmatrix} lm' \\ r' \end{smallmatrix}\right)$ can be derived from the equation for the original matrix $\rho(\boldsymbol{r}, \sigma; \boldsymbol{r}', \sigma')$. It is, however, more appropriate to use, instead of the density matrix ρ, a function whose classical limit (that is, whose limit as $\hbar \to 0$) is the classical phase space distribution function. Such a function is the *Wigner distribution function*, which is defined as

$$f_{lm,\,lm'}(\boldsymbol{r}, \boldsymbol{p}, t) = \frac{1}{(\pi\hbar)^3} \int \exp(2i\boldsymbol{p}\cdot\boldsymbol{R}/\hbar)\, \rho \left(\begin{matrix} lm \\ \boldsymbol{r}+\boldsymbol{R} \end{matrix} \middle| \begin{matrix} lm' \\ \boldsymbol{r}-\boldsymbol{R} \end{matrix} \right) d^3R. \qquad (11.4\text{-}6)$$

(Note that, in the present section, momentum ($\boldsymbol{p}$) is used as the independent variable, in place of velocity ($\boldsymbol{c}$). The time dependence of f comes from the time-dependence of ρ, which has not been displayed explicitly.)

The Wigner distribution function is a density matrix in the space of internal quantum numbers, diagonal in l, but not in general diagonal in m. Its normalization is such that

$$n(\boldsymbol{r}, t) = \mathrm{tr} \int f\, d^3p, \qquad (11.4\text{-}7)$$

where the trace is over internal state space and $n(\boldsymbol{r}, t)$ is the number density of the gas at $(\boldsymbol{r}, t)$.

In the absence of external forces, the equation satisfied by the Wigner distribution function reads

$$\left(\frac{\partial}{\partial t} + \frac{\boldsymbol{p}}{m}\cdot\nabla_r\right) f_{lm,\,lm'}(\boldsymbol{r}, \boldsymbol{p}, t)$$

$$= (2\pi\hbar)^2 \sum_{\substack{L, L_1, l_1, m_1 \\ n, n', n_1, n'_1}} \iint \left\langle \begin{matrix} lm\, l_1\, m_1 \\ (\boldsymbol{p}, \boldsymbol{p}_1) \end{matrix} \middle| t \middle| \begin{matrix} Ln\, L_1\, n_1 \\ (\boldsymbol{p}', \boldsymbol{p}'_1) \end{matrix} \right\rangle f_{Ln,\,Ln'}(\boldsymbol{r}, \boldsymbol{p}', t) \times$$

* If the molecule were in a definite m-state with respect to some particular axis, it would not be in such a state with respect to any other axis. Since the choice of coordinates is arbitrary, we cannot assign a particular m to a state.

$$\times f_{L_1 n_1, L_1 n'_1}(\boldsymbol{r}, \boldsymbol{p}'_1, t) \left\langle \begin{matrix} Ln'\, L_1\, n'_1 \\ (\boldsymbol{p}', \boldsymbol{p}'_1) \end{matrix} \right| \delta(E) t^\dagger \left| \begin{matrix} lm'\, l_1\, m_1 \\ (\boldsymbol{p}, \boldsymbol{p}_1) \end{matrix} \right\rangle \mathrm{d}^3 p_1 \, \mathrm{d}^3 p'$$

$$+ \frac{2\pi h^2}{\mathrm{i}} \sum_{\substack{l_1, m_1 \\ n_1, n}} \int \left[\left\langle \begin{matrix} lm\, l_1\, m_1 \\ (\boldsymbol{p}, \boldsymbol{p}_1) \end{matrix} \right| t \left| \begin{matrix} ln\, l_1\, n_1 \\ (\boldsymbol{p}, \boldsymbol{p}_1) \end{matrix} \right\rangle f_{ln, lm'}(\boldsymbol{r}, \boldsymbol{p}, t) f_{l_1 n_1, l_1 m_1}(\boldsymbol{r}, \boldsymbol{p}_1, t) \right.$$

$$\left. - f_{lm, ln}(\boldsymbol{r}, \boldsymbol{p}, t) f_{l_1 m_1, l_1 n_1}(\boldsymbol{r}, \boldsymbol{p}_1, t) \left\langle \begin{matrix} ln\, l_1\, n_1 \\ (\boldsymbol{p}, \boldsymbol{p}_1) \end{matrix} \right| t^\dagger \left| \begin{matrix} lm'\, l_1\, m_1 \\ (\boldsymbol{p}, \boldsymbol{p}_1) \end{matrix} \right\rangle \right] \mathrm{d}^3 p_1 . \tag{11.4-8}$$

Here, $\boldsymbol{p}'_1$ is understood to mean $\boldsymbol{p}+\boldsymbol{p}_1-\boldsymbol{p}'$, and $(\boldsymbol{p}, \boldsymbol{p}') = \frac{1}{2}(\boldsymbol{p}_1 - \boldsymbol{p})$ is half the relative linear momentum of the colliding molecules. The quantum mechanical operator t describes the scattering process; it is related to the cross section in the following way,

$$\sigma_{\mathrm{a}\to\mathrm{b}} = (\pi m h)^2 (p_{\mathrm{b}}/p_{\mathrm{a}}) |\langle b|t|a\rangle|^2 . \tag{11.4-9}$$

Thus, the first term on the right-hand side of Eq. (11.4-8) is the contribution of scattering to the states contributing to $f_{lm, lm'}$, while the second term represents the rate of removal from such states due to scattering.

The equation (11.4-8) is known as the *Waldmann-Snider equation*. It reduces to the Wang Chang-Uhlenbeck equation in the absence of degenerate rotational states.

Snider [1964b] has verified that the Waldmann-Snider equation gives rise to an H-theorem, and that its collision term possesses all of the necessary conservation properties – i.e., it conserves mass, linear momentum, angular momentum, and energy.

The equation was used as the basis for a theory of transport properties of polyatomic gases by McCourt – cf. McCourt [1966]; McCourt and Snider [1964, 1965]. The essential difference between this theory and the one presented in the previous section is that, following the suggestion of Kagan and Afanas'ev [1961], in the zero-order solution the possibility of a non-zero local angular momentum density is taken into account. In the previous section it was noted that $\log f^{(0)}$ is a linear combination of the summational invariants, but the coefficient of the angular momentum was set equal to zero. However, there is no reason why a local angular momentum density cannot exist; such an angular momentum density will be small, to be sure, but it can be quite important.

As before, the theory begins with the presentation of the macroscopic conservation equations. With the normalization (11.4-7), the *mass density*,

ρ, the *hydrodynamic velocity*, $\boldsymbol{v}$, and the *internal energy* per unit mass, u, are given by

$$\rho(\boldsymbol{r}, t) = n\langle m\rangle, \tag{11.4-10}$$

$$\rho\boldsymbol{v}(\boldsymbol{r}, t) = n\langle \boldsymbol{p}\rangle, \tag{11.4-11}$$

$$\rho u(\boldsymbol{r}, t) = n\langle \tfrac{1}{2}mC^2 + H'\rangle, \tag{11.4-12}$$

where $\boldsymbol{C}$ is the peculiar velocity, $\boldsymbol{C} = (\boldsymbol{p}/m) - \boldsymbol{v}$. The operator H' is the Hamiltonian (total energy) less the kinetic energy contribution, $H' = H - p^2/2m$; in other words, the operator H' is the Hamiltonian for internal degrees of freedom. The brackets $\langle\ \rangle$ are used to denote an average over momentum and internal state spaces – i.e., for any molecular property φ, $\langle\varphi\rangle$ is defined by

$$n\langle\varphi\rangle = \mathrm{tr}\int \varphi f\, \mathrm{d}^3p, \tag{11.4-13}$$

where the trace is over internal state space.

The conservation equations for mass, momentum and energy are precisely the equations (11.3-6), (11.3-7) and (11.3-8) of the previous section,

$$\frac{1}{\rho}\frac{\mathrm{d}\rho}{\mathrm{d}t} = -\nabla\cdot\boldsymbol{v}, \tag{11.4-14}$$

$$\rho\frac{\mathrm{d}\boldsymbol{v}}{\mathrm{d}t} = -\nabla\cdot\mathsf{P}, \tag{11.4-15}$$

$$\rho\frac{\mathrm{d}u}{\mathrm{d}t} = -\nabla\cdot\boldsymbol{q} - \mathsf{P} : \nabla\boldsymbol{v}, \tag{11.4-16}$$

where, now, the pressure tensor, P, and the heat flow vector, $\boldsymbol{q}$, are defined by

$$\mathsf{P} = n\langle m\boldsymbol{C}\boldsymbol{C}\rangle \tag{11.4-17}$$

and

$$\boldsymbol{q} = n\langle(\tfrac{1}{2}mC^2 + H')\boldsymbol{C}\rangle, \tag{11.4-18}$$

respectively. Again, Eq. (11.4-16) may be written in terms of the temperature T, since $\mathrm{d}u/\mathrm{d}t = c_v(\mathrm{d}T/\mathrm{d}t)$, where c_v is the specific heat of the gas.

In addition, each molecule is characterized by its angular momentum operator, $\boldsymbol{M}$, which is the sum of $\boldsymbol{r}\times\boldsymbol{p}$ and the internal angular momentum, $\boldsymbol{J}$,

$$\boldsymbol{M} = \boldsymbol{r}\times\boldsymbol{p} + \boldsymbol{J}. \tag{11.4-19}$$

The molecular property $\boldsymbol{J}$ gives rise to a new macroscopic observable of the gas, $\boldsymbol{J}_o$,

$$\boldsymbol{J}_o = \langle \boldsymbol{J} \rangle = \frac{1}{n} \operatorname{tr} \int \boldsymbol{J} f \, \mathrm{d}^3 p, \tag{11.4-20}$$

which is the *average internal angular momentum*. The conservation equation for $\boldsymbol{J}_o$ is readily found by operating on the Waldmann-Snider equation by the angular momentum operator $\boldsymbol{M}$, integrating the resulting equation over momentum space and summing over internal states. As the collision operator conserves angular momentum, one finds

$$\frac{\mathrm{d}\boldsymbol{J}_o}{\mathrm{d}t} = -\nabla \cdot \mathbf{L}, \tag{11.4-21}$$

where $\mathbf{L}$ is the angular momentum flux tensor,

$$\mathbf{L} = n\langle \boldsymbol{C}\boldsymbol{J} \rangle. \tag{11.4-22}$$

The role of $\mathbf{L}$ is analogous to that of the pressure tensor $\mathbf{P}$ for linear momentum. Eq. (11.4-21) actually contains no contribution from the $\boldsymbol{r} \times \boldsymbol{p}$ term in the angular momentum operator; it is not difficult to see that this is so because the term $\boldsymbol{r} \times \boldsymbol{p}$ simply yields terms which are the cross product of $\boldsymbol{r}$ with the momentum conservation equation, Eq. (11.4-15). Eq. (11.4-21) states that the change in angular momentum of an element moving with the stream is the divergence of the angular momentum flux tensor.

The zero-order solution (i.e., equilibrium solution) of the Waldmann-Snider equation (11.4-8) can be written as

$$f^{(0)}(\boldsymbol{p}) = \frac{n}{(2\pi mkT)^{\frac{3}{2}}} \exp\left(-p^2/2mkT\right) \frac{\exp\left(-H'/kT\right) \exp\left(\boldsymbol{\alpha} \cdot \boldsymbol{J}/kT\right)}{Q}. \tag{11.4-23}$$

Here, the function

$$Q = \operatorname{tr} \exp\left(-H'/kT\right) \exp\left(\boldsymbol{\alpha} \cdot \boldsymbol{J}/kT\right) \tag{11.4-24}$$

is the internal state partition function; the parameter $\boldsymbol{\alpha}$ is related to the angular momentum density $\boldsymbol{J}_o$ by

$$\boldsymbol{J}_o = kT\nabla_\alpha \log Q. \tag{11.4-25}$$

In this approximation the pressure tensor, heat flow vector and angular momentum flux tensor become

$$\mathsf{P} = p\mathsf{I} \quad \text{with } p = nkT, \tag{11.4-26}$$

$$\boldsymbol{q} = 0, \tag{11.4-27}$$

$$\mathsf{L} = 0, \tag{11.4-28}$$

so that the conservation equations yield the Euler equations of fluid dynamics, together with the equation of angular momentum,

$$\frac{\mathrm{d}\boldsymbol{J}_\mathrm{o}}{\mathrm{d}t} = 0. \tag{11.4-29}$$

One can now proceed in a manner similar to the usual Chapman-Enskog development provided that proper care is taken of the fact that one is dealing with quantum mechanical functions and operators and, consequently, that certain commutation operations may not be admissible. McCourt [1966] and McCourt and Snider [1964, 1965] have, in fact, carried out this program, although using slightly different language. The development is involved and will be presented here only in summary; for the details the reader is referred to the works cited above.

In addition to the usual Chapman-Enskog hypotheses one assumes that the angular momentum density – i.e., $\boldsymbol{\alpha}$ – is small and that all quantities can be expanded in powers of $\boldsymbol{\alpha}$, Thus, to terms quadratic in $\boldsymbol{\alpha}$,

$$Q = Q_\mathrm{o}\{1+[\alpha^2/6(kT)^2]\langle J^2\rangle\}, \tag{11.4-30}$$

where

$$Q_\mathrm{o} = \operatorname{tr} \exp(-H'/kT) \tag{11.4-31}$$

and

$$\langle J^2\rangle = (Q_\mathrm{o})^{-1} \operatorname{tr} J^2 \exp(-H'/kT). \tag{11.4-32}$$

Furthermore, to terms linear in $\boldsymbol{\alpha}$,

$$\boldsymbol{J}_\mathrm{o} = (\langle J^2\rangle/3kT)\boldsymbol{\alpha}. \tag{11.4-33}$$

If the equilibrium distribution is expanded in powers of $\boldsymbol{\alpha}$ and only terms linear in $\boldsymbol{\alpha}$ are retained, one has

$$f^{(0)}(\boldsymbol{p}) = f_\mathrm{o}^{(0)}(\boldsymbol{p})(1+\boldsymbol{\alpha}\cdot\boldsymbol{J}/kT), \tag{11.4-34}$$

where

$$f_\mathrm{o}^{(0)}(\boldsymbol{p}) = \frac{n}{(2\pi mkT)^{\frac{3}{2}}} \exp\left(-\frac{p^2}{2mkT}\right) \frac{\exp(-H'/kT)}{Q_\mathrm{o}}. \tag{11.4-35}$$

The expansion (11.4-34) for $f^{(0)}$ is used in computing the streaming term in the first Chapman-Enskog approximation; however, it is *not* used in the collision term where its use would destroy the important conservation properties of the operator.

The linearization of f is carried out in a manner that makes the perturbation ϕ Hermitian:

$$f = f^{(0)} + \tfrac{1}{2}(f^{(0)}\phi + \phi f^{(0)}), \tag{11.4-36}$$

which is necessary because $f^{(0)}$ and ϕ, being operators, do not commute. (Basically the difficulty is caused by the non-commutivity of $\boldsymbol{\alpha}$ and $\boldsymbol{J}$.) When the Chapman-Enskog program is carried out and only terms linear in $\boldsymbol{\alpha}$, but including $\boldsymbol{\alpha}\nabla\boldsymbol{\alpha}$, are retained, one obtains the following equation for ϕ,

$$\begin{aligned}
n^2 I(\phi) = -f_0^{(0)} \Bigg\{ & \left(\mathscr{C}^2 - \tfrac{5}{2} + \frac{H' - \langle H' \rangle}{kT}\right) \boldsymbol{C} \cdot \nabla \log T + 2(\boldsymbol{\mathscr{C}}\boldsymbol{\mathscr{C}} - \tfrac{1}{3}\mathscr{C}^2 \boldsymbol{I}) : \nabla \boldsymbol{v} \\
& + \left[\frac{2}{3}\frac{c_{v,\,\mathrm{int}}}{c_v}(\mathscr{C}^2 - \tfrac{3}{2}) - \frac{k/m}{c_v}\frac{H' - \langle H' \rangle}{kT}\right] \nabla \cdot \boldsymbol{v} \\
& + \frac{1}{kT}\boldsymbol{J}\boldsymbol{C} : \nabla\boldsymbol{\alpha} + \frac{1}{kT}\left(\mathscr{C}^2 - \tfrac{7}{2} + \frac{H' - \langle H' \rangle}{kT}\right) \boldsymbol{J}\boldsymbol{C} : (\nabla \log T)\boldsymbol{\alpha} \\
& + \frac{2}{kT}\boldsymbol{J}(\boldsymbol{\mathscr{C}}\boldsymbol{\mathscr{C}} - \tfrac{1}{3}\mathscr{C}^2\boldsymbol{I}) \vdots (\nabla\boldsymbol{v})\boldsymbol{\alpha} \\
& + \left[\frac{2}{3}\frac{c_{v,\,\mathrm{int}}}{c_v}(\mathscr{C}^2 - \tfrac{3}{2}) - \frac{k/m}{c_v}\frac{H' - \langle H' \rangle}{kT}\right. \\
& \left. + \frac{k/m}{c_v}\frac{\langle J^2 H' \rangle - \langle J^2 \rangle \langle H' \rangle}{kT\langle J^2 \rangle}\right] \boldsymbol{J} \cdot (\nabla \cdot \boldsymbol{v})\boldsymbol{\alpha} \\
& + \frac{1}{(kT)^2}(\boldsymbol{J}\boldsymbol{J})^* \boldsymbol{C} \vdots (\nabla\boldsymbol{\alpha})\boldsymbol{\alpha} + \frac{J^2 - \langle J^2 \rangle}{3(kT)^2}\boldsymbol{C} \cdot (\nabla\boldsymbol{\alpha}) \cdot \boldsymbol{\alpha} \Bigg\}.
\end{aligned} \tag{11.4-37}$$

Here, I is the linearized collision operator, $\boldsymbol{\mathscr{C}}$ is the usual dimensionless velocity variable, $\boldsymbol{\mathscr{C}} = (m/2kT)^{\frac{1}{2}}\boldsymbol{C}$, and $(\boldsymbol{J}\boldsymbol{J})^*$ is the symmetric traceless part of the tensor $\boldsymbol{J}\boldsymbol{J}$. Furthermore, c_v and $c_{v,\,\mathrm{int}}$ are the coefficients of specific heat, defined by

$$c_v = \partial u / \partial T, \tag{11.4-38a}$$

$$c_{v,\,\mathrm{int}} = c_v - (3k/2m). \tag{11.4-38b}$$

In the classical case, and in the absence of a local angular momentum density – i.e., if $\boldsymbol{\alpha} = 0$ – Eq. (11.4-37) is identical with Eq. (11.3-21) of the previous section.

From Eq. (11.4-37) it is readily seen that the perturbation function ϕ must have the form

$$\phi = -\frac{1}{n}\boldsymbol{A}\cdot\nabla\log T - \frac{1}{n}\mathbf{B}:\nabla\boldsymbol{v} - \frac{1}{n}\mathbf{E}:\nabla\boldsymbol{\alpha}$$

$$-\frac{1}{n}\mathbf{A}':(\nabla\log T)\boldsymbol{\alpha} - \frac{1}{n}\mathbf{B}'\vdots(\nabla\boldsymbol{v})\boldsymbol{\alpha} - \frac{1}{n}\mathbf{E}'\vdots(\nabla\boldsymbol{\alpha})\boldsymbol{\alpha}, \qquad (11.4\text{-}39)$$

where $\boldsymbol{A}$, $\mathbf{B}$, $\mathbf{A}'$, $\mathbf{B}'$, $\mathbf{E}$ and $\mathbf{E}'$ are all tensor functions (of the proper rank) of the peculiar velocity and the internal coordinates of the molecules. (The terms involving $\nabla\boldsymbol{v}$ and $\nabla\cdot\boldsymbol{v}$ have been written as a single term involving $\nabla\boldsymbol{v}$; similarly, the terms involving $(\nabla\boldsymbol{\alpha})\boldsymbol{\alpha}$ and $(\nabla\boldsymbol{\alpha})\cdot\boldsymbol{\alpha}$ have been written as a single term involving $(\nabla\boldsymbol{\alpha})\boldsymbol{\alpha}$.) Strictly speaking, there is a different set of tensors for each set of indices (lm, lm'), but we use a symbolic notation in order to avoid proliferation of subscripts. Substitution of the expression (11.4-39) into Eq. (11.4-37) yields a set of six equations for the tensors enumerated above,

$$I(\boldsymbol{A}) = \frac{1}{n}f_o^{(0)}\left(\mathscr{C}^2 - \tfrac{5}{2} + \frac{H' - \langle H'\rangle}{kT}\right)\boldsymbol{C}, \qquad (11.4\text{-}40)$$

$$I(\mathbf{B}) = \frac{1}{n}f_o^{(0)}\left\{(\mathscr{C}\mathscr{C} - \tfrac{1}{3}\mathscr{C}^2\mathbf{I}) + \left[\frac{2}{3}\frac{c_{v,\,\mathrm{int}}}{c_v}(\mathscr{C}^2 - \tfrac{3}{2}) - \frac{k/m}{c_v}\frac{H' - \langle H'\rangle}{kT}\right]\mathbf{I}\right\}, \qquad (11.4\text{-}41)$$

$$I(\mathbf{E}) = \frac{1}{nkT}f_o^{(0)}\boldsymbol{J}\boldsymbol{C}, \qquad (11.4\text{-}42)$$

$$I(\mathbf{A}') = \frac{1}{nkT}f_o^{(0)}\left(\mathscr{C}^2 - \tfrac{7}{2} + \frac{H' - \langle H'\rangle}{kT}\right)\boldsymbol{J}\boldsymbol{C}, \qquad (11.4\text{-}43)$$

$$I(\mathbf{B}') = \frac{1}{n}f_o^{(0)}\left\{\frac{2}{kT}\boldsymbol{J}(\mathscr{C}\mathscr{C} - \tfrac{1}{3}\mathscr{C}^2\mathbf{I})\right.$$
$$\left. + \left[\frac{2}{3}\frac{c_{v,\,\mathrm{int}}}{c_v}(\mathscr{C}^2 - \tfrac{3}{2}) - \frac{k/m}{c_v}\frac{H' - \langle H'\rangle}{kT} + \frac{k/m}{c_v}\frac{\langle J^2H'\rangle - \langle J^2\rangle\langle H'\rangle}{kT\langle J^2\rangle}\right]\boldsymbol{J}\mathbf{I}\right\}, \qquad (11.4\text{-}44)$$

$$I(\mathbf{E}') = \frac{1}{n(kT)^2}f_o^{(0)}\{(\boldsymbol{J}\boldsymbol{J})^*\boldsymbol{C} + \tfrac{1}{3}(J^2 - \langle J^2\rangle)\mathbf{I}\boldsymbol{C}\}. \qquad (11.4\text{-}45)$$

The next task is to write the fluxes in terms of the tensors appearing in the expression (11.4-39). Since $\boldsymbol{C}$ is an ordinary (polar) vector while $\boldsymbol{J}$ is an axial or pseudovector (i.e., $\boldsymbol{J}$ does not change sign under a reflection), $\boldsymbol{q}$, $\boldsymbol{P}$ and $\boldsymbol{L}$ are a vector, a tensor and a pseudotensor, respectively; also, they will certainly be linear combinations of the gradients appearing in Eq. (11.4-39). Furthermore, the expressions for the flux quantities must be written so as to be invariant under coordinate transformations and, in particular, under rotations. Otherwise, the heat flux, for example, would take on different forms in different coordinate systems. Thus, in the expressions for the flux quantities, there may appear only those combinations of the gradients with isotropic tensors (i.e., tensors whose form is rotationally invariant) which have the proper tensor character. Not all of the gradients can be fashioned into tensors of all the various characters. To take a particularly simple example, we note that from $\nabla \log T$ one can obtain only a vector, namely $\nabla \log T$ itself, and a pseudotensor $\boldsymbol{\epsilon} \cdot \nabla \log T$, where $\boldsymbol{\epsilon}$ is the alternating tensor (i.e., the completely antisymmetric tensor of rank three). Hence, the pressure tensor, which is a symmetric tensor of rank two, cannot contain terms proportional to $\nabla \log T$. Similar arguments are used in the development of phenomenological theories; see De Groot and Mazur [1962].

The essentials of expressing the transport coefficients in terms of generalized bracket integrals, as well as of finding approximate solutions to the integral equations (11.4-40) through (11.4-45) may be found in or extracted from McCourt and Snider [1964, 1965] or from McCourt's thesis [1966]. In fact, completion of the Chapman-Enskog program is almost entirely a matter of lengthy, but relatively straight-forward computation.

The procedure is based upon splitting Eq. (11.4-37) into two parts, involving terms independent of $\boldsymbol{\alpha}$ and terms linear in $\boldsymbol{\alpha}$, respectively. Since $|\boldsymbol{\alpha}|$ is assumed small, it is not necessary to carry the calculation further than to the first order. In the left-hand member of Eq. (11.4-37), the collision operator itself is thereby also split in the same way by expanding $f^{(0)}$. Thus, one finds that, to zero order (i.e., if $\boldsymbol{\alpha} = 0$), the transport properties are exactly those obtained from the Wang Chang-Uhlenheck approach given in the previous section. This is as expected, since $\boldsymbol{\alpha} = 0$ implies an isotropic angular momentum distribution and one would then expect any effects due to the angular momentum to average out. The angular momentum flux pseudotensor is not, however, zero in this case.

Since the complete first-order theory is not yet available and, even if it were, would be too lengthy to present in its entirety, we shall simply sketch the analysis. The major interest lies in the new effects that arise, particular cross effects such as the effect of non-zero angular momentum density on the pressure tensor, and we shall emphasize the form of the flux quantities rather than explicit results for the transport coefficients which occur in the expressions for the fluxes.

Turning our attention first to the pressure tensor, we find that, due to tensor properties mentioned above, only the second and fifth terms of the expression (11.4-39) make any contribution, so that the pressure tensor is a linear combination of $\nabla \boldsymbol{v}$ and $(\nabla \boldsymbol{v})\boldsymbol{\alpha}$. In contrast to the case in which the local angular momentum density is assumed zero, the coefficient tensors cannot be reduced to the product of a scalar and a known tensor of the proper character. Consequently, the transport properties are no longer scalar quantities or, stated otherwise, several transport properties are needed to relate the pressure tensor to the phenomenological gradients. This also means that the expressions for the unknown tensors take a much more complicated form than in the previous cases. For the tensors required for the computation of the pressure tensor we must write

$$B_{ij} = \sum_{p,q,\{k\}} B^{pq}_{k_1 \ldots k_{p+q} ij} [\mathscr{C}]^p_{k_1 \ldots k_p} [J]^q_{k_{p+1} \ldots k_{p+q}}, \tag{11.4-46}$$

$$B'_{ijl} = \sum_{p,q,\{k\}} B'^{pq}_{k_1 \ldots k_{p+q} ijl} [\mathscr{C}]^p_{k_1 \ldots k_p} [J]^q_{k_{p+1} \ldots k_{p+q}}. \tag{11.4-47}$$

In these expansions, the quantities

$$[\mathscr{C}]^p_{k_1 \ldots k_p} \equiv [\mathscr{C}_{k_1} \ldots \mathscr{C}_{k_p}]^p \tag{11.4-48}$$

are the components of the irreducible tensor of rank p and weight p generated by the vector $\mathscr{C}$. Similarly, the quantities

$$[J]^q_{k_{p+1} \ldots k_{p+q}} \equiv [J_{k_{p+1}} \ldots J_{k_{p+q}}]^q \tag{11.4-49}$$

are the components of the irreducible tensor of rank q and weight q generated by the vector $\boldsymbol{J}$. An irreducible tensor whose weight is equal to its rank, is completely symmetric and completely contractionless. For a given rank, n, such a tensor is uniquely determined; it has $2n+1$ independent components. For example, for $n = 1, 2, 3$, the components of the irreducible tensors of rank n and weight n generated by the vector $\boldsymbol{a} = (a_1, a_2, a_3)$ are

$$[a]^1_i = a_i \quad \text{(trace undefined)},$$

$$[a]^2_{ij} = a_i a_j - \tfrac{1}{3} a^2 \delta_{ij},$$

$$[a]^3_{ijk} = a_i a_j a_k - \tfrac{1}{5} a^2 (a_i \delta_{jk} + a_j \delta_{ki} + a_k \delta_{ij}).$$

The tensorial coefficients in the expansions (11.4-46) and (11.4-47) may be functions of the scalars $\mathscr{C}$ and J and, furthermore, they must be linear combinations of isotropic (i.e., rotationally invariant) tensors of the proper rank in order to yield the correct invariance properties. Of course, in practice only a few terms in the expansions (11.4-46) and (11.4-47) are retained.*

It is natural now to expand the tensorial coefficients in terms of Sonine polynomials in velocity ($\mathscr{C}^2$) and in polynomials in H' as well. For the latter, we introduce sets of polynomials $R_q^{(t)}(H'/kT)$ of order t and index q, which satisfy orthogonality relations of the following type,**

$$Q_0^{-1} \sum_j e^{-E_j/kT}(2j+1)\,g(q)\,R_q^{(t)}(E_j/kT)\,R_q^{(t')}(E_j/kT) = N_q^{(t)}\delta_{tt'}, \quad (11.4\text{-}50)$$

where the sum extends over all internal states j; E_j is the energy of internal state j and is, therefore, an eigenvalue of the internal energy operator, H'. The first few weights $g(q)$ are

$$g(0) = 1, \quad (11.4\text{-}51a)$$

$$g(1) = j(j+1), \quad (11.4\text{-}51b)$$

$$g(2) = \tfrac{1}{20}j(j+1)(4j^2+4j-3), \quad (11.4\text{-}51c)$$

* If the internal degrees of freedom are neglected altogether, the expansion (11.4-46) reduces to one single term,

$$B^{pq}_{k_1\ldots k_{p+q}ij} = 0, \quad \text{unless } p = 2 \text{ and } q = 0.$$

Moreover,

$$B^{20}_{k_1k_2ij} = (m/2kT)B(C)\delta_{k_1i}\delta_{k_2j},$$

where $B(C)$ is identical with the function $B(C)$ introduced in Eq. (5.4-20) for the case of the monatomic gas (the factor $m/2kT$ is due to the fact that in Eq. (5.4-20) the irreducible tensor is $[C]^2_{k_1k_2}$, rather than $[\mathscr{C}]^2_{k_1k_2}$). Because of the symmetry of the tensor $[\mathscr{C}]^2_{k_1k_2}$ we may also write

$$B^{20}_{k_1k_2ij} = (m/2kT)B(C)\tfrac{1}{2}(\delta_{k_1i}\delta_{k_2j}+\delta_{k_1j}\delta_{k_2i}).$$

The expansion (11.4-47) is not required if internal degrees of freedom are neglected.

** The polynomial $R_q^{(t)}$ corresponds to the polynomial $R_t^{(q)}$ of McCourt and Snider [1965]. Our notation is consistent with the nomenclature for Sonine polynomials, the superscript denoting the *order* of the polynomial, the subscript its *index* (cf. Section 5.5).

and the first few normalization constants $N_q^{(t)}$ are

$$t = 0: \quad N_q^{(0)} = 1, \quad q = 0, 1, 2, \ldots \tag{11.4-51d}$$

$$t = 1: \quad N_0^{(1)} = \langle (H'/kT)^2 \rangle - \langle H'/kT \rangle^2 = \frac{c_{v,\,\mathrm{int}}}{k/m}, \tag{11.4-51e}$$

$$N_1^{(1)} = \langle (H'/kT)^2 J^2 \rangle - \langle (H'/kT) J^2 \rangle^2 / \langle J^2 \rangle. \tag{11.4-51f}$$

(Higher values of q and t are not needed for the following theory.) Using the quantum mechanical expression for the magnitude of the angular momentum of a rotator, $J^2 = j(j+1)$ (units of $\hbar^2$), one readily evaluates the first few polynomials $R_q^{(t)}$:

$$R_0^{(0)}(H'/kT) = 1, \tag{11.4-52a}$$

$$R_0^{(1)}(H'/kT) = (H'/kT) - \langle H'/kT \rangle, \tag{11.4-52b}$$

$$R_1^{(0)}(H'/kT) = \langle J^2 \rangle^{-\frac{1}{2}}, \tag{11.4-52c}$$

$$R_1^{(1)}(H'/kT) = (H'/kT) - \langle (H'/kT) J^2 \rangle / \langle J^2 \rangle, \tag{11.4-52d}$$

$$R_2^{(0)}(H'/kT) = (\tfrac{4}{20} \langle J^4 \rangle - \tfrac{3}{20} \langle J^2 \rangle^2)^{-\frac{1}{2}}. \tag{11.4-52e}$$

Thus, we expand the tensorial coefficients in Eqs. (11.4-46) and (11.4-47) in the following way,*

$$B^{pq}_{k_1 \ldots k_{p+q} ij} = \sum_{st} B^{pqst}_{k_1 \ldots k_{p+q} ij} S^{(s)}_{p+\frac{1}{2}}(\mathscr{C}^2) R_q^{(t)}(H'/kT), \tag{11.4-53}$$

$$B'^{pq}_{k_1 \ldots k_{p+q} ijl} = \sum_{st} B'^{pqst}_{k_1 \ldots k_{p+q} ijl} S^{(s)}_{p+\frac{1}{2}}(\mathscr{C}^2) R_q^{(t)}(H'/kT). \tag{11.4-54}$$

Now, the tensors $\mathbf{B}^{pqst}$ and $\mathbf{B}'^{pqst}$, whose components are $B^{pqst}_{k_1 \ldots k_{p+q} ij}$ and $B'^{pqst}_{k_1 \ldots k_{p+q} ijl}$, respectively, can be written as sums of *isotropic* tensors of appropriate rank ($p+q+2$ for $\mathbf{B}^{pqst}$, $p+q+3$ for $\mathbf{B}'^{pqst}$). Complete sets of isotropic tensors of ranks up to five are given in Table 11.4. Such sets

* If the internal degrees of freedom are neglected (cf. footnote on p. 342) one has

$$B^{pqst}_{k_1 \ldots k_{p+q} ij} = 0, \text{ unless } p = 2,\ q = 0 \text{ and } t = 0,$$

and $\mathbf{B}'$ plays no role in the theory. Moreover,

$$B^{20s0}_{k_1 k_2 ij} = \tfrac{1}{2} b_s (\delta_{k_1 i} \delta_{k_2 j} + \delta_{k_1 j} \delta_{k_2 i}),$$

where b_s is identical with the coefficient b_s introduced in Eq. (5.6-13) for the case of a monatomic gas.

TABLE 11.4

Complete sets of independent irreducible isotropic tensors of ranks two through five

Rank		Isotropic tensors
2		$T_{ij}^{(2)} = \delta_{ij}$
3		$T_{ijk}^{(3)} = -\varepsilon_{ijk}$
4		$T_{ijkl}^{(4,1)} = \frac{1}{2}(\delta_{ik}\delta_{jl}+\delta_{il}\delta_{jk})-\frac{1}{3}\delta_{ij}\delta_{kl}$
		$T_{ijkl}^{(4,2)} = \frac{1}{2}(\delta_{il}\delta_{jk}-\delta_{ik}\delta_{jl})$
		$T_{ijkl}^{(4,3)} = \delta_{ij}\delta_{kl}$
	or	$\hat{T}_{ijkl}^{(4,1)} = \frac{1}{2}(\delta_{ik}\delta_{jl}+\delta_{ij}\delta_{kl})-\frac{1}{3}\delta_{il}\delta_{jk}$
		$\hat{T}_{ijkl}^{(4,2)} = \frac{1}{2}(\delta_{ik}\delta_{jl}-\delta_{ij}\delta_{kl})$
		$\hat{T}_{ijkl}^{(4,3)} = \delta_{il}\delta_{jk}$
5		$T_{ijklm}^{(5,1)} = -[\frac{1}{2}(\varepsilon_{ilm}\delta_{jk}+\varepsilon_{jlm}\delta_{ik})-\frac{1}{3}\varepsilon_{klm}\delta_{ij}]$
		$T_{ijklm}^{(5,2)} = -\frac{1}{4}(\varepsilon_{ikl}\delta_{jm}+\varepsilon_{jkl}\delta_{im}+\varepsilon_{ikm}\delta_{jl}+\varepsilon_{jkm}\delta_{il})$
		$T_{ijklm}^{(5,3)} = -\frac{1}{2}(\varepsilon_{ijl}\delta_{km}-\varepsilon_{ijm}\delta_{kl})$
		$T_{ijklm}^{(5,4)} = -\varepsilon_{ijk}\delta_{lm}$
		$T_{ijklm}^{(5,5)} = -[\frac{1}{2}(\varepsilon_{ijl}\delta_{km}+\varepsilon_{ijm}\delta_{kl})-\frac{1}{3}\varepsilon_{ijk}\delta_{lm}]$
		$T_{ijklm}^{(5,6)} = -\varepsilon_{klm}\delta_{ij}$
	or	$\hat{T}_{ijklm}^{(5,1)} = -[\frac{1}{2}(\varepsilon_{jkl}\delta_{im}+\varepsilon_{jkm}\delta_{il})-\frac{1}{3}\varepsilon_{ijk}\delta_{lm}]$
		$\hat{T}_{ijklm}^{(5,2)} = -\frac{1}{4}(\varepsilon_{ijm}\delta_{kl}+\varepsilon_{ijl}\delta_{km}+\varepsilon_{ikm}\delta_{jl}+\varepsilon_{ikl}\delta_{jm})$
		$\hat{T}_{ijklm}^{(5,3)} = -\frac{1}{2}(\varepsilon_{klm}\delta_{ij}-\varepsilon_{jlm}\delta_{ik})$
		$\hat{T}_{ijklm}^{(5,4)} = -\varepsilon_{ilm}\delta_{jk}$
		$\hat{T}_{ijklm}^{(5,5)} = -\frac{1}{2}(\varepsilon_{jlm}\delta_{ik}+\varepsilon_{klm}\delta_{ij})$
		$\hat{T}_{ijklm}^{(5,6)} = -\varepsilon_{ijk}\delta_{lm}$

are not unique and two different sets are, in fact, given for ranks four and five.

Thus, we expand

$$\mathbf{B}^{pqst} = \sum_h B_h^{pqst} \mathbf{T}^{(p+q+2,\,h)}, \tag{11.4-55}$$

$$\mathbf{B}'^{pqst} = \sum_h B_h'^{pqst} \mathbf{T}^{(p+q+3,\,h)}, \tag{11.4-56}$$

where the coefficients B_h^{pqst} and $B_h'^{pqst}$ are scalars. In those cases in which there is only one istropic tensor (ranks 2 and 3) we shall drop the subscript on the expansion coefficient.

It is not difficult (although not particularly simple either) to express the pressure tensor in terms of the scalars B_h^{pqst} and $B_h'^{pqst}$. One finds

$$\mathbf{P} = (nkT - \kappa \nabla \cdot \boldsymbol{v} - \kappa_1 \boldsymbol{\alpha} \cdot \nabla \times \boldsymbol{v})\mathbf{I} - 2\eta \mathbf{S} - 2\eta_1 \mathbf{S}_1 - 2\eta_2 \mathbf{S}_2, \tag{11.4-57}$$

where $\mathbf{S}$ is the usual rate-of-shear tensor,

$$\mathbf{S} = \tfrac{1}{2}[\nabla \boldsymbol{v} + (\nabla \boldsymbol{v})^T] - \tfrac{1}{3}\nabla \cdot \boldsymbol{v}\mathbf{I}, \tag{11.4-58a}$$

and $\mathbf{S}_1$ and $\mathbf{S}_2$ are defined by

$$\mathbf{S}_1 = \tfrac{1}{2}[\boldsymbol{\alpha}(\nabla \times \boldsymbol{v}) + (\nabla \times \boldsymbol{v})\boldsymbol{\alpha}] - \tfrac{1}{3}\boldsymbol{\alpha} \cdot (\nabla \times \boldsymbol{v})\mathbf{I}, \tag{11.4-58b}$$

$$\mathbf{S}_2 = \tfrac{1}{2}[\boldsymbol{\alpha} \times \mathbf{S} - \mathbf{S} \times \boldsymbol{\alpha}]. \tag{11.4-58c}$$

The coefficients appearing in Eq. (11.4-57) are given by*

$$\eta = \tfrac{1}{2}kT B_1^{2000}, \tag{11.4-59a}$$

$$\eta_1 = \tfrac{1}{2}kT[B_1'^{2000} + (1/3kT)B_1^{2100}], \tag{11.4-59b}$$

$$\eta_2 = \tfrac{1}{2}kT[B_2'^{2000} + (1/3kT)B_2^{2100}], \tag{11.4-59c}$$

and

$$\kappa = -kT B^{0010}, \tag{11.4-60a}$$

$$\kappa_1 = -kT[B'^{0010} + (1/3kT)B^{0110}]. \tag{11.4-60b}$$

* From the footnote on p. 343 we see that, in the case of no internal degrees of freedom, the only non-zero coefficient in the expansions (11.4-55) and (11.4-56) is $B_1^{20s0} = b_s$. Moreover, one verifies that the coefficient of $\frac{1}{2}(\delta_{k_1 i}\delta_{k_2 j} + \delta_{k_1 j}\delta_{k_2 i})$ in the expansion (11.4-49) of $B_{k_1 k_2 ij}^{20s0}$ is equal to B_1^{20s0}. Thus, comparing with the expression for $B_{k_1 k_2 ij}^{20s0}$ given in the footnote on p. 343 we conclude that $B_1^{20s0} = b_s$. Thus, B_1^{2000} is the generalization of the coefficient b_0 in the expansion (5.6-13) and the result (11.4-59a) generalizes the result (5.6-23) for monatomic gases to polyatomic gases.

Again, in the case $\boldsymbol{\alpha} = 0$, Eq. (11.4-57) reduces to Eq. (11.3-33) with η, the shear viscosity, and κ, the bulk viscosity, reducing to those of the Wang Chang-Uhlenbeck theory.

Now, turning to the angular momentum flux pseudotensor, we find that only the terms which did not contribute to the pressure tensor, i.e., the first, third, fourth and sixth terms of the right member of Eq. (11.4-39) contribute. We assume expansions of the form,

$$A_i = -\left(\frac{m}{2kT}\right)^{\frac{1}{2}} \sum_{pqst} A^{pqst}_{k_1\ldots k_{p+q}i} [\mathscr{C}]^p [J]^q S^{(s)}_{p+\frac{1}{2}}(\mathscr{C}^2) R^{(t)}_q(H'/kT), \tag{11.4-61}$$

$$A'_{ij} = -\left(\frac{m}{2kT}\right)^{\frac{1}{2}} \sum_{pqst} A'^{pqst}_{k_1\ldots k_{p+q}ij} [\mathscr{C}]^p [J]^q S^{(s)}_{p+\frac{1}{2}}(\mathscr{C}^2) R^{(t)}_q(H'/kT), \tag{11.4-62}$$

$$E_{ij} = -\left(\frac{m}{2kT}\right)^{\frac{1}{2}} \sum_{pqst} E^{pqst}_{k_1\ldots k_{p+q}ij} [\mathscr{C}]^p [J]^q S^{(s)}_{p+\frac{1}{2}}(\mathscr{C}^2) R^{(t)}_q(H'/kT), \tag{11.4-63}$$

$$E'_{ijl} = -\left(\frac{m}{2kT}\right)^{\frac{1}{2}} \sum_{pqst} E'^{pqst}_{k_1\ldots k_{p+q}ijl} [\mathscr{C}]^p [J]^q S^{(s)}_{p+\frac{1}{2}}(\mathscr{C}^2) R^{(t)}_q(H'/kT). \tag{11.4-64}$$

Then, if the tensors $\mathbf{A}^{pqst}$, $\mathbf{A}'^{pqst}$, $\mathbf{E}^{pqst}$ and $\mathbf{E}'^{pqst}$, whose components are $A^{pqst}_{k_1\ldots k_{p+q}i}$, $A'^{pqst}_{k_1\ldots k_{p+q}ij}$, $E^{pqst}_{k_1\ldots k_{p+q}ij}$ and $E'^{pqst}_{k_1\ldots k_{p+q}ijl}$, respectively, are, in turn, expanded in terms of the isotropic tensors of proper rank,

$$\mathbf{A}^{pqst} = \sum_h A^{pqst}_h \mathbf{T}^{(p+q+1,\,h)}, \tag{11.4-65}$$

$$\mathbf{A}'^{pqst} = \sum_h A'^{pqst}_h \mathbf{T}^{(p+q+2,\,h)}, \tag{11.4-66}$$

$$\mathbf{E}^{pqst} = \sum_h E^{pqst}_h \mathbf{T}^{(p+q+2,\,h)}, \tag{11.4-67}$$

$$\mathbf{E}'^{pqst} = \sum_h E'^{pqst}_h \mathbf{T}^{(p+q+3,\,h)}, \tag{11.4-68}$$

one finds that the angular momentum flux pseudotensor takes the form

$$\begin{aligned}\mathbf{L} = {} & -2\zeta_1 \boldsymbol{\Omega}^{(2)} - 2\zeta_2 \boldsymbol{\Omega}^{(1)} - \zeta_3 (\nabla\cdot\boldsymbol{\alpha})\mathbf{I} - 2\xi_1 \boldsymbol{\Omega}^{(2)}_1 \\ & -2\xi_2 \boldsymbol{\Omega}^{(1)}_1 - \xi_3 \boldsymbol{\alpha}\cdot(\nabla\times\boldsymbol{\alpha})\mathbf{I} - 2\xi_4 \boldsymbol{\Omega}^{(2)}_2 - 2\xi_5\, \boldsymbol{\epsilon}\cdot\boldsymbol{\Omega}^{(2)}\cdot\boldsymbol{\alpha} - \xi_6\, \boldsymbol{\epsilon}\cdot(\nabla\cdot\boldsymbol{\alpha})\boldsymbol{\alpha} \\ & -\Lambda\boldsymbol{\epsilon}\cdot\nabla\log T - 2\Lambda_1 \mathbf{K}^{(2)} - 2\Lambda_2 \mathbf{K}^{(1)} - \Lambda_3 \boldsymbol{\alpha}\cdot(\nabla\log T)\mathbf{I},\end{aligned} \tag{11.4-69}$$

where $\boldsymbol{\Omega}^{(2)}$ and $\boldsymbol{\Omega}^{(1)}$ are the symmetric traceless and anti-symmetric parts of the tensor $\nabla\boldsymbol{\alpha}$,*

$$\boldsymbol{\Omega}^{(2)} = \tfrac{1}{2}[\nabla\boldsymbol{\alpha}+(\nabla\boldsymbol{\alpha})^T]-\tfrac{1}{3}\nabla\cdot\boldsymbol{\alpha}\,\mathbf{I}, \tag{11.4-70a}$$

$$\boldsymbol{\Omega}^{(1)} = \tfrac{1}{2}[\nabla\boldsymbol{\alpha}-(\nabla\boldsymbol{\alpha})^T], \tag{11.4-70b}$$

$\boldsymbol{\Omega}_1^{(2)}$ and $\boldsymbol{\Omega}_1^{(1)}$ are the symmetric traceless and anti-symmetric parts of the tensor $(\nabla\times\boldsymbol{\alpha})\boldsymbol{\alpha}$,

$$\boldsymbol{\Omega}_1^{(2)} = \tfrac{1}{2}[(\nabla\times\boldsymbol{\alpha})\boldsymbol{\alpha}+\boldsymbol{\alpha}(\nabla\times\boldsymbol{\alpha})]-\tfrac{1}{3}\boldsymbol{\alpha}\cdot(\nabla\times\boldsymbol{\alpha})\mathbf{I}, \tag{11.4-71a}$$

$$\boldsymbol{\Omega}_1^{(1)} = \tfrac{1}{2}[(\nabla\times\boldsymbol{\alpha})\boldsymbol{\alpha}-\boldsymbol{\alpha}(\nabla\times\boldsymbol{\alpha})], \tag{11.4-71b}$$

and $\boldsymbol{\Omega}_2^{(2)}$ is defined by

$$\boldsymbol{\Omega}_2^{(2)} = \tfrac{1}{2}[\boldsymbol{\alpha}\times\boldsymbol{\Omega}^{(2)}-\boldsymbol{\Omega}^{(2)}\times\boldsymbol{\alpha}]. \tag{11.4-71c}$$

Finally, $\mathbf{K}^{(2)}$ and $\mathbf{K}^{(1)}$ are the symmetric traceless and anti-symmetric parts of the tensor $(\nabla \log T)\boldsymbol{\alpha}$,

$$\mathbf{K}^{(2)} = \tfrac{1}{2}[(\nabla\log T)\boldsymbol{\alpha}+\boldsymbol{\alpha}(\nabla\log T)]-\tfrac{1}{3}\boldsymbol{\alpha}\cdot(\nabla\log T)\mathbf{I}, \tag{11.4-72a}$$

$$\mathbf{K}^{(1)} = \tfrac{1}{2}(\nabla\log T)\boldsymbol{\alpha}-\boldsymbol{\alpha}(\nabla\log T)]. \tag{11.4-72b}$$

The transport coefficients may be written in terms of the expansion coefficients,**

$$\zeta_1 = -\tfrac{1}{12}E_1^{1100}, \tag{11.4-73a}$$

$$\zeta_2 = -\tfrac{1}{12}E_2^{1100}, \tag{11.4-73b}$$

$$\zeta_3 = -\tfrac{1}{6}E_3^{1100}, \tag{11.4-73c}$$

$$\xi_1 = -\frac{1}{12}\left[E_1'^{1100}+3I\,\frac{c_{v,\,\text{int}}}{3k/2m}\,E^{1001}+\frac{1}{3kT}\,E_1^{1200}\right], \tag{11.4-74a}$$

* The superscript refers to the weight of the irreducible tensor.

** If the internal degrees of freedom are neglected, one has $A^{pqst}_{k_1\ldots k_{p+q}l}=0$ unless $p=1$, $q=0$ and $t=0$, and $\mathbf{A}'$, $\mathbf{E}$ and $\mathbf{E}'$ play no role in the theory. Moreover, $A^{10s0}_{k_1 i} = a_s\delta_{k_1 i}$, where a_s is identical with the coefficient a_s introduced in the expansion (5.6-1) for the case of a monatomic gas, and one verifies that the coefficient of $\delta_{k_1 i}$ in the expansion (11.4-42) of $A^{10s0}_{k_1 i}$ is equal to A_1^{10s0} (which may be denoted by A^{10s0} since there is only one isotropic tensor of rank two). Thus, we have $A^{10s0}=a_s$ and all other coefficients vanish. From Eqs. (11.4-73), (11.4-74) and (11.4-75) one verifies that the angular momentum flux pseudotensor vanishes identically if the internal degrees of freedom are neglected.

$$\xi_2 = -\frac{1}{12}\left[E_3'^{1100} + 3I\frac{c_{v,\,\mathrm{int}}}{3k/2m}E^{1001} - \frac{5}{3kT}E_1^{1200}\right], \qquad (11.4\text{-}74b)$$

$$\xi_3 = -\frac{1}{6}\left[E_6'^{1100} + I\frac{c_{v,\,\mathrm{int}}}{3k/2m}E^{1001} + \frac{10}{9kT}E_1^{1200}\right], \qquad (11.4\text{-}74c)$$

$$\xi_4 = \frac{1}{12}\left[E_2'^{1100} - \frac{1}{kT}E_2^{1200}\right], \qquad (11.4\text{-}74d)$$

$$\xi_5 = \frac{1}{12}\left[E_5'^{1100} - \frac{3}{2kT}E_2^{1200}\right], \qquad (11.4\text{-}74e)$$

$$\xi_6 = \frac{1}{6}\left[E_4'^{1100} - \frac{1}{3kT}E_2^{1200}\right], \qquad (11.4\text{-}74f)$$

$$\Lambda = \tfrac{1}{6}A^{1100}, \qquad (11.4\text{-}75a)$$

$$\Lambda_1 = -\frac{1}{12}\left[A_1'^{1100} + 3I\frac{c_{v,\,\mathrm{int}}}{3k/2m}A^{1001} + \frac{1}{3kT}A_1^{1200}\right], \qquad (11.4\text{-}75b)$$

$$\Lambda_2 = -\frac{1}{12}\left[A_2'^{1100} + 3I\frac{c_{v,\,\mathrm{int}}}{3k/2m}A^{1001} - \frac{5}{3kT}A_1^{1200}\right], \qquad (11.4\text{-}75c)$$

$$\Lambda_3 = -\frac{1}{6}\left[A_3'^{1100} + I\frac{c_{v,\,\mathrm{int}}}{3k/2m}A^{1001} + \frac{10}{9kT}A_1^{1200}\right]. \qquad (11.4\text{-}75d)$$

In these expressions the quantity I corresponds to the moment of inertia of the molecule defined such that

$$E_j = J^2/2I \qquad \text{(in units } \hbar^2\text{)}, \qquad (11.4\text{-}76)$$

if all internal energy is rotational.

A very similar procedure yields the heat flux vector. Again, only the first, third, fourth, and sixth terms of the right member of Eq. (11.4-39) contribute. With the aid of the expansions (11.4-55) until (11.4-62), one finds

$$\boldsymbol{q} = -\lambda\nabla T - \lambda'\boldsymbol{\alpha}\times\nabla T - \nu\nabla\times\boldsymbol{\alpha} - 2\nu_1\boldsymbol{\Omega}^{(2)}\cdot\boldsymbol{\alpha} - 2\nu_2\boldsymbol{\Omega}^{(1)}\cdot\boldsymbol{\alpha} - \nu_3\boldsymbol{\alpha}\nabla\cdot\boldsymbol{\alpha}, \quad (11.4\text{-}77)$$

with*

$$\lambda = \tfrac{5}{4}k\left[A^{1010} - \frac{3}{5}\frac{c_{v,\,\mathrm{int}}}{3k/2m}A^{1001}\right], \tag{11.4-78a}$$

$$\lambda' = -\tfrac{5}{4}k\left[A'^{1010} - \frac{3}{5}\frac{c_{v,\,\mathrm{int}}}{3k/2m}A'^{1001} - \frac{2}{5}\frac{\langle H'\rangle\langle J^2\rangle - \langle H'J^2\rangle}{kT\langle J^2\rangle}A'^{1000} - \frac{1}{3kT}A^{1110} + \frac{2}{15}\frac{\langle H'J^2\rangle\langle J^2\rangle - \langle H'J^2\rangle^2}{(kT)^3\langle J^2\rangle}A^{1101}\right], \tag{11.4-78b}$$

and

$$\nu = \tfrac{5}{4}kT\left[E^{1010} - \frac{3}{5}\frac{c_{v,\,\mathrm{int}}}{3k/2m}E^{1001}\right], \tag{11.4-79a}$$

$$\nu_i = \tfrac{5}{8}kT\left[E_i'^{1010} - \frac{3}{5}\frac{c_{v,\,\mathrm{int}}}{3k/2m}E_i'^{1001} - \frac{2}{5}\frac{\langle H'\rangle\langle J^2\rangle - \langle H'J^2\rangle}{kT\langle J^2\rangle}E_i'^{1000} + \frac{1}{3kT}E_i^{1110} - \frac{2}{15}\frac{\langle H'J^2\rangle\langle J^2\rangle - \langle H'J^2\rangle^2}{(kT)^3\langle J^2\rangle}E_i^{1101}\right], \quad i = 1, 2, 3. \tag{11.4-79b}$$

The constraints that always appear in the Chapman-Enskog approach and that place some restrictions on the expansion coefficients have been utilized in writing down the expressions for the transport coefficients.

As might be expected whenever there are "cross-effects" – such as the angular momentum flow due to a temperature gradient and a heat flow due to an angular momentum density gradient – there are Onsager relations connecting the various transport coefficients. These may be proven directly by examining the expressions for the transport coefficients provided by the theory or they may be obtained by methods of non-equilibrium thermodynamics; see De Groot and Mazur [1962]. In the present case, there are six such relations:

$$\nu = -\Lambda, \tag{11.4-80}$$

$$\nu_i = \Lambda_i, \qquad i = 1, 2, 3, \tag{11.4-81}$$

$$\xi_1 = -\xi_5, \tag{11.4-82}$$

$$\xi_3 = -\xi_6. \tag{11.4-83}$$

* Since, according to the footnote on p. 347, A^{1010} is the generalization of the coefficient a_1 in the expansion (5.6-1), the result (11.4-78a) generalizes the result (5.6-22) for monatomic gases to polyatomic gases.

The final stage in the calculation is, of course, the computation of the various expansion coefficients and, thence, the transport coefficients. McCourt [1966] has obtained variational principles from which the coefficients may be computed but did not complete the computation. He did carry out the first iteration of the $\boldsymbol{\alpha}$ expansion described above and obtained expressions for the transport coefficients to this order. It was found that the expressions for shear viscosity, bulk viscosity and thermal conductivity are those of the Wang Chang-Uhlenbeck theory. New expressions were obtained for the rotational diffusion coefficients ξ_i $(i = 1, 2, 3)$ and Λ. All of the other coefficients are zero to this order of approximation. The interesting feature of this calculation is that the integrals occurring in the expressions for the new coefficients cannot be reduced to integrals involving the cross section defined by Eq. (11.4-8). Rather, they contain combinations of the t-matrix and the angular momentum operator $\boldsymbol{J}$. The occurrence of these new cross sections is of considerable potential importance. If the rotational diffusion coefficients could be measured, their analysis could provide much more information about the nature of the intermolecular interaction than is presently attainable from transport property measurements. In fact, just the simple introduction of a whole set of new properties considerably enlarges the potential for obtaining information from transport property measurements. Unfortunately, at present no experimental technique for the measurement of the angular momentum density is known so that it is not clear that this approach is feasible. However, very similar effects are obtained when a magnetic field is applied to the gas and one can, from the measurement of transport properties under these circumstances, obtain information comparable to that just described.

Before passing on to a consideration of the effects of magnetic fields, we mention that the addition of the angular momentum equation of motion to the usual equations of fluid dynamics can be expected to produce a great enrichment in the phenomena inherent in the equations. Undoubtedly, in many situations the additional equation of motion is of no particular importance; the success of present-day fluid dynamics gives evidence of this. But investigation of the phenomena inherent in the new equations is only beginning and it is too early to make sweeping generalizations; a review of this work and further references are given in Dahler [1965]. Further discussion of these phenomena is outside the scope of this book and therefore not given here.

11.5. The Senftleben-Beenakker effect

It was mentioned in the previous section that a magnetic field may affect the transport properties of a gas. The effect was observed in paramagnetic gases by Senftleben [1930] and a qualitative explanation of it was first given by Gorter [1938]. Interest in the effect was rather minimal in the period 1940–1960 but in the early 1960's Beenakker and his group at Leiden undertook a systematic study of the effect and discovered that it exists in diamagnetic as well as in paramagnetic gases. Hence, it is now known as the *Senftleben-Beenakker effect* and has become an important tool for studying the properties of polyatomic gases.

Some of the characteristics of the effect are readily explained in a qualitative manner. Consider, for example, a paramagnetic gas, i.e., a gas the molecules of which possess a permanent magnetic dipole moment. An applied magnetic field will, of course, interact with this magnetic moment. As a consequence, the rotation axis will precess about the magnetic field direction with the Larmor frequency. Since the collision cross section depends, in general, on the orientation, the result of the interaction is a periodic variation of the cross section. Finally, since the transport properties depend on certain weighted averages of the cross section and not simply on the cross section averaged over orientations, the transport properties of such a gas depend on the strength of the applied magnetic field.

From this picture, some of the essential features of the effect are readily deduced. The frequency of precession is $\omega = (\gamma\mu_0/\hbar)H$, where γ is the gyromagnetic ratio (ratio of magnetic dipole moment to angular momentum), μ_0 is the Bohr magneton ($\mu_0 = e\hbar/2m_\mathrm{p}c$ where m_p is the mass of a proton) and H the magnetic field strength. Now, the effect must depend upon the ratio of the oscillation time, ω^{-1}, to the average time between collisions, τ. Clearly, if $\omega\tau \ll 1$, the precession is unimportant while, if $\omega\tau \gg 1$, the molecule precesses many times between collisions and complete averaging over orientations will obtain. Since τ is inversely proportional to the density or, at fixed temperature, to the pressure, p, $\omega\tau$ is proportional to the ratio of the magnetic field strength, H, and the pressure, p. Therefore, the effect of the magnetic field on the transport properties is largely dependent upon the parameter H/p, saturation being achieved for large values of H/p. Furthermore, the effect of precession will tend to give more collisions in those orientations for which the cross section is highest, simply because these are more probable. Thus, we expect the averaged cross sections that enter into the transport property calculations to be larger than in the zero field case. As the transport properties are always inversely proportional to the averaged cross sections, the result is a reduction in the transport properties. To be sure, the effect is small (typically 0.1–1 %), but it is measurable and its measurement does provide useful information about the molecules.

For diatomic molecules, which possess very small magnetic moments, the effect is very similar, but much higher fields are required to achieve saturation.

In an analogous manner electric fields may influence transport properties. For molecules with permanent electric dipole moments, the important parameter is E/p, where E is the electric field strength. On the other hand, in non-polar molecules the effect depends on an induced dipole moment and the characteristic parameter becomes E^2/p.

Measurements concerning the effect of fields on transport properties have been reported at an increasing rate over the past few years and it is reasonable to expect a further increase of the work done in this area, especially when it becomes possible to relate the results to the properties of molecular interactions. A review of the status of measurements as of mid-1968 has been given by Beenakker [1969].

Since the only experimentally observable quantities are the temperature and velocity gradients, it is convenient to express the heat flow vector and the pressure tensor in terms of them by means of the relations

$$\boldsymbol{q} = -\boldsymbol{\lambda} \cdot \nabla T, \tag{11.5-1}$$

$$\boldsymbol{P} = nkT\boldsymbol{I} - 2\boldsymbol{\eta} : \boldsymbol{S} - \boldsymbol{\zeta}\nabla \cdot \boldsymbol{v}, \tag{11.5-2}$$

where $\boldsymbol{\lambda}$ and $\boldsymbol{\eta}$ are now tensors of rank two and four, respectively, and $\boldsymbol{\zeta}$ is a new tensor of rank two. Again, the rate-of-shear tensor $\boldsymbol{S}$ makes no contribution to the heat flow vector, and the temperature gradient does not affect the pressure tensor. (We have assumed that the anti-symmetric part of the pressure tensor can be neglected entirely; this assumption is justified in most cases). There are, of course, additional terms due to the angular momentum density that should be added to the right-hand sides of Eqs. (11.5-1) and (11.5-2), but, as these are not measurable by present technique we shall ignore them.

The methods of irreversible thermodynamics can be used to determine the forms that the tensors $\boldsymbol{\lambda}$, $\boldsymbol{\eta}$ and $\boldsymbol{\zeta}$ must take, see De Groot and Mazur [1962]. If the magnetic field is assumed to be in the x-direction, the results may be expressed schematically as in Tables 11.5 and 11.6. Of course, the same re-

TABLE 11.5

Form of the thermal conductivity tensor

	$\partial T/\partial x$	$\partial T/\partial y$	$\partial T/\partial z$
q_x	$-\lambda_{\|\|}$	0	0
q_y	0	$-\lambda_{\perp}$	λ_{tr}
q_z	0	$-\lambda_{tr}$	$-\lambda_{\perp}$

Note: $\lambda_{\|\|}$ and $\lambda_{\perp}$ are even functions of $\boldsymbol{H}$, λ_{tr} is an odd function of $\boldsymbol{H}$. For $\boldsymbol{H} = 0$, $\lambda_{\|\|} = \lambda_{\perp}$; $\lambda_{tr} = 0$.

TABLE 11.6

Form of the viscosity tensor

	S_{xx}	S_{yy}	S_{zz}	S_{xy}	S_{xz}	S_{yz}	$\nabla \cdot v$
P_{xx}	$-2\eta_1$	0	0	0	0	0	-2ζ
P_{yy}	0	$-2\eta_2$	$2(\eta_2-\eta_1)$	0	0	$-2\eta_4$	ζ
P_{zz}	0	$2(\eta_2-\eta_1)$	$-2\eta_2$	0	0	$2\eta_4$	ζ
P_{xy}	0	0	0	$-2\eta_3$	$2\eta_5$	0	0
P_{xz}	0	0	0	$-2\eta_5$	$-2\eta_3$	0	0
P_{yz}	0	η_4	$-\eta_4$	0	0	$2(\eta_1-2\eta_2)$	0
p	-2ζ	ζ	ζ	0	0	0	$-\kappa$

Note: η_1, η_2, η_3, ζ and κ are all even functions of $\boldsymbol{H}$; η_4 and η_5 are odd functions of $\boldsymbol{H}$. For $\boldsymbol{H} = 0$, $\eta_1 = \eta_2 = \eta_3$; $\eta_4 = \eta_5 = \zeta = 0$.

sults must come out of the theory, but the tables are useful in predicting how many quantities must be computed. One sees that there are three coefficients of thermal conductivity and seven coefficients of viscosity. The coefficients $\eta_1, \eta_2, \ldots, \eta_5$ connect the components of the pressure tensor to the components of $\boldsymbol{S}$; they can therefore be called coefficients of shear viscosity. The coefficient κ connects the traces p and $\nabla \cdot \boldsymbol{v}$ and is therefore the coefficient of bulk viscosity. The seventh coefficient, ξ, describes a cross-effect between shear and bulk viscosity. Further application of irreversible thermodynamics predicts the odd or even character of the various quantities with respect to the magnetic field. In principle, all of the coefficients could be measured (and many of them have been) so that there are now available ten experimental quantities in place of the three found in the absence of the magnetic field.

Since work on predicting these properties is in progress (references may be found in Beenakker [1969]) and far from complete, we shall only sketch the theory. First, we must look at the modifications in the Boltzmann equation made necessary by the presence of a magnetic field. The important effect, mentioned above, is the interaction with the magnetic moment of the nucleus and the precession caused thereby. Furthermore, the magnetic field causes a shift in the energy levels of the molecules (the Zeeman effect) and, in fact, causes the degenerate levels to split. For high magnetic fields, the Zeeman splitting is so great that it is appropriate to treat the energy levels as completely distinct and one thus arrives at the surprising conclusion that the Wang Chang-Uhlenbeck equation is valid in the high-field limit. However, these effects also obtain for the collision operator used in the previous section and thus it seems reasonable to adopt this operator *mutatis mutandis*. Waldmann [1968] has given a more rigorous argument for using this operator; Hess [1967] has given the modification of the operator necessary to account for the Zeeman splitting, but this modified operator has apparently not been used to date. Thus, one has to deal only with the modifications occurring in the streaming term of the Boltzmann equation and, fortunately, this is not a difficult task. The Hamiltonian of a single particle may be written as

$$H = \frac{p^2}{2m} + H' - \gamma \boldsymbol{H} \cdot \boldsymbol{J} = \frac{p^2}{2m} + H_{\text{int}} \tag{11.5-3}$$

where H' is the Hamiltonian for the internal states of the molecule in the absence of the field; the term $\gamma \boldsymbol{H} \cdot \boldsymbol{J}$ represents the effect of the magnetic field (γ is the gyromagnetic ratio). In Eq. (11.5-3) we have omitted a term quadratic in the field strength which is normally small. In fact, Eq. (11.5-3) is the Hamiltonian normally used in studying the Zeeman effect in quantum mechanics. The desired modified Boltzmann equation is obtained by replacing the left-hand side of Eq. (11.4-8) by

$$\frac{\partial f}{\partial t} + \frac{\boldsymbol{p}}{m} \cdot \nabla_r f + \frac{\mathrm{i}}{\hbar}(-\gamma \boldsymbol{H} \cdot \boldsymbol{J} f + f \gamma \boldsymbol{H} \cdot \boldsymbol{J}). \tag{11.5-4}$$

Another term is required to account for the possibility of an inhomogeneous magnetic field but is of no interest in the present case.

Since the collision term is unaltered by the addition of the magnetic field, the linearized equations for the perturbation are easily obtained and can be solved by methods analogous to those used in the previous section. Again, one finds that the results cannot be expressed simply in terms of the scattering cross section, and various new types of cross sections enter. The transport properties may be expressed in terms of bracket integrals; the latter are now twelve-fold integrals in which six of the integrations may be carried out without knowledge of the molecular interaction. We are thus left with the extremely difficult task of evaluating six-fold integrals instead of the double integrals of the monatomic gas theory. Evaluation of these integrals requires the establishment of some model for the interaction, but realistic models have not yet been developed. Thus, it may be expected that several years of further work will be required before useful results as to the molecular interaction can be obtained from the theory.

Dense gases - Enskog's theory 12

In the preceding chapters we have dealt with the kinetic theory of dilute gases and derived relations between the microscopic properties of the molecules and the transport coefficients. The analysis was based entirely upon the Boltzmann equation (or appropriate modifications thereof). From the discussion of Chapter 3, however, we know that the use of the Boltzmann equation must be restricted to gases which are sufficiently dilute that only binary collisions need to be taken into account, and in which the molecular dimensions are small in comparison with the mean distance between the molecules. Obviously, these restrictions are not met in dense gases and liquids, so the results of the previous theory cannot be applied. In this chapter we turn our attention to the study of transport phenomena in dense systems.

A first attempt towards a kinetic theory of dense gases is due to Enskog [1922b]. It is an *ad hoc* extension of the kinetic theory of gases at normal densities given in Chapter 5, which has been developed for rigid spherical molecules only. The advantage of the rigid sphere model in this connection is that collisions are instantaneous and the probability of multiple simultaneous encounters is negligible. Enskog's extension involves the introduction of corrections that account for the fact that the molecular diameter is no longer small compared with the average intermolecular distance. A major consequence of this is that a mechanism of momentum and energy transfer which is negligible at normal densities and, thus, has not been considered so far, becomes important – viz., during a collision momentum and energy are transferred over a distance equal to the separation of the molecules. In the case of rigid spherical molecules this *collisional transfer* of momentum and energy takes place instantaneously and results in a transfer over the distance σ_{12} between their centers. In very dense gases the collisional transfer is indeed the principal transport mechanism since the molecules are almost "locked in place" by their neighbors and transport by molecular flow becomes very difficult.

Although Enskog's theory is valid for rigid spheres only, its results may be applied to real gases with considerable success if an appropriately chosen effective collision diameter is used. It is also an aid in understanding the physics of the problem and the method can be extended to other types of rigid molecules, e.g., rough spheres.

12.1. Enskog's equation

Consider a simple gas composed of rigid spherical molecules of diameter σ. In Section 3.1 the Boltzmann equation for the distribution function $f(\boldsymbol{r}, \boldsymbol{c}, t)$ has been derived using heuristic arguments. Similarly, we may derive an integro-differential equation for the velocity distribution function in the case of a dense gas. It is clear that the streaming part of such an equation will have the same form as before. The collision term, however, will be different. In Section 3.1 it was found that, in a gas at normal pressures, the average number of collisions per unit time at time t, such that the center of the first molecule lies in the volume element $\mathrm{d}^3 r$, the velocities of the two molecules before the collision lie in the velocity ranges $\mathrm{d}^3 c$, $\mathrm{d}^3 c_1$, and the geometrical collision variables lie in the range $\mathrm{d}b$, $\mathrm{d}\varepsilon$ about b, ε, is equal to

$$f(\boldsymbol{r}, \boldsymbol{c}, t) f(\boldsymbol{r}, \boldsymbol{c}_1, t) g b \,\mathrm{d}b \,\mathrm{d}\varepsilon \,\mathrm{d}^3 c \,\mathrm{d}^3 c_1 \,\mathrm{d}^3 r, \tag{12.1-1}$$

where g is the magnitude of the relative velocity, $\boldsymbol{g} = \boldsymbol{c}_1 - \boldsymbol{c}$. It is convenient to introduce the unit vector $\boldsymbol{k}$ along the apse-line which, in the case of rigid spherical molecules, is the line joining the centers of the two molecules at the instant of contact, cf. Fig. 12.1. Since $\boldsymbol{k}$ makes an angle $\psi = \frac{1}{2}(\pi - \chi)$ with $\boldsymbol{g}$, and the plane through $\boldsymbol{k}$ and $\boldsymbol{g}$ makes an angle ε with a reference plane through $\boldsymbol{g}$, ψ and ε are polar coordinates specifying the direction of $\boldsymbol{k}$ on the unit sphere. Then, since $b = \sigma \sin \psi$,

$$\mathrm{d}^2 k = \sin \psi \,\mathrm{d}\psi \,\mathrm{d}\varepsilon = (\sigma^2 \cos \psi)^{-1} b \,\mathrm{d}b \,\mathrm{d}\varepsilon, \tag{12.1-2}$$

whence it follows that the infinitesimal solid angle $g b \,\mathrm{d}b \,\mathrm{d}\varepsilon$ is equal to

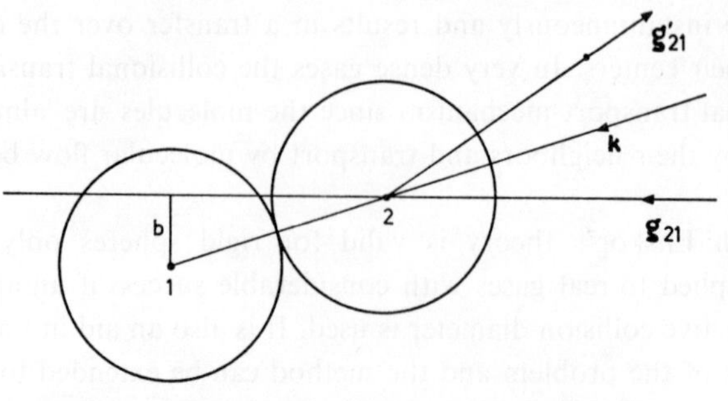

Fig. 12.1. Geometry of a collision of two rigid spheres.

$\sigma^2(\boldsymbol{g} \cdot \boldsymbol{k})\mathrm{d}^2k$, so that, instead of Eq. (12.1-1), we may write

$$f(\boldsymbol{r}, \boldsymbol{c}, t)f(\boldsymbol{r}, \boldsymbol{c}_1, t)\sigma^2(\boldsymbol{g} \cdot \boldsymbol{k})\mathrm{d}^2k\,\mathrm{d}^3c\,\mathrm{d}^3c_1\,\mathrm{d}^3r. \tag{12.1-3}$$

This expression requires correction when the gas is dense:

(i) Because of the finite size of the molecules, the centers of two colliding molecules are not at the same point. If, at the instant of collision, the center of the first molecule is at $\boldsymbol{r}$, that of the second is at $\boldsymbol{r}-\sigma\boldsymbol{k}$, so that $f(\boldsymbol{r}, \boldsymbol{c}_1, t)$ must be replaced by $f(\boldsymbol{r}-\sigma\boldsymbol{k}, \boldsymbol{c}_1, t)$.

(ii) In a dense gas, the volume per molecule – $1/n$, where n is the number density of the gas – becomes comparable with the volume of a molecule – viz., $\frac{4}{3}\pi(\frac{1}{2}\sigma)^3$, in the case of rigid spherical molecules of diameter σ. Thus, the volume in which any one of the molecules can lie (i.e., the "free volume") is reduced and, as we will see, the probability of a collision is increased. We will account for this effect by multiplication of the expression (12.1-3) by a factor χ, which may be a function of the number density and, therefore, of position and time. The function χ should be evaluated at the point of contact, $\boldsymbol{r}-\frac{1}{2}\sigma\boldsymbol{k}$, of the two colliding molecules. Hence, the corrected form of the expression (12.1-3) is

$$\chi(\boldsymbol{r}-\tfrac{1}{2}\sigma\boldsymbol{k}, t)f(\boldsymbol{r}, \boldsymbol{c}, t)f(\boldsymbol{r}-\sigma\boldsymbol{k}, \boldsymbol{c}_1, t)\,\sigma^2\boldsymbol{g}\cdot\boldsymbol{k}\,\mathrm{d}^2k\,\mathrm{d}^3c\,\mathrm{d}^3c_1\,\mathrm{d}^3r$$
$$= \chi(\boldsymbol{r}-\tfrac{1}{2}\sigma\boldsymbol{k})f(\boldsymbol{r})f_1(\boldsymbol{r}-\sigma\boldsymbol{k})\,\sigma^2\boldsymbol{g}\cdot\boldsymbol{k}\,\mathrm{d}^2k\,\mathrm{d}^3c\,\mathrm{d}^3c_1\,\mathrm{d}^3r, \tag{12.1-4}$$

using a notation similar to that of Section 3.1.

Corresponding to any direct collision specified by the variables $\boldsymbol{c}$, $\boldsymbol{c}_1$, $\boldsymbol{k}$ there is an analogous inverse collision in which $\boldsymbol{c}$, $\boldsymbol{c}_1$ are the velocities of the molecules after the collision, while $-\boldsymbol{k}$ is the direction of the apse-line. In such a collision the center of the second molecule is at $\boldsymbol{r}+\sigma\boldsymbol{k}$, while the point of contact is at $\boldsymbol{r}+\frac{1}{2}\sigma\boldsymbol{k}$. Hence, the average number of inverse collisions per unit time at time t, such that the center of the first molecule lies in the volume element d^3r, the velocities of the two molecules after collision lie in the ranges d^3c, d^3c_1 while the direction of the line of centers is $-\boldsymbol{k}$, where $\boldsymbol{k}$ lies in the range d^2k, will be

$$\begin{aligned}\chi(\boldsymbol{r}+\tfrac{1}{2}\sigma\boldsymbol{k},\, t)\, f(\boldsymbol{r},\, \boldsymbol{c}',\, t)\, f(\boldsymbol{r}+\sigma\boldsymbol{k},\, \boldsymbol{c}_1',\, t)\,\sigma^2\boldsymbol{g}\cdot\boldsymbol{k}\,\mathrm{d}^2k\,\mathrm{d}^3c\,\mathrm{d}^3c_1\,\mathrm{d}^3r \\ = \chi(\boldsymbol{r}+\tfrac{1}{2}\sigma\boldsymbol{k})\, f'(\boldsymbol{r})\, f_1'(\boldsymbol{r}+\sigma\boldsymbol{k})\,\sigma^2\boldsymbol{g}\cdot\boldsymbol{k}\,\mathrm{d}^2k\,\mathrm{d}^3c\,\mathrm{d}^3c_1\,\mathrm{d}^3r, \quad (12.1\text{-}5)\end{aligned}$$

where $\boldsymbol{c}'$ and $\boldsymbol{c}_1'$ denote the velocities of the two molecules before the collision. It is easily verified that $\boldsymbol{c}'$ and $\boldsymbol{c}_1'$ are related to $\boldsymbol{c}$ and $\boldsymbol{c}_1$ through

$$\boldsymbol{c}' = \boldsymbol{c}+\boldsymbol{k}(\boldsymbol{g}\cdot\boldsymbol{k}), \quad \boldsymbol{c}_1' = \boldsymbol{c}_1-\boldsymbol{k}(\boldsymbol{g}\cdot\boldsymbol{k}). \qquad (12.1\text{-}6)$$

Thus we find the following integro-differential equation for the velocity distribution function in the case of a dense gas of rigid spherical molecules,

$$\begin{aligned}\frac{\partial f}{\partial t}+\boldsymbol{c}\cdot\nabla_r f+\boldsymbol{F}\cdot\nabla_c f = \iint \{\chi(\boldsymbol{r}+\tfrac{1}{2}\sigma\boldsymbol{k})\, f'(\boldsymbol{r})\, f_1'(\boldsymbol{r}+\sigma\boldsymbol{k}) \\ -\chi(\boldsymbol{r}-\tfrac{1}{2}\sigma\boldsymbol{k})\, f(\boldsymbol{r})\, f_1(\boldsymbol{r}-\sigma\boldsymbol{k})\}\,\sigma^2(\boldsymbol{g}\cdot\boldsymbol{k})\,\mathrm{d}^2k\,\mathrm{d}^3c_1. \quad (12.1\text{-}7)\end{aligned}$$

This equation is called *Enskog's equation*. Assuming that the conditions in the gas are sufficiently smooth we expand each of the functions $\chi(\boldsymbol{r}\pm\frac{1}{2}\sigma\boldsymbol{k})$ and $f_1(\boldsymbol{r}\pm\sigma\boldsymbol{k})$ in a Taylor series near $\boldsymbol{r}$ and neglect third- and higher-order terms. Then, Enskog's equation may be written as

$$\mathfrak{D}f = J_0(ff)+J_1(ff)+J_2(ff), \qquad (12.1\text{-}8)$$

where $\mathfrak{D}f$ is the streaming part of the equation, and

$$J_0(ff) = \chi\iint (f'f_1'-ff_1)\sigma^2(\boldsymbol{g}\cdot\boldsymbol{k})\,\mathrm{d}^2k\,\mathrm{d}^3c_1, \qquad (12.1\text{-}9)$$

$$\begin{aligned}J_1(ff) = \sigma\iint [\boldsymbol{k}\chi\cdot(f'\nabla_r f_1'+f\nabla_r f_1) \\ +\tfrac{1}{2}(\boldsymbol{k}\cdot\nabla_r\chi)(f'f_1'+ff_1)]\sigma^2(\boldsymbol{g}\cdot\boldsymbol{k})\mathrm{d}^2k\,\mathrm{d}^3c_1, \quad (12.1\text{-}10)\end{aligned}$$

$$\begin{aligned}J_2(ff) = \tfrac{1}{2}\sigma^2\iint [\boldsymbol{k}\boldsymbol{k}\chi : (f'\nabla_r\nabla_r f_1'-f\nabla_r\nabla_r f_1) \\ +(\boldsymbol{k}\cdot\nabla_r\chi)\boldsymbol{k}\cdot(f'\nabla_r f_1'-f\nabla_r f_1) \\ +\tfrac{1}{4}(\boldsymbol{k}\boldsymbol{k} : \nabla_r\nabla_r\chi)(f'f_1'-ff_1)]\,\sigma^2(\boldsymbol{g}\cdot\boldsymbol{k})\,\mathrm{d}^2k\,\mathrm{d}^3c_1, \quad (12.1\text{-}11)\end{aligned}$$

where all quantities are evaluated at $\boldsymbol{r}$.

12.2. Flux vectors

In Section 2.2 we studied the flux of a molecular property φ across a small surface element d^2S in a gas at low pressure. There, the molecules could be considered to be point masses. In dense gases, however, the finite size of the molecules must be taken into account and, consequently, *collisional transfer* of the molecular property φ across d^2S – i.e., the (instantaneous) transfer due to collisions between molecules whose centers lie on opposite sides of d^2S while the line joining their centers cuts d^2S – must be taken into account.

Let $\boldsymbol{n}$ be the unit vector normal to d^2S pointing from the negative to the positive side; d^2S is located at position $\boldsymbol{r}$. Consider a collision between a molecule having velocity $\boldsymbol{c}$ and a molecule having velocity $\boldsymbol{c}_1$. If the first molecule lies on the positive side of d^2S and the second on the negative side, the inner product $\boldsymbol{k} \cdot \boldsymbol{n}$ is positive. At collision, the line joining the centers of the molecules must cut d^2S, so the center of the first molecule must lie within a cylinder having d^2S as base and whose generators are parallel to $\boldsymbol{k}$ and of length σ. Thus, the volume of this cylinder is $\sigma(\boldsymbol{k} \cdot \boldsymbol{n})d^2S$. The mean positions of the centers of the two molecules at the instant of the collision are $\boldsymbol{r}+\frac{1}{2}\sigma\boldsymbol{k}$ and $\boldsymbol{r}-\frac{1}{2}\sigma\boldsymbol{k}$, and the mean position of the point of contact is $\boldsymbol{r}$. Hence, the average number of such collisions per unit time at t, in which $\boldsymbol{c}$, $\boldsymbol{c}_1$ and $\boldsymbol{k}$ lie in the ranges d^3c, d^3c_1 and d^2k, respectively, is, by analogy with Eq. (12.1-4)

$$\chi(\boldsymbol{r})f(\boldsymbol{r}+\tfrac{1}{2}\sigma\boldsymbol{k})f_1(\boldsymbol{r}-\tfrac{1}{2}\sigma\boldsymbol{k})\,\sigma^3\boldsymbol{g}\cdot\boldsymbol{k}\,d^2k\,d^3c\,d^3c_1(\boldsymbol{k}\cdot\boldsymbol{n})\,d^2S. \quad (12.2\text{-}1)$$

Each such collision causes the transfer of a quantity $(\varphi'-\varphi)$ of the molecular property φ across d^2S, so the total transfer by collisions of this type, per unit time and per unit area, is

$$\sigma\chi(\boldsymbol{r})\iiint(\varphi'-\varphi)f(\boldsymbol{r}+\tfrac{1}{2}\sigma\boldsymbol{k})f_1(\boldsymbol{r}-\tfrac{1}{2}\sigma\boldsymbol{k})\,\sigma^2(\boldsymbol{g}\cdot\boldsymbol{k})(\boldsymbol{k}\cdot\boldsymbol{n})\,d^2k\,d^3c\,d^3c_1, \quad (12.2\text{-}2)$$

the integration being over all values of the variables such that $\boldsymbol{g}\cdot\boldsymbol{k}$ and $\boldsymbol{k}\cdot\boldsymbol{n}$ are positive (the condition $\boldsymbol{g}\cdot\boldsymbol{k}>0$ ensures that the molecules collide, the condition $\boldsymbol{k}\cdot\boldsymbol{n}>0$ is a consequence of the definition of $\boldsymbol{k}$ and $\boldsymbol{n}$). The vector

$$\boldsymbol{\Phi} = \sigma\chi(\boldsymbol{r})\iiint_{\substack{\boldsymbol{g}\cdot\boldsymbol{k}>0\\ \boldsymbol{k}\cdot\boldsymbol{n}>0}} (\varphi'-\varphi)\, f(\boldsymbol{r}+\tfrac{1}{2}\sigma\boldsymbol{k})\, f_1(\boldsymbol{r}-\tfrac{1}{2}\sigma\boldsymbol{k})\,\sigma^2(\boldsymbol{g}\cdot\boldsymbol{k})\,\boldsymbol{k}\,\mathrm{d}^2k\,\mathrm{d}^3c\,\mathrm{d}^3c_1 \tag{12.2-3}$$

is the *flux vector* associated with the collisional transfer of property φ. We note in passing that Eq. (12.2-3) is a special case of a more general statistical mechanical formula.

Of particular interest are the flux vectors related to the transfer of each of the summational invariants – mass, momentum and energy. Let φ be a summational invariant, ψ, so that $\psi+\psi_1 = \psi'+\psi_1'$. In Eq. (12.2-3), let the variables of integration $\boldsymbol{c}$ and $\boldsymbol{c}_1$ be interchanged; this is equivalent to interchanging the roles of the two colliding molecules, so that $\boldsymbol{k}$ is replaced by $-\boldsymbol{k}$, $\boldsymbol{g}$ by $-\boldsymbol{g}$, and $\psi'-\psi$ by $\psi_1'-\psi_1$ which is equal to $-(\psi'-\psi)$. Consequently, the integrand of Eq. (12.2-3) is unchanged but the region of integration is now restricted by the conditions $\boldsymbol{g}\cdot\boldsymbol{k} > 0$ and $\boldsymbol{k}\cdot\boldsymbol{n} < 0$. The flux vector associated with the collisional transfer of a summational invariant ψ is, therefore, equal to

$$\boldsymbol{\Psi}_{\Phi} = \tfrac{1}{2}\sigma\chi(\boldsymbol{r})\iiint_{\boldsymbol{g}\cdot\boldsymbol{k}>0} (\psi'-\psi)\, f(\boldsymbol{r}+\tfrac{1}{2}\sigma\boldsymbol{k})\, f_1(\boldsymbol{r}-\tfrac{1}{2}\sigma\boldsymbol{k})\,\sigma^2(\boldsymbol{g}\cdot\boldsymbol{k})\,\boldsymbol{k}\,\mathrm{d}^2k\,\mathrm{d}^3c\,\mathrm{d}^3c_1\,. \tag{12.2-4}$$

In addition to the collisional transfer of the summational invariant, ψ, there is also the contribution due to the flow of molecules across the surface as given by Eq. (2.2-1),

$$\boldsymbol{\Psi}_K = \int \psi f \boldsymbol{C}\,\mathrm{d}^3c. \tag{12.2-5}$$

The total flux of the property ψ is then the sum of the two contributions,

$$\boldsymbol{\Psi} = \boldsymbol{\Psi}_K + \boldsymbol{\Psi}_{\Phi}. \tag{12.2-6}$$

The subscripts K and Φ indicate the contributions due to the kinetic energy (flow of molecules) and the potential energy (collisional transfer). The latter is negligible if the density of the gas is low; at high densities the principal contribution to the flux vector comes from the second term. The reason for this is easily seen: from Eq. (12.2-5) we see that $\boldsymbol{\Psi}_K$ is a linear functional of f and, therefore, proportional to the density, while from Eq. (12.2-4) we see that $\boldsymbol{\Psi}_{\Phi}$ is quadratic in f and, therefore, proportional to the square of the density.

If the conditions in the gas are sufficiently smooth, $f(\boldsymbol{r}\pm\tfrac{1}{2}\sigma\boldsymbol{k})$ may be expanded in a Taylor series around $\boldsymbol{r}$, so that $\boldsymbol{\Psi}_{\Phi}$ may be taken as

$$\boldsymbol{\Psi}_{\Phi} = \tfrac{1}{2}\sigma\chi \iiint\limits_{\boldsymbol{g}\cdot\boldsymbol{k}>0} (\psi'-\psi) f f_1 \, \sigma^2 (\boldsymbol{g}\cdot\boldsymbol{k}) \boldsymbol{k} \, \mathrm{d}^2 k \, \mathrm{d}^3 c \, \mathrm{d}^3 c_1$$

$$+\tfrac{1}{4}\sigma^2\chi \iiint\limits_{\boldsymbol{g}\cdot\boldsymbol{k}>0} (\psi'-\psi)\left(\boldsymbol{k}\cdot f f_1 \nabla_r \log\frac{f}{f_1}\right) \sigma^2(\boldsymbol{g}\cdot\boldsymbol{k})\boldsymbol{k}\,\mathrm{d}^2 k\,\mathrm{d}^3 c\,\mathrm{d}^3 c_1, \quad (12.2\text{-}7)$$

where all quantities are evaluated at $\boldsymbol{r}$. (It has been assumed throughout that, if the summational invariant ψ is space-dependent, it is evaluated at the *same* point $\boldsymbol{r}$ for *both* molecules involved in a collision.)

12.3. Macroscopic conservation equations

The macroscopic conservation equations may be derived in a manner which is entirely analogous to the one used in Section 4.1. When Enskog's equation (12.1-8) is multiplied by a summational invariant, ψ, and integrated over $\boldsymbol{c}$, the result is

$$\frac{\partial}{\partial t}\int \psi f \,\mathrm{d}^3 c + \nabla_r \cdot \int \psi \boldsymbol{c} f \,\mathrm{d}^3 c - \boldsymbol{F}\cdot\int f \nabla_c \psi \,\mathrm{d}^3 c = I_0 + I_1 + I_2, \quad (12.3\text{-}1)$$

where

$$I_i = \int \psi J_i(ff)\,\mathrm{d}^3 c, \qquad i = 0, 1, 2. \quad (12.3\text{-}2)$$

As in Section 4.1 one proves that the integral I_0 vanishes identically. By invoking the equivalence of collisions and their inverses one may prove that the integral I_1, which is defined as

$$I_1 = \sigma \iiint \psi[\boldsymbol{k}\chi\cdot(f'\nabla_r f_1' + f\nabla_r f_1) + \tfrac{1}{2}(\boldsymbol{k}\cdot\nabla_r\chi)(f'f_1' + ff_1)] \times \sigma^2(\boldsymbol{g}\cdot\boldsymbol{k})\,\mathrm{d}^2 k\,\mathrm{d}^3 c\,\mathrm{d}^3 c_1, \quad (12.3\text{-}3)$$

is equal to

$$I_1 = -\sigma \iiint \psi'[\boldsymbol{k}\chi\cdot(f'\nabla_r f_1' + f\nabla_r f_1) + \tfrac{1}{2}(\boldsymbol{k}\cdot\nabla_r\chi)(f'f_1' + ff_1)] \times \sigma^2(\boldsymbol{g}\cdot\boldsymbol{k})\,\mathrm{d}^2 k\,\mathrm{d}^3 c\,\mathrm{d}^3 c_1, \quad (12.3\text{-}4)$$

so, instead of Eq. (12.3-3), we may write

$$I_1 = \tfrac{1}{2}\sigma \iiint (\psi-\psi')[\boldsymbol{k}\chi\cdot(f'\nabla_r f_1' + f\nabla_r f_1) + \tfrac{1}{2}(\boldsymbol{k}\cdot\nabla_r\chi)(f'f_1' + ff_1)] \times \sigma^2(\boldsymbol{g}\cdot\boldsymbol{k})\,\mathrm{d}^2 k\,\mathrm{d}^3 c\,\mathrm{d}^3 c_1. \quad (12.3\text{-}5)$$

Again using the same arguments one may prove that the latter expression is, in turn, equivalent to

$$I_1 = \sigma \iiint (\psi - \psi')[\boldsymbol{k}\chi \cdot f \boldsymbol{\nabla}_r f_1 + \tfrac{1}{2}(\boldsymbol{k} \cdot \boldsymbol{\nabla}_r \chi) f f_1] \sigma^2 (\boldsymbol{g} \cdot \boldsymbol{k}) \, \mathrm{d}^2 k \, \mathrm{d}^3 c \, \mathrm{d}^3 c_1 \tag{12.3-6}$$

which, by interchanging the roles of the two molecules as described in connection with Eq. (12.2-4) can be shown to be equivalent to

$$I_1 = \sigma \iiint (\psi - \psi')[\boldsymbol{k}\chi \cdot f_1 \boldsymbol{\nabla}_r f + \tfrac{1}{2}(\boldsymbol{k} \cdot \boldsymbol{\nabla}_r \chi) f_1 f] \sigma^2 (\boldsymbol{g} \cdot \boldsymbol{k}) \, \mathrm{d}^2 k \, \mathrm{d}^3 c \, \mathrm{d}^3 c_1 . \tag{12.3-7}$$

Combining the expressions (12.3-6) and (12.3-7) we obtain

$$I_1 = \tfrac{1}{2}\sigma \iiint (\psi - \psi')[\boldsymbol{k}\chi \cdot \boldsymbol{\nabla}_r f f_1 + (\boldsymbol{k} \cdot \boldsymbol{\nabla}_r \chi) f f_1] \sigma^2 (\boldsymbol{g} \cdot \boldsymbol{k}) \, \mathrm{d}^2 k \, \mathrm{d}^3 c \, \mathrm{d}^3 c_1 \tag{12.3-8}$$

or

$$I_1 = \tfrac{1}{2}\sigma \boldsymbol{\nabla}_r \cdot \chi \iiint (\psi - \psi') f f_1 \sigma^2 (\boldsymbol{g} \cdot \boldsymbol{k}) \boldsymbol{k} \, \mathrm{d}^2 k \, \mathrm{d}^3 c \, \mathrm{d}^3 c_1 . \tag{12.3-9}$$

The last part of $J_2(ff)$, i.e., the term involving $\boldsymbol{kk} : \boldsymbol{\nabla}_r \boldsymbol{\nabla}_r \chi$ in Eq. (12.1-11), gives a zero contribution to the integral I_2 (for the same reason that I_0 vanishes). The remaining two parts yield

$$I_2 = \tfrac{1}{4}\sigma^2 \boldsymbol{\nabla}_r \cdot \chi \iiint (\psi - \psi') \left(\boldsymbol{k} \cdot f f_1 \boldsymbol{\nabla}_r \log \frac{f}{f_1} \right) \sigma^2 (\boldsymbol{g} \cdot \boldsymbol{k}) \boldsymbol{k} \, \mathrm{d}^2 k \, \mathrm{d}^3 c \, \mathrm{d}^3 c_1 , \tag{12.3-10}$$

as may be shown by manipulations that are similar to the ones used above to derive Eq. (12.3-9). A comparison of these results with Eq. (12.2-7) indicates that

$$I_0 + I_1 + I_2 = -\boldsymbol{\nabla}_r \cdot \boldsymbol{\Psi}_\Phi , \tag{12.3-11}$$

so that the *general conservation equation* (12.3-1) may be written as

$$\frac{\partial}{\partial t} \int \psi f \, \mathrm{d}^3 c + \boldsymbol{\nabla}_r \cdot \int \psi \boldsymbol{c} f \, \mathrm{d}^3 c - \boldsymbol{F} \cdot \int f \boldsymbol{\nabla}_c \psi \, \mathrm{d}^3 c = -\boldsymbol{\nabla}_r \cdot \boldsymbol{\Psi}_\Phi . \tag{12.3-12}$$

This equation is the generalization of Eq. (4.1-19) to a dense gas of identical, rigid spherical molecules. When we let ψ be, in turn, one of the summational invariants as was done in Eq. (4.1-20), we obtain the usual equation of continuity, Eq. (4.1-33), the equation of motion, Eq. (4.1-34), and the equation of energy, Eq. (4.1-35). The only difference in the final result is that the pressure tensor, $\mathbf{P}$, and heat flow vector, $\boldsymbol{q}$, are made up of two parts (kinetic and potential) as indicated in Eq. (12.2-6),

$$\mathbf{P} = \mathbf{P}_K + \mathbf{P}_\Phi, \qquad \boldsymbol{q} = \boldsymbol{q}_K + \boldsymbol{q}_\Phi. \tag{12.3-13}$$

12.4. Zero-order approximation

Enskog's equation (12.1-8) may be solved by a method quite analogous to that used for the solution of the Boltzmann equation in Section 5.2. The distribution function may be expanded in a series as in Eq. (5.2-3),

$$f = f^{(0)} + f^{(1)} + f^{(2)} + \ldots. \tag{12.4-1}$$

If the gas is uniform, the expressions χ and f do not depend on $\boldsymbol{r}$; hence in this case only $J_0(ff)$ is different from zero and the collision term of Enskog's equation reduces to the collision term of Boltzmann's equation except for the factor χ. Consequently, the arguments of Section 5.2 can be used to show that in the uniform steady state f takes the Maxwellian form,

$$f = f^{(0)} = n(m/2\pi kT)^{\frac{3}{2}} \exp(-mC^2/2kT), \tag{12.4-2}$$

with $\boldsymbol{C} = \boldsymbol{c} - \boldsymbol{v}$.

Next, we calculate the flux vectors to this order of approximation. These vectors are, as we have seen, composed of a kinetic part and a collisional part. Of course, the kinetic parts of the pressure tensor and heat flow vector are identical with the total pressure tensor and heat flow vector, respectively, in ordinary gases. So, with the aid of Eqs. (5.3-2) and (5.3-3), we immediately find

$$\mathbf{P}_K^{(0)} = nkT\,\mathbf{I}, \tag{12.4-3}$$

$$\boldsymbol{q}_K^{(0)} = 0. \tag{12.4-4}$$

Notice that, now, the product nkT is *not* equal to the hydrostatic pressure, p, since the collisional contribution to the latter is non-zero. With $f = f^{(0)}$ we have from Eq. (12.2-7), taking $\psi = m\boldsymbol{c}$,

$$\mathbf{P}_\Phi^{(0)} = \tfrac{1}{2}\sigma\chi \iiint m(\boldsymbol{c}' - \boldsymbol{c})\, f^{(0)} f_1^{(0)} \sigma^2 (\boldsymbol{g}\cdot\boldsymbol{k})\boldsymbol{k}\, \mathrm{d}^2 k\, \mathrm{d}^3 c\, \mathrm{d}^3 c_1. \tag{12.4-5}$$

The integral can be evaluated analytically, as shown in Appendix D. The result is

$$\mathbf{P}_{\Phi}^{(0)} = \tfrac{2}{3}\pi n\sigma^3\chi nkT\mathbf{I}. \tag{12.4-6}$$

Thus, combining the results (12.4-3) and (12.4-6) we conclude that in a zero-order theory, the pressure tensor assumes the form

$$\mathbf{P}^{(0)} = (1+\tfrac{2}{3}\pi n\sigma^3\chi)nkT\mathbf{I}. \tag{12.4-7}$$

Similarly, the potential part of the heat flow vector, $\boldsymbol{q}_{\Phi}$, is found from Eq. (12.2-7) by taking $\psi = \frac{1}{2}mC^2$, where $\boldsymbol{C}$ is taken as the velocity of a molecule relative to the mass velocity of the gas at the point $\boldsymbol{r}$ under consideration, cf. the remark after Eq. (12.2-7). Thus, with $f = f^{(0)}$,

$$\boldsymbol{q}_{\Phi}^{(0)} = \tfrac{1}{2}\sigma\chi\iiint \tfrac{1}{2}m(C'^2-C^2)f^{(0)}f_1^{(0)}\sigma^2(\boldsymbol{g}\cdot\boldsymbol{k})\boldsymbol{k}\,\mathrm{d}^2k\,\mathrm{d}^3c\,\mathrm{d}^3c_1. \tag{12.4-8}$$

This integral vanishes identically, as is shown in Appendix D. Thus, in a zero-order theory, the heat flow vector is zero,

$$\boldsymbol{q}^{(0)} = 0. \tag{12.4-9}$$

From the definition (2.2-6) and the result (12.4-7) it follows that the kinetic theory gives for the hydrostatic pressure,

$$p = \tfrac{1}{3}\mathbf{P}:\mathbf{I} = (1+\tfrac{2}{3}\pi n\sigma^3\chi)nkT. \tag{12.4-10}$$

Since the hydrostatic pressure in equilibrium is related to the temperature and the density of the gas through the equation of state, Eq. (12.4-10) may be used to evaluate the quantity χ from equilibrium measurements or computations. We return to this point in Section 12.6.

12.5. First-order approximation

When the gas is not in the uniform steady state, a first approximation to f is given by $f = f^{(0)}+f^{(1)}$ or, with $f^{(1)} = f^{(0)}\phi^{(1)}$,

$$f = f^{(0)}(1+\phi^{(1)}), \tag{12.5-1}$$

where $\phi^{(1)}$ is linear in the first-order derivatives of the macroscopic variables. The equation satisfied by $\phi^{(1)}$ is obtained from Eq. (12.1-8) by neglecting all terms involving products of derivatives of these quantities or derivatives of higher order than the first. Thus, in substituting from Eq. (12.5-1) into the right member of Eq. (12.1-8), J_2 may be neglected altogether and, in J_1, all terms involving $\phi^{(1)}$ may be neglected so that there we can simply write $f^{(0)}$ in place of f. Hence, $\phi^{(1)}$ must satisfy the integral equation

$$-n^2\chi I(\phi^{(1)}) = (\mathfrak{D}f)^{(0)} - J_1(f^{(0)}f^{(0)}), \tag{12.5-2}$$

where the integral operator I has been defined in Eq. (4.7-1) and $(\mathfrak{D}f)^{(0)}$ has been defined in Eq. (5.4-3). From the discussion of Section 5.4 we deduce

$$(\mathfrak{D}f)^{(0)} = f^{(0)}\left\{\frac{1}{n}\frac{\mathrm{d}_0 n}{\mathrm{d}t} + \left(\frac{mC^2}{2kT} - \frac{3}{2}\right)\frac{1}{T}\frac{\mathrm{d}_0 T}{\mathrm{d}t} + \boldsymbol{C}\cdot\nabla\log n \right.$$
$$+\left(\frac{mC^2}{2kT} - \frac{3}{2}\right)\boldsymbol{C}\cdot\nabla\log T$$
$$\left. -\frac{m}{kT}\left(\boldsymbol{F} - \frac{\mathrm{d}_0\boldsymbol{v}}{\mathrm{d}t}\right)\cdot\boldsymbol{C} + \frac{m}{kT}\boldsymbol{CC}:\nabla\boldsymbol{v}\right\}, \tag{12.5-3}$$

where, now, the time derivatives of the macroscopic variables must be taken from the results of the preceeding section,

$$\frac{\mathrm{d}_0 n}{\mathrm{d}t} = -n\nabla\cdot\boldsymbol{v}, \tag{12.5-4}$$

$$\frac{\mathrm{d}_0\boldsymbol{v}}{\mathrm{d}t} = \boldsymbol{F} - \frac{1}{nm}\nabla(1+\tfrac{2}{3}\pi n\sigma^3\chi)nkT, \tag{12.5-5}$$

$$\frac{\mathrm{d}_0 T}{\mathrm{d}t} = -\tfrac{2}{3}(1+\tfrac{2}{3}\pi n\sigma^3\chi)T\nabla\cdot\boldsymbol{v}. \tag{12.5-6}$$

The integral J_1 becomes, upon substitution,

$$J_1(f^{(0)}f^{(0)}) = \sigma\iint[\boldsymbol{k}\chi\cdot(f^{(0)\prime}\nabla_r f_1^{(0)\prime} + f^{(0)}\nabla_r f_1^{(0)})$$
$$+\tfrac{1}{2}(\boldsymbol{k}\cdot\nabla_r\chi)(f^{(0)\prime}f_1^{(0)\prime} + f^{(0)}f_1^{(0)})]\sigma^2(\boldsymbol{g}\cdot\boldsymbol{k})\,\mathrm{d}^2k\,\mathrm{d}^3c_1 \tag{12.5-7}$$

or, since $f^{(0)}f_1^{(0)} = f^{(0)\prime}f_1^{(0)\prime}$,

$$J_1(f^{(0)}f^{(0)}) = \sigma f^{(0)}\iint f_1^{(0)}[\chi\boldsymbol{k}\cdot\nabla_r\log f_1^{(0)}f_1^{(0)\prime} + \boldsymbol{k}\cdot\nabla_r\chi]\sigma^2(\boldsymbol{g}\cdot\boldsymbol{k})\,\mathrm{d}^2k\,\mathrm{d}^3c_1. \tag{12.5-8}$$

The integral can be evaluated analytically, as is shown in Appendix D. The result is

$$J_1(f^{(0)}f^{(0)}) = -\tfrac{2}{3}\pi n\sigma^3 f^{(0)}\left\{\chi\left[2\boldsymbol{C}\cdot\nabla\log n + \left(\frac{3mC^2}{10kT}-\frac{1}{2}\right)\boldsymbol{C}\cdot\nabla\log T\right.\right.$$

$$\left.\left.+\frac{2m}{5kT}\boldsymbol{CC}:\nabla\boldsymbol{v}+\left(\frac{mC^2}{5kT}-1\right)\nabla\cdot\boldsymbol{v}\right]+\boldsymbol{C}\cdot\nabla\chi\right\}. \qquad (12.5\text{-}9)$$

Hence, finally, Eq. (12.5-2) is equivalent to

$$n^2 I(\phi^{(1)}) = -\frac{f^{(0)}}{\chi}\left\{(1+\tfrac{2}{5}\pi n\sigma^3\chi)\left(\frac{mC^2}{2kT}-\frac{5}{2}\right)\boldsymbol{C}\cdot\nabla\log T\right.$$

$$\left.+(1+\tfrac{4}{15}\pi n\sigma^3\chi)\frac{m}{kT}(\boldsymbol{CC}-\tfrac{1}{3}C^2\mathbf{I}):\nabla\boldsymbol{v}\right\} \qquad (12.5\text{-}10)$$

or, in terms of the dimensionless velocity variable $\mathscr{C} = (m/2kT)^{\frac{1}{2}}\boldsymbol{C}$,

$$n^2 I(\phi^{(1)}) = -\frac{f^{(0)}}{\chi}\{(1+\tfrac{2}{5}\pi n\sigma^3\chi)(\mathscr{C}^2-\tfrac{5}{2})\boldsymbol{C}\cdot\nabla\log T$$

$$+2(1+\tfrac{4}{15}\pi n\sigma^3\chi)(\mathscr{C}\mathscr{C}-\tfrac{1}{3}\mathscr{C}^2\mathbf{I}):\nabla\boldsymbol{v}\}. \qquad (12.5\text{-}11)$$

This equation differs from the corresponding equation (5.4-13) for a dilute gas only in that the term involving ∇ log T is multiplied by $(1+\frac{2}{5}\pi n\sigma^3\chi)/\chi$ and the term involving $\nabla\boldsymbol{v}$ by $(1+\frac{4}{15}\pi n\sigma^3\chi)/\chi$. Hence, its solution can be written down in terms of the vector function $\boldsymbol{A}$ and the tensor function $\mathbf{B}$ of Section 5.4; it is

$$\phi^{(1)} = -\frac{1}{n\chi}\{(1+\tfrac{2}{5}\pi n\sigma^3\chi)\boldsymbol{A}\cdot\nabla\log T+(1+\tfrac{4}{15}\pi n\sigma^3\chi)\mathbf{B}:\nabla\boldsymbol{v}\}. \quad (12.5\text{-}12)$$

The final task now consists of computing the transport coefficients to the present order of approximation. For this, the flux vectors must be evaluated; they are expressed as integrals over the distribution function f in Eqs. (12.2-5), (12.2-6) and (12.2-7). In the first-order theory, f is expressed in terms of the equilibrium distribution $f^{(0)}$ and a perturbation function $\phi^{(1)}$ which are both known explicitly and are given by the formulae (12.4-2) and (12.5-12), respectively. So we have everything we need for the evaluation of the transport coefficients.

The kinetic contributions to the pressure tensor and the heat flow vector may be obtained from Eq. (12.2-5). Since $\phi^{(1)}$ is the same as that for the dilute gas except for simple multiplicative factors, using the result (12.5-12) and the analysis of Section 5.4, we may write down at once the following results,

$$\mathbf{P}_K^{(1)} = -\frac{2}{\chi}(1+\tfrac{4}{15}\pi n\sigma^3\chi)\eta^{(0)}\mathbf{S}, \tag{12.5-13}$$

$$\boldsymbol{q}_K^{(1)} = -\frac{1}{\chi}(1+\tfrac{2}{5}\pi n\sigma^3\chi)\lambda^{(0)}\nabla T, \tag{12.5-14}$$

where $\eta^{(0)}$ and $\lambda^{(0)}$ are the coefficients of viscosity and thermal conductivity of the gas at ordinary densities, cf. Eqs. (5.4-36) and (5.4-42). If these corrections are added to the zero-order expressions, Eqs. (12.4-3) and (12.4-4) respectively, we obtain the kinetic contributions to the pressure tensor and heat flow vector correct to this order of approximation,

$$\mathbf{P}_K = nkT\mathbf{I} - \frac{2}{\chi}(1+\tfrac{4}{15}\pi n\sigma^3\chi)\eta^{(0)}\mathbf{S}, \tag{12.5-15}$$

$$\boldsymbol{q}_K = -\frac{1}{\chi}(1+\tfrac{2}{5}\pi n\sigma^3\chi)\lambda^{(0)}\nabla T. \tag{12.5-16}$$

To these kinetic contributions must be added the potential contributions, $\mathbf{P}_\Phi$ and $\boldsymbol{q}_\Phi$, respectively. These will now be evaluated using Eq. (12.2-7). Consider first the collisional transfer of momentum. Let $\psi = mc$, then Eq. (12.2-7) gives, to first order,

$$\mathbf{P}_\Phi = \tfrac{1}{2}\sigma\chi\iiint m(\boldsymbol{c}'-\boldsymbol{c})ff_1\,\sigma^2(\boldsymbol{g}\cdot\boldsymbol{k})\,\boldsymbol{k}\,\mathrm{d}^2k\,\mathrm{d}^3c\,\mathrm{d}^3c_1$$
$$+\tfrac{1}{4}\sigma^2\chi\iiint m(\boldsymbol{c}'-\boldsymbol{c})f^{(0)}f_1^{(0)}\left(\boldsymbol{k}\cdot\nabla_r\log\frac{f^{(0)}}{f_1^{(0)}}\right)\sigma^2(\boldsymbol{g}\cdot\boldsymbol{k})\,\boldsymbol{k}\,\mathrm{d}^2k\,\mathrm{d}^3c\,\mathrm{d}^3c_1\,. \tag{12.5-17}$$

These integrals are further discussed in Appendix D, where it is shown that

$$\mathbf{P}_\Phi = \tfrac{2}{3}\pi n\sigma^3\chi nkT\mathbf{I} - \tfrac{8}{15}\pi n\sigma^3(1+\tfrac{4}{15}\pi n\sigma^3\chi)\eta^{(0)}\mathbf{S} - \kappa[\tfrac{6}{5}\mathbf{S}+(\nabla\cdot\boldsymbol{v})\mathbf{I}], \tag{12.5-18}$$

in which

$$\kappa = \tfrac{4}{9}n^2\sigma^4\chi(\pi mkT)^{\frac{1}{2}}. \tag{12.5-19}$$

Next, consider the collisional transfer of energy. Let $\psi = \frac{1}{2}mC^2$. From Eq. (12.2-7) we have

$$\boldsymbol{q}_\Phi = \tfrac{1}{2}\sigma\chi\iiint \tfrac{1}{2}m(C'^2-C^2)ff_1\,\sigma^2(\boldsymbol{g}\cdot\boldsymbol{k})\,\boldsymbol{k}\,\mathrm{d}^2k\,\mathrm{d}^3c\,\mathrm{d}^3c_1$$

$$+\tfrac{1}{4}\sigma^2\chi\iiint \tfrac{1}{2}m(C'^2-C^2)f^{(0)}f_1^{(0)}\left(\boldsymbol{k}\cdot\nabla_r\log\frac{f^{(0)}}{f_1^{(0)}}\right)\sigma^2(\boldsymbol{g}\cdot\boldsymbol{k})\,\boldsymbol{k}\,\mathrm{d}^2k\,\mathrm{d}^3c\,\mathrm{d}^3c_1. \tag{12.5-20}$$

The integrals are evaluated in Appendix D. The result is

$$\boldsymbol{q}_\Phi = -[\tfrac{2}{5}\pi n\sigma^3(1+\tfrac{2}{5}\pi n\sigma^3\chi)\lambda^{(0)}+c_v\kappa]\nabla T, \tag{12.5-21}$$

where the coefficient κ has been defined in Eq. (12.5-19) and c_v is the specific heat, $c_v = 3k/2m$.

Adding the contributions (12.5-15) and (12.5-18) we finally obtain for the total pressure tensor,

$$\mathsf{P} = p\boldsymbol{I}-\kappa(\nabla\cdot\boldsymbol{v})\boldsymbol{I}-2\left[\frac{1}{\chi}(1+\tfrac{4}{15}\pi n\sigma^3\chi)^2\eta^{(0)}+\tfrac{3}{5}\kappa\right]\mathsf{S}. \tag{12.5-22}$$

Similarly, adding the contributions (12.5-16) and (12.5-21) we obtain for the heat flow vector,

$$\boldsymbol{q} = -\left[\frac{1}{\chi}(1+\tfrac{2}{5}\pi n\sigma^3\chi)^2\lambda^{(0)}+c_v\kappa\right]\nabla T. \tag{12.5-23}$$

The transport coefficients for a moderately dense gas can be readily extracted from these results. A comparison of Eqs. (5.4-38), (11.3-33)and (12.5-22) shows that the *coefficient of shear viscosity*, η, is given by the expression

$$\eta = \frac{1}{\chi}(1+\tfrac{4}{15}\pi n\sigma^3\chi)^2\eta^{(0)}+\tfrac{3}{5}\kappa, \tag{12.5-24}$$

and the *coefficient of bulk viscosity* is simply κ. A comparison of Eqs. (5.4-43) and (12.5-23) finally shows that the *coefficient of thermal conductivity*, λ, is given by the expression

$$\lambda = \frac{1}{\chi}(1+\tfrac{2}{5}\pi n\sigma^3\chi)^2\lambda^{(0)}+c_v\kappa. \tag{12.5-25}$$

The expressions for η and λ may be put in a more lucid form if the coefficient κ is expressed in terms of the low-density values $\eta^{(0)}$ and $\lambda^{(0)}$, respectively. Since, for a gas of hard spheres,

$$\eta^{(0)} = 1.0160\times\frac{5}{16}\left(\frac{mkT}{\pi}\right)^{\frac{1}{2}}\sigma^2, \tag{12.5-26}$$

$$\lambda^{(0)} = 2.522c_v\eta^{(0)}, \tag{12.5-27}$$

we have

$$\kappa = \frac{16}{1.0160\times 5\pi}(\tfrac{2}{3}\pi n\sigma^3)^2\chi\eta^{(0)} = 1.002(\tfrac{2}{3}\pi n\sigma^3)^2\chi\eta^{(0)} \qquad (12.5\text{-}28)$$

and

$$c_v\kappa = \frac{1.002}{2.522}(\tfrac{2}{3}\pi n\sigma^3)^2\chi\lambda^{(0)}. \qquad (12.5\text{-}29)$$

Thus, Enskog's theory predicts the following expressions for the true coefficients of shear viscosity and thermal conductivity of a moderately dense gas in terms of the same coefficients at ordinary densities (at the *same* temperature),

$$\eta = (1/\chi)[1+\tfrac{4}{5}(\tfrac{2}{3}\pi n\sigma^3\chi)+0.7614(\tfrac{2}{3}\pi n\sigma^3\chi)^2]\eta^{(0)}, \qquad (12.5\text{-}30)$$

$$\lambda = (1/\chi)[1+\tfrac{6}{5}(\tfrac{2}{3}\pi n\sigma^3\chi)+0.7574(\tfrac{2}{3}\pi n\sigma^3\chi)^2]\lambda^{(0)}. \qquad (12.5\text{-}31)$$

These formulae will receive further attention in the next section.

We conclude this section by mentioning two immediate generalizations of Enskog's theory. First, the extension to the case of a binary mixture of rigid spherical molecules which was carried out by Thorne, whose results are quoted by Chapman and Cowling [1952]. Second, the extension to the case of a dense gas of perfectly rough spheres by McCoy, Sandler and Dahler [1966].

12.6. Application of Enskog's results

In the previous section we have shown that, in a moderately dense gas of rigid spherical molecules the transport coefficients may be expressed in terms of their values at ordinary densities (at the same temperature) with the aid of two parameters – viz., χ and $\frac{2}{3}\pi n\sigma^3$; the latter parameter is commonly denoted by $b\rho$, where b is defined as

$$b = \tfrac{2}{3}\pi\sigma^3/m \qquad (12.6\text{-}1)$$

and ρ is the mass density of the gas; $b\rho$ is called the *covolume* of the molecules. In fact, from Eqs. (12.5-30), (12.5-28) and (12.5-31) we have,

shear viscosity:

$$(\eta/\eta^{(0)})/b\rho = (b\rho\chi)^{-1}+0.8+0.7614\, b\rho\chi, \qquad (12.6\text{-}2)$$

bulk viscosity:

$$(\kappa/\eta^{(0)})/b\rho = 1.002\, b\rho\chi, \tag{12.6-3}$$

thermal conductivity:

$$(\lambda/\lambda^{(0)})/b\rho = (b\rho\chi)^{-1}+1.2+0.7574\, b\rho\chi. \tag{12.6-4}$$

Thus, as ρ increases, $(\eta/\eta^{(0)})/b\rho$ has a minimum value of 2.545 corresponding to $b\rho\chi = (b\rho\chi)_{\min,\eta}$, where

$$(b\rho\chi)_{\min,\eta} = (0.7614)^{-\frac{1}{2}} = 1.146. \tag{12.6-5}$$

Similarly, as ρ increases, $(\lambda/\lambda^{(0)})/b\rho$ has a minimum value of 2.938 corresponding to $b\rho\chi = (b\rho\chi)_{\min,\lambda}$, where

$$(b\rho\chi)_{\min,\lambda} = (0.7574)^{-\frac{1}{2}} = 1.151. \tag{12.6-6}$$

The variation of the various transport coefficients with the parameter $b\rho\chi$ is shown graphically in Fig. 12.2.

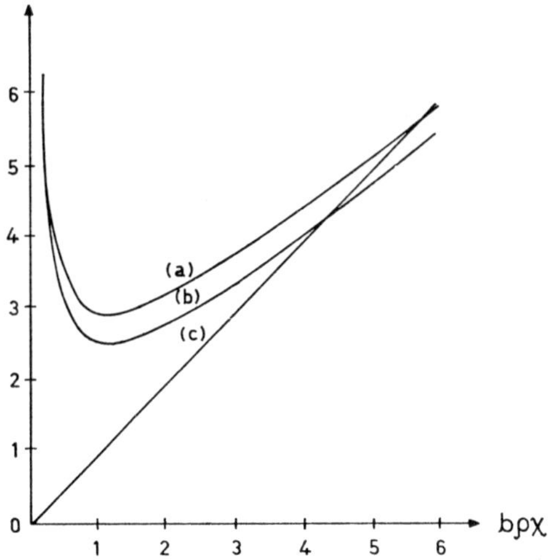

Fig. 12.2. The transport coefficients of a gas of rigid spherical molecules according to Enskog's theory; (a) $(\lambda/\lambda^{(0)})/b\rho$; (b) $(\eta/\eta^{(0)})/b\rho$; (c) $(\kappa/\eta^{(0)})/b\rho$. (From Hirschfelder, Curtiss and Bird [1954].)

The results of Enskog's theory may be applied to real gases if the parameters χ and b are properly adjusted. The expression (12.4-10) for the hydrostatic pressure of a dense gas at equilibrium provides a means for evaluating χ. From equilibrium thermodynamics it is known that, in the case of a gas of rigid spherical molecules, the virial coefficients do not vary with the temperature. In fact, the equation of state in that case reads

$$pV/RT = 1+b\rho+0.6250(b\rho)^2+0.2869(b\rho)^3+0.115(b\rho)^4+\ \ldots \qquad (12.6\text{-}7)$$

if b is defined as in Eq. (12.6-1) – see e.g., Hirschfelder, Curtiss and Bird [1954]. On the other hand, the kinetic theory result (12.4-10) may be written as

$$pV/RT = 1+b\rho\chi, \qquad (12.6\text{-}8)$$

whence, comparing the last two formulae, we have the following virial expansion of χ,

$$\chi = 1+0.6250\, b\rho+0.2869(b\rho)^2+0.115(b\rho)^3+\ \ldots . \qquad (12.6\text{-}9)$$

The sole problem for the application of Enskog's theory to real gases is, therefore, the calculation of an effective diameter σ and, consequently, an effective value of the covolume $b\rho$. This may be done in a number of ways. One possibility is to take the experimental data for $\eta^{(0)}$ and $\lambda^{(0)}$ and use the Chapman-Enskog results for the rigid sphere model to compute an effective molecular diameter σ as a function of temperature; χ can then be obtained from the expansion (12.6-9). This procedure was carried out for the noble gases by Sengers [1965]. Comparison of the variation of $\eta/\eta^{(0)}$ and $\lambda/\lambda^{(0)}$ with the density thus predicted with the experimental data shows that the data follow the behavior predicted reasonably well up to a density of $b\rho = 0.4$; at higher densities, however, the increase predicted is much too steep. (If the density of a gas is expressed in amagat units – one amagat unit is the density of the substance at 0^0 C and one atmosphere – the density $b\rho = 0.4$ corresponds to a density of about 650 amagat for He, to about 400 amagat for Ne, to about 160 amagat for Ar and to about 65 amagat for Xe.)

As was suggested by Enskog, an attempt can be made to represent the transport coefficients of a real gas over a larger density range by evaluating the parameters b and χ from observations of the variation of pressure with temperature at constant volume, i.e., from the compressibility isotherms of the gas. In this way one partially accounts for the influence of attractive forces between the molecules. The procedure most commonly used is the one introduced by Michels and Gibson [1931]. In this procedure effective values of $b\rho\chi$ are calculated by requiring that the pressure p in the equation of state for rigid spheres, Eq. (12.6-8), be identified with the thermal pressure $T(\partial p/\partial T)_V$ of the real gas. Then Eq. (12.6-8) reads $(V/R)(\partial p/\partial T)_V = 1+b\rho\chi$, from which one deduces the expression

$$b\rho\chi = \frac{1}{R}\left(\frac{\partial pV}{\partial T}\right)_V - 1. \tag{12.6-10}$$

A consistent value of b may then be found from the following argument. In order that $\eta^{(0)}$ and $\lambda^{(0)}$ be the low-density limits of η and λ, respectively, it is necessary that $\chi \to 1$ as $\rho \to 0$. Now, from the equation of state,

$$pV/RT = 1+B(T)/V+C(T)/V^2+ \dots \tag{12.6-11}$$

and the definition (12.6-10) we have

$$b\chi = \frac{1}{\rho V}\frac{\mathrm{d}}{\mathrm{d}T}TB(T)+\frac{\rho}{(\rho V)^2}\frac{\mathrm{d}}{\mathrm{d}T}TC(T)+\dots. \tag{12.6-12}$$

Hence, upon passing to the limit $\rho = 0$, we find (since $V = 1/\rho$) that b must satisfy the relation

$$b = \frac{\mathrm{d}}{\mathrm{d}T}[TB(T)]. \tag{12.6-13}$$

In other words, if $b\rho\chi$ is determined from the compressibility isotherms according to Eq. (12.6-10), then a consistent evaluation of the covolume, $b\rho$, based upon the temperature variation of the second virial coefficient, is given by Eq. (12.6-13).

The latter procedure has also been used by Sengers [1965] to calculate η and λ for the noble gases. The experimental data for the shear viscosity of He, Ne, Ar and Xe, and for the thermal conductivity of Ne and Ar were compared with the behavior predicted by Enskog's theory. The results are shown graphically in Figs. 12.3 and 12.4. It is seen that the density dependence of η and λ is described fairly well over a large density range up to a density corresponding to $b\rho = 0.6$. Similar comparisons were carried out for the shear viscosity of H_2 and N_2 as well as the thermal conductivity of N_2. Here, the results were somewhat less satisfactory, so that one may expect Enskog's theory to lead to larger deviations from the experimental data in the case of polyatomic gases. However, at least for the simple gases, Enskog's theory does contain the essential transport mechanism for a dilute as well as for a strongly compressed gas and, provided that proper values of b and χ are used, does account for the major trends of the density dependence of viscosity and thermal conductivity.

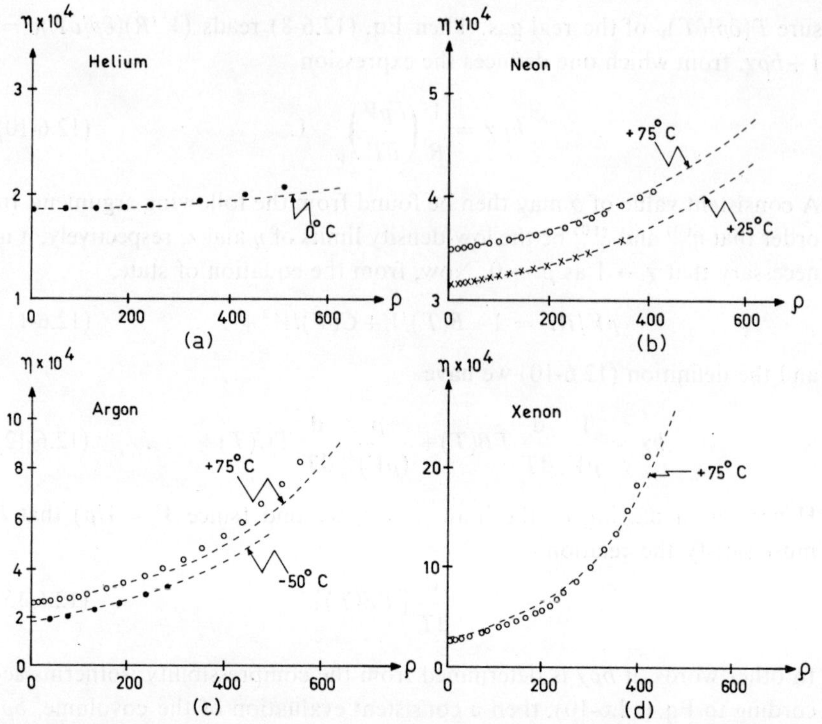

Fig. 12.3. Density dependence of the viscosity of the noble gases compared with the results of Enskog's theory. Units: [ρ] amagat, [η] g cm^{-1} sec^{-1}. (a) Helium, ● 0° C; (b) Neon, ○ +75° C, × +25° C; (c) Argon, ○ +75 °C, ● −50 °C; (d) Xenon, ○ +75 °C. (From Sengers [1965].)

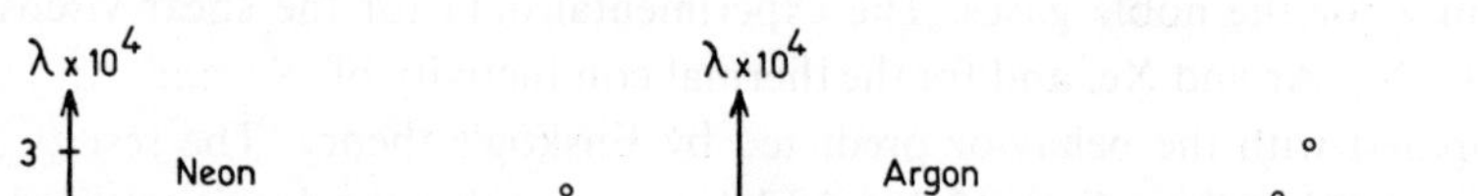

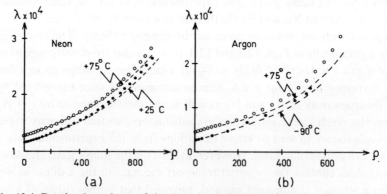

Fig. 12.4. Density dependence of the therma conductivity of neon and argon compared with the results of Enskog's theory. Units: [ρ] amagat, [λ] cal cm^{-1} sec^{-1} deg^{-1}. (a) Neon, ○ +75 °C, ● +25 °C; (b) Argon, ○ +75 °C, ● −90 °C. (From Sengers [1965].)

It is remarked that a study of the temperature dependence of η and λ requires a more subtle adaption of Enskog's theory. For instance, Kim and Ross [1965] observed that according to Enskog's theory the initial slope $(\partial\eta/\partial\rho)_T$ is proportional to $\eta^{(0)}$ and, therefore, to $T^{\frac{1}{2}}$, which does not seem to apply at high temperature. An empirical formula for the molecular diameter which represents the temperature dependence of the transport coefficients at least to some degree of accuracy has been given by Coremans and Beenakker [1966].

Dense gases - General theory 13

In the previous chapter we have shown that Enskog's theory may be used successfully in the kinetic theory of moderately dense gases – at least, of the simplest of them. The foundation of Enskog's theory, however, is far from satisfactory. Clearly, to develop a rigorous theory of dense gases one must have a generalized Boltzmann equation properly derived from first principles, i.e., from the Liouville equation. Until now, only the first steps have been made in this direction. The results, though, have been unexpected insofar as they contradict the intuitive idea that the density dependence of the transport coefficients can be represented by means of power series expansions in n and that the low-density results correspond to the lowest-order terms of these expansions. It has been found that the assumption of molecular chaos may not always be valid and that the particles become, in fact, more correlated as the density of the gas increases.

At the risk of becoming obsolete in a very short time we feel that at least a discussion of the difficulties encountered in this area of kinetic theory is appropriate. In this perspective we have written the present chapter on dense gases. The reader should, therefore, be aware of the fact that the validity of some of the arguments may be open to discussion and that, certainly, much work remains to be done in this field.

The plan of the chapter is as follows. In the first section we recall certain results from Chapter 3 and define a generalized Boltzmann equation. Furthermore, we establish the relation between the kinetic theory of dense gases and fluid mechanics by deriving the macroscopic conservation equations and defining the flux vectors (i.e., the pressure tensor and the heat flow vector).

In Section 13.2 we develop a Chapman-Enskog type theory to solve the generalized Boltzmann equation to first order in the spatial gradients, and derive the expressions for the transport coefficients. The results of this section are still general, as no specific form of the functional dependence of the two-particle distribution function upon the one-particle distribution function is assumed. In the next section, Section 13.3, the results are worked out for the special model of Choh and Uhlenbeck for a moderately dense gas. Finally, in Section 13.4, we discuss some of the difficulties which are encountered in the verification of Bogoliubov's functional assumption.

13.1. The generalized Boltzmann equation and the macroscopic conservation equations

The first step in the development of a general kinetic theory of dense gases involves the definition of a generalized Boltzmann equation. To avoid complications we restrict ourselves to the case of an intermolecular potential which is *monotonically repulsive* and of *finite range*, r_0.

From the analysis of Section 3.2 we recall that the reduced distribution functions F_s satisfy the BBGKY-hierarchy of equations (3.2-27),

$$\frac{\partial F_s}{\partial t} = -\mathfrak{H}_s F_s + n\mathfrak{L}_s F_{s+1}, \qquad s = 1, 2, \ldots \tag{13.1-1}$$

where $\mathfrak{H}_s$ is the Hamiltonian operator associated with a system of s particles,

$$\mathfrak{H}_s \equiv \sum_{i=1}^{s} \left[\frac{\boldsymbol{p}_i}{m} \cdot \frac{\partial}{\partial \boldsymbol{r}_i} + \mathfrak{F}_i \cdot \frac{\partial}{\partial \boldsymbol{p}_i}\right] - \sum_{1 \leqq i < j \leqq s} \Theta_{ij}, \tag{13.1-2}$$

Θ_{ij} is the molecular interaction operator,

$$\Theta_{ij} \equiv \frac{\partial \varphi_{ij}}{\partial \boldsymbol{r}_i} \cdot \frac{\partial}{\partial \boldsymbol{p}_i} + \frac{\partial \varphi_{ij}}{\partial \boldsymbol{r}_j} \cdot \frac{\partial}{\partial \boldsymbol{p}_j}, \tag{13.1-3}$$

and $\mathfrak{L}_s$ is the phase mixing operator,

$$\mathfrak{L}_s \equiv \sum_{i=1}^{s} \int \mathrm{d}^6 x_{s+1}\, \Theta_{i,s+1}; \tag{13.1-4}$$

$\mathfrak{F}_i$ is the external force acting on particle i; the symbol x represents the pair of vectors $(\boldsymbol{r}, \boldsymbol{p})$. We shall be especially interested in the cases $s = 1$ and $s = 2$, which read, explicitly,

$$\left(\frac{\partial}{\partial t}+\frac{\boldsymbol{p}_1}{m}\cdot\nabla_{r_1}+\mathfrak{F}_1\cdot\nabla_{p_1}\right)F_1(x_1;t)=n\int\Theta_{12}F_2(x_1x_2;t)\mathrm{d}^6x_2, \tag{13.1-5}$$

$$\left(\frac{\partial}{\partial t}+\frac{\boldsymbol{p}_1}{m}\cdot\nabla_{r_1}+\frac{\boldsymbol{p}_2}{m}\cdot\nabla_{r_2}+\mathfrak{F}_1\cdot\nabla_{p_1}+\mathfrak{F}_2\cdot\nabla_{p_2}-\Theta_{12}\right)F_2(x_1x_2;t)$$
$$=n\int(\Theta_{13}+\Theta_{23})F_3(x_1x_2x_3;t)\mathrm{d}^6x_3. \tag{13.1-6}$$

Now, the possibility of deriving a generalized Boltzmann equation depends on whether or not the two-particle distribution function, F_2, can be expressed as a time-independent functional of the one-particle distribution function, F_1. Only then is Eq. (13.1-5) self-contained in terms of F_1 and only then does its right member – which represents the interaction between the single particle and all remaining particles – have the main characteristics of the collision operator of Boltzmann's equation. In Sections 3.3 and 3.4 it was shown that, at any time t, F_2 can be written in terms of F_1 by means of the streaming operators $\tilde{\mathfrak{S}}_t^{(s)}$, $s = 2, 3\ldots$:

$$F_2(x_1x_2;t)=F_2^{(0)}(x_1x_2;t)+\frac{n}{1!}F_2^{(1)}(x_1x_2;t)+\ldots, \tag{13.1-7}$$

cf. Eq. (3.4-15), where

$$F_2^{(0)}(x_1x_2;t)=\tilde{\mathfrak{S}}_t^{(2)}(x_1x_2)\prod_{i=1}^{2}F_1(x_i;t) \tag{13.1-8}$$

and, for $s = 1, 2, \ldots$,

$$F_2^{(s)}(x_1x_2;t)=\int\ldots\int\tilde{\mathfrak{T}}_t^{(s+2)}(x_1x_2|x_3\ldots x_{s+2})\prod_{i=1}^{s+2}F_1(x_i;t)\mathrm{d}^6x_3\ldots\mathrm{d}^6x_{s+2}. \tag{13.1-9}$$

The operators $\tilde{\mathfrak{T}}_t^{(s+2)}$ are (known) combinations of the streaming operators $\tilde{\mathfrak{S}}_t^{(2)}\ldots\tilde{\mathfrak{S}}_t^{(s+2)}$. The expansion (13.1-7) was established using the method of cluster expansions, which was adopted from equilibrium statistical mechanics. The density n was thereby used as an ordering parameter, no claim was made about the convergence of the expansion. In Section 3.5 it was subsequently shown that the zero-order term of the expansion, $F_2^{(0)}$, reduces to the collision term of Boltzmann's equation if one makes the assumption of molecular chaos and if, furthermore, one neglects spatial inhomogeneities over the range of the intermolecular force. Thus, one is led to believe that, by retaining more terms of the expansion (13.1-7) and using the ideas of Section 3.5, one should be able to generalize the collision term of Boltzmann's equation. Furthermore, the fact that successive terms of the expansion contain successively higher powers of the density suggests that such a generalization may be valid for a gas at higher densities. One expects, of course, that the spatial inhomogeneities over the range of the intermolecular force will not present any significant difficulties: one should, at least, be able to correct for these inhomogeneities by including terms linear in the spatial gradients. (We recall that the present discussion is limited to short-range potentials.) Therefore, the fundamental difficulty in this approach is related to the assumption of molecular chaos: Is the assumption of molecular chaos valid or not? From the discussion of Chapter 3 we infer that, to answer this question, we must verify two hypotheses:

(H_1) At the time t of interest ($t \gg t_{\text{coll}}$) any correlation present in the initial state ($t = 0$) can be ignored.

(H_2) For each $s (s = 2, 3, \ldots)$, the s-particle streaming operator $\mathfrak{S}_t^{(s)}$ tends to a bounded operator $\mathfrak{S}_\infty^{(s)}$ as $t \to \infty$ and, at the time t of interest, $\mathfrak{S}_t^{(s)}$ can be replaced by $\mathfrak{S}_\infty^{(s)}$ to a sufficiently good approximation.

The hypothesis (H_1) renders the streaming operators Markovian and implies that, for each s, the operator $\tilde{\mathfrak{S}}_t^{(s)}$ reduces to the operator $\mathfrak{S}_t^{(s)}$, where $\mathfrak{S}_t^{(s)}$ is defined by

$$\mathfrak{S}_t^{(s)}(x_1 \ldots x_s) = S_{-t}^{(s)}(x_1 \ldots x_s) \prod_{i=1}^{s} S_t^{(1)}(x_i), \qquad (13.1\text{-}10)$$

cf. Eqs. (3.4-8) and (3.5-2). The hypothesis (H_2) reduces each term of the expansion (13.1-17) – and, therefore, the two-particle distribution function F_2 – to a time-independent functional of the one-particle distribution function, F_1.

Since the streaming operators describe the motion of isolated groups of particles under the influence of their mutual interactions only, a discussion of the dynamics of isolated groups of 2, 3, . . . particles is necessary in order to check the validity of the hypotheses (H_1) and (H_2). We will postpone such a discussion until Section 13.4 and assume, for the time being, that both hypotheses can be verified. That is, we assume that, on the collision time scale, F_1 can be considered as a constant of motion and that, on the kinetic time scale, a contracted description of the state of the gas is possible, in which the temporal development is determined by F_1 alone. We express the assumption formally by the identity

$$F_2(x_1 x_2; t) \equiv F_2(x_1 x_2 | F_1).$$

Similarly, we assume that all higher-order reduced distribution functions F_s, $s = 3, 4, \ldots$, can be written as time-independent functionals of F_1, so that, in general,

$$F_s(x_1 \ldots x_s; t) \equiv F_s(x_1 \ldots x_s | F_1), \qquad s = 2, 3, \ldots. \tag{13.1-11}$$

This is Bogoliubov's *functional assumption*; its validity will be analyzed in Section 13.4.

To facilitate the comparison of the following discussion with the corresponding discussion of the Boltzmann equation for a dilute gas we will use the velocity variable $\boldsymbol{c}$, instead of the momentum variable $\boldsymbol{p}$, and the velocity distribution functions $f_1, f_2, \ldots$, instead of the reduced distribution functions $F_1, F_2, \ldots$. The connection between these functions is given by

$$f_1(\boldsymbol{r}, \boldsymbol{c}, t)\mathrm{d}^3c = nF_1(\boldsymbol{r}, \boldsymbol{p}, t)\mathrm{d}^3p,$$

$$f_2(\boldsymbol{r}_1, \boldsymbol{c}_1, \boldsymbol{r}_2, \boldsymbol{c}_2, t)\mathrm{d}^3c_1\mathrm{d}^3c_2 = n^2F_2(\boldsymbol{r}_1, \boldsymbol{p}_1, \boldsymbol{r}_2, \boldsymbol{p}_2, t)\mathrm{d}^3p_1\mathrm{d}^3p_2,$$

etc. The first two equations of the BBGKY-hierarchy, Eqs. (13.1-5) and (13.1-6), thus become,

$$\left(\frac{\partial}{\partial t} + \boldsymbol{c} \cdot \nabla_r + \boldsymbol{F} \cdot \nabla_c\right) f_1(x; t) = \int \Theta_{12} f_2(x_1 x_2; t)\mathrm{d}^6x_2|_{x_1 = x}, \tag{13.1-12}$$

$$\left(\frac{\partial}{\partial t} + \boldsymbol{c}_1 \cdot \nabla_{r_1} + \boldsymbol{c}_2 \cdot \nabla_{r_2} + \boldsymbol{F}_1 \cdot \nabla_{c_1} + \boldsymbol{F}_2 \cdot \nabla_{c_2} - \Theta_{12}\right) f_2(x_1 x_2; t)$$
$$= \int (\Theta_{13} + \Theta_{23}) f_3(x_1 x_2 x_3; t)\mathrm{d}^6x_3, \tag{13.1-13}$$

where, now, the symbol x stands for the pair of vector variables $(\boldsymbol{r}, \boldsymbol{c})$. When the functional assumption (13.1-11) is made, Eq. (13.1-12) describes the temporal development of the one-particle distribution function, which is a characteristic of Boltzmann's equation. Therefore, we define the *generalized Boltzmann equation* as

$$\left(\frac{\partial}{\partial t} + \boldsymbol{c} \cdot \nabla_r + \boldsymbol{F} \cdot \nabla_c\right) f_1(x; t) = \int \Theta_{12} f_2(x_1, x_2 | f_1)\mathrm{d}^6x_2|_{x_1 = x}, \tag{13.1-14}$$

where no specific model for the functional dependence of f_2 upon f_1 is assumed.

Before we attempt to solve the generalized Boltzmann equation we will establish the formal relation between the kinetic theory of dense gases and fluid mechanics. We will do so on the basis of the equations of the BBGKY-hierarchy, in particular Eqs. (13.1-12) and (13.1-13), i.e., without reference to the functional assumption.

The macroscopic quantities which describe the state of the gas in the hydrodynamical stage are the *mass density* $\rho = nm$, where

$$n(\boldsymbol{r}, t) = \int f_1(\boldsymbol{r}, \boldsymbol{c}, t)\,\mathrm{d}^3c, \tag{13.1-15}$$

the *hydrodynamic velocity* $\boldsymbol{v}$, defined by

$$\rho\boldsymbol{v}(\boldsymbol{r}, t) = \int m\boldsymbol{c}\, f_1(\boldsymbol{r}, \boldsymbol{c}, t)\,\mathrm{d}^3c, \tag{13.1-16}$$

and the *internal energy* per unit mass, u, which is the sum of the kinetic energy associated with the peculiar motion of the molecules and the average intermolecular potential energy,

$$u = u_K + u_\Phi, \tag{13.1-17a}$$

where

$$\rho u_K(\boldsymbol{r}, t) = \int \tfrac{1}{2}mC^2 f_1(\boldsymbol{r}, \boldsymbol{c}, t)\,\mathrm{d}^3c, \qquad \boldsymbol{C} = \boldsymbol{c} - \boldsymbol{v}, \tag{13.1-17b}$$

and

$$\rho u_\Phi(\boldsymbol{r}, t) = \tfrac{1}{2}\int \varphi(r_{12})\, n_2(\boldsymbol{r}_1, \boldsymbol{r}_2, t)\,\mathrm{d}^3r_2|_{\boldsymbol{r}_1=\boldsymbol{r}}. \tag{13.1-17c}$$

Here, n_2 is the *pair distribution function* which is defined by

$$n_2(\boldsymbol{r}_1, \boldsymbol{r}_2, t) = \iint f_2(\boldsymbol{r}_1, \boldsymbol{c}_1, \boldsymbol{r}_2, \boldsymbol{c}_2, t)\,\mathrm{d}^3c_1\,\mathrm{d}^3c_2. \tag{13.1-18}$$

Note that u_K corresponds to the quantity u of Chapter 5. Furthermore, if it is assumed that f_2 is a time-independent functional of f_1, u_Φ and, therefore, u is a time-independent functional of f_1, i.e., $u(\boldsymbol{r}, t) = u(\boldsymbol{r}|f_1)$.

In terms of these macroscopic quantities the general conservation equations may be obtained in exactly the same form as in Chaper 4 – provided the definitions of the pressure tensor and the heat flow vector are appropriately generalized – as we will now show.

Integration of the equation for f_1, Eq. (13.1-12), with respect to $\boldsymbol{c}$, leads immediately to the *equation of continuity*,

$$\frac{1}{\rho}\frac{\mathrm{d}\rho}{\mathrm{d}t} = -\nabla\cdot\boldsymbol{v}, \tag{13.1-19}$$

since the right-hand side of Eq. (13.1-12) vanishes upon integration. Eq. (13.1-19) is indeed identical with the equation of continuity for a dilute gas, Eq. (4.1-33).

If we multiply Eq. (13.1-12) by $m\boldsymbol{c}$ before integrating over $\boldsymbol{c}$ we obtain, using the equation of continuity and the definition (13.1-18) of the pair distribution function, n_2,

$$\rho\frac{\mathrm{d}\boldsymbol{v}}{\mathrm{d}t} = \rho\boldsymbol{F} - \nabla_r\cdot\int m\boldsymbol{CC}\,f_1(\boldsymbol{r},\boldsymbol{c},t)\,\mathrm{d}^3c - \int\frac{\partial\varphi(|\boldsymbol{r}_1-\boldsymbol{r}_2|)}{\partial\boldsymbol{r}_1}\,n_2(\boldsymbol{r}_1,\boldsymbol{r}_2,t)\,\mathrm{d}^3r_2|_{\boldsymbol{r}_1=\boldsymbol{r}}. \tag{13.1-20}$$

We will now demonstrate that the last term of this equation can be written as the divergence of a second-rank tensor. Since $\boldsymbol{r}_{12} = \boldsymbol{r}_2-\boldsymbol{r}_1$ we have the identity

$$\int\frac{\partial\varphi(|\boldsymbol{r}_1-\boldsymbol{r}_2|)}{\partial\boldsymbol{r}_1}\,n_2(\boldsymbol{r}_1,\boldsymbol{r}_2,t)\,\mathrm{d}^3r_2 = -\int\varphi'(r_{12})\frac{\boldsymbol{r}_{12}}{r_{12}}\,n_2(\boldsymbol{r}_1,\boldsymbol{r}_1+\boldsymbol{r}_{12},t)\,\mathrm{d}^3r_{12}. \tag{13.1-21a}$$

Also, since n_2 is invariant under the interchange of the particles 1 and 2,

$$\int\frac{\partial\varphi(|\boldsymbol{r}_1-\boldsymbol{r}_2|)}{\partial\boldsymbol{r}_1}\,n_2(\boldsymbol{r}_1,\boldsymbol{r}_2,t)\,\mathrm{d}^3r_2 = -\int\varphi'(r_{12})\frac{\boldsymbol{r}_{12}}{r_{12}}\,n_2(\boldsymbol{r}_1+\boldsymbol{r}_{12},\boldsymbol{r}_1,t)\,\mathrm{d}^3r_{12},$$

or, changing the variable of integration from $\boldsymbol{r}_{12}$ to $-\boldsymbol{r}_{12}$,

$$\int\frac{\partial\varphi(|\boldsymbol{r}_1-\boldsymbol{r}_2|)}{\partial\boldsymbol{r}_1}\,n_2(\boldsymbol{r}_1,\boldsymbol{r}_2,t)\,\mathrm{d}^3r_2 = \int\varphi'(r_{12})\frac{\boldsymbol{r}_{12}}{r_{12}}\,n_2(\boldsymbol{r}_1-\boldsymbol{r}_{12},\boldsymbol{r}_1,t)\,\mathrm{d}^3r_{12}. \tag{13.1-21b}$$

Thus, combining Eqs. (13.1-21a) and (13.1-21b) we obtain

$$\int\frac{\partial\varphi(|\boldsymbol{r}_1-\boldsymbol{r}_2|)}{\partial\boldsymbol{r}_1}\,n_2(\boldsymbol{r}_1,\boldsymbol{r}_2,t)\,\mathrm{d}^3r_2$$

$$= -\tfrac{1}{2}\int\varphi'(r_{12})\frac{\boldsymbol{r}_{12}}{r_{12}}\{n_2(\boldsymbol{r}_1,\boldsymbol{r}_1+\boldsymbol{r}_{12},t)-n_2(\boldsymbol{r}_1-\boldsymbol{r}_{12},\boldsymbol{r}_1,t)\}\,\mathrm{d}^3r_{12}. \tag{13.1-22}$$

Next, we observe that

$$n_2(\boldsymbol{r}_1, \boldsymbol{r}_1+\boldsymbol{r}_{12}, t)-n_2(\boldsymbol{r}_1-\boldsymbol{r}_{12}, \boldsymbol{r}_1, t)$$
$$= \int_0^1 \frac{\partial}{\partial\mu} n_2(\boldsymbol{r}_1-(1-\mu)\boldsymbol{r}_{12}, \boldsymbol{r}_1+\mu\boldsymbol{r}_{12}, t)\mathrm{d}\mu$$
$$= \int_0^1 \boldsymbol{r}_{12}\cdot\nabla_{\boldsymbol{r}_1} n_2(\boldsymbol{r}_1-(1-\mu)\boldsymbol{r}_{12}, \boldsymbol{r}_1+\mu\boldsymbol{r}_{12}, t)\mathrm{d}\mu.$$

Upon substitution of this result into Eq. (13.1-22) we find that

$$\int \frac{\partial\varphi(|\boldsymbol{r}_1-\boldsymbol{r}_2|)}{\partial\boldsymbol{r}_1} n_2(\boldsymbol{r}_1, \boldsymbol{r}_2, t)\mathrm{d}^3r_2$$
$$= -\nabla_{\boldsymbol{r}_1}\cdot\tfrac{1}{2}\int \boldsymbol{r}_{12}\boldsymbol{r}_{12}\frac{\varphi'(r_{12})}{r_{12}}\int_0^1 n_2(\boldsymbol{r}_1-(1-\mu)\boldsymbol{r}_{12}, \boldsymbol{r}_1+\mu\boldsymbol{r}_{12}, t)\mathrm{d}\mu\,\mathrm{d}^3r_{12}, \tag{13.1-23}$$

which shows that the last term of Eq. (13.1-20) is indeed the divergence of a second-rank tensor. Hence, the *equation of motion*, Eq. (13.1-20), can be written in the form

$$\rho\frac{\mathrm{d}\boldsymbol{v}}{\mathrm{d}t} = \rho\boldsymbol{F}-\nabla\cdot\mathsf{P}, \tag{13.1-24}$$

where the *pressure tensor* P is defined by

$$\mathsf{P} = \mathsf{P}_K+\mathsf{P}_\Phi \tag{13.1-25a}$$

with

$$\mathsf{P}_K(\boldsymbol{r}, t) = \int mCCf_1(\boldsymbol{r}, \boldsymbol{c}, t)\,\mathrm{d}^3c \tag{13.1-25b}$$

and

$$\mathsf{P}_\Phi(\boldsymbol{r}, t) = -\tfrac{1}{2}\int \boldsymbol{r}_{12}\boldsymbol{r}_{12}\frac{\varphi'(r_{12})}{r_{12}}$$
$$\times\int_0^1 n_2(\boldsymbol{r}_1-(1-\mu)\boldsymbol{r}_{12}, \boldsymbol{r}_1+\mu\boldsymbol{r}_{12}, t)\mathrm{d}\mu\,\mathrm{d}^3r_{12}|_{\boldsymbol{r}_1=\boldsymbol{r}}. \tag{13.1-25c}$$

The equation (13.1-24) is identical with the equation of motion for a dilute gas, Eq. (4.1-34). The pressure tensor is now composed of two parts – viz., a kinetic part, P_K, and a potential part, P_Φ. The former corresponds to the pressure tensor of a dilute gas and represents the transfer of momentum due to the flow of the molecules, the latter represents the transfer of momentum between molecules by the intermolecular force. The latter is positive (negative) when the repulsive (attractive) part of the potential dominates. The tensor $\mathbf{P}_\Phi$ generalizes the pressure tensor associated with collisional transfer of momentum of Enskog's theory. Notice that the pressure tensor is again symmetric.

To derive the equation of transfer of kinetic energy we multiply Eq. (13.1-12) by $\frac{1}{2}mC^2$ and integrate over $\mathbf{c}$. Using the equation of continuity (13.1-19) and the definition (13.1-25) of the pressure tensor we find

$$\rho \frac{\mathrm{d}u_K}{\mathrm{d}t} = -\mathbf{P}_K : \nabla \mathbf{v} - \nabla \cdot \int \tfrac{1}{2}mC^2 \mathbf{C} f_1(\mathbf{r}, \mathbf{c}, t)\, \mathrm{d}^3c - \int \frac{\partial \varphi(|\mathbf{r}_1 - \mathbf{r}_2|)}{\partial \mathbf{r}_1} \cdot \mathbf{C}_1 f_2(x_1 x_2; t)\, \mathrm{d}^6x_2\, \mathrm{d}^3c_1|_{\mathbf{r}_1 = \mathbf{r}} . \tag{13.1-26}$$

The equation of transfer of potential energy must be added to this equation to obtain the equation of energy. From the definition (13.1-17c) we see that such an equation can be derived only from the equation for f_2. Hence, at this point we draw upon the second equation of the BBGKY-hierarchy. Multiplying Eq. (13.1-13) by $\frac{1}{2}\varphi(|\mathbf{r}_1 - \mathbf{r}_2|)$ and integrating over $\mathbf{c}_1$, $\mathbf{r}_2$ and $\mathbf{c}_2$ we find

$$\rho \frac{\mathrm{d}u_\Phi}{\mathrm{d}t} = -\nabla_{\mathbf{r}_1} \cdot \tfrac{1}{2} \iint \varphi(|\mathbf{r}_1 - \mathbf{r}_2|) \mathbf{C}_1 f_2(x_1 x_2; t)\, \mathrm{d}^6x_2\, \mathrm{d}^3c_1|_{\mathbf{r}_1 = \mathbf{r}} + \tfrac{1}{2} \iint \frac{\partial \varphi(|\mathbf{r}_1 - \mathbf{r}_2|)}{\partial \mathbf{r}_1} \cdot (\mathbf{C}_1 - \mathbf{C}_2) f_2(x_1 x_2; t)\, \mathrm{d}^6x_2\, \mathrm{d}^3c_1|_{\mathbf{r}_1 = \mathbf{r}} . \tag{13.1-27}$$

Here, the three-particle distribution function f_3 does not enter, as all terms involving f_3 vanish upon integration. Adding Eqs. (13.1-26) and (13.1-27) we obtain the energy equation in the form

$$\rho \frac{\mathrm{d}u}{\mathrm{d}t} = -\mathbf{P}_K : \nabla \mathbf{v} - \nabla \cdot \int \tfrac{1}{2}mC^2 \mathbf{C} f_1(\mathbf{r}, \mathbf{c}, t)\, \mathrm{d}^3c - \nabla_{\mathbf{r}_1} \cdot \tfrac{1}{2} \iint \varphi(|\mathbf{r}_1 - \mathbf{r}_2|) \mathbf{C}_1 f_2(x_1 x_2; t)\, \mathrm{d}^6x_2\, \mathrm{d}^3c_1|_{\mathbf{r}_1 = \mathbf{r}} - \tfrac{1}{2} \iint \frac{\partial \varphi(|\mathbf{r}_1 - \mathbf{r}_2|)}{\partial \mathbf{r}_1} \cdot (\mathbf{C}_1 + \mathbf{C}_2) f_2(x_1 x_2; t)\, \mathrm{d}^6x_2\, \mathrm{d}^3c_1|_{\mathbf{r}_1 = \mathbf{r}} . \tag{13.1-28}$$

Now, consider the last term of this equation. Performing similar transformations as in Eqs. (13.1-21) and (13.1-22), we find

$$\tfrac{1}{2}\iint \frac{\partial\varphi(|\boldsymbol{r}_1-\boldsymbol{r}_2|)}{\partial\boldsymbol{r}_1}\cdot(\boldsymbol{C}_1+\boldsymbol{C}_2)\,f_2(x_1\,x_2;\,t)\,\mathrm{d}^6x_2\,\mathrm{d}^3c_1$$

$$= -\tfrac{1}{4}\iint \varphi'\,(r_{12})\frac{\boldsymbol{r}_{12}}{r_{12}}\cdot(\boldsymbol{C}_1+\boldsymbol{C}_2)$$

$$\times\,\boldsymbol{r}_{12}\cdot\boldsymbol{\nabla}_{r_1}\int_0^1 f_2(\boldsymbol{r}_1-(1-\mu)\boldsymbol{r}_{12},\,\boldsymbol{c}_1,\,\boldsymbol{r}_1+\mu\boldsymbol{r}_{12},\,\boldsymbol{c}_2;\,t)\mathrm{d}\mu\,\mathrm{d}^6x_2\,\mathrm{d}^3c_1\,.$$

The right member of this identity is equal to

$$-\boldsymbol{\nabla}_{r_1}\cdot\tfrac{1}{4}\iint \boldsymbol{r}_{12}\boldsymbol{r}_{12}\frac{\varphi'(r_{12})}{r_{12}}\cdot(\boldsymbol{C}_1+\boldsymbol{C}_2)$$

$$\times\int_0^1 f_2(\boldsymbol{r}_1-(1-\mu)\boldsymbol{r}_{12},\,\boldsymbol{c}_1,\,\boldsymbol{r}_1+\mu\boldsymbol{r}_{12},\,\boldsymbol{c}_2,\,t)\mathrm{d}\mu\,\mathrm{d}^6x_2\,\mathrm{d}^3c_1$$

$$+\tfrac{1}{4}\iint \boldsymbol{r}_{12}\boldsymbol{r}_{12}\frac{\varphi'(r_{12})}{r_{12}}:(\boldsymbol{\nabla}_{r_1}(\boldsymbol{C}_1+\boldsymbol{C}_2))$$

$$\times\int_0^1 f_2(\boldsymbol{r}_1-(1-\mu)\boldsymbol{r}_{12},\,\boldsymbol{c}_1,\,\boldsymbol{r}_1+\mu\boldsymbol{r}_{12},\,\boldsymbol{c}_2,\,t)\mathrm{d}\mu\,\mathrm{d}^6x_2\,\mathrm{d}^3c_1\,.$$

The last term of this expression can be simplified considerably. Since $\boldsymbol{\nabla}_{r_1}(\boldsymbol{C}_1+\boldsymbol{C}_2) = -2\boldsymbol{\nabla}\boldsymbol{v}$, which is independent of the variables of integration, we have

$$\tfrac{1}{4}\iint \boldsymbol{r}_{12}\boldsymbol{r}_{12}\frac{\varphi'(r_{12})}{r_{12}}:(\boldsymbol{\nabla}_{r_1}(\boldsymbol{C}_1+\boldsymbol{C}_2))$$

$$\times\int_0^1 f_2(\boldsymbol{r}_1-(1-\mu)\boldsymbol{r}_{12},\,\boldsymbol{c}_1,\,\boldsymbol{r}_1+\mu\boldsymbol{r}_{12},\,\boldsymbol{c}_2,\,t)\mathrm{d}\mu\,\mathrm{d}^6x_2\,\mathrm{d}^3c_1$$

$$= -\tfrac{1}{2}\int \boldsymbol{r}_{12}\boldsymbol{r}_{12}\frac{\varphi'(r_{12})}{r_{12}}\int_0^1 n_2(\boldsymbol{r}_1-(1-\mu)\boldsymbol{r}_{12},\,\boldsymbol{r}_1+\mu\boldsymbol{r}_{12},\,t)\mathrm{d}\mu\,\mathrm{d}^3r_2:\boldsymbol{\nabla}\boldsymbol{v}$$

$$= \mathsf{P}_\Phi:\boldsymbol{\nabla}\boldsymbol{v},$$

where we have used the definition (13.1-25c) of the potential part of the pressure tensor, P_Φ. Thus, the *equation of energy*, Eq. (13.1-28), can be written in the form

$$\rho\,\frac{\mathrm{d}u}{\mathrm{d}t} = -(\boldsymbol{\nabla}\cdot\boldsymbol{q}+\mathsf{P}:\boldsymbol{\nabla}\boldsymbol{v}), \qquad (13.1\text{-}29)$$

where the *heat flow vector* $\boldsymbol{q}$ is defined by

$$\boldsymbol{q} = \boldsymbol{q}_K + \boldsymbol{q}_{\Phi 1} + \boldsymbol{q}_{\Phi 2} \tag{13.1-30a}$$

with

$$\boldsymbol{q}_K(\boldsymbol{r}, t) = \int \tfrac{1}{2} m C^2 \boldsymbol{C} f_1(\boldsymbol{r}, \boldsymbol{c}, t) \mathrm{d}^3 c, \tag{13.1-30b}$$

$$\boldsymbol{q}_{\Phi 1}(\boldsymbol{r}, t) = \tfrac{1}{2} \int \varphi(r_{12}) \iint \boldsymbol{C}_1 f_2(\boldsymbol{r}_1, \boldsymbol{c}_1, \boldsymbol{r}_2, \boldsymbol{c}_2, t) \mathrm{d}^3 c_1 \mathrm{d}^3 c_2 \mathrm{d}^3 r_2 |_{\boldsymbol{r}_1 = \boldsymbol{r}}, \tag{13.1-30c}$$

and

$$\boldsymbol{q}_{\Phi 2}(\boldsymbol{r}, t) = -\tfrac{1}{4} \int \boldsymbol{r}_{12} \boldsymbol{r}_{12} \frac{\varphi'(r_{12})}{r_{12}} \iint (\boldsymbol{C}_1 + \boldsymbol{C}_2)$$
$$\times \int_0^1 f_2(\boldsymbol{r}_1 - (1-\mu)\boldsymbol{r}_{12}, \boldsymbol{c}_1, \boldsymbol{r}_1 + \mu \boldsymbol{r}_{12}, \boldsymbol{c}_2, t) \mathrm{d}\mu \, \mathrm{d}^3 c_1 \mathrm{d}^3 c_2 \mathrm{d}^3 r_2 |_{\boldsymbol{r}_1 = \boldsymbol{r}}. \tag{13.1-30d}$$

The equation (13.1-29) is identical with the equation of energy for a dilute gas, Eq. (4.1-35). The heat flow vector is now composed of three parts – viz., a kinetic part $\boldsymbol{q}_K$ and two potential parts, $\boldsymbol{q}_{\Phi 1}$ and $\boldsymbol{q}_{\Phi 2}$. $\boldsymbol{q}_K$ corresponds to the heat flow vector of a dilute gas and represents the transfer of thermal energy due to the flow of the molecules; $\boldsymbol{q}_{\Phi 1}$ represents the flow of potential energy carried by the molecules; the scalar product of the vector $\boldsymbol{q}_{\Phi 2}$ with the normal $\boldsymbol{n}$ on a surface element $\mathrm{d}^2 S$ represents the work per unit time done on molecules on one side of $\mathrm{d}^2 S$ by molecules on the other side (work is done on the former as they move due to their peculiar motion through the force field of the latter). Thus, the vector $\boldsymbol{q}_{\Phi 2}$ generalizes the heat flow vector associated with collisional transfer of energy of Enskog's theory; the vector $\boldsymbol{q}_{\Phi 1}$ has no counterpart in Enskog's theory. Notice that both $\boldsymbol{q}_{\Phi 1}$ and $\boldsymbol{q}_{\Phi 2}$ depend on the peculiar velocities of the molecules only.

As in the theory of dilute gases, the formal agreement between the conservation equations and the equations of mass, momentum and energy of fluid mechanics provides the hope that the constitutive relations needed to close the set of equations may be derived from first principles.

The question now is, of course, whether we can solve the generalized Boltzmann equation. Obviously, there is no hope that we will ever to able to solve the equation exactly. In fact, we cannot even prove an *H*-theorem and, hence, we cannot demonstrate that the generalized Boltzmann equation describes the irreversible approach of a gas towards equilibrium. This, of course, is a very serious objection. Nevertheless, in the next section we will show that it is possible to develop a solution technique for the generalized Boltzmann equation which closely resembles the Chapman-Enskog solution technique for the Boltzmann equation in a dilute gas. By formally applying this technique we will obtain an approximate solution for the velocity distribution function f_1 and, thence, the expressions for the flux vectors of a dense gas in terms of the intermolecular potential and the functional $f_2(|f_1)$.

13.2. Solution of the generalized Boltzmann equation

In this section we turn to the solution of the generalized Boltzmann equation following the investigations of García-Colín, Green and Chaos [1966]. We will write the equation in the form

$$\mathfrak{D} f_1 = \Phi(|f_1), \tag{13.2-1}$$

where $\mathfrak{D}$ is the streaming operator,

$$\mathfrak{D} f_1(\boldsymbol{r}, \boldsymbol{c}, t) = \left(\frac{\partial}{\partial t} + \boldsymbol{c} \cdot \nabla_r + \boldsymbol{F} \cdot \nabla_c\right) f_1(\boldsymbol{r}, \boldsymbol{c}, t), \tag{13.2-2}$$

cf. Eq. (4.1-2), and Φ represents the binary interaction functional,

$$\Phi(x|f_1) = \int \Theta_{12} f_2(x_1\, x_2|f_1) \mathrm{d}^6 x_2|_{x_1 = x}, \quad x = (\boldsymbol{r}, \boldsymbol{c}). \tag{13.2-3}$$

First, we observe that $\Phi(|f_1)$ is a *nonlocal* functional, i.e., its value at any particular point depends upon the values of f_1 throughout the entire system. Hence, Eq. (13.2-1) is valid to all orders in the gradients of the system. However, we shall restrict ourselves to non-local effects which are, at most, linear in the gradients. To this end, we define a function $\bar{f}_1(; \boldsymbol{r})$ whose value at time t at any point $x' = (\boldsymbol{r}', \boldsymbol{c}')$ of phase space is equal to $f_1(\boldsymbol{r}, \boldsymbol{c}', t)$. In other words, the velocity distribution represented by $\bar{f}_1(; \boldsymbol{r})$ is spatially homogeneous and coincides with the actual velocity distribution at the point $\boldsymbol{r}$. In the following we shall also use the notation $\bar{f}_1$ instead of $\bar{f}_1(; \boldsymbol{r})$ if there is no cause of confusion. Then we may write

$$\Phi(x|f_1) = \Phi(x|\bar{f}_1 + \delta f_1), \tag{13.2-4}$$

with $\delta f_1 = f_1 - \bar{f}_1$. Now, the functional Φ involves the molecular interaction operator, Θ, which, in turn, involves the gradient of the intermolecular potential, φ. Since we have assumed that the latter is monotonically repulsive and of finite range, r_0, the domain of influence of the interaction operator Θ is limited to a small spatial region of radius r_0 around the point of interest. Hence, the value of the functional $\Phi(|f_1)$ at the point $\boldsymbol{r}$ depends only on the values of f_1 in a small neighborhood of $\boldsymbol{r}$. Thus we are led to expand the functional (13.2-4) at $\boldsymbol{r}$ around the spatially homogeneous function $\bar{f}_1$, retaining only the terms up to the first order in the spatial inhomogeneities:

$$\Phi(x|f_1) = \Phi(x|\bar{f}_1(;\boldsymbol{r})) + \int \Phi'(x|\bar{f}_1(;\boldsymbol{r}); x')\,\delta f_1(x')\,\mathrm{d}^6x', \tag{13.2-5}$$

where $\Phi'(x|\bar{f}_1(;\boldsymbol{r}); x')\delta f_1\,\mathrm{d}^6x'$ is the variation of Φ at x due to a variation $\Delta f_1 = \delta f_1\,\mathrm{d}^6x' = (f_1 - \bar{f}_1(;\boldsymbol{r}))\mathrm{d}^6x'$ of f_1 in an infinitesimal element d^6x' around the point x' of phase space. Now,

$$\delta f_1(x') = f_1(\boldsymbol{r}', \boldsymbol{c}', t) - f_1(\boldsymbol{r}, \boldsymbol{c}', t), \tag{13.2-6}$$

so, retaining only terms up to first order in the gradients we may substitute

$$\delta f_1(x') = (\nabla_{\boldsymbol{r}'} f_1(x'))_{\boldsymbol{r}'=\boldsymbol{r}} \cdot (\boldsymbol{r}' - \boldsymbol{r}). \tag{13.2-7}$$

Thus, combining Eqs. (13.2-1), (13.2-5) and (13.2-7) we obtain the generalized Boltzmann equation to first order in the spatial gradients,

$$\mathfrak{D} f_1 = \Phi(x|f_1) + \int \Phi'(x|f_1; x')(\nabla_{\boldsymbol{r}'} f_1(x'; t))_{\boldsymbol{r}'=\boldsymbol{r}} \cdot (\boldsymbol{r}' - \boldsymbol{r})\,\mathrm{d}^6x', \tag{13.2-8}$$

where, from now on, it is implicitly understood that the functionals Φ and Φ' are evaluated for the spatially homogeneous function $f_1 = \bar{f}_1(;\boldsymbol{r})$.

To obtain a solution of this equation we use a method analogous to the method of Chapman and Enskog. The essential idea is to expand f_1 around its local equilibrium value, $f_1^{(0)}$, in a power series expansion with respect to a *uniformity parameter*, μ, which is a measure of the gradients of the macroscopic variables, and to require that time is not an explicit argument of f_1, but only appears implicitly through the mass density, ρ, the hydrodynamic velocity, $\boldsymbol{v}$, and the internal energy, u. If $\boldsymbol{\beta}$ is the vector whose components are ρ, the three components of $\boldsymbol{v}$, and u, the latter requirement is expressed by the relation

$$f_1(\boldsymbol{r}, \boldsymbol{c}, t) \equiv f_1(\boldsymbol{r}, \boldsymbol{c}|\boldsymbol{\beta}(\boldsymbol{r}, t), \nabla_{\boldsymbol{r}}\boldsymbol{\beta}(\boldsymbol{r}, t), \ldots). \tag{13.2-9}$$

Furthermore, it is required that the time evolution of $\boldsymbol{\beta}$ is given by the hydrodynamic equations for this system.

Therefore, we expand f_1 in the following way,

$$f_1 = f_1^{(0)} + \mu f_1^{(1)} + \mu^2 f_1^{(2)} + \ldots, \tag{13.2-10}$$

requiring f_1 and each coefficient of the expansion to satisfy Eq. (13.2-9). To be consistent with the fact that the generalized Boltzmann equation (13.2-8) contains terms only to first order in the spatial gradients, we shall not go beyond terms linear in μ.

To zero order, the variation of the velocity distribution function over an intermolecular distance may be ignored. We obtain the equation

$$\Phi(x|f_1^{(0)}) \equiv \int \Theta_{12} f_2(\boldsymbol{r}_1, \boldsymbol{c}_1, \boldsymbol{r}_2, \boldsymbol{c}_2 | f_1^{(0)}) \mathrm{d}^3 r_2 \, \mathrm{d}^3 c_2 |_{x_1 = x} = 0 \tag{13.2-14}$$

or, taking into account the definition of Θ_{12},

$$\int \frac{\partial \varphi(r_{12})}{\partial \boldsymbol{r}_{12}} \cdot \frac{\partial}{\partial \boldsymbol{c}_1} f_1(\boldsymbol{r}_1, \boldsymbol{c}_1, \boldsymbol{r}_2, \boldsymbol{c}_2 | f_1^{(0)}) \mathrm{d}^3 r_2 \, \mathrm{d}^3 c_2 = 0. \tag{13.2-12}$$

Clearly, this equation is satisfied by the function

$$f_2(\boldsymbol{r}_1, \boldsymbol{c}_1, \boldsymbol{r}_2, \boldsymbol{c}_2 | f_1^{(0)}) = f_1^{(0)}(\boldsymbol{r}_1, \boldsymbol{c}_1, t) f_1^{(0)}(\boldsymbol{r}_1, \boldsymbol{c}_2, t) g(r_{12}), \tag{13.2-13}$$

where g is *any* function of $r_{12} = |\boldsymbol{r}_1 - \boldsymbol{r}_2|$, because in that case the integrand in Eq. (13.2-12) is an odd function of the components of the vector $\boldsymbol{r}_{12}$. Since the spatial variations of the macroscopic variables are neglected in this order of approximation, g may depend upon the variables n, $\boldsymbol{v}$ and u The reason why the function g is arbitrary is, of course, that the result (13.2-13) has been obtained from the first equation of the BBGKY-hierarchy alone. To determine the function g one must draw upon the second equation of the hierarchy, which determines the function f_2. On the other hand, if the physics has been put in correctly, the functional f_2 should be such that, at equilibrium, the expression (13.2-13) reduces to the expression for f_2 obtained from equilibrium statistical mechanics and g must be the *pair correlation function*. Also, note that nothing in the theory determines $f_1^{(0)}$. The reason for this is obscure and we are forced to *assume* $f_1^{(0)}$ to be a local Maxwellian. This is consistent with all of the special cases of the present theory (of which the Boltzmann equation is one) and with equilibrium statistical mechanics.

There still remains the problem of determining the arbitrary parameters in $f_1^{(0)}$, since we can take *any* local Maxwellian in Eq. (13.2-13). Because we have required $f_1^{(0)}$ to describe the local equilibrium state of the gas, we choose the parameters to be the true local values of the number density, the hydrodynamic velocity and the temperature. Therefore, if we write

$$f_1^{(0)} = n(m/2\pi\theta)^{\frac{3}{2}} \exp(-mC^2/2\theta), \tag{13.2-14}$$

where $\boldsymbol{C} = \boldsymbol{c} - \boldsymbol{v}$, Eqs. (13.1-15) and (13.1-16) identify n and $\boldsymbol{v}$ with the true local averages of the particle density and velocity, respectively. Furthermore, Eq. (13.1-17) gives a relation between the true average of the local internal energy per unit mass, u, and n and θ. A trivial calculation yields,

$$\rho u(\boldsymbol{r}, t) = \tfrac{3}{2}n\theta + \tfrac{1}{2}n^2 \int \varphi(r')g(r')\mathrm{d}^3r', \tag{13.2-15}$$

from which we derive the following expression for the internal energy per mole, U,

$$U = \tfrac{3}{2}N\theta + \tfrac{1}{2}(N^2/V) \int \varphi(r')g(r')\mathrm{d}^3r', \tag{13.2-16}$$

where V is the volume of one mole of the gas and N is Avogadro's number.

The interpretation of the parameter θ requires some more care, as it is intimately connected with the definition of a local temperature. The latter quantity has an unambiguous meaning only at equilibrium, in which case it is defined by the second law of thermodynamics. For dilute gases, a simple extension of the definition is possible, as we have seen in Chapter 2. However, in dense gases various definitions are possible, see Ernst [1966]. Green, García-Colín and Chaos [1966] require that the zeroth order solution be one of local thermodynamic equilibrium and, thus, the local temperature must be related to other properties by the relations of equilibrium thermodynamics. Since the one- and two-particle distribution functions which appear in Eq. (13.2-13) are the equilibrium functions, the equation (13.2-16) is nothing else than the equilibrium thermodynamic relationship giving the energy density as a function of the particle density and the temperature. Hence, the parameter θ corresponds to the quantity kT at equilibrium and we can interpret the ratio θ/k as the temperature of a gas in non-equilibrium.

The pressure tensor and heat flow vector are easily calculated for the distribution function (13.2-13). One has

$$\mathsf{P}^{(0)} = p\mathsf{I} \tag{13.2-17}$$

with

$$p = n\theta - \tfrac{1}{6}n^2 \int \varphi'(r)\, r\, g(r)\, \mathrm{d}^3 r, \tag{13.2-18}$$

and

$$\boldsymbol{q}^{(0)} = 0. \tag{13.2-19}$$

Equation (13.2-17) expresses the fact that the pressure system is hydrostatic. From Eq. (13.2-18) it follows that the *equation of state* is

$$pV = N\theta - \tfrac{1}{6}(N^2/V) \int \varphi'(r)\, r\, g(r)\, \mathrm{d}^3 r. \tag{13.2-20}$$

Equation (13.2-19) implies that, for local equilibrium, entropy must be conserved along the streamlines.

The caloric equation (13.2-16) and the equation of state (13.2-20) may be transformed to the virial form in the usual way by developing the pair distribution function as a power series in the density. Using the Mayer notation,

$$f_{ij} = [\exp(-\varphi_{ij}/\theta) - 1], \tag{13.2-21}$$

where $\varphi_{ij} = \varphi(|\boldsymbol{r}_i - \boldsymbol{r}_j|)$, and defining the functions

$$\beta_1(\theta) = \frac{1}{1!} \int f_{12}\, \mathrm{d}^3 r_2, \tag{13.2-22}$$

$$\beta_2(\theta) = \frac{1}{2!} \int\!\!\int f_{12} f_{23} f_{31}\, \mathrm{d}^3 r_2\, \mathrm{d}^3 r_3, \tag{13.2-23}$$

etc., we find

$$pV = N\theta[1 - \tfrac{1}{2}(N/V)\beta_1(\theta) - \tfrac{2}{3}(N/V)^2 \beta_2(\theta) - \ldots] \tag{13.2-24}$$

and

$$U = N\theta[\tfrac{3}{2} + \tfrac{1}{2}(N/V)\theta\beta_1'(\theta) + \tfrac{1}{3}(N/V)^2 \theta\beta_2'(\theta) + \ldots]. \tag{13.2-25}$$

When the results (13.2-17) and (13.2-19) of the zero-order approximation are substituted into the general conservation equations we obtain the *Euler equations* of hydrodynamics,

$$\frac{1}{\rho}\frac{\mathrm{d}\rho}{\mathrm{d}t} = -\nabla \cdot \boldsymbol{v}, \tag{13.2-26}$$

$$\rho \frac{\mathrm{d}\boldsymbol{v}}{\mathrm{d}t} = \rho \boldsymbol{F} - \nabla p, \tag{13.2-27}$$

$$\rho \frac{\mathrm{d}u}{\mathrm{d}t} = -p\nabla \cdot \boldsymbol{v}. \tag{13.2-28}$$

Before turning to a discussion of the first-order approximation we emphasize once more that no H-theorem establishing the monotonic approach to equilibrium can be proven. Hence it is also not known whether the local Maxwellian (13.2-14) is the *unique* solution to Eq. (13.2-11). Therefore, the results of the present theory are necessarily of a somewhat speculative nature.

Next, we turn to the first-order approximation, i.e., to the determination of the coefficient $f_1^{(1)}$ in the expansion (13.2-10). Substituting $f_1 = f_1^{(0)} + \mu f_1^{(1)}$ into Eq. (13.2-8) and collecting terms linear in μ we obtain the following equation for $f_1^{(1)}$,

$$(\mathfrak{D}f_1)^{(0)} = \int \Phi'(x|f_1^{(0)}; \boldsymbol{c}') f_1^{(1)}(\boldsymbol{r}, \boldsymbol{c}')\,\mathrm{d}^3c' + \int \Phi'(x|f_1^{(0)}; x')(\nabla_{r'} f_1^{(0)}(x'))_{\boldsymbol{r}'=\boldsymbol{r}} \cdot (\boldsymbol{r}'-\boldsymbol{r})\,\mathrm{d}^6x'. \quad (13.2\text{-}29)$$

Here, the first term of the right member arises from the spatially homogeneous term $\Phi(x_1|f_1)$ of Eq. (13.2-8); it contains the unknown function $f_1^{(1)}$. The other terms in Eq. (13.2-29) depend entirely upon $f_1^{(0)}$; $(\mathfrak{D}f_1)^{(0)}$ is defined as in Chapter 5, cf. Eqs. (5.4-3) and (5.4-5),

$$\begin{aligned}(\mathfrak{D}f_1)^{(0)} &= \partial_0 f_1^{(0)}/\partial t + \boldsymbol{c} \cdot \nabla_r f_1^{(0)} + \boldsymbol{F} \cdot \nabla_c f_1^{(0)} \\ &= f_1^{(0)} \left\{ \frac{\mathrm{d}_0 \log f_1^{(0)}}{\mathrm{d}t} + \boldsymbol{C} \cdot \nabla_r \log f_1^{(0)} + \left(\boldsymbol{F} - \frac{\mathrm{d}_0 \boldsymbol{v}}{\mathrm{d}t}\right) \cdot \nabla_C \log f_1^{(0)} \right. \\ &\left. -(\nabla_C \log f_1^{(0)})\boldsymbol{C} : \nabla \boldsymbol{v} \right\}, \end{aligned} \quad (13.2\text{-}30)$$

where the zero-order time derivatives $\mathrm{d}_0/\mathrm{d}t$ are to be calculated from the Euler equations. From Eq. (13.2-26) we have

$$\frac{\mathrm{d}_0 n}{\mathrm{d}t} = -n\nabla \cdot \boldsymbol{v}. \quad (13.2\text{-}31)$$

Since p is a function of ρ and θ it follows from Eq. (13.2-27) that

$$\frac{\mathrm{d}_0 \boldsymbol{v}}{\mathrm{d}t} = \boldsymbol{F} - \frac{1}{\rho}\nabla p = \boldsymbol{F} - \frac{1}{\rho}\left[\left(\frac{\partial p}{\partial \rho}\right)_\theta \nabla\rho + \left(\frac{\partial p}{\partial \theta}\right)_\rho \nabla\theta\right]. \quad (13.2\text{-}32)$$

Finally, since u is a function of ρ and θ we have

$$\frac{\mathrm{d}_0 u}{\mathrm{d}t} = \left(\frac{\partial u}{\partial \rho}\right)_\theta \frac{\mathrm{d}_0 \rho}{\mathrm{d}t} + \left(\frac{\partial u}{\partial \theta}\right)_\rho \frac{\mathrm{d}_0 \theta}{\mathrm{d}t} \quad (13.2\text{-}33)$$

and, therefore, from Eqs. (13.2-28) and (13.2-31),

$$\frac{\mathrm{d}_0 \theta}{\mathrm{d}t} = -\frac{p-\rho^2(\partial u/\partial \rho)_\theta}{\rho(\partial u/\partial \theta)_\rho} \nabla \cdot \boldsymbol{v}. \tag{13.2-34}$$

Now, from the second law of thermodynamics we have the identity $(\partial U/\partial V)_T = T(\partial p/\partial T)_V - p$, which gives the variation of the internal energy per mole, U, with volume as a function of the pressure and the temperature of the gas. In terms of the variables u, ρ, p and θ the identity assumes the following form,

$$-\rho^2\left(\frac{\partial u}{\partial \rho}\right)_\theta = \theta\left(\frac{\partial p}{\partial \theta}\right)_\rho - p. \tag{13.2-35}$$

Combining Eqs. (13.2-34) and (13.2-35) we arrive at the following expression for $\mathrm{d}_0\theta/\mathrm{d}t$,

$$\frac{\mathrm{d}_0 \theta}{\mathrm{d}t} = -\theta \frac{(\partial p/\partial \theta)_\rho}{\rho(\partial u/\partial \theta)_\rho} \nabla \cdot \boldsymbol{v}. \tag{13.2-36}$$

Thus, if we introduce the *specific heat* (heat capacity per unit mass), c_v, the *thermal expansion coefficient*, β, and the *isothermal compressibility*, κ, defined by the relations

$$c_v = \left(\frac{\partial u}{\partial(\theta/k)}\right)_\rho, \quad \beta = \kappa\left(\frac{\partial p}{\partial(\theta/k)}\right)_\rho, \quad \kappa^{-1} = \rho\left(\frac{\partial p}{\partial \rho}\right)_\theta, \tag{13.2-37}$$

respectively, we obtain after some trivial algebra the following expression for $(\mathfrak{D}f_1)^{(0)}$,

$$(\mathfrak{D}f_1)^{(0)} = f_1^{(0)}\left\{\left(1-\frac{1}{n\kappa\theta}\right)\boldsymbol{C}\cdot\nabla\log n + \left[\frac{mC^2}{2\theta} - \tfrac{5}{2} + \left(1-\frac{\beta}{n\kappa k}\right)\right] \times \boldsymbol{C}\cdot\nabla\log\theta \right.$$

$$\left. + \left[\frac{m}{\theta}(\boldsymbol{CC}-\tfrac{1}{3}C^2\boldsymbol{I}) + \left(\frac{mC^2}{3\theta}-1\right)\left(1-\frac{3}{2}\frac{\beta}{\rho\kappa c_v}\right)\boldsymbol{I}\right] : \nabla\boldsymbol{v}\right\}, \tag{13.2-38}$$

which is a generalization of the expression (5.4-12). From the virial expansions (13.2-24) and (13.2-25) and the definitions (13.2-37) it is readily verified that, for small n,

$$1 - \frac{1}{n\kappa\theta} = n\beta_1(\theta) + O(n^2), \tag{13.2-39a}$$

$$1 - \frac{\beta}{n\kappa k} = \tfrac{1}{2}n[\beta_1(\theta) + \theta\,\beta_1'(\theta)] + O(n^2), \tag{13.2-39b}$$

$$1 - \frac{3}{2}\frac{\beta}{\rho\kappa c_v} = n[\tfrac{1}{2}\beta_1(\theta) + \tfrac{7}{6}\theta\beta_1'(\theta) + \tfrac{1}{3}\theta^2\beta_1''(\theta)] + O(n^2). \tag{13.2-39c}$$

The second term of the right member of Eq. (13.2-29) may be written in a form similar to the one obtained for $(\mathfrak{D}f_1)^{(0)}$ in Eq. (13.2-38). To this end, it is sufficient to notice that the term $(\nabla_{r'} f_1^{(0)}(x'))_{r'=r}$ may be expanded in the following way,

$$(\nabla_{r'} f_1^{(0)}(x'))_{r'=r} = f_1^{(0)}(\boldsymbol{r}, \boldsymbol{c}', t) \times \left\{\nabla_r \log n + \left(\frac{mC'^2}{2\theta} - \frac{3}{2}\right)\nabla_r \log\theta + \frac{m}{\theta}\nabla_r \boldsymbol{v}\cdot\boldsymbol{C}'\right\}, \tag{13.2-40}$$

where the vectors $\nabla\log n$, $\nabla\log\theta$ and the tensor $\nabla\boldsymbol{v}$ do not depend upon the variable of integration, $x' = (\boldsymbol{r}', \boldsymbol{c}')$. Thus, writing

$$f_1^{(1)} = f_1^{(0)}\phi, \tag{13.2-41}$$

we find that ϕ satisfies an integral equation of the following form,

$$\begin{aligned} n^2 I(\phi) = -\Bigg\{&\left[f_1^{(0)}\left(1 - \frac{1}{n\kappa\theta}\right)\boldsymbol{C} - \int \Phi'(x|f_1^{(0)}; x')(\boldsymbol{r}'-\boldsymbol{r})\, f_1^{(0)}(\boldsymbol{c}')\,\mathrm{d}^6x'\right]\cdot\nabla\log n \\ &+ \left[f_1^{(0)}\left(\frac{mC^2}{2\theta} - \tfrac{5}{2} + \left(1 - \frac{\beta}{nk\kappa}\right)\right)\boldsymbol{C}\right. \\ &\left. - \int \Phi'(x|f_1^{(0)}; x')(\boldsymbol{r}'-\boldsymbol{r})\left(\frac{mC'^2}{2\theta} - \frac{3}{2}\right) f_1^{(0)}(\boldsymbol{c}')\,\mathrm{d}^6x'\right]\cdot\nabla\log\theta \\ &+ \left[f_1^{(0)}\left(\frac{m}{\theta}(\boldsymbol{CC} - \tfrac{1}{3}C^2\mathsf{I}) + \left(\frac{mC^2}{3\theta} - 1\right)\left(1 - \frac{3}{2}\frac{\beta}{\rho\kappa c_v}\right)\mathsf{I}\right)\right. \\ &\left. - \frac{m}{\theta}\int \Phi'(x|f_1^{(0)}; x')\boldsymbol{C}'(\boldsymbol{r}'-\boldsymbol{r})\, f_1^{(0)}(\boldsymbol{c}')\,\mathrm{d}^6x'\right] : \nabla\boldsymbol{v}\Bigg\}, \end{aligned} \tag{13.2-42}$$

where the linear integral operator I is defined by the relation

$$\int \Phi'(x|f_1^{(0)}; \boldsymbol{c}') f_1^{(0)}(\boldsymbol{c}')F(\boldsymbol{c}')\,\mathrm{d}^3c' = -n^2 I(F) \tag{13.2-43}$$

for any function F of the molecular velocity.

In the integral equation (13.2-42) two more reductions can be performed. The first one is that the coefficient of the gradient of the density is equal to zero; as the proof is too long to present here, the reader is referred to the original article by García-Colín, Green and Chaos [1966]. The second reduction concerns the structure of the tensor multiplying $\nabla \boldsymbol{v}$. Any tensor may be decomposed uniquely into a sum of a symmetric traceless part, a diagonal part and an antisymmetric part and, furthermore, upon integration, the tensorial character does not change. If we decompose the tensor $\boldsymbol{C}'(\boldsymbol{r}'-\boldsymbol{r})$ and perform the integration over x' in the coefficient of $\nabla \boldsymbol{v}$ in Eq. (13.2-42), we get a tensor whose components are functions of $\boldsymbol{C}$ only. However, since the most general such tensor of rank two is the sum of a scalar multiple of $\boldsymbol{CC}-\frac{1}{3}C^2\boldsymbol{I}$ and a scalar multiple of $\boldsymbol{I}$, we conclude that the integral of the antisymmetric part of $\boldsymbol{C}'(\boldsymbol{r}'-\boldsymbol{r})$ must be equal to zero. Defining the symmetric traceless tensor $\boldsymbol{T}$,

$$\boldsymbol{T} = \tfrac{1}{2}[(\boldsymbol{\mathscr{C}}'(\boldsymbol{r}'-\boldsymbol{r})+(\boldsymbol{r}'-\boldsymbol{r})\boldsymbol{\mathscr{C}}']-\tfrac{1}{3}\boldsymbol{\mathscr{C}}'\cdot(\boldsymbol{r}'-\boldsymbol{r})\boldsymbol{I}, \qquad (13.2\text{-}44)$$

where $\boldsymbol{\mathscr{C}}$ is the dimensionless velocity variable,

$$\boldsymbol{\mathscr{C}} = (m/2\theta)^{\frac{1}{2}}\boldsymbol{C}, \qquad (13.2\text{-}45)$$

we obtain the following integral equation for the unknown function ϕ,

$$\begin{aligned} n^2 I(\phi) = &-\Bigg\{\Bigg[f_1^{(0)}\left(\mathscr{C}^2-\tfrac{5}{2}+\left(1-\frac{\beta}{nk\kappa}\right)\right)\boldsymbol{C} \\ &-\int \Phi'(x|f_1^{(0)};x')(\boldsymbol{r}'-\boldsymbol{r})(\mathscr{C}'^2-\tfrac{3}{2})f_1^{(0)}(\boldsymbol{c}')\,\mathrm{d}^6x'\Bigg]\cdot\nabla\log\theta \\ &+2\Bigg[f_1^{(0)}(\boldsymbol{\mathscr{C}\mathscr{C}}-\tfrac{1}{3}\mathscr{C}^2\boldsymbol{I}) \\ &-\left(\frac{m}{2\theta}\right)^{\frac{1}{2}}\int \Phi'(x|f_1^{(0)};x')\,\boldsymbol{T}f_1^{(0)}(\boldsymbol{c}')\,\mathrm{d}^6x'\Bigg]:\nabla\boldsymbol{v} \\ &+\frac{2}{3}\Bigg[f_1^{(0)}(\mathscr{C}^2-\tfrac{3}{2})\left(1-\frac{3}{2}\frac{\beta}{\rho\kappa c_v}\right) \\ &-\left(\frac{m}{2\theta}\right)^{\frac{1}{2}}\int \Phi'(x|f_1^{(0)};x')\boldsymbol{\mathscr{C}}'\cdot(\boldsymbol{r}'-\boldsymbol{r})\,f_1^{(0)}(\boldsymbol{c}')\,\mathrm{d}^6x'\Bigg]\,\nabla\cdot\boldsymbol{v}\Bigg\}. \end{aligned} \qquad (13.2\text{-}46)$$

The next step consists of the solution of Eq. (13.2-46) for ϕ. Notice that the structure of the integral equation is similar to that of the corresponding equation (5.4-13) which was found in the dilute gas case. However, there is one basic difference regarding the kernel of the integral operator. In fact, García-Colín, Green and Chaos [1966] have shown that

(i) the right eigenfunctions with zero eigenvalue, i.e., the solutions to the homogeneous equation

$$\int \Phi'(x|f_1^{(0)}; \boldsymbol{c}')\, f_1^{(0)}(\boldsymbol{c}')\, \phi_r(\boldsymbol{c}')\, \mathrm{d}^3c' = 0, \tag{13.2-47}$$

are 1, $m\boldsymbol{c}$ and $\frac{1}{2}mC^2$;

(ii) the left eigenfunctions with zero eigenvalue, i.e., the solutions to the homogeneous equation

$$\int \phi_l(\boldsymbol{c}')\, \Phi'(x|f_1^{(0)}; \boldsymbol{c}')\, f_1^{(0)}(\boldsymbol{c}')\, \mathrm{d}^3c' = 0, \tag{13.2-48}$$

are 1, $m\boldsymbol{c}$ and $U'(x'|f_1^{(0)})$, where the latter quantity stands for the functional derivative of the total energy evaluated for the local Maxwellian.

The results (i) and (ii) imply that the kernel is not symmetric. It may be shown that the solutions of Eq. (13.2-48) are orthogonal to the inhomogeneous part of Eq. (13.2-45) so that the existence of a nontrivial solution to the integral equation is indeed guaranteed. Of course, the solution is not unique but it is determined up to a linear combination of the solutions of Eq. (13.2-47). Thus, using similar arguments as in Section 5.4 one may verify that the solution ϕ to Eq. (13.2-45) is of the form

$$\phi = -\frac{1}{n}\boldsymbol{A}\cdot\nabla\log\theta - \frac{1}{n}\mathsf{B}:\nabla\boldsymbol{v} - \frac{1}{n}\Gamma\nabla\cdot\boldsymbol{v} + \boldsymbol{\alpha}^{(1)}\cdot\boldsymbol{\psi}, \tag{13.2-49}$$

where $\boldsymbol{A}$, B and Γ are a vector, a tensor and, scalar function of $\boldsymbol{C}$, respectively, while $\boldsymbol{\alpha}^{(1)}$ is a vector independent of the velocity variable (which may, however, be a function of $\boldsymbol{r}$ and t) and $\boldsymbol{\psi}$ is the vector whose components are 1, the three components of $m\boldsymbol{c}$ and $\frac{1}{2}mC^2$, The unknowns $\boldsymbol{A}$, B, and Γ satisfy the integral equations

$$n\,I(\boldsymbol{A}) = f_1^{(0)}\left(\mathscr{C}^2 - \tfrac{5}{2} + \left(1 - \frac{\beta}{nk\kappa}\right)\right)\boldsymbol{C}$$

$$-\int \Phi'(x|f_1^{(0)}; x')(\boldsymbol{r}'-\boldsymbol{r})(\mathscr{C}'^2 - \tfrac{3}{2})f_1^{(0)}(\boldsymbol{c}')\,\mathrm{d}^6x', \tag{13.2-50}$$

$$n\,I(\mathbf{B}) = 2f_1^{(0)}(\mathscr{C}\mathscr{C}-\tfrac{1}{3}\mathscr{C}^2\mathbf{I})$$
$$-2\left(\frac{m}{2\theta}\right)^{\frac{1}{2}}\int \Phi'(x|f_1^{(0)};x')\,\mathbf{T}\,f_1^{(0)}(\boldsymbol{c}')\,\mathrm{d}^6x', \qquad (13.2\text{-}51)$$

$$n\,I(\Gamma) = \tfrac{2}{3}f_1^{(0)}(\mathscr{C}^2-\tfrac{3}{2})\left(1-\frac{3}{2}\,\frac{\beta}{\rho\kappa c_v}\right)$$
$$-\frac{2}{3}\left(\frac{m}{2\theta}\right)^{\frac{1}{2}}\int \Phi'(x|f_1^{(0)};x')\,\mathscr{C}'\cdot(\boldsymbol{r}'-\boldsymbol{r})\,f_1^{(0)}(\boldsymbol{c}')\,\mathrm{d}^6x', \qquad (13.2\text{-}52)$$

respectively. From these equations it follows that the vector $\boldsymbol{A}$ and the tensor $\mathbf{B}$ must be of the form

$$\boldsymbol{A} = A(\mathscr{C})\boldsymbol{C},\quad \mathbf{B} = B(\mathscr{C})(\mathscr{C}\mathscr{C}-\tfrac{1}{3}\mathscr{C}^2\mathbf{I}), \qquad (13.2\text{-}53)$$

where A, B and, similarly, Γ are scalar functions of n, θ and C (or $\mathscr{C}$).

Uniqueness of the solution (13.2-49) is obtained by the requirement that, in accordance with Eq. (13.2-14), the true values of the local macroscopic variables ρ, $\boldsymbol{v}$ and u be determined by $f_1^{(0)}$ alone. This immediately gives five subsidiary conditions for ϕ, viz.,

$$\int f_1^{(0)}(\boldsymbol{c})\,\phi(\boldsymbol{c})\,\mathrm{d}^3c = 0, \qquad (13.2\text{-}54a)$$

$$\int f_1^{(0)}(\boldsymbol{c})\,\phi(\boldsymbol{c})\,\boldsymbol{c}\,\mathrm{d}^3c = 0, \qquad (13.2\text{-}54b)$$

$$\int u'(\boldsymbol{r}|f_1^{(0)};\boldsymbol{c}')\,f_1^{(0)}(\boldsymbol{c}')\,\phi(\boldsymbol{c}')\,\mathrm{d}^3c'$$
$$+\int u'(\boldsymbol{r}|f_1^{(0)};x')(\nabla_{\boldsymbol{r}'}\,f_1^{(0)}(x'))_{\boldsymbol{r}'=\boldsymbol{r}}\cdot(\boldsymbol{r}'-\boldsymbol{r})\,\mathrm{d}^6x' = 0. \qquad (13.2\text{-}54c)$$

If we substitute ϕ according to Eq. (13.2-49) into these conditions we find that the vector $\boldsymbol{\alpha}^{(1)}$ may be set equal to zero provided that the functions A and Γ satisfy the constraints

$$\int f_1^{(0)}A(\mathscr{C})\,\mathscr{C}^2\,\mathrm{d}^3c = 0, \qquad (13.2\text{-}55)$$

and

$$\int f_1^{(0)}\,\Gamma(\mathscr{C})\,\mathrm{d}^3c = 0, \qquad (13.2\text{-}56a)$$

$$\frac{1}{n}\int u'(r|f_1^{(0)};\,c')\,f_1^{(0)}(c')\,\Gamma(\mathscr{C}')\,\mathrm{d}^3c'$$
$$=\frac{1}{3}\left(\frac{m}{2\theta}\right)^{\frac{1}{2}}\iiint \varphi(|r-r_1|)\,f_2'(x,\,x_1|f_1^{(0)};\,x')\,\mathscr{C}'\cdot(r'-r)\,f_1^{(0)}(c')\,\mathrm{d}^6x_1\,\mathrm{d}^3c\,\mathrm{d}^6x', \tag{13.2-56b}$$

respectively. Therefore, the solution ϕ is now uniquely determined; its explicit form is

$$\phi = -\frac{1}{n}\,A\cdot\nabla\log\theta-\frac{1}{n}\,\mathbf{B}:\nabla v-\frac{1}{n}\,\Gamma\nabla\cdot v. \tag{13.2-57}$$

Thus, Eq. (13.2-57) can be used to compute the transport coefficients to first order in the gradients for a gas composed of a single kind of molecules which interact between themselves with short-range repulsive forces. The results will, of course, depend on the functions A, $\mathbf{B}$ and Γ which are, in principle, determined by Eqs. (13.2-50), (13.2-51) and (13.2-52), respectively, subject to the constraints (13.2-55) and (13.2-56).

The pressure tensor and heat flow vector can be evaluated from Eqs. (13.1-25) and (13.1-30), respectively, upon substitution of the explicit form of f_1, viz.

$$f_1 = f_1^{(0)}(1+\phi), \tag{13.2-58}$$

where, now, ϕ is known. The calculation is rather straightforward and has been performed by García-Colín, Green and Chaos [1966]. The following considerations may be helpful.

(i) In an isotropic system the flux vectors are represented by the product of an *isotropic* tensor function of the variables specifying the state of the gas times the thermodynamic forces.

(ii) The expressions for $\mathbf{P}_\Phi$ and $q_{\Phi 1}$ contain the quantity $f_2(r_1+(\mu-1)\,r_{12},\ c_1,\ r_1+\mu r_{12},\ c_2|f_1)$, a nonlocal functional of f_1. Hence, several expansions must be made to take care of the spatial inhomogeneities. First, one deals with the spatial arguments of f_2 by expanding in a Taylor series around $\mu = 1$. Thus, the quantity $f_2(r_1+(\mu-1)r_{12},\ c_1,\ r_1+\mu r_{12},\ c_2|f_1)$ gives rise to two terms, one involving $f_2(x_1\,x_2|f_1)$ and another involving the gradients of f_2 with respect to its spatial arguments. However, the latter gradients are proportional to r_{12} and, hence, the second term vanishes upon integration with respect to r_2. Now, consider the first term, which involves the (nonlocal) functional $f_2(x_1\,x_2|f_1)$. To reduce it to a local functional one expands around the function $\bar{f}_1 = \bar{f}_1(;\boldsymbol{r}_1)$. One obtains two terms, one involving $f_2(x_1 x_2|\bar{f}_1)$, and another involving the variation of the functional, $\int f_2'(x_1 x_2|\bar{f}_1; x')\,\delta f_1(x')\mathrm{d}^6x'$, as in Eq. (13.2-5). In the latter, $\delta f_1(x')$ may be replaced by $(\nabla_{r'} f_1(x'))_{r'=r} \cdot (\boldsymbol{r}'-\boldsymbol{r})$, as in Eq. (13.2-7). All functionals have now been localized and one can omit the bar above f_1, as in Eq. (13.2-8). The final step consists of the substitution $f_1 = f_1^{(0)}(1+\phi)$ followed by an expansion around the equilibrium solution, $f_1^{(0)}$. The first term, $f_2(x_1 x_2|f_1)$, gives rise to two terms: $f_2(x_1 x_2|f_1^{(0)})$, and a second term involving the variation of the functional. In evaluating the latter one must remember that, in Eq. (13.2-58), f_1 and $f_1^{(0)}$ are both evaluated at the same point. Hence, the variation of the functional is affected only by the variation of f_1 in the velocity domain; it is given by $\int f_2'(x_1 x_2|f_1^{(0)}; \boldsymbol{c}')f_1^{(0)}\phi(\boldsymbol{r}_1, \boldsymbol{c}', t) \times \mathrm{d}^3c'$. Finally, in the term $\int f_2'(x_1 x_2|f_1; x')(\nabla_{r'} f_1(x'))_{r'=r} \cdot (\boldsymbol{r}'-\boldsymbol{r})\mathrm{d}^6x'$ one may simply replace f_1 by $f_1^{(0)}$, as this term is already linear in the gradients of f_1. Thus, collecting all contributions one has

$$
\begin{aligned}
f_2(x_1 x_2|f_1) = {} & f_2(x_1 x_2|f_1^{(0)}) \\
& + \int f_2'(x_1 x_2|f_1^{(0)}; \boldsymbol{c}')\, f_1^{(0)}\phi(\boldsymbol{r}_1, \boldsymbol{c}', t)\,\mathrm{d}^3c' \\
& + \int f_2'(x_1 x_2|f_1^{(0)}; x')(\nabla_{r'} f_1^{(0)}(x'))_{r'=r_1}\cdot(\boldsymbol{r}'-\boldsymbol{r}_1)\,\mathrm{d}^6x'. \quad (13.2\text{-}59)
\end{aligned}
$$

Referring to the Appendix E for a detailed exposition of the methods of calculating the flux vectors, we only give the final results. For the heat flow vector, $\boldsymbol{q}$, we obtain *Fourier's law*,

$$\boldsymbol{q} = -\lambda\nabla(\theta/k), \qquad (13.2\text{-}60)$$

where λ, the *thermal conductivity*, is given by

$$\lambda = \lambda_K + \lambda_{\Phi 1,1} + \lambda_{\Phi 2,1} + \lambda_{\Phi 1,2} + \lambda_{\Phi 2,2}, \qquad (13.2\text{-}61a)$$

λ_K is the kinetic contribution,

$$\lambda_K = \frac{2k\theta}{3mn}\int f_1^{(0)}(\boldsymbol{c})\,A(\mathscr{C})\,\mathscr{C}^4\mathrm{d}^3c, \qquad (13.2\text{-}61b)$$

$\lambda_{\Phi 1,1}$ and $\lambda_{\Phi 2,1}$ are the contributions arising from the macroscopic gradients,

$$\lambda_{\Phi 1,1} = \frac{k}{3mn}\int \varphi(r_{12})\left\{\iint \mathscr{C}_1\cdot\left[\int \mathscr{C}'\right.\right.$$

$$\left.\left.\times f_2'(x_1\,x_2|f_1^{(0)};\boldsymbol{c}')f_1^{(0)}(\boldsymbol{c}')A(\mathscr{C}')\mathrm{d}^3c'\right]\mathrm{d}^3c_1\,\mathrm{d}^3c_2\right\}\mathrm{d}^3r_2|_{\boldsymbol{r}_1=\boldsymbol{r}}, \quad (13.2\text{-}61c)$$

$$\lambda_{\Phi 2,1} = -\frac{k}{6mn}\int \frac{\varphi'(r_{12})}{r_{12}}\boldsymbol{r}_{12}\cdot\left\{\iint(\mathscr{C}_1+\mathscr{C}_2)\right.$$

$$\left.\times\left[\int f_2'(x_1\,x_2|f_1^{(0)};\boldsymbol{c}')f_1^{(0)}(\boldsymbol{c}')A(\mathscr{C}')(\boldsymbol{r}_{12}\cdot\mathscr{C}')\mathrm{d}^3c'\right]\mathrm{d}^3c_1\,\mathrm{d}^3c_2\right\}\mathrm{d}^3r_2|_{\boldsymbol{r}_1=\boldsymbol{r}},$$

(13.2-61d)

and $\lambda_{\Phi 1,2}$ and $\lambda_{\Phi 2,2}$ are the contributions arising from spatial inhomogeneities over the range of the intermolecular force,

$$\lambda_{\Phi 1,2} = -\frac{k}{6\theta}\left(\frac{2\theta}{m}\right)^{\frac{1}{2}}\int \varphi(r_{12})\left\{\iint \mathscr{C}_1\cdot\left[\int(\boldsymbol{r}'-\boldsymbol{r}_1)\right.\right.$$

$$\left.\left.\times f_2'(x_1\,x_2|f_1^{(0)};x')f_1^{(0)}(\boldsymbol{c}')\mathscr{C}'^2\mathrm{d}^6x'\right]\mathrm{d}^3c_1\,\mathrm{d}^3c_2\right\}\mathrm{d}^3r_2|_{\boldsymbol{r}_1=\boldsymbol{r}}, \quad (13.2\text{-}61e)$$

$$\lambda_{\Phi 2,2} = \frac{k}{12\theta}\left(\frac{2\theta}{m}\right)^{\frac{1}{2}}\int \frac{\varphi'(r_{12})}{r_{12}}\boldsymbol{r}_{12}\cdot\left\{\iint(\mathscr{C}_1+\mathscr{C}_2)\right.$$

$$\left.\times\left[\int f_2'(x_1\,x_2|f_1^{(0)};x')f_1^{(0)}(\boldsymbol{c}')\mathscr{C}'^2\boldsymbol{r}_{12}\cdot(\boldsymbol{r}'-\boldsymbol{r}_1)\mathrm{d}^6x'\right]\mathrm{d}^3c_1\,\mathrm{d}^3c_2\right\}\mathrm{d}^3r_2|_{\boldsymbol{r}_1=\boldsymbol{r}}.$$

(13.2-61f)

For the pressure tensor, **P**, we obtain *Newton's law*,

$$\mathbf{P} = p\mathbf{I} - 2\eta\mathbf{S} - \kappa\nabla\cdot\boldsymbol{v}, \quad (13.2\text{-}62)$$

where the hydrostatic pressure p is given by Eq. (13.2-18) and **S** is the rate-of-shear tensor. The *shear viscosity*, η, is given by

$$\eta = \eta_K + \eta_{\Phi,1} + \eta_{\Phi,2}, \quad (13.2\text{-}63a)$$

where η_K is the kinetic contribution,

$$\eta_K = \frac{2\theta}{15n}\int f_1^{(0)}(\boldsymbol{c})B(\mathscr{C})\mathscr{C}^4\,\mathrm{d}^3c, \quad (13.2\text{-}63b)$$

$\eta_{\Phi,1}$ is the contribution arising from the macroscopic gradients,

$$\eta_{\Phi,1} = -\frac{1}{20n}\int \frac{\varphi'(r_{12})}{r_{12}} \left\{ \iint \left[\int f_2'(x_1 x_2 | f_1^{(0)}; \boldsymbol{c}') f_1^{(0)}(\boldsymbol{c}') B(\mathscr{C}') \right.\right.$$
$$\left.\left. \times [(\boldsymbol{r}_{12} \cdot \mathscr{C}')^2 - \tfrac{1}{3} r_{12}^2 \mathscr{C}'^2] \mathrm{d}^3 c' \right] \mathrm{d}^3 c_1 \, \mathrm{d}^3 c_2 \right\} \mathrm{d}^3 r_2 |_{\boldsymbol{r}_1 = \boldsymbol{r}} \qquad (13.2\text{-}63c)$$

and $\eta_{\Phi,2}$ is the contribution arising from the spatial inhomogeneities over the range of the intermolecular force,

$$\eta_{\Phi,2} = \frac{1}{10}\left(\frac{m}{2\theta}\right)^{\frac{1}{2}} \int \frac{\varphi'(r_{12})}{r_{12}} \left\{ \iint \left[\int f_2'(x_1 x_2 | f_1^{(0)}; x') f_1^{(0)}(\boldsymbol{c}') \right.\right.$$
$$\left.\left. \times [\boldsymbol{r}_{12} \cdot (\boldsymbol{r}' - \boldsymbol{r}_1) \boldsymbol{r}_{12} \cdot \mathscr{C}' - \tfrac{1}{3} r_{12}^2 (\boldsymbol{r}' - \boldsymbol{r}_1) \cdot \mathscr{C}'] \mathrm{d}^6 x' \right] \mathrm{d}^3 c_1 \, \mathrm{d}^3 c_2 \right\} \mathrm{d}^3 r_2 |_{\boldsymbol{r}_1 = \boldsymbol{r}} . \qquad (13.2\text{-}63d)$$

Similarly, the *bulk viscosity*, κ, is given by

$$\kappa = \kappa_K + \kappa_{\Phi,1} + \kappa_{\Phi,2}, \qquad (13.2\text{-}64a)$$

where κ_K is the kinetic contribution,

$$\kappa_K = \frac{2\theta}{3n}\int f_1^{(0)}(\boldsymbol{c}) \Gamma(\mathscr{C}) \mathscr{C}^2 \, \mathrm{d}^3 c, \qquad (13.2\text{-}64b)$$

$\kappa_{\Phi,1}$ is the contribution arising from the macroscopic gradients,

$$\kappa_{\Phi,1} = -\frac{1}{6n}\int r_{12} \varphi'(r_{12})$$
$$\times \left\{ \iint \left[\int f_2'(x_1 x_2 | f_1^{(0)}; \boldsymbol{c}') f_1^{(0)}(\boldsymbol{c}') \Gamma(\mathscr{C}') \mathrm{d}^3 c' \right] \mathrm{d}^3 c_1 \, \mathrm{d}^3 c_2 \right\} \mathrm{d}^3 r_2 |_{\boldsymbol{r}_1 = \boldsymbol{r}} \qquad (13.2\text{-}64c)$$

and $\kappa_{\Phi,2}$ is the contribution arising from the spatial inhomogeneities over the range of the intermolecular force,

$$\kappa_{\Phi,2} = \frac{1}{9}\left(\frac{m}{2\theta}\right)^{\frac{1}{2}} \int r_{12} \varphi'(r_{12}) \left\{ \iint \left[\int f_2'(x_1 x_2 | f_1^{(0)}; x') f_1^{(0)}(\boldsymbol{c}') \right.\right.$$
$$\left.\left. \times \mathscr{C}' \cdot (\boldsymbol{r}' - \boldsymbol{r}_1) \mathrm{d}^6 x' \right] \mathrm{d}^3 c_1 \, \mathrm{d}^3 c_2 \right\} \mathrm{d}^3 r_2 |_{\boldsymbol{r}_1 = \boldsymbol{r}} . \qquad (13.2\text{-}64d)$$

When the expressions (13.2-60) and (13.2-62) are substituted into the general macroscopic conservation equations we obtain the Navier-Stokes equations of fluid dynamics with expressions for the transport coefficients given by Eqs. (13.2-61), (13.2-63) and (13.2-64).

13.3. The case of a moderately dense gas – The Choh-Uhlenbeck theory

The expressions for the transport coefficients that were obtained from the generalized Boltzmann equation in the previous section are still formal in two respects: it has been assumed that f_2 may be written as a time-independent functional of f_1 (Bogoliubov's *functional assumption*), and, moreover, the functional dependence of f_2 upon f_1 – if such exists – must still be prescribed. We now turn to a discussion of these two points.

The verification of Bogoliubov's functional assumption is the more difficult and, in fact, this point is far from settled at the time of this writing. The difficulties encountered here are intimately related to the occurrence of divergences in the cluster expansions to be discussed later. A discussion of this point will therefore be deferred until the next section.

Some progress has been made on the resolution of the second point mentioned above, namely the functional dependence of f_2 upon f_1. However, it is well to remember that the significance of the results cannot be truly evaluated until the question of the validity of Bogoliubov's functional assumption is settled.

With this *caveat* we turn our attention to the problem of explicitly evaluating the right-hand side of Eq. (13.1-5). Specifically, the task is one of writing each term in the expansion (13.1-7) in terms of F_1 alone. As we saw in Chapter 3, most configurations $(x_1 x_2)$ of two particles at time t turn out to give no contribution to $F_2^{(0)}$ and the same is true for all higher order terms $F_2^{(n)}$. Location of the portions of phase space which do contribute to $F_2^{(n)}$, and the evaluation of those contributions, is thus the task before us. In Chapter 3 we saw that the nontrivial portion of $F_2^{(0)}$ results from configurations in which the two particles under consideration suffered a collision in the interval $(0, t)$. On evaluating that term explicitly, we found that a modified Boltzmann equation resulted and that with a few additional and readily apparent approximations the Boltzmann equation could be obtained. Thus, as a next step, we turn our attention to $F_2^{(1)}$ with a view to locating the contributing portions of phase space and evaluating $F_2^{(1)}$ as an explicit functional of F_1. This problem is considerably more difficult than the corresponding one in the evaluation of $F_2^{(0)}$. It is not hard to trace the source of the difficulty. Whereas the two-body problem possesses an explicit solution whose employment rendered the evaluation of $F_2^{(0)}$ almost trivial, the three-body problem has no explicit solution, and even some of the qualitative features have only recently been elucidated. For example, it was shown only recently by T. J. Murphy –

see Appendix II of Cohen [1966a] – that three hard spheres may collide at most four times, and it is not yet known whether this is the case for all purely repulsive force laws. The dynamical problem for three particles has been studied in detail by Green [1956], Cohen [1966a], Sengers [1966a, 1967], and Dorfman and Cohen [1967].

As stated above, in the case of two particles, the phase points $(x_1 x_2)$ at time t were distinguished by whether or not an interaction between the two particles took place in the interval $(0, t)$. We must expect that, for three particles, the situation will be similar and that the phase points should be classified according to what events transpired among them in the time interval $(0, t)$. The simplest (and most prevalent) case is that of phases that result from no interaction, but, as this case is easily seen to yield no contribution to $F_2^{(1)}$, we may ignore it from the start. The situations that must be considered are illustrated in the diagrams of Fig. 13.1, where the actual positions (at time t) of the particles are indicated by the black dots, the direction of the corresponding velocities by the arrows. The lines represent particle trajectories and are labelled with the particle numbers. The circles represent

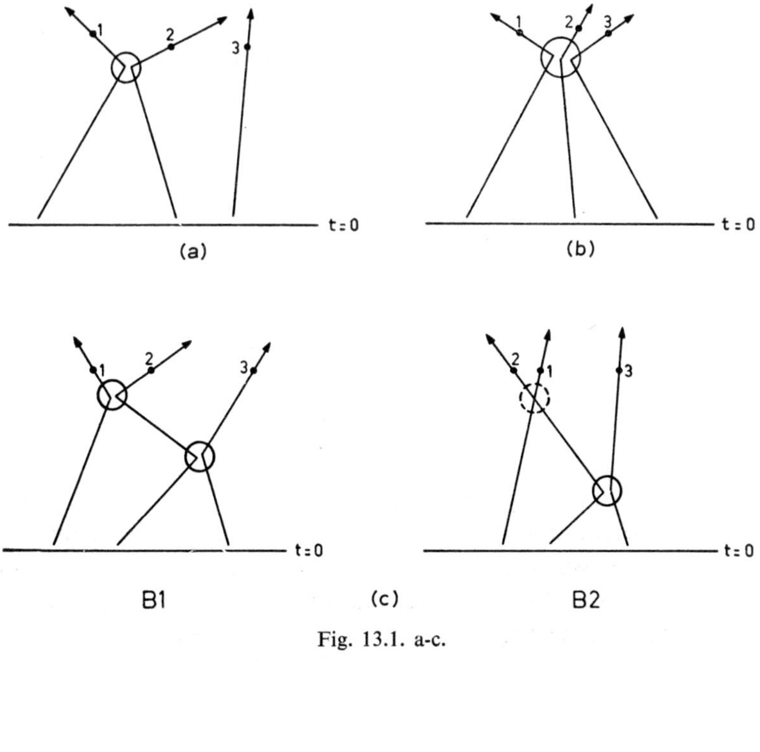

Fig. 13.1. a-c.

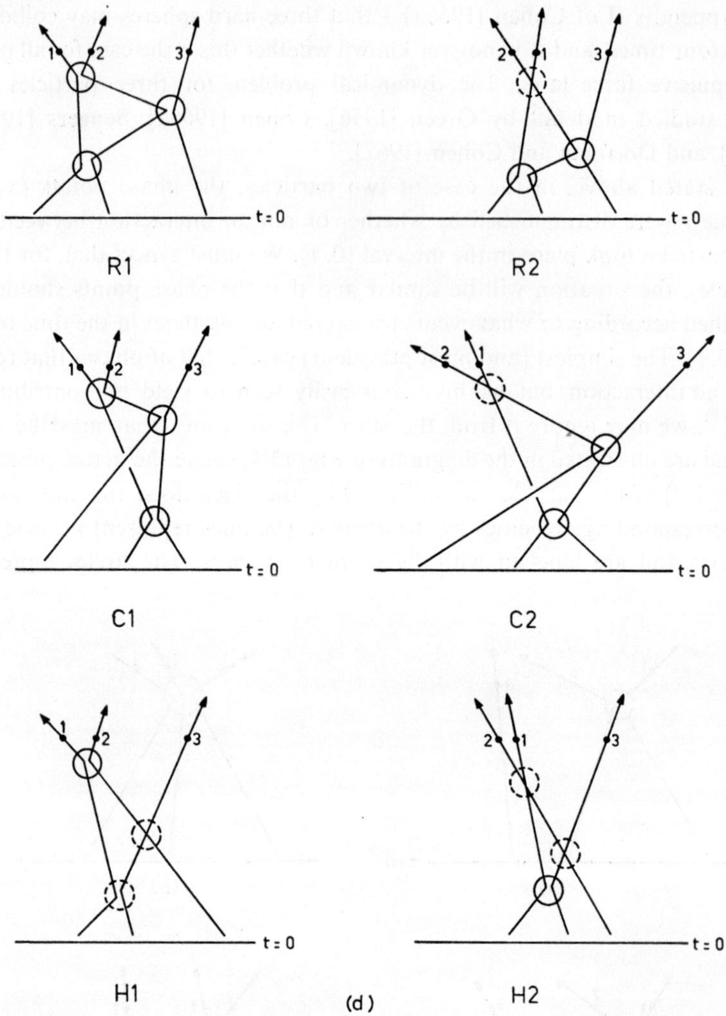

Fig. 13.1. d.

collisions between the particles whose trajectories are enclosed. The streaming operators S_t transform the present phases of the particles into the phases that the particles would have had in the past. Accordingly, the diagrams should be read from top to bottom and sequences of successive collisions will be ordered in this way.

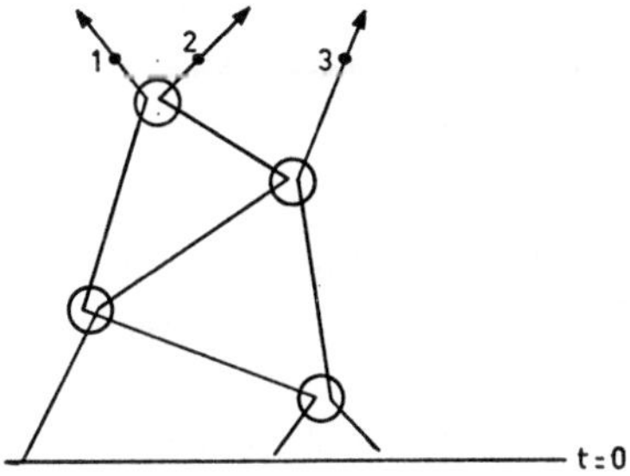

Fig. 13.1. Basic collision events of three particles. (a) One binary collision; (b) One triple collision; (c) Two binary collisions; (d) Three binary collisions; (e) Four binary collisions.

Phases resulting from actual sequences of collisions certainly contribute importantly to $F_2^{(1)}$. However, there are also other phases that give contributions, although their past histories do not contain only real physical interactions. Accordingly we introduce the following definitions.

A binary collision which is encountered when the actual trajectories of the particles are retraced into the past, is called a *real* collision. A binary collision which is encountered only when the effect of some other collision on the trajectories of the particles is disregarded, i.e., when the potential between two of the particles is turned off, is called a *hypothetical* collision. The latter will be distinguished by a bar; an illustration is given by comparing the two parts of Fig. 13.1(c). In addition, a distinction is made between *interacting* and *non-interacting* collisions: In a non-interacting collision the trajectories cross, but are continued without deflection – the interaction potential is ignored. Non-interacting collisions are indicated by a dotted circle and provided with the subscript "ni". Evidently, from these definitions it follows that any collision encountered when the trajectories are retraced from a non-interacting collision is a hypothetical collision. Thus, the possible dynamical events between time 0 and time t in which the last event was a 1–2 collision may be cataloged as follows (always reading the corresponding diagrams from the top to the bottom):

(a) One binary collision – see Fig. 13.1(a);

(b) A (genuine) triple collision – see Fig. 13.1(b).

Note that a triple collision cannot be followed by binary collisions, unless the latter take place within a time interval t_{coll} following the triple collision; in that case, the whole sequence of events is considered as an "extended" triple collision. This assumption is certainly justified if the intermolecular potential is strongly repulsive.

(c) Two successive binary collisions – the binary collision (12) is followed by another binary collision. There are two basic types – see Fig. 13.1(c),

B1 – a sequence of the first kind – (12)(23),
B2 – a sequence of the second kind – $(12)(\overline{23})$.

(d) Three successive binary collisions – the binary collision (12) is followed by two binary collisions. There are six basic types – see Fig. 13.1(d),

R1 – recollision of the first kind – (12)(23)(12),
R2 – recollision of the second kind – $(12)_{\text{ni}}(\overline{23})(\overline{12})$,
C1 – cyclic collision of the first kind – (12)(23)(31),
C2 – cyclic collision of the second kind – $(12)_{\text{ni}}(\overline{23})(\overline{31})$,
H1 – hypothetical cyclic collision of the first kind – $(12)(23)_{\text{ni}}(\overline{31})$,
H2 – hypothetical cyclic collision of the second kind – $(12)_{\text{ni}}(\overline{23})_{\text{ni}}(\overline{31})$.

The difference between recollisions and cyclic collisions is that, in the former, the same pair of particles is involved in the first and the third collision while, in the latter, different pairs are involved in these collisions; the ordering is essential to the argument. The dynamical histories under (d) and those obtained by permutations of the particles are the central ones in this theory. In general, it is possible that the dynamical history of a particular phase point can be represented by more than one basic dynamical history; an example of this is given by Sengers [1966a].

(e) Four or more successive binary collisions – see Fig. 13.1(e). In the dynamical histories R1 and C1 under (d) a fourth real collision may occur after the last event. Four is also the maximum number of real successive binary collisions between three rigid spheres.

The above classification may be used to discuss the long-time behavior of the coefficient $F_2^{(1)}$ in the expansion (13.1-7) of F_2, and the decay of initial correlations with time. It is convenient to write $F_2^{(1)}$ as the sum of two terms, one of which contains the contribution of the initial state,

$$F_2^{(1)}(x_1 x_2; t) = \int \mathfrak{T}_t^{(3)}(x_1 x_2 | x_3) \prod_{i=1}^{3} F_1(x_i; t) \, \mathrm{d}^6 x_3 + C_2^{(1)}(x_1 x_2; t), \qquad (13.3\text{-}1)$$

where the operator $\mathfrak{T}_t^{(3)}(x_1 x_2 | x_3)$ is obtained from the operator $\tilde{\mathfrak{T}}_t^{(3)}$ by setting all $a_s(x_1 \ldots x_s; 0)$ equal to unity. Thus, Eq. (13.3-1) defines the quantity $C_2^{(1)}(x_1 x_2;\ t)$.

As has been shown by Cohen [1966a], the structure of the operator $\mathfrak{T}_t^{(3)}(x_1 x_2|x_3)$ is such that, among the dynamical events listed above only a (genuine) triple collision or a sequence of three or more successive binary collisions contributes to $F_2^{(1)}$. Moreover, the structure of the term $C_2^{(1)}$ is such that the only contributions come from dynamical histories which involve a sequence of three or more successive binary collisions and in which, apart from the (12)-collision at time t, another binary collision occurs at time 0. Dorfman and Cohen [1967] have shown that, for $t \gg t_{\text{coll}}$,

$$C_2^{(1)}(x_1\, x_2;\, t) = \mathrm{O}((t_{\text{coll}}/t)^2), \tag{13.3-2}$$

$$\int \mathfrak{T}_t^{(3)}(x_1\, x_2|x_3) \prod_{i=1}^{3} F_1(x_i;\, t)\, \mathrm{d}^6 x_3 = \left[1+\mathrm{O}\left(\frac{t_{\text{coll}}}{t}\right)\right] \int \mathfrak{T}_\infty^{(3)}(x_1\, x_2|x_3) \prod_{i=1}^{3} F_1(x_i;\, t)\, \mathrm{d}^6 x_3, \tag{13.3-3}$$

where $\mathfrak{T}_\infty^{(3)}\ (x_1 x_2|x_3)$ represents the limiting form of the operator $\mathfrak{T}_t^{(3)}(x_1 x_2|x_3)$ as $t \to \infty$,

$$\mathfrak{T}_\infty^{(3)}(x_1\, x_2|x_3) = \mathfrak{S}_\infty^{(3)}(x_1\, x_2|x_3) - \mathfrak{S}_\infty^{(2)}(x_1\, x_2)\, \mathfrak{S}_\infty^{(2)}(x_1\, x_3) - \mathfrak{S}_\infty^{(2)}(x_1\, x_2)\, \mathfrak{S}_\infty^{(2)}(x_2\, x_3) + \mathfrak{S}_\infty^{(2)}(x_1\, x_2). \tag{13.3-4}$$

Now, suppose that the expansion (13.1-7) may be truncated after the first two terms. Then it follows from the foregoing analysis that, if we assume $t \gg t_{\text{coll}}$ and neglect terms of the order t_{coll}/t, we may write

$$\begin{aligned} F_2(x_1 x_2;\, t) &= F_2(x_1 x_2|F_1) \\ &= F_2^{(0)}(x_1\, x_2|F_1) + \frac{n}{1!}\, F_2^{(1)}(x_1\, x_2|F_1) \\ &= \mathfrak{S}_\infty^{(2)}(x_1\, x_2) \prod_{i=1}^{2} F_1(x_i;\, t) + \frac{n}{1!} \int \mathfrak{T}_\infty^{(3)}(x_1\, x_2|x_3) \prod_{i=1}^{3} F_1(x_i;\, t)\, \mathrm{d}^6 x_3, \end{aligned} \tag{13.3-5}$$

and the theory of the previous section is applicable. Thus, we obtain the results first derived by Choh and Uhlenbeck [1958] who calculated the first corrections to the density-independent transport coefficients of the Chapman-Enskog theory due to collisional transfer and three-body interactions.

The formulae of the previous section are most conveniently evaluated if the density dependence of all the quantities involved is written out explicitly. Thus, in terms of the velocity distribution functions f_1 and f_2, we write the functional $\Phi(x|f_1)$ as a sum of two terms,

$$\Phi(x|f_1) = \Phi^{(0)}(x|f_1)+n\Phi^{(1)}(x|f_1), \tag{13.3-6}$$

where

$$\Phi^{(0)}(x|f_1) = \int \Theta_{12} f_2^{(0)}(x_1 x_2|f_1)\, d^6x_2|_{x_1=x}, \tag{13.3-7}$$

$$\Phi^{(1)}(x|f_1) = \int \Theta_{12} f_2^{(1)}(x_1 x_2|f_1)\, d^6x_2|_{x_1=x}. \tag{13.3-8}$$

Similarly, we write the coefficients $\boldsymbol{A}$, $\mathbf{B}$ and Γ of the perturbation function ϕ, cf. Eq. (13.2-57), as a sum of two terms,

$$\boldsymbol{A} = \boldsymbol{A}^{(0)}+n\boldsymbol{A}^{(1)}, \tag{13.3-9}$$

$$\mathbf{B} = \mathbf{B}^{(0)}+n\mathbf{B}^{(1)}, \tag{13.3-10}$$

$$\Gamma = \Gamma^{(0)}+n\Gamma^{(1)}, \tag{13.3-11}$$

and, likewise, the function ϕ itself as

$$\phi = \phi^{(0)}+n\phi^{(1)}, \tag{13.3-12a}$$

where, for $i = 0, 1$,

$$\phi^{(i)} = -\frac{1}{n}\boldsymbol{A}^{(i)}\cdot\nabla\log\theta-\frac{1}{n}\mathbf{B}^{(i)} : \nabla\boldsymbol{v}-\frac{1}{n}\Gamma^{(i)}\nabla\cdot\boldsymbol{v}. \tag{13.3-12b}$$

From the explicit form of the functionals $f_2^{(0)}(x_1 x_2|f_1)$ and $f_2^{(1)}(x_1 x_2|f_1)$ it is immediately obvious that the functional derivatives $\Phi^{(0)\prime}(x|f_1^{(0)};\, \boldsymbol{c}')$ and $\Phi^{(0)\prime}(x|f_1^{(0)};\, x')$ are linear in $f_1^{(0)}$ and, hence, proportional to the density, while the functional derivatives $\Phi^{(1)\prime}(x|f_1^{(0)};\, \boldsymbol{c}')$ and $\Phi^{(1)\prime}(x|f_1^{(0)};\, x')$ are quadratic in $f_1^{(0)}$ and, hence, proportional to the density squared. Thus if we decompose the integral operator I, defined in Eq. (13.2-43), into a sum of two integral operators,

$$I = I_2+nI_3, \tag{13.3-13a}$$

and define I_2 by the relation

$$\int \Phi^{(0)\prime}(x|f_1^{(0)};\, \boldsymbol{c}')\, f_1^{(0)}(\boldsymbol{c}')\, F(\boldsymbol{c}')\, d^3c' = -n^2 I_2(F), \tag{13.3-13b}$$

and I_3 by the relation

$$\int \Phi^{(1)\prime}(x|f_1^{(0)}; \boldsymbol{c}')\, f_1^{(0)}(\boldsymbol{c}')\, F(\boldsymbol{c}')\, \mathrm{d}^3c' = -n^3\, I_3(F), \qquad (13.3\text{-}13c)$$

then – for any function F of the molecular velocity – F, $I_2(F)$, and $I_3(F)$ are all of the same order in the density. The subscripts 2 and 3 on the integral operators refer to the fact that I_2 and I_3 involve two-particle and three-particle interactions, respectively.

The integral operator I_2 may be further evaluated. First, from the definition of the functional $\Phi^{(0)}(x|f_1)$ it is readily verified that the functional derivative $\Phi^{(0)\prime}(x|f_1^{(0)}; \boldsymbol{c}')$ is given by

$$\Phi^{(0)\prime}(x|f_1^{(0)}; \boldsymbol{c}')$$
$$= \int \Theta_{12}\, \mathfrak{S}_\infty^{(2)}(x_1\, x_2)\, \{\delta(\boldsymbol{c}'-\boldsymbol{c}_1)\, f_1^{(0)}(\boldsymbol{c}_2) + \delta(\boldsymbol{c}'-\boldsymbol{c}_2)\, f_1^{(0)}(\boldsymbol{c}_1)\}\, \mathrm{d}^6x_2|_{x_1=x}, \qquad (13.3\text{-}14)$$

where it is understood that, under the integral sign, the local parameters in each $f_1^{(0)}$ are taken at the *same* position, $\boldsymbol{r}_1$, and that the operator $\mathfrak{S}_\infty^{(2)}(x_1\, x_2)$ acts on the velocity variables only. Then, the integral over $\boldsymbol{c}'$ in Eq. (13.3-13b) may be evaluated and we find the relation

$$\int \Theta_{12}\, \mathfrak{S}_\infty^{(2)}(x_1\, x_2)\, f_1^{(0)}(\boldsymbol{c}_1)\, f_1^{(0)}(\boldsymbol{c}_2)\, \{F(\boldsymbol{c}_1)+F(\boldsymbol{c}_2)\}\, \mathrm{d}^6x_2|_{x_1=x} = -n^2\, I_2(F), \qquad (13.3\text{-}15)$$

defining I_2. Now, the integral in the left member may be treated by the procedure which was used in Section 3.5 to derive the binary collision operator. The procedure is even simpler here since all the quantities under the integral sign are evaluated at the *same* position, $\boldsymbol{r}_1$. As a result we find that I_2 is defined by the relation

$$I_2(F) = \frac{1}{n^2} \iiint f_1^{(0)}(\boldsymbol{c}_1)\, f_1^{(0)}(\boldsymbol{c}_2)\, \{F(\boldsymbol{c}_1)+F(\boldsymbol{c}_2)-F(\boldsymbol{c}_1')-F(\boldsymbol{c}_2')\}$$
$$\times\, g b\, \mathrm{d}b\, \mathrm{d}\varepsilon\, \mathrm{d}^3c_2, \qquad (13.3\text{-}16)$$

where $\boldsymbol{c}_1'$, $\boldsymbol{c}_2'$ and $\boldsymbol{c}_1$, $\boldsymbol{c}_2$ are the velocities of the particles "before" and "after" the collision, respectively, and $g = |\boldsymbol{c}_1 - \boldsymbol{c}_2|$. Comparing this result with Eq. (4.6-4) we conclude that I_2 is identical with the *linearized Boltzmann* (*binary*) *collision operator*.

The integral operator I_3 may be treated by similar methods. From the definition of the functional $\Phi^{(1)}(x|f_1)$ it follows that the derivative $\Phi^{(1)\prime}(x|f_1^{(0)}; \boldsymbol{c}')$ is given by

$$\Phi^{(1)\prime}(x|f_1^{(0)}; \boldsymbol{c}')$$
$$= \int \Theta_{12} \int \mathfrak{T}_\infty^{(3)}(x_1 x_2|x_3) \sum_{(1,2,3)} \delta(\boldsymbol{c}'-\boldsymbol{c}_1) f_1^{(0)}(\boldsymbol{c}_2) f_1^{(0)}(\boldsymbol{c}_3) \mathrm{d}^6 x_3 \mathrm{d}^6 x_2|_{x_1=x}, \tag{13.3-17}$$

where $\sum_{(1,2,3)}$ indicates that the sum must be taken over all permutations of the indices 1, 2, and 3 of the velocity variables and where, as in Eq. (13.3-14), it is understood that, under the integral sign, the local parameters in each $f_1^{(0)}$ are taken at the same position, $\boldsymbol{r}_1$. Also, the operator $\mathfrak{T}_\infty^{(3)}(x_1 x_2|x_3)$ acts only on the velocity variables. Then, the integral over $\boldsymbol{c}'$ in Eq. (13.3-13c) may be evaluated and we obtain the following relation defining I_3,

$$I_3(F) = -\frac{1}{n^3} \int \Theta_{12} \int \mathfrak{T}_\infty^{(3)}(x_1 x_2|x_3) f_1^{(0)}(\boldsymbol{c}_1) f_1^{(0)}(\boldsymbol{c}_2) f_1^{(0)}(\boldsymbol{c}_3)$$
$$\times \{F(\boldsymbol{c}_1)+F(\boldsymbol{c}_2)+F(\boldsymbol{c}_3)\} \mathrm{d}^6 x_3 \mathrm{d}^6 x_2|_{x_1=x}. \tag{13.3-18}$$

By analogy, I_3 may be called the *linearized ternary collision operator.*

Thus, collecting terms proportional to the density in Eqs. (13.2-50), (13.2-51) and (13.2-52), we arrive at the following set of integral equations for the vector function $\boldsymbol{A}^{(0)} = A^{(0)}(\mathscr{C})\boldsymbol{C}$, the tensor function $\mathbf{B}^{(0)} = B^{(0)}(\mathscr{C})(\mathscr{C}\mathscr{C}-\frac{1}{3}\mathscr{C}^2\mathbf{I})$ and the scalar function $\Gamma^{(0)} = \Gamma^{(0)}(\mathscr{C})$,

$$n I_2(\boldsymbol{A}^{(0)}) = f_1^{(0)}(\mathscr{C}^2-\tfrac{5}{2})\boldsymbol{C}, \tag{13.3-19}$$

$$n I_2(\mathbf{B}^{(0)}) = 2f_1^{(0)}(\mathscr{C}\mathscr{C}-\tfrac{1}{3}\mathscr{C}^2\mathbf{I}), \tag{13.3-20}$$

$$n I_2(\Gamma^{(0)}) = 0, \tag{13.3-21}$$

together with the constraints

$$\int f_1^{(0)} A^{(0)}(\mathscr{C})\mathscr{C}^2 \mathrm{d}^3 c = 0, \tag{13.3-22}$$

$$\int f_1^{(0)} \Gamma^{(0)}(\mathscr{C}) \mathrm{d}^3 c = 0, \tag{13.3-23a}$$

$$\int f_1^{(0)} \Gamma^{(0)}(\mathscr{C})\mathscr{C}^2 \mathrm{d}^3 c = 0. \tag{13.3-23b}$$

From Eqs. (13.3-21) and (13.3-23) we immediately conclude that the only solution to Eq. (13.3-21) which is not one of the summational invariants, is the zero solution, so $\Gamma^{(0)} = 0$. Since Eqs. (13.3-19), (13.3-20) and (13.3-22) are identical with Eqs. (5.4-15), (5.4-16) and (5.4-30), respectively, we have, to the present order of approximation, exact agreement with the Chapman-Enskog theory of dilute gases.

To the next order in the density, Eqs. (13.2-50), (13.2-51) and (13.2-52) yield a set of integral equations for the vector functions $\boldsymbol{A}^{(1)} = A^{(1)}(\mathscr{C})\boldsymbol{C}$, the tensor function $\mathbf{B}^{(1)} = B^{(1)}(\mathscr{C})(\mathscr{C}\mathscr{C}-\frac{1}{3}\mathscr{C}^2\boldsymbol{I})$ and the scalar function $\Gamma^{(1)} = \Gamma^{(1)}(\mathscr{C})$, viz.,

$$n^2 I_2(\boldsymbol{A}^{(1)}) = \tfrac{1}{2}n[\beta_1(\theta)+\theta\,\beta_1'(\theta)]\, f_1^{(0)}\boldsymbol{C}-n^2 I_3(\boldsymbol{A}^{(0)})$$

$$-\int \Phi^{(0)\prime}(x|f_1^{(0)};\, x')(\boldsymbol{r}'-\boldsymbol{r})(\mathscr{C}'^2-\tfrac{3}{2})\, f_1^{(0)}(\boldsymbol{c}')\,\mathrm{d}^6x', \qquad (13.3\text{-}24)$$

$$n^2 I_2(\mathbf{B}^{(1)}) = -n^2 I_3(\mathbf{B}^{(0)})$$

$$-2\left(\frac{m}{2\theta}\right)^{\frac{1}{2}}\int \Phi^{(0)\prime}(x|f_1^{(0)};\, x')\,\mathbf{T}\, f_1^{(0)}(\boldsymbol{c}')\,\mathrm{d}^6x', \qquad (13.3\text{-}25)$$

$$n^2 I_2(\Gamma^{(1)}) = \tfrac{2}{3}n[\tfrac{1}{2}\beta_1(\theta)+\tfrac{7}{6}\theta\,\beta_1'(\theta)+\tfrac{1}{3}\theta^2\,\beta_1''(\theta)]\, f_1^{(0)}(\mathscr{C}^2-\tfrac{3}{2})$$

$$-\tfrac{2}{3}\left(\frac{m}{2\theta}\right)^{\frac{1}{2}}\int \Phi^{(0)\prime}(x|f_1^{(0)};\, x')\,\mathscr{C}'\cdot(\boldsymbol{r}'-\boldsymbol{r})\, f_1^{(0)}(\boldsymbol{c}')\,\mathrm{d}^6x', \quad (13.3\text{-}26)$$

together with the constraints

$$\int f_1^{(0)} A^{(1)}(\mathscr{C})\mathscr{C}^2\,\mathrm{d}^3c = 0, \qquad (13.3\text{-}27)$$

$$\int f_1^{(0)}\,\Gamma^{(1)}(\mathscr{C})\,\mathrm{d}^3c = 0, \qquad (13.3\text{-}28a)$$

$$\int f_1^{(0)}\,\Gamma^{(1)}(\mathscr{C})\mathscr{C}^2\,\mathrm{d}^3c = 0. \qquad (13.3\text{-}28b)$$

The integrals occurring in the right member of Eqs. (13.3-24), (13.3-25) and (13.3-26) may be simplified by carrying out the integration over the primed variables. Consider first the integral of Eq. (13.3-24). Using the explicit form of the functional $f_2^{(0)}(x_1 x_2|f_1)$ one readily verifies the relation

$$-\int \Phi^{(0)'}(x|f_1^{(0)};x')(\mathscr{C}'^2-\tfrac{3}{2})(\boldsymbol{r}'-\boldsymbol{r})\,f_1^{(0)}(\boldsymbol{c}')\,\mathrm{d}^6x'$$

$$= -\iint \Theta_{12}\,\mathfrak{S}_\infty^{(2)}(x_1x_2)\,f_1^{(0)}(\boldsymbol{c}_1)\,f_1^{(0)}(\boldsymbol{c}_2)$$

$$\times(\mathscr{C}'^2-\tfrac{3}{2})(\boldsymbol{r}'-\boldsymbol{r})\{\delta(x'-x_1)+\delta(x'-x_2)\}\,\mathrm{d}^6x_2\,\mathrm{d}^6x'. \qquad (13.3\text{-}29)$$

If we decompose the sum of the two δ-functions,

$$\delta(x'-x_1)+\delta(x'-x_2) = \delta(\boldsymbol{c}'-\boldsymbol{c}_1)\delta(\boldsymbol{r}'-\boldsymbol{r}_1)+\delta(\boldsymbol{c}'-\boldsymbol{c}_2)\delta(\boldsymbol{r}'-\boldsymbol{r}_2)$$

$$= \tfrac{1}{2}\{\delta(\boldsymbol{c}'-\boldsymbol{c}_1)+\delta(\boldsymbol{c}'-\boldsymbol{c}_2)\}\{\delta(\boldsymbol{r}'-\boldsymbol{r}_1)+\delta(\boldsymbol{r}'-\boldsymbol{r}_2)\}$$

$$+\tfrac{1}{2}\{\delta(\boldsymbol{c}'-\boldsymbol{c}_1)-\delta(\boldsymbol{c}'-\boldsymbol{c}_2)\}\{\delta(\boldsymbol{r}'-\boldsymbol{r}_1)-\delta(\boldsymbol{r}'-\boldsymbol{r}_2)\}$$

$$(13.3\text{-}30)$$

and perform the integration over $x' = (\boldsymbol{r}', \boldsymbol{c}')$, we obtain from Eq. (13.3-29)

$$-\int \Phi^{(0)'}(x|f_1^{(0)};x')(\mathscr{C}'^2-\tfrac{3}{2})(\boldsymbol{r}'-\boldsymbol{r}_1)\,f_1^{(0)}(\boldsymbol{c}')\,\mathrm{d}^6x'$$

$$= -\int \Theta_{12}\,\mathfrak{S}_\infty^{(2)}(x_1x_2)\,f_1^{(0)}(\boldsymbol{c}_1)\,f_1^{(0)}(\boldsymbol{c}_2)$$

$$\times\{\tfrac{1}{2}(\mathscr{C}_1^2+\mathscr{C}_2^2-3)(\boldsymbol{r}_2-\boldsymbol{r}_1)-\tfrac{1}{2}(\mathscr{C}_1^2-\mathscr{C}_2^2)(\boldsymbol{r}_2-\boldsymbol{r}_1)\}\,\mathrm{d}^6x_2. \qquad (13.3\text{-}31)$$

Now, we recall from the discussion of Section 3.5 that $\mathfrak{S}_\infty^{(2)}(x_1x_2)$ – which, in the present case, operates on the velocity variables only – transforms the actual momenta $\boldsymbol{p}_1$ and $\boldsymbol{p}_2$ into $\boldsymbol{P}_1^{(2)}$ and $\boldsymbol{P}_2^{(2)}$, respectively, where $\boldsymbol{P}_1^{(2)}$ and $\boldsymbol{P}_2^{(2)}$ are the (constant) initial momenta of the particles 1 and 2 that lead to the phase (x_1x_2) at time t. Since the total energy is invariant under such a transformation it follows that we must have the identity

$$\frac{1}{2m}(P_1^{(2)^2}+P_2^{(2)^2}) = \frac{1}{2m}(p_1^2+p_2^2)+\varphi(r_{12}) \qquad (13.3\text{-}32a)$$

or, in terms of the dimensionless velocity variable $\mathscr{C}$,

$$\mathfrak{S}_\infty^{(2)}(x_1x_2)(\mathscr{C}_1^2+\mathscr{C}_2^2) = (\mathscr{C}_1^2+\mathscr{C}_2^2)+\varphi(r_{12})/\theta, \qquad (13.3\text{-}32b)$$

where $r_{12} = |\boldsymbol{r}_2-\boldsymbol{r}_1|$. Hence, Eq. (13.3-31) is equivalent to

$$-\int \Phi^{(0)\prime}(x|f_1^{(0)};x')(\mathscr{C}'^2-\tfrac{3}{2})(\boldsymbol{r}'-\boldsymbol{r})\,f_1^{(0)}(\boldsymbol{c}')\,\mathrm{d}^6x'$$

$$=-\tfrac{1}{2}\int \Theta_{12}\,f_1^{(0)}(\boldsymbol{c}_1)\,f_1^{(0)}(\boldsymbol{c}_2)\,\mathrm{e}^{-\varphi(r_{12})/\theta}\left(\mathscr{C}_1^2+\mathscr{C}_2^2+\frac{\varphi(r_{12})}{\theta}-3\right)(\boldsymbol{r}_2-\boldsymbol{r}_1)\,\mathrm{d}^6x_2$$

$$+\tfrac{1}{2}\int\!\!\int \Theta_{12}\,\mathfrak{S}_\infty^{(2)}(x_1\,x_2)\,f_1^{(0)}(\boldsymbol{c}_1)\,f_1^{(0)}(\boldsymbol{c}_2)(\mathscr{C}_1^2-\mathscr{C}_2^2)(\boldsymbol{r}_2-\boldsymbol{r}_1)\,\mathrm{d}^6x_2. \quad (13.3\text{-}33)$$

Substituting the molecular interaction operator Θ_{12} and leaving out vanishing integrals we see that the first term of the right member of Eq. (13.3-33) may, in turn, be rewritten as

$$-\tfrac{1}{2}\int \Theta_{12}\,f_1^{(0)}(\boldsymbol{c}_1)\,f_1^{(0)}(\boldsymbol{c}_2)\,\mathrm{e}^{-\varphi(r_{12})/\theta}\left(\mathscr{C}_1^2+\mathscr{C}_2^2+\frac{\varphi(r_{12})}{\theta}-3\right)$$

$$\times(\boldsymbol{r}_2-\boldsymbol{r}_1)\,\mathrm{d}^6x_2|_{x_1=x}$$

$$=-\frac{1}{2\theta}\int\!\!\int f_1^{(0)}(\boldsymbol{c}_2)\,\mathrm{e}^{-\varphi(r_{12})/\theta}\,\frac{\varphi'(r_{12})}{r_{12}}\left(\mathscr{C}_1^2+\mathscr{C}_2^2+\frac{\varphi(r_{12})}{\theta}-4\right)$$

$$\times\boldsymbol{r}_{12}\,\boldsymbol{r}_{12}\,\mathrm{d}^3r_{12}\,\mathrm{d}^3c_2|_{x_1=x}\cdot f_1^{(0)}(\boldsymbol{c})\,\boldsymbol{C}$$

$$=-\frac{1}{6\theta}\int \mathrm{e}^{-\varphi(r_{12})/\theta}\varphi'(r_{12})r_{12}\left\{\int f_1^{(0)}(\boldsymbol{c}_2)\,\mathrm{d}^3c_2\left(\mathscr{C}^2+\frac{\varphi(r_{12})}{\theta}-4\right)\right.$$

$$\left.+\int f_1^{(0)}(\boldsymbol{c}_2)\,\mathscr{C}_2^2\,\mathrm{d}^3c_2\right\}\mathrm{d}^3r_{12}|_{\boldsymbol{r}_1=\boldsymbol{r}}f_1^{(0)}(\boldsymbol{c})\,\boldsymbol{C}$$

$$=-\frac{n}{6\theta}\left[(\mathscr{C}^2-\tfrac{5}{2})\int \mathrm{e}^{-\varphi(r_{12})/\theta}\varphi'(r_{12})\,r_{12}\,\mathrm{d}^3r_{12}\right.$$

$$\left.+\frac{1}{\theta}\int \mathrm{e}^{-\varphi(r_{12})/\theta}\varphi'(r_{12})\,\varphi(r_{12})\,r_{12}\,\mathrm{d}^3r_{12}\right]_{\boldsymbol{r}_1=\boldsymbol{r}} f_1^{(0)}(\boldsymbol{c})\,\boldsymbol{C}. \quad (13.3\text{-}34)$$

By integration by parts, using the definition (13.2-22) of the first virial coefficient, $\beta_1(\theta)$, one may verify that Eq. (13.3-34) yields the relation

$$-\tfrac{1}{2}\int \Theta_{12}\,f_1^{(0)}(\boldsymbol{c}_1)\,f_1^{(0)}(\boldsymbol{c}_2)\,\mathrm{e}^{-\varphi(r_{12})/\theta}\left(\mathscr{C}_1^2+\mathscr{C}_2^2+\frac{\varphi(r_{12})}{\theta}-3\right)$$

$$\times(\boldsymbol{r}_2-\boldsymbol{r}_1)\,\mathrm{d}^6x_2|_{x_1=x}$$

$$=-\tfrac{1}{2}n\left[(\mathscr{C}^2-\tfrac{5}{2})\beta_1(\theta)+\frac{\mathrm{d}}{\mathrm{d}\theta}\,\theta\,\beta_1(\theta)\right]f_1^{(0)}(\boldsymbol{c})\,\boldsymbol{C}. \quad (13.3\text{-}35)$$

Thus, combining Eqs. (13.3-33) and (13.3-35) we find that the integral occurring in the right member of Eq. (13.3-24) can indeed be simplified and that the result can be written as

$$-\int \Phi^{(0)\prime}(x|f_1^{(0)};x')(\mathscr{C}'^2-\tfrac{3}{2})(\boldsymbol{r}'-\boldsymbol{r})\,f_1^{(0)}(\boldsymbol{c}')\,\mathrm{d}^6x'|_{x_1=x}$$

$$= -n\left[\tfrac{1}{2}(\mathscr{C}^2-\tfrac{5}{2})\beta_1(\theta)+\frac{1}{2}\frac{\mathrm{d}}{\mathrm{d}\theta}\,\theta\,\beta_1(\theta)+\bar{b}(\mathscr{C})\right]f_1^{(0)}(\boldsymbol{c})\,\boldsymbol{C}, \qquad (13.3\text{-}36)$$

where the quantity $\bar{b}$ is defined by the relation

$$\tfrac{1}{2}\int \Theta_{12}\,\mathfrak{S}_\infty^{(2)}(x_1\,x_2)\,f_1^{(0)}(\boldsymbol{c}_1)\,f_1^{(0)}(\boldsymbol{c}_2)(\mathscr{C}_1^2-\mathscr{C}_2^2)(\boldsymbol{r}_2-\boldsymbol{r}_1)\,\mathrm{d}^6x_2|_{x_1=x}$$

$$= -n\,\bar{b}(\mathscr{C})\,f_1^{(0)}(\boldsymbol{c})\,\boldsymbol{C}. \qquad (13.3\text{-}37)$$

Similarly, one may show that the integrals occurring in the right members of Eqs. (13.3-25) and (13.3-26) can be simplified and that the results are given by

$$-2\left(\frac{m}{2\theta}\right)^{\frac{1}{2}}\int \Phi^{(0)\prime}(x|f_1^{(0)};x')\,\mathsf{T}\,f_1^{(0)}(\boldsymbol{c}')\,\mathrm{d}^6x'$$

$$= -n[\beta_1(\theta)+\bar{C}_1(\mathscr{C})]\,f_1^{(0)}(\boldsymbol{c})(\boldsymbol{\mathscr{C}}\boldsymbol{\mathscr{C}}-\tfrac{1}{3}\mathscr{C}^2\boldsymbol{I}), \qquad (13.3\text{-}38)$$

where the quantity $\bar{C}_1$ is defined by the relation

$$\left(\frac{m}{2\theta}\right)^{\frac{1}{2}}\int \Theta_{12}\,\mathfrak{S}_\infty^{(2)}(x_1\,x_2)\,f_1^{(0)}(\boldsymbol{c}_1)\,f_1^{(0)}(\boldsymbol{c}_2)\,\{\tfrac{1}{2}[(\boldsymbol{\mathscr{C}}_1-\boldsymbol{\mathscr{C}}_2)(\boldsymbol{r}_2-\boldsymbol{r}_1)$$

$$+(\boldsymbol{r}_2-\boldsymbol{r}_1)(\boldsymbol{\mathscr{C}}_1-\boldsymbol{\mathscr{C}}_2)]-\tfrac{1}{3}(\boldsymbol{\mathscr{C}}_1-\boldsymbol{\mathscr{C}}_2)\cdot(\boldsymbol{r}_2-\boldsymbol{r}_1)\boldsymbol{I}\}\mathrm{d}^6x_2|_{x_1=x}$$

$$= -n\bar{C}_1(\mathscr{C})f_1^{(0)}(\boldsymbol{c})(\boldsymbol{\mathscr{C}}\boldsymbol{\mathscr{C}}-\tfrac{1}{3}\mathscr{C}^2\boldsymbol{I}), \qquad (13.3\text{-}39)$$

and

$$-\tfrac{2}{3}\left(\frac{m}{2\theta}\right)^{\frac{1}{2}}\int \Phi^{(0)\prime}(x|f_1^{(0)};x')\boldsymbol{\mathscr{C}}'\cdot(\boldsymbol{r}'-\boldsymbol{r})\,f_1^{(0)}(\boldsymbol{c}')\,\mathrm{d}^6x'$$

$$= -n[\tfrac{1}{3}\beta_1(\theta)(\mathscr{C}_1^2-\tfrac{3}{2})+\bar{C}_2(\mathscr{C})]f_1^{(0)}(\boldsymbol{c}), \qquad (13.3\text{-}40)$$

where the quantity $\bar{C}_2$ is defined by the relation

$$\tfrac{1}{3}\left(\frac{m}{2\theta}\right)^{\frac{1}{2}}\int \Theta_{12}\,\mathfrak{S}_\infty^{(2)}(x_1\,x_2)\,f_1^{(0)}(\boldsymbol{c}_1)\,f_1^{(0)}(\boldsymbol{c}_2)(\boldsymbol{\mathscr{C}}_1-\boldsymbol{\mathscr{C}}_2)\cdot(\boldsymbol{r}_2-\boldsymbol{r}_1)\,\mathrm{d}^6x_2|_{x_1=x}$$

$$= -n\bar{C}_2(\mathscr{C})\,f_1^{(0)}(\boldsymbol{c}). \qquad (13.3\text{-}41)$$

Finally, when we substitute these results into Eqs. (13.3-24), (13.3-25) and (13.3-26) we obtain the following integral equations for the unknown functions $\boldsymbol{A}^{(1)}$, $\mathbf{B}^{(1)}$ and $\Gamma^{(1)}$:

$$I_2(\boldsymbol{A}^{(1)}) = -I_3(\boldsymbol{A}^{(0)}) - \frac{f_1^{(0)}}{n}\left[\tfrac{1}{2}(\mathscr{C}^2-\tfrac{5}{2})\beta_1(\theta)+\bar{b}(\mathscr{C})\right]\boldsymbol{C}, \tag{13.3-42}$$

$$I_2(\mathbf{B}^{(1)}) = -I_3(\mathbf{B}^{(0)}) - \frac{f_1^{(0)}}{n}\left[\beta_1(\theta)+\bar{C}_1(\mathscr{C})\right](\boldsymbol{\mathscr{C}}\boldsymbol{\mathscr{C}}-\tfrac{1}{3}\mathscr{C}^2\mathbf{I}), \tag{13.3-43}$$

$$I_2(\Gamma^{(1)}) = \frac{f_1^{(0)}}{n}\left[\tfrac{2}{3}(\mathscr{C}^2-\tfrac{3}{2})(\tfrac{7}{6}\theta\,\beta_1'(\theta)+\tfrac{1}{3}\theta^2\,\beta_1''(\theta))+\bar{C}_2(\mathscr{C})\right], \tag{13.3-44}$$

respectively. These equations must be solved subject to the auxiliary conditions (13.3-27) and (13.3-28) on $A^{(1)}$ and $\Gamma^{(1)}$, respectively.

The results for the transport coefficients follow in a straight-forward manner. It is easily seen by inspection of Eqs. (13.2-61), (13.2-62) and (13.2-63) that the contributions arising from spatial inhomogeneities are at least of second order in the density so that they do not contribute to the present order of approximation. For the thermal conductivity one obtains

$$\lambda = \lambda^{(0)} + n\lambda^{(1)}, \tag{13.3-45a}$$

where $\lambda^{(0)}$ is the low-density limit which has been found in Chapter 5,

$$\lambda^{(0)} = \frac{2k\theta}{3m}\int \frac{f_1^{(0)}(\boldsymbol{c})}{n} A^{(0)}(\mathscr{C})\mathscr{C}^4\, \mathrm{d}^3c, \tag{13.3-45b}$$

and $\lambda^{(1)}$ is the correction due to the effect of triple collisions and of macroscopic gradients,

$$\lambda^{(1)} = \lambda_K^{(1)} + \lambda_{\Phi 1}^{(1)} + \lambda_{\Phi 2}^{(1)} \tag{13.3-45c}$$

with

$$\lambda_K^{(1)} = \frac{2k\theta}{3m}\int \frac{f_1^{(0)}(\boldsymbol{c})}{n} A^{(1)}(\mathscr{C})\mathscr{C}^4\, \mathrm{d}^3c, \tag{13.3-45d}$$

$$\lambda_{\Phi 1}^{(1)} = \frac{k}{3m}\int \varphi(r_{12})\left\{\iint \boldsymbol{\mathscr{C}}_2\cdot \mathfrak{S}_\infty^{(2)}(x_1 x_2)\,\frac{f_1^{(0)}(\boldsymbol{c}_1)}{n}\,\frac{f_1^{(0)}(\boldsymbol{c}_2)}{n}\right.$$
$$\left.\times\left[A^{(0)}(\mathscr{C}_1)\boldsymbol{\mathscr{C}}_1 + A^{(0)}(\mathscr{C}_2)\boldsymbol{\mathscr{C}}_2\right]\mathrm{d}^3c_1\,\mathrm{d}^3c_2\right\}\mathrm{d}^3r_2\big|_{\boldsymbol{r}_1=\boldsymbol{r}}, \tag{13.3-45e}$$

$$\lambda_{\Phi 2}^{(1)} = -\frac{k}{6m}\int \frac{\varphi'(r_{12})}{r_{12}} \boldsymbol{r}_{12} \cdot \left\{ \iint (\mathscr{C}_1 + \mathscr{C}_2)\, \mathfrak{S}_{\infty}^{(2)}(x_1 x_2) \frac{f_1^{(0)}(\boldsymbol{c}_1)}{n} \frac{f_1^{(0)}(\boldsymbol{c}_2)}{n} \right.$$

$$\left. \times \boldsymbol{r}_{12} \cdot [A^{(0)}(\mathscr{C}_1)\mathscr{C}_1 + A^{(0)}(\mathscr{C}_2)\mathscr{C}_2]\, \mathrm{d}^3c_1\, \mathrm{d}^3c_2 \right\} \mathrm{d}^3r_2|_{\boldsymbol{r}_1=\boldsymbol{r}} . \qquad (13.3\text{-}45\text{f})$$

In the expressions (13.3-45e) and (13.3-45f) the streaming operator $\mathfrak{S}_{\infty}^{(2)}(x_1 x_2)$ is understood to act upon the velocity variables only.

For the shear viscosity we obtain

$$\eta = \eta^{(0)} + n\eta^{(1)}, \qquad (13.3\text{-}46\text{a})$$

where $\eta^{(0)}$ is the low-density limit,

$$\eta^{(0)} = \tfrac{2}{15}\theta \int \frac{f_1^{(0)}(\boldsymbol{c})}{n} B^{(0)}(\mathscr{C})\mathscr{C}^4\, \mathrm{d}^3c, \qquad (13.3\text{-}46\text{b})$$

and $\eta^{(1)}$ is the correction due to the effect of triple collisions and of macro scopic gradients,

$$\eta^{(1)} = \eta_K^{(1)} + \eta_\Phi^{(1)} \qquad (13.3\text{-}46\text{c})$$

with

$$\eta_K^{(1)} = \tfrac{2}{15}\theta \int \frac{f_1^{(0)}(\boldsymbol{c})}{n} B^{(1)}(\mathscr{C})\mathscr{C}^4\, \mathrm{d}^3c, \qquad (13.3\text{-}46\text{d})$$

$$\eta_\Phi^{(1)} = -\frac{1}{20n}\int \frac{\varphi'(r_{12})}{r_{12}} \left\{ \iint \mathfrak{S}_{\infty}^{(2)}(x_1 x_2) \frac{f_1^{(0)}(\boldsymbol{c}_1)}{n} \frac{f_1^{(0)}(\boldsymbol{c}_2)}{n} \right.$$

$$\times [B^{(0)}(\mathscr{C}_1)((\boldsymbol{r}_{12}\cdot \mathscr{C}_1)^2 - \tfrac{1}{3}r_{12}^2\mathscr{C}_1^2)$$

$$\left. + B^{(0)}(\mathscr{C}_2)((\boldsymbol{r}_{12}\cdot \mathscr{C}_2)^2 - \tfrac{1}{3}r_{12}^2\mathscr{C}_2^2)]\mathrm{d}^3c_1\, \mathrm{d}^3c_2 \right\} \mathrm{d}^3r_2|_{\boldsymbol{r}_1=\boldsymbol{r}}, \qquad (13.3\text{-}46\text{e})$$

with the same comment about $\mathfrak{S}_{\infty}^{(2)}(x_1 x_2)$ as under Eq. (13.3-45).

Finally, the bulk viscosity, which is zero in the low-density limit, also vanishes in the present approximation: the kinetic part is zero on account of the constraint (13.3-28b) on $\Gamma^{(1)}$, the potential part is zero as it depends only upon the solution $\Gamma^{(0)}$ which is zero. Hence, for the bulk viscosity we obtain

$$\kappa = 0. \qquad (13.3\text{-}47)$$

For the special case of hard spheres, Sengers [1966, 1967] has computed the contributions to the thermal conductivity and shear viscosity due to triple collisions, i.e., $\lambda_K^{(1)}$ and $\eta_K^{(1)}$, using a modified Monte Carlo quadrature technique. From the preliminary results based on a numerical analysis of approximately 128 000 triple collision events he found

$$\lambda_K^{(1)}/\lambda^{(0)} = -0.593 \pm 0.011, \tag{13.3-48}$$

$$\eta_K^{(1)}/\eta^{(0)} = -0.601 \pm 0.008. \tag{13.3-49}$$

The Enskog theory predicts a value of -0.625 for each of these ratios. The difference is therefore only four percent in the case of the thermal conductivity and five percent in the case of the viscosity.

13.4. Difficulties in the theory of dense gases

On the basis of the analysis of the foregoing sections one is tempted to assume the validity of the hypotheses (H_1) and (H_2) formulated in Section 13.1 for *all* terms in the (formally exact) expansion (13.1-7) for F_2. Yet, we will now demonstrate that the two hypotheses cannot in general be true for the terms beyond the second one in the expansion and that, therefore, it is impossible to obtain an expression of the form (13.1-11) for F_2 by a termwise evaluation of the expansion (13.1-7).

For example, consider the third term of the expansion, which is proportional to $F_2^{(2)}(x_1 x_2;\, t)$ and, therefore, requires the analysis of the dynamics of four particles. A complete classification of all possible dynamical events among four particles is not available at present. However, such a classification is not needed for a qualitative argument. First, we write $F_2^{(2)}$ as the sum of two terms, one of which contains the contribution of the initial state,

$$F_2^{(2)}(x_1 x_2;\, t) = \iint \mathfrak{T}_t^{(4)}(x_1 x_2|x_3 x_4) \prod_{i=1}^{4} F_1(x_i;\, t)\, \mathrm{d}^6 x_3\, \mathrm{d}^6 x_4 + C_2^{(2)}(x_1 x_2;\, t). \tag{13.4-1}$$

Here the operator $\mathfrak{T}_t^{(4)}(x_1 x_2|x_3 x_4)$ is obtained from the operator $\tilde{\mathfrak{T}}_t^{(4)}(x_1 x_2|x_3 x_4)$ by setting all $a_s(x_1 \ldots x_s;\, 0)$ equal to unity; $C_2^{(2)}$ represents the influence of initial correlations upon $F_2^{(2)}$ at time t. As has been shown by Cohen [1966a], the structure of $\mathfrak{T}_t^{(4)}$ is such that the only contributions to $F_2^{(2)}$ come from the following dynamical events:

(a) a (genuine) quadruple collision at t,
(b) a sequence of a binary and at least one (genuine) triple collision,
(c) four or more successive binary collisions.

Of course, sequences of the types (b) and (c) which contain non-interacting and/or hypothetical collisions must be taken into account. Examples of these dynamical events are given in the diagrams of Fig. 13.2. Also, the structure of the term $C_2^{(2)}$ in Eq. (13.4-1) is such that the only contributions come from those dynamical events listed under (b) and (c).

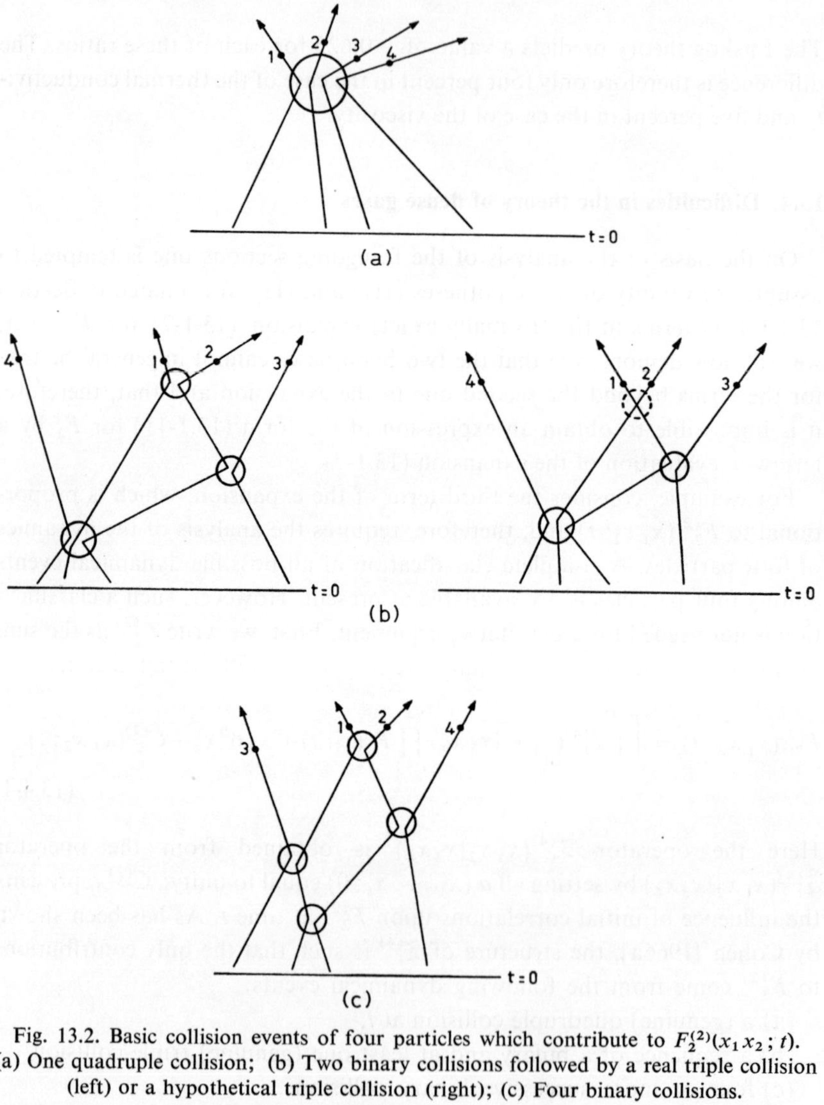

Fig. 13.2. Basic collision events of four particles which contribute to $F_2^{(2)}(x_1 x_2\,; t)$. (a) One quadruple collision; (b) Two binary collisions followed by a real triple collision (left) or a hypothetical triple collision (right); (c) Four binary collisions.

On the basis of these considerations Dorfman and Cohen [1967] have obtained the following estimates for $t \gg t_{\text{coll}}$,

$$C_2^{(2)}(x_1\, x_2;\, t) = \mathrm{O}(t_{\text{coll}}/t), \tag{13.4-2}$$

$$\iint \mathfrak{T}_t^{(4)}(x_1\, x_2|x_3\, x_4) \prod_{i=1}^{4} F_1(x_i;\, t)\,\mathrm{d}^6 x_3\,\mathrm{d}^6 x_4 = \mathrm{O}\left(\log \frac{t}{t_{\text{coll}}}\right). \tag{13.4-3}$$

Thus it is seen that for $t \gg t_{\text{coll}}$, if terms of the order of t_{coll}/t are neglected, the initial state is "forgotten" to $\mathrm{O}(n^2)$, i.e., $C_2^{(2)}(x_1\, x_2;\, t)$ may be ignored. However, $F_2^{(2)}(x_1\, x_2;\, t)$ contains contributions which grow with time as $\log\,(t/t_{\text{coll}})$. Consequently, as $t \to \infty$, $F_2^{(2)}(x_1\, x_2;\, t)$ does *not* tend to a finite limit, which implies a breakdown of the theory – either a time-independent functional $F_2(x_1\, x_2|F_1)$ does not exist, or the density expansion is improper.

Furthermore, the considerations of Dorfman and Cohen may be generalized to higher order terms in the expansion (13.1-7) of F_2. In fact, one finds that, to the next order, the initial conditions give rise to contributions which diverge logarithmically as $t \to \infty$,

$$C_2^{(3)}(x_1\, x_2;\, t) = \mathrm{O}\left(\log \frac{t}{t_{\text{coll}}}\right), \tag{13.4-4}$$

while

$$\iiint \mathfrak{T}_t^{(5)}(x_1\, x_2|x_3\, x_4\, x_5) \prod_{i=1}^{5} F_1(x_i;\, t)\,\mathrm{d}^6 x_3\,\mathrm{d}^6 x_4\,\mathrm{d}^6 x_5 = \mathrm{O}\left(\frac{t}{t_{\text{coll}}}\right). \tag{13.4-5}$$

Also, for $s = 4, 5, \ldots$

$$C_2^{(s)}(x_1\, x_2;\, t) = \mathrm{O}((t/t_{\text{coll}})^{s-3}) \tag{13.4-6}$$

and

$$\int \cdots \int \mathfrak{T}_t^{(s+2)}(x_1\, x_2|x_3 \cdots x_{s+2}) \prod_{i=1}^{s+2} F_1(x_i;\, t)\mathrm{d}^6 x_3 \cdots \mathrm{d}^6 x_{s+2} = \mathrm{O}\left(\left(\frac{t}{t_{\text{coll}}}\right)^{s-2}\right). \tag{13.4-7}$$

These results may be understood on the basis of dynamical events like those of Fig. 13.3, where the (12)-collision is triggered by a sequence of successive binary collisions $(s, s-1)(s-1, s-2) \ldots (32)$, see Dorfman and Cohen [1967].

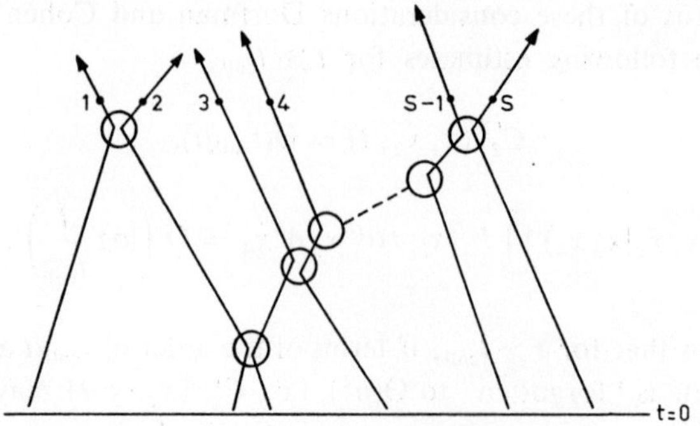

Fig. 13.3. s-particle event contributing to $F_2(x_1 x_2; t)$.

Therefore, the expansion (13.1-7) for F_2 contains terms which *diverge* as $t \to \infty$ and, in fact, the divergence becomes stronger for successively higher-order terms. Hence, the existence of a kinetic stage for $t \gg t_{\text{coll}}$, in which for a large class of initial conditions F_2 may be written as a time-independent functional of F_1, cannot be established on the basis of the simple density expansion, Eq. (13.1-7). Obviously, due to these difficulties, the significance of the first two terms and, hence, of the theory of Choh and Uhlenbeck presented in Section 13.3, also becomes open to a critical re-evaluation (see below).

The reason for the occurrence of divergences is that the expansion (13.1-7) is one in terms of the dynamics of *isolated* groups of particles in infinite space. Their contributions to F_2 are essentially proportional to the volume of phase space the particles must occupy at $t = 0$ in order to produce the specified sequences of events by time t. It turns out that this phase space volume increases with the number of particles involved. Although this can be shown only by detailed calculation, the increase results from particles enjoying long free flight trajectories before "joining the action". In an actual system, with a large number of particles, no single member of a small set of particles can travel for very long without meeting a particle from outside the set. On the average, flights are restricted to a time of order t_{mfp} and this effect will cut off the phase volume available for specified types of events. What this means in terms of our cluster expansion is, that there is some contribution in the $(n+1)$-particle term that just cancels the divergence in the n-particle term. Stated otherwise, portions of the various terms must be re-combined in such a way as to cause the divergences to disappear, i.e., a *resummation* is required. A less rigorous procedure is to cut off the time at t_{mfp}. This problem has been considered by Weinstock [1965], Kawasaki and Oppenheim [1965], Haines, Dorfman and Ernst [1966], Frieman and Goldman [1966], Goldman and Frieman [1967], Dorfman and Cohen [1967], Van Leeuwen and Weyland [1967a, 1967b].

Only very recently has some insight been gained in regard to the consequences of such a resummation. A study of the analogous problem in two dimensions has proven to be of considerable help in appreciating the nature of the difficulties since, in a two-dimensional gas, similar divergences occur one term earlier in the expansion (13.1-7). For the details of the computation in two-dimensional systems the reader may be referred to two papers by Sengers [1965b, 1966b] who studied the case of a gas of hard discs, and to an article by Cohen [1968] who studied the case of a two-dimensional Lorentz gas. For three-dimensional systems Cohen [1967] and Dorfman [1967] have indicated how such a resummation may be performed. For example, for the four-particle term $F_2^{(2)}$ a convergent contribution to F_2 as $t \to \infty$ is obtained if a summation is performed over those sequences of events that lead to interruptions of the "free path" of a particle by imaginary collisions. As might be expected, the behavior of the contribution as $t \to \infty$ is not as $\log (t/t_{\text{coll}})$ as in Eq. (13.4-3) but, instead, as $\log (t_{\text{mfp}}/t_{\text{coll}})$ or, since $t_{\text{mfp}}/t_{\text{coll}}$ is of the order nr_0^3, as $\log (nr_0^3)$, i.e., the contribution remains finite. Thus, one sees that the resummation introduces a new time, viz., t_{mfp}, in addition to t_{coll}. It is claimed that, by similar resummations, all divergent contributions in the expansion (13.1-7) can be made finite and that, at least in first approximation, the result would be a replacement of t by t_{mfp} in the asymptotic estimates (13.3-2), (13.3-3) and (13.4-2) through (13.4-7). Such a program has yet to be carried out, though.

There is strong evidence that the logarithmic divergence in time of the four-particle term in F_2 leads to a logarithmic dependence on density of the four-particle contributions to the transport coefficients. The easiest way to see this is to add the (time-dependent) four-particle term, $\frac{1}{2}n^2F_2^{(2)}(x_1 x_2; t)$ to the expansion (13.3-5) and treat the resulting kinetic equation with the method of Chapman and Enskog. The result would be a contribution to the transport coefficients which would behave as $(nr_0^3)^2 \log (t/t_{\text{coll}})$ and, therefore, diverge logarithmically as $t \to \infty$, just as the four-particle contribution to F_2 does. However, if one now accounts for the resummation by replacing t by t_{mfp}, then a contribution to the transport coefficients would be obtained which would behave as $(nr_0^3)^2 \log (t_{\text{mfp}}/t_{\text{coll}})$ or, equivalently, as $(nr_0^3)^2 \times \log (nr_0^3)$. This logarithmic density dependence of the transport coefficients follows under the same assumption also directly from the binary collision expansion technique, see Dorfman [1967].

Of course, in addition to terms which depend logarithmically upon density and which are due to the contributions from recollisions to F_2 (i.e., events which incorporate memory effects over a mean free path), the transport coefficients also contain terms which depend upon powers of the density alone and which are due to convergent contributions from small groups of isolated particles to F_2. Thus, one would expect that the transport coefficients can be written not as power series in the density, but in the form of a hybrid power series expansion with respect to the density and the logarithm of the density, e.g., for the thermal conductivity and shear viscosity,

$$\lambda = \lambda^{(0)} + n\lambda^{(1)} + n^2(\lambda_1^{(2)} \log n + \lambda^{(2)}) + \cdots, \tag{13.4-8}$$

$$\eta = \eta^{(0)} + n\eta^{(1)} + n^2(\eta_1^{(2)} \log n + \eta^{(2)}) + \ldots, \tag{13.4-9}$$

and similarly for the other transport coefficients. It may be shown that, while the values of the coefficients $\lambda_1^{(2)}$, $\eta_1^{(2)}$, . . . depend only weakly on the way the resummation of the expansion (13.1-2) is performed, the values of the coefficients $\lambda^{(2)}$, $\eta^{(2)}$, . . . depend very sensitively on it. Obviously, the resummation will also affect the convergent three-particle term since all free paths of the particles 1, 2 and 3 will be cut off at a few mean free paths. As far as one can see, however, this will not affect the contributions to the transport coefficients which are proportional to the density, but it will affect the coefficient of the logarithmic term. On this basis, one would expect that the Choh-Uhlenbeck expressions for $\eta^{(1)}$ and $\lambda^{(1)}$ still give the correct coefficients for the corrections to the Chapman-Enskog results $\eta^{(0)}$ and $\lambda^{(0)}$, respectively.

Ionized gases 14

It is well known that, at high temperatures, gases ionize and a whole range of new phenomena may arise. Ignoring the effects of magnetic fields for the moment, we note that the most significant difference between an ionized gas and a neutral gas is due to the Coulomb portion of the intermolecular force, which occurs only in the former. In fact, the decay of the Coulomb potential at large interparticle separations ($\sim r^{-1}$) is sufficiently slow that one can no longer consider collisions as isolated, short duration, events and, if one attempts to apply the Chapman-Enskog theory to an ionized gas, one finds that the Ω-integrals from which transport properties are calculated, do not converge. Clearly, then, one cannot treat an ionized gas simply as a gas mixture with a particular intermolecular potential. When the further complications of a magnetic field and the associated Lorentz force are taken into consideration, one sees that the range of phenomena that are possible in ionized gases is much richer than in neutral gases. This, in turn, is responsible for the enormous growth that *plasma physics* has enjoyed in the last decade or two.

As plasma physics can scarcely be treated in a single volume, it is certainly not our intent to cover this field here. Rather, we shall seek those areas of plasma physics, particularly those regimes of temperature and density, in which theories similar to those presented earlier in the present text, may be applied. Essentially, we are seeking what may be regarded as the region of overlap between kinetic theory and plasma physics. This necessitates a short qualitative discussion of some of the topics covered by the term "plasma physics".

A neutral gas may be characterized by three length parameters – viz., the range of the intermolecular force, r_0, the mean free path, l, and the scale length for macroscopic properties, L. (The mean distance between particles, $n^{-\frac{1}{3}}$, could be added to the list but is not truly independent, since $l \sim (nr_0^2)^{-1}$.) Furthermore, in normal gases $r_0 \ll l \ll L$ and the entire theory of Chapters 2–11 is based on this ordering; Chapters 12, 13 and 15 show that substantial modifications are necessary when r_0/l and l/L cannot be treated as small parameters. In plasma physics at least two new length parameters come into play: the Debye length, d (cf. Eq. (14.1-6), and the cyclotron radius, r_c (cf. Eq. (14.2-2b)). This is the minimum set of length scales; in actuality, the large mass difference between ions and electrons requires that two cyclotron radii, one for electrons and one for ions, be introduced and there are scale lengths (i.e., wavelengths) associated with electromagnetic phenomena that may also enter the theory. The very richness of characteristic lengths is what makes plasma physics simultaneously interesting and difficult. Theories valid for one ordering of the parameters will not be valid for other orderings; many of the possible cases are realized in physical systems and this has led to various sub-branches of plasma physics.

One of the first problems in plasma physics is that of finding the set of equations with which to describe the system in the regime under consideration. In particular, the Coulomb potential must be handled very carefully. It can be treated within the framework of a Boltzmann-like equation either through a collision term or through the inclusion of a force due to electric and magnetic fields which are, in turn, generated by the charges and motions of the particles as described by Maxwell's equations. The problem with the former approach is that simultaneous interactions of many particles must be considered, while the latter would require accounting for the rapid local field variations caused by individual particles. In any case, one must be careful that the same interaction does not enter the theory twice. This problem has been considered by various authors, including Rostocker and Rosenbluth [1960], Guernsey [1960], Simon and Harris [1960] and Lenard [1960]; Wu [1966] and Montgomery [1967] have given reviews of this work. These investigations are along the lines of Chapter 3 with increased difficulty arising from the phenomena discussed above. Although these investigations are of great interest and importance, a proper treatment of them would require far more space than can be devoted to them here; the interested reader is referred to the reviews cited above. We shall cite only a few results of direct interest to us.

Probably the most important single observation is that even small deviations from local charge neutrality in an ionized gas produce very intense electric fields which cause the plasma to quickly return to a charge-neutral state; the time required for charge neutrality is approximately the inverse of the "plasma frequency" and is equal to $(4\pi ne^2/m)^{-1}$. For the plasmas of interest here this time is short enough to be neglected. (It is typical of plasma physics that there are interesting cases for which this is not so.) The consequence of this is that each charged particle tends to attract a slight excess of opposite charge to its neighborhood and acts as if it carried a "shield" around with it. The Debye length is a measure of the size of this shield. The phenomenon, known as *Debye shielding*, is discussed in Section 14.1.

Use of the concept of shielding considerably reduces the difficulty of the problem: Interactions between particles at distances greater than the Debye length can be considered to be mediated by the electric and magnetic fields, while those at shorter distances may be considered true collisions. Thus, one of the major difficulties of the theory of ionized gases is surmounted.

There remains the question of finding an appropriate operator to represent interactions at distances shorter than the Debye length, i.e., a collision operator. Not surprisingly, the form of the collision operator depends on the relative values of some of the parameters. For a sufficiently rarefied plasma (such as occur in many fusion devices), the short-range interactions may be neglected entirely and we are left with the *collisionless Boltzmann*, or *Vlasov, equation*. On the other extreme, in which the Coulomb potential is unimportant relative to the other components of the intermolecular force (this is not a realistic physical model), the usual Boltzmann equation is satisfactory. In the most common regime simultaneous interactions of many particles must be considered. However, due to the long-range nature of the Coulomb potential, the cross section $\sigma(g, \chi) \sim \sin^{-4}\frac{1}{2}\chi$ and the preponderant number of collisions result in small deflections. In fact, despite the small deflections, these collisions are more important than the much rarer large deflection collisions. Small deflection collisions produce small changes in the velocities of the colliding molecules. One way of deriving the collision operator representing this type of behavior is to start with the Boltzmann operator, note that $\boldsymbol{c}' \approx \boldsymbol{c}$ and $\boldsymbol{c}_1' \approx \boldsymbol{c}_1$ and expand $f(\boldsymbol{c}')f(\boldsymbol{c}_1')$ in a Taylor series about $f(\boldsymbol{c})f(\boldsymbol{c}_1)$, keeping terms up to second order. There results a second-order differential operator and the resulting equation is known as the *Fokker-Planck equation*. In all fairness, it should be noted that the above recipe provides one of the least satisfactory ways of deriving the Fokker-

Planck equation. Finally, it should be noted that modifications of the Fokker-Planck equation are required in some regimes.

Our interest naturally lies in the regime which may be treated by something akin to hydrodynamics. Since the equations must be modified at least to account for electromagnetic phenomena, this subject is commonly known as magnetohydrodynamics (MHD) or magnetogasdynamics. Clearly, this regime must possess the characteristics that permit application of hydrodynamics to a neutral gas, i.e., $r_0 \ll l \ll L$. Additionally, we must have $nd^3 \gg 1$, i.e., the number of particles in a Debye sphere must be large. Furthermore, r_c, the cyclotron radius must be much larger than d, and finally, the wavelength of any electromagnetic wave must be larger than d. Physically, such a regime is realized at pressures approximately atmospheric and at temperatures on the order 10^3–10^5 °K. From the point of view of normal engineering these are low pressure, high temperature gases. A plasma physicist accustomed to working with fusion devices would, on the other hand, regard this as a high pressure, low temperature plasma. The regime we are discussing is appropriate to engineering devices such as magnetohydrodynamic power generators, to certain laboratory discharge plasmas and to stellar interiors, among others.

It should be clear by now that we shall be dealing with the application of the Chapman-Enskog theory to ionized gases. In this application it turns out that identical results are obtained (within a few percent) whether the Boltzmann or Fokker-Planck equations are used. The former retains a few difficulties that are removed in the derivation of the latter, but, as the Boltzmann equation is more familiar to the readers of this text, we shall adopt the former approach. Having made that choice, we find that the treatment of an ionized gas in the absence of a magnetic field becomes a special case of a gas mixture and that special handling is required only in the calculation of the transport properties. As the problems are very similar in the presence of a magnetic field, we proceed in Section 14.2 directly to the treatment of the more general case. Here, features not present in the neutral gas mixture problem come into play and substantial modifications of the theory are necessary as we shall see. Explicit formulas for the transport properties are obtained in Section 14.3.

Although the results of the first three sections are appropriate to a large number of potential applications, there are numerous situations to which they do not apply. For example, since the ratio of the mass of a positively charged ion to the mass of an electron is very large, the rate of energy transfer between the two species is very low and, should one of the species become hotter than the other, the non-equilibrium situation will persist for a significant length of time. Consequently, the energy conservation equations for the different species must be treated separately, and a number of new transport properties arise. This situation is discussed in Section 14.4.

Still other transport phenomena may arise in non-equilibrium situations in ionized gases. We only mention those situations in which the characteristic frequency of a disturbance is shorter than the duration of a collision. In such circumstances, the collision term in the Boltzmann equation must be modified and new methods of solving the resulting equation are required. Another interesting situation arises in the study of certain devices for the production of controlled thermonuclear fusion, in which the velocity distribution function may differ greatly from the Maxwellian and, again, a new approach is necessary. These phenomena properly belong to plasma physics and will not be studied here. The reader who desires a survey of the various problems posed by the electrodynamics of plasmas is referred to the monograph by Jancel and Kahan [1966].

14.1. Debye shielding

Shielding, mentioned in the introduction to this chapter, was first discovered by Debye in connection with his work on ionic solutions and is now known by his name. The problem is to find the charge distribution that develops in the neighborhood of any charged particle.

Of particular importance is the size of the shield region. Assume, for convenience, that a positive ion of charge Ze is located at the origin. The problem is to compute the charge distribution and potential in the neighborhood of the ion. Since the electrons are very light and hence move much more rapidly than the ions, they will very quickly approach equilibrium with the central ion. In accordance with Eq. (4.5-36), the electron density in this neighborhood will be given by

$$n_e = n_{e0} \exp(eV/kT). \qquad (14.1\text{-}1)$$

Here n_{e0} is the average electron density, V the electric potential, and e the electronic charge. The ion density, on the other hand, can be considered to be essentially constant in this neighborhood because the ions, being much heavier, do not normally have time to come to equilibrium. Thus, the local charge density is

$$Q = Ze\,n_{i0} - en_{e0} \exp(eV/kT) = en_{e0}[1 - \exp(eV/kT)], \quad (14.1\text{-}2)$$

since $Zn_{i0} = n_{e0}$. As the ratio (eV/kT) is normally small, we can expand the exponential to obtain

$$Q \approx -(e^2 n_{e0}/kT)V. \quad (14.1\text{-}3)$$

Finally, from Poisson's equation, we have the equation for the electric potential, V,

$$\nabla^2 V = -4\pi \left[Ze\,\delta(r) - \left(\frac{e^2 n_{e0}}{kT}\right) V \right], \quad (14.1\text{-}4)$$

the solution of which is given by

$$V(r) = -\frac{Ze}{r} \exp\left(-\frac{r}{d}\right), \quad (14.1\text{-}5)$$

where

$$d = \left(\frac{kT}{4\pi n_{e0}\, e^2}\right)^{\frac{1}{2}}. \quad (14.1\text{-}6)$$

The length d is called the *Debye length* and is one of the most important parameters of plasma physics.

At gas densities appropriate to the regime discussed in the introduction to this chapter, the Debye length is much larger than the distance between particles. Also, since the electrostatic forces set up by charge imbalance are very strong, the potential distribution (14.1-5) is set up on a time scale very much shorter than any other time of interest. Hence it is proper to regard an ion (or electron) as carrying with it a charge cloud described by Eqs. (14.1-5) and (14.1-1) and one can thus treat the gas by means of the Boltzmann equation, provided the potential (14.1-5) is used in place of the unshielded Coulomb potential. A rigorous proof of this statement may be found in the works cited earlier. The necessary calculations have been carried out, but it is found that almost equally good results may be obtained by using the unshielded Coulomb potential with a radial cut-off at $r = d$, i.e., by taking

$$V(r) = \begin{cases} -Ze/r & r < d, \\ 0 & r > d. \end{cases} \quad (14.1\text{-}7)$$

Since the integrals needed for computing the transport properties can be evaluated in closed form to very good approximation with the truncated Coulomb potential (14.1-7), it has been favored in the literature and will be adopted here.

14.2. The Chapman-Enskog method for an ionized gas

As stated earlier, the case of an ionized gas in the absence of a magnetic field is relatively straight-forward and may, in any event, be obtained as the zero magnetic field limit of the general case. Thus, we proceed directly to the study of an ionized gas in a magnetic field.

The presence of a magnetic field has a profound effect on the behavior of a gas. Not only is the force exerted on a charged particle by a magnetic field velocity dependent, it is also perpendicular to both the velocity and the magnetic field direction. Thus, the force exerted on a particle of electric charge e (e.s.u.), moving with the velocity $\boldsymbol{c}$ in a magnetic field $\boldsymbol{H}$(e.m.u.) is given by

$$\boldsymbol{F} = (e/\mathrm{c})\boldsymbol{c} \times \boldsymbol{H}, \tag{14.2-1}$$

where c is the speed of light in vacuum; this is the well-known *Lorentz force*.

The presence of the Lorentz force causes a particle whose velocity is normal to the field direction to move in a circle, whose radius (r_c) is determined by equating the Lorentz force with the centrifugal force,

$$(e/\mathrm{c})cH = mc^2/r_c. \tag{14.2-2a}$$

Thus we find that r_c is given by

$$r_c = \frac{mc}{(e/\mathrm{c})H}. \tag{14.2-2b}$$

The radius r_c is called the *gyromagnetic radius* (*gyroradius*) or *cyclotron radius*. The frequency of rotation is

$$\omega = c/r_c = \frac{(e/\mathrm{c})H}{m} \tag{14.2-3}$$

and is known as the *cyclotron frequency*. (It is sometimes called the Larmor frequency, but, as there is some ambiguity in this term, we shall avoid it.) A particle which has velocity components both parallel and perpendicular to the magnetic field will execute a helical motion (assuming no electric field is present); its velocity component parallel to the field will remain unaffected while its velocity component perpendicular to the field will undergo the gyration described above. A further discussion of the motion of a particle in fields, e.g., combined electric and magnetic fields, inhomogeneous fields, etc., is not required here and the interested reader is referred to any introductory plasma physics text, e.g., Spitzer [1962], Longmire [1963].

The admission of a magnetic field thus introduces a new characteristic length, the gyromagnetic radius, into the theory. There now exist four distinct characteristic lengths; in addition to the gyroradius we have the range of the intermolecular force, the mean free path, and the Debye length. There may be various other scale lengths of the type $X^{-1}|\nabla X|$, where X is any property of importance. The abundance of characteristic parameters is, in fact, what makes plasma physics so difficult. Approximations appropriate to one regime generally fail in the other regimes. As a discussion of the whole of plasma physics is not intended here, we shall restrict ourselves to the regime mentioned in the introduction – temperatures less than 10^4 °K and pressures greater that 10^{-4} atm. Included in this regime are most atmospheric phenomena, high-speed gas flows, magnetohydrodynamic power generation and a number of interesting laboratory plasmas.

To proceed rigorously, one ought to again derive a Boltzmann equation. However, as the equation retains essentially its previous form, we shall simply quote the result. For particles of species i (mass m_i, charge e_i) the Boltzmann equation assumes the form

$$\frac{\partial f_i}{\partial t} + \boldsymbol{c}_i \cdot \nabla_r f_i + \left(\boldsymbol{F}_i + \frac{e_i}{m_i}\left(\boldsymbol{E} + \frac{1}{c}\,\boldsymbol{c}_i \times \boldsymbol{H}\right)\right) \cdot \nabla_{c_i} f_i = \sum_j J(f_i f_j). \quad (14.2\text{-}4)$$

Here we have included the Lorentz force and the electric field explicitly; $\boldsymbol{F}_i$ represents any non-electromagnetic forces; J is the usual collision operator. It can be shown that this equation is correct only if the gyroradius is very large compared to the Debye length, but, as this condition is satisfied in the regime described above, no difficulty arises.

We may now proceed to the Chapman–Enskog development; the development presented here follows that of Marshall [1960] very closely. We shall, of course, also need the macroscopic conservation equations. As in the case of a mixture of uncharged particles, the equations required are those of conservation of particles of each species, total momentum and total energy. In the derivation of these equations from Eq. (14.2-4), the only new feature is the presence of the Lorentz force term and only this term will receive special attention.

The conservation equation for particles of species i is obtained by integrating the Boltzmann equation over velocity. The new term is

$$\int (\boldsymbol{c}_i \times \boldsymbol{H} \cdot \nabla_{ci} f_i) \mathrm{d}^3 c_i,$$

which can be integrated by parts and yields zero, because $\nabla_c \cdot \boldsymbol{c} \times \boldsymbol{H} = 0$. Thus, the conservation equation for particles of species i remains unchanged. In terms of the mass density ρ_i we have

$$\frac{\mathrm{d}\rho_i}{\mathrm{d}t} = -\rho_i \nabla \cdot \boldsymbol{v} - \nabla \cdot \rho_i \boldsymbol{V}_i, \tag{14.2-5}$$

cf. Eq. (4.1-32), and the combined continuity equation for all species is

$$\frac{1}{\rho} \frac{\mathrm{d}\rho}{\mathrm{d}t} = -\nabla \cdot \boldsymbol{v}, \tag{14.2-6}$$

cf. Eq. (4.1-28).

The equation of motion, or momentum conservation equation, is obtained by multiplying the Boltzmann equation by $m_i \boldsymbol{c}_i$, integrating over $\boldsymbol{c}_i$ and summing the resulting equations. Here, the new term is

$$\sum_i \frac{e_i}{\mathrm{c}} \int \boldsymbol{c}_i \boldsymbol{c}_i \times \boldsymbol{H} \cdot \nabla_{ci} f_i \mathrm{d}^3 c_i,$$

which is equal to

$$\sum_i \frac{e_i}{\mathrm{c}} \int \boldsymbol{c}_i \nabla_{ci} \cdot (f_i \boldsymbol{c}_i \times \boldsymbol{H}) \mathrm{d}^3 c_i .$$

Integrating by parts we find that the latter expression is, in turn, equal to

$$-\sum_i \frac{e_i}{\mathrm{c}} \int f_i \boldsymbol{c}_i \times \boldsymbol{H} \mathrm{d}^3 c_i$$

or, in terms of the hydrodynamic velocity $\boldsymbol{v}$ and the diffusion velocities $\boldsymbol{V}_i$,

$$-\sum_i \frac{n_i e_i}{\mathrm{c}} (\boldsymbol{v} + \boldsymbol{V}_i) \times \boldsymbol{H}.$$

Thus, if we introduce the *charge density*, Q,

$$Q = \sum_i n_i e_i \tag{14.2-7a}$$

and the *current density*, $\boldsymbol{j}$, measured by an observer moving with the gas,

$$\boldsymbol{j} = \sum_i n_i e_i \boldsymbol{V}_i, \tag{14.2-7b}$$

we have

$$\sum_i \frac{e_i}{\mathrm{c}} \int \boldsymbol{c}_i \boldsymbol{c}_i \times \boldsymbol{H} \cdot \nabla_{c_i} f_i \mathrm{d}^3 c_i = - \frac{1}{\mathrm{c}} (Q\boldsymbol{v} + \boldsymbol{j}) \times \boldsymbol{H}.$$

Note that the vector $Q\boldsymbol{v}+\boldsymbol{j}$ is the current density observed by a stationary observer.

The equation of motion can then be written

$$\rho \frac{\mathrm{d}\boldsymbol{v}}{\mathrm{d}t} = -\nabla \cdot \mathsf{P} + Q\left(\boldsymbol{E} + \frac{1}{\mathrm{c}} \boldsymbol{v} \times \boldsymbol{H}\right) + \frac{1}{\mathrm{c}} \boldsymbol{j} \times \boldsymbol{H} + \sum_i \rho_i \boldsymbol{F}_i, \tag{14.2-8}$$

cf. Eq. (4.1-29). The new terms may be interpreted as the effect of the Lorentz force acting on the gas as a whole.

Finally, the energy conservation equation is obtained by multiplying the Boltzmann equation by $\frac{1}{2} m_i C_i^2$, integrating over velocity and summing over species. Here, the new term is

$$\sum_i \frac{1}{2} \frac{e_i}{\mathrm{c}} \int C_i^2 (\boldsymbol{c}_i \times \boldsymbol{H}) \cdot \nabla_{c_i} f_i \mathrm{d}^3 c_i,$$

which is equal to

$$\sum_i \frac{1}{2} \frac{e_i}{\mathrm{c}} \int C_i^2 \nabla_{c_i} \cdot (f_i \boldsymbol{c}_i \times \boldsymbol{H}) \mathrm{d}^3 c_i.$$

Integrating by parts we find that the latter expression, in turn, is equal to

$$-\sum_i \frac{e_i}{\mathrm{c}} \int \boldsymbol{C}_i \cdot f_i \boldsymbol{c}_i \times \boldsymbol{H} \, \mathrm{d}^3 c_i$$

or, in terms of the hydrodynamic velocity $\boldsymbol{v}$ and the diffusion velocities $\boldsymbol{V}_i$,

$$-\sum_i \frac{n_i e_i}{\mathrm{c}} \boldsymbol{V}_i \cdot (\boldsymbol{v} \times \boldsymbol{H}).$$

Thus we have

$$\sum_i \frac{1}{2} \frac{e_i}{\mathrm{c}} \int C_i^2 (\boldsymbol{c}_i \times \boldsymbol{H}) \cdot \nabla_{c_i} f_i \mathrm{d}^3 c_i = - \frac{1}{\mathrm{c}} \boldsymbol{j} \cdot (\boldsymbol{v} \times \boldsymbol{H}),$$

where $\boldsymbol{j}$ is the current density introduced by Eq. (14.2-7b).

Using the equation of continuity (14.2-6) to eliminate ($\mathrm{d}\rho/\mathrm{d}t$), we obtain the following equation for the rate of change of the internal energy per unit mass, u,

$$\rho \frac{\mathrm{d}u}{\mathrm{d}t} = -\mathsf{P} : \nabla \boldsymbol{v} - \nabla \cdot \boldsymbol{q} + \boldsymbol{j} \cdot \left(\boldsymbol{E} + \frac{1}{\mathrm{c}} \boldsymbol{v} \times \boldsymbol{H} \right) + \sum_i \rho_i \boldsymbol{V}_i \cdot \boldsymbol{F}_i, \qquad (14.2\text{-}9)$$

cf. Eq. (4.1-30). As $\boldsymbol{E} + \boldsymbol{v} \times \boldsymbol{H}/\mathrm{c}$ is the electric field seen by the moving fluid the new term simply represents the work done by the electric field, i.e., the Ohmic heating. The equation for the temperature T can be obtained from Eq. (14.2-9) by using the definition (2.3-15), $\rho u = \frac{3}{2} nkT$.

Once the flow quantities P, $\boldsymbol{q}$ and $\boldsymbol{j}$ are specified in terms of the other quantities in these equations, Eqs. (14.2-6), (14.2-8) and (14.2-9), together with Maxwell's equations (which relate the fields to $\boldsymbol{j}$ and Q) form a closed set of equations governing the behavior of an ionized gas – the *magnetohydrodynamic* (MHD) or, more correctly, the *magnetogasdynamic equations.*

The stage is now set for the Chapman-Enskog development and, as the procedure follows that of Chapter 6 rather closely, some of the details will be omitted. A little care is needed before proceeding, however. Recall that in Chapter 5 it was argued that the streaming term of the Boltzmann equation produces changes in the distribution function on a time scale long compared with the time between collisions, τ, and consequently, the effect of the collision term was to maintain the gas in a state which is nearly a local Maxwellian. This idea is central to the Chapman-Enskog approach. The same argument applies in the present case, except to the term $(e_i/m_i \mathrm{c}) \boldsymbol{c}_i \times \boldsymbol{H} \cdot \nabla_{c_i} f_i$, whose natural time sale of variation is the inverse of the gyrofrequency. As $\omega\tau$ may or may not be small compared to unity, it is not apparent that one is permitted to ignore this term in the zeroth order calculation. On the other hand,

$$\boldsymbol{E} + \frac{1}{\mathrm{c}} \boldsymbol{c}_i \times \boldsymbol{H} = \frac{1}{\mathrm{c}} \boldsymbol{C}_i \times \boldsymbol{H} + \left(\frac{1}{\mathrm{c}} \boldsymbol{v} \times \boldsymbol{H} + \boldsymbol{E} \right) \qquad (14.2\text{-}10)$$

and the last term is just the electric field measured in the frame moving with the gas. If the electrical conductivity is high – as it is if the gas is ionized to any appreciable degree – we must expect the electric field

$$\boldsymbol{E}' = \boldsymbol{E} + \frac{1}{\mathrm{c}} \boldsymbol{v} \times \boldsymbol{H} \qquad (14.2\text{-}11)$$

to be small. Thus, it is more correct to consider $\boldsymbol{E}'$ to be small (rather than $\boldsymbol{E}$) and to retain only the term involving $\boldsymbol{C}_i \times \boldsymbol{H}$ in the zeroth order approximation. Thus, in the zeroth order approximation, we must solve the system of equations

$$\frac{e_i}{m_i \mathrm{c}} \boldsymbol{C}_i \times \boldsymbol{H} \cdot \boldsymbol{\nabla}_{c_i} f_i^{(0)} = \sum_j J(f_i^{(0)} f_j^{(0)}). \tag{14.2-12}$$

The solution is again a set of Maxwellian velocity distributions,

$$f_i^{(0)} = n_i \left(\frac{m_i}{2\pi kT}\right)^{\frac{3}{2}} \exp\left(-m_i C_i^2/2kT\right) \tag{14.2-13}$$

with $\boldsymbol{C}_i = \boldsymbol{c}_i - \boldsymbol{v}$, as may be verified by direct substitution ($\boldsymbol{\nabla}_{c_i} f_i^{(0)}$ is proportional to $\boldsymbol{C}_i$ and thus orthogonal to $\boldsymbol{C}_i \times \boldsymbol{H}$). As before, the Maxwellian leads to a hydrostatic pressure tensor, zero heat flow vector, zero diffusion velocities and, therefore, a zero electric current density.

Proceeding to the first order in the Chapman-Enskog development, assuming each f_i to be of the form

$$f_i = f_i^{(0)}(1+\phi_i), \tag{14.2-14}$$

we find the following set of equations for the unknowns ϕ_i,

$$-\sum_j n_i n_j I_{ij}(\phi) - \frac{e_i}{m_i \mathrm{c}} \boldsymbol{C}_i \times \boldsymbol{H} \cdot \boldsymbol{\nabla}_{c_i} f_i^{(0)} \phi_i$$
$$= \frac{\partial_0 f_i^{(0)}}{\partial t} + \boldsymbol{c}_i \cdot \boldsymbol{\nabla}_r f_i^{(0)} + \left(\boldsymbol{F}_i + \frac{e_i}{m_i} \boldsymbol{E}'\right) \cdot \boldsymbol{\nabla}_{c_i} f_i^{(0)} \tag{14.2-15}$$

or, if $f_i^{(0)}$ is regarded as a function of $\boldsymbol{r}$, $\boldsymbol{C}_i$ and t, instead of $\boldsymbol{r}$, $\boldsymbol{c}_i$ and t,

$$-\sum_j n_i n_j I_{ij}(\phi) - \frac{e_i}{m_i \mathrm{c}} \boldsymbol{C}_i \times \boldsymbol{H} \cdot \boldsymbol{\nabla}_{C_i} f_i^{(0)} \phi_i$$
$$= f_i^{(0)} \left\{ \frac{\mathrm{d}_0 \log f_i^{(0)}}{\mathrm{d}t} + \boldsymbol{C}_i \cdot \boldsymbol{\nabla}_r \log f_i^{(0)} \right.$$
$$\left. + \left(\boldsymbol{F}_i + \frac{e_i}{m_i} \boldsymbol{E}' - \frac{\mathrm{d}_0 \boldsymbol{v}}{\mathrm{d}t}\right) \cdot \boldsymbol{\nabla}_{C_i} \log f_i^{(0)} - (\boldsymbol{\nabla}_{C_i} \log f_i^{(0)}) \boldsymbol{C}_i : \boldsymbol{\nabla} \boldsymbol{v} \right\}. \tag{14.2-16}$$

The term $(\mathrm{d}_0 \log f_i^{(0)}/\mathrm{d}t)$ is obtained in a manner similar to that described in Chapters 5 and 6. In fact, the evaluation of $(\mathrm{d}_0 n_i/\mathrm{d}t)$ and $(\mathrm{d}_0 T/\mathrm{d}t)$

is exactly analogous to that of the earlier chapters and we have

$$\frac{\mathrm{d}_0 n_i}{\mathrm{d}t} = -n_i \nabla \cdot \boldsymbol{v}, \tag{14.2-17}$$

$$\frac{\mathrm{d}_0 T}{\mathrm{d}t} = -\tfrac{2}{3} T \nabla \cdot \boldsymbol{v}. \tag{14.2-18}$$

In evaluating $(\mathrm{d}_0 \boldsymbol{v}/\mathrm{d}t)$, however, some extra care is necessary. Although $\boldsymbol{j} = 0$ in the zeroth order approximation, the term

$$\frac{1}{\mathrm{c}} \boldsymbol{j} \times \boldsymbol{H} = \sum_i \frac{e_i}{\mathrm{c}} \int \boldsymbol{C}_i \times \boldsymbol{H} f_i \mathrm{d}^3 c_i$$

$$= \sum_i \frac{e_i}{\mathrm{c}} \int \boldsymbol{C}_i \times \boldsymbol{H} f_i^{(0)} \phi_i \mathrm{d}^3 c_i$$

is of the same order as the term retained on the left-hand side of the Boltzmann equation in zeroth order. Consistency therefore demands that, although this term is apparently of first, rather than zeroth order, it be retained in the evaluation of $(\mathrm{d}_0 \boldsymbol{v}/\mathrm{d}t)$. Thus,

$$\frac{\mathrm{d}_0 \boldsymbol{v}}{\mathrm{d}t} = -\frac{1}{\rho} \nabla p + \frac{Q}{\rho} \boldsymbol{E}' + \frac{1}{\rho \mathrm{c}} \sum_i e_i \int f_i^{(0)} \phi_i (\boldsymbol{C}_i \times \boldsymbol{H}) \mathrm{d}^3 c_i + \frac{1}{\rho} \sum_i \rho_i \boldsymbol{F}_i. \tag{14.2-19}$$

Substituting these expressions into Eq. (14.2-15), we have the first-order Chapman-Enskog equation

$$\sum_j n_i n_j I_{ij}(\phi) + f_i^{(0)} \left[\frac{m_i}{\rho \mathrm{c} k T} \boldsymbol{C}_i \cdot \sum_j e_j \int \boldsymbol{C}_j \times \boldsymbol{H} f_j^{(0)} \phi_j \mathrm{d}^3 c_j \right.$$

$$\left. + \frac{e_i}{m_i \mathrm{c}} \boldsymbol{C}_i \times \boldsymbol{H} \cdot \nabla_{\boldsymbol{C}_i} \phi_i \right]$$

$$= -f_i^{(0)} \left[2(\mathscr{C}_i \mathscr{C}_i - \tfrac{1}{3} \mathscr{C}_i^2 \mathbf{I}) : \nabla \boldsymbol{v} + (\mathscr{C}_i^2 - \tfrac{5}{2}) \boldsymbol{C}_i \cdot \nabla \log T + \frac{n}{n_i} \boldsymbol{C}_i \cdot \boldsymbol{d}_i \right], \tag{14.2-20}$$

where $\boldsymbol{d}_i$, the diffusion driving force for species i, has the same meaning as in Chapter 6, cf. Eq. (6.3-10), except that the electric field term is explicit,

$$\boldsymbol{d}_i = \nabla \left(\frac{n_i}{n} \right) + \left(\frac{n_i}{n} - \frac{\rho_i}{\rho} \right) \nabla \log p - \frac{\rho_i}{p} \left(\boldsymbol{F}_i - \sum_j \frac{\rho_j}{\rho} \boldsymbol{F}_j \right)$$

$$- \frac{\rho_i}{p} \left[\frac{e_i}{m_i} - \sum_j \frac{n_j e_j}{\rho} \right] \boldsymbol{E}'. \tag{14.2-21}$$

As in the case of a neutral gas, $\sum_i \boldsymbol{d}_i = 0$. The solution to Eq. (14.2-20) may be expressed in the following form,

$$\phi_i = -\frac{1}{n}\mathbf{B}_i : \nabla \boldsymbol{v} - \frac{1}{n}\boldsymbol{A}_i \cdot \nabla \log T - \frac{1}{n}\sum_k \boldsymbol{D}_i^k \cdot \boldsymbol{d}_k. \qquad (14.2\text{-}22)$$

As in the neutral gas case, this ansatz permits the splitting of Eq. (14.2-20) into separate equations for the unknowns $\boldsymbol{A}$, $\mathbf{B}$ and $\boldsymbol{D}^k$. Again, some difficulty arises out of the linear dependence of the $\boldsymbol{d}_i$. This difficulty may be circumvented exactly as in Chapter 6, by introducing a set of $\boldsymbol{d}_i^*$, defined by Eq. (6.3-13). One then finds a set of equations for the functions $\boldsymbol{D}^k$,

$$\sum_j n_i n_j I_{ij}(\boldsymbol{D}^k) + \frac{m_i}{\rho c k T} f_i^{(0)} \boldsymbol{C}_i \cdot \sum_j e_j \int (\boldsymbol{C}_j \times \boldsymbol{H}) f_j^{(0)} \boldsymbol{D}_j^k \mathrm{d}^3 c_j$$

$$+ \frac{e_i}{m_i \mathrm{c}} f_i^{(0)} (\boldsymbol{C}_i \times \boldsymbol{H}) \cdot \nabla_{C_i} \boldsymbol{D}_i^k = \frac{n^2}{n_i} f_i^{(0)} \left(\delta_{ik} - \frac{\rho_i}{\rho}\right) \boldsymbol{C}_i, \qquad (14.2\text{-}23a)$$

together with the constraint

$$\sum_l (\rho_l/\rho) \boldsymbol{D}^l = 0. \qquad (14.2\text{-}23b)$$

The equations for the tensor $\mathbf{B}$ and the vector $\boldsymbol{A}$ are easily found to be

$$\sum_j n_i n_j I_{ij}(\mathbf{B}) + \frac{m_i}{\rho c k T} f_i^{(0)} \boldsymbol{C}_i \cdot \sum_j e_j \int (\boldsymbol{C}_j \times \boldsymbol{H}) f_j^{(0)} \mathbf{B}_j \mathrm{d}^3 c_j$$

$$+ \frac{e_i}{m_i \mathrm{c}} f_i^{(0)} (\boldsymbol{C}_i \times \boldsymbol{H}) \cdot \nabla_{C_i} \mathbf{B}_i = 2 n f_i^{(0)} (\mathscr{C}_i \mathscr{C}_i - \tfrac{1}{3} \mathscr{C}_i^2 \mathbf{I}) \qquad (14.2\text{-}24)$$

and

$$\sum_j n_i n_j I_{ij}(\boldsymbol{A}) + \frac{m_i}{\rho c k T} f_i^{(0)} \boldsymbol{C}_i \cdot \sum_j e_j \int (\boldsymbol{C}_j \times \boldsymbol{H}) f_j^{(0)} \boldsymbol{A}_j \mathrm{d}^3 c_j$$

$$+ \frac{e_i}{m_i \mathrm{c}} f_i^{(0)} (\boldsymbol{C}_i \times \boldsymbol{H}) \cdot \nabla_{C_i} \boldsymbol{A}_i = n f_i^{(0)} (\mathscr{C}_i^2 - \tfrac{5}{2}) \boldsymbol{C}_i, \qquad (14.2\text{-}25)$$

respectively.

The solution of these equations is more difficult than in the neutral gas case. We shall consider Eqs. (14.2-23a) and (14.2-25) first. Before proceeding, we note that the magnetic field vector, $\boldsymbol{H}$, is actually a pseudo-vector (it changes sign under coordinate inversion) and the vector product of two vectors is likewise a pseudovector. However, all the terms in Eq. (14.2-23a) and (14.2-25) are true (polar) vectors and, thus, $\boldsymbol{A}$ and $\boldsymbol{D}^k$ are true vectors and must be composed from the various true vectors that can be constructed from $\boldsymbol{C}$ and $\boldsymbol{H}$. The only such vectors are

$$\boldsymbol{C},\ \boldsymbol{C}\times\boldsymbol{H},\ (\boldsymbol{C}\cdot\boldsymbol{H})\boldsymbol{H}, \tag{14.2-26}$$

all other vectors can be expressed as linear combinations of these vectors. The expansion for $\boldsymbol{A}$ must, therefore, take the form

$$\boldsymbol{A} = \boldsymbol{A}^{(1)}+\boldsymbol{A}^{(2)}+\boldsymbol{A}^{(3)}, \tag{14.2-27a}$$

where

$$\boldsymbol{A}^{(1)} = A^{(1)}\boldsymbol{C},\ \boldsymbol{A}^{(2)} = A^{(2)}\boldsymbol{C}\times\boldsymbol{H},\ \boldsymbol{A}^{(3)} = A^{(3)}(\boldsymbol{C}\cdot\boldsymbol{H})\boldsymbol{H}, \tag{14.2-27b}$$

and, similarly, for $\boldsymbol{D}^k$,

$$\boldsymbol{D}^k = \boldsymbol{D}^{k(1)}+\boldsymbol{D}^{k(2)}+\boldsymbol{D}^{k(3)}, \tag{14.2-28a}$$

where

$$\boldsymbol{D}^{k(1)} = D^{k(1)}\boldsymbol{C},\quad \boldsymbol{D}^{k(2)} = D^{k(2)}\boldsymbol{C}\times\boldsymbol{H},\quad \boldsymbol{D}^{k(3)} = D^{k(3)}(\boldsymbol{C}\cdot\boldsymbol{H})\boldsymbol{H}. \tag{14.2-28b}$$

(The subscript i on the velocity variable is suppressed wherever its occurrence is obvious.) The coefficients in Eqs. (14.2-27) and (14.2-28) must be (scalar) functions of the scalars that can be created from $\boldsymbol{C}$ and $\boldsymbol{H}$. The only such true scalars are

$$C^2,\ H^2 \text{ and } (\boldsymbol{C}\cdot\boldsymbol{H})^2. \tag{14.2-29}$$

As the operators appearing on the left-hand sides of Eqs. (14.2-23a) and (14.2-25) are identical, the treatment of each is the same and we shall give the details for Eq. (14.2-25) only. From Eq. (14.2-27) we have, after some calculation,

$$(\boldsymbol{C}\times\boldsymbol{H})\cdot\nabla_C\boldsymbol{A} = A^{(1)}\boldsymbol{C}\times\boldsymbol{H}+A^{(2)}[(\boldsymbol{C}\cdot\boldsymbol{H})\boldsymbol{H}-H^2\boldsymbol{C}], \tag{14.2-30}$$

and with some further manipulation,

$$\begin{aligned}
\boldsymbol{C}_i\cdot\sum_j e_j&\int(\boldsymbol{C}_j\times\boldsymbol{H})f_j^{(0)}\boldsymbol{A}_j\,\mathrm{d}^3c_j\\
&= -(\boldsymbol{C}_i\times\boldsymbol{H})\cdot\sum_j e_j\int f_j^{(0)}\boldsymbol{C}_j\boldsymbol{A}_j\,\mathrm{d}^3c_j\\
&= \tfrac{1}{3}(\boldsymbol{C}_i\times\boldsymbol{H})\sum_j e_j\int f_j^{(0)}\left[C_j^2-\frac{1}{H^2}(\boldsymbol{C}_j\cdot\boldsymbol{H})^2\right]A_j^{(1)}\,\mathrm{d}^3c_j\\
&\quad+\tfrac{1}{2}[H^2\boldsymbol{C}_i-(\boldsymbol{C}_i\cdot\boldsymbol{H})\boldsymbol{H}]\sum_j e_j\int f_j^{(0)}\left[C_j^2-\frac{1}{H^2}(\boldsymbol{C}_j\cdot\boldsymbol{H})^2\right]A_j^{(2)}\mathrm{d}^3c_j.
\end{aligned} \tag{14.2-31}$$

When Eq. (14.2-27) is substituted into Eq. (14.2-25) and use is made of Eqs. (14.2-30) and (14.2-31), there result terms proportional to each of the vectors $\boldsymbol{C}_i$, $\boldsymbol{C}_i \times \boldsymbol{H}$ and $(\boldsymbol{C}_i \cdot \boldsymbol{H})\boldsymbol{H}$. As these vectors are linearly independent, the equation (14.2-25) splits into three separate equations – coupled integral equations for the functions $A_i^{(1)}$, $A_i^{(2)}$ and $A_i^{(3)}$ – viz.,

$$\sum_j n_i n_j I_{ij}(\boldsymbol{A}^{(1)}) + \frac{m_i}{2\rho c k T} f_i^{(0)} H^2 \boldsymbol{C}_i \sum_j e_j \int f_j^{(0)} \left[C_j^2 - \frac{1}{H^2}(\boldsymbol{C}_j \cdot \boldsymbol{H})^2 \right] A_j^{(2)} \mathrm{d}^3 c_j$$
$$- \frac{e_i}{m_i \mathrm{c}} f_i^{(0)} A_i^{(2)} H^2 \boldsymbol{C}_i = n f_i^{(0)} (\mathscr{C}_i^2 - \tfrac{5}{2}) \boldsymbol{C}_i, \qquad (14.2\text{-}32\text{a})$$

$$\sum_j n_i n_j I_{ij}(\boldsymbol{A}^{(2)}) - \frac{m_i}{2\rho c k T} f_i^{(0)} (\boldsymbol{C}_i \times \boldsymbol{H}) \sum_j e_j \int f_j^{(0)} \left[C_j^2 - \frac{1}{H^2}(\boldsymbol{C}_j \cdot \boldsymbol{H})^2 \right]$$
$$\times A_j^{(1)} \mathrm{d}^3 c_j + \frac{e_i}{m_i \mathrm{c}} f_i^{(0)} A_i^{(1)} \boldsymbol{C}_i \times \boldsymbol{H} = 0, \qquad (14.2\text{-}32\text{b})$$

$$\sum_j n_i n_j I_{ij}(\boldsymbol{A}^{(3)}) - \frac{m_i}{2\rho c k T} f_i^{(0)} (\boldsymbol{C}_i \cdot \boldsymbol{H}) \boldsymbol{H} \sum_j e_j \int f_j^{(0)} \left[C_j^2 - \frac{1}{H^2}(\boldsymbol{C}_j \cdot \boldsymbol{H})^2 \right]$$
$$\times A_j^{(2)} \mathrm{d}^3 c_j + \frac{e_i}{m_i \mathrm{c}} f_i^{(0)} A_i^{(2)} (\boldsymbol{C}_i \cdot \boldsymbol{H}) \boldsymbol{H} = 0. \qquad (14.2\text{-}32\text{c})$$

Further simplification is obtained if, instead of three real quantities $A^{(1)}$, $A^{(2)}$ and $A^{(3)}$, we introduce one real and one complex quantity, thus:

$$\mathfrak{A}^{(1)} = A^{(1)} + H^2 A^{(3)},$$
$$\mathfrak{A}^{(2)} = A^{(1)} + \mathrm{i} H A^{(2)}. \qquad (14.2\text{-}33)$$

By taking the scalar product of Eqs. (14.2-32a) and (14.2-32c) with $\boldsymbol{H}$ and adding the resulting equations, one finds a set of integral equations for the unknown vector $\boldsymbol{\mathfrak{A}}^{(1)} \equiv \mathfrak{A}^{(1)} \boldsymbol{C}$, viz.,

$$\sum_j n_i n_j I_{ij}(\boldsymbol{\mathfrak{A}}^{(1)}) = n f_i^{(0)} (\mathscr{C}_i^2 - \tfrac{5}{2}) \boldsymbol{C}_i. \qquad (14.2\text{-}34)$$

The latter equation is precisely the equation (6.3-19) for the vector $\boldsymbol{A} \equiv A\boldsymbol{C}$ in the case of neutral gases and thus requires no further discussion. The equations for the complex functions $\mathfrak{A}^{(2)}$ are obtained by taking the vector product of Eq. (14.2-32a) with $\boldsymbol{H}$ and multiplying Eq. (14.2-32b) by $\mathrm{i}H$, and then adding the two resulting equations. In terms of the unknown vector $\boldsymbol{\mathfrak{A}}^{(2)} \equiv \mathfrak{A}^{(2)} \boldsymbol{C}$ one has

$$\sum_j n_i n_j I_{ij}(\mathfrak{A}^{(2)}) - \mathrm{i}H \frac{m_i}{2\rho c k T} f_i^{(0)} \boldsymbol{C}_i \sum_j e_j \int f_j^{(0)} \left[C_j^2 - \frac{1}{H^2} (\boldsymbol{C}_j \cdot \boldsymbol{H})^2 \right] \mathfrak{A}_j^{(2)} \mathrm{d}^3 c_j$$
$$+ \mathrm{i}H \frac{e_i}{m_i \mathrm{c}} f_i^{(0)} \mathfrak{A}_i^{(2)} = n f_i^{(0)} (\mathscr{C}_i^2 - \tfrac{5}{2}) \boldsymbol{C}_i . \qquad (14.2\text{-}35)$$

Thus, instead of a system of three coupled integral equations for the unknowns $A^{(1)}$, $A^{(2)}$ and $A^{(3)}$, we now have two uncoupled integral equations for the unknowns $\mathfrak{A}^{(1)}$ and $\mathfrak{A}^{(2)}$. Once Eqs. (14.2-34) and (14.2-35) have been solved, it is an easy matter to obtain the original functions $A^{(1)}$, $A^{(2)}$ and $A^{(3)}$. Alternatively, we may express all transport quantities of interest in terms of the $\mathfrak{A}^{(1)}$ and $\mathfrak{A}^{(2)}$ directly and avoid further reference to the $A^{(1)}$, $A^{(2)}$ and $A^{(3)}$. Note also that, when $H = 0$, Eq. (14.2-35) reduces to (14.2-34) so that $\mathfrak{A}^{(1)} = \mathfrak{A}^{(2)}$, which implies that $A^{(1)} = A$, i.e., the neutral gas A, and furthermore, $A^{(2)} = A^{(3)} = 0$.

As was stated earlier, as a consequence of the similarity between Eqs. (14.2-23a) and (14.2-25) the development above can be followed through for $\boldsymbol{D}^k$ as well as $\boldsymbol{A}$. In fact, the functions $\boldsymbol{\mathfrak{D}}^{k(1)} \equiv \mathfrak{D}^{k(1)}\boldsymbol{C}$ and $\boldsymbol{\mathfrak{D}}^{k(2)} \equiv \mathfrak{D}^{k(2)}\boldsymbol{C}$, with

$$\begin{aligned} \mathfrak{D}^{k(1)} &= D^{k(1)} + H^2 D^{k(3)}, \\ \mathfrak{D}^{k(2)} &= D^{k(1)} + \mathrm{i}HD^{k(2)}, \end{aligned} \qquad (14.2\text{-}36)$$

obey equations identical to Eqs. (14.2-34) and (14.2-35), with the single change that, on the right-hand sides, the factors $(\mathscr{C}_i^2 - \frac{5}{2})$ are replaced by $(n/n_i)(\delta_{ik} - \rho_i/\rho)$. The solution of the equations for $\boldsymbol{\mathfrak{D}}^{k(1)}$ is thus identical to that of Eq. (6.3-18) for the functions $\boldsymbol{D}^k$ in the neutral gas case, and the solution of the equations for the $\mathfrak{D}^{k(2)}$ can be obtained in a manner analogous to that which will be used for Eq. (14.2-35).

As in the neutral gas case, the first-order Chapman-Enskog solutions must satisfy subsidiary conditions similar to Eqs. (6.3-21) and (6.3-22),

$$\sum_i m_i \int f_i^{(0)} C_i^2 \mathfrak{A}_i^{(j)} \mathrm{d}^3 c_i = 0, \quad j = 1, 2, \qquad (14.2\text{-}37a)$$

$$\sum_i m_i \int f_i^{(0)} C_i^2 \mathfrak{D}_i^{k(j)} \mathrm{d}^3 c_i = 0, \quad j = 1, 2; \text{all } k. \qquad (14.2\text{-}37b)$$

It is important to note that, because I_{ij} is an isotropic operator, i.e., it converts a tensor constructed from $\mathscr{C}$ into another tensor of the same type, $\mathfrak{A}^{(2)}$ cannot be a function of $(\mathscr{C} \cdot \boldsymbol{H})^2$. Then it is permissible to make the expansions

$$\mathfrak{A}_i^{(j)} \equiv \mathfrak{A}_i^{(j)} \boldsymbol{C}_i = -\left(\frac{m_i}{2kT}\right)^{\frac{1}{2}} \sum_{p=0}^{\infty} a_{i,p}^{(j)} S_{\frac{3}{2}}^{(p)}(\mathscr{C}_i^2)\mathscr{C}_i, \quad j = 1, 2, \tag{14.2-38a}$$

$$\mathfrak{D}_i^{k(j)} \equiv \mathfrak{D}_i^{k(j)} \boldsymbol{C}_i = \left(\frac{m_i}{2kT}\right)^{\frac{1}{2}} \sum_{p=0}^{\infty} d_{i,p}^{k(j)} S_{\frac{3}{2}}^{(p)}(\mathscr{C}_i^2)\mathscr{C}_i, \quad j = 1, 2; \text{all } k, \tag{14.2-38b}$$

where the coefficients are, of course, functions of the magnetic field strength, H.

In terms of the expansion coefficients the auxiliary conditions (14.2-37) become

$$\sum_i (\rho_i/\rho) a_{i,0}^{(j)} = 0, \qquad j = 1, 2, \tag{14.2-39a}$$

$$\sum_i (\rho_i/\rho) d_{i,0}^{k(j)} = 0, \qquad j = 1, 2; \text{all } k. \tag{14.2-39b}$$

Furthermore, the second term of Eq. (14.2-35) may be evaluated in terms of the expansion coefficients as follows:

$$\begin{aligned} &\int f_j^{(0)} \left[C_j^2 - \frac{1}{H^2}(\boldsymbol{C}_j \cdot \boldsymbol{H})^2 \right] \mathfrak{A}_j^{(2)} \mathrm{d}^3 c_j \\ &= -\frac{n_j}{\pi^{\frac{3}{2}}} \sum_{p=0}^{\infty} a_{j,p}^{(2)} \int \exp(-\mathscr{C}_j^2) \left[\mathscr{C}_j^2 - \frac{1}{H^2}(\mathscr{C}_j \cdot \boldsymbol{H})^2 \right] S_{\frac{3}{2}}^{(p)}(\mathscr{C}_j^2) \mathrm{d}^3 \mathscr{C}_j \\ &= -n_j a_{j,0}^{(2)}. \end{aligned} \tag{14.2-40}$$

The variational approach used in Chapter 6 may be applied to this problem; this technique has been worked out in detail by Marshall [1960] for the case of a binary mixture and is the most elegant and rigorous approach. Since the method followed in Chapter 6 need be changed in only minor details, we shall follow a short-cut approach similar to that used in Chapter 11. We simply truncate the expansions (14.2-38) after n terms, substitute them in Eq. (14.2-35) and the analogous equation for $\mathfrak{D}^{k(2)}$, take the scalar product of the resulting equations with $S_{\frac{3}{2}}^{(q)}(\mathscr{C}_i^2)\mathscr{C}_i$ for $q = 0, 1, \ldots, n-1$, and integrate over $\mathscr{C}_i$. This procedure yields the same equations for the expansion coefficients as the more elaborate variational method. Recall also that Eq. (14.2-34) requires no special treatment here, as it is identical to Eq. (6.3-19), which has already been studied. By the procedure outlined above, one has

$$\sum_j \sum_{q=0}^{n-1} \Lambda_{ij}^{pq} a_{j,q}^{(2)} + \mathrm{i} \frac{4H}{25cn^2k^2T} \frac{4}{3\sqrt{\pi}} \frac{\Gamma(p+\frac{5}{2})}{\Gamma(p+1)} \sum_j \left(\delta_{ij} - \delta_{p0} \frac{\rho_i}{\rho}\right) n_j e_j a_{j,p}^{(2)}$$
$$= \frac{4}{5k} \frac{n_i}{n} \delta_{p1}, \quad p = 0, \ldots, n-1; \text{ all } i; \qquad (14.2\text{-}41)$$

$$\sum_j \sum_{q=0}^{n-1} \Lambda_{ij}^{pq} d_{j,q}^{k(2)} + \mathrm{i} \frac{4H}{25cn^2k^2T} \frac{4}{3\sqrt{\pi}} \frac{\Gamma(p+\frac{5}{2})}{\Gamma(p+1)} \sum_j \left(\delta_{ij} - \delta_{p0} \frac{\rho_i}{\rho}\right) n_j e_j d_{j,p}^{(2)}$$
$$= \frac{8}{25k} (\delta_{ik} - \rho_i/\rho)\delta_{p0}, \quad p = 0, \ldots, n-1; \text{ all } i \text{ and } k. \qquad (14.2\text{-}42)$$

The coefficients Λ_{ij}^{pq} are identical with those defined in Chapter 6, Eq. (6.4-15). For $p = 0$, the equations (14.2-41) are linearly dependent. An additional equation is provided by the constraint (14.2-39a). Similarly, for $p = 0$, the system of equations (14.2-42) must be supplemented by Eq. (14.2-39b).

We turn now to the computation of the diffusion velocities of the various species and the heat flow vector. From the former, we readily obtain the electric current and, thereby, the electrical conductivity, which turns out to be a second-rank tensor. From the heat flow vector we obtain the thermal conductivity, which also turns out to be a second-rank tensor. As in all previous cases, the velocity gradient term in Eq. (14.2-22) makes no contribution to $\boldsymbol{V}_i$ or $\boldsymbol{q}$ and there remain only terms involving ∇T and the $\boldsymbol{d}_i$. In fact, from the form of the vectors $\boldsymbol{A}$ and $\boldsymbol{D}^k$ displayed in Eqs. (14.2-27) and (14.2-28), one sees that there will be terms proportional to the vectors

$$\nabla T, \boldsymbol{H} \times \nabla T, \boldsymbol{H}(\boldsymbol{H} \cdot \nabla T); \boldsymbol{d}_i, \boldsymbol{H} \times \boldsymbol{d}_i, \boldsymbol{H}(\boldsymbol{H} \cdot \boldsymbol{d}_i) \qquad (14.2\text{-}43)$$

in each of the flux vectors. This formulation is certainly valid, but, as the vectors $\boldsymbol{H}$ and ∇T (or $\boldsymbol{H}$ and $\boldsymbol{d}_i$) are not generally orthogonal, the first three vectors of the set (14.2-43) (and the last three, as well) do not form an orthogonal set. It is therefore preferable to express the flux vectors in terms of the following orthogonal sets:

$$(\nabla T)_\parallel = \boldsymbol{H}(\boldsymbol{H} \cdot \nabla T)/H^2, (\nabla T)_\perp = \nabla T - (\nabla T)_\parallel, (\nabla T)_t = \boldsymbol{H} \times \nabla T/H; \qquad (14.2\text{-}44a)$$

$$\boldsymbol{d}_{i_\parallel} = \boldsymbol{H}(\boldsymbol{H} \cdot \boldsymbol{d}_i)/H^2, \quad \boldsymbol{d}_{i_\perp} = \boldsymbol{d}_i - \boldsymbol{d}_{i_\parallel}, \quad \boldsymbol{d}_{i_t} = \boldsymbol{H} \times \boldsymbol{d}_i/H. \qquad (14.2\text{-}44b)$$

By a straight-forward direct calculation, one now finds

$$V_i = -[D_{Ti}^{\parallel}(\nabla \log T)_{\parallel} + D_{Ti}^{\perp}(\nabla \log T)_{\perp} + D_{Ti}^{t}(\nabla \log T)_{t}] - \sum_j [D_{ij}^{\parallel} d_{j_{\parallel}} + D_{ij}^{\perp} d_{j_{\perp}} + D_{ij}^{t} d_{j_t}], \quad (14.2\text{-}45)$$

where the *coefficients of diffusion and thermal diffusion* are defined in a manner analogous to that of Chapter 6,

$$D_{ij}^{\parallel} = \frac{1}{3n}[\mathfrak{D}^{i(1)}, \mathfrak{D}^{j(1)}] = \frac{1}{2n} d_{i,0}^{j(1)} = \frac{1}{2n} d_{j,0}^{i(1)}, \quad (14.2\text{-}46a)$$

$$D_{ij}^{\perp} + \mathrm{i}D_{ij}^{t} = \frac{1}{3n}[\mathfrak{D}^{i(2)}, \mathfrak{D}^{j(2)}] = \frac{1}{2n} d_{i,0}^{j(2)} = \frac{1}{2n} d_{j,0}^{i(2)}, \quad (14.2\text{-}46b)$$

$$D_{Ti}^{\parallel} = \frac{1}{3n}[\mathfrak{D}^{i(1)}, \mathfrak{A}^{(1)}] = -\frac{1}{2n} a_{i,0}^{(1)} = -\frac{5}{4n}\sum_j \frac{n_j}{n} d_{j,1}^{i(1)}, \quad (14.2\text{-}47a)$$

$$D_{Ti}^{\perp} + \mathrm{i}D_{Ti}^{t} = \frac{1}{3n}[\mathfrak{D}^{i(2)}, \mathfrak{A}^{(2)}] = -\frac{1}{2n} a_{i,0}^{(2)} = -\frac{5}{4n}\sum_j \frac{n_j}{n} d_{j,1}^{i(2)}. \quad (14.2\text{-}47b)$$

From the result (14.2-45) and the expression (14.2-21) for the diffusion driving forces we see that the diffusion velocity contains terms proportional to the concentration gradients, the pressure gradient, the difference between the external (non-electrical) forces acting on the various particles, the electric field (measured in the frame moving with the gas) and the temperature gradient. Except for the electric field term, these terms are the same as for diffusion in ordinary (neutral) gases. The important thing to note, however, is that in an ionized gas, in the presence of a magnetic field, particle fluxes perpendicular to both $\boldsymbol{H}$ and the ordinary diffusion and thermal diffusion flows occur. These transverse particle fluxes are characterized by the transverse diffusion and thermal diffusion coefficients, D_{ij}^{t} and D_{Ti}^{t}.

The electric current density is obtained almost trivially from Eqs. (14.2-7b) and (14.2-45),

$$\begin{aligned} j &= \sum_i n_i e_i V_i \\ &= -\sum_{ij} [\sigma_{i_{\parallel}}^{j} d'_{j_{\parallel}} + \sigma_{i_{\perp}}^{j} d'_{j_{\perp}} + \sigma_{i_t}^{j} d'_{j_t}] \\ &\quad - \sum_i [\varphi_{i_{\parallel}}(\nabla T)_{\parallel} + \varphi_{i_{\perp}}(\nabla T)_{\perp} + \varphi_{i_t}(\nabla T)_t]. \end{aligned} \quad (14.2\text{-}48)$$

Here, the $\boldsymbol{d}'$ are related to the $\boldsymbol{d}$ by

$$\begin{aligned}\boldsymbol{d}_i' &= \frac{p}{\rho_i}\left[\frac{e_i}{m_i} - \sum_j \frac{n_j e_j}{\rho}\right]^{-1} \boldsymbol{d}_i \\ &= \frac{p}{\rho_i}\left[\frac{e_i}{m_i} - \sum_j \frac{n_j e_j}{\rho}\right]^{-1}\left[\nabla\left(\frac{n_i}{n}\right) + \left(\frac{n_i}{n} - \frac{\rho_i}{\rho}\right)\nabla \log p\right. \\ &\quad \left. - \frac{\rho_i}{p}\left(\boldsymbol{F}_i - \sum_j \frac{\rho_j}{\rho}\boldsymbol{F}_j\right)\right] - \boldsymbol{E}', \qquad (14.2\text{-}49)\end{aligned}$$

while the coefficients σ and φ are related to the coefficients of diffusion and thermal diffusion, respectively,

$$\sigma_i^j = \frac{n_i e_i \rho_j}{p}\left[\frac{e_j}{m_j} - \sum_k \frac{n_k e_k}{\rho}\right] D_{ij}, \qquad (14.2\text{-}50a)$$

$$\varphi_i = \frac{n_i e_i}{T} D_{Ti}. \qquad (14.2\text{-}50b)$$

(Each of the relations (14.2-49) and (14.2-50) is understood to hold for the parallel, perpendicular and transverse quantities separately.) The change from the $\boldsymbol{d}_i$ to the $\boldsymbol{d}_i'$ is made in order to eliminate the coefficient appearing in $\boldsymbol{d}_i$ multiplying the electric field term, thus clearing the way for the interpretation of the expression (14.2-48) as *Ohm's law* for an ionized gas. The coefficients σ_i^j are called the *partial electrical conductivities*; the coefficients φ_i are the *partial electro-thermal* (*inverse thermo-electric*) *coefficients*. The elements of the total electrical conductivity tensor and the total electro-thermal conductivity tensor are found from the partial coefficients (14.2-50) by summing over the indices, i.e.,

$$\sigma = \sum_{ij} \sigma_i^j, \qquad (14.2\text{-}51a)$$

$$\varphi = \sum_i \varphi_i. \qquad (14.2\text{-}51b)$$

In the present case the diffusivities, conductivities and thermo-electric coefficients are all tensors; this is a consequence of the nature of the Lorentz force. By using the coordinate systems defined by Eqs. (14.2-44) we have brought the tensors to their principal axes, i.e., we have chosen the coordinate systems in which the tensors are diagonal. The transport coefficient tensors in other coordinate systems may be obtained by the normal rules for the rotation of vectors and tensors.

Observe that the presence of an electric field $\boldsymbol{E}'$ produces an electric current which has a component perpendicular to both the magnetic field and the electric field, characterized by the transverse electrical conductivity, σ_t. This transverse current is commonly known as the *Hall current* and is due to the fact that in crossed electric and magnetic fields a charged particle drifts with a velocity proportional to $\boldsymbol{E}' \times \boldsymbol{H}$.

From Eq. (14.2-34) for $\mathfrak{A}^{(1)}$ and the analogous equation for $\mathfrak{D}^{k(1)}$ we see that these quantities are independent of the magnetic field strength H. Thus the electrical conductivity and the thermo-electric coefficients parallel to the magnetic field are unaffected by the field. This is a natural consequence of the fact that the magnetic field has no effect on particle motions parallel to itself.

In the limit of vanishing magnetic field $\mathfrak{A}^{(2)} = \mathfrak{A}^{(1)}$ and $\mathfrak{D}^{k(2)} = \mathfrak{D}^{k(1)}$, so that the electrical conductivity and thermo-electric coefficients perpendicular to the magnetic field become equal to their counterparts parallel to the field in this limit. As the field strength increases, however, the perpendicular transport coefficients decrease while the transverse coefficients increase. This may be traced to the turning of particles with velocities perpendicular to the magnetic field. Explicit formulae displaying these effects will be given in the next section.

Next, we compute the heat flux vector. Again expressing the results in terms of the vector sets (14.2-44), we find

$$\begin{aligned} \boldsymbol{q} = &-[\lambda'_{\parallel}(\nabla T)_{\parallel} + \lambda'_{\perp}(\nabla T)_{\perp} + \lambda'_{t}(\nabla T)_{t}] \\ &- p \sum_i [D^{\parallel}_{Ti}\boldsymbol{d}_{i\parallel} + D^{\perp}_{Ti}\boldsymbol{d}_{i\perp} + D^{t}_{Ti}\boldsymbol{d}_{it}] + \tfrac{5}{2}kT \sum_i n_i \boldsymbol{V}_i, \end{aligned} \qquad (14.2\text{-}52)$$

where the *partial coefficients of thermal conductivity*, λ', are given by

$$\lambda'_{\parallel} = \tfrac{1}{3}k[\mathfrak{A}^{(1)}, \mathfrak{A}^{(1)}] = \tfrac{5}{4}\sum_i \frac{n_i}{n} a^{(1)}_{i,1}, \qquad (14.2\text{-}53a)$$

$$\lambda'_{\perp} + i\lambda'_{t} = \tfrac{1}{3}k[\mathfrak{A}^{(2)}, \mathfrak{A}^{(2)}] = \tfrac{5}{4}\sum_i \frac{n_i}{n} a^{(2)}_{i,1}, \qquad (14.2\text{-}53b)$$

and the D_{Ti} are the thermal diffusion coefficients defined in Eq. (14.2-47). As in the neutral gas case, the λ' are not the true thermal conductivities. Again, this is the case because thermal conductivities are usually measured in situations in which the diffusion velocities of all species (and, of course, the electric current) are zero. To obtain the true conductivities one must again eliminate the $\boldsymbol{d}_j$ from Eq. (14.2-52). This may be done for each component separately following a procedure analogous to that used in Chapter 6. Thus, introducing the parallel, perpendicular and transverse thermal diffusion ratios, $k_{Ti}^{\parallel}$, $k_{Ti}^{\perp}$ and k_{Ti}^{t}, respectively, for each component of the mixture, by the definition

$$\sum_j D_{ij} k_{Tj} = D_{Ti}, \tag{14.2-54a}$$

$$\sum_i k_{Ti} = 0, \tag{14.2-54b}$$

and using Eq. (6.3-47) for each of the three directions, we obtain the result

$$\begin{aligned} \boldsymbol{q} = & -[\lambda_{\parallel}(\nabla T)_{\parallel} + \lambda_{\perp}(\nabla T)_{\perp} + \lambda_{\mathrm{t}}(\nabla T)_{\mathrm{t}}] \\ & + p \sum_i \left[\left(k_{Ti}^{\parallel} + \frac{5}{2}\frac{n_i}{n} \right) V_{i_{\parallel}} + \left(k_{Ti}^{\perp} + \frac{5}{2}\frac{n_i}{n} \right) V_{i_{\perp}} + \left(k_{Ti}^{\mathrm{t}} + \frac{5}{2}\frac{n_i}{n} \right) V_{i_{\mathrm{t}}} \right], \end{aligned} \tag{14.2-55}$$

where the λ's are the *coefficients of thermal conductivity*,

$$\lambda = \lambda' - nk \sum_i k_{Ti} D_{Ti}. \tag{14.2-56}$$

(This relation is understood to hold for the parallel, perpendicular and transverse coefficients of thermal conductivity separately.)

The contributions to the heat flux vector behave in a manner analogous to the contributions to the electric current. In the absence of the field the parallel and perpendicular thermal conductivities are equal and the transverse conductivity is zero. As the field strength is increased, the parallel conductivity remains constant, the perpendicular conductivity decreases and the transverse conductivity increases.

Finally, we consider the solution of Eq. (14.2-24) and evaluate the pressure tensor. In terms of the dimensionless velocity variable $\mathscr{C}$, Eq. (14.2-24) reads

$$\begin{aligned} & \sum_j n_i n_j I_{ij}(\mathbf{B}) + \frac{2}{\rho c} f_i^{(0)} \mathscr{C}_i \cdot \sum_j \left(\frac{m_i}{m_j}\right)^{\frac{1}{2}} \frac{n_j e_j}{\pi^{\frac{3}{2}}} \int \exp(-\mathscr{C}_j^2)(\mathscr{C}_j \times \boldsymbol{H}) \mathbf{B}_j \, \mathrm{d}^3\mathscr{C}_j \\ & \quad + \frac{e_i}{m_i c} f_i^{(0)} (\mathscr{C}_i \times \boldsymbol{H}) \cdot \nabla_{\mathscr{C}_i} \mathbf{B}_i = 2n f_i^{(0)} (\mathscr{C}_i \mathscr{C}_i - \tfrac{1}{3} \mathscr{C}_i^2 \mathbf{I}). \end{aligned} \tag{14.2-57}$$

The tensor **B** must be constructed from all the true symmetric, traceless second-rank tensors that can be created from the vector $\mathscr{C}$ and the pseudovector $\boldsymbol{H}$. The choice is again arbitrary, since any complete, linearly independent set will suffice. The simplest approach is to adopt the dyadics that are formed by taking the vectors $\mathscr{C}$, $\mathscr{C}\times\boldsymbol{H}$ and $(\mathscr{C}\cdot\boldsymbol{H})\boldsymbol{H}$ in pairs (after symmetrizing and removing the trace). Thus we have

$$\mathbf{T}^1 = \mathscr{C}\mathscr{C}-\tfrac{1}{3}\mathscr{C}^2\mathbf{I}, \tag{14.2-58a}$$

$$\mathbf{T}^2 = \tfrac{1}{2}[\mathscr{C}(\mathscr{C}\times\boldsymbol{H})+(\mathscr{C}\times\boldsymbol{H})\mathscr{C}], \tag{14.2-58b}$$

$$\mathbf{T}^3 = (\mathscr{C}\times\boldsymbol{H})(\mathscr{C}\times\boldsymbol{H})-\tfrac{1}{3}[\mathscr{C}^2H^2-(\mathscr{C}\cdot\boldsymbol{H})^2]\mathbf{I}, \tag{14.2-58c}$$

$$\mathbf{T}^4 = \tfrac{1}{2}[\mathscr{C}\boldsymbol{H}+\boldsymbol{H}\mathscr{C}]-\tfrac{1}{3}(\mathscr{C}\cdot\boldsymbol{H})^2\mathbf{I}, \tag{14.2-58d}$$

$$\mathbf{T}^5 = \tfrac{1}{2}[\boldsymbol{H}(\mathscr{C}\times\boldsymbol{H})+(\mathscr{C}\times\boldsymbol{H})\boldsymbol{H}]\mathscr{C}\cdot\boldsymbol{H}, \tag{14.2-58e}$$

$$\mathbf{T}^6 = \boldsymbol{H}\boldsymbol{H}(\mathscr{C}\cdot\boldsymbol{H})^2-\tfrac{1}{3}H^2(\mathscr{C}\cdot\boldsymbol{H})^2\mathbf{I}. \tag{14.2-58f}$$

It should be noted that not all of these tensors are independent, since there can be at most five independent symmetric traceless tensors of rank two. In fact it may be shown that

$$\mathscr{C}^2\mathbf{T}^6+(\mathscr{C}\cdot\boldsymbol{H})^2[\mathbf{T}^3+H^2\mathbf{T}^1-2\mathbf{T}^4] = 0. \tag{14.2-59}$$

However, it is simpler to work with all six tensors and doing so is justified by the fact that the various tensors are of different powers in the components of the vectors $\mathscr{C}$ and $\boldsymbol{H}$. Thus, using all six tensors is equivalent to rearranging the Sonine polynomial expansions used later. It also has the important advantage of removing all dependence of the expansion coefficients on $(\mathscr{C}\cdot\boldsymbol{H})^2$. Next, we expand $\mathbf{B}$ in terms of these tensors,

$$\mathbf{B} = \sum_{n=1}^{6} B^{(n)}\mathbf{T}^n, \tag{14.2-60}$$

where the $B^{(n)}$ are scalar functions of $\mathscr{C}^2$, H^2 and $(\mathscr{C}\cdot\boldsymbol{H})^2$.

From the form of the $\mathbf{T}^n$ and the fact that $B^{(n)}$ depends only on $\mathscr{C}^2$, $(\mathscr{C}\cdot\boldsymbol{H})^2$ and H^2, we see that all terms in the integrand of the second term in Eq. (14.2-57) are odd in at least one component of $\mathscr{C}$ and, therefore, the integral vanishes identically. Next, we turn attention to the third term. From the nature of the $B^{(n)}$ one readily shows that

$$\mathscr{C}\times\boldsymbol{H}\cdot\nabla_{\mathscr{C}}B^{(n)} = 0, \qquad n = 1, 2, \ldots, 6 \tag{14.2-61}$$

so that this operator effectively acts only on the tensors $\mathbf{T}^n$. Its action produces:

$$(\mathscr{C}\times H)\cdot\nabla_{\mathscr{C}}\mathbf{T}^1 = 2\mathbf{T}^2, \qquad (14.2\text{-}62a)$$

$$(\mathscr{C}\times H)\cdot\nabla_{\mathscr{C}}\mathbf{T}^2 = -H^2\mathbf{T}^1+\mathbf{T}^3+\mathbf{T}^4, \qquad (14.2\text{-}62b)$$

$$(\mathscr{C}\times H)\cdot\nabla_{\mathscr{C}}\mathbf{T}^3 = -2H^2\mathbf{T}^2+2\mathbf{T}^5, \qquad (14.2\text{-}62c)$$

$$(\mathscr{C}\times H)\cdot\nabla_{\mathscr{C}}\mathbf{T}^4 = \mathbf{T}^5, \qquad (14.2\text{-}62d)$$

$$(\mathscr{C}\times H)\cdot\nabla_{\mathscr{C}}\mathbf{T}^5 = -H^2\mathbf{T}^4+\mathbf{T}^6, \qquad (14.2\text{-}62e)$$

$$(\mathscr{C}\times H)\cdot\nabla_{\mathscr{C}}\mathbf{T}^6 = 0. \qquad (14.2\text{-}62f)$$

As the collision operator I_{ij} is a linear, isotropic operator in velocity space, its action on any one of the tensors constructed from $\mathscr{C}$ produces a tensor of precisely the same type multiplied by a scalar. Then, substituting the expansion (14.2-60) into Eq. (14.2-57) and equating the coefficients of the various tensors, one obtains six equations for the $B^{(n)}$. Note that this procedure is permissible even though the tensors $\mathbf{T}^n$ are not linearly independent, because they contain different powers of the components of $\boldsymbol{H}$. The results are

$$\sum_j n_i n_j I_{ij}(\mathbf{B}^{(1)}) - \frac{e_i}{m_i c} f_i^{(0)} H^2 \mathbf{B}_i^{(2)} = 2n f_i^{(0)}(\mathscr{C}_i\mathscr{C}_i - \tfrac{1}{3}\mathscr{C}_i^2\mathbf{I}), \qquad (14.2\text{-}63a)$$

$$\sum_j n_i n_j I_{ij}(\mathbf{B}^{(2)}) + \frac{2e_i}{m_i c} f_i^{(0)}(\mathbf{B}_i^{(1)} - H^2\mathbf{B}_i^{(3)}) = 0, \qquad (14.2\text{-}63b)$$

$$\sum_j n_i n_j I_{ij}(\mathbf{B}^{(3)}) + \frac{e_i}{m_i c} f_i^{(0)}\mathbf{B}_i^{(2)} = 0, \qquad (14.2\text{-}63c)$$

$$\sum_j n_i n_j I_{ij}(\mathbf{B}^{(4)}) + \frac{e_i}{m_i c} f_i^{(0)}(\mathbf{B}_i^{(2)} - H^2\mathbf{B}_i^{(5)}) = 0, \qquad (14.2\text{-}63d)$$

$$\sum_j n_i n_j I_{ij}(\mathbf{B}^{(5)}) + \frac{e_i}{m_i c} f_i^{(0)}(2\mathbf{B}_i^{(3)} + \mathbf{B}_i^{(4)}) = 0, \qquad (14.2\text{-}63e)$$

$$\sum_j n_i n_j I_{ij}(\mathbf{B}^{(6)}) + \frac{e_i}{m_i c} f_i^{(0)}\mathbf{B}_i^{(5)} = 0, \qquad (14.2\text{-}63f)$$

where we have abbreviated

$$\mathbf{B}^{(n)} = B^{(n)}(\mathscr{C}\mathscr{C} - \tfrac{1}{3}\mathscr{C}^2\mathbf{I}). \qquad (14.2\text{-}64)$$

We note immediately that by combining Eqs. (14.2-63c, d, f) we find

$$\sum_j n_i n_j I_{ij}(\mathbf{B}^{(3)} - \mathbf{B}^{(4)} - H^2\mathbf{B}^{(6)}) = 0 \qquad (14.2\text{-}65)$$

and, as this equation possesses only the trivial solution, we must have

$$H^2 B^{(6)} = B^{(3)} - B^{(4)}. \tag{14.2-66}$$

Thus, it is sufficient to solve Eqs. (14.2-63a–e) and then determine $B^{(6)}$ from Eq. (14.2-66). As was the case for the $A^{(i)}$, considerable simplification results from the use of linear combinations of the $B^{(i)}$. In particular, we define

$$\mathfrak{B}_j^{(1)} = B_j^{(1)} + H^2 B_j^{(3)}, \tag{14.2-67a}$$

$$\mathfrak{B}_j^{(2)} = B_j^{(1)} + iHB_j^{(2)} - H^2 B_j^{(3)}, \tag{14.2-67b}$$

$$\mathfrak{B}_j^{(3)} = B_j^{(1)} + \tfrac{1}{2} iHB_j^{(2)} + \tfrac{1}{2} H^2 B_j^{(4)} + \tfrac{1}{2} iH^3 B_j^{(5)}. \tag{14.2-67c}$$

Again, as in the case of the $A^{(i)}$, we can obtain the $B^{(i)}$ once the $\mathfrak{B}^{(i)}$ are determined, or we can simply express the pressure tensor in terms of the $\mathfrak{B}^{(i)}$ and avoid further reference to the $B^{(i)}$. The equations for the $\mathfrak{B}^{(i)}$ are readily obtained by forming the appropriate linear combinations of Eqs. (14.2-63),

$$\sum_j n_i n_j I_{ij}(\boldsymbol{\mathfrak{B}}^{(1)}) = 2n f_i^{(0)}(\mathscr{C}_i \mathscr{C}_i - \tfrac{1}{3}\mathscr{C}_i^2 \mathbf{I}), \tag{14.2-68a}$$

$$\sum_j n_i n_j I_{ij}(\boldsymbol{\mathfrak{B}}^{(2)}) + \frac{2iHe_i}{m_i c} f_i^{(0)} \boldsymbol{\mathfrak{B}}_i^{(2)} = 2n f_i^{(0)}(\mathscr{C}_i \mathscr{C}_i - \tfrac{1}{3}\mathscr{C}_i^2 \mathbf{I}), \tag{14.2-68b}$$

$$\sum_j n_i n_j I_{ij}(\boldsymbol{\mathfrak{B}}^{(3)}) + \frac{iHe_i}{m_i c} f_i^{(0)} \boldsymbol{\mathfrak{B}}_i^{(3)} = 2n f_i^{(0)}(\mathscr{C}_i \mathscr{C}_i - \tfrac{1}{3}\mathscr{C}_i^2 \mathbf{I}), \tag{14.2-68c}$$

where $\boldsymbol{\mathfrak{B}} = \mathfrak{B}\,(\mathscr{C}\,\mathscr{C} - \tfrac{1}{3}\mathscr{C}^2\,\mathbf{I})$, and we are left with three uncoupled integral equations which must be solved. The first of these, Eq. (14.2-68a), is just the equation (6.3-20) for the tensor $\mathbf{B}$ of the neutral gas problem. It is also clear that, as H approaches zero, all the $\mathfrak{B}^{(i)}$ become identical. Furthermore, if $H \neq 0$, the three equations (14.2-68) differ only in the coefficient of the second term on the left-hand side so that, if we denote

$$\mathfrak{B}^{(2)} \equiv \mathfrak{B}(\mathscr{C}, H), \tag{14.2-69a}$$

we shall then have

$$\mathfrak{B}^{(1)} \equiv \mathfrak{B}(\mathscr{C}, 0), \tag{14.2-69b}$$

$$\mathfrak{B}^{(3)} \equiv \mathfrak{B}(\mathscr{C}, \tfrac{1}{2}H). \tag{14.2-69c}$$

Thus, it is actually sufficient to solve just the single equation (14.2-68b), the solutions of the others then being determined via Eqs. (14.2-69).

The method of solution again follows the techniques established in Chapter 6. First we note that $\mathfrak{B}$ may be a function of the scalars $\mathscr{C}^2$,

H^2, $(\mathscr{C}\cdot\boldsymbol{H})^2$. But the structure of Eq. (14.2-68b) forbids $(\mathscr{C}\cdot\boldsymbol{H})^2$ from being an argument of $\mathfrak{B}$. Thus we may expand

$$\mathfrak{B}_i(\mathscr{C}_i, H) = \sum_{p=0}^{\infty} b_{i,p}(H) S_{\frac{5}{2}}^{(p)}(\mathscr{C}_i^2), \tag{14.2-70}$$

where we have indicated explicitly that the coefficients are functions of H only. Following the technique used for the $\mathfrak{A}^{(2)}$, we truncate the series (14.2-70) at n terms, substitute the truncated series into Eq. (14.2-68b), take the scalar product with $S_{\frac{5}{2}}^{(q)}(\mathscr{C}_i^2)(\mathscr{C}_i\mathscr{C}_i - \frac{1}{3}\mathscr{C}_i^2\mathsf{I})$, $q = 0, 1, \ldots, n-1$, and integrate over $\mathscr{C}_i$ to obtain

$$\sum_j \sum_{q=0}^{n-1} H_{ij}^{pq} b_{j,q} + \frac{2iH}{cn^2kT}\frac{8}{15\sqrt{\pi}}\frac{\Gamma(p+\frac{7}{2})}{\Gamma(p+1)}\frac{n_i e_i}{m_i} b_{i,p} = \frac{2}{kT}\frac{n_i}{n}\delta_{p0},$$
$$p = 0, 1, \ldots, n-1; \tag{14.2-71}$$

the coefficients H_{ij}^{pq} are identical with those defined in Chapter 6, Eq. (6.4-36).

All that remains is to compute the pressure tensor and the viscosity, which now becomes a fourth-rank tensor. Again, as in the neutral gas case, the last two terms of Eq. (14.2-22) make no contribution and we have

$$\mathsf{P} = p\mathsf{I} - \sum_i m_i \int \boldsymbol{C}_i \boldsymbol{C}_i f_i^{(0)} \mathbf{B}_i : \nabla \boldsymbol{v} \mathrm{d}^3 c_i. \tag{14.2-72}$$

The working out of the integral in terms of the expansion coefficients has been done by Marshall by the straight-forward method of using a coordinate system in which $\boldsymbol{H}$ lies along one of the coordinate axes. It is somewhat simpler, however, to use tensor index notation and this approach has the advantage of expressing P in a form which is not tied to the choice of coordinate system. Using the relations (14.2-66), (14.2-67) and (14.2-69) and the expansion (14.2-70) one can express the components of P in the following form

$$P_{\alpha\beta} = p\delta_{\alpha\beta} - \tfrac{1}{2}kT \sum_i n_i \Big\{ [b_{i,0}(0) + \tfrac{1}{2}(b_{i,0}(H) + b_{i,0}^*(H))] S_{\alpha\beta}$$
$$+ \frac{1}{2iH}[b_{i,0}(H) - b_{i,0}^*(H)][\varepsilon_{\alpha\zeta\eta}\delta_{\beta\gamma}H_\zeta + \varepsilon_{\beta\zeta\eta}\delta_{\alpha\gamma}H_\zeta] S_{\gamma\eta}$$
$$+ \frac{1}{H^2}[b_{i,0}(0) - \tfrac{1}{2}(b_{i,0}(H) + b_{i,0}^*(H))][\varepsilon_{\alpha\zeta\eta}\varepsilon_{\beta\theta\gamma}H_\zeta H_\theta + H^{-2}H_\alpha H_\beta H_\gamma H_\eta] S_{\gamma\eta} +$$

$$+\frac{1}{H^2}[(b_{i,0}(\tfrac{1}{2}H)+b^*_{i,0}(\tfrac{1}{2}H))-b_{i,0}(0)-\tfrac{1}{2}(b_{i,0}(H)+b^*_{i,0}(H))]$$

$$\times[\delta_{\alpha\gamma}H_\beta H_\eta+\delta_{\beta\gamma}H_\alpha H_\eta-2H^{-2}H_\alpha H_\beta H_\gamma H_\eta]S_{\gamma\eta}$$

$$+\frac{1}{iH^3}[(b_{i,0}(\tfrac{1}{2}H)-b^*_{i,0}(\tfrac{1}{2}H))-\tfrac{1}{2}(b_{i,0}(H)-b^*_{i,0}(H))]$$

$$\left.\times[\varepsilon_{\alpha\zeta\eta}H_\beta H_\zeta H_\gamma+\varepsilon_{\beta\zeta\eta}H_\alpha H_\zeta H_\gamma]S_{\gamma\eta}\right\}. \qquad (14.2\text{-}73)$$

In this expression the asterisk (*) represents complex conjugation and we have used the tensor sum convention, i.e., it is understood that any index which is repeated is to be summed over. Also, ϵ represents the alternating tensor, i.e., the completely antisymmetric tensor of rank three; its presence arises because $(\boldsymbol{A}\times\boldsymbol{B})_i=\varepsilon_{ijk}A_jB_k$. See Appendix A for further details on tensor analysis.

Note that Eq. (14.2-73) can be written symbolically as

$$P_{\alpha\beta}=p\delta_{\alpha\beta}-2\eta_{\alpha\beta\gamma\eta}S_{\gamma\eta}, \qquad (14.2\text{-}74)$$

where $\eta_{\alpha\beta\gamma\eta}$ is a fourth-rank tensor, which may be identified with the *coefficient of viscosity*. Although such a tensor may have as many as 81 independent components, a large number of its elements are zero and the tensor possesses a considerable degree of symmetry. It is symmetric, for example, under the interchange of α and β and under the interchange of γ and η. In fact, since $\eta_{\alpha\beta\gamma\eta}$ is determined by just five numbers – viz., $\sum_i b_{i,0}(H)$, $\sum_i b^*_{i,0}(H)$, $\sum_i b_{i,0}(\tfrac{1}{2}H)$, $\sum_i b^*_{i,0}(\tfrac{1}{2}H)$ and $\sum_i b_{i,0}(0)$ – only five of its components are truly independent. De Groot and Mazur [1962] show, on very general grounds, that there may, in principle, be as many as seven independent coefficients. Evidently, even in the presence of a magnetic field an ionized gas of molecules with no internal structure possesses symmetry beyond what is required by the general principles of non-equilibrium thermodynamics. As opposed to the thermal and electrical conductivities, however, the viscosity coefficients have no standard nomenclature.

In the next section we shall complete the computation of the transport coefficients for ionized gases. The only job remaining is to actually calculate the coefficients from the systems of linear algebraic equations (14.2-41), (14.2-42) and (14.2-71), solve the resulting equations for the expansion coefficients $a^{(2)}_{i,p}$, $d^{k(2)}_{i,p}$ and $b_{i,p}$ and substitute the appropriate coefficients into the expressions for the various transport coefficients. Apart from the calculation of the bracket integrals, the procedure is very similar to what has been done in Chapters 7 and 9 for neutral gases. We shall, therefore, be brief in our presentation and focus upon the difficulties encountered in computing the matrix elements (or Ω-integrals) for the case of the truncated Coulomb potential (14.1-17) and present results for a binary mixture of singly ionized atoms and electrons.

14.3. Results for binary mixtures of ionized gases

With the aid of the techniques of Chapter 7, the expressions for the transport coefficients of an ionized gas developed in the previous section may be reduced to the ratios of determinants whose elements are linear combinations of Ω-integrals and the mole fractions of the various constituents. As the results are very similar to those for neutral gases, we restrict ourselves to the presentation of explicit results for binary mixtures. In particular, we consider a fully ionized gas consisting of singly ionized atoms of mass m and number density n and electrons whose mass is m_e and whose number density is also n. As m_e is always small compared to m (for protons, $m_e/m = 1/1836$), we neglect the ratio m_e/m wherever it occurs.

Our first task is to reduce the bracket integrals that occur in the coefficients Λ and H, cf. Eqs. (14.2-41), (14.2-42) and (14.2-70). As was stated earlier, difficulty arises out of the long range of the Coulomb potential. The bracket integrals can indeed be reduced to Ω-integrals, as in Chapter 7, and the latter can be computed as was done for point centers of repulsion in Section 9.2(b) using $\nu = 1$ and $\sigma_{ij} = e_i e_j$. In fact, for $\nu = 1$, the integral (9.2-5) can be evaluated analytically to yield

$$\cos \chi = \frac{4z^2 - 1}{4z^2 + 1}, \tag{14.3-1}$$

where z is given by Eq. (9.2-6) with the substitutions given above, i.e.,

$$z = (\tfrac{1}{2} m_{ij} g^2 / e_i e_j) b. \tag{14.3-2}$$

Thus, the factor $A_l(1)$ of Eq. (9.2-9) becomes

$$A_l(1) = \int_0^\infty (1 - \cos^l \chi) z \, dz = \int_0^\infty \left[1 - \left(\frac{4z^2 - 1}{4z^2 + 1}\right)^l\right] z \, dz. \tag{14.3-3}$$

It is here that the difficulty arises since, for large z the integrand falls off only as z^{-1}. The integral is, therefore, logarithmically divergent. As z is just a non-dimensional impact parameter (as may be seen from Eq. (14.3-2)) the divergence clearly arises from the long-range nature of the $1/r$ potential. On the other hand, from the discussion of Section 14.1, we know that at large distances the effective potential actually falls off more rapidly than $1/r$ due to Debye shielding and it is appropriate to use the truncated potential (14.1-7) in place of the unmodified Coulomb potential. The truncation means that an impact parameter b larger than the Debye length d, does not result in collisions ($\chi = 0$), and the integral (9.2-5) should be cut off with an upper limit d. However, the latter change does not result in a large change in the integral, except when the impact parameter b is very close to d. Thus, as a good approximation, we may take

$$\cos\chi = \begin{cases} \dfrac{4z^2-1}{4z^2+1} & \text{if} \quad z < z_0, \\ 1 & \text{if} \quad z > z_0, \end{cases} \tag{14.3-4}$$

where

$$z_0 = (\tfrac{1}{2}m_{ij}g^2/e_i e_j)d. \tag{14.3-5}$$

Then the integral defining $A_l(1)$ in Eq. (14.3-3) is also cut off (at the upper limit z_0) and one has

$$A_1(1) \approx \tfrac{1}{4}\log(1+4z_0^2), \tag{14.3-6a}$$

$$A_2(1) \approx \frac{1}{2}\left[\log(1+4z_0^2) - \frac{4z_0^2}{4z_0^2+1}\right]. \tag{14.3-6b}$$

Note that the effect of cutting off the potential is to make the $A_l(1)$ velocity dependent. Consequently, the evaluation of

$$\Omega_{ij}^{(l,r)} = \left(\frac{2\pi kT}{m_{ij}}\right)^{\frac{1}{2}}(e_i e_j)^2\int_0^\infty \mathscr{g}^{2r+3}A_l(1)(\tfrac{1}{2}m_{ij}g^2)^{-2}\exp(-\mathscr{g}^2)d\mathscr{g} \tag{14.3-7}$$

is no longer simple and a further approximation is required.

To find the appropriate approximation we remark that, when two particles of relative velocity g which interact with a Coulomb potential meet in a head-on collision, the distance of closest approach is

$$r_0 = e_i e_j/(\tfrac{1}{2}m_{ij}g^2), \tag{14.3-8}$$

so that z_0 is the ratio of the Debye length to the distance of closest approach. Replacing $\frac{1}{2}m_{ij}g^2$ by its average value, kT, we have $r_0 \approx (e_i e_j/kT)$ and, therefore, $r_0 \approx (nd^2)^{-1}$ – cf. Eq. (14.1-6). Thus we find that $z_0 \approx nd^3$, the number of particles in a sphere of radius equal to the Debye length. The number of particles in the "Debye sphere" is normally very large – in the range 10^7–10^{10} – so that, except for particles with very low relative speeds, z_0 is very large. Now, for large z_0, $A_l(1)$ is approximately equal to $\frac{1}{2}l \log z_0$, which is a rather slowly varying function of g over the range of values of g that contribute significantly to the integral (14.3-7), and, to a good approximation, we may replace z_0 by its average value with respect to the weight $g^3 \exp(-g^2)$ (corresponding to $r = 2$ in Eq. (14.3-7)), i.e., $\hat{z}_0 = (2kT/e_i e_j)d$. (The precise average value that should be used is somewhat unclear, but as the results are not very sensitive to the value chosen, the choice is not of critical importance.) Thus we are led to approximate the function $A_l(1)$ under the integral sign in Eq. (14.3-7) by the constant value $\hat{A}_l(1)$,

$$A_l(1) \approx \hat{A}_l(1) = \int_0^{\hat{z}_0} \left[1 - \left(\frac{4z^2-1}{4z^2+1}\right)^l\right] z\,dz \quad \text{with } \hat{z}_0 = (2kT/e_i e_j)d. \tag{14.3-9}$$

Then the integral (14.3-7) can be evaluated trivially and we have the following (approximate) result

$$\Omega_{ij}^{(l,r)} \approx \left(\frac{2\pi kT}{m_{ij}}\right)^{\frac{1}{2}} \hat{A}_l(1) \frac{(r-1)!}{2} \left(\frac{e_i e_j}{kT}\right)^2. \tag{14.3-10}$$

Now we turn to explicit results for the transport coefficients of an electrically neutral gas composed of singly ionized atoms and electrons. As mentioned earlier, terms of order m_e/m will be neglected, as the resulting error is certainly smaller than the errors introduced in approximating the Ω-integrals.

It turns out that the results can be expressed almost entirely in terms of two parameters, viz.,

$$\omega = \frac{eH}{m_e c} \tag{14.3-11}$$

and

$$\tau = \frac{3}{\sqrt{2\pi}} \frac{m_e^{\frac{1}{2}}(kT)^{\frac{3}{2}}}{ne^4 \log(1+4\hat{z}_0^2)}. \tag{14.3-12}$$

ω is, of course, just the electron gyrofrequency; τ, whose presence is due to the form (14.3-10) of the Ω-integrals, is essentially an average inverse collision frequency or a mean free time. As was noted in the introduction to this chapter, we may expect the dependence of the transport coefficients on the magnetic field strength to be governed by the dimensionless quantity $\omega\tau$. Unless stated otherwise the results are for the second Chapman-Cowling approximation, i.e., $n = 2$ in Eqs. (14.2-41), (14.2-42) and (14.2-70).

The electrical conductivities are given by

$$\sigma_{\parallel} = 1.931 \frac{ne^2\tau}{2m_e}, \tag{14.3-13a}$$

$$\sigma_{\perp} = \frac{ne^2\tau}{2m_e} \frac{\omega^2\tau^2+1.802}{\omega^4\tau^4+6.282\omega^2\tau^2+0.933}, \tag{14.3-13b}$$

$$\sigma_t = -\frac{ne^2\tau}{2m_e} \frac{\omega\tau(\omega^2\tau^2+4.382)}{\omega^4\tau^4+6.282\omega^2\tau^2+0.933}. \tag{14.3-13c}$$

They display the expected behavior – $\sigma_{\parallel}$ is independent of the magnetic field strength and is identical to the conductivity in the absence of a magnetic field; $\sigma_{\perp}$ does depend on the field strength but reduces to $\sigma_{\parallel}$ when $\omega \to 0$; finally, σ_t vanishes in the limit of no magnetic field. We also note that the first Chapman-Cowling approximation for $\sigma_{\parallel}$ yields Eq. (14.3-13a) with the numerical factor 1.931 deleted; higher-order approximations raise the value of this constant to approximately 1.97. For $\sigma_{\perp}$ and σ_t, the first approximation is obtained by replacing the fractions in Eqs. (14.3-13b, c) by $(1+\omega^2\tau^2)^{-1}$ and $\omega\tau(1+\omega^2\tau^2)^{-1}$, respectively. Thus, at low frequency (all frequencies for $\sigma_{\parallel}$) the first approximation produces a considerable error and the second approximation is necessary for reasonable accuracy.

The thermal conductivities are

$$\lambda_{\parallel} = 1.020 \frac{n\tau k^2 T}{m_e}, \tag{14.3-14a}$$

$$\lambda_{\perp} = \frac{n\tau k^2 T}{m_e} \frac{6.79\omega^2\tau^2+45.18}{\omega^4\tau^4+16.2\omega^2\tau^2+44.3}, \tag{14.3-14b}$$

$$\lambda_t = -\frac{n\tau k^2 T}{m_e} \frac{11.6\omega\tau}{\omega^4\tau^4+16.2\omega^2\tau^2+44.3}, \tag{14.3-14c}$$

and the electro-thermal coefficients are

$$\varphi_{\parallel} = 0.777 \frac{kne\tau}{m_e}, \tag{14.3-15a}$$

$$\varphi_{\mathrm{l}} = -0.75 \frac{kne\tau}{m_{\mathrm{e}}} \frac{\omega^2\tau^2 - 0.966}{\omega^4\tau^4 + 6.282\omega^2\tau^2 + 0.933}, \tag{14.3-15b}$$

$$\varphi_{\mathrm{t}} = -0.75 \frac{kne\tau}{m_{\mathrm{e}}} \frac{2.866\omega\tau}{\omega^4\tau^4 + 6.282\omega^2\tau^2 + 0.933}. \tag{14.3-15c}$$

It should be noted that in these equations all of the quantities depend only on the electron mass – the ion mass appears nowhere in the formulae. This is a consequence of the fact that, on the average, electrons have the same energy as the ions and, because they are also much lighter, their thermal velocities are much larger. The more rapid motion of the electrons causes them to contribute far more to the electric current and heat flow than the ions.

On the other hand, due to their larger mass, for a given flow velocity, most of the momentum will reside in the ions. As a result, the viscosity coefficients are essentially just properties of the ion gas and do not depend on the electron mass. For this reason it is convenient to introduce two new quantities,

$$\omega_{\mathrm{i}} = \frac{eH}{mc} = \frac{m_{\mathrm{e}}}{m}\,\omega, \tag{14.3-16a}$$

$$\tau_{\mathrm{i}} = \frac{5}{2\sqrt{2}} \left(\frac{m}{m_{\mathrm{e}}}\right)^{\frac{1}{2}} \tau. \tag{14.3-16b}$$

These are, of course, the ion gyrofrequency and an effective inverse ion collision frequency. The coefficients $b_{\mathrm{i},0}(H)$ and $b_{\mathrm{e},0}(H)$ are, in the first Chapman-Cowling approximation,

$$b_{\mathrm{i},0}(H) = \frac{4\tau_{\mathrm{i}}(3 - 4i\omega_{\mathrm{i}}\tau_{\mathrm{i}})}{9 + 16\omega_{\mathrm{i}}^2\tau_{\mathrm{i}}^2}, \tag{14.3-17a}$$

$$b_{\mathrm{e},0}(H) = \frac{5\tau(3(2+\sqrt{2}) - 10i\omega\tau)}{3(3+2\sqrt{2}) + 50\omega^2\tau^2}. \tag{14.3-17b}$$

Since τ_{i} is larger than τ by a factor of more than $(m/m_{\mathrm{e}})^{\frac{1}{2}} \approx 40$, we see that the electron contribution to the pressure tensor is indeed very small.

Measurements of plasma properties are extremely difficult to make. The high temperatures rule out the possibility of using most of the instruments commonly used to measure physical properties and indirect means, such as scattering or emission of electromagnetic radiation by the plasma, must be employed. Unfortunately, such measurements do not generally give values of properties at a point. These must be inferred indirectly and, as a result, the experimental errors tend to be large. Thus, a comparison of theory and experiment is not as meaningful as it was in the neutral gas case and will not be given. It suffices to say that the theory is, at least, consistent with experiment.

14.4. Non-equilibrium ionized gases

A consequence of the large ion–to–electron mass ratio is the small energy exchange that occurs in collisions between ions and electrons. The situation may be visualized as follows: In equilibrium, at a given temperature, the average electron speed is much greater than that of the ions, so we may imagine the ions to be essentially stationary. Now, an electron hitting a stationary ion will rebound with only a small change in speed; its direction will, however, be greatly altered and we may, for purposes of argument, assume the electron to be scattered with equal probability in all directions, i.e., isotropically, in a frame moving with the gas. We thus see that, if the electron gas possesses any drift velocity with respect to the ion gas, it will quickly lose it in collisions with the ions. On the other hand, the temperature of the electron gas will be changed only slowly.

There are situations in which one of the species will attain a temperature that is different from that of the other species. In such an instance the lack of equilibrium will persist for a time long compared with the mean free time τ, and, quite possibly, for a time as long as one of the characteristic macroscopic times of the gas. In fact, if energy is being introduced into the system by electric fields, a quasi-steady state may arise in which the difference in the temperatures of the species is essentially permanent. As the transport properties of an ionized gas depend almost entirely on the properties of one species or the other and are highly temperature dependent (e.g., $\sigma \sim T^{\frac{3}{2}}$) the non-equilibrium may have an important effect on the behavior of the gas.

Specific situations in which non-equilibrium plays an important role are gas flows in magnetohydrodynamic (MHD) electric generators and in certain types of plasma accelerators. In MHD-systems the electric field does more work on the electrons than on the ions (the force on each species is the same, but the electrons have a higher average speed, so the power input $\boldsymbol{F} \cdot \boldsymbol{c}$ is higher) and, as a result, the electrons tend to be considerably hotter than the ions. Such a situation is desirable, as the electrical conductivity depends mainly on the electron temperature and a high conductivity is conducive to good performance. The inverse situation occurs in plasma accelerators (guns) in which a beam of high-speed plasma is produced and then slowed down to convert the energy of directed motion into internal thermal energy. As both species are accelerated to the same speed, the ions have by far the greater fraction of the energy and, therefore, are much hotter.

Clearly, a gas of the type described above cannot be legitimately treated by means of hydrodynamic equations. However, as the major effect is a difference between the temperatures of the ion and electron gases one might hope to describe such a gas by means of a set of equations consisting of

a) a conservation equation for the number of particles of each species,
b) an equation for the conservation of total momentum,
c) an energy conservation equation for the ions,
d) an energy conservation equation for the electrons,
e) Maxwell's equations,
f) various constitutive equations – Newton's, Fourier's and Ohm's laws, etc., or their equivalents.

Thus, the only change from the magnetogasdynamic equations of the previous section is the presence of two energy equations rather than one. The problem is then one of finding the coefficients that occur in the various constitutive equations.

The basic equations are again the Boltzmann equations (14.2-4); likewise, the conservation equation for each species is (14.2-5), and the momentum equation is (14.2-8). The energy equation for the ions is derived by multiplying the Boltzmann equation by $\frac{1}{2}m_i C_i^2$, integrating over velocity and summing the resulting equations for the heavy species (ions and neutral atoms). One then has

$$\rho_h \frac{du_h}{dt} = -\nabla \cdot \boldsymbol{q}_h - \mathbf{P}_h : \nabla \boldsymbol{v} + \boldsymbol{j}_h \cdot \boldsymbol{E}' + \sum_{j\in\{h\}} \rho_j \boldsymbol{V}_j \cdot \boldsymbol{F}_j - \sum_{j\in\{h\}} \Delta E_{je}. \tag{14.4-1}$$

Similarly, the energy conservation equation for the electrons is

$$\rho_e \frac{du_e}{dt} = -\nabla \cdot \boldsymbol{q}_e - \mathbf{P}_e : \nabla \boldsymbol{v} + \boldsymbol{j}_e \cdot \boldsymbol{E}' + \rho_e \boldsymbol{V}_e \cdot \boldsymbol{F}_e - \sum_{j\in\{h\}} \Delta E_{ej}. \tag{14.4-2}$$

In these equations the symbol $\sum_{j\in\{h\}}$ denotes a sum over all of the "heavy" species, i.e., ions and neutral atoms. The macroscopic variables ($\boldsymbol{q}$, $\mathbf{P}$, $\boldsymbol{j}$) have their usual definitions; the index h(e) indicates that only the contributions due to the heavy particles (electrons) are considered.

Furthermore, ΔE_{ij} is defined by

$$\Delta E_{ij} = \int \tfrac{1}{2} m_i C_i^2 J(f_i f_j) \, d^3c_i, \tag{14.4-3}$$

and represents the thermal energy transferred from species j to species i by collisions per unit time per unit volume. Clearly, adding Eqs. (14.4-1) and (14.4-2) produces the usual overall energy conservation equation.

Now, the Chapman-Enskog approach cannot be used to relate the transport quantities to macroscopic variables, because it requires all such species to have identical temperatures. A modification of the method which circumvents this difficulty was given by Chmieleski [1966]; see also Chmieleski and Ferziger [1967a, 1967b]. The essence of the method lies in the realization that there now exist two small parameters – the usual Chapman-Enskog ε and the mass ratio m_e/m_h, where m_h is a typical heavy particle mass. In the calculation, it is convenient to regard ε and $(m_e/m_h)^{\frac{1}{2}}$ as being of the same order; Chmieleski [1967] showed that the identification of ε with any power of (m_e/m_h) up to $(m_e/m_h)^{\frac{3}{4}}$ produces substantially identical results, but that this choice is most convenient. In fact, it appears that the resulting equations are valid even when the relation $\varepsilon \sim (m_e/m_h)^{\frac{3}{2}}$ fails to hold, because the electron energy equation would then reduce to $T_e \approx T_h$ and the equations become those for a one-temperature gas.

In the Boltzmann equation for any of the heavy species, the term representing collisions with electrons may be estimated by approximating the distributions of the two species involved by Maxwellians with different temperatures. One then finds that this term is of order $(m_e/m_h)^{\frac{3}{2}} \sim \varepsilon^3$ with respect to the terms representing collisions with heavy species. In the equations for heavy species the terms representing collisions with electrons may therefore be neglected entirely if the calculation is only taken to second order. Consequently, the Boltzmann equations for the heavy species are completely uncoupled from the Boltzmann equation for the electrons. As the presence of the energy exchange term in Eq. (14.4-1) has no effect through the first order in the Chapman-Enskog analysis, it turns out that the calculation for the heavy species is identical in all respects to the one presented in Section 14.2, and we need not carry out the analysis anew.

The treatment of the Boltzmann equation describing the electrons does, however, present some new difficulties related to the treatment of the electron–heavy particle collision terms. Clearly, these terms can neither be neglected entirely nor retained without modification. Neglecting them would make the electron distribution function entirely independent of the presence of the heavy species, while retaining them without modification would require the electrons to have the same temperature at equilibrium as the heavy species. The required modification is obtained by splitting the electron–heavy species collision term into two parts. In the first part the ions are treated as if they had *infinite mass*; the second part, which may be treated as being of order ε, is then simply the difference between the exact collision term and the first term.

To zeroth order, the distribution function for the (heavy) species j is Maxwellian, i.e.,

$$f_j^{(0)} = n_j \left(\frac{m_j}{2\pi k T_h}\right)^{\frac{3}{2}} \exp\left(-m_j C_j^2/2kT_h\right), \tag{14.4-4}$$

where $\boldsymbol{C}_j = \boldsymbol{c}_j - \boldsymbol{v}$, and T_h is the equilibrium temperature of the heavy species. In the limit as m_j becomes infinite, $f_j^{(0)}$ becomes

$$\lim_{m_j \to \infty} f_j^{(0)} = n_j \delta(\boldsymbol{C}), \tag{14.4-5}$$

where we have abbreviated $\delta(\boldsymbol{C}) \equiv (4\pi C^2)^{-1}\delta(C)$. Thus, we write

$$f_j^{(0)} = n_j \delta(\boldsymbol{C}) + \varepsilon n_j [f_j^{(0)} - \delta(\boldsymbol{C})] \tag{14.4-6}$$

in the electron–heavy particle collision term. In the zeroth order approximation the electron distribution function obeys the equation

$$J(f_e^{(0)} f_e^{(0)}) + \sum_{j\in\{h\}} n_j J(f_e^{(0)}, \delta(\boldsymbol{C})) = -\frac{e}{m_e c}(\boldsymbol{C}_e \times \boldsymbol{H}) \cdot \nabla_{C_e} f_e^{(0)}. \tag{14.4-7}$$

This equation has the solution

$$f_e^{(0)} = n_e \left(\frac{m_e}{2\pi k T_e}\right)^{\frac{3}{2}} \exp\left(-m_e C_e^2/2kT_e\right), \tag{14.4-8}$$

where $\boldsymbol{C}_e = \boldsymbol{c}_e - \boldsymbol{v}$. It is important to note that $\boldsymbol{v}$ is the same hydrodynamic velocity that appears in the heavy particle Maxwellian (14.4-4), but T_e is an independent parameter. This is what was anticipated on the basis of the physical model discussed at the beginning of this section.

The first-order Chapman-Enskog equation for the electrons is obtained in the standard manner, except for the handling of the collision terms. The collision terms of order ε are

$$J(f_e^{(0)}, \phi_e f_e^{(0)}) + J(\phi_e f_e^{(0)}, f_e^{(0)}) + \sum_{j\in\{h\}} [J(f_e^{(0)} f_j^{(0)}) - J(f_e^{(0)}, n_j \delta(\boldsymbol{C}))$$
$$+ J(\phi_e f_e^{(0)}, n_j \delta(\boldsymbol{C})) + J(f_e^{(0)}, n_j \phi_j \delta(\boldsymbol{C}))]. \tag{14.4-9}$$

Of these, the first two terms are immediately recognized as $-n_e^2 I_{ee}(\phi)$; the second and fourth terms within the summation are zero; the first term in the summation contains only known functions as its arguments and can therefore be lumped with the inhomogeneous terms; finally, the third term will be designated by

$$-n_e n_j I'_{ej}(\phi) \equiv J(\phi_e f_e^{(0)}, n_j \delta(\boldsymbol{C})), \tag{14.4-10}$$

i.e., it is the usual linearized operator for a mixture with the mass of the heavy species made infinite. The first-order Chapman-Enskog equation for the electrons is thus

$$\begin{aligned} n^2 \mathfrak{H}_e(\phi) &\equiv n_e^2 I_{ee}(\phi) + \sum_{j\in\{h\}} n_e n_j I'_{ej}(\phi) - f_e^{(0)} \left[\frac{e}{m_e c} (\boldsymbol{C}_e \times \boldsymbol{H}) \cdot \boldsymbol{\nabla}_{C_e} \phi_e \right. \\ &\quad \left. + \frac{m_e e}{\rho c k T_e} \boldsymbol{C}_e \cdot \int (\boldsymbol{C}_e \times \boldsymbol{H}) f_e^{(0)} \phi_e \, d^3 c_e \right] \\ &= -f_e^{(0)} \left[(\mathscr{C}_e^2 - \tfrac{5}{2}) \boldsymbol{C}_e \cdot \boldsymbol{\nabla} \log T_e + 2(\mathscr{C}_e \mathscr{C}_e - \tfrac{1}{3} \mathscr{C}_e^2 \boldsymbol{I}) : \boldsymbol{\nabla} \boldsymbol{v} + \frac{n}{n_e} \boldsymbol{C}_e \cdot \boldsymbol{d}_e \right] \\ &\quad + \frac{2 f_e^{(0)}}{3 n_e k T_e} (\mathscr{C}_e^2 - \tfrac{3}{2}) \sum_{j\in\{h\}} \Delta E_{ej}^0 - \sum_{j\in\{h\}} J(f_e^{(0)} f_j^{(0)}), \end{aligned} \tag{14.4-11}$$

where $\boldsymbol{d}_e$ is the electron diffusion driving force,

$$\boldsymbol{d}_e = \frac{\rho_e}{\rho n k T_e} \left\{ \left[\sum_{j\in\{h\}} n_j e_j - \frac{\rho_h e}{m_e} \right] \boldsymbol{E}' + \frac{\rho_h}{\rho_e} \boldsymbol{\nabla} n_e k T_e - \boldsymbol{\nabla} n_h k T_h \right\}. \tag{14.4-12}$$

In the expression (14.4-12), ρ_h is the mass density of the heavy species, $\rho_h = \sum_{j\in\{h\}} \rho_j$.

Equation (14.4-11) is very similar to Eq. (14.2-20) and the procedure for its solution is therefore very similar. The differences between this equation and the one encountered earlier are: the modified collision operator occurring in the left-hand side, a change in the definition of the diffusion driving force for electrons, and, finally, the occurrence of two new terms on the right-hand side. The modified expression for the diffusion driving force has no effect on the method of solution or on the form of the solution. The modification of the collision operator, however, means that the homogeneous equation now has only two independent solutions, viz., 1 and $\frac{1}{2} m_e C_e^2$. As an immediate consequence, we find that only two conditions need be imposed on the second-order solution,

$$\int f_e^{(0)} \phi_e \, d^3 c_e = 0, \tag{14.4-13a}$$

and

$$\int f_e^{(0)} \phi_e C_e^2 \mathrm{d}^3 c_e = 0. \tag{14.4-13b}$$

We also note that the last term on the left hand side of Eq. (14.4-11) is of order m_e/m_i with respect to the term preceding it and may be neglected.

From the form of the right-hand side of Eq. (14.4-11) we conclude that ϕ_e must have the form

$$\phi_e = -\frac{1}{n} \boldsymbol{A}_e \cdot \nabla \log T - \frac{1}{n} \boldsymbol{D}_e \cdot \boldsymbol{d}_e - \frac{1}{n} \mathbf{B}_e : \nabla \boldsymbol{v} - \frac{1}{n} \psi_e(\mathscr{C}_e^2) + \alpha_{1e} + \tfrac{1}{2}\alpha_{2e} m_e C_e^2, \tag{14.4-14}$$

and the functions $\boldsymbol{A}$, $\mathbf{B}$, $\boldsymbol{D}$ and ψ satisfy

$$\mathfrak{H}_e(\boldsymbol{A}_e) = \frac{1}{n} f_e^{(0)} (\mathscr{C}_e^2 - \tfrac{5}{2}) \boldsymbol{C}_e, \tag{14.4-15a}$$

$$\mathfrak{H}_e(\mathbf{B}_e) = \frac{2}{n} f_e^{(0)} (\boldsymbol{\mathscr{C}}_e \boldsymbol{\mathscr{C}}_e - \tfrac{1}{3}\mathscr{C}_e^2 \mathbf{I}), \tag{14.4-15b}$$

$$\mathfrak{H}_e(\boldsymbol{D}_e) = \frac{1}{n_e} f_e^{(0)} \boldsymbol{C}_e, \tag{14.4-15c}$$

$$\mathfrak{H}_e(\psi_e) = -\frac{2 f_e^{(0)}}{3 n n_e k T_e} (\mathscr{C}_e^2 - \tfrac{3}{2}) \sum_{j \in \{h\}} \Delta E_{ej}^0 + \frac{1}{n^2} \sum_{j \in \{h\}} J(f_e^{(0)} f_j^{(0)}). \tag{14.4-15d}$$

From this point on, the treatment of Eqs. (14.4-15) is quite similar to that of Eqs. (14.2-32). The vectors $\boldsymbol{A}_e$ and $\boldsymbol{D}_e$ are expanded in terms of the vectors (14.2-26) as in Eq. (14.2-27) and (14.2-28), the coefficient functions are re-arranged as in Eqs. (14.2-33) and (14.2-36), and the modified coefficient functions are expanded in Sonine polynomials as in Eqs. (14.2-38). The theory is, in fact, simplified a bit with respect to the earlier theory by the neglect of the last term on the left-hand side of Eq. (14.4-11). We thus arrive at a set of equations for the Sonine expansion coefficients,

$$\sum_{q=0}^{n-1} \Lambda_{ee}'^{pq} a_{e,q}^{(k)} - \frac{4 i H n_e e}{25 c n^2 k^2 T_e} \frac{4}{3\sqrt{\pi}} \frac{\Gamma(p+\frac{5}{2})}{\Gamma(p+1)} a_{e,p}^{(k)} \delta_{k2} = \frac{4}{5k} \frac{n_e}{n} \delta_{p1}, \quad p = 0, 1, \ldots, n-1, \tag{14.4-16a}$$

$$\sum_{q=0}^{n-1} \Lambda_{ee}'^{pq} d_{e,q}^{(k)} - \frac{4 i H n_e e}{25 c n^2 k T_e} \frac{4}{3\sqrt{\pi}} \frac{\Gamma(p+\frac{5}{2})}{\Gamma(p+1)} d_{e,p}^{(k)} \delta_{k2} = \frac{8}{25k} \delta_{p0}, \quad p = 0, 1, \ldots, n-1, \tag{14.4-16b}$$

where

$$\Lambda'^{pq}_{ee} = \frac{8m_e}{75k^2T_e}\left\{\sum_{j\in\{h\}} \frac{n_e n_j}{n^2} \int S^{(p)}_{\frac{3}{2}}(\mathscr{C}_e^2)\mathscr{C}_e \cdot I'_{ej}(S^{(q)}_{\frac{3}{2}}(\mathscr{C}_e^2)\mathscr{C}_e)\mathrm{d}^3\mathscr{C}_e \right.$$

$$\left. + \frac{n_e^2}{n^2}[S^{(p)}_{\frac{3}{2}}(\mathscr{C}^2)\mathscr{C}, S^{(q)}_{\frac{3}{2}}(\mathscr{C}^2)\mathscr{C}]'\right\}. \quad (14.4\text{-}17)$$

Note that these equations do not contain the first-order solution for the heavy species and, for this reason, it is a relatively straight-forward task to solve them.

Equation (14.4-15b) may likewise be solved in a manner analogous to the one employed for the solution of Eq. (14.2-24). One expands $\boldsymbol{B}_e$ in terms of the tensors (14.2-57) as in Eq. (14.2-59) and rearranges the coefficient functions as in Eqs. (14.2-66). Again, all three of the modified coefficient functions are readily determined once one of them ($\mathfrak{B}_e^{(2)}$) is found. Then $\mathfrak{B}_e^{(2)}$ is expanded in a series of Sonine polynomials as in Eq. (14.2-69) and one obtains the following set of algebraic equations:

$$\sum_{q=0}^{n-1} H'^{pq}_{ee} b_{e,q} - \frac{2iHn_e e}{cn^2kTm_e}\frac{8}{15\sqrt{\pi}}\frac{\Gamma(p+\frac{7}{2})}{\Gamma(p+1)} b_{e,p} = \frac{2}{kT}\frac{n_i}{n}\delta_{p0},$$

$$p = 0, \ldots, n-1, \quad (14.4\text{-}18)$$

where

$$H'^{pq}_{ee} = \frac{2}{5kT_e}\left\{\sum_{j\in\{h\}} \int S^{(p)}_{\frac{5}{2}}(\mathscr{C}_e^2)(\mathscr{C}_e\mathscr{C}_e - \tfrac{1}{3}\mathscr{C}_e^2\mathbf{I}) : I'_{ej}(S^{(q)}_{\frac{5}{2}}(\mathscr{C}^2)(\mathscr{C}\mathscr{C} - \tfrac{1}{3}\mathscr{C}^2\mathbf{I}))\mathrm{d}^3\mathscr{C}_e \right.$$

$$\left. + [S^{(p)}_{\frac{5}{2}}(\mathscr{C}^2)(\mathscr{C}\mathscr{C} - \tfrac{1}{3}\mathscr{C}^2\mathbf{I}), S^{(q)}_{\frac{5}{2}}(\mathscr{C}^2)(\mathscr{C}\mathscr{C} - \tfrac{1}{3}\mathscr{C}^2\mathbf{I})]'_{ee}\right\}. \quad (14.4\text{-}19)$$

Finally, we must consider the solution of Eq. (14.4-15d). The right-hand side of this equation is a scalar function of $\mathscr{C}_e^2$ and, because the linearized collision operator is rotationally invariant, the ψ_e is also a scalar function of $\mathscr{C}_e^2$. Consequently, ψ_e makes no contribution to the transfer quantities and, thus, has no effect on the transport properties. The only way in which ψ_e enters the MHD-equations is through the term ΔE_{ej} but, as its contribution to this term is small compared to the zeroth order contribution, we may neglect it here as well. Thus, we need not consider Eq. (14.4-15d) further.

The energy transfer between the electrons and a typical heavy species is easily evaluated,

$$\Delta E_{ej} = \int \tfrac{1}{2} m_e C_e^2 J(f_e^{(0)} f_j^{(0)}) \mathrm{d}^3 c_e$$

$$= \frac{8(T_h - T_e)}{(2k)^{\frac{1}{2}} (m_e T_h + m_j T_e)^{\frac{3}{2}}} n_e n_j (m_e m_j)^{\frac{1}{2}} e^4 \log \Lambda, \qquad (14.4\text{-}20)$$

where the approximations used in Section 14.3 have been used.

The remaining task is that of finding the flux quantities and the associated transport coefficients. This again is similar to what has been done earlier, and the details will not be given. The new feature is that, wherever the terms involving ∇T appeared in the expressions in Section 14.2, there now appear terms involving both ∇T_e and ∇T_h. Thus, we have, for example, instead of Eq. (14.2-48), the following result for the electric current density,

$$\begin{aligned} \boldsymbol{j} = &- \sum_{i,j \in \{h\}} [\sigma_{i\parallel}^{j} \boldsymbol{d}'_{j\parallel} + \sigma_{i\perp}^{j} \boldsymbol{d}'_{j\perp} + \sigma_{i_t}^{j} \boldsymbol{d}'_{j_t}] - [\sigma_{\parallel}^{e} \boldsymbol{d}'_{e\parallel} + \sigma_{\perp}^{e} \boldsymbol{d}'_{e\perp} + \sigma_{t}^{e} \boldsymbol{d}'_{e_t}] \\ &- \sum_{i \in \{h\}} [\varphi_{i\parallel} (\nabla T_i)_{\parallel} + \varphi_{i\perp} (\nabla T_i)_{\perp} + \varphi_{i_t} (\nabla T_i)_t] \\ &- [\varphi_{e\parallel} (\nabla T_e)_{\parallel} + \varphi_{e\perp} (\nabla T_e)_{\perp} + \varphi_{e_t} (\nabla T_e)_t], \end{aligned} \qquad (14.4\text{-}21)$$

where the transport coefficients are related to the Sonine expansion coefficients exactly as in Eqs. (14.2-50), and $\boldsymbol{d}'_i$ is defined by Eq. (14.2-49). Note that, because $\boldsymbol{d}_e$ contains a term proportional to ∇T_h while $\boldsymbol{d}_j$ for $j \neq$ e does not contain ∇T_e at all, the thermo-electric coefficient that appears in front of ∇T_h contains a contribution from the electron gas, but the coefficient of ∇T_e has no contribution arising from the heavy species. We shall, however, omit the rather lengthy list of formulae for the transport coefficients.

The computation of the pressure tensor is somewhat simpler as it is just the sum of electron and heavy species contributions and these may be computed exactly as in Section 14.2.

Dynamics of rarefied gases 15

The Chapman-Enskog theory of Chapters 5 and 6, and its modification in Chapters 11 through 14, have, at their heart, the assumption that the time between collisions is much smaller than any other time scale of interest and, consequently, that the gas is locally very nearly in equilibrium. In fact, the Chapman-Enskog parameter ε can be identified with the ratio of the mean free time to a macroscopic time, or with the ratio of the mean free path l to some characteristic macroscopic dimension L. The ratio l/L, usually called the *Knudsen number*, Kn, measures the relative importance of collisions and particle streaming in determining the distribution function and must be small if the Chapman-Enskog theory is to be valid.

Now, if the gas density is reduced, the mean free path – which is inversely proportional to the density – will obviously increase and, given a sufficiently large decrease in the density, the Knudsen number need no longer be small. In the extreme limit in which the gas is so rarefied that collisions may be neglected entirely, the problem becomes one of tracing the paths of the particles as they fly through space and interact with any boundaries that may be present. Then we speak of *free molecule flow*. Similarly, if the characteristic dimension L is reduced to the order of a mean free path, the Knudsen number again becomes large. In either case, we must expect that the Chapman-Enskog expansion and the hydrodynamic, i.e., Navier-Stokes, equations that follow from it are no longer valid. Note, however, that the assumptions underlying the Boltzmann equation remain valid. One is therefore led to the necessity of solving the Boltzmann equation together with boundary and/or initial conditions appropriate to the particular physical situation. In general, this is a far more formidable task than the solution of the analogous hydrodynamic problem, but considerable progress has been made in this area over the last decade. Much of the recent progress is summarized in the monographs by Cercignani [1969a], Kogan [1969] and Williams [1971]. This chapter is intended only as an introduction to some of the ideas and methods of rarefied gas dynamics; the interested reader is referred to the above references for a more thorough discussion of the subject.

15.1. The linearized Boltzmann equation – Models and boundary conditions

Near a boundary, the appropriate Knudsen number is the mean free path divided by the distance from the boundary. Thus, within a few mean free paths of the wall, the Knudsen number is not small and the underlying assumptions of the Chapman-Enskog method become invalid. In a number of interesting situations, particularly those involving low-speed flows, the distribution function may still be nearly Maxwellian and linearization of the Boltzmann equation may be appropriate. However, the deviation of the distribution function from the Maxwellian is not of the Chapman-Enskog character. As the theory of linear equations is much more advanced than that of non-linear equations, the use of a linearized theory produces a significant increase in the number of tractable problems.

The linearized Boltzmann equation was derived in Section 4.6, cf. Eqs. (4.6-5) and (4.6-27). In terms of the function $\varphi \equiv \varphi(\boldsymbol{r}, \boldsymbol{c}, t)$, it may be written

$$\frac{\partial \varphi}{\partial t} + \boldsymbol{c} \cdot \boldsymbol{\nabla}_r \varphi = -\nu(c)\varphi(\boldsymbol{c}) + \int K(\boldsymbol{c}, \boldsymbol{c}')\,\varphi(\boldsymbol{c}')\,\mathrm{d}^3c' \equiv L(\varphi), \qquad (15.1\text{-}1)$$

where we have written $\nu(c)$ in place of $K_0(\boldsymbol{c})/f_\mathrm{M}(\boldsymbol{c})$, and $K(\boldsymbol{c}, \boldsymbol{c}')$ in place of $K(\boldsymbol{c}, \boldsymbol{c}')/f_\mathrm{M}(\boldsymbol{c})$. Thus, $L(\varphi) \equiv -n^2 I(\varphi)/f_\mathrm{M}$. Note that the Maxwellian employed in the linearization is spatially constant and has zero hydrodynamic velocity. This accords with the assumption of low-speed flow. Both the nature and method of finding solutions of Eq. (15.1-1) depend to a large extent on the properties of the linearized collision operator and we therefore give a listing of some of its properties before proceeding.

(i) From Eq. (4.6-4) we see immediately that

$$L(1) = L(\boldsymbol{c}) = L(c^2) = 0. \qquad (15.1\text{-}2)$$

In other words, the summational invariants are all eigenfunctions of the linearized collision operator with eigenvalue zero.

(ii) By multiplying both members of (4.6-4) by $F(\mathbf{c})$ and integrating over $\mathbf{c}$, and using the identities of Section 4.7, we see that

$$\int F(\mathbf{c})\, L(F)\, \mathrm{d}^3 c = -\tfrac{1}{4}\int \mathrm{d}^3 c \int \mathrm{d}^3 c_1 (F+F_1-F'-F_1')^2 f_{\mathrm{M}}(c_1) g b\, \mathrm{d}b\, \mathrm{d}\varepsilon \leqq 0, \tag{15.1-3}$$

with equality holding if and only if F is a linear combination of the summational invariants. Thus, L is a negative semi-definite operator.

(iii) L is rotationally invariant. That is, if a rotation of the coordinate system is made, L does not change. This property, which has been used in the preceding chapters, means that the eigenfunctions of L may be represented as products of spherical harmonics in the direction of $\mathbf{c}$ and functions of c^2. Alternatively, they may be represented in terms of isotropic tensors such as were employed in Chapter 11. Thus, if we define the eigenfunctions of L by

$$L(\psi_{rlm}) = \lambda_{rlm}\psi_{rlm}, \tag{15.1-4}$$

then

$$\psi_{rlm}(\mathbf{c}) = R_{rl}(c^2)\, Y_l^m(\mathbf{c}/c), \tag{15.1-5}$$

where the functions Y_l^m are the spherical harmonics in velocity space. Note that three indices are required because the problem is three dimensional. The "radial" eigenfunctions $R_{rl}(c^2)$ have been found for only one particular molecular model – Maxwell molecules, for which $R_{rl}(c^2) = c^l S_{l+\frac{1}{2}}^{(r)}(mc^2/2kT)$, see Wang Chang and Uhlenbeck [1952]. For this model the eigenvalues can also be computed explicitly.

(iv) Except for Maxwell molecules, for which the collision frequency ν is independent of c, the operator L has as eigenvalues all of the values that ν takes on as c varies from 0 to ∞. There may, additionally, be discrete eigenvalues, but little is known about them except in the case of a hard sphere gas (Kuščer and Williams [1967], Yan [1969]).

(v) As was discussed in Chapter 4, ν is not finite if the intermolecular potential has infinite range. This is due to a preponderance of small-angle collisions and is compensated for by the integral term in such a way that L operating on a bounded function of $\mathbf{c}$ will produce another bounded function. Physically, small-angle collisions have a negligible effect on the distribution function and nothing of significance is lost if ν is assumed bounded.

(The single exception is the Coulomb potential for which special treatment is needed; see Chapter 14.)

Thus we see that L is generally a rather complicated operator and that, except for a few special cases, only a few of its properties are known. Fortunately, it turns out that most of the properties of the gas as a whole are not strongly dependent on the precise form of L. Strictly speaking, we can only verify this after the fact, but some indication of the truth of this statement is found in the near-insensitivity of the transport properties to the potential. For these reasons, calculations are generally carried out with L replaced by a relatively simply approximate operator, i.e., a *model.*

We shall not go into model construction in great detail, but will merely display some of the more commonly used models. For this purpose it is most convenient to work in terms of the non-dimensional variable $\mathscr{C} = (m/2kT)^{\frac{1}{2}}\mathbf{c}$. Thus,

$$L(\varphi(\mathbf{c})) \rightarrow \mathfrak{L}(\varphi(\mathscr{C})) \equiv -\nu(\mathscr{C})\varphi(\mathscr{C}) + \int K(\mathscr{C}, \mathscr{C}')\varphi(\mathscr{C}')\mathrm{d}^3\mathscr{C}'. \tag{15.1-6}$$

The simplest models are obtained by considering the case of Maxwell molecules. Then it follows from the Hilbert-Schmidt theory of integral equations (see Courant and Hilbert [1953]) that $K(\mathscr{C}, \mathscr{C}')$ can be written

$$K(\mathscr{C}, \mathscr{C}') = \sum_{r,l,m} (\lambda_{rlm} + \nu)\psi_{rlm}(\mathscr{C})\psi_{rlm}(\mathscr{C}')\mathrm{e}^{-\mathscr{C}'^2}, \tag{15.1-7}$$

where the sum runs over all r from 0 to ∞, all l from 0 to ∞, and $m = -l, \ldots, +l$. Simple models are obtained by truncating the series after a small number of terms, or, equivalently, setting all but a few of the eigenvalues equal to $-\nu$. To retain the important conservation properties it is essential that the terms corresponding to the summational invariants – $(r, l, m) = (0, 0, 0)$, $(0, 1, -1)$, $(0, 1, 0)$, $(0, 1, 1)$ and $(1, 0, 0)$ – be left undisturbed. Thus, the simplest sensible model for K is (in vector form)

$$K(\mathscr{C}, \mathscr{C}') = \pi^{-\frac{3}{2}}\nu\,\mathrm{e}^{-\mathscr{C}'^2}[1 + \mathscr{C}\cdot\mathscr{C}' + (\mathscr{C}^2 - \tfrac{3}{2})(\mathscr{C}'^2 - \tfrac{3}{2})] \tag{15.1-8}$$

and is known as the *linearized BGK model* (Bhatnagar, Gross and Krook [1954]); the non-linear version of this model will be discussed later. Higher-order models, in which further eigenvalues are left unchanged, were first discussed by Gross and Jackson [1959]. Clearly, by allowing the number of terms in Eq. (15.1-7) which are retained to increase, it is possible to approximate the exact linearized Maxwell molecule collision operator with great accuracy.

Since real molecules are not very well approximated by the Maxwell molecule model, it is natural to seek generalizations of these models. The simplest procedure is to assume that a model with constant collision frequency can be made to represent the actual molecule and then adjust whatever constants are available to fit properties of the true molecules. One then finds models similar to the Gross-Jackson models, but with the coefficients replaced by matrices. As this is essentially a special case of the models we will discuss below, no details will be given.

It was found in the related area of neutron transport theory,* in which many detailed calculations of the distribution function have been made, that the major effect of the collision operator on the distribution function is caused by the speed dependence of ν, rather than by the detailed structure of $K(\mathscr{C}, \mathscr{C}')$. Physically, this is due to the fact that $\nu(\mathscr{C})$ governs the rate at which particles are removed from the vicinity of velocity $\mathscr{C}$, while the rate of arrival, which is governed by $K(\mathscr{C}, \mathscr{C}')$, is nearly model independent. This idea led Loyalka and Ferziger [1967] to adapt models proposed by Shapiro and Corngold [1965] in neutron transport theory for use in kinetic theory.

The models are derived by assuming that they have the form:

$$\tilde{\mathfrak{L}}(\varphi) = -\nu(\mathscr{C})\varphi(\mathscr{C}) + \sum_{i,j=1}^{N} (\beta^{-1})_{ij}\gamma_i(\mathscr{C}) \int e^{-\mathscr{C}'^2}\gamma_j(\mathscr{C}')\varphi(\mathscr{C}')\,d^3\mathscr{C}', \tag{15.1-9}$$

where

$$\gamma_i(\mathscr{C}) = \int e^{-\mathscr{C}'^2} K(\mathscr{C}, \mathscr{C}')\psi_i(\mathscr{C}')\,d^3\mathscr{C}', \tag{15.1-10}$$

and where the functions $\psi_i(\mathscr{C})$ are any convenient complete set of functions (usually chosen to be the Maxwell molecule eigenfunctions; i represents an ordered set of three indices). The constants β_{ij} ($(\beta^{-1})_{ij}$ is an element of the inverse matrix) are determined by requiring that the model retain some properties of the exact linearized collision operator. Loyalka and Ferziger

* Neutron transport theory is actually a special case of linearized kinetic theory. In neutron transport problems the neutrons are generally so dilute that their interactions with each other may be neglected so that the neutrons act like a "foreign gas". Consequently, the neutron transport equation is nothing more than the linearized Boltzmann equation with a collision operator that conserves particle number but not momentum or energy.

chose

$$\mathfrak{L}(\psi_i) = \tilde{\mathfrak{L}}(\psi_i), \qquad i = 1, \ldots, N \tag{15.1-11}$$

as the conditions to be satisfied and found

$$\beta_{ij} = \int e^{-\mathscr{C}^2} \nu(\mathscr{C}) \psi_i(\mathscr{C}) \psi_j(\mathscr{C}) \, d^3\mathscr{C}. \tag{15.1-12}$$

Other choices in place of (15.1-11) lead to similar approximate kernels. If the γ_i were chosen to be constants, the generalized Gross-Jackson models obtained earlier would be obtained.

Next, we turn to consideration of the boundary conditions. Some of the general properties that must be possessed by these have already been discussed in Section 4.4. We now look in detail at models for the function $W(\boldsymbol{c}, \boldsymbol{c}')$ introduced there. This function plays a role similar to the one $K(\mathscr{C}, \mathscr{C}')$ plays in the bulk of the gas. Although W is in principle amenable to experimental measurement, almost no data are available and one must rely on models.

The simplest models, which have been most often employed, are those in which a molecule encountering the surface is presumed either to bounce off as a billiard ball (*specular reflection*) or to stick to the surface momentarily and then be emitted with a Maxwellian distribution corresponding to the wall temperature (*diffuse reflection*). In the former case,

$$W_s(\boldsymbol{c}, \boldsymbol{c}') = \delta(\boldsymbol{c}' - \boldsymbol{c} + 2(\boldsymbol{n} \cdot \boldsymbol{c})\boldsymbol{n}), \tag{15.1-13}$$

whereas, in the latter case,

$$W_d(\boldsymbol{c}, \boldsymbol{c}') = \frac{1}{2\pi} \left(\frac{m}{kT_w} \right)^2 (\boldsymbol{c} \cdot \boldsymbol{n}) \exp(-mc^2/2kT_w), \tag{15.1-14}$$

where the wall is assumed to be stationary (T_w is the temperature of the wall).

A model which has been frequently used is simply a constant linear combination of the two preceding models,

$$W(\boldsymbol{c}, \boldsymbol{c}') = \alpha W_d(\boldsymbol{c}, \boldsymbol{c}') + (1-\alpha) W_s(\boldsymbol{c}, \boldsymbol{c}'), \tag{15.1-15}$$

where α, called the *accommodation coefficient*, is usually chosen to fit whatever sparse experimental data are available.

Although, as we have said, little data are available, it is clear that this model cannot be correct. A fast molecule striking the surface should have a smaller probability of sticking than should a slower molecule. Furthermore, molecules hitting the surface normally might be expected to stick more easily than those which merely graze the surface. Thus, the accommodation coefficient should be velocity dependent and a better model might be

$$W(\boldsymbol{c}, \boldsymbol{c}') = P(\boldsymbol{c}')F^{+}(\boldsymbol{c})+(1-P(\boldsymbol{c}'))\delta(\boldsymbol{c}'-\boldsymbol{c}+2(\boldsymbol{n}\cdot\boldsymbol{c})\boldsymbol{n}). \quad (15.1\text{-}16)$$

If we require the model to satisfy the relations (4.4-3) and (4.4-4), the function F^{+} is uniquely determined in terms of P and the wall temperature T_{w},

$$F^{+}(\boldsymbol{c}) = \frac{(\boldsymbol{c}\cdot\boldsymbol{n})\,P(\boldsymbol{c}-2(\boldsymbol{c}\cdot\boldsymbol{n})\boldsymbol{n})\exp\left(-mc^{2}/2kT_{\mathrm{w}}\right)}{\int_{\boldsymbol{c}'\cdot\boldsymbol{n}<0} -(\boldsymbol{c}'\cdot\boldsymbol{n})\,P(\boldsymbol{c}')\exp\left(-mc'^{2}/2kT_{\mathrm{w}}\right)\mathrm{d}^{3}c'}, \quad (15.1\text{-}17)$$

so that we arrive at the following model of the function W:

$$W(\boldsymbol{c}, \boldsymbol{c}') = \frac{(\boldsymbol{c}\cdot\boldsymbol{n})\,P(\boldsymbol{c}-2(\boldsymbol{c}\cdot\boldsymbol{n})\boldsymbol{n})\exp\left(-mc^{2}/2kT_{\mathrm{w}}\right)}{\int_{\boldsymbol{c}''\cdot\boldsymbol{n}<0} -(\boldsymbol{c}''\cdot\boldsymbol{n})\,P(\boldsymbol{c}'')\exp\left(-mc''^{2}/2kT_{\mathrm{w}}\right)\mathrm{d}^{3}c''}\,P(\boldsymbol{c}')$$
$$+(1-P(\boldsymbol{c}'))\delta(\boldsymbol{c}'-\boldsymbol{c}+2(\boldsymbol{c}\cdot\boldsymbol{n})\boldsymbol{n}), \quad (15.1\text{-}18)$$

where $\boldsymbol{c}'\cdot\boldsymbol{n} < 0$, $\boldsymbol{c}\cdot\boldsymbol{n} > 0$. The constant accommodation model is recovered when $P(\boldsymbol{c}) = \alpha$, and specular and diffuse reflection obtain for $P(\boldsymbol{c}) = 0$ and $P(\boldsymbol{c}) = 1$, respectively.

Of course, one still needs a model for $P(\boldsymbol{c})$ which is more accurate than these simple special cases. This question was considered by Epstein [1967]. Based on experimental data for thermal accommodation coefficients, Epstein proposed the following model of P:

$$P(\boldsymbol{c}') = \exp\left(-mc'^{2}/2k\theta_{1}\right)+B[1-\exp\left(-mc'^{2}/2k\theta_{2}\right)], \quad (15.1\text{-}19)$$

where B, θ_1 and θ_2 are constants to be determined below. If $\theta_1 < \theta_2$, the first term controls the low-energy behavior of P (where there is almost complete accommodation), whereas the second term controls the high-energy behavior of P (which tends towards B). With this model one obtains

$$W(\boldsymbol{c}, \boldsymbol{c}') = \frac{m^{2}}{2\pi k^{2}T_{\mathrm{w}}^{2}}(\boldsymbol{c}\cdot\boldsymbol{n})\exp\left(-mc^{2}/2kT_{\mathrm{w}}\right)$$
$$\times\frac{P(\boldsymbol{c})P(\boldsymbol{c}')}{(1+T_{\mathrm{w}}/\theta_{1})^{-2}+B[1-(1+T_{\mathrm{w}}/\theta_{2})^{-2}]}+(1-P(\boldsymbol{c}'))\,\delta(\boldsymbol{c}'-\boldsymbol{c}+2(\boldsymbol{n}\cdot\boldsymbol{c})\boldsymbol{n}).$$
$$(15.1\text{-}20)$$

Numerical values of the constants B, θ_1 and θ_2 are best determined from experimental data which, usually, refer to the energy accommodation coefficient. The latter is defined by

$$\alpha_E = (Q^- - Q^+)/(Q^- - Q_w^+) \tag{15.1-21}$$

with

$$Q^- = -\tfrac{1}{2}m \int (c' \cdot n)c'^2 f^-(c')\,d^3c', \tag{15.1-22a}$$

$$Q^+ = \tfrac{1}{2}m \int (c \cdot n)c^2 f^+(c)\,d^3c, \tag{15.1-22b}$$

$$Q_w^+ = \lim_{\substack{\theta_1, \theta_2 \to 0 \\ B \to 1}} Q^+. \tag{15.1-22c}$$

These quantities are the energy transfer to and from the wall, and the energy that would be carried away from a perfectly accommodating wall, respectively. If the incoming distribution function f^- is taken as a Maxwellian with temperature T_∞,

$$f^-(c') = n(m/2\pi kT_\infty)^{\frac{3}{2}} \exp(-mc'^2/2kT_\infty), \tag{15.1-23}$$

the quantities Q defined by Eqs. (15.1-22) become,

$$Q^- = n(2/\pi m)^{\frac{1}{2}}(kT_\infty)^{\frac{3}{2}}, \tag{15.1-24a}$$

$$\begin{aligned} Q^+ = n(2/\pi m)^{\frac{1}{2}}(kT_\infty)^{\frac{3}{2}}\{&A_2(T_w/T_\infty)[(\theta_1/(\theta_1+T_w))^3 + B[1-(\theta_2/(\theta_2+T_w))^3]] \\ &+1-(\theta_1/(\theta_1+T_\infty))^3 - B[1-(\theta_2/(\theta_2+T_\infty))^3]\}, \end{aligned} \tag{15.1-24b}$$

$$Q_w^+ = n(2/\pi m)^{\frac{1}{2}}(kT_\infty)^{\frac{3}{2}}(T_w/T_\infty), \tag{15.1-24c}$$

so that, in this case, the following expression for the energy accommocation coefficient α_E is obtained:

$$\alpha_E = \left\{\left(\frac{\theta_1}{\theta_1+T_\infty}\right)^3 + B\left[1-\left(\frac{\theta_2}{\theta_2+T_\infty}\right)^3\right]\right\} \frac{1-(A_2/A_3)(T_w/T_\infty)}{1-(T_w/T_\infty)}. \tag{15.1-25}$$

In these expressions the quantities A_r ($r = 2, 3$) are defined by

$$A_r = \frac{(\theta_1/(\theta_1+T_\infty))^r + B[1-(\theta_2/(\theta_2+T_\infty))^r]}{(\theta_1/(\theta_1+T_w))^r + B[1-(\theta_2/(\theta_2+T_w))^r]}. \tag{15.1-26}$$

The coefficients B, θ_1 and θ_2 may now be determined by applying a least-squares fitting of Eq. (15.1-25) to appropriate experimental data.

In the reference cited before, Epstein has shown that, for several gases, experimental data of the energy accommodation coefficient could be fit accurately over an energy range large enough for α_E to vary by a factor of 2, thus showing the efficacy of the model. Generalizations of the model, which are still tractable analytically, are obtained if each exponential in Eq. (15.1-19) is multiplied by a polynomial in the velocity components. Also, to allow for anisotropy in the energy accommodation one can replace each of the exponential functions in the model (15.1-19) by

$$\exp(-mc^2/2k\theta) \rightarrow \exp\left[-\frac{m(\boldsymbol{c}\cdot\boldsymbol{n})^2}{2k\theta_\perp} - \frac{m(c^2-(\boldsymbol{c}\cdot\boldsymbol{n})^2)}{2k\theta_\parallel}\right]. \tag{15.1-27}$$

Thus, although the proposed model is entirely empirical, it has the advantages of combining great flexibility with analytical tractability and may, therefore, prove useful in the study of boundary value problems of kinetic theory.

A number of aspects of atom-wall interactions and a model in which the particles leaving the wall may be described by means of different temperatures parallel and normal to the wall are discussed in a recent paper by Cercignani and Lampis [1970].

15.2. The slip flow problem

In this section we shall treat a problem which may be regarded as the prototype for a great number of more complicated problems, viz., the *slip flow problem* (*Kramer's problem*). This problem, which is an abstraction from physical reality, is that of computing the steady-state velocity field of a gas flowing past an infinite flat plate lying in the plane $x = 0$; flow is in the z-direction. As there is momentum transfer to the wall, but no change in the velocity of the gas along a streamline, there must be a constant flow of z-momentum per unit area toward the plate. Far from the plate the Chapman -Enskog solution is valid and, in order to have a uniform momentum flow, it follows from Eqs. (5.4-33) and (5.4-46) that the velocity must be increasing linearly with x at large distances,

$$v_z \sim \alpha x, \qquad x \rightarrow \infty. \tag{15.2-1}$$

The constant α is in fact related to the momentum flow by

$$P_{xz} = -\eta \frac{\mathrm{d}v_z}{\mathrm{d}x} = -\alpha\eta. \tag{15.2-2}$$

Physically, this problem may be viewed as the infinite separation limit of *Couette flow*, that is, the flow between two infinite parallel plates in uniform relative motion, when the separation between the plates becomes infinite. Or, in terms of current ideas in boundary layer theory, we can say that we are seeking the "inner" solution (the solution valid in the region near the plate) and that the condition at infinity is simply a representation of the "outer" flow. Again, we stress that this problem is presented here mainly to illustrate a commonly used method and to display some of the important physical effects.

Mathematically, the problem is one of solving the Boltzmann equation with an appropriate boundary condition at the plate. We expect that far (i.e., several mean free paths) from the plate any peculiarities in the distribution function induced by the plate will have died out and that the hydrodynamic solution (15.2-1) will be valid. Thus, the velocity distribution function should assume the Chapman-Enskog form far from the plate.

Since the slip flow problem is presented mainly for illustrative purposes, we adopt the simplest model of the linearized collision operator – viz., the linearized BGK model. Thus, we deal with the equation

$$\mathscr{C}\cdot\nabla_r\varphi = \nu\Big[\pi^{-\frac{3}{2}}\int e^{-\mathscr{C}'^2}\varphi(\boldsymbol{r},\mathscr{C}')\,d^3\mathscr{C}' + \pi^{-\frac{3}{2}}\mathscr{C}\cdot\int e^{-\mathscr{C}'^2}\mathscr{C}'\varphi(\boldsymbol{r},\mathscr{C}')\,d^3\mathscr{C}'$$

$$+\pi^{-\frac{3}{2}}(\mathscr{C}^2-\tfrac{3}{2})\int e^{-\mathscr{C}'^2}(\mathscr{C}'^2-\tfrac{3}{2})\varphi(\boldsymbol{r},\mathscr{C}')\,d^3\mathscr{C}'-\varphi(\boldsymbol{r},\mathscr{C})\Big]. \qquad (15.2\text{-}3)$$

The integrals appearing on the right-hand side of this equation are easily recognized to be the perturbations in the density, hydrodynamic velocity, and temperature from their Maxwellian values, respectively. In the particular problem at hand, however, the state of the gas varies only in the x-direction, the hydrodynamic velocity is entirely in the z-direction, and variations in density and temperature may be ignored, so that Eq. (15.2-3) reduces to

$$\mathscr{C}_x\frac{\partial\varphi(x,\mathscr{C})}{\partial x} = \nu\Big[\pi^{-\frac{3}{2}}\mathscr{C}_z\int \mathscr{C}'_z e^{-\mathscr{C}'^2}\varphi(x,\mathscr{C}')\,d^3\mathscr{C}'-\varphi(x,\mathscr{C})\Big]. \qquad (15.2\text{-}4)$$

For the boundary condition at the plate we shall again adopt the simplest possibility – viz., a diffuse reflector at the temperature of the gas. Since this means that the molecules coming from the plate have a Maxwellian velocity distribution with zero hydrodynamic velocity, we have

$$\varphi(0, \mathscr{C}) = 0 \quad \text{for} \quad \mathscr{C}_x > 0. \tag{15.2-5}$$

Note that only half the distribution function is specified at the boundary – this is a peculiarity of the Boltzmann equation which has been discussed earlier; see Sections 4.4 and 4.8. To complete the specification of the problem we need only insist that φ grow no faster than linearly as $x \to \infty$; this is in accord with the condition that the hydrodynamic velocity grows linearly at large distances from the plate. Thus,

$$\varphi \sim \nu x \qquad (x \to \infty), \tag{15.2-6}$$

where the proportionality constant has been set equal to ν for later convenience (this acts as a normalization, since the entire problem is homogeneous).

The problem posed by Eq. (15.2-4) and boundary conditions (15.2-5) and (15.2-6) is the simplest non-trivial problem in kinetic theory and may be solved by a variety of methods. Because it is one of the few problems for which exact solution is possible, it provides an excellent means of testing approximate methods. Furthermore, as we have mentioned earlier, it possesses considerable intrinsic interest of its own.

A method of solution due to Van Kampen [1955] and Case [1960] and first applied to kinetic theory by Cercignani [1962] will be used here. To begin, the problem is simplified by introducing the function

$$g(x, \mathscr{C}_x) = \int \mathrm{d}\mathscr{C}_z \int \mathrm{d}\mathscr{C}_y \mathrm{e}^{-\mathscr{C}_z{}^2} \mathrm{e}^{-\mathscr{C}_y{}^2} \mathscr{C}_z \varphi(x, \mathscr{C}). \tag{15.2-7}$$

Then, multiplying both members of Eq. (15.2-4) and the boundary conditions (15.2-5) and (15.2-6) by $\mathscr{C}_z \exp(-\mathscr{C}_z^2) \exp(-\mathscr{C}_y^2)$, integrating over $\mathscr{C}_y$ and $\mathscr{C}_z$, and defining a non-dimensional coordinate $\tilde{x} = \nu x$, we obtain

$$\mathscr{C}_x \frac{\partial g(\tilde{x}, \mathscr{C}_x)}{\partial \tilde{x}} + g(\tilde{x}, \mathscr{C}_x) = \pi^{-\frac{1}{2}} \int_{-\infty}^{\infty} g(\tilde{x}, \mathscr{C}'_x) \mathrm{e}^{-\mathscr{C}'_x{}^2} \mathrm{d}\mathscr{C}'_x, \tag{15.2-8}$$

$$g(\tilde{x}, \mathscr{C}_x) \sim \tilde{x} \qquad (\tilde{x} \to \infty), \tag{15.2-9}$$

$$g(0, \mathscr{C}_x) = 0 \quad \text{for} \quad \mathscr{C}_x > 0. \tag{15.2-10}$$

(In the following, the tilde on x will be dropped as no confusion should arise.)

The method is basically just the well-known separation of variables method of the theory of partial differential equations, i.e., we seek solutions of Eq. (15.2-8) that can be written as a product of a function of x alone and a function of $\mathscr{C}_x$ alone, show that the general solution is a superposition of such solutions, and expand the solution of our problem in terms of these "normal modes". Noting that Eq. (15.2-8) is indeed separable and that the solutions of the separated x-equations are exponentials, we write

$$g(x, \mathscr{C}_x) = \mathrm{e}^{-x/\lambda} g_\lambda(\mathscr{C}_x) \tag{15.2-11}$$

and find that g_λ must satisfy the equation

$$(\lambda - \mathscr{C}_x) g_\lambda(\mathscr{C}_x) = \pi^{-\frac{1}{2}} \lambda \int_{-\infty}^{\infty} g_\lambda(\mathscr{C}'_x) \mathrm{e}^{-\mathscr{C}'_x{}^2} \mathrm{d}\mathscr{C}'_x. \tag{15.2-12}$$

Now, since this equation is homogeneous in g_λ, we may choose the normalization such that

$$\pi^{-\frac{1}{2}} \int_{-\infty}^{\infty} g_\lambda(\mathscr{C}_x) \mathrm{e}^{-\mathscr{C}_x{}^2} \mathrm{d}\mathscr{C}_x = 1. \tag{15.2-13}$$

Thus, the functions g_λ satisfy the equation

$$(\lambda - \mathscr{C}_x) g_\lambda(\mathscr{C}_x) = \lambda. \tag{15.2-14}$$

The crux of the method lies in the observation that the solutions (15.2-11) do not in themselves represent any physical quantity and, hence, there is no reason to require that they have the smoothness properties that one requires of observable quantities. Thus, while

$$g_\lambda(\mathscr{C}_x) = \lambda/(\lambda - \mathscr{C}_x) \tag{15.2-15}$$

would be the only solution if one admitted only measurable functions, by allowing generalized functions as solutions a great many more solutions can be found. If generalized functions are admitted, the homogeneous equation

$$(\lambda - \mathscr{C}_x) g_\lambda(\mathscr{C}_x) = 0 \tag{15.2-16}$$

has the solution

$$g_\lambda(\mathscr{C}_x) = \delta(\lambda - \mathscr{C}_x) \tag{15.2-17}$$

and an arbitrary multiple of this solution may be added to the solution (15.2-15), so that

$$g_\lambda(\mathscr{C}_x) = \frac{\lambda}{\lambda - \mathscr{C}_x} + p(\lambda)\,\delta(\lambda - \mathscr{C}_x) \tag{15.2-18}$$

is the general solution of Eq. (15.2-14). Since $p(\lambda)$ is arbitrary, there is a solution of this type for every λ which is a possible value of $\mathscr{C}_x$, i.e., for every real λ. For non-real λ, the delta function is of no significance and the solution is again of the form (15.2-15).

Recall, however, that the normalization condition (15.2-13) must be satisfied. Treating first the case of non-real λ we find that λ must be such that

$$\pi^{-\frac{1}{2}}\lambda \int_{-\infty}^{\infty} \frac{e^{-\mathscr{C}_x^2}}{\lambda - \mathscr{C}_x}\, d\mathscr{C}_x = 1. \tag{15.2-19}$$

One can readily show (either directly, or by use of the argument principle) that the only root of this equation is a double root at infinity. The normal modes associated with the double root at infinity are simply given by

$$g_+(x, \mathscr{C}_x) = 1, \tag{15.2-20a}$$

$$g_-(x, \mathscr{C}_x) = x - \mathscr{C}_x. \tag{15.2-20b}$$

That these functions satisfy Eq. (15.2-8) is readily verified by direct substitution.

In applying the normalization condition to the case of real λ, one must first decide on a definition of the singular integral arising from the first term in Eq. (15.2-18). In fact, any consistent choice is allowable; a difference in the choice of definition results in a different $p(\lambda)$, but no change in the final results. As a matter of convenience we shall adopt the principal value prescription,

$$\int_{-\infty}^{\infty} \frac{f(x)}{x-y}\, dx = \lim_{\substack{N\to\infty \\ \varepsilon\to 0}} \left[\int_{-N}^{y-\varepsilon} \frac{f(x)}{x-y}\, dx + \int_{y+\varepsilon}^{N} \frac{f(x)}{x-y}\, dx \right].$$

Then the normalization condition applied to g_λ simply determines the function p,

$$e^{-\lambda^2} p(\lambda) = \sqrt{\pi} - \lambda \int_{-\infty}^{\infty} \frac{e^{-\mathscr{C}_x^2}}{\lambda - \mathscr{C}_x}\, d\mathscr{C}_x. \tag{15.2-21}$$

We note that the definition of $p(\lambda)$ can be extended into the complex λ-plane by simply removing the principal value requirement. Thus we have a continuum of solutions of Eq. (15.2-8),

$$g_\lambda(x, \mathscr{C}_x) = e^{-x/\lambda}\left[\frac{\lambda}{\lambda-\mathscr{C}_x} + p(\lambda)\delta(\lambda-\mathscr{C}_x)\right]. \qquad (15.2\text{-}22)$$

A number of interesting properties of the normal modes can be proven. Of these, the most important one is that, for any x, the functions g_+, g_- and $g_\lambda(-\infty < \lambda < \infty)$ form a "complete set" in the sense that any function f of $\mathscr{C}_x$ defined on the real axis and satisfying the condition

$$\int_{-\infty}^{\infty} \mathscr{C}_x^2 e^{-\mathscr{C}_x^2} f(\mathscr{C}_x)\, d\mathscr{C}_x < \infty \qquad (15.2\text{-}23)$$

can be uniquely expanded in terms of these functions. In particular, if one takes $x = 0$, the completeness theorem states that, for any value of $\mathscr{C}_x$, the representation

$$f(\mathscr{C}_x) = a_+ - a_- \mathscr{C}_x + \int_{-\infty}^{\infty} A(\lambda) g_\lambda(\mathscr{C}_x)\, d\lambda \qquad (15.2\text{-}24)$$

is valid, provided the expansion coefficients a_+, a_- and $A(\lambda)$ are properly chosen. The theorem is thus the strong statement that the expansion represents the function exactly at every point. The proof of the completeness theorem is performed in two steps. The first step consists of the construction of the expansion coefficients in terms of the function f. The second step consists of the actual verification of the representation (15.2-24) and is performed by proving a so-called "closure relation", i.e., by substituting the expansion coefficients in the right member of Eq. (15.2-24) and showing that the right member thus reduces to $f(\mathscr{C}_x)$. The second step of the proof is almost trivial, as it can be accomplished by merely reversing the arguments of the first step. As the construction of the expansion coefficients, which constitutes the first step of the proof, is precisely the same type of mathematical problem as the particular boundary value problem (15.2-8) through (15.2-10), we shall solve the latter and then outline how the method of solution is adapted in the proof of the completeness theorem.

The proof of the completeness theorem as outlined here requires the function f to be a Hölder continuous function of its argument*. However, by for-

* A function $f(x)$ is said to be Hölder continuous if there exist two constants M and $\alpha(0 < \alpha \leq 1)$ such that, for every pair (x, x'), the inequality

$$|f(x) - f(x')| < M|x-x'|^\alpha$$

is satisfied. The condition of Hölder continuity is a necessary condition for the existence of the principal value integral of f. See Gakhov [1966].

mally applying it to the generalized function (distribution) $\delta(\mathscr{C}_x-\mathscr{C}_{x0})$, where $\mathscr{C}_{x0}$ is some real number, one can show that the representation (15.2-24) holds also in that case. Although no mathematically rigorous proof has been given yet, we shall make use of this result in the following discussion.

The most general solution to Eq. (15.2-8) may be written

$$g(x, \mathscr{C}_x) = a_+ g_+(x, \mathscr{C}_x) + a_- g_-(x, \mathscr{C}_x) + \int_{-\infty}^{\infty} A(\lambda)\, g_\lambda(x, \mathscr{C}_x)\, \mathrm{d}\lambda. \qquad (15.2\text{-}25)$$

In particular, the solution to the slip flow problem has this representation. However, the boundary condition (15.2-9) requires that $A(\lambda) = 0$ for $\lambda < 0$ and that $a_- = 1$. Thus, inserting the explicit form of the normal modes, we have

$$g(\lambda, \mathscr{C}_x) = (x-\mathscr{C}_x) + a_+ + \int_0^{\infty} A(\lambda)\, \mathrm{e}^{-x/\lambda} \left[\frac{\lambda}{\lambda-\mathscr{C}_x} + p(\lambda)\, \delta(\lambda-\mathscr{C}_x)\right] \mathrm{d}\lambda. \qquad (15.2\text{-}26)$$

This function has been constructed to satisfy the Boltzmann equation and the boundary condition at infinity. All that remains is to determine $a_+$and $A(\lambda)$ for $\lambda \geqq 0$ such that the boundary condition (15.2-10) is satisfied. Requiring the expansion (15.2-26) to satisfy this boundary condition, we have

$$\mathscr{C}_x - a_+ = \int_0^{\infty} \frac{\lambda A(\lambda)}{\lambda-\mathscr{C}_x}\, \mathrm{d}\lambda + p(\mathscr{C}_x)\, A(\mathscr{C}_x) \quad \text{for } \mathscr{C}_x > 0, \qquad (15.2\text{-}27)$$

that is, a singular integral equation of the Cauchy type for the unknown function $A(\lambda)$. The theory of such equations is well developed (see Muskhelishvili [1953]; Gakhov [1966]).

One begins by assuming that a solution to Eq. (15.2-27) exists. Then, one can define the function

$$N(z) = \frac{1}{2\pi \mathrm{i}} \int_0^{\infty} \frac{\lambda A(\lambda)}{\lambda - z}\, \mathrm{d}\lambda, \qquad (15.2\text{-}28)$$

which is analytic in the entire z-plane cut along the positive real axis. Also since $p(\lambda) \sim \sqrt{\pi}\, \exp(\lambda^2)/2\lambda^2$ as $\lambda \to \infty$, $A(\lambda) \sim \exp(-\lambda^2)$ if Eq. (15.2-26) is to make sense. Then $\int_0^\infty \lambda A(\lambda) \mathrm{d}\lambda$ certainly exists and we see that $N(z)$ vanishes at least as fast as z^{-1} as $z \to \infty$.

Now, the Plemelj formulae (see Muskhelishvili [1953]) state that, as z

approaches the positive real axis, $N(z)$ takes the values

$$N^+(t) = \lim_{\varepsilon\to 0} N(t+\mathrm{i}\varepsilon) = \frac{1}{2\pi\mathrm{i}}\int_0^\infty \frac{\lambda A(\lambda)}{\lambda - t}\,\mathrm{d}\lambda + \tfrac{1}{2}t\,A(t), \tag{15.2-29a}$$

$$N^-(t) = \lim_{\varepsilon\to 0} N(t-\mathrm{i}\varepsilon) = \frac{1}{2\pi\mathrm{i}}\int_0^\infty \frac{\lambda A(\lambda)}{\lambda - t}\,\mathrm{d}\lambda - \tfrac{1}{2}t\,A(t). \tag{15.2-29b}$$

The Plemelj formulae allow us to replace the integral and $A(\mathscr{C}_x)$ in Eq. (15.2-27) by the boundary values of N thus:

$$\mathscr{C}_x(\mathscr{C}_x - a_+) = (p(\mathscr{C}_x) + \pi\mathrm{i}\mathscr{C}_x)N^+(\mathscr{C}_x) + (p(\mathscr{C}_x) - \pi\mathrm{i}\mathscr{C}_x)\,N^-(\mathscr{C}_x). \tag{15.2-30}$$

Thus, if A solves Eq. (15.2-27), the boundary values of the related function N satisfy the relation (15.2-30). Now, the latter problem, known as the Riemann-Hilbert boundary value problem, possesses a solution which vanishes at infinity only for a particular value of a_+, and this solution is unique. Thus, N is determined uniquely and, once it is found, $A(\lambda)$ is obtained by taking the difference of Eqs. (15.2-29a) and (15.2-29b),

$$\lambda A(\lambda) = N^+(\lambda) - N^-(\lambda). \tag{15.2-31}$$

To construct N, we introduce the auxiliary function

$$X(z) = z^{-1}\exp\left\{\frac{1}{2\pi\mathrm{i}}\int_0^\infty \log\left(\frac{p(t)+\pi\mathrm{i}t}{p(t)-\pi\mathrm{i}t}\right)\frac{\mathrm{d}t}{t-z}\right\}, \tag{15.2-32}$$

where the branch of the logarithm which vanishes at infinity is chosen. The function X is also analytic in the z-plane cut along the positive real axis and vanishes as z^{-1} as $z \to \infty$. Furthermore, from the Plemelj formulae,

$$\frac{X^+(t)}{X^-(t)} = \frac{p(t)+\pi\mathrm{i}t}{p(t)-\pi\mathrm{i}t}. \tag{15.2-33}$$

Thus, the boundary value problem (15.2-30) may be written

$$X^+(\mathscr{C}_x)\,N^+(\mathscr{C}_x) - X^-(\mathscr{C}_x)\,N^-(\mathscr{C}_x) = \frac{X^-(\mathscr{C}_x)}{p(\mathscr{C}_x) - \pi\mathrm{i}\mathscr{C}_x}\,\mathscr{C}_x(\mathscr{C}_x - a_+). \tag{15.2-34}$$

The left-hand side of this equation contains only the boundary values of functions analytic in the complex plane cut along the positive real axis. Thus, from the Plemelj formulae, we see that

$$X(z)\,N(z) = \frac{1}{2\pi\mathrm{i}}\int_0^\infty \frac{X^-(t)}{p(t)-\pi\mathrm{i}t}\,\frac{t(t-a_+)}{t-z}\,\mathrm{d}t \tag{15.2-35}$$

is a solution to Eq. (15.2-34). Then

$$N(z) = \frac{X^{-1}(z)}{2\pi i} \int_0^{\infty} \frac{X^{-}(t)}{p(t)-\pi i t} \frac{t(t-a_+)}{t-z} \mathrm{d}t \tag{15.2-36}$$

is certainly a function satisfying Eq. (15.2-30), which is analytic in the complex plane cut along the positive real axis. This function may be identified with the function defined by Eq. (15.2-28) if only we can show that it vanishes as fast as z^{-1} as $z \to \infty$. An inspection of the right-hand side of Eq. (15.2-36) shows that, in general, it becomes a constant at infinity. If, however,

$$a_+ \int_0^{\infty} \frac{tX^{-}(t)}{p(t)-\pi i t} \mathrm{d}t = \int_0^{\infty} \frac{t^2 X^{-}(t)}{p(t)-\pi i t} \mathrm{d}t, \tag{15.2-37}$$

the function defined by (15.2-36) will indeed vanish as z^{-1} as $z \to \infty$.

Thus, provided that a_+ takes the value assigned by Eq. (15.2-37), we have constructed the correct function N and $A(\lambda)$ is obtained from Eq. (15.2-31). In other words, an explicit solution for $A(\lambda)$ and a_+ is obtained.

Before discussing the results, we point out that the completeness theorem given earlier is proven in precisely the same manner. The only difference is that we deal with an equation whose range is $(-\infty, \infty)$ instead of $(0, \infty)$. It turns out in this case that the function X is identical with the function p, and the analysis is thereby simplified. Condition (15.2-23) is required simply for the convergence of integrals.

Returning to the solution of the slip flow problem, we see that the non-dimensional hydrodynamic velocity is given by

$$\begin{aligned} v_z(x) &= \int_{-\infty}^{\infty} \mathrm{e}^{-\mathscr{C}_x^2} g(x, \mathscr{C}_x) \mathrm{d}\mathscr{C}_x \\ &= x + a_+ + \int_0^{\infty} A(\lambda) \mathrm{e}^{-x/\lambda} \mathrm{d}\lambda, \end{aligned} \tag{15.2-38}$$

and note that the first term is just the hydrodynamic solution given at the beginning of this section and is indeed the dominant term far from the plate. The last term dies out within a few mean free paths of the plate and happens to be small near the plate, so that the second term is the most important correction.

Physically, the significant item of note in this solution is that the velocity of the gas is non-zero at the plate; this is called *slip* and is the origin of the name of the problem. The explanation for this is simple – although the molecules leaving the plate have no net momentum parallel to the plate, the molecules which are about to strike the plate do, so that the gas near the plate does have a (small) net momentum content. That this should be the case was, in fact, recognized by Maxwell, who used mean free path arguments to compute the slip. Crudely, the average momentum of the molecules about to strike the plate should be approximately that corresponding to the gas displaced one mean free path from the plate. However, only half the molecules have this momentum, since those leaving the plate have none. Thus, Maxwell obtained

$$v_z(0) \approx \tfrac{1}{2} v_z(l) \approx \tfrac{1}{2} l \left. \frac{dv_z}{dx} \right|_{x=0} . \tag{15.2-39}$$

In non-dimensional units the coefficient is just $\frac{1}{2}$. More careful analysis based on the mean free path model gives a similar result with a different numerical coefficient.

In gas flows under ordinary conditions the bulk of the gas is many mean free paths in thickness (recall $l \approx 10^{-5}$ cm) so that the slip correction is entirely negligible, but for rarefied gas flows the effect may be significant. The effect of slip can be at least partially accounted for by modifying the usual zero slip boundary condition, i.e., $v_z(0) = 0$, used in conjunction with continuum gas dynamics, to one of the form (15.2-39),

$$v_z(0) = a_+ \left. \frac{dv_z}{dx} \right|_{x=0} \tag{15.2-40}$$

or, in dimensional form,

$$v_z(0) = a_+ l \left. \frac{dv_z}{dx} \right|_{x=0} = a_+ \frac{2\eta}{\rho} \left(\frac{m}{2kT} \right)^{\frac{1}{2}} \left. \frac{dv_z}{dx} \right|_{x=0} , \tag{15.2-41}$$

where the relation between the mean free path and the viscosity appropriate to the BGK model has been used to write the expression in terms of observable properties of the gas.

All that remains is the evaluation of the numerical coefficient a_+. This can be obtained by carrying out the quadratures indicated in Eq. (15.2-35); some analytical simplification is possible before carrying out the quadrature, but is not of concern here. The numerical value is found to be 1.0161. Also, in Table 15.1 numerical values of the non-asymptotic portion of the velocity field (the integral term in Eq. (15.2-38)) are given and are seen to be small with respect to the numerical values of the asymptotic portion, $x + a_+$.

TABLE 15.1
Deviation of the velocity from its asymptotic value*

x	$v_z(x)-(x+a_+)$	x	$v_z(x)-(x+a_+)$
0.00	0.3090	2.00	0.0313
0.05	0.2665	2.50	0.0216
0.10	0.2360	3.00	0.0152
0.20	0.1948	3.50	0.0116
0.50	0.1268	4.00	0.0080
1.00	0.0736	4.50	0.0059
1.50	0.0468	5.00	0.0040

* From Loyalka [1967].

As was stated earlier, the slip flow problem serves as a standard against which approximations may be tested and has been solved by various other methods. Among these are:

a) Polynomial expansion methods. Historically, this was the first method used and it bears a strong similarity to the Chapman-Cowling method for solving the Chapman-Enskog integral equations. The distribution function is approximated by a finite series of functions (usually the Maxwell molecule eigenfunctions) and the moments of the Boltzmann equation are taken with the same eigenfunctions to yield a finite set of equations for the expansion coefficients. Likewise, moments of the boundary conditions (with the distribution function replaced by its finite series approximation) are taken with the same set of functions to yield a set of boundary conditions to be applied to the expansion coefficients. This method tends to converge rather slowly because it has difficulty in matching the peculiarities of the distribution function near the plate (the distribution function has a discontinuity at $\mathscr{C}_x = 0$ at the plate).

b) Double polynomial expansions. One means of overcoming the problem arising from the discontinuity mentioned above is to use different expansions for $\mathscr{C}_x > 0$ and $\mathscr{C}_x < 0$. This technique gives improved convergence, but, for a given order, one must carry twice the number of coefficients. However, when the double expansion is compared with the simple expansion of twice the order, there is still some improvement, although this improvement is obtained at the cost of having to solve a more complicated set of equations.

c) Variational methods. Solution of the Boltzmann equation and its associated boundary conditions can be shown to be equivalent to finding the extrema of particular functionals of the distribution function; in fact, there exist numerous such functionals (called variational principles) and the choice is one of convenience. In particular, one can find variational principles whose extrema are quantities of interest, e.g., the slip coefficient. The variational character of the problem means that accurate results for the quantity of interest are obtained even with poor estimates of the distribution function so that this method is well suited to the computation of selected coefficients but not of the distribution function. A good example of this approach is the variational calculation of the transport coefficients in Chapter 5.

d) Moment methods. In these methods one supplements the usual macroscopic conservation equations with equations obtained by taking moments of the Boltzmann equation with other functions of the molecular velocity. Since a closed set of equations is never obtained in this manner, the set must be closed by arbitrarily setting some of the terms in these equations equal to zero. For the linearized equation this method produces equations similar to those obtained by the polynomial expansion methods. An extensive treatment of moments methods may be found in Burgers [1969].

e) Wiener-Hopf method. In this method one converts the problem to an integral equation for the hydrodynamic velocity and solves the latter by a Fourier transform technique. This method is completely equivalent to the normal mode expansion technique outlined above.

The solution of the slip flow problem presented above represents the prototype for a whole class of solvable problems in linearized kinetic theory and its publication by Cercignani in 1962 gave rise to a great deal of activity in this area. Most of the work has concentrated on finding solutions of the slip flow problem with fewer assumptions – use of better models of the collision operator, more complicated boundary conditions, time-dependent problems – and the application of the method to other physical situations – Couette flow, Poiseuille flow, sound wave propagation, heat flow, etc. Space does not permit coverage of this work here and the interested reader is referred to Cercignani's book for further details.

15.3. Non-linear problems

A great many rarefied gas-flow problems of physical and technological interest do not permit linearization. The prime example is the problem of computing the flow field in the initial stages of the entry of a body from space (a meteor or a man-made object). In these flows, a large part of the gas flows at supersonic speeds, the flow in the vicinity of the body is characterized by very steep gradients in the properties of the gas, and shock waves will occur. Within a shock wave the state of a gas is so far displaced from equilibrium and varying so rapidly that the only realistic treatment of the problem must utilize the non-linear Boltzmann equation. The simplest problem of non-linear rarefied gas dynamics – that of predicting the distribution function within a plane shock wave – has played the role of a prototype problem. Unfortunately, despite the large amount of attention devoted to this problem, many of the approaches used have produced rather unsatisfactory results.

In this section, we shall avoid the approaches which have met with limited success and discuss those which appear to have the most promise and which have produced the best results; these number just two. The first approach involves solving the Boltzmann equation numerically, but since this is a formidable task, even on the largest computers, the collision operator is generally modeled. In the second approach, the Boltzmann equation is bypassed entirely and the physics of molecular motion is modeled directly; this is known as the *Monte Carlo* method and has produced some of the most useful results obtained to date.

The first approach requires the selection of a model for the collision operator. By far the most popular model is the BGK model, whose linearization was discussed in the previous section. This model may be developed by considering the exact collision operator,

$$J(ff) = \int (f(\boldsymbol{c}')f(\boldsymbol{c}_1') - f(\boldsymbol{c})f(\boldsymbol{c}_1))\, gb\, \mathrm{d}b\, \mathrm{d}\varepsilon\, \mathrm{d}^3c_1 . \tag{15.3-1}$$

Beginning with the second term, which represents the removal of particles from velocities near $\boldsymbol{c}$, one adopts the simplest possible approximation to it,

$$-f(\boldsymbol{c}) \int f(\boldsymbol{c}_1) gb\, \mathrm{d}b\, \mathrm{d}\varepsilon\, \mathrm{d}^3c_1 \approx -n\nu f(\boldsymbol{c}), \tag{15.3-2}$$

i.e., the collision frequency is approximated by a constant times the density (cf. Eq. (2.4-6)). For the first term in the integral (15.3-1), which represents the particles arriving at velocities near $\boldsymbol{c}$, one assumes that, independent of what the distribution function of the colliding particles may be, the result is a Maxwellian distribution,

$$\int f(\boldsymbol{c}')f(\boldsymbol{c}_1') gb\, \mathrm{d}b\, \mathrm{d}\varepsilon\, \mathrm{d}^3c_1 \approx n\nu f^{(0)}(\boldsymbol{c}). \tag{15.3-3}$$

Thus,

$$J_{\mathrm{BGK}}(ff) = n\nu(f^{(0)} - f). \tag{15.3-4}$$

Since $f^{(0)}$ is a Maxwellian whose parameters are the *local* density, hydrodynamic velocity and temperature, one easily sees that this operator possesses the correct moments with respect to the summational invariants:

$$\int \psi_i J_{\text{BGK}}(ff)\,\mathrm{d}^3c = 0, \quad i = 1, 2, \ldots, 5, \tag{15.3-5}$$

where $\psi_1 = 1$, $\psi_2 = c_x$, $\psi_3 = c_y$, $\psi_4 = c_z$, $\psi_5 = \frac{1}{2}mc^2$. It should be noted that one can allow the collision frequency ν to be a function of the macroscopic gas properties without affecting these results. In particular, the freedom to set $\nu = \nu(T)$ allows one to fit the viscosity of a real gas. However, for the BGK model the Eucken ratio is always $\frac{3}{2}$, instead of the correct $\frac{5}{2}$, and this cannot be remedied.

The appearance of linearity in this model is deceptive; it is highly non-linear due to the dependence of the parameters in $f^{(0)}$ on f. To study the structure of a plane shock, the Boltzmann equation with the BGK model,

$$c_z \frac{\partial f}{\partial z} = \nu n(f^{(0)} - f), \tag{15.3-6}$$

must be solved together with the boundary conditions

$$f \sim n_1 \left(\frac{m}{2\pi k T_1}\right)^{\frac{3}{2}} \exp\left(\frac{-m(\boldsymbol{c} - v_1 \hat{\boldsymbol{z}})^2}{2kT_1}\right) \text{ as } z \to -\infty, \tag{15.3-7a}$$

$$f \sim n_2 \left(\frac{m}{2\pi k T_2}\right)^{\frac{3}{2}} \exp\left(\frac{-m(\boldsymbol{c} - v_2 \hat{\boldsymbol{z}})^2}{2kT_2}\right) \text{ as } z \to \infty, \tag{15.3-7b}$$

where $\hat{\boldsymbol{z}}$ is a unit vector in the direction of the flow. The parameter sets n_1, v_1, T_1 and n_2, v_2, T_2 must be connected by the Rankine-Hugoniot conditions, which are obtained by integrating the conservation equations from a position well upstream of the shock to one well downstream from the shock (see, e.g., Shapiro [1953]),

$$\rho_1 v_1 = \rho_2 v_2, \tag{15.3-8a}$$

$$\rho_1 v_1^2 + p_1 = \rho_2 v_2^2 + p_2, \tag{15.3-8b}$$

$$\rho_1 v_1(u_1 + p_1/\rho_1 + \tfrac{1}{2}v_1^2) = \rho_2 v_2(u_2 + p_2/\rho_2 + \tfrac{1}{2}v_2^2), \tag{15.3-8c}$$

which together with the equations of state $u = 3kT/2m$ and $p = nkT$ give the state of the gas on one side of the shock when the state on the other side is known. Various other forms of these equations have been given but are not needed here.

A straight-forward numerical integration of this problem is rendered extremely difficult by the non-linearity. The procedure adopted by Liepmann, Narasimha and Chahine [1962] and Anderson [1966] is to convert the problem into a system of integral equations for the macroscopic properties of the gas and solve these resulting equations numerically. This is done as follows. First, let us write Eq. (15.3-6) in the form

$$c_z \frac{\partial f}{\partial z} + \nu n f = \nu n f^{(0)}. \tag{15.3-9}$$

Now, treating $f^{(0)}$ and n as known functions for the moment, we can integrate this equation to yield

$$f(z, \boldsymbol{c}) = \begin{cases} \displaystyle\int_{-\infty}^{z} \frac{\nu n(z')}{c_z} f^{(0)}(z', \boldsymbol{c}) \exp\left(-\int_{z'}^{z} \frac{\nu n(z'')}{c_z} \mathrm{d}z''\right) \mathrm{d}z', & \text{for } c_z > 0, \\ \displaystyle -\int_{z}^{\infty} \frac{\nu n(z')}{c_z} f^{(0)}(z', \boldsymbol{c}) \exp\left(-\int_{z}^{z'} \frac{\nu n(z'')}{c_z} \mathrm{d}z''\right) \mathrm{d}z', & \text{for } c_z < 0. \end{cases} \tag{15.3-10}$$

For convenience, define a new coordinate $\tilde{z}$,

$$\tilde{z} = \int_0^z \nu(z') n(z') \mathrm{d}z'. \tag{15.3-11}$$

Then, on taking moments of Eq. (15.3-10) with 1, $\boldsymbol{c}$ and $\frac{1}{2}c^2$, respectively, one obtains (dropping the tilde from $\tilde{z}$)

$$n(z) = \int_{-\infty}^{\infty} \mathrm{d}z' \, n(z') \left(\frac{m}{kT(z')}\right)^{\frac{1}{2}} H_1 \left(\left(\frac{m}{kT(z')}\right)^{\frac{1}{2}} v(z') \operatorname{sgn}(z-z'), \left(\frac{m}{kT(z')}\right)^{\frac{1}{2}} |z-z'|\right), \tag{15.3-12a}$$

$$n(z)v(z) = n(-\infty)v(-\infty), \tag{15.3-12b}$$

$$\begin{aligned} \tfrac{1}{2}n(z)\left[\frac{3kT(z)}{m} + v^2(z)\right] = \int_{-\infty}^{\infty} \mathrm{d}z' \, n(z') \left(\frac{kT(z')}{m}\right)^{\frac{1}{2}} \\ \times H_1 \left(\left(\frac{m}{kT(z')}\right)^{\frac{1}{2}} v(z') \operatorname{sgn}(z-z'), \left(\frac{m}{kT(z')}\right)^{\frac{1}{2}} |z-z'|\right) \\ + \tfrac{1}{2}n(-\infty)\left[\frac{kT(-\infty)}{m} + v^2(-\infty)\right], \end{aligned} \tag{15.3-12c}$$

where we have defined

$$H_n(p, q) = (2\pi)^{-\frac{1}{2}} \int_0^\infty \mathrm{d}c\, c^{n-2} \mathrm{e}^{-\frac{1}{2}(c-p)^2} \mathrm{e}^{-q/c}. \qquad (15.3\text{-}13)$$

In writing Eqs. (15.3-12b) and (15.3-12c), we have made use of the facts that the $\boldsymbol{c}$ and c_z^2 moments of f are invariants of a one-dimensional flow, to eliminate H_2 and H_3 which would occur in Eqs. (15.3-12b, c) and also to reduce the three equations to two coupled non-linear integral equations. H_n may be evaluated numerically without great difficulty. Then the two cou-

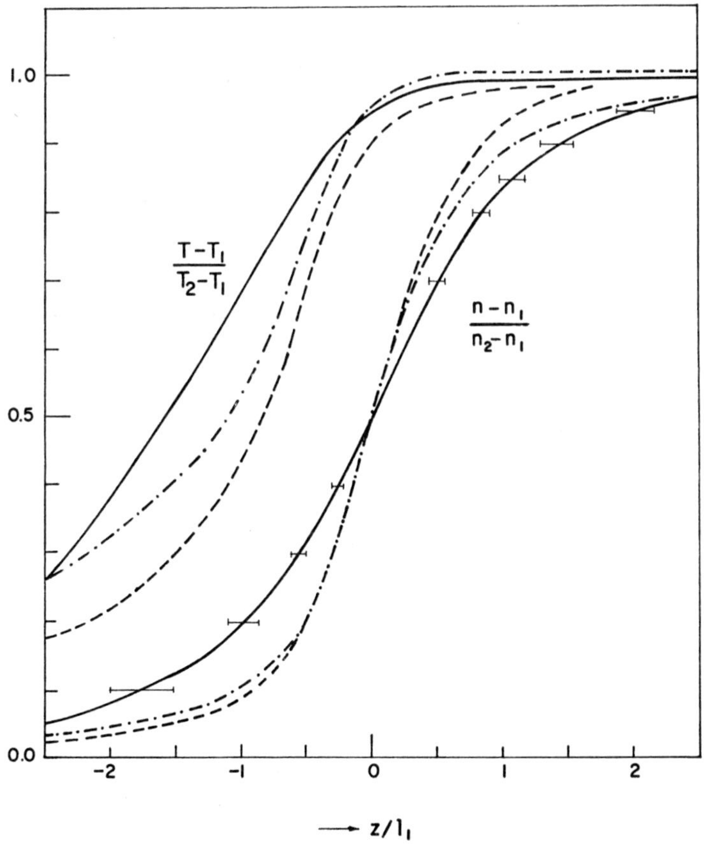

Fig. 15.1. Reduced density and temperature profiles for a plane shock wave in a monatomic gas of hard sphere molecules (Mach number $M = 10$); l_1 is the upstream mean free path. ——— Monte Carlo, – – – – BGK-model, –.–.–.– ellipsoidal model (Holway). (From Segal [1971].)

pled non-linear integral equations are solved by any of a variety of iterative numerical techniques.

From the results (cf. Fig. 15.1), one finds that the BGK model predicts a shock that is too thin. This result is not surprising since this model produces a Maxwellian distribution after a single collision, while real molecules require a few collisions to attain this distribution. It clearly displays the need for better models, but such few have been proposed. The simplest advance on the BGK model is due to Holway [1963]. The model was originally developed using statistical arguments, but can be derived by more heuristic arguments. The molecules moving in a direction perpendicular to a shock have, on the average, made their previous collisions a mean free path removed from their current positions. As large changes occur over a distance of one mean free path in a shock, the distribution function for these molecules will be different from that for molecules traveling transverse to the shock. As this situation will persist after collision, it is reasonable to write

$$J(ff) \approx n\nu(\Phi - f), \tag{15.3-14}$$

where, for Φ, we adopt the simplest function that reflects the anisotropy:

$$\Phi = n\pi^{-\frac{3}{2}}(\det|\alpha|)^{\frac{1}{2}} \exp\left[-\sum_{i,j=1}^{3} \alpha_{ij} C_i C_j\right]. \tag{15.3-15}$$

The matrix elements α_{ij} are obtained by requiring the model to yield the correct Eucken ratio and conservation properties (mass and momentum conservation are already built in). One finds

$$\alpha_{ij} = \frac{2kT}{m}(\mathrm{f}/\gamma)\delta_{ij} - 2\frac{\mathrm{f}-\gamma}{\gamma}\frac{P_{ij}}{\rho}, \tag{15.3-16}$$

where f is the Eucken ratio, $\lambda/\eta c_v$, and γ is the ratio of specific heats, $\gamma = c_p/c_v$. For $\mathrm{f} = \gamma = \frac{5}{3}$, this model reduces to the BGK model. Calculations indicate that the model (15.3-14) does give some improvement in the predicted shock structure versus the BGK model, but still does not reproduce the results obtained by the Monte Carlo method (see below). Thus, the problem of constructing appropriate models for the non-linear collision operator is one which is far from solved. Recently, Segal [1971] and Segal and Ferziger [1970] have given models especially tailored to the problem and have obtained good results for shock structure.

A method which circumvents these difficulties is the Monte Carlo approach developed by Bird [1968] and others. In this approach one models the behavior of molecules rather than the Boltzmann equation. We shall give only a short description of this method. Actually, its application to the shock structure problem is complicated by the fact that the location of the shock is arbitrary (because there is no fixed point to which a coordinate system may be attached). To avoid this difficulty, one must either treat a moving shock (which is what Bird did) or introduce a body which will fix the shock position (which is the principal method used in experimental investigations).

Suppose then we start with a system in which all the molecules at positions $x < 0$ have the distribution characteristic of the gas far upstream from the shock and those at positions $x > 0$ have the characteristic downstream distribution. Physically, a shock will develop which will propagate to the right. This situation is modeled by dividing the portion of space within a few mean free paths of the discontinuity into a number of cells. Within each cell one chooses a small number of molecules for study (modeling of all the molecules would be impossible even in a highly rarefied gas). The velocities of the molecules are chosen by a random sampling method in accordance with the known distribution function.* Then the molecules are allowed to fly freely for a short length of time Δt and their new positions are computed. Next, from among the molecules in each cell, pairs are selected at random and their relative velocities are computed. If the collision frequency is known as a function of relative velocity, one can determine the probability that a collision between the two molecules of a given pair will take place. Note, however, that because each model "molecule" represents a much larger number of physical molecules the probabilities must be modified to account for this. Then, again by choosing a random number, it is determined whether a collision actually takes place and, if such is the case, two further random numbers suffice to completely fix the nature of the collision, i.e., they determine χ and ε and new velocities are assigned to the molecules. When all pairs in each cell have been treated in this way, the molecules are again allowed to fly

* For a one-dimensional distribution this is done as follows. Let

$$x(c) = \int_{-\infty}^{c} f(c)\,\mathrm{d}c \bigg/ \int_{-\infty}^{\infty} f(c)\,\mathrm{d}c.$$

Clearly, x is a number between zero and one and stands in a one-to-one relation to c. Now choose a random number between zero and one (there are many recipes for doing this on a computer). Then the molecule is assigned this value of x, or by inversion of the above relation, the corresponding value of c. It is easy to show that, when a sufficiently large number of molecules has been assigned velocities in this way, their distribution function will be $f(c)$. In three dimensions one simply applies this method to each of the three velocity components.

freely and all of the above is repeated. In the shock problem, the cell structure is moved at the velocity of the shock and eventually a steady, propagating shock is found. One can then study any desired property of the gas by straight-forward computation.

There are, of course, disadvantages in this method. First of all, it is not a completely faithful representation of the physical situation because it separates the free flight and collision processes arbitrarily and because the collision model does not depend on the position of a molecule within a cell; these difficulties disappear if the cell size and Δt are made sufficiently small. Secondly, because the method is statistical in nature, a large number of trials is needed to produce accurate results. This means that a great deal of computation time is necessary if one wants to study small effects.

On the other hand, the method possesses a great deal of flexibility and is readily applied to more complex problems. It thus appears to hold considerable potential for studying realistic flows.

Appendix A
Vectors and tensors

Throughout the text we have made extensive use of vectors and Cartesian tensors. This appendix contains a summary of the vector and tensor algebra, as well as some integral relations involving vectors and tensors. For a detailed account of vector and tensor analysis the reader is referred, for example, to the monograph by Jeffreys [1931].

Light-face italic type is used for scalar quantities.

Bold-face italic type is used for vector quantities.

Sans serif italic type is used for tensor quantities.

A.1. *Definition of a vector.* A *vector* is a quantity that possesses both a magnitude and a direction and obeys the parallelogram law of addition.

A.2. *Representation of a vector.* Consider an orthogonal right-handed Cartesian coordinate system. Let the three axes be labeled with the indices 1, 2 and 3 (or, alternately, with the indices x, y and z) and let $\boldsymbol{e}_1$, $\boldsymbol{e}_2$, $\boldsymbol{e}_3$ be the reference unit vectors (i.e., the vectors of unit length in the direction of the axis 1, 2, 3, respectively).

A vector $\boldsymbol{a}$ is uniquely determined by a set of three numbers which form the *components* of the vector. Thus, if a_1, a_2, a_3 are the components of $\boldsymbol{a}$, then the vector $\boldsymbol{a}$ is expressed by

$$\boldsymbol{a} = a_1 \boldsymbol{e}_1 + a_2 \boldsymbol{e}_2 + a_3 \boldsymbol{e}_3 .$$

We will often write the expression symbolically in the form

$$\boldsymbol{a} = (a_1, a_2, a_3) \text{ or, simply, } \boldsymbol{a} = (a_i).$$

As we shall see, in many situations simplicity of notation is achieved if we adopt *Einstein's summation convention*: whenever an index is repeated in a term, such a term is to be summed over all values of that index. With this convention we may also write

$$\boldsymbol{a} = a_i \boldsymbol{e}_i.$$

The nonnegative real number

$$a = (a_1^2 + a_2^2 + a_3^2)^{\frac{1}{2}} = (a_i a_i)^{\frac{1}{2}}$$

is called the *magnitude* of the vector $\boldsymbol{a}$.

A.3. *Products of vectors.* The *scalar product* of two vectors $\boldsymbol{a} = (a_i)$ and $\boldsymbol{b} = (b_i)$ is a scalar quantity defined by

$$\boldsymbol{a} \cdot \boldsymbol{b} = a_1 b_1 + a_2 b_2 + a_3 b_3 = a_i b_i.$$

Of particular importance are the scalar products between the unit vectors $\boldsymbol{e}_1$, $\boldsymbol{e}_2$ and $\boldsymbol{e}_3$. There are nine such products, expressed by $\boldsymbol{e}_i \cdot \boldsymbol{e}_j$ where each i and j is given independently all the values 1, 2, 3. Then,

$$\boldsymbol{e}_i \cdot \boldsymbol{e}_j = \delta_{ij},$$

where δ_{ij} is the *Kronecker delta* or the *Kronecker symbol* ($\delta_{ij} = 1$ if $i = j$, $\delta_{ij} = 0$ if $i \neq j$).

The *vector product* of two vectors $\boldsymbol{a} = (a_i)$ and $\boldsymbol{b} = (b_i)$ is a vector quantity defined by

$$\boldsymbol{a} \times \boldsymbol{b} = (a_2 b_3 - a_3 b_2)\boldsymbol{e}_1 + (a_3 b_1 - a_1 b_3)\boldsymbol{e}_2 + (a_1 b_2 - a_2 b_1)\boldsymbol{e}_3.$$

Consider the various vector products between the unit vectors $\boldsymbol{e}_1$, $\boldsymbol{e}_2$ and $\boldsymbol{e}_3$. There are nine such products, expressed by $\boldsymbol{e}_j \times \boldsymbol{e}_k$ where each j and k is given independently all the values 1, 2, 3. The component of the vector $\boldsymbol{e}_j \times \boldsymbol{e}_k$ in the direction of $\boldsymbol{e}_i$ is given by

$$\boldsymbol{e}_i \cdot \boldsymbol{e}_j \times \boldsymbol{e}_k = \begin{cases} 0 & \text{if any two of } i, j \text{ and } k \text{ are equal,} \\ +1 & \text{if } i, j \text{ and } k \text{ are unequal and in cyclic order,} \\ -1 & \text{if } i, j \text{ and } k \text{ are unequal and not in cyclic order.} \end{cases}$$

The result can be represented compactly with the aid of a symbol ε_{ijk} defined by the properties

$$\varepsilon_{ijk} = \begin{cases} 0 & \text{if any two of the indices are equal,} \\ +1 & \text{if the indices are unequal and in cyclic order,} \\ -1 & \text{if the indices are unequal and not in cyclic order.} \end{cases}$$

The symbol ε_{ijk} is known as the *alternating symbol* or the *permutation symbol*. We see that

$$\varepsilon_{ijk} = \boldsymbol{e}_i \cdot \boldsymbol{e}_j \times \boldsymbol{e}_k.$$

Thus,

$$\boldsymbol{e}_j \times \boldsymbol{e}_k = \varepsilon_{ijk} \boldsymbol{e}_i$$

and, consequently,

$$\boldsymbol{a} \times \boldsymbol{b} = \varepsilon_{ijk} a_j b_k \boldsymbol{e}_i.$$

A.4. *Multiple products of vectors.* Products between three vectors are called triple products. The scalar triple product of three vectors $\boldsymbol{a} = (a_i)$, $\boldsymbol{b} = (b_i)$ and $\boldsymbol{c} = (c_i)$ is a scalar quantity given by

$$\boldsymbol{a} \cdot (\boldsymbol{b} \times \boldsymbol{c}) = a_i \boldsymbol{e}_i \cdot b_j \boldsymbol{e}_j \times c_k \boldsymbol{e}_k = \varepsilon_{ijk} a_i b_j c_k,$$

the *vector triple product* is a vector quantity given by

$$\boldsymbol{a} \times (\boldsymbol{b} \times \boldsymbol{c}) = \varepsilon_{ijk} \varepsilon_{klm} a_j b_l c_m \boldsymbol{e}_i.$$

It is easily verified that

$$\boldsymbol{a} \times (\boldsymbol{b} \times \boldsymbol{c}) = (\boldsymbol{a} \cdot \boldsymbol{c})\boldsymbol{b} - (\boldsymbol{a} \cdot \boldsymbol{b})\boldsymbol{c}.$$

A.5. *Transformation properties of the position vector.* A special vector is the position vector $\boldsymbol{r} = (x_i)$. Let us see how the components and products of the components of the position vector transform under a rotation of the coordinate axes. The new components x'_j are related to the original components x_i by

$$x'_j = \alpha_{ji} x_i$$

and, conversely,

$$x_i = \alpha_{ij} x'_j.$$

For any of the nine products $x_i x_k$, where each i and k takes independently all the values 1, 2, 3, we have

$$x'_j x'_l = \alpha_{ji} \alpha_{lk} x_i x_k$$

and, conversely,

$$x_i x_k = \alpha_{ij} \alpha_{kl} x'_j x'_l.$$

In general, for any of the 3^n products $x_{i_1} \ldots x_{i_n}$, where each $i_1, \ldots, i_n$

takes independently the values 1, 2, 3, we have

$$x'_{j_1} \ldots x'_{j_n} = \alpha_{j_1 i_1} \ldots \alpha_{j_n i_n} x_{i_1} \ldots x_{i_n}$$

and, conversely,

$$x_{i_1} \ldots x_{i_n} = \alpha_{i_1 j_1} \ldots \alpha_{i_n j_n} x'_{j_1} \ldots x'_{j_n}.$$

A.6. *Definition of a tensor.* An *nth-rank Cartesian tensor* is a set of 3^n numbers which are characterized by n indices and which transform in the same manner as a product of n components of the position vector (see the previous paragraph). Thus, if $t_{i_1 \ldots i_n}$, where each of the indices $i_1, \ldots, i_n$ takes independently the values 1, 2, 3, denotes the general component of a tensor **t** with respect to the axes X_1, X_2, X_3, then the general component with respect to the axes X'_1, X'_2, X'_3 is given by

$$t'_{j_1 \ldots j_n} = \alpha_{j_1 i_1} \ldots \alpha_{j_n i_n} t_{i_1 \ldots i_n} \tag{1}$$

and, conversely,

$$t_{i_1 \ldots i_n} = \alpha_{i_1 j_1} \ldots \alpha_{i_n j_n} t'_{j_1 \ldots j_n}.$$

The number of indices characterizing the general component of a tensor is called the *rank* or *order* of that tensor. Some examples are:

(i) All scalars are zero-rank tensors.

(ii) All vectors are first-rank tensors. In particular, if ϕ is a position-dependent scalar, $\phi = \phi(\boldsymbol{r})$, the gradient of ϕ, $\nabla\phi = (\partial\phi/\partial x_i)$, is a vector or a first-rank tensor.

(iii) Let $\boldsymbol{a}$ be a position-dependent vector, $\boldsymbol{a} = \boldsymbol{a}(\boldsymbol{r})$. The set of nine numbers $(\partial a_i/\partial x_k)$ transforms according to Eq. (1) with $n = 2$. These numbers are therefore the components of a second-rank tensor. This tensor is called the *tensor gradient* of the vector $\boldsymbol{a}$ and is denoted by $\nabla\boldsymbol{a}$. In particular, taking $\boldsymbol{a} = \boldsymbol{r}$ we find that the Kronecker delta forms a second-rank tensor.

(iv) The products $a_i b_k c_m$ of the components of the vectors $\boldsymbol{a}$, $\boldsymbol{b}$ and $\boldsymbol{c}$ generate a third-rank tensor $\mathbf{t} = \boldsymbol{abc}$. The alternating symbol ε_{ijk} constitutes a third-rank tensor, $\boldsymbol{\epsilon}$, which is called the *alternating tensor* or the *completely antisymmetric tensor of rank three.*

A.7. *Tensor algebra.* Tensors can be added and multiplied by a scalar in the obvious manner (i.e., by termwise addition and multiplication, respectively). The *tensor product* of a tensor **a** of rank m and a tensor **b** of rank n is a tensor **c** of rank $m+n$ whose components are given by

$$c_{i_1 \ldots i_m j_1 \ldots j_n} = a_{i_1 \ldots i_m} b_{j_1 \ldots j_n}.$$

A.8. *Contraction.* A typical operation with tensors is the process of *contraction*, which consists of putting two indices in a tensor equal and summing over that index. In general, contraction of a tensor results in a tensor whose rank is two less than that of the original tensor.

The following contraction formulae prove useful in the derivation of certain expressions involving the Kronecker delta tensor and the alternating tensor:

$$\delta_{ii} = 3,$$

$$\varepsilon_{ikm}\varepsilon_{ikm} = 6,$$

$$\varepsilon_{iks}\varepsilon_{mps} = \varepsilon_{sik}\varepsilon_{smp} = \varepsilon_{ksi}\varepsilon_{psm},$$

$$\delta_{ik}\varepsilon_{ikm} = 0,$$

$$\varepsilon_{iks}\varepsilon_{mps} = \delta_{im}\delta_{kp} - \delta_{ip}\delta_{km},$$

$$\varepsilon_{iks}\varepsilon_{mks} = 2\delta_{im}.$$

A.9. *Isotropic tensors.* Due to the rotational invariance of the collision operator a particularly important role in kinetic theory is played by those Cartesian tensors which are isotropic. A tensor is called *isotropic* if its components retain the same value however the axes of the coordinate system are rotated.

There is no isotropic tensor of rank one other than the zero tensor. The only isotropic tensors of rank two are scalar multiples of the Kronecker delta tensor. Similarly, the only isotropic tensors of rank three are scalar multiples of the alternating tensor. Finally, the general isotropic tensor of rank four is

$$\lambda\delta_{ik}\delta_{mp} + \mu(\delta_{im}\delta_{kp} + \delta_{ip}\delta_{km}) + \nu(\delta_{im}\delta_{kp} - \delta_{ip}\delta_{km}),$$

where λ, μ and ν are scalars. See also Table 11.4.

A.10. *Second-rank tensors.* Throughout most of the text we are dealing with tensors of rank zero (scalars), one (vectors) or two. In this paragraph we discuss some specific properties of second-rank tensors. Also, we establish a notation for various algebraic operations involving second-rank tensors in which the index representation is suppressed.

Consider a second-rank tensor $\mathbf{w}$ with general component w_{ij}; $\mathbf{w}$ may be represented by the array

$$\begin{pmatrix} w_{11} & w_{12} & w_{13} \\ w_{21} & w_{22} & w_{23} \\ w_{31} & w_{32} & w_{33} \end{pmatrix}.$$

The *transpose* of $\boldsymbol{w}$, obtained from $\boldsymbol{w}$ by interchanging rows and columns, is denoted by $\boldsymbol{w}^{\mathrm{T}}$. When $\boldsymbol{w}^{\mathrm{T}}$ is identical with $\boldsymbol{w}$, $\boldsymbol{w}$ is said to be a *symmetric tensor*; the condition of symmetry is $w_{ij} = w_{ji}$. If $\boldsymbol{w}^{\mathrm{T}} = -\boldsymbol{w}$, $\boldsymbol{w}$ is said to be an *antisymmetric tensor*; the condition of antisymmetry is $w_{ij} = -w_{ji}$. Any tensor of rank two may be written as a sum of two parts, one symmetric and the other antisymmetric,

$$\boldsymbol{w} = \tfrac{1}{2}(\boldsymbol{w}+\boldsymbol{w}^{\mathrm{T}})+\tfrac{1}{2}(\boldsymbol{w}-\boldsymbol{w}^{\mathrm{T}}).$$

This decomposition is invariant under rotation.

The simplest example of a symmetric tensor is the Kronecker delta δ_{ij} with the array

$$\begin{pmatrix} 1 & 0 & 0 \\ 0 & 1 & 0 \\ 0 & 0 & 1 \end{pmatrix}.$$

This tensor is known as the *unit tensor* and usually denoted by the symbol $\boldsymbol{I}$.

The sum of the diagonal terms of a tensor is known as the *trace* of the tensor,

$$\operatorname{tr} \boldsymbol{w} = w_{ii}.$$

The trace is a scalar and, therefore, invariant under a coordinate transformation. If the trace is zero, the tensor is said to be *traceless*. The trace of the unit tensor is equal to three.

The tensor product of a vector $\boldsymbol{a}$ and a tensor $\boldsymbol{w}$ is a tensor of rank three, $\boldsymbol{a}\,\boldsymbol{w}$, whose elements are $a_i w_{jk}$. If we contract this product we obtain a tensor of rank one, i.e., a vector. We denote this contracted product by inserting a dot between $\boldsymbol{a}$ and $\boldsymbol{w}$,

$$\boldsymbol{a} \cdot \boldsymbol{w} = (a_i w_{ij}).$$

Such a product is known as a *simple product* of a vector and a tensor. (The use of the dot is consistent with the usage in the scalar product of two vectors; it represents the presence, in the index notation, of two adjacent, equal, and therefore dummy indices.)

Consider now the tensor product of two tensors $\boldsymbol{w}$ and $\boldsymbol{w}'$. The result is a tensor of rank four, $\boldsymbol{w}\boldsymbol{w}' = (w_{ij} w'_{kl})$. If we contract this product once, we obtain a second-rank tensor; if we contract it twice, we obtain a scalar. We denote a single contraction by a single dot and a double contraction by two dots. We call the once-contracted product a *simple product* and the twice-contracted product a *double product* or *scalar product* of the tensors. Thus, for the simple product,

$$\mathbf{w} \cdot \mathbf{w}' = (w_{ij} w'_{jk}) \qquad \text{(tensor of rank two)},$$

and for the double product,

$$\mathbf{w} : \mathbf{w}' = w_{ij} w'_{ji} \qquad \text{(scalar)}.$$

It is readily verified that, for any vector $\boldsymbol{a}$,

$$\boldsymbol{I} \cdot \boldsymbol{a} = \boldsymbol{a} \cdot \boldsymbol{I} = \boldsymbol{a}$$

and that, for any tensor $\mathbf{w}$,

$$\boldsymbol{I} : \mathbf{w} = \mathbf{w} : \boldsymbol{I} = \operatorname{tr} \mathbf{w}.$$

As a special case of the latter foımula one has

$$\boldsymbol{I} : \boldsymbol{I} = 3.$$

A.11. *Dyadics.* A second-rank tensor that is a tensor product of two vectors is called a *dyadic*. The dyadic $\boldsymbol{ab}$ of two vectors $\boldsymbol{a}$ and $\boldsymbol{b}$ is represented by the array $(a_i b_j)$. A particular example of a dyadic is provided by the unit tensor, which may be written in the form,

$$\boldsymbol{I} = \boldsymbol{e}_1 \boldsymbol{e}_1 + \boldsymbol{e}_2 \boldsymbol{e}_2 + \boldsymbol{e}_3 \boldsymbol{e}_3 = \boldsymbol{e}_i \boldsymbol{e}_i.$$

A.12. *Pseudovectors.* In three dimensions, the number of numerically independent components of an antisymmetric tensor of rank two is equal to three, i.e. equal to the number of components of a vector. In fact, with any vector $\boldsymbol{a} = (a_i)$ we can associate an antisymmetric tensor of rank two, $\mathbf{a} = (a_{ik})$, by multiplying the vector by the alternating tensor and contracting once:

$$a_{ik} = \varepsilon_{ikm} a_m = \begin{pmatrix} 0 & a_3 & -a_2 \\ -a_3 & 0 & a_1 \\ a_2 & -a_1 & 0 \end{pmatrix}.$$

Conversely, if $\mathbf{a} = (a_{ik})$ is an antisymmetric tensor of rank two, then we can form a vector $\boldsymbol{a} = (a_i)$ by taking

$$a_i = \varepsilon_{ikm} a_{km},$$

whose components are numerically twice those of $\mathbf{a}$.

Thus, we can associate an antisymmetric tensor of rank two with a vector and *vice versa.* Vectors obtained in this way are called *pseudovectors* or *axial vectors.* To bring out the distinction, an ordinary vector may be called a *polar vector*. The difference between an axial vector and a polar vector is unimportant for most, but not all purposes. Indeed, many quantities usually thought of as vectors are pseudovectors. For example, the vector product of two vectors $\boldsymbol{a}$ and $\boldsymbol{b}$ is a pseudovector: $(\boldsymbol{a} \times \boldsymbol{b})_i = \varepsilon_{ikm} a_k b_m$. Under inversion, the components of both $\boldsymbol{a}$ and $\boldsymbol{b}$ change sign, hence those of $\boldsymbol{a} \times \boldsymbol{b}$ do not, indicating that it must be a pseudovector.

The scalar product of a pseudovector and a polar vector is sometimes called a *pseudoscalar*.

A.13. *Differentiation.* The *differential operator* with respect to the vector $\boldsymbol{a}$ is defined as $\boldsymbol{\nabla}_a = \boldsymbol{e}_i(\partial/\partial a_i)$. A special case is the *nabla operator*, with $\boldsymbol{a} = \boldsymbol{r}$. If ϕ is any scalar, the vector $\boldsymbol{\nabla}_a \phi$ is called the *gradient* of ϕ with respect to $\boldsymbol{a}$. If ϕ is a function of the magnitude a of $\boldsymbol{a}$ alone, one has the identity

$$\boldsymbol{\nabla}_a \phi = \phi'(a) \frac{\boldsymbol{a}}{a}, \quad \text{where } \phi'(a) = \frac{\mathrm{d}\phi(a)}{\mathrm{d}a}.$$

The differential operator $\boldsymbol{\nabla}_a$ may be one of the vectors in a dyadic. When it appears in the product of two tensors, attention must be paid to the order in which the terms occur. For example,

$$\boldsymbol{\nabla}_r \cdot \boldsymbol{ab} = (\boldsymbol{a} \cdot \boldsymbol{\nabla}_r)\boldsymbol{b} + \boldsymbol{b}(\boldsymbol{\nabla}_r \cdot \boldsymbol{a}),$$

$$\boldsymbol{ab} : \boldsymbol{\nabla}_r \boldsymbol{c} = \boldsymbol{a} \cdot (\boldsymbol{b} \cdot \boldsymbol{\nabla}_r)\boldsymbol{c}.$$

A.14. *Some integral relations.* In this final paragraph we derive some integral relations involving vectors and tensors that are used in the text.

Let $\boldsymbol{C}$ be a vector with components C_x, C_y, C_z. Consider integrals of the type

$$\int \phi(\boldsymbol{C})\,\mathrm{d}^3C,$$

taken over the whole $\boldsymbol{C}$-space.

If ϕ is a function of odd degree in any of the components of $\boldsymbol{C}$, the integral vanishes.

If $\phi(\boldsymbol{C}) = C_i^2 F(C)$, then, by symmetry,

$$\int C_i^2 F(C)\,\mathrm{d}^3C = \tfrac{1}{3}\int C^2 F(C)\,\mathrm{d}^3C,$$

so, for any function $F(C)$,

$$\int F(C)\boldsymbol{CC}\,\mathrm{d}^3C = \tfrac{1}{3}\int F(C)C^2\mathrm{d}^3C\,\boldsymbol{I}. \tag{1}$$

Hence, if $\boldsymbol{a}$ is any vector independent of $\boldsymbol{C}$,

$$\int F(C)\boldsymbol{C}(\boldsymbol{C}\cdot\boldsymbol{a})\mathrm{d}^3C = \tfrac{1}{3}\int F(C)C^2\mathrm{d}^3C\,\boldsymbol{a}. \tag{2}$$

Also, if $\mathbf{w}$ is any second-rank tensor independent of $\boldsymbol{C}$,

$$\int F(C)\boldsymbol{CC}\{(\boldsymbol{CC}-\tfrac{1}{3}C^2\boldsymbol{I}):\mathbf{w}\}\,\mathrm{d}^3C$$

$$= \int F(C)(\boldsymbol{CC}-\tfrac{1}{3}C^2\boldsymbol{I})\{(\boldsymbol{CC}-\tfrac{1}{3}C^2\boldsymbol{I}):\mathbf{w}\}\,\mathrm{d}^3C$$

$$= \int F(C)(\boldsymbol{CC}-\tfrac{1}{3}C^2\boldsymbol{I})(\boldsymbol{CC}:\mathbf{w})\mathrm{d}^3C. \tag{3}$$

Again, let $\phi(\boldsymbol{C}) = C_i^4 F(C)$. Using polar coordinates one finds

$$\int C_i^4 F(C)\mathrm{d}^3C = \iiint (C\cos\theta)^4 F(C)C^2\mathrm{d}C\sin\theta\,\mathrm{d}\theta\,\mathrm{d}\phi$$

$$= \tfrac{1}{5}\int C^4 F(C)\mathrm{d}^3C. \tag{4}$$

Similarly, if $i \neq j$,

$$\int C_i^2C_j^2 F(C)\mathrm{d}^3C = \iiint (C\cos\theta)^2(C\sin\theta\cos\phi)^2 F(C)C^2\mathrm{d}C\sin\theta\,\mathrm{d}\theta\,\mathrm{d}\phi$$

$$= \tfrac{1}{15}\int C^4 F(C)\mathrm{d}^3C. \tag{5}$$

Now, consider the integral in the first member of the equation (3). Since, for any tensor $\mathbf{w}$,

$$(\boldsymbol{CC}-\tfrac{1}{3}C^2\boldsymbol{I}):\mathbf{w} = (\boldsymbol{CC}-\tfrac{1}{3}C^2\boldsymbol{I}):\tfrac{1}{2}(\mathbf{w}+\mathbf{w}^{\mathrm{T}})$$

and, also,

$$(\boldsymbol{CC}-\tfrac{1}{3}C^2\boldsymbol{I}):\mathbf{w} = \boldsymbol{CC}:(\mathbf{w}-\tfrac{1}{3}(\mathbf{w}:\boldsymbol{I})\boldsymbol{I}),$$

one may replace the double product $(\boldsymbol{CC}-\tfrac{1}{3}C^2\boldsymbol{I}):\mathbf{w}$ by the double product

$\boldsymbol{CC} : \boldsymbol{S}$, where $\boldsymbol{S}$ is the symmetric traceless part of $\boldsymbol{w}$, i.e.,

$$(\boldsymbol{CC}-\tfrac{1}{3}C^2\boldsymbol{I}) : \boldsymbol{w} = \boldsymbol{CC} : \boldsymbol{S},$$

where

$$\boldsymbol{S} = \tfrac{1}{2}(\boldsymbol{w}+\boldsymbol{w}^{\mathrm{T}})-\tfrac{1}{3}(\boldsymbol{w} : \boldsymbol{I})\boldsymbol{I}.$$

Hence, the integral in the first member of equation (3) is identical with the integral

$$\int F(C)\boldsymbol{CC}(\boldsymbol{CC} : \boldsymbol{S})\,\mathrm{d}^3C.$$

The diagonal and nondiagonal terms of the latter tensor may be evaluated with the aid of the equations (4) and (5), respectively. The result is

$$\int F(C)\boldsymbol{CC}\{(\boldsymbol{CC}-\tfrac{1}{3}C^2\boldsymbol{I}) : \boldsymbol{w}\}\,\mathrm{d}^3C = \tfrac{2}{15}\int F(C)\,C^4\,\mathrm{d}^3C\,\boldsymbol{S}$$

or, equivalently,

$$\int F(C)(\boldsymbol{CC}-\tfrac{1}{3}C^2\boldsymbol{I})\{(\boldsymbol{CC}-\tfrac{1}{3}C^2\boldsymbol{I}) : \boldsymbol{w}\}\,\mathrm{d}^3C$$
$$= \tfrac{1}{5}\int F(C)(\boldsymbol{CC}-\tfrac{1}{3}C^2\boldsymbol{I}) : (\boldsymbol{CC}-\tfrac{1}{3}C^2\boldsymbol{I})\,\mathrm{d}^3C\,\boldsymbol{S}. \quad (6)$$

Using similar methods one may prove that, for any function $F(C)$,

$$\int F(C)\boldsymbol{CC}\{(\boldsymbol{CC}-\tfrac{1}{3}C^2\boldsymbol{I}) : \boldsymbol{w}\}\{(\boldsymbol{CC}-\tfrac{1}{3}C^2\boldsymbol{I}) : \boldsymbol{w}\}\,\mathrm{d}^3C$$
$$= \tfrac{8}{105}\int F(C)\,C^6\,\mathrm{d}^3C(\boldsymbol{S}\cdot\boldsymbol{S}-\tfrac{1}{3}(\boldsymbol{I} : \boldsymbol{S}\cdot\boldsymbol{S})\boldsymbol{I}), \quad (7)$$

where, again, $\boldsymbol{S}$ is the symmetric traceless part of the tensor $\boldsymbol{w}$.

Appendix B
Reduction of bracket integrals

In Chapter 7 the reduction of two of the simpler bracket integrals for a simple gas was carried out. Here we give the reduction of the bracket integrals for mixtures; the bracket integrals for a simple gas may be obtained as special cases. We use the indices 1 and 2 to distinguish the molecular species involved in the collision process.

B.1. *Kinematic relations*

We begin by listing the kinetic relations for a binary collision that were discussed in Chapter 3. The center-of-mass and relative velocities are defined by

$$\boldsymbol{G} = \mu_1 \boldsymbol{c}_1 + \mu_2 \boldsymbol{c}_2 \text{ and } \boldsymbol{g} = \boldsymbol{c}_2 - \boldsymbol{c}_1,$$

respectively, where

$$\mu_1 = \frac{m_1}{m_1 + m_2} \quad \text{and} \quad \mu_2 = \frac{m_2}{m_1 + m_2}.$$

The inverse relations are

$$\boldsymbol{c}_1 = \boldsymbol{G} - \mu_2 \boldsymbol{g} \text{ and } \boldsymbol{c}_2 = \boldsymbol{G} + \mu_1 \boldsymbol{g}.$$

Similar relations hold for the post-collision velocities; one simply replaces $\boldsymbol{c}_1$, $\boldsymbol{c}_2$, and $\boldsymbol{g}$ by $\boldsymbol{c}_1'$, $\boldsymbol{c}_2'$, and $\boldsymbol{g}'$, respectively.

The peculiar velocities are, of course,

$$\boldsymbol{C}_1 = \boldsymbol{c}_1 - \boldsymbol{v} \text{ and } \boldsymbol{C}_2 = \boldsymbol{c}_2 - \boldsymbol{v}.$$

Their dimensionless counterparts are $\mathscr{C}_1$ and $\mathscr{C}_2$, where

$$\mathscr{C}_1 = \left(\frac{m_1}{2kT}\right)^{\frac{1}{2}} \boldsymbol{C}_1 \quad \text{and} \quad \mathscr{C}_2 = \left(\frac{m_2}{2kT}\right)^{\frac{1}{2}} \boldsymbol{C}_2 .$$

It is also convenient, as in Chapter 7, to introduce a center-of-mass velocity relative to the moving gas stream,

$$\boldsymbol{G}_0 = \boldsymbol{G} - \boldsymbol{v}.$$

Non-dimensional center-of-mass and relative velocities are most conveniently defined by

$$\mathscr{G}_0 = \left(\frac{m_1+m_2}{2kT}\right)^{\frac{1}{2}} \boldsymbol{G}_0 \quad \text{and} \quad \mathscr{g} = \left(\frac{m_{12}}{2kT}\right)^{\frac{1}{2}} \boldsymbol{g},$$

respectively, where m_{12} is the reduced mass defined by

$$m_{12}^{-1} = m_1^{-1} + m_2^{-1}.$$

From these relations it follows that

$$\mathscr{C}_1 = \sqrt{\mu_1}\,\mathscr{G}_0 - \sqrt{\mu_2}\,\mathscr{g} \text{ and } \mathscr{C}_2 = \sqrt{\mu_2}\,\mathscr{G}_0 + \sqrt{\mu_1}\,\mathscr{g}.$$

It is easily verified that

$$\mathscr{C}_1^2 + \mathscr{C}_2^2 = \mathscr{G}_0^2 + \mathscr{g}^2,$$

and

$$\frac{\partial(\mathscr{G}_0, \mathscr{g})}{\partial(\mathscr{C}_1, \mathscr{C}_2)} = 1.$$

Furthermore,

$$\mathscr{g}^2 = \mathscr{g}'^2 \text{ and } \mathscr{g} \cdot \mathscr{g}' = \mathscr{g}^2 \cos \chi.$$

B.2. *Evaluation of* $[S_{\frac{3}{2}}^{(p)}(\mathscr{C}^2)\mathscr{C}, S_{\frac{3}{2}}^{(q)}(\mathscr{C}^2)\mathscr{C}]_{12}''$

According to the definition (4.7-10b) and (4.7-3b), we have

$$[S_{\frac{3}{2}}^{(p)}(\mathscr{C}^2)\mathscr{C}, S_{\frac{3}{2}}^{(q)}(\mathscr{C}^2)\mathscr{C}]_{12}'' = \frac{1}{n_1 n_2} \int \mathrm{d}^3 c_1 \int \mathrm{d}^3 c_2\, f_{\mathrm{M}}(\boldsymbol{c}_1) f_{\mathrm{M}}(\boldsymbol{c}_2) \int g b \,\mathrm{d}b \int \mathrm{d}\varepsilon$$
$$\times (S_{\frac{3}{2}}^{(p)}(\mathscr{C}_1^2)\mathscr{C}_1 - S_{\frac{3}{2}}^{(p)}(\mathscr{C}_1'^2)\mathscr{C}_1') \cdot \mathscr{C}_2 S_{\frac{3}{2}}^{(q)}(\mathscr{C}_2^2).$$

In order to carry out the integration for arbitrary p and q we introduce the generating function for Sonine polynomials,

$$(1-s)^{-\nu-1} \mathrm{e}^{-xs/(1-s)} = \sum_{n=0}^{\infty} s^n S_\nu^{(n)}(x).$$

Thus the bracket integral is equal to the coefficient of $s^p t^q$ in the power series expansion of

$$\frac{(1-s)^{-\frac{5}{2}}(1-t)^{-\frac{5}{2}}}{n_1 n_2}\int \mathrm{d}^3 c_1 \int \mathrm{d}^3 c_2\, f_{\mathrm{M}}(\boldsymbol{c}_1)\, f_{\mathrm{M}}(\boldsymbol{c}_2) \int g b\, \mathrm{d}b \int \mathrm{d}\varepsilon$$
$$\times (\mathscr{C}_1 \exp\{-s\mathscr{C}_1^2/(1-s)\} - \mathscr{C}_1' \exp\{-s\mathscr{C}_1'^2/(1-s)\}) \cdot \mathscr{C}_2 \exp\{-t\mathscr{C}_2^2/(1-t)\}.$$

Next, we substitute the expressions for $f_{\mathrm{M}}(\boldsymbol{c}_1)$ and $f_{\mathrm{M}}(\boldsymbol{c}_2)$, use the factors $(m_1/2kT)^{\frac{3}{2}}$, $(m_2/2kT)^{\frac{3}{2}}$ to non-dimensionalize $\mathrm{d}^3 c_1 \mathrm{d}^3 c_2$ and change the variables of integration to the center-of-mass and relative coordinates. There results

$$(1-s)^{-\frac{5}{2}}(1-t)^{-\frac{5}{2}}\pi^{-3}\int \mathrm{d}^3 \mathscr{g} \int g b\, \mathrm{d}b \int \mathrm{d}\varepsilon \int \mathrm{d}^3 \mathscr{G}_0 \exp\{-(\mathscr{C}_1^2+\mathscr{C}_2^2)\}$$
$$\times [\mathscr{C}_1 \exp\{-s\mathscr{C}_1^2/(1-s)\} - \mathscr{C}_1' \exp\{-s\mathscr{C}_1'^2/(1-s)\}] \cdot \mathscr{C}_2 \exp\{-t\mathscr{C}_2^2/(1-t)\}. \tag{1}$$

The integration over $\mathscr{G}_0$ is the simplest and will be carried out first. To avoid the carrying of unnecessary factors we define

$$H_{12}(\chi) = \int \mathrm{d}^3 \mathscr{G}_0 \mathscr{C}_2 \cdot \mathscr{C}_1' \exp\left[-\left(\mathscr{C}_1^2+\mathscr{C}_2^2+\frac{s}{1-s}\mathscr{C}_1'^2+\frac{t}{1-t}\mathscr{C}_2^2\right)\right],$$

which is a function of the scattering angle $\chi(b, g)$ and the magnitude of the center-of-mass velocity. The first $\mathscr{G}_0$-integral occurring in the expression (1) is obtained from the second $\mathscr{G}_0$-integral simply by replacing $\mathscr{C}_1'$ by $\mathscr{C}_1$. Since $\mathscr{C}_1$ is just the value of $\mathscr{C}_1'$ when the scattering angle χ vanishes, the latter integral can be obtained from the integral $H_{12}(\chi)$ by setting $\chi = 0$. Thus the expression (1) can be written as

$$(1-s)^{-\frac{5}{2}}(1-t)^{-\frac{5}{2}}\pi^{-3}\int \mathrm{d}^3 \mathscr{g} \int g b\, \mathrm{d}b \int \mathrm{d}\varepsilon\, [H_{12}(0)-H_{12}(\chi)]. \tag{2}$$

The computation of $H_{12}(\chi)$ is straight-forward. First, all of the velocities are replaced by their representations in terms of $\mathscr{G}_0$, $\mathscr{g}$ and $\mathscr{g}'$. Thus,

$$\mathscr{C}_1^2+\mathscr{C}_2^2+\frac{s}{1-s}\mathscr{C}_1'^2+\frac{t}{1-t}\mathscr{C}_2^2$$
$$= \alpha\mathscr{G}_0^2+\beta\mathscr{g}^2+2(\mu_1\mu_2)^{\frac{1}{2}}\mathscr{G}_0 \cdot \left[-\frac{t}{1-t}\mathscr{g}+\frac{s}{1-s}\mathscr{g}'\right],$$

where we have abbreviated

$$\alpha \equiv \alpha(s, t) = 1 + \frac{\mu_1 s}{1-s} + \frac{\mu_2 t}{1-t},$$

and

$$\beta \equiv \beta(s, t) = 1 + \frac{\mu_2 s}{1-s} + \frac{\mu_1 t}{1-t}.$$

Furthermore,

$$\mathscr{C}'_1 \cdot \mathscr{C}_2 = (\mu_1 \mu_2)^{\frac{1}{2}} \mathscr{G}_0^2 + \mu_2 \boldsymbol{g}' \cdot \mathscr{G}_0 - \mu_1 \boldsymbol{g} \cdot \mathscr{G}_0 - (\mu_1 \mu_2)^{\frac{1}{2}} g^2.$$

The integration is now carried out by completing the square with respect to $\mathscr{G}_0$ in the exponent. In terms of the new vector variable

$$\boldsymbol{\Gamma} \equiv \boldsymbol{\Gamma}(s, t) = \mathscr{G}_0 + (\mu_1 \mu_2)^{\frac{1}{2}} \alpha^{-1} \left(-\frac{t}{1-t} \boldsymbol{g} + \frac{s}{1-s} \boldsymbol{g}' \right)$$

we have

$$\mathscr{C}_1^2 + \mathscr{C}_2^2 + \frac{s}{1-s} \mathscr{C}_1'^2 + \frac{t}{1-t} \mathscr{C}_2^2 = \alpha \Gamma^2 + \gamma g^2,$$

where we have abbreviated,

$$\gamma \equiv \gamma(s, t; \chi) = \beta - \mu_1 \mu_2 \alpha^{-1} \left[\left(\frac{t}{1-t} \right)^2 + \left(\frac{s}{1-s} \right)^2 - 2 \left(\frac{t}{1-t} \right) \left(\frac{s}{1-s} \right) \cos \chi \right]$$

$$= \frac{1 - 2\mu_1 \mu_2 st(1 - \cos \chi)}{1 - (\mu_2 s + \mu_1 t)}.$$

In terms of $\boldsymbol{\Gamma}$, the product $\mathscr{C}'_1 \cdot \mathscr{C}_2$ becomes, after some algebraic manipulation,

$$\mathscr{C}'_1 \cdot \mathscr{C}_2 = (\mu_1 \mu_2)^{\frac{1}{2}} \{ \Gamma^2 + (g^2/\alpha)(1 - \gamma - \cos \chi) + \boldsymbol{\Gamma} \cdot (\ldots) \},$$

where the factor multiplying $\boldsymbol{\Gamma}$ in the last term may be neglected, as it will drop out when the integration is performed. Thus we have

$$H_{12}(\chi) = (\mu_1 \mu_2)^{\frac{1}{2}} \exp(-\gamma g^2) \int e^{-\alpha \Gamma^2} \{ \Gamma^2 + (g^2/\alpha)(1 - \gamma - \cos \chi) \} d^3\Gamma$$

$$= \pi^{\frac{3}{2}} (\mu_1 \mu_2)^{\frac{1}{2}} \exp(-\gamma g^2) \alpha^{-\frac{5}{2}} [\tfrac{3}{2} + g^2(1 - \gamma - \cos \chi)].$$

When this result is substituted into the expression (2), it is possible to carry out the angular integrals over $\boldsymbol{g}$ and ε trivially. The result must then be expanded in a double power series in s and t to obtain the required expressions for the bracket integrals. However, since the expression (2) contains the difference $H_{12}(\chi)-H_{12}(0)$ it is easier to proceed by first expanding $H_{12}(\chi)$. Thus, we consider

$$\pi^{-\frac{3}{2}}(\mu_1\mu_2)^{-\frac{1}{2}}(1-s)^{-\frac{5}{2}}(1-t)^{-\frac{5}{2}}H_{12}(\chi)$$
$$= \exp(-g^2)\exp\{(1-\gamma)g^2\}(1-(\mu_1 t+\mu_2 s))^{-\frac{5}{2}}[\tfrac{3}{2}+g^2(1-\gamma-\cos\chi)].$$

One notes by examining this equation and the expression defining γ, that s and t and the mass ratios μ_1 and μ_2 occur only in the combinations $\mu_2 s+\mu_1 t$ and $\mu_2 s\mu_1 t$. Thus a power series in s and t automatically produces a power series in μ_1 and μ_2 as well. Now,

$$\exp\{(1-\gamma)g^2\} = \sum_{n=0}^{\infty}\frac{(1-\gamma)^n}{n!}g^{2n} = \sum_{n=0}^{\infty}\frac{[2\mu_1\mu_2(1-\cos\chi)-(\mu_2 s+\mu_1 t)]^n}{n![1-(\mu_2 s+\mu_1 t)]^n}g^{2n},$$

from which it can be seen that

$$\pi^{-\frac{3}{2}}(\mu_1\mu_2)^{-\frac{1}{2}}(1-s)^{-\frac{5}{2}}(1-t)^{-\frac{5}{2}}H_{12}(\chi)$$
$$= \exp(-g^2)\sum_{pqlr=0}^{\infty}A_{pqlr}(\mu_2 s)^p(\mu_1 t)^q g^{2r}\cos^l\chi \tag{3}$$

where the A_{pqlr} are pure numbers not depending on any of the molecular properties. The A_{pqlr} may be obtained by carrying out the power series development but, as only the bracket integrals for small values of p and q (which involve only small values of r and l) are required, it is as simple or simpler to obtain the values by differentiating $H_{12}(\chi)$ p times with respect to s and q times with respect to t and setting $s = t = 0$. Substituting the expansion (3) into the expansion (2), and carrying out the trivial integrals over the directions of $\boldsymbol{g}$ and over ε, one obtains the desired result,

$$[S_{\frac{3}{2}}^{(p)}(\mathscr{C}^2)\mathscr{C},\, S_{\frac{3}{2}}^{(q)}(\mathscr{C}^2)\mathscr{C}]_{12}'' = 8\mu_2^{p+\frac{1}{2}}\mu_1^{q+\frac{1}{2}}\sum_{l,r=0}^{\infty}A_{pqlr}\,\Omega_{12}^{(l,r)}$$

where, in accordance with Eqs. (7.1-21) and (7.1-22),

$$\Omega_{12}^{(l,r)} = \left(\frac{kT}{2\pi m_{12}}\right)^{\frac{1}{2}}\int_0^\infty \exp(-g^2)\,g^{2r+3}\left\{2\pi\int_0^\infty(1-\cos^l\chi(b,g))b\,\mathrm{d}b\right\}\mathrm{d}g.$$

Explicit expressions for some values of p and q are given in Chapter 7 (Table 7.4) of the text.

B.3. *Evaluation of* $[S_{\frac{5}{2}}^{(p)}(\mathscr{C}^2)(\mathscr{C}\mathscr{C}-\frac{1}{3}\mathsf{I}\mathscr{C}^2),\ S_{\frac{5}{2}}^{(q)}(\mathscr{C}^2)(\mathscr{C}\mathscr{C}-\frac{1}{3}\mathsf{I}\mathscr{C}^2)]_{12}''$

Many of the tricks involved in the reduction of this partial bracket integral are identical to those used in the previous section. Thus, a less detailed treatment will be given. By using the definition of the bracket integral and the generating function for the Sonine polynomials it is easily shown that the required integral is the coefficient of $s^p t^q$ in the expansion of

$$(1-s)^{-\frac{7}{2}}(1-t)^{-\frac{7}{2}}\pi^{-3}\int \mathrm{d}^3 g \int g b\,\mathrm{d}b \int \mathrm{d}\varepsilon\,[L_{12}(\chi)-L_{12}(0)],$$

where

$$L_{12}(\chi) = \int \mathrm{d}^3\mathscr{G}_0[\mathscr{C}_1'\mathscr{C}_1'-\tfrac{1}{3}\mathsf{I}\mathscr{C}_1'^2] : [\mathscr{C}_2\mathscr{C}_2-\tfrac{1}{3}\mathsf{I}\mathscr{C}_2^2] \times \exp\left[-\left(\mathscr{C}_1^2+\mathscr{C}_2^2+\frac{s}{1-s}\mathscr{C}_1'^2+\frac{t}{1-t}\mathscr{C}_2^2\right)\right].$$

The function L_{12} plays a role similar to that played by H_{12} in the previous section.

Again, a transformation to center-of-mass and relative coordinates is made. The exponent in $L_{12}(\chi)$ is the same as that in $H_{12}(\chi)$; thus, it can be written as $\exp(-\alpha\Gamma^2-\gamma g^2)$. The scalar product of the two tensors can be transformed via a great deal of straight-forward, but difficult algebraic manipulation,

$$\begin{aligned}[\mathscr{C}_1'\mathscr{C}_1'-\tfrac{1}{3}\mathsf{I}\mathscr{C}_1'^2] &: [\mathscr{C}_2\mathscr{C}_2-\tfrac{1}{3}\mathsf{I}\mathscr{C}_2^2] \\ &= (\mathscr{C}_1'\cdot\mathscr{C}_2)^2-\tfrac{1}{3}\mathscr{C}_1'^2\mathscr{C}_2^2 \\ &= \tfrac{2}{3}\Gamma^4+\tfrac{20}{9}\Gamma^2(g^2/\alpha)[1-\gamma-\cos\chi] \\ &\quad +\tfrac{1}{3}(g^4/\alpha^2)[2(1-\gamma-\cos\chi)^2-(1-\cos^2\chi)].\end{aligned}$$

After carrying out the integration over $\boldsymbol{\Gamma}$ one finds

$$\begin{aligned}\pi^{-\frac{3}{2}}(\mu_1\mu_2)^{-\frac{5}{2}}(1-s)^{-\frac{7}{2}}(1-t)^{-\frac{7}{2}}L_{12}(\chi) &= \tfrac{2}{3}\exp(-g^2)\exp\{(1-\gamma)g^2\}(1-(\mu_1 t+\mu_2 s))^{-\frac{7}{2}}\{\tfrac{15}{4}+5(1-\gamma-\cos\chi)g^2 \\ &\quad +[(1-\gamma-\cos\chi)-\tfrac{1}{2}(1-\cos^2\chi)]g^4\},\end{aligned}$$

which again contains μ_1, μ_2, s and t only in the combinations $\mu_2 s+\mu_1 t$ and $\mu_2 s\mu_1 t$. Thus, $L_{12}(\chi)$ has the same type of power series expansion as $H_{12}(\chi)$

and we find

$$\pi^{-\frac{3}{2}}(\mu_1\mu_2)^{-\frac{1}{2}}(1-s)^{-\frac{7}{2}}(1-t)^{-\frac{7}{2}}L_{12}(\chi)$$

$$= \exp(-g^2)\sum_{pqlr=0}^{\infty} B_{pqlr}(\mu_2 s)^p(\mu_1 t)^q g^{2r}\cos^l\chi,$$

where, again, the B_{pqlr} are universal numbers.

Carrying out the directional integrations over $\boldsymbol{g}$ and the integration over ε, one obtains

$$[S_{\frac{5}{2}}^{(p)}(\mathscr{C}^2)(\mathscr{C}\mathscr{C}-\tfrac{1}{3}I\mathscr{C}^2),\ S_{\frac{5}{2}}^{(q)}(\mathscr{C}^2)(\mathscr{C}\mathscr{C}-\tfrac{1}{3}I\mathscr{C}^2)]_{12}''$$

$$= 8\mu_2^{p+\frac{1}{2}}\mu_1^{q+\frac{1}{2}}\sum_{l,r=0}^{\infty} B_{pqlr}\,\Omega_{12}^{(l,r)}$$

with $\Omega_{12}^{(l,r)}$ given by Eqs. (7.1-21) and (7.1-22). Explicit expressions for some values of p and q are given in Chapter 7 (Table 7.6) of the text.

B.4. *Evaluation of* $[S_{\frac{3}{2}}^{(p)}(\mathscr{C}^2)\mathscr{C},\ S_{\frac{3}{2}}^{(q)}(\mathscr{C}^2)\mathscr{C}]_{12}'$ *and* $[S_{\frac{5}{2}}^{(p)}(\mathscr{C}^2)(\mathscr{C}\mathscr{C}-\frac{1}{3}I\mathscr{C}^2),\ S_{\frac{5}{2}}^{(q)}(\mathscr{C}^2)(\mathscr{C}\mathscr{C}-\frac{1}{3}I\mathscr{C}^2)]_{12}'$

These partial bracket integrals can be reduced by following the methods of the previous two sections. The only difference between these integrals and those considered earlier is that $\mathscr{C}_2$ is replaced by $\mathscr{C}_1$ in the second argument of the bracket. In turn, this entails only a change in the sign of $\boldsymbol{g}$ and an interchange of m_1 and m_2 in certain terms. Otherwise, the reductions are identical to those already given. For this reason only the results are given.

In place of α, β and γ, the coefficients defined in Section B.2, the following coefficients occur:

$$\alpha' \equiv \alpha'(s,t) = 1+\mu_1\left(\frac{s}{1-s}+\frac{t}{1-t}\right),$$

$$\beta' \equiv \beta'(s,t) = 1+\mu_2\left(\frac{s}{1-s}+\frac{t}{1-t}\right),$$

$$\gamma' \equiv \gamma'(s,t;\chi)$$

$$= \beta'-\mu_1\mu_2\,\alpha^{-1}\left[\left(\frac{s}{1-s}\right)^2+\left(\frac{t}{1-t}\right)^2+2\left(\frac{s}{1-s}\right)\left(\frac{t}{1-t}\right)\cos\chi\right].$$

The integral $[S_{\frac{3}{2}}^{(p)}(\mathscr{C}^2)\mathscr{C},\ S_{\frac{3}{2}}^{(q)}(\mathscr{C}^2)\mathscr{C}]_{12}'$ is the coefficient of $s^p t^q$ in the expansion of

$$(1-s)^{-\frac{5}{2}}(1-t)^{-\frac{5}{2}}\pi^{-3}\int d^3g\int gb\,db\int d\varepsilon[H_1(0)-H_1(\chi)],$$

where

$$H_1(\chi) = \int \mathrm{d}^3\mathscr{G}_0 \mathscr{C}_1' \cdot \mathscr{C}_1 \exp\left[-\left(\mathscr{G}_0^2 + g^2 + \frac{s}{1-s}\mathscr{C}_1'^2 + \frac{t}{1-t}\mathscr{C}_1^2\right)\right]$$

$$= \pi^{\frac{3}{2}}\alpha'^{-\frac{5}{2}}\exp(-\gamma' g^2)[\tfrac{2}{3}\mu_1 + \{\mu_1(1-\gamma') + \mu_2 \cos\chi\}g^2].$$

Likewise, $[S_{\frac{5}{2}}^{(p)}(\mathscr{C}^2)(\mathscr{C}\mathscr{C}-\frac{1}{3}I\mathscr{C}^2), S_{\frac{5}{2}}^{(q)}(\mathscr{C}^2)(\mathscr{C}\mathscr{C}-\frac{1}{3}I\mathscr{C}^2)]_{12}'$ is the coefficient of $s^p t^q$ in the expansion of

$$(1-s)^{-\frac{7}{2}}(1-t)^{-\frac{7}{2}}\pi^{-3}\int \mathrm{d}^3 g \int g b \,\mathrm{d}b \int \mathrm{d}\varepsilon\, [L_1(0) - L_1(\chi)],$$

where

$$L_1(\chi) = \int \mathrm{d}^3\mathscr{G}_0 [(\mathscr{C}_1 \cdot \mathscr{C}_1')^2 - \tfrac{1}{3}\mathscr{C}_1^2\mathscr{C}_1'^2]$$

$$\times \exp\left[-\left(\mathscr{G}_0^2 + g^2 + \frac{s}{1-s}\mathscr{C}_1^2 + \frac{t}{1-t}\mathscr{C}_1'^2\right)\right]$$

$$= \tfrac{2}{3}\pi^{\frac{3}{2}}\alpha'^{-\frac{7}{2}}\exp(-\gamma' g^2)\{\tfrac{15}{4}\mu_1^2 + 5\mu_1[\mu_1(1-\gamma') + \mu_2\cos\chi]g^2$$

$$+ [\mu_1(1-\gamma') + \mu_2 \cos\chi]^2 g^4 - \tfrac{1}{2}\mu_2^2(1-\cos^2\chi)g^4\}.$$

In view of the similarity of these expressions to those obtained earlier, $H_1(\chi)$ and $L_1(\chi)$ possess the same type of power series expansion as $H_{12}(\chi)$ and $L_{12}(\chi)$, respectively. Thus, we have

$$[S_{\frac{3}{2}}^{(p)}(\mathscr{C}^2)\mathscr{C}, S_{\frac{3}{2}}^{(q)}(\mathscr{C}^2)\mathscr{C}]_{12}' = 8 \sum_{l,r=0}^{\infty} A'_{pqlr}\Omega_{12}^{(l,r)},$$

and

$$[S_{\frac{5}{2}}^{(p)}(\mathscr{C}^2)(\mathscr{C}\mathscr{C}-\tfrac{1}{3}I\mathscr{C}^2), S_{\frac{5}{2}}^{(q)}(\mathscr{C}^2)(\mathscr{C}\mathscr{C}-\tfrac{1}{3}I\mathscr{C}^2)]_{12}' = 8 \sum_{l,r=0}^{\infty} B'_{pqlr}\Omega_{12}^{(l,r)}.$$

However, in this case the A'_{pqlr} and B'_{pqlr} may depend on the mass ratios μ_1 and μ_2. This is a consequence of the fact that, in $H_1(\chi)$ and $L_1(\chi)$, the mass ratios do not occur in unified combinations with s and t.

Explicit expressions for some values of p and q are given in Chapter 7 (Tables 7.3 and 7.5) of the text.

B.5. *Reduction of bracket integrals for a simple gas*

It follows from relations (4.7-1), (4.7-2) and (4.7-3) that

$$[S_{\frac{3}{2}}^{(p)}(\mathscr{C}^2)\mathscr{C}, S_{\frac{3}{2}}^{(q)}(\mathscr{C}^2)\mathscr{C}] = [S_{\frac{3}{2}}^{(p)}(\mathscr{C}^2)\mathscr{C}, S_{\frac{3}{2}}^{(q)}(\mathscr{C}^2)\mathscr{C}]_{1=2}'$$

$$+ [S_{\frac{3}{2}}^{(p)}(\mathscr{C}^2)\mathscr{C}, S_{\frac{3}{2}}^{(q)}(\mathscr{C}^2)\mathscr{C}]_{1=2}''$$

and

$$[S_{\frac{5}{2}}^{(p)}(\mathscr{C}^2)(\mathscr{C}\mathscr{C}-\tfrac{1}{3}\mathsf{I}\mathscr{C}^2), S_{\frac{5}{2}}^{(q)}(\mathscr{C}^2)(\mathscr{C}\mathscr{C}-\tfrac{1}{3}\mathsf{I}\mathscr{C}^2)]$$
$$= [S_{\frac{5}{2}}^{(p)}(\mathscr{C}^2)(\mathscr{C}\mathscr{C}-\tfrac{1}{3}\mathsf{I}\mathscr{C}^2), S_{\frac{5}{2}}^{(q)}(\mathscr{C}^2)(\mathscr{C}\mathscr{C}-\tfrac{1}{3}\mathsf{I}\mathscr{C}^2)]'_{1=2}$$
$$+[S_{\frac{5}{2}}^{(p)}(\mathscr{C}^2)(\mathscr{C}\mathscr{C}-\tfrac{1}{3}\mathsf{I}\mathscr{C}^2), S_{\frac{5}{2}}^{(q)}(\mathscr{C}^2)(\mathscr{C}\mathscr{C}-\tfrac{1}{3}\mathsf{I}\mathscr{C}^2)]''_{1=2},$$

where it is understood that in the right-hand sides of these expressions m_2 is to be set equal m_1 and the interaction law between molecules of type 1 and those of type 2 is to be taken as that between a pair of 1-molecules. With these changes $\mu_1 = \mu_2 = \frac{1}{2}$ and $H_{12}(\chi)$ and $L_{12}(\chi)$ become identical to $H_1(\chi)$ and $L_1(\chi)$ except that $\cos\chi$ is everywhere replaced by $-\cos\chi$.

Thus, to express the bracket integrals on the left-hand sides in terms of Ω-integrals, one can simply take the Ω-integral representations of the first terms on the right-hand sides in which all terms with even l are doubled and all terms with odd l are suppressed. It is also clear that in this case $\Omega_{12}^{(l,r)}$ simply reduces to $\Omega^{(l,r)}$.

Appendix C
Tables of transport integrals

The transport integrals $\Omega^{(l,r)}$ are defined by Eqs. (7.1-17) and (7.1-18),

$$\Omega^{(l,r)} = \left(\frac{kT}{\pi m}\right)^{\frac{1}{2}} \int_0^\infty \exp(-g^2) g^{2r+3} \left\{2\pi \int_0^\infty [1-\cos^l \chi(b,g)] b \, \mathrm{d}b\right\} \mathrm{d}g$$

and the reduced transport integrals $\Omega^{(l,r)\star}$ by Eqs. (7.1-29) and (7.1-30),

$$\Omega^{(l,r)\star} = \frac{2}{(r+1)!\pi\sigma^2}\left[1-\frac{1+(-1)^l}{2(l+1)}\right]^{-1} \times \int_0^\infty \exp(-g^2) g^{2r+3} \left\{2\pi \int_0^\infty [1-\cos^l \chi(b,g)] b \, \mathrm{d}b\right\} \mathrm{d}g.$$

It was shown in Chapter 9 that, for the two-parameter potentials, the latter are functions of the single variable $T^* = kT/\epsilon$. In this appendix are tabulated the reduced transport integrals and/or the related functions

$$\mathrm{A}^\star = \Omega^{(2,2)\star}/\Omega^{(1,1)\star},$$

$$\mathrm{B}^\star = [5\Omega^{(1,2)\star} - 4\Omega^{(1,3)\star}]/\Omega^{(1,1)\star},$$

$$\mathrm{C}^\star = \Omega^{(1,2)\star}/\Omega^{(1,1)\star},$$

$$\mathrm{E}^\star = \Omega^{(2,3)\star}/\Omega^{(2,2)\star},$$

$$\mathrm{F}^\star = \Omega^{(3,3)\star}/\Omega^{(1,1)\star},$$

for some of the more important model potentials.

TABLE C.1

Transport integrals for the modified Buckingham (6-exp) potential, Eq. (8.4-6)*

$$Z^{(l,r)}(T^*) = [T^*(1-6/\alpha)]^{\frac{1}{3}}\Omega^{(l,r)*}(T^*),\ T^* = kT/\epsilon$$

$\alpha = 12$ T^*	$Z^{(1,1)}$	$Z^{(1,2)}$	$Z^{(1,3)}$	$Z^{(1,4)}$	$Z^{(1,5)}$	$Z^{(3,3)}$
0	1.1870	1.0551	0.9672	0.9027	0.8572	1.1178
0.1	1.1911	1.0523	0.9584	0.8883	0.8320	1.1079
0.2	1.1662	1.0142	0.9025	0.8113	0.7338	1.0704
0.3	1.1243	0.9515	0.8232	0.7239	0.6475	1.0161
0.4	1.0750	0.8900	0.7597	0.6667	0.6005	0.9641
0.5	1.0282	0.8402	0.7160	0.6328	0.5764	0.9194
0.6	0.9873	0.8025	0.6870	0.6131	0.5643	0.8828
0.7	0.9530	0.7745	0.6679	0.6019	0.5589	0.8536
0.8	0.9248	0.7538	0.6555	0.5959	0.5572	0.8305
0.9	0.9016	0.7387	0.6475	0.5930	0.5576	0.8125
1.0	0.8825	0.7272	0.6425	0.5923	0.5593	0.7985
1.2	0.8541	0.7130	0.6385	0.5942	0.5646	0.7793
1.4	0.8350	0.7058	0.6388	0.5985	0.5711	0.7684
1.6	0.8221	0.7028	0.6414	0.6039	0.5779	0.7625
1.8	0.8135	0.7025	0.6451	0.6096	0.5845	0.7601
2.0	0.8080	0.7037	0.6496	0.6155	0.5910	0.7600
2.5	0.8023	0.7104	0.6616	0.6296	0.6059	0.7654
3.0	0.8031	0.7195	0.6735	0.6425	0.6180	0.7746
3.5	0.8070	0.7288	0.6847	0.6541	0.6305	0.7853
4	0.8126	0.7382	0.6950	0.6644	0.6404	0.7963
5	0.8253	0.7556	0.7129	0.6819	0.6572	0.8174
6	0.8384	0.7709	0.7280	0.6962	0.6705	0.8364
7	0.8509	0.7842	0.7409	0.7081	0.6817	0.8533
8	0.8623	0.7961	0.7520	0.7185	0.6912	0.8680
9	0.8730	0.8067	0.7618	0.7272	0.6993	0.8813
10	0.8829	0.8160	0.7702	0.7351	0.7066	0.8929
12	0.9002	0.8322	0.7848	0.7485	0.7189	0.9133
14	0.9150	0.8457	0.7969	0.7594	0.7292	0.9301
16	0.9280	0.8572	0.8074	0.7690	0.7380	0.9444
18	0.9394	0.8674	0.8164	0.7773	0.7459	0.9568
20	0.9497	0.8762	0.8245	0.7846	0.7530	0.9680
25	0.9710	0.8951	0.8415	0.8007	0.7682	0.9910
30	0.9885	0.9103	0.8555	0.8138	0.7808	1.0094
35	1.0032	0.9232	0.8674	0.8253	0.7921	1.0250
40	1.0157	0.9346	0.8781	0.8355	0.8021	1.0385
45	1.0270	0.9446	0.8876	0.8450	0.8114	1.0508
50	1.0371	0.9538	0.8965	0.8535	0.8199	1.0614
60	1.0549	0.9701	0.9123	0.8691	0.8355	1.0807
70	1.0700	0.9846	0.9264	0.8827	0.8495	1.0978
80	1.0834	0.9973	0.9388	0.8957	0.8622	1.1129
90	1.0959	1.0091	0.9504	0.9074	0.8743	1.1265
100	1.1071	1.0201	0.9615	0.9184	0.8853	1.1391
200	1.1906	1.1038	1.0462	1.0049	0.9729	1.2356

* This table has been adopted from Mason [1954] and is based entirely on classical mechanics.

TABLE C.1 (continued)

$$Z^{(l,r)}(T^*) = [T^*(1-6/\alpha)]^{\frac{1}{3}}\Omega^{(l,r)*}(T^*),\ T^* = kT/\epsilon$$

$\alpha = 12$

T^*	$Z^{(2,2)}$	$Z^{(2,3)}$	$Z^{(2,4)}$	$Z^{(2,5)}$	$Z^{(2,6)}$	$Z^{(4,4)}$
0	1.1947	1.0951	1.0221	0.9706	0.9193	1.0927
0.1	1.1983	1.1002	1.0291	0.9750	0.9323	1.1062
0.2	1.2098	1.1094	1.0401	0.9770	0.9188	1.1119
0.3	1.2065	1.0936	0.9978	0.9129	0.8381	1.0843
0.4	1.1799	1.0471	0.9367	0.8453	0.7711	1.0360
0.5	1.1421	0.9974	0.8835	0.7953	0.7280	0.9870
0.6	1.1028	0.9538	0.8428	0.7614	0.7017	0.9450
0.7	1.0667	0.9185	0.8133	0.7391	0.6862	0.9111
0.8	1.0352	0.8907	0.7923	0.7249	0.6775	0.8847
0.9	1.0084	0.8692	0.7775	0.7160	0.6731	0.8642
1.0	0.9859	0.8528	0.7673	0.7108	0.6716	0.8488
1.2	0.9520	0.8309	0.7561	0.7075	0.6736	0.8282
1.4	0.9290	0.8189	0.7525	0.7093	0.6790	0.8170
1.6	0.9137	0.8130	0.7531	0.7139	0.6860	0.8120
1.8	0.9039	0.8111	0.7560	0.7198	0.6936	0.8107
2.0	0.8978	0.8116	0.7604	0.7263	0.7014	0.8119
2.5	0.8930	0.8189	0.7742	0.7435	0.7202	0.8207
3.0	0.8963	0.8300	0.7891	0.7600	0.7374	0.8331
3.5	0.9033	0.8423	0.8035	0.7752	0.7529	0.8466
4	0.9122	0.8546	0.8171	0.7892	0.7668	0.8599
5	0.9313	0.8778	0.8415	0.8135	0.7906	0.8849
6	0.9503	0.8987	0.8623	0.8339	0.8105	0.9072
7	0.9681	0.9172	0.8806	0.8516	0.8275	0.9269
8	0.9845	0.9339	0.8967	0.8670	0.8424	0.9445
9	0.9995	0.9488	0.9110	0.8808	0.8559	0.9602
10	1.0133	0.9624	0.9241	0.8935	0.8678	0.9745
12	1.0381	0.9862	0.9469	0.9153	0.8891	0.9998
14	1.0598	1.0068	0.9664	0.9341	0.9073	1.0213
16	1.0788	1.0248	0.9838	0.9508	0.9236	1.0404
18	1.0958	1.0409	0.9993	0.9658	0.9383	1.0575
20	1.1113	1.0557	1.0134	0.9796	0.9516	1.0731
25	1.1448	1.0875	1.0441	1.0095	0.9812	1.1070
30	1.1729	1.1145	1.0703	1.0352	1.0067	1.1357
35	1.1976	1.1382	1.0935	1.0582	1.0294	1.1608
40	1.2193	1.1593	1.1143	1.0787	1.0499	1.1832
45	1.2393	1.1786	1.1332	1.0976	1.0688	1.2041
50	1.2573	1.1962	1.1509	1.1152	1.0863	1.2228
60	1.2901	1.2283	1.1826	1.1469	1.1180	1.2572
70	1.3189	1.2571	1.2113	1.7156	1.1468	1.2878
80	1.3451	1.2828	1.2370	1.2014	1.1730	1.3157
90	1.3690	1.3068	1.2609	1.2257	1.1973	1.3413
100	1.3915	1.3292	1.2835	1.2481	1.2198	1.3649
200	1.5600	1.4988	1.4547	1.4208	1.3939	1.5470

TABLE C.1 (continued)

$$Z^{(l,r)}(T^*) = [T^*(1-6/\alpha)]^{\frac{1}{3}}\Omega^{(l,r)*}(T^*),\ T^* = kT/\epsilon$$

$\alpha = 13$ T^*	$Z^{(1,1)}$	$Z^{(1,2)}$	$Z^{(1,3)}$	$Z^{(1,4)}$	$Z^{(1,5)}$	$Z^{(3,3)}$
0	1.1870	1.0551	0.9672	0.9027	0.8572	1.1178
0.1	1.1813	1.0447	0.9528	0.8846	0.8305	1.1020
0.2	1.1618	1.0150	0.9084	0.8217	0.7479	1.0718
0.3	1.1275	0.9619	0.8389	0.7429	0.6681	1.0250
0.4	1.0851	0.9071	0.7807	0.6893	0.6236	0.9787
0.5	1.0436	0.8617	0.7400	0.6575	0.6011	0.9383
0.6	1.0068	0.8268	0.7129	0.6393	0.5904	0.9048
0.7	0.9756	0.8009	0.6953	0.6294	0.5862	0.8780
0.8	0.9498	0.7818	0.6841	0.6244	0.5855	0.8569
0.9	0.9285	0.7678	0.6772	0.6226	0.5871	0.8405
1.0	0.9112	0.7577	0.6732	0.6229	0.5897	0.8279
1.2	0.8854	0.7454	0.6710	0.6266	0.5970	0.8111
1.4	0.8684	0.7400	0.6729	0.6326	0.6052	0.8022
1.6	0.8573	0.7386	0.6770	0.6395	0.6136	0.7981
1.8	0.8503	0.7396	0.6823	0.6468	0.6219	0.7974
2.0	0.8461	0.7421	0.6881	0.6542	0.6299	0.7987
2.5	0.8436	0.7519	0.7032	0.6715	0.6483	0.8073
3.0	0.8472	0.7637	0.7181	0.6875	0.6643	0.8193
3.5	0.8535	0.7757	0.7320	0.7017	0.6785	0.8323
4	0.8614	0.7874	0.7446	0.7145	0.6910	0.8454
5	0.8782	0.8091	0.7671	0.7365	0.7121	0.8700
6	0.8950	0.8282	0.7862	0.7547	0.7296	0.8922
7	0.9107	0.8451	0.8024	0.7702	0.7441	0.9118
8	0.9253	0.8601	0.8166	0.7836	0.7568	0.9294
9	0.9389	0.8734	0.8292	0.7954	0.7678	0.9453
10	0.9514	0.8855	0.8404	0.8059	0.7777	0.9596
12	0.9736	0.9065	0.8599	0.8240	0.7948	0.9847
14	0.9927	0.9245	0.8765	0.8394	0.8094	1.0063
16	1.0096	0.9399	0.8907	0.8528	0.8223	1.0250
18	1.0247	0.9537	0.9034	0.8648	0.8337	1.0417
20	1.0384	0.9659	0.9147	0.8754	0.8440	1.0569
25	1.0674	0.9922	0.9392	0.8987	0.8664	1.0895
30	1.0913	1.0139	0.9596	0.9184	0.8855	1.1166
35	1.1118	1.0328	0.9775	0.9357	0.9027	1.1403
40	1.1296	1.0492	0.9933	0.9510	0.9179	1.1616
45	1.1456	1.0643	1.0076	0.9651	0.9318	1.1806
50	1.1602	1.0778	1.0208	0.9783	0.9447	1.1983
60	1.1861	1.1023	1.0447	1.0017	0.9682	1.2297
70	1.2087	1.1239	1.0659	1.0230	0.9894	1.2577
80	1.2287	1.1431	1.0849	1.0422	1.0089	1.2830
90	1.2469	1.1608	1.1025	1.0598	1.0267	1.3063
100	1.2639	1.1774	1.1193	1.0762	1.0434	1.3277
200	1.3892	1.3027	1.2456	1.2042	1.1723	1.4915

TABLE C.1 (continued)

$$Z^{(l,r)}(T^*) = [T^*(1-6/\alpha)]^{\frac{1}{3}}\Omega^{(l,r)*}(T^*),\ T^* = kT/\epsilon$$

$\alpha = 13$

T^*	$Z^{(2,2)}$	$Z^{(2,3)}$	$Z^{(2,4)}$	$Z^{(2,5)}$	$Z^{(2,6)}$	$Z^{(4,4)}$
0	1.1947	1.0951	1.0221	0.9706	0.9193	1.0927
0.1	1.1985	1.0994	1.0269	0.9709	0.9265	1.0986
0.2	1.2052	1.1090	1.0376	0.9783	0.9247	1.1116
0.3	1.2056	1.0977	1.0077	0.9278	0.8562	1.0944
0.4	1.1862	1.0602	0.9549	0.8664	0.7934	1.0526
0.5	1.1549	1.0164	0.9059	0.8190	0.7519	1.0076
0.6	1.1207	0.9766	0.8676	0.7866	0.7267	0.9681
0.7	1.0882	0.9436	0.8395	0.7653	0.7120	0.9360
0.8	1.0594	0.9175	0.8196	0.7519	0.7043	0.9110
0.9	1.0346	0.8972	0.8057	0.7439	0.7009	0.8917
1.0	1.0138	0.8820	0.7965	0.7398	0.7005	0.8775
1.2	0.9824	0.8618	0.7870	0.7382	0.7044	0.8590
1.4	0.9613	0.8514	0.7850	0.7418	0.7118	0.8499
1.6	0.9475	0.8471	0.7871	0.7481	0.7206	0.8466
1.8	0.9391	0.8465	0.7916	0.7557	0.7299	0.8469
2.0	0.9344	0.8484	0.7976	0.7638	0.7392	0.8496
2.5	0.9327	0.8590	0.8148	0.7844	0.7615	0.8617
3.0	0.9388	0.8731	0.8326	0.8040	0.7817	0.8771
3.5	0.9485	0.8880	0.8497	0.8219	0.7998	0.8931
4	0.9597	0.9027	0.8658	0.8381	0.8161	0.9087
5	0.9831	0.9303	0.8944	0.8668	0.8443	0.9378
6	1.0058	0.9548	0.9191	0.8911	0.8679	0.9637
7	1.0270	0.9769	0.9408	0.9121	0.8885	0.9867
8	1.0464	0.9964	0.9600	0.9309	0.9066	1.0072
9	1.0643	1.0143	0.9773	0.9477	0.9230	1.0257
10	1.0809	1.0306	0.9929	0.9628	0.9377	1.0425
12	1.1105	1.0592	1.0207	0.9898	0.9641	1.0725
14	1.1363	1.0843	1.0449	1.0133	0.9871	1.0984
16	1.1595	1.1064	1.0662	1.0342	1.0076	1.1216
18	1.1801	1.1264	1.0857	1.0530	1.0262	1.1422
20	1.1991	1.1448	1.1033	1.0704	1.0432	1.1611
25	1.2406	1.1847	1.1423	1.1085	1.0812	1.2027
30	1.2758	1.2187	1.1757	1.1416	1.1138	1.2382
35	1.3066	1.2486	1.2052	1.1709	1.1430	1.2694
40	1.3344	1.2757	1.2320	1.1975	1.1694	1.2974
45	1.3595	1.3004	1.2565	1.2217	1.1937	1.3230
50	1.3826	1.3232	1.2792	1.2444	1.2165	1.3466
60	1.4243	1.3644	1.3202	1.2854	1.2577	1.3896
70	1.4615	1.4015	1.3569	1.3224	1.2945	1.4280
80	1.4951	1.4348	1.3906	1.3559	1.3282	1.4628
90	1.5262	1.4656	1.4215	1.3869	1.3595	1.4948
100	1.5546	1.4942	1.4500	1.4161	1.3885	1.5248
200	1.7718	1.7128	1.6699	1.6376	1.6114	1.7518

TABLE C.1 (continued)

$$Z^{(l,r)}(T^*) = [T^*(1-6/\alpha)]^{\frac{1}{3}}\Omega^{(l,r)*}(T^*),\ T^* = kT/\epsilon$$

$\alpha = 14$ T^*	$Z^{(1,1)}$	$Z^{(1,2)}$	$Z^{(1,3)}$	$Z^{(1,4)}$	$Z^{(1,5)}$	$Z^{(3,3)}$
0	1.1870	1.0551	0.9672	0.9027	0.8572	1.1178
0.1	1.1727	1.0382	0.9481	0.8814	0.8287	1.0987
0.2	1.1575	1.0146	0.9115	0.8281	0.7571	1.0729
0.3	1.1286	0.9684	0.8497	0.7566	0.6836	1.0308
0.4	1.0914	0.9191	0.7963	0.7069	0.6420	0.9891
0.5	1.0542	0.8776	0.7585	0.6773	0.6214	0.9524
0.6	1.0208	0.8455	0.7336	0.6608	0.6121	0.9218
0.7	0.9924	0.8217	0.7176	0.6521	0.6091	0.8974
0.8	0.9688	0.8042	0.7076	0.6483	0.6096	0.8781
0.9	0.9495	0.7917	0.7018	0.6476	0.6121	0.8633
1.0	0.9337	0.7827	0.6989	0.6488	0.6157	0.8519
1.2	0.9106	0.7724	0.6984	0.6542	0.6246	0.8373
1.4	0.8956	0.7686	0.7019	0.6617	0.6342	0.8300
1.6	0.8863	0.7687	0.7074	0.6700	0.6440	0.8273
1.8	0.8808	0.7710	0.7139	0.6784	0.6534	0.8277
2.0	0.8780	0.7747	0.7208	0.6868	0.6625	0.8300
2.5	0.8784	0.7871	0.7385	0.7068	0.6836	0.8406
3.0	0.8843	0.8011	0.7557	0.7252	0.7022	0.8545
3.5	0.8928	0.8152	0.7716	0.7416	0.7188	0.8692
4	0.9025	0.8288	0.7863	0.7565	0.7335	0.8840
5	0.9228	0.8539	0.8125	0.7824	0.7587	0.9121
6	0.9426	0.8764	0.8349	0.8042	0.7796	0.9376
7	0.9612	0.8962	0.8543	0.8229	0.7974	0.9605
8	0.9784	0.9140	0.8715	0.8391	0.8129	0.9812
9	0.9944	0.9300	0.8867	0.8536	0.8266	0.9999
10	1.0092	0.9455	0.9005	0.8666	0.8390	1.0169
12	1.0358	0.9701	0.9245	0.8893	0.8606	1.0468
14	1.0588	0.9918	0.9450	0.9086	0.8792	1.0726
16	1.0794	1.0110	0.9629	0.9255	0.8956	1.0953
18	1.0978	1.0280	0.9789	0.9408	0.9101	1.1157
20	1.1143	1.0436	0.9931	0.9546	0.9235	1.1339
25	1.1501	1.0766	1.0244	0.9844	0.9526	1.1737
30	1.1802	1.1044	1.0507	1.0097	0.9772	1.2070
35	1.2060	1.1281	1.0736	1.0320	0.9992	1.2359
40	1.2286	1.1494	1.0940	1.0520	1.0191	1.2615
45	1.2491	1.1685	1.1125	1.0703	1.0372	1.2848
50	1.2678	1.1862	1.1296	1.0871	1.0538	1.3063
60	1.3008	1.2176	1.1604	1.1179	1.0844	1.3446
70	1.3297	1.2455	1.1877	1.1450	1.1115	1.3789
80	1.3555	1.2708	1.2125	1.1699	1.1363	1.4095
90	1.3789	1.2938	1.2354	1.1926	1.1595	1.4377
100	1.4009	1.3150	1.2568	1.2141	1.1809	1.4637
200	1.5626	1.4758	1.4185	1.3768	1.3452	1.6611

TABLE C.1 (continued)

$$Z^{(l,r)}(T^*) = [T^*(1-6/\alpha)]^{\frac{1}{3}}\Omega^{(l,r)*}(T^*),\ T^* = kT/\epsilon$$

$\alpha = 14$

T^*	$Z^{(2,2)}$	$Z^{(2,3)}$	$Z^{(2,4)}$	$Z^{(2,5)}$	$Z^{(2,6)}$	$Z^{(4,4)}$
0	1.1947	1.0951	1.0221	0.9706	0.9193	1.0927
0.1	1.2017	1.1027	1.0301	0.9742	0.9296	1.0990
0.2	1.2090	1.1128	1.0416	0.9831	0.9309	1.1098
0.3	1.2109	1.1046	1.0168	0.9391	0.8695	1.0965
0.4	1.1947	1.0716	0.9690	0.8825	0.8109	1.0607
0.5	1.1670	1.0317	0.9235	0.8382	0.7718	1.0206
0.6	1.1359	0.9948	0.8878	0.8076	0.7481	0.9846
0.7	1.1060	0.9641	0.8615	0.7877	0.7346	0.9549
0.8	1.0794	0.9397	0.8428	0.7754	0.7277	0.9315
0.9	1.0564	0.9209	0.8300	0.7682	0.7249	0.9135
1.0	1.0370	0.9066	0.8215	0.7647	0.7250	0.9001
1.2	1.0078	0.8882	0.8132	0.7641	0.7299	0.8829
1.4	0.9885	0.8790	0.8122	0.7686	0.7380	0.8746
1.6	0.9761	0.8756	0.8151	0.7756	0.7476	0.8720
1.8	0.9688	0.8759	0.8205	0.7840	0.7580	0.8731
2.0	0.9649	0.8785	0.8271	0.7930	0.7683	0.8764
2.5	0.9651	0.8910	0.8464	0.8162	0.7937	0.8905
3.0	0.9729	0.9069	0.8666	0.8385	0.8172	0.9081
3.5	0.9844	0.9239	0.8862	0.8594	0.8384	0.9267
4	0.9973	0.9409	0.9048	0.8785	0.8575	0.9449
5	1.0244	0.9729	0.9384	0.9121	0.8908	0.9790
6	1.0507	1.0015	0.9765	0.9408	0.9188	1.0092
7	1.0755	1.0274	0.9882	0.9656	0.9428	1.0361
8	1.0982	1.0504	1.0154	0.9875	0.9639	1.0600
9	1.1192	1.0712	1.0357	1.0068	0.9827	1.0816
10	1.1385	1.0902	1.0539	1.0244	0.9996	1.1013
12	1.1731	1.1237	1.0859	1.0554	1.0293	1.1359
14	1.2034	1.1526	1.1136	1.0818	1.0550	1.1656
16	1.2302	1.1781	1.1379	1.1051	1.0777	1.1917
18	1.2540	1.2007	1.1596	1.1261	1.0980	1.2153
20	1.2758	1.2213	1.1792	1.1452	1.1168	1.2364
25	1.3229	1.2659	1.2222	1.1868	1.1576	1.2826
30	1.3622	1.3034	1.2583	1.2222	1.1925	1.3217
35	1.3963	1.3363	1.2902	1.2535	1.2231	1.3559
40	1.4266	1.3653	1.3185	1.2816	1.2510	1.3863
45	1.4543	1.3917	1.3445	1.3070	1.2766	1.4141
50	1.4793	1.4161	1.3684	1.3308	1.2999	1.4396
60	1.5239	1.4596	1.4115	1.3735	1.3427	1.4856
70	1.5636	1.4983	1.4497	1.4118	1.3806	1.5263
80	1.5990	1.5332	1.4846	1.4463	1.4156	1.5633
90	1.6314	1.5656	1.5165	1.4782	1.4474	1.5972
100	1.6612	1.5950	1.5461	1.5079	1.4771	1.6289
200	1.8858	1.8198	1.7713	1.7344	1.7043	1.8674

TABLE C.1 (continued)

$$Z^{(l,r)}(T^*) = [T^*(1-6/\alpha)]^{\frac{1}{3}}\Omega^{(l,r)*}(T^*),\ T^* = kT/\epsilon$$

$\alpha = 15$ T^*	$Z^{(1,1)}$	$Z^{(1,2)}$	$Z^{(1,3)}$	$Z^{(1,4)}$	$Z^{(1,5)}$	$Z^{(3,3)}$
0	1.1870	1.0551	0.9672	0.9027	0.8572	1.1178
0.1	1.1722	1.0378	0.9477	0.8810	0.8285	1.1007
0.2	1.1577	1.0163	0.9153	0.8343	0.7655	1.0765
0.3	1.1320	0.9750	0.8594	0.7687	0.6970	1.0392
0.4	1.0984	0.9300	0.8099	0.7221	0.6579	1.0012
0.5	1.0645	0.8916	0.7747	0.6944	0.6388	0.9672
0.6	1.0337	0.8618	0.7514	0.6792	0.6306	0.9387
0.7	1.0075	0.8397	0.7367	0.6715	0.6285	0.9158
0.8	0.9857	0.8235	0.7278	0.6687	0.6299	0.8978
0.9	0.9678	0.8121	0.7230	0.6687	0.6333	0.8840
1.0	0.9533	0.8040	0.7208	0.6707	0.6378	0.8735
1.2	0.9323	0.7954	0.7218	0.6777	0.6484	0.8605
1.4	0.9191	0.7931	0.7266	0.6867	0.6596	0.8548
1.6	0.9112	0.7944	0.7335	0.6964	0.6708	0.8535
1.8	0.9070	0.7979	0.7412	0.7063	0.6817	0.8554
2.0	0.9054	0.8028	0.7494	0.7159	0.6921	0.8593
2.5	0.9083	0.8179	0.7699	0.7389	0.7160	0.8734
3.0	0.9167	0.8344	0.7896	0.7597	0.7372	0.8903
3.5	0.9273	0.8507	0.8078	0.7783	0.7558	0.9077
4	0.9389	0.8662	0.8245	0.7953	0.7725	0.9245
5	0.9627	0.8951	0.8541	0.8245	0.8012	0.9558
6	0.9856	0.9203	0.8796	0.8494	0.8250	0.9836
7	1.0069	0.9430	0.9017	0.8708	0.8458	1.0086
8	1.0268	0.9632	0.9214	0.8895	0.8638	1.0308
9	1.0449	0.9816	0.9390	0.9063	0.8800	1.0514
10	1.0619	0.9981	0.9549	0.9216	0.8946	1.0699
12	1.0922	1.0277	0.9829	0.9483	0.9205	1.1032
14	1.1188	1.0530	1.0070	0.9715	0.9428	1.1321
16	1.1426	1.0756	1.0284	0.9921	0.9628	1.1579
18	1.1640	1.0957	1.0475	1.0104	0.9805	1.1812
20	1.1834	1.1140	1.0648	1.0273	0.9970	1.2026
25	1.2257	1.1537	1.1029	1.0642	1.0333	1.2494
30	1.2614	1.1872	1.1353	1.0957	1.0642	1.2897
35	1.2923	1.2167	1.1637	1.1234	1.0920	1.3254
40	1.3199	1.2429	1.1890	1.1486	1.1169	1.3574
45	1.3446	1.2666	1.2123	1.1718	1.1397	1.3869
50	1.3675	1.2886	1.2339	1.1929	1.1609	1.4141
60	1.4083	1.3280	1.2726	1.2316	1.1996	1.4631
70	1.4443	1.3630	1.3073	1.2660	1.2340	1.5072
80	1.4766	1.3945	1.3389	1.2974	1.2654	1.5471
90	1.5059	1.4235	1.3675	1.3267	1.2946	1.5837
100	1.5331	1.4505	1.3945	1.3534	1.3217	1.6180
200	1.7372	1.6538	1.5991	1.5596	1.5291	1.8773

TABLE C.1 (continued)

$$Z^{(l,r)}(T^*) = [T^*(1-6/\alpha)]^{\frac{1}{3}}\Omega^{(l,r)*}(T^*),\ T^* = kT/\epsilon$$

$\alpha = 15$

T^*	$Z^{(2,2)}$	$Z^{(2,3)}$	$Z^{(2,4)}$	$Z^{(2,5)}$	$Z^{(2,6)}$	$Z^{(4,4)}$
0	1.1947	1.0951	1.0221	0.9706	0.9193	1.0927
0.1	1.2016	1.1025	1.0306	0.9736	0.9290	1.1061
0.2	1.2081	1.1114	1.0403	0.9828	0.9325	1.1150
0.3	1.2107	1.1061	1.0211	0.9465	0.8796	1.1029
0.4	1.1978	1.0783	0.9791	0.8952	0.8252	1.0705
0.5	1.1739	1.0426	0.9374	0.8536	0.7879	1.0333
0.6	1.1462	1.0088	0.9039	0.8247	0.7654	0.9996
0.7	1.1191	0.9803	0.8791	0.8060	0.7529	0.9718
0.8	1.0945	0.9575	0.8617	0.7946	0.7470	0.9499
0.9	1.0734	0.9400	0.8498	0.7883	0.7454	0.9331
1.0	1.0555	0.9268	0.8423	0.7858	0.7466	0.9209
1.2	1.0285	0.9102	0.8359	0.7872	0.7535	0.9057
1.4	1.0110	0.9027	0.8365	0.7936	0.7637	0.8992
1.6	1.0003	0.9009	0.8412	0.8024	0.7751	0.8982
1.8	0.9944	0.9027	0.8481	0.8124	0.7868	0.9007
2.0	0.9920	0.9067	0.8562	0.8227	0.7986	0.9054
2.5	0.9953	0.9223	0.8786	0.8488	0.8266	0.9222
3.0	1.0059	0.9409	0.9013	0.8734	0.8521	0.9419
3.5	1.0196	0.9602	0.9229	0.8960	0.8750	0.9622
4	1.0345	0.9790	0.9431	0.9169	0.8958	0.9819
5	1.0651	1.0139	0.9796	0.9533	0.9318	1.0187
6	1.0943	1.0452	1.0111	0.9844	0.9623	1.0514
7	1.1213	1.0733	1.0389	1.0115	0.9887	1.0805
8	1.1462	1.0985	1.0634	1.0356	1.0121	1.1064
9	1.1691	1.1212	1.0856	1.0570	1.0332	1.1300
10	1.1904	1.1421	1.1059	1.0766	1.0523	1.1515
12	1.2283	1.1791	1.1416	1.1115	1.0860	1.1895
14	1.2616	1.2111	1.1727	1.1416	1.1154	1.2225
16	1.2911	1.2397	1.2002	1.1683	1.1417	1.2518
18	1.3179	1.2655	1.2250	1.1925	1.1656	1.2781
20	1.3423	1.2887	1.2477	1.2148	1.1873	1.3020
25	1.3954	1.3399	1.2975	1.2634	1.2353	1.3544
30	1.4406	1.3835	1.3400	1.3054	1.2768	1.3992
35	1.4802	1.4219	1.3775	1.3425	1.3138	1.4382
40	1.5152	1.4561	1.4114	1.3759	1.3471	1.4737
45	1.5474	1.4874	1.4424	1.4067	1.3776	1.5057
50	1.5769	1.5163	1.4710	1.4352	1.4060	1.5353
60	1.6298	1.5684	1.5225	1.4865	1.4575	1.5889
70	1.6768	1.6150	1.5687	1.5326	1.5034	1.6365
80	1.7190	1.6569	1.6107	1.5747	1.5453	1.6797
90	1.7580	1.6952	1.6491	1.6132	1.5841	1.7194
100	1.7938	1.7312	1.6850	1.6489	1.6200	1.7562
200	2.0647	2.0026	1.9572	1.9227	1.8945	2.0351

TABLE C.2

Tables of the transport integral ratios A*, B*, C* for the modified Buckingham (6-exp) potential*

T^*	$\alpha = 12$	The function A* $\alpha = 13$	$\alpha = 14$	$\alpha = 15$
0	1.0065	1.0065	1.0065	1.0065
0.1	1.0060	1.0146	1.0247	1.0251
0.2	1.0374	1.0374	1.0445	1.0435
0.3	1.0731	1.0693	1.0729	1.0695
0.4	1.0976	1.0932	1.0946	1.0905
0.5	1.1108	1.1067	1.1070	1.1028
0.6	1.1170	1.1131	1.1128	1.1088
0.7	1.1193	1.1154	1.1145	1.1108
0.8	1.1194	1.1154	1.1142	1.1104
0.9	1.1185	1.1143	1.1126	1.1091
1.0	1.1173	1.1126	1.1106	1.1071
1.2	1.1146	1.1096	1.1067	1.1032
1.4	1.1126	1.1070	1.1037	1.1000
1.6	1.1114	1.1052	1.1013	1.0978
1.8	1.1111	1.1044	1.0999	1.0964
2.0	1.1111	1.1044	1.0990	1.0956
2.5	1.1130	1.1056	1.0987	1.0958
3.0	1.1161	1.1081	1.1002	1.0973
3.5	1.1193	1.1113	1.1026	1.0995
4	1.1226	1.1141	1.1050	1.1018
5	1.1284	1.1194	1.1101	1.1064
6	1.1335	1.1238	1.1147	1.1103
7	1.1377	1.1277	1.1189	1.1136
8	1.1417	1.1309	1.1224	1.1163
9	1.1494	1.1336	1.1255	1.1189
10	1.1477	1.1361	1.1281	1.1210
12	1.1532	1.1406	1.1326	1.1246
14	1.1583	1.1447	1.1366	1.1276
16	1.1625	1.1485	1.1397	1.1300
18	1.1665	1.1517	1.1423	1.1322
20	1.1702	1.1548	1.1449	1.1343
25	1.1790	1.1623	1.1502	1.1385
30	1.1865	1.1691	1.1542	1.1421
35	1.1938	1.1752	1.1578	1.1454
40	1.2005	1.1813	1.1612	1.1480
45	1.2067	1.1867	1.1643	1.1508
50	1.2123	1.1917	1.1668	1.1531
60	1.2230	1.2008	1.1715	1.1573
70	1.2326	1.2092	1.1759	1.1610
80	1.2416	1.2168	1.1796	1.1642
90	1.2492	1.2240	1.1831	1.1674
100	1.2569	1.2300	1.1858	1.1700
200	1.3103	1.2754	1.2068	1.1885

* From Mason [1954].

TABLE C.2 (continued)

The function B*

T^*	$\alpha = 12$	$\alpha = 13$	$\alpha = 14$	$\alpha = 15$
0	1.1851	1.1851	1.1851	1.1851
0.1	1.1988	1.1955	1.1926	1.1928
0.2	1.2528	1.2407	1.2328	1.2268
0.3	1.3028	1.2895	1.2788	1.2698
0.4	1.3127	1.3019	1.2922	1.2840
0.5	1.3003	1.2922	1.2844	1.2768
0.6	1.2808	1.2737	1.2668	1.2609
0.7	1.2601	1.2539	1.2476	1.2424
0.8	1.2403	1.2346	1.2289	1.2238
0.9	1.2239	1.2172	1.2125	1.2074
1.0	1.2079	1.2025	1.1973	1.1930
1.2	1.1837	1.1780	1.1733	1.1689
1.4	1.1662	1.1612	1.1561	1.1523
1.6	1.1536	1.1490	1.1440	1.1392
1.8	1.1458	1.1394	1.1347	1.1298
2.0	1.1387	1.1324	1.1279	1.1226
2.5	1.1288	1.1222	1.1174	1.1119
3.0	1.1250	1.1167	1.1113	1.1057
3.5	1.1217	1.1136	1.1084	1.1024
4	1.1211	1.1128	1.1067	1.1002
5	1.1225	1.1126	1.1048	1.1001
6	1.1242	1.1131	1.1059	1.0989
7	1.1252	1.1155	1.1067	1.1006
8	1.1278	1.1176	1.1079	1.1009
9	1.1298	1.1185	1.1094	1.1025
10	1.1317	1.1203	1.1103	1.1026
12	1.1351	1.1225	1.1127	1.1050
14	1.1376	1.1247	1.1135	1.1056
16	1.1384	1.1259	1.1149	1.1066
18	1.1405	1.1271	1.1153	1.1070
20	1.1404	1.1274	1.1178	1.1077
25	1.1426	1.1282	1.1176	1.1070
30	1.1426	1.1281	1.1178	1.1058
35	1.1427	1.1279	1.1162	1.1055
40	1.1427	1.1268	1.1159	1.1050
45	1.1418	1.1270	1.1148	1.1035
50	1.1407	1.1255	1.1142	1.1023
60	1.1388	1.1236	1.1119	1.1003
70	1.1378	1.1218	1.1106	1.0980
80	1.1365	1.1198	1.1096	1.0950
90	1.1350	1.1180	1.1077	1.0940
100	1.1331	1.1154	1.1049	1.0922
200	1.1206	1.1021	1.0911	1.0779

TABLE C.2 (continued)

The function c*

T^*	$\alpha = 12$	$\alpha = 13$	$\alpha = 14$	$\alpha = 15$
0	0.8889	0.8889	0.8889	0.8889
0.1	0.8835	0.8844	0.8853	0.8853
0.2	0.8697	0.8736	0.8765	0.8779
0.3	0.8463	0.8531	0.8581	0.8613
0.4	0.8279	0.8360	0.8421	0.8467
0.5	0.8172	0.8257	0.8325	0.8376
0.6	0.8128	0.8212	0.8283	0.8337
0.7	0.8127	0.8209	0.8280	0.8334
0.8	0.8151	0.8231	0.8301	0.8354
0.9	0.8193	0.8269	0.8338	0.8391
1.0	0.8240	0.8315	0.8383	0.8434
1.2	0.8348	0.8419	0.8482	0.8532
1.4	0.8453	0.8521	0.8582	0.8629
1.6	0.8549	0.8615	0.8673	0.8718
1.8	0.8636	0.8698	0.8753	0.8797
2.0	0.8709	0.8771	0.8823	0.8867
2.5	0.8855	0.8913	0.8961	0.9005
3.0	0.8959	0.9014	0.9059	0.9102
3.5	0.9031	0.9088	0.9131	0.9174
4	0.9084	0.9141	0.9183	0.9226
5	0.9155	0.9213	0.9253	0.9298
6	0.9195	0.9254	0.9298	0.9337
7	0.9216	0.9280	0.9324	0.9365
8	0.9232	0.9295	0.9342	0.9381
9	0.9241	0.9302	0.9352	0.9394
10	0.9242	0.9307	0.9359	0.9399
12	0.9245	0.9311	0.9366	0.9409
14	0.9243	0.9313	0.9367	0.9412
16	0.9237	0.9310	0.9366	0.9414
18	0.9234	0.9307	0.9364	0.9413
20	0.9226	0.9302	0.9366	0.9414
25	0.9218	0.9295	0.9361	0.9413
30	0.9209	0.9291	0.9358	0.9412
35	0.9203	0.9289	0.9354	0.9415
40	0.9202	0.9288	0.9355	0.9417
45	0.9198	0.9290	0.9355	0.9420
50	0.9197	0.9290	0.9356	0.9423
60	0.9196	0.9293	0.9360	0.9430
70	0.9202	0.9298	0.9367	0.9437
80	0.9205	0.9303	0.9375	0.9444
90	0.9208	0.9309	0.9383	0.9453
100	0.9214	0.9316	0.9387	0.9461
200	0.9271	0.9377	0.9445	0.9520

TABLE C.3

Transport integrals $\Omega^{(1,1)*}$ and $\Omega^{(2,2)*}$ for the Stockmayer potential, Eq. (8.4-11)*

(a) $\Omega^{(1,1)*}$

T^* \ δ	−2.5	−2.0	−1.5	−1.0	−0.5	0	0.25	0.5	0.75	1.0	1.5	2.0	2.5
0.1	32.491	27.796	22.860	17.337	11.034	4.0079	2.5030	4.8195	7.0752	8.7831	11.683	14.223	16.042
0.2	20.231	17.334	14.250	10.858	7.0967	3.1300	1.8648	2.5558	4.1729	5.3665	7.2732	8.9026	10.378
0.3	15.325	13.141	10.813	8.2836	5.5257	2.6494	1.6049	1.8216	3.0004	3.9671	5.4825	6.7496	7.8875
0.4	12.581	10.796	8.8974	6.8508	4.6309	2.3144	1.4442	1.4953	2.3752	3.1823	4.4702	5.5351	6.4844
0.5	10.796	9.2718	7.6546	5.9160	4.0268	2.0661	1.3343	1.3208	1.9970	2.6795	3.8070	4.7390	5.5655
0.6	9.5286	8.1904	6.7712	5.2434	3.5795	1.8767	1.2556	1.2156	1.7489	2.3320	3.3346	4.1700	4.9086
0.7	8.5753	7.3764	6.1031	4.7278	3.2304	1.7293	1.1966	1.1462	1.5766	2.0795	2.9798	3.7400	4.4118
0.8	7.8277	6.7363	5.5738	4.3143	2.9491	1.6122	1.1509	1.0973	1.4514	1.8890	2.7031	3.4019	4.0205
0.9	7.2226	6.2159	5.1402	3.9730	2.7172	1.5175	1.1144	1.0609	1.3571	1.7411	2.4815	3.1284	3.7032
1.0	6.7201	5.7816	4.7758	3.6850	2.5231	1.4398	1.0845	1.0329	1.2839	1.6235	2.3001	2.9023	3.4399
1.2	5.9263	5.0909	4.1924	3.2244	2.2173	1.3204	1.0382	0.99162	1.1787	1.4496	2.0215	2.5495	3.0269
1.4	5.3195	4.5588	3.7421	2.8717	1.9888	1.2336	1.0038	0.96264	1.1070	1.3283	1.8185	2.2869	2.7165
1.6	4.8352	4.1326	3.3823	2.5938	1.8131	1.1679	0.97690	0.94058	1.0553	1.2394	1.6646	2.0838	2.4742
1.8	4.4363	3.7816	3.0880	2.3694	1.6745	1.1166	0.95503	0.92286	1.0161	1.1719	1.5445	1.9225	2.2797
2.0	4.1006	3.4873	2.8428	2.1852	1.5626	1.0753	0.93686	0.90797	0.98537	1.1189	1.4484	1.7912	2.1199
2.5	3.4515	2.9229	2.3808	1.8463	1.3614	1.0006	0.90147	0.87918	0.93080	1.0261	1.2763	1.5509	1.8232
3.0	2.9827	2.5221	2.0594	1.6168	1.2286	0.95003	0.87520	0.85747	0.89424	0.96585	1.1630	1.3887	1.6188
3.5	2.6294	2.2243	1.8258	1.4531	1.1353	0.91311	0.85440	0.83989	0.86740	0.92332	1.0831	1.2742	1.4701

4.0	2.3546	1.9960	1.6495	1.3318	1.0662	0.88453	0.83720	0.82506	0.84643	0.89140	1.0239	1.1852	1.3573
5.0	1.9592	1.6729	1.4042	1.1650	0.97130	0.84277	0.80977	0.80087	0.81486	0.84594	0.94178	1.0637	1.1982
6.0	1.6916	1.4582	1.2435	1.0567	0.90894	0.81287	0.78837	0.78154	0.79143	0.81432	0.88729	0.98319	1.0916
7.0	1.5008	1.3070	1.1318	0.98147	0.86449	0.78976	0.77086	0.76543	0.77282	0.79045	0.84814	0.92586	1.0155
8.0	1.3595	1.1958	1.0494	0.92618	0.83109	0.77111	0.75604	0.75162	0.75737	0.77142	0.81835	0.88283	0.95847
9.0	1.2507	1.1110	0.98706	0.88364	0.80494	0.75553	0.74322	0.73954	0.74416	0.75566	0.79471	0.84923	0.91405
10.0	1.1656	1.0441	0.93815	0.84990	0.78360	0.74220	0.73191	0.72881	0.73262	0.74224	0.77533	0.82214	0.87844
12.0	1.0403	0.94699	0.86604	0.79981	0.75055	0.72022	0.71269	0.71042	0.71316	0.72023	0.74504	0.78084	0.82468
14.0	0.95392	0.87951	0.81546	0.76397	0.72575	0.70254	0.69678	0.69505	0.69714	0.70261	0.72202	0.75047	0.78576
16.0	0.89087	0.82989	0.77819	0.73667	0.70615	0.68776	0.68322	0.68187	0.68354	0.68793	0.70362	0.72688	0.75603
18.0	0.84280	0.79194	0.74929	0.71501	0.69007	0.67510	0.67144	0.67036	0.67174	0.67536	0.68838	0.70782	0.73240
20.0	0.80498	0.76209	0.72605	0.69726	0.67649	0.66405	0.66103	0.66015	0.66133	0.66438	0.67539	0.69194	0.71301
25.0	0.73865	0.70865	0.68355	0.66381	0.64973	0.64136	0.63938	0.63884	0.63970	0.64183	0.64958	0.66134	0.67647
30.0	0.69505	0.67267	0.65416	0.63977	0.62950	0.62350	0.62211	0.62175	0.62244	0.62404	0.62987	0.63875	0.65026
35.0	0.66372	0.64638	0.63216	0.62115	0.61331	0.60882	0.60780	0.60755	0.60814	0.60939	0.61398	0.62099	0.63011
40.0	0.63977	0.62598	0.61475	0.60603	0.59988	0.59640	0.59563	0.59544	0.59597	0.59698	0.60072	0.60642	0.61386
50.0	0.60473	0.59563	0.58819	0.58246	0.57848	0.57626	0.57578	0.57666	0.57612	0.57681	0.57946	0.58347	0.58871
75.0	0.55113	0.54806	0.54560	0.54375	0.54254	0.54146	0.54128	0.54138	0.54158	0.54206	0.54338	0.54532	0.54782
100.0	0.51718	0.51777	0.51829	0.51875	0.51915	0.51803	0.51795	0.51860	0.51820	0.51901	0.51964	0.52049	0.52153

* These integrals are for a particular molecular orientation, i.e., for fixed $\theta_1, \theta_2, \phi_1, \phi_2$. They may also be regarded as the transport integrals for the average Stockmayer potential, Eq. (11.1-3). For $\delta = 0$, these integrals reduce to those for the Lennard-Jones potential, Eq. (8.4-5). Table C.3 has been adapted from Monchick and Mason [1961].

TABLE C.3 (continued)

(b) $\Omega^{(2,2)*}$

T^* \ δ	−2.5	−2.0	−1.5	−1.0	−0.5	0	0.25	0.5	0.75	1.0	1.5	2.0	2.5
0.1	29.139	25.148	20.830	16.015	10.451	4.1005	3.0142	6.4500	9.4870	11.699	15.508	18.860	21.923
0.2	18.420	15.917	13.211	10.225	6.8596	3.2626	2.1702	3.2515	5.6362	7.1946	9.6786	11.820	13.765
0.3	14.097	12.196	10.146	7.9029	5.4303	2.8399	1.8408	2.2186	4.0126	5.3309	7.3133	8.9737	10.471
0.4	11.668	10.106	8.4275	6.6083	4.6384	2.5310	1.6348	1.7708	3.1216	4.2632	5.9726	7.3682	8.6157
0.5	10.081	8.7414	7.3096	5.7737	4.1155	2.2837	1.4952	1.5375	2.5764	3.5654	5.0888	6.3144	7.4001
0.6	8.9498	7.7708	6.5189	5.1848	3.7272	2.0838	1.3966	1.3997	2.2184	3.0764	4.4541	5.5594	6.5307
0.7	8.0965	7.0416	5.9275	4.7397	3.4170	1.9220	1.3241	1.3105	1.9705	2.7178	3.9732	4.9867	5.8723
0.8	7.4276	6.4722	5.4661	4.3847	3.1588	1.7902	1.2691	1.2487	1.7914	2.4458	3.5951	4.5346	5.3528
0.9	6.8882	6.0142	5.0923	4.0903	2.9388	1.6823	1.2259	1.2034	1.6576	2.2341	3.2899	4.1672	4.9306
1.0	6.4427	5.6362	4.7810	3.8383	2.7488	1.5929	1.1913	1.1688	1.5544	2.0657	3.0384	3.8618	4.5794
1.2	5.7488	5.0427	4.2814	3.4239	2.4382	1.4551	1.1391	1.1192	1.4076	1.8169	2.6495	3.3822	4.0259
1.4	5.2280	4.5900	3.8897	3.0928	2.1967	1.3551	1.1012	1.0846	1.3091	1.6440	2.3638	3.0221	3.6074
1.6	4.8159	4.2254	3.5677	2.8214	2.0057	1.2800	1.0722	1.0587	1.2390	1.5182	2.1464	2.7416	3.2786
1.8	4.4774	3.9202	3.2965	2.5953	1.8522	1.2219	1.0491	1.0382	1.1865	1.4232	1.9762	2.5173	3.0131
2.0	4.1898	3.6589	3.0638	2.4045	1.7270	1.1757	1.0300	1.0212	1.1458	1.3492	1.8399	2.3341	2.7939
2.5	3.6210	3.1378	2.6058	2.0415	1.4993	1.0933	0.99337	0.98859	1.0748	1.2210	1.5960	1.9969	2.3838

3.0	3.1902	2.7466	2.2710	1.7881	1.3486	1.0388	0.96645	0.96415	1.0281	1.1391	1.4361	1.7683	2.0992
3.5	2.8502	2.4433	2.0194	1.6045	1.2430	0.99963	0.94519	0.95453	0.99448	1.0822	1.3240	1.6042	1.8910
4.0	2.5756	2.2032	1.8255	1.4672	1.1655	0.96988	0.92764	0.92808	0.96854	1.0400	1.2414	1.4814	1.7325
5.0	2.1635	1.8526	1.5510	1.2784	1.0605	0.92676	0.89963	0.90136	0.93011	0.98086	1.1281	1.3108	1.5089
6.0	1.8736	1.6134	1.3694	1.1566	0.99273	0.89616	0.87766	0.88005	0.90203	0.94049	1.0539	1.1985	1.3594
7.0	1.6622	1.4429	1.2426	1.0727	0.94537	0.87272	0.85956	0.86230	0.87996	0.91048	1.0012	1.1191	1.2529
8.0	1.5033	1.3169	1.1501	1.0115	0.91013	0.85379	0.84417	0.84709	0.86180	0.88684	0.96160	1.0600	1.1734
9.0	1.3806	1.2208	1.0801	0.96522	0.88269	0.83795	0.83079	0.83379	0.84637	0.86745	0.93046	1.0142	1.1118
10.0	1.2839	1.1456	1.0256	0.92892	0.86055	0.82435	0.81895	0.82198	0.83296	0.85108	0.90517	0.97754	1.0627
12.0	1.1424	1.0364	0.94629	0.87547	0.82661	0.80184	0.79874	0.80172	0.81047	0.82449	0.86609	0.92218	0.98904
14.0	1.0452	0.96153	0.89171	0.83772	0.80132	0.78363	0.78189	0.78478	0.79207	0.80342	0.83680	0.88197	0.93627
16.0	0.97494	0.90740	0.85178	0.80938	0.78137	0.76834	0.76747	0.77024	0.77650	0.78600	0.81365	0.85108	0.89633
18.0	0.92214	0.86654	0.82120	0.78711	0.76498	0.75518	0.75488	0.75753	0.76303	0.77117	0.79464	0.82637	0.86485
20.0	0.88115	0.83457	0.79693	0.76897	0.75112	0.74364	0.74372	0.74626	0.75117	0.75828	0.77859	0.80596	0.83923
25.0	0.81004	0.77827	0.75314	0.73492	0.72377	0.71982	0.72045	0.72270	0.72662	0.73198	0.74700	0.76707	0.79150
30.0	0.76376	0.74088	0.72306	0.71041	0.70302	0.70097	0.70183	0.70381	0.70713	0.71137	0.72317	0.73876	0.75770
35.0	0.73046	0.71348	0.70044	0.69140	0.68638	0.68545	0.68641	0.68813	0.69107	0.69450	0.70412	0.71669	0.73192
40.0	0.70459	0.69195	0.68238	0.67591	0.67256	0.67232	0.67331	0.67480	0.67746	0.68028	0.68832	0.69870	0.71121
50.0	0.66536	0.65902	0.65445	0.65164	0.65061	0.65099	0.65195	0.65314	0.65538	0.65733	0.66314	0.67045	0.67917
75.0	0.60030	0.60470	0.60849	0.61171	0.61438	0.61397	0.61475	0.61624	0.61708	0.61832	0.62067	0.62330	0.62620
100.0	0.55518	0.56801	0.57833	0.58620	0.59170	0.58870	0.58931	0.59240	0.59105	0.59291	0.59278	0.59207	0.59082

TABLE C.4

Orientation averaged transport integrals for the Stockmayer potential, Eq. (11.1-4)*

(a) $\langle \Omega^{(1,1)*} \rangle$

T^* \ δ	0	0.25	0.50	0.75	1.0	1.5	2.0	2.5
0.1	4.0079	4.002	4.655	5.521	6.454	8.214	9.824	11.31
0.2	3.1300	3.164	3.355	3.721	4.198	5.230	6.225	7.160
0.3	2.6494	2.657	2.770	3.002	3.319	4.054	4.785	5.483
0.4	2.3144	2.320	2.402	2.572	2.812	3.386	3.972	4.539
0.5	2.0661	2.073	2.140	2.278	2.472	2.946	3.437	3.918
0.6	1.8767	1.885	1.944	2.060	2.225	2.628	3.054	3.474
0.7	1.7293	1.738	1.791	1.893	2.036	2.388	2.763	3.137
0.8	1.6122	1.622	1.670	1.760	1.886	2.198	2.535	2.872
0.9	1.5175	1.527	1.572	1.653	1.765	2.044	2.349	2.657
1.0	1.4398	1.450	1.490	1.564	1.665	1.917	2.196	2.478
1.2	1.3204	1.330	1.364	1.425	1.509	1.720	1.956	2.199
1.4	1.2336	1.242	1.272	1.324	1.394	1.573	1.777	1.990
1.6	1.1679	1.176	1.202	1.246	1.306	1.461	1.639	1.827
1.8	1.1166	1.124	1.146	1.185	1.237	1.372	1.530	1.698
2.0	1.0753	1.082	1.102	1.135	1.181	1.300	1.441	1.592
2.5	1.0006	1.005	1.020	1.046	1.080	1.170	1.278	1.397
3.0	0.95003	0.9538	0.9656	0.9852	1.012	1.082	1.168	1.265
3.5	0.91311	0.9162	0.9256	0.9413	0.9626	1.019	1.090	1.170
4.0	0.88453	0.8871	0.8948	0.9076	0.9252	0.9721	1.031	1.098
5.0	0.84277	0.8446	0.8501	0.8592	0.8716	0.9053	0.9483	0.9984
6.0	0.81287	0.8142	0.8183	0.8251	0.8344	0.8598	0.8927	0.9316
7.0	0.78976	0.7908	0.7940	0.7993	0.8066	0.8265	0.8526	0.8836
8.0	0.77111	0.7720	0.7745	0.7788	0.7846	0.8007	0.8219	0.8474
9.0	0.75553	0.7562	0.7584	0.7619	0.7667	0.7800	0.7976	0.8189
10.0	0.74220	0.7428	0.7446	0.7475	0.7515	0.7627	0.7776	0.7957
12.0	0.72022	0.7206	0.7220	0.7241	0.7271	0.7354	0.7464	0.7600
14.0	0.70254	0.7029	0.7039	0.7055	0.7078	0.7142	0.7228	0.7334
16.0	0.68776	0.6880	0.6888	0.6901	0.6919	0.6970	0.7040	0.7125
18.0	0.67510	0.6753	0.6760	0.6770	0.6785	0.6827	0.6884	0.6955
20.0	0.66405	0.6642	0.6648	0.6657	0.6669	0.6704	0.6752	0.6811
25.0	0.64136	0.6415	0.6418	0.6425	0.6433	0.6457	0.6490	0.6531
30.0	0.62350	0.6236	0.6239	0.6243	0.6249	0.6267	0.6291	0.6321
35.0	0.60882	0.6089	0.6091	0.6094	0.6099	0.6112	0.6131	0.6154
40.0	0.59640	0.5964	0.5966	0.5969	0.5972	0.5983	0.5998	0.6017
50.0	0.57626	0.5763	0.5764	0.5766	0.5768	0.5775	0.5785	0.5798
75.0	0.54146	0.5415	0.5416	0.5416	0.5418	0.5421	0.5424	0.5429
100.0	0.51803	0.5181	0.5182	0.5184	0.5184	0.5185	0.5186	0.5187

* These integrals are obtained from those in Table C.3 by performing the averaging process defined by Eq. (11.1-4). This table has been adapted from Monchick and Mason [1961].

TABLE C.4 (continued)

(b) $\langle \Omega^{(2,2)*} \rangle$

T^* \ δ	0	0.25	0.50	0.75	1.0	1.5	2.0	2.5
0.1	4.1005	4.266	4.833	5.742	6.729	8.624	10.34	11.89
0.2	3.2626	3.305	3.516	3.914	4.433	5.570	6.637	7.618
0.3	2.8399	2.836	2.936	3.168	3.511	4.329	5.126	5.874
0.4	2.5310	2.522	2.586	2.749	3.004	3.640	4.282	4.895
0.5	2.2837	2.277	2.329	2.460	2.665	3.187	3.727	4.249
0.6	2.0838	2.081	2.130	2.243	2.417	2.862	3.329	3.786
0.7	1.9220	1.924	1.970	2.072	2.225	2.614	3.028	3.435
0.8	1.7902	1.795	1.840	1.934	2.070	2.417	2.788	3.156
0.9	1.6823	1.689	1.733	1.820	1.944	2.258	2.596	2.933
1.0	1.5929	1.601	1.644	1.725	1.838	2.124	2.435	2.746
1.2	1.4551	1.465	1.504	1.574	1.670	1.913	2.181	2.451
1.4	1.3551	1.365	1.400	1.461	1.544	1.754	1.989	2.228
1.6	1.2800	1.289	1.321	1.374	1.447	1.630	1.838	2.053
1.8	1.2219	1.231	1.259	1.306	1.370	1.532	1.718	1.912
2.0	1.1757	1.184	1.209	1.251	1.307	1.451	1.618	1.795
2.5	1.0933	1.100	1.119	1.150	1.193	1.304	1.435	1.578
3.0	1.0388	1.044	1.059	1.083	1.117	1.204	1.310	1.428
3.5	0.99963	1.004	1.016	1.035	1.062	1.133	1.220	1.319
4.0	0.96988	0.9732	0.9830	0.9991	1.021	1.079	1.153	1.236
5.0	0.92676	0.9291	0.9360	0.9473	0.9628	1.005	1.058	1.121
6.0	0.89616	0.8979	0.9030	0.9114	0.9230	0.9545	0.9955	1.044
7.0	0.87272	0.8741	0.8780	0.8845	0.8935	0.9181	0.9505	0.9893
8.0	0.85379	0.8549	0.8580	0.8632	0.8703	0.8901	0.9164	0.9482
9.0	0.83795	0.8388	0.8414	0.8456	0.8515	0.8678	0.8895	0.9160
10.0	0.82435	0.8251	0.8273	0.8308	0.8356	0.8493	0.8676	0.8901
12.0	0.80184	0.8024	0.8039	0.8065	0.8101	0.8201	0.8337	0.8504
14.0	0.78363	0.7840	0.7852	0.7872	0.7899	0.7976	0.8081	0.8212
16.0	0.76834	0.7687	0.7696	0.7712	0.7733	0.7794	0.7878	0.7983
18.0	0.75518	0.7554	0.7562	0.7575	0.7592	0.7642	0.7711	0.7797
20.0	0.74364	0.7438	0.7445	0.7455	0.7470	0.7512	0.7569	0.7642
25.0	0.71982	0.7200	0.7204	0.7211	0.7221	0.7250	0.7289	0.7339
30.0	0.70097	0.7011	0.7014	0.7019	0.7026	0.7047	0.7076	0.7112
35.0	0.68545	0.6855	0.6858	0.6861	0.6867	0.6883	0.6905	0.6932
40.0	0.67232	0.6724	0.6726	0.6728	0.6733	0.6745	0.6762	0.6784
50.0	0.65099	0.6510	0.6512	0.6513	0.6516	0.6524	0.6534	0.6546
75.0	0.61397	0.6141	0.6143	0.6145	0.6147	0.6148	0.6148	0.6147
100.0	0.58870	0.5889	0.5894	0.5900	0.5903	0.5901	0.5895	0.5885

TABLE C.5

Orientation averaged transport integral ratios for the Stockmayer potential*

(a) $\langle A^* \rangle$

T^* \ δ	0	0.25	0.50	0.75	1.0	1.5	2.0	2.5
0	1.0065	1.084	1.084	1.084	1.084	1.084	1.084	1.084
0.1	1.0231	1.066	1.038	1.040	1.043	1.050	1.052	1.051
0.2	1.0424	1.045	1.048	1.052	1.056	1.065	1.066	1.064
0.3	1.0719	1.067	1.060	1.055	1.058	1.068	1.071	1.071
0.4	1.0936	1.087	1.077	1.069	1.068	1.075	1.078	1.078
0.5	1.1053	1.098	1.088	1.080	1.078	1.082	1.084	1.084
0.6	1.1104	1.104	1.096	1.089	1.086	1.089	1.090	1.090
0.7	1.1114	1.107	1.100	1.095	1.093	1.095	1.096	1.095
0.8	1.1104	1.107	1.102	1.099	1.098	1.100	1.100	1.099
0.9	1.1086	1.106	1.102	1.101	1.101	1.105	1.105	1.104
1.0	1.1063	1.104	1.103	1.103	1.104	1.108	1.109	1.108
1.2	1.1020	1.102	1.103	1.105	1.107	1.112	1.115	1.115
1.4	1.0985	1.099	1.101	1.104	1.108	1.115	1.119	1.120
1.6	1.0960	1.096	1.099	1.103	1.108	1.116	1.121	1.124
1.8	1.0943	1.095	1.099	1.102	1.108	1.117	1.123	1.126
2.0	1.0934	1.094	1.097	1.102	1.107	1.116	1.123	1.128
2.5	1.0926	1.094	1.097	1.099	1.105	1.115	1.123	1.130
3.0	1.0934	1.095	1.097	1.099	1.104	1.113	1.122	1.129
3.5	1.0948	1.096	1.098	1.100	1.103	1.112	1.119	1.127
4.0	1.0965	1.097	1.099	1.101	1.104	1.110	1.118	1.126
5.0	1.0997	1.100	1.101	1.102	1.105	1.110	1.116	1.123
6.0	1.1025	1.103	1.104	1.105	1.106	1.110	1.115	1.121
7.0	1.1050	1.105	1.106	1.107	1.108	1.111	1.115	1.120
8.0	1.1072	1.107	1.108	1.108	1.109	1.112	1.115	1.119
9.0	1.1091	1.109	1.109	1.110	1.111	1.113	1.115	1.119
10.0	1.1107	1.111	1.111	1.111	1.112	1.114	1.116	1.119
12.0	1.1133	1.114	1.113	1.114	1.114	1.115	1.117	1.119
14.0	1.1154	1.115	1.116	1.116	1.116	1.117	1.118	1.120
16.0	1.1172	1.117	1.117	1.118	1.118	1.118	1.119	1.120
18.0	1.1186	1.119	1.119	1.119	1.119	1.119	1.120	1.121
20.0	1.1199	1.120	1.120	1.120	1.120	1.121	1.121	1.122
25.0	1.1223	1.122	1.122	1.122	1.122	1.123	1.123	1.124
30.0	1.1243	1.124	1.124	1.124	1.124	1.124	1.125	1.125
35.0	1.1259	1.126	1.126	1.126	1.126	1.126	1.126	1.126
40.0	1.1273	1.127	1.127	1.127	1.127	1.127	1.127	1.128
50.0	1.1297	1.130	1.130	1.130	1.130	1.130	1.130	1.129
75.0	1.1339	1.134	1.134	1.135	1.135	1.134	1.134	1.132
100.0	1.1364	1.137	1.137	1.138	1.139	1.138	1.137	1.135
∞	1.14187	1.14187	1.14187	1.14187	1.14187	1.14187	1.14187	1.14187

* These are ratios of orientation averaged transport integrals similar to those presented in Table C.4. Thus,

$$\langle A^* \rangle = \langle \Omega^{(2,2)*} \rangle / \langle \Omega^{(1,1)*} \rangle, \qquad \langle E^* \rangle = \langle \Omega^{(2,3)*} \rangle / \langle \Omega^{(2,2)*} \rangle,$$

$$\langle B^* \rangle = [5\langle \Omega^{(1,2)*} \rangle - 4\langle \Omega^{(1,3)*} \rangle] / \langle \Omega^{(1,1)*} \rangle, \qquad \langle F^* \rangle = \langle \Omega^{(3,3)*} \rangle / \langle \Omega^{(1,1)*} \rangle.$$

$$\langle C^* \rangle = \langle \Omega^{(1,2)*} \rangle / \langle \Omega^{(1,1)*} \rangle,$$

For $\delta = 0$, they reduce to the transport integral ratios for the Lennard-Jones potential. This table has been adapted from Monchick and Mason [1961].

TABLE C.5 (continued)

(b) ⟨B*⟩

T^* \ δ	0	0.25	0.50	0.75	1.0	1.5	2.0	2.5
0	1.1852	1.2963	1.2963	1.2963	1.2963	1.2963	1.2963	1.2963
0.1	1.1960	1.216	1.237	1.269	1.285	1.290	1.297	1.294
0.2	1.2451	1.257	1.340	1.389	1.366	1.327	1.314	1.278
0.3	1.2900	1.294	1.272	1.258	1.262	1.282	1.290	1.299
0.4	1.2986	1.291	1.284	1.278	1.277	1.288	1.294	1.297
0.5	1.2865	1.281	1.276	1.272	1.277	1.286	1.292	1.298
0.6	1.2665	1.264	1.261	1.263	1.269	1.284	1.292	1.298
0.7	1.2455	1.244	1.248	1.255	1.262	1.278	1.289	1.296
0.8	1.2253	1.225	1.234	1.240	1.252	1.271	1.284	1.295
0.9	1.2078	1.210	1.216	1.227	1.242	1.264	1.281	1.292
1.0	1.1919	1.192	1.205	1.216	1.230	1.256	1.273	1.287
1.2	1.1678	1.172	1.181	1.195	1.209	1.237	1.261	1.277
1.4	1.1496	1.155	1.161	1.174	1.189	1.221	1.246	1.266
1.6	1.1366	1.141	1.147	1.159	1.174	1.202	1.231	1.256
1.8	1.1270	1.130	1.138	1.148	1.162	1.191	1.218	1.242
2.0	1.1197	1.122	1.129	1.140	1.149	1.178	1.205	1.231
2.5	1.1080	1.110	1.116	1.122	1.132	1.154	1.180	1.205
3.0	1.1016	1.103	1.107	1.112	1.120	1.138	1.160	1.183
3.5	1.0980	1.099	1.102	1.106	1.112	1.127	1.145	1.165
4.0	1.0958	1.097	1.099	1.102	1.107	1.119	1.135	1.153
5.0	1.0935	1.094	1.095	1.097	1.100	1.109	1.120	1.134
6.0	1.0925	1.092	1.094	1.095	1.098	1.104	1.112	1.122
7.0	1.0922	1.092	1.093	1.094	1.096	1.100	1.106	1.115
8.0	1.0922	1.092	1.093	1.093	1.095	1.098	1.103	1.110
9.0	1.0923	1.092	1.093	1.093	1.094	1.097	1.101	1.106
10.0	1.0923	1.092	1.092	1.093	1.094	1.096	1.099	1.103
12.0	1.0927	1.093	1.093	1.093	1.094	1.095	1.098	1.101
14.0	1.0930	1.093	1.093	1.093	1.094	1.094	1.096	1.099
16.0	1.0933	1.094	1.093	1.094	1.094	1.095	1.096	1.098
18.0	1.0937	1.093	1.094	1.094	1.094	1.094	1.096	1.097
20.0	1.0939	1.094	1.094	1.094	1.094	1.095	1.095	1.097
25.0	1.0943	1.094	1.094	1.094	1.095	1.095	1.096	1.096
30.0	1.0944	1.095	1.094	1.094	1.094	1.095	1.095	1.096
35.0	1.0944	1.094	1.095	1.094	1.094	1.095	1.096	1.096
40.0	1.0943	1.095	1.094	1.094	1.095	1.095	1.095	1.095
50.0	1.0941	1.094	1.094	1.094	1.094	1.094	1.094	1.096
75.0	1.0947	1.095	1.094	1.094	1.093	1.093	1.094	1.095
100.0	1.0957	1.095	1.094	1.093	1.092	1.093	1.093	1.094
∞	1.10185	1.10185	1.10185	1.10185	1.10185	1.10185	1.10185	1.10185

TABLE C.5 (continued)

(c) ⟨c*⟩

T \ δ	0	0.25	0.50	0.75	1.0	1.5	2.0	2.5
0	0.8889	0.77778	0.77778	0.77778	0.77778	0.77778	0.77778	0.77778
0.1	0.88575	0.8988	0.8378	0.8029	0.7876	0.7805	0.7799	0.7801
0.2	0.87268	0.8692	0.8647	0.8479	0.8237	0.7975	0.7881	0.7784
0.3	0.85182	0.8525	0.8386	0.8198	0.8054	0.7903	0.7839	0.7820
0.4	0.83542	0.8362	0.8306	0.8196	0.8076	0.7918	0.7842	0.7808
0.5	0.82629	0.8278	0.8252	0.8169	0.8074	0.7916	0.7838	0.7802
0.6	0.82299	0.8249	0.8230	0.8165	0.8072	0.7922	0.7839	0.7798
0.7	0.82357	0.8257	0.8241	0.8178	0.8084	0.7927	0.7839	0.7794
0.8	0.82657	0.8280	0.8264	0.8199	0.8107	0.7939	0.7842	0.7796
0.9	0.83110	0.8324	0.8295	0.8228	0.8136	0.7960	0.7854	0.7798
1.0	0.83630	0.8366	0.8342	0.8267	0.8168	0.7986	0.7864	0.7805
1.2	0.84762	0.8474	0.8438	0.8358	0.8250	0.8041	0.7904	0.7822
1.4	0.85846	0.8583	0.8530	0.8444	0.8336	0.8118	0.7957	0.7854
1.6	0.86840	0.8674	0.8619	0.8531	0.8423	0.8186	0.8011	0.7998
1.8	0.87713	0.8755	0.8709	0.8616	0.8504	0.8265	0.8072	0.7939
2.0	0.88479	0.8831	0.8779	0.8695	0.8578	0.8338	0.8133	0.7990
2.5	0.89972	0.8986	0.8936	0.8846	0.8742	0.8504	0.8294	0.8125
3.0	0.91028	0.9089	0.9043	0.8967	0.8869	0.8649	0.8438	0.8253
3.5	0.91793	0.9166	0.9125	0.9058	0.8970	0.8768	0.8557	0.8372
4.0	0.92371	0.9226	0.9189	0.9128	0.9050	0.8861	0.8664	0.8484
5.0	0.93135	0.9304	0.9274	0.9226	0.9164	0.9006	0.8833	0.8662
6.0	0.93607	0.9353	0.9329	0.9291	0.9240	0.9109	0.8958	0.8802
7.0	0.93927	0.9387	0.9366	0.9334	0.9292	0.9182	0.9050	0.8911
8.0	0.94149	0.9409	0.9393	0.9366	0.9331	0.9236	0.9122	0.8997
9.0	0.94306	0.9426	0.9412	0.9388	0.9357	0.9276	0.9175	0.9065
10.0	0.94419	0.9437	0.9425	0.9406	0.9380	0.9308	0.9219	0.9119
12.0	0.94571	0.9455	0.9445	0.9430	0.9409	0.9353	0.9283	0.9201
14.0	0.94662	0.9464	0.9456	0.9444	0.9428	0.9382	0.9325	0.9258
16.0	0.94723	0.9471	0.9464	0.9455	0.9442	0.9405	0.9355	0.9298
18.0	0.94764	0.9474	0.9469	0.9462	0.9450	0.9418	0.9378	0.9328
20.0	0.94790	0.9478	0.9474	0.9465	0.9457	0.9430	0.9394	0.9352
25.0	0.94827	0.9481	0.9480	0.9472	0.9467	0.9447	0.9422	0.9391
30.0	0.94842	0.9484	0.9481	0.9478	0.9472	0.9458	0.9437	0.9415
35.0	0.94852	0.9484	0.9483	0.9480	0.9475	0.9465	0.9449	0.9430
40.0	0.94861	0.9487	0.9484	0.9481	0.9479	0.9468	0.9455	0.9438
50.0	0.94872	0.9486	0.9486	0.9483	0.9482	0.9475	0.9464	0.9452
75.0	0.94881	0.9488	0.9489	0.9490	0.8487	0.9482	0.9476	0.9468
100.0	0.94863	0.9487	0.9489	0.9491	0.9493	0.9491	0.9483	0.9476
∞	0.94444	0.94444	0.94444	0.94444	0.94444	0.94444	0.94444	0.94444

TABLE C.5 (continued)

(d) $\langle E^* \rangle$

T^* \ δ	0	0.25	0.50	0.75	1.0	1.5	2.0	2.5
0	0.91667	0.83333	0.83333	0.83333	0.83333	0.83333	0.83333	0.83333
0.1	0.91652	0.8669	0.8258	0.8015	0.7970	0.8081	0.8182	0.8247
0.2	0.91829	0.9129	0.8877	0.8643	0.8511	0.8445	0.8397	0.8409
0.3	0.90774	0.9073	0.8910	0.8715	0.8590	0.8457	0.8426	0.8418
0.4	0.89143	0.8906	0.8859	0.8763	0.8652	0.8500	0.8440	0.8413
0.5	0.87857	0.8810	0.8789	0.8728	0.8649	0.8510	0.8446	0.8416
0.6	0.87081	0.8741	0.8751	0.8725	0.8659	0.8529	0.8456	0.8418
0.7	0.86743	0.8706	0.8731	0.8711	0.8656	0.8535	0.8461	0.8422
0.8	0.86722	0.8702	0.8728	0.8707	0.8662	0.8544	0.8476	0.8435
0.9	0.86899	0.8721	0.8736	0.8714	0.8663	0.8547	0.8475	0.8435
1.0	0.87212	0.8745	0.8753	0.8725	0.8672	0.8555	0.8480	0.8438
1.2	0.88028	0.8812	0.8810	0.8767	0.8707	0.8578	0.8496	0.8450
1.4	0.88909	0.8894	0.8879	0.8830	0.8756	0.8616	0.8517	0.8461
1.6	0.89758	0.8976	0.8940	0.8886	0.8804	0.8650	0.8542	0.8475
1.8	0.90515	0.9041	0.9007	0.8943	0.8861	0.8695	0.8568	0.8494
2.0	0.91188	0.9105	0.9065	0.9001	0.8914	0.8739	0.8609	0.8524
2.5	0.92509	0.9236	0.9205	0.9130	0.9044	0.8857	0.8704	0.8593
3.0	0.93431	0.9329	0.9289	0.9230	0.9141	0.8962	0.8802	0.8669
3.5	0.94095	0.9394	0.9358	0.9305	0.9228	0.9056	0.8893	0.8749
4.0	0.94573	0.9446	0.9414	0.9361	0.9293	0.9139	0.8977	0.8835
5.0	0.95202	0.9511	0.9485	0.9454	0.9388	0.9252	0.9112	0.8974
6.0	0.95577	0.9550	0.9529	0.9496	0.9451	0.9338	0.9210	0.9082
7.0	0.95812	0.9574	0.9558	0.9531	0.9493	0.9399	0.9287	0.9169
8.0	0.95970	0.9592	0.9578	0.9555	0.9524	0.9443	0.9345	0.9240
9.0	0.96077	0.9604	0.9591	0.9572	0.9546	0.9476	0.9391	0.9296
10.0	0.96153	0.9611	0.9602	0.9585	0.9563	0.9502	0.9426	0.9341
12.0	0.96246	0.9621	0.9614	0.9602	0.9584	0.9537	0.9477	0.9409
14.0	0.96297	0.9628	0.9622	0.9611	0.9599	0.9561	0.9511	0.9453
16.0	0.96327	0.9631	0.9626	0.9617	0.9607	0.9577	0.9535	0.9488
18.0	0.96340	0.9633	0.9628	0.9621	0.9613	0.9588	0.9553	0.9513
20.0	0.96348	0.9634	0.9631	0.9626	0.9617	0.9595	0.9567	0.9530
25.0	0.96357	0.9635	0.9632	0.9628	0.9623	0.9607	0.9587	0.9561
30.0	0.96365	0.9635	0.9634	0.9631	0.9627	0.9615	0.9599	0.9580
35.0	0.96373	0.9637	0.9635	0.9633	0.9629	0.9619	0.9606	0.9590
40.0	0.96383	0.9637	0.9636	0.9636	0.9632	0.9623	0.9611	0.9596
50.0	0.96395	0.9641	0.9639	0.9639	0.9636	0.9629	0.9619	0.9606
75.0	0.96373	0.9638	0.9640	0.9645	0.9644	0.9641	0.9631	0.9618
100.0	0.96316	0.9635	0.9642	0.9649	0.9651	0.9649	0.9639	0.9624
∞	0.95833	0.95833	0.95833	0.95833	0.95833	0.95833	0.95833	0.95833

TABLE C.5 (continued)

(e) $\langle F^* \rangle$

T^*	0	0.25	0.50	0.75	1.0	1.5	2.0	2.5
0	0.9417	0.8812	0.8812	0.8812	0.8812	0.8812	0.8812	0.8812
0.1	0.9361	0.9500	0.8928	0.8696	0.8602	0.8594	0.8594	0.8592
0.2	0.9211	0.9134	0.9004	0.8861	0.8761	0.8702	0.8668	0.8645
0.3	0.9063	0.9040	0.8949	0.8831	0.8756	0.8703	0.8679	0.8665
0.4	0.8992	0.8974	0.8909	0.8814	0.8748	0.8698	0.8678	0.8669
0.5	0.8969	0.8953	0.8897	0.8815	0.8754	0.8693	0.8676	0.8668
0.6	0.8972	0.8960	0.8909	0.8840	0.8773	0.8706	0.8677	0.8667
0.7	0.8990	0.8982	0.8939	0.8870	0.8806	0.8719	0.8686	0.8668
0.8	0.9018	0.9007	0.8970	0.8915	0.8844	0.8744	0.8694	0.8673
0.9	0.9053	0.9050	0.9008	0.8953	0.8884	0.8777	0.8714	0.8683
1.0	0.9091	0.9083	0.9054	0.8996	0.8925	0.8811	0.8734	0.8697
1.2	0.9171	0.9165	0.9142	0.9088	0.9019	0.8884	0.8788	0.8731
1.4	0.9250	0.9251	0.9222	0.9169	0.9103	0.8964	0.8852	0.8779
1.6	0.9324	0.9320	0.9293	0.9246	0.9181	0.9035	0.8914	0.8834
1.8	0.9392	0.9386	0.9363	0.9308	0.9248	0.9103	0.8974	0.8881
2.0	0.9454	0.9445	0.9419	0.9374	0.9314	0.9169	0.9035	0.8932
2.5	0.9580	0.9579	0.9554	0.9501	0.9444	0.9299	0.9171	0.9062
3.0	0.9677	0.9671	0.9646	0.9603	0.9545	0.9418	0.9289	0.9170
3.5	0.9751	0.9745	0.9721	0.9682	0.9631	0.9513	0.9385	0.9274
4.0	0.9810	0.9804	0.9783	0.9748	0.9700	0.9586	0.9468	0.9362
5.0	0.9894	0.9889	0.9871	0.9842	0.9803	0.9705	0.9600	0.9496
6.0	0.9951	0.9947	0.9932	0.9908	0.9875	0.9793	0.9700	0.9604
7.0	0.9993	0.9990	0.9977	0.9956	0.9928	0.9857	0.9775	0.9689
8.0	1.0025	1.002	1.001	0.9992	0.9969	0.9908	0.9835	0.9756
9.0	1.0049	1.005	1.004	1.002	1.000	0.9946	0.9881	0.9810
10.0	1.0068	1.007	1.006	1.004	1.003	0.9979	0.9920	0.9855
12.0	1.0097	1.010	1.009	1.008	1.006	1.002	0.9979	0.9925
14.0	1.0117	1.012	1.011	1.010	1.009	1.006	1.002	0.9975
16.0	1.0133	1.013	1.013	1.012	1.011	1.009	1.005	1.001
18.0	1.0145	1.014	1.014	1.013	1.013	1.010	1.008	1.004
20.0	1.0155	1.015	1.015	1.015	1.014	1.012	1.010	1.007
25.0	1.0174	1.017	1.017	1.017	1.016	1.015	1.013	1.011
30.0	1.0188	1.019	1.019	1.018	1.018	1.017	1.015	1.014
35.0	1.0201	1.020	1.020	1.020	1.019	1.018	1.017	1.016
40.0	1.0213	1.021	1.021	1.021	1.021	1.020	1.018	1.017
50.0	1.0231	1.023	1.023	1.023	1.023	1.022	1.021	1.019
75.0	1.0261	1.026	1.027	1.027	1.027	1.026	1.024	1.022
100.0	1.0274	1.028	1.029	1.030	1.031	1.030	1.027	1.023
∞	1.02580	1.02580	1.02580	1.02580	1.02580	1.02580	1.02580	1.02580

TABLE C.6

Transport integrals and ratios for the Morse potential, Eq. (8.4-9)*

T^*	$\Omega^{(1,1)*}$	$\Omega^{(2,2)*}$	A*	B*	C*	E*	F*
			$C = 1.0$				
0.004	81.543	68.632	0.8417	1.1212	0.9458	0.9421	0.9558
0.010	68.217	54.950	0.8055	1.1592	0.9240	0.9358	0.9169
0.020	57.238	45.519	0.7953	1.1819	0.9074	0.9280	0.8954
0.040	46.596	36.871	0.7913	1.1900	0.8962	0.9213	0.8850
0.100	32.674	28.463	0.8711	1.4321	0.8197	0.9284	0.7885
0.200	19.597	20.628	1.0526	1.5130	0.6898	0.8227	0.7446
0.400	9.169	10.731	1.1704	1.4219	0.5864	0.7109	0.7366
			$C = 2.0$				
0.004	21.3303	21.6425	1.0146	1.0932	0.9523	0.9480	1.0147
0.010	18.5383	17.7060	0.9551	1.1151	0.9443	0.9423	0.9870
0.020	16.3252	14.9869	0.9180	1.1430	0.9324	0.9375	0.9606
0.040	13.9429	12.5350	0.8990	1.1736	0.9153	0.9335	0.9325
0.100	10.6486	9.7379	0.9145	1.2757	0.8811	0.9327	0.8855
0.200	7.6296	7.8154	1.0243	1.4478	0.7904	0.8912	0.8205
0.400	4.4535	5.0400	1.1317	1.4268	0.6923	0.7953	0.7965
1.000	1.8140	2.1403	1.1799	1.2793	0.6776	0.7637	0.8205
2.000	0.9790	1.1794	1.2047	1.2398	0.7285	0.8104	0.8761
4.000	0.5830	0.7362	1.2627	1.2800	0.7687	0.8431	0.9365
			$C = 4.0$				
0.004	7.0039	7.9153	1.1301	1.0956	0.9497	0.9570	1.0395
0.010	6.0782	6.7197	1.1055	1.0991	0.9473	0.9537	1.0308
0.020	5.4385	5.8879	1.0826	1.1040	0.9455	0.9508	1.0223
0.040	4.8363	5.1122	1.0570	1.1168	0.9411	0.9472	1.0097
0.100	4.0509	4.1757	1.0308	1.1497	0.9284	0.9424	0.9866
0.200	3.3969	3.5466	1.0441	1.2611	0.8948	0.9377	0.9405
0.400	2.5430	2.8232	1.1102	1.3457	0.8262	0.8888	0.8955
1.000	1.4859	1.6903	1.1376	1.2354	0.8046	0.8504	0.8956
2.000	1.0321	1.1708	1.1344	1.1633	0.8491	0.8891	0.9348
4.000	0.7889	0.9084	1.1514	1.1498	0.8867	0.9239	0.9761
10.000	0.5940	0.7071	1.1904	1.1713	0.9010	0.9343	1.0080
20.000	0.4808	0.5856	1.2180	1.1944	0.8941	0.9286	1.0164
40.000	0.3806	0.4740	1.2453	1.2204	0.8802	0.9182	1.0193
100.000	0.2645	0.3394	1.2831	1.2637	0.8534	0.8984	1.0168
200.000	0.1896	0.2491	1.3136	1.3052	0.8252	0.8775	1.0094

* These calculations have been done quantum mechanically; for large T^* they are equivalent to the classical results. In this table, $C = \alpha\sigma$, where α is the parameter appearing in Eq. (8.4-9) and a/σ has been set equal to $1+(\ln 2)/C$. This table has been adapted from Smith and Munn [1964 .

TABLE C.6 (continued)

T^*	$\Omega^{(1,1)*}$	$\Omega^{(2,2)*}$	A^*	B^*	C^*	E^*	F^*
			$C = 6.0$				
0.004	4.2323	4.8452	1.1448	1.0905	0.9521	0.9637	1.0390
0.010	3.7030	4.2226	1.1403	1.0933	0.9506	0.9613	1.0381
0.020	3.3374	3.7824	1.1333	1.0951	0.9494	0.9592	1.0362
0.040	3.0005	3.3665	1.1220	1.0979	0.9481	0.9567	1.0323
0.100	2.5911	2.8514	1.1005	1.1096	0.9443	0.9525	1.0206
0.200	2.2879	2.4906	1.0886	1.1500	0.9335	0.9502	1.0004
0.400	1.9261	2.1349	1.1084	1.2434	0.8963	0.9317	0.9617
1.000	1.3598	1.5224	1.1195	1.2001	0.8649	0.8934	0.9374
2.000	1.0522	1.1611	1.1035	1.1303	0.8917	0.9171	0.9549
4.000	0.8685	0.9629	1.1087	1.1068	0.9221	0.9452	0.9843
10.000	0.7193	0.8122	1.1292	1.1104	0.9368	0.9579	1.0083
20.000	0.6310	0.7228	1.1456	1.1196	0.9363	0.9572	1.0168
40.000	0.5504	0.6389	1.1609	1.1304	0.9318	0.9535	1.0211
100.000	0.4512	0.5326	1.1804	1.1464	0.9232	0.9469	1.0240
200.000	0.38157	0.4561	1.1954	1.1601	0.9153	0.9411	1.0251
			$C = 8.0$				
0.004	3.1590	3.5914	1.1368	1.0828	0.9571	0.9682	1.0383
0.010	2.7997	3.1867	1.1382	1.0855	0.9552	0.9666	1.0376
0.020	2.5475	2.8992	1.1381	1.0868	0.9541	0.9652	1.0369
0.040	2.3136	2.6269	1.1354	1.0876	0.9534	0.9637	1.0359
0.100	2.0331	2.2896	1.1262	1.0897	0.9525	0.9612	1.0328
0.200	1.8382	2.0481	1.1142	1.1023	0.9497	0.9584	1.0247
0.400	1.6330	1.8131	1.1103	1.1650	0.9322	0.9509	0.9982
1.000	1.2870	1.4217	1.1047	1.1721	0.9008	0.9202	0.9624
2.000	1.0572	1.1513	1.0890	1.1121	0.9158	0.9322	0.9677
4.000	0.9097	0.9874	1.0854	1.0849	0.9389	0.9555	0.9871
10.000	0.7872	0.8638	1.0973	1.0818	0.9526	0.9683	1.0062
20.000	0.7149	0.7928	1.1090	1.0866	0.9541	0.9691	1.0142
40.000	0.6488	0.7262	1.1193	1.0925	0.9522	0.9673	1.0182
100.000	0.5659	0.6404	1.1316	1.1009	0.9481	0.9639	1.0213
200.000	0.5060	0.5771	1.1406	1.1076	0.9443	0.9611	1.0227
			$C = 20.0$				
0.004	1.6882	1.8187	1.0772	1.0492	0.9748	0.9817	1.0183
0.010	1.5739	1.6986	1.0793	1.0504	0.9742	0.9811	1.0188
0.020	1.4908	1.6105	1.0803	1.0512	0.9737	0.9804	1.0187
0.040	1.4109	1.5246	1.0805	1.0512	0.9734	0.9801	1.0185
0.100	1.3117	1.4171	1.0803	1.0498	0.9737	0.9800	1.0190
0.200	1.2428	1.3410	1.0790	1.0476	0.9745	0.9801	1.0192
0.400	1.1800	1.2692	1.0756	1.0450	0.9757	0.9801	1.0184
1.000	1.1030	1.1741	1.0644	1.0601	0.9734	0.9755	1.0069
2.000	1.0364	1.0821	1.0440	1.0679	0.9674	0.9649	0.9829
4.000	0.9673	0.9783	1.0114	1.0481	0.9679	0.9709	0.9631
10.000	0.9061	0.9359	1.0329	1.0232	0.9838	0.9955	1.0054
20.000	0.8783	0.9174	1.0446	1.0320	0.9845	0.9901	1.0096
40.000	0.8488	0.8892	1.0476	1.0343	0.9834	0.9883	1.0098
100.000	0.8095	0.8502	1.0503	1.0354	0.9823	0.9874	1.0105
200.000	0.7798	0.8205	1.0522	1.0362	0.9817	0.9870	1.0112

Appendix D
Evaluation of certain integrals in Enskog's theory of dense gases

In this appendix we evaluate various integrals which were encountered in the presentation of Enskog's theory of dense gases. The notation is the same as in Chapter 12.

D.1. *Auxiliary results.* From Chapter 12 we recall that if two rigid spherical molecules with the velocities $\boldsymbol{c}$ and $\boldsymbol{c}_1$ collide and emerge from the collision with the velocities $\boldsymbol{c}'$ and $\boldsymbol{c}_1'$, respectively, then

$$\boldsymbol{c}' = \boldsymbol{c}+\boldsymbol{k}(\boldsymbol{g}\cdot\boldsymbol{k}) \quad \text{and} \quad \boldsymbol{c}_1' = \boldsymbol{c}_1-\boldsymbol{k}(\boldsymbol{g}\cdot\boldsymbol{k}),$$

where $\boldsymbol{g}$ is the relative velocity, $\boldsymbol{g} = \boldsymbol{c}_1-\boldsymbol{c}$, and $\boldsymbol{k}$ is the unit vector along the apse line, i.e., the line joining the centers of the two molecules. Since the peculiar velocities of the molecules involved in a collision are taken relative to the hydrodynamic velocity of the gas at the special point $\boldsymbol{r}$ under consideration – which, therefore, is the same for both molecules – we have also

$$\boldsymbol{C}' = \boldsymbol{C}+\boldsymbol{k}(\boldsymbol{g}\cdot\boldsymbol{k}) \quad \text{and} \quad \boldsymbol{C}_1' = \boldsymbol{C}_1-\boldsymbol{k}(\boldsymbol{g}\cdot\boldsymbol{k}).$$

In the following we shall frequently work in terms of the center-of-mass and relative velocities in a coordinate system moving with the gas, $\boldsymbol{G}_0$ and $\boldsymbol{g}$, respectively – i.e., $\boldsymbol{G}_0 = \frac{1}{2}(\boldsymbol{C}+\boldsymbol{C}_1)$ and $\boldsymbol{g} = \boldsymbol{C}_1-\boldsymbol{C}$. It is readily verified that the Jacobian of the transformation of variables $(\boldsymbol{c}, \boldsymbol{c}_1) \Rightarrow (\boldsymbol{G}_0, \boldsymbol{g})$ is equal to unity.

We will need the following integrals with respect to $\boldsymbol{k}$, evaluated over the hemisphere for which $\boldsymbol{g}\cdot\boldsymbol{k}$ is positive.

$$\int (\boldsymbol{g}\cdot\boldsymbol{k})\boldsymbol{k}\,\mathrm{d}^2k = \tfrac{2}{3}\pi\boldsymbol{g},$$

$$\int (\boldsymbol{g}\cdot\boldsymbol{k})^2\boldsymbol{kk}\,\mathrm{d}^2k = \tfrac{2}{15}\pi(2\boldsymbol{gg}+g^2\mathsf{I}),$$

$$\int (\boldsymbol{g}\cdot\boldsymbol{k})^3\boldsymbol{k}\,\mathrm{d}^2k = \tfrac{2}{5}\pi g^2\boldsymbol{g},$$

$$\int (\boldsymbol{a}\cdot\boldsymbol{k})(\boldsymbol{g}\cdot\boldsymbol{k})^2\boldsymbol{kk}\,\mathrm{d}^2k = \tfrac{1}{12}\pi[(\boldsymbol{a}\cdot\boldsymbol{g})(\boldsymbol{gg}+g^2\mathsf{I})/g+g(\boldsymbol{ag}+\boldsymbol{ga})].$$

A derivation of the first, second and fourth identity may be found in Chapman and Cowling [1952], p. 280; the third identity follows from the second one simply by taking the scalar product with $\boldsymbol{g}$.

D.2. *The integral* $\mathsf{P}_\Phi^{(0)}$

From the definition (12.4-5) we have

$$\mathsf{P}_\Phi^{(0)} = \tfrac{1}{2}\sigma^3\chi\iiint m(\boldsymbol{c}'-\boldsymbol{c})\,f^{(0)}f_1^{(0)}(\boldsymbol{g}\cdot\boldsymbol{k})\boldsymbol{k}\,\mathrm{d}^2k\,\mathrm{d}^3c\,\mathrm{d}^3c_1\,.$$

Expressing $\boldsymbol{c}'$ in terms of the variables of integration we have

$$\mathsf{P}_\Phi^{(0)} = \tfrac{1}{2}\sigma^3\chi\iiint mf^{(0)}f_1^{(0)}(\boldsymbol{g}\cdot\boldsymbol{k})^2\boldsymbol{kk}\,\mathrm{d}^2k\,\mathrm{d}^3c\,\mathrm{d}^3c_1$$

or, evaluating the integral over $\boldsymbol{k}$,

$$\mathsf{P}_\Phi^{(0)} = \tfrac{1}{15}\pi\sigma^3\chi m\iint f^{(0)}f_1^{(0)}(2\boldsymbol{gg}+g^2\mathsf{I})\,\mathrm{d}^3c\,\mathrm{d}^3c_1\,.$$

Then, substituting the explicit form of $f^{(0)}$ and $f_1^{(0)}$ and changing the variables of integration from $(\boldsymbol{c}, \boldsymbol{c}_1)$ to $(\boldsymbol{G}_0, \boldsymbol{g})$, we have

$$\mathsf{P}_\Phi^{(0)} = \tfrac{1}{15}\pi\sigma^3\chi mn^2\left(\frac{m}{2\pi kT}\right)^3\int\exp\left(-\frac{mG_0^2}{kT}\right)\mathrm{d}^3G_0 \times\int(2\boldsymbol{gg}+g^2\mathsf{I})\exp\left(-\frac{mg^2}{4kT}\right)\mathrm{d}^3g$$

which can be integrated directly. The result is

$$\mathsf{P}_\Phi^{(0)} = \tfrac{2}{3}\pi\sigma^3\chi n^2kT\mathsf{I}.$$

D.3. *The integral* $\boldsymbol{q}_\Phi^{(0)}$

From the definition (12.4-8) we have

$$\boldsymbol{q}_\Phi^{(0)} = \tfrac{1}{2}\sigma^3\chi \iiint \tfrac{1}{2}m(C'^2 - C^2) f^{(0)} f_1^{(0)} (\boldsymbol{g}\cdot\boldsymbol{k})\boldsymbol{k}\, \mathrm{d}^2k\, \mathrm{d}^3c\, \mathrm{d}^3c_1 .$$

Again, we first express C'^2 in terms of the variables of integration. One verifies easily that $C'^2 = C^2 + 2(\boldsymbol{C}\cdot\boldsymbol{k})(\boldsymbol{g}\cdot\boldsymbol{k}) + (\boldsymbol{g}\cdot\boldsymbol{k})^2$, so

$$\boldsymbol{q}_\Phi^{(0)} = \tfrac{1}{2}\sigma^3\chi \iiint \tfrac{1}{2}m f^{(0)} f_1^{(0)} (2\boldsymbol{C}+\boldsymbol{g})\cdot\boldsymbol{k}(\boldsymbol{g}\cdot\boldsymbol{k})^2 \boldsymbol{k}\, \mathrm{d}^2k\, \mathrm{d}^3c\, \mathrm{d}^3c_1$$

or, in terms of the center-of-mass velocity $\boldsymbol{G}_0$ and the relative velocity $\boldsymbol{g}$,

$$\boldsymbol{q}_\Phi^{(0)} = \tfrac{1}{4}\sigma^3\chi m n^2 \left(\frac{m}{2\pi kT}\right)^3 \iiint \exp\{-mG_0^2/kT - mg^2/4kT\}$$
$$\times 2(\boldsymbol{G}_0\cdot\boldsymbol{k})(\boldsymbol{g}\cdot\boldsymbol{k})^2 \boldsymbol{k}\, \mathrm{d}^2k\, \mathrm{d}^3G_0\, \mathrm{d}^3g\cdot$$

Here, the integrand is odd in the components of $\boldsymbol{G}_0$. The integral therefore vanishes upon integration and we have

$$\boldsymbol{q}_\Phi^{(0)} = 0.$$

D.4. *The integral* $J_1(f^{(0)}f^{(0)})$

From the definition (12.5-8) we have

$$J_1(f^{(0)}f^{(0)})$$
$$= \sigma^3 f^{(0)} \iint f_1^{(0)} [\chi \boldsymbol{k}\cdot\nabla_r \log f_1^{(0)} f_1^{(0)\prime} + \boldsymbol{k}\cdot\nabla_r\chi](\boldsymbol{g}\cdot\boldsymbol{k})\, \mathrm{d}^2k\, \mathrm{d}^3c_1,$$

or, after substituting the values of $f_1^{(0)}$ and $f_1^{(0)\prime}$ under the logarithm,

$$J_1(f^{(0)}f^{(0)}) = \sigma^3 f^{(0)} \iint f_1^{(0)} \{\chi[2\nabla \log n - 3\nabla \log T$$
$$+ \frac{m}{2kT}(C_1^2 + C_1'^2)\nabla \log T + \frac{m}{kT}(\boldsymbol{C}_1 + \boldsymbol{C}_1')\cdot\nabla\boldsymbol{v}] + \nabla\chi\}\cdot\boldsymbol{k}(\boldsymbol{g}\cdot\boldsymbol{k})\, \mathrm{d}^2k\, \mathrm{d}^3c_1 .$$

Expressing $\boldsymbol{C}_1'$ and $C_1'^2$ in terms of the variables of integration we see that

$$J_1(f^{(0)}f^{(0)}) = \sigma^3 f^{(0)} \iint f_1^{(0)} \{\chi[2\nabla \log n - 3\nabla \log T$$
$$+ \frac{m}{2kT}(2C_1^2 - 2(\boldsymbol{C}_1\cdot\boldsymbol{k})(\boldsymbol{g}\cdot\boldsymbol{k}) + (\boldsymbol{g}\cdot\boldsymbol{k})^2)\nabla \log T$$
$$+ \frac{m}{kT}(2\boldsymbol{C}_1 - (\boldsymbol{g}\cdot\boldsymbol{k})\boldsymbol{k})\cdot\nabla\boldsymbol{v}] + \nabla\chi\}\cdot\boldsymbol{k}(\boldsymbol{g}\cdot\boldsymbol{k})\, \mathrm{d}^2k\, \mathrm{d}^3c_1 .$$

Then, performing the integration over $\boldsymbol{k}$, we obtain

$$J_1(f^{(0)}f^{(0)}) = \tfrac{2}{3}\pi\sigma^3 f^{(0)} \int f_1^{(0)}\{\chi[2\boldsymbol{g}\cdot\nabla\log n - 3\boldsymbol{g}\cdot\nabla\log T$$

$$+\frac{m}{2kT}(2C_1^2\boldsymbol{g} - \tfrac{2}{5}\boldsymbol{C}_1\cdot(2\boldsymbol{g}\boldsymbol{g}+g^2\mathbf{I})+\tfrac{3}{5}g^2\boldsymbol{g})\cdot\nabla\log T$$

$$+\frac{m}{kT}(2\boldsymbol{C}_1\boldsymbol{g} - \tfrac{1}{5}(2\boldsymbol{g}\boldsymbol{g}+g^2\mathbf{I})) : \nabla\boldsymbol{v}] + \boldsymbol{g}\cdot\nabla\chi\}\,\mathrm{d}^3c_1 .$$

Since $\boldsymbol{g} = \boldsymbol{C}_1 - \boldsymbol{C}$, the evaluation of the integral over $\boldsymbol{c}_1$ is now straightforward. The result is

$$J_1(f^{(0)}f^{(0)}) = -\tfrac{2}{3}\pi n\sigma^3 f^{(0)}\left\{\chi\left[2\boldsymbol{C}\cdot\nabla\log n + \left(\frac{3mC^2}{10kT} - \tfrac{1}{2}\right)\boldsymbol{C}\cdot\nabla\log T\right.\right.$$

$$\left.\left. + \frac{2m}{5kT}\boldsymbol{C}\boldsymbol{C} : \nabla\boldsymbol{v} + \left(\frac{mC^2}{5kT} - 1\right)\nabla\cdot\boldsymbol{v}\right] + \boldsymbol{C}\cdot\nabla\chi\right\} .$$

D.5. *The integral* P_Φ

From the definition (12.5-17) we have

$$\mathsf{P}_\Phi = \tfrac{1}{2}\sigma^3\chi\iiint m(\boldsymbol{c}'-\boldsymbol{c})ff_1(\boldsymbol{g}\cdot\boldsymbol{k})\boldsymbol{k}\,\mathrm{d}^2k\,\mathrm{d}^3c\,\mathrm{d}^3c_1$$

$$+\tfrac{1}{4}\sigma^4\chi\iiint m(\boldsymbol{c}'-\boldsymbol{c})\,f^{(0)}f_1^{(0)}\left(\boldsymbol{k}\cdot\nabla_r\log\frac{f^{(0)}}{f_1^{(0)}}\right)(\boldsymbol{g}\cdot\boldsymbol{k})\boldsymbol{k}\,\mathrm{d}^2k\,\mathrm{d}^3c\,\mathrm{d}^3c_1 .$$

The first term is treated in the following way. First, $\boldsymbol{c}'-\boldsymbol{c}$ is replaced by $(\boldsymbol{g}\cdot\boldsymbol{k})\boldsymbol{k}$, then the integral over $\boldsymbol{k}$ is evaluated, next $\boldsymbol{g}$ is replaced by $\boldsymbol{C}_1 - \boldsymbol{C}$ and finally the integrations over $\boldsymbol{c}$ and $\boldsymbol{c}_1$ are performed. As the first three steps are trivial, we illustrate only the final step. One has

$$\tfrac{1}{2}\sigma^3\chi\iiint m(\boldsymbol{c}'-\boldsymbol{c})ff_1(\boldsymbol{g}\cdot\boldsymbol{k})\boldsymbol{k}\,\mathrm{d}^2k\,\mathrm{d}^3c\,\mathrm{d}^3c_1$$

$$= \tfrac{1}{15}\pi\sigma^3\chi m\iint ff_1[2\boldsymbol{C}\boldsymbol{C} - 2\boldsymbol{C}\boldsymbol{C}_1 - 2\boldsymbol{C}_1\boldsymbol{C} + 2\boldsymbol{C}_1\boldsymbol{C}_1$$

$$+(C^2 - 2\boldsymbol{C}\cdot\boldsymbol{C}_1 + C_1^2)\mathbf{I}]\mathrm{d}^3c\,\mathrm{d}^3c_1$$

$$= \tfrac{2}{15}\pi\sigma^3\chi m\iint ff_1[2\boldsymbol{C}\boldsymbol{C} - 2\boldsymbol{C}\boldsymbol{C}_1 + (C^2 - \boldsymbol{C}\cdot\boldsymbol{C}_1)\mathbf{I}]\mathrm{d}^3c\,\mathrm{d}^3c_1$$

$$= \tfrac{2}{15}\pi\sigma^3\chi m\iint ff_1(2\boldsymbol{C}\boldsymbol{C} + C^2\mathbf{I})\,\mathrm{d}^3c\,\mathrm{d}^3c_1 .$$

Here we have used the fact that the integrals of the terms involving $\boldsymbol{C}\boldsymbol{C}_1$ and $\boldsymbol{C}\cdot\boldsymbol{C}_1$ are proportional to the tensors $\overline{\boldsymbol{C}}\,\overline{\boldsymbol{C}}_1$ and $(\overline{\boldsymbol{C}}\cdot\overline{\boldsymbol{C}}_1)\mathbf{I}$, respectively; therefore, they vanish identically (both $\overline{\boldsymbol{C}}$ and $\overline{\boldsymbol{C}}_1$ are zero, by definition). Now, the integral over $\boldsymbol{C}_1$ can be evaluated trivially. The resulting expression is

$$\tfrac{2}{15}\pi\sigma^3\chi nm\int f(2\boldsymbol{C}\boldsymbol{C}+C^2\mathbf{I})\,\mathrm{d}^3c.$$

Here, we substitute the expression for f,

$$f = f^{(0)}\left\{1-\frac{1}{n\chi}\,[(1+\tfrac{2}{5}\pi n\sigma^3\chi)\boldsymbol{A}\cdot\nabla\log T+(1+\tfrac{4}{15}\pi n\sigma^3\chi)\mathbf{B}:\nabla\boldsymbol{v}]\right\},$$

where the vector $\boldsymbol{A}$ and the tensor $\mathbf{B}$ are the same functions of the velocity as in the case of a dilute gas, cf. Section 5.4. Since $\boldsymbol{A} = A(C)\boldsymbol{C}$, the term proportional to $\nabla\log T$ is odd in the components of $\boldsymbol{C}$ and its integral over $\boldsymbol{c}$ is zero. Hence, the first integral in the definition of $\mathbf{P}_\Phi$ is equal to

$$\tfrac{2}{15}\pi\sigma^3\chi nm\int f^{(0)}\left\{1-\frac{1}{n\chi}\left(1+\frac{4}{15}\pi n\sigma^3\chi\right)\mathbf{B}:\nabla\boldsymbol{v}\right\}(2\boldsymbol{C}\boldsymbol{C}+C^2\mathbf{I})\,\mathrm{d}^3c$$
$$= \tfrac{2}{3}\pi\sigma^3\chi n^2kT\mathbf{I}-\tfrac{8}{15}\pi n\sigma^3(1+\tfrac{4}{15}\pi n\sigma^3\chi)\eta^{(0)}\mathbf{S},$$

where $\mathbf{S}$ is the symmetric traceless part of the tensor $\nabla\boldsymbol{v}$ (i.e., $\mathbf{S}$ is the rate-of-shear tensor) and $\eta^{(0)} = \tfrac{1}{10}kT[\mathbf{B},\mathbf{B}]$ (i.e., $\eta^{(0)}$ is the viscosity coefficient for a gas at low density, cf. Eq. (5.4-36)).

The second integral in the definition of $\mathbf{P}_\Phi$ is treated in the following manner. First, $\boldsymbol{c}'-\boldsymbol{c}$ is replaced by $(\boldsymbol{g}\cdot\boldsymbol{k})\boldsymbol{k}$. Then, the integral over $\boldsymbol{k}$ is evaluated. Thus we obtain

$$\tfrac{1}{4}\sigma^4\chi\iiint m(\boldsymbol{c}'-\boldsymbol{c})f^{(0)}f_1^{(0)}\left(\boldsymbol{k}\cdot\nabla_r\log\frac{f^{(0)}}{f_1^{(0)}}\right)(\boldsymbol{g}\cdot\boldsymbol{k})\boldsymbol{k}\,\mathrm{d}^2k\,\mathrm{d}^3c\,\mathrm{d}^3c_1$$
$$= \tfrac{1}{48}\pi\sigma^4\chi m\iint f^{(0)}f_1^{(0)}\left\{\left(\boldsymbol{g}\cdot\nabla_r\log\frac{f^{(0)}}{f_1^{(0)}}\right)(\boldsymbol{g}\boldsymbol{g}+g^2\mathbf{I})/g\right.$$
$$\left.+g\left[\left(\nabla_r\log\frac{f^{(0)}}{f_1^{(0)}}\right)\boldsymbol{g}+\boldsymbol{g}\left(\nabla_r\log\frac{f^{(0)}}{f_1^{(0)}}\right)\right]\right\}\mathrm{d}^3c\,\mathrm{d}^3c_1.$$

Now,

$$\nabla_r\log\frac{f^{(0)}}{f_1^{(0)}} = \frac{m}{2kT}(C^2-C_1^2)\nabla\log T+\frac{m}{kT}(\nabla\boldsymbol{v})\cdot(\boldsymbol{C}-\boldsymbol{C}_1),$$

and, since $\boldsymbol{g} = \boldsymbol{C}_1 - \boldsymbol{C}$, the terms involving $\boldsymbol{\nabla} \log T$ are odd in the components of $\boldsymbol{C}$ and $\boldsymbol{C}_1$ and vanish upon integration. Thus, the second integral in $\mathbf{P}_\Phi$ is equal to

$$-\tfrac{1}{48}\pi\sigma^4\chi \frac{m^2}{kT} \iint f^{(0)} f_1^{(0)} \{(\boldsymbol{gg} : \boldsymbol{\nabla v})(\boldsymbol{gg} + g^2 \mathbf{I})/g + g[(\boldsymbol{\nabla v}) \cdot \boldsymbol{gg} + \boldsymbol{gg} \cdot (\boldsymbol{\nabla v})]\} \mathrm{d}^3 c \, \mathrm{d}^3 c_1 .$$

These integrals are most easily evaluated after a transformation of variables from $(\boldsymbol{c}, \boldsymbol{c}_1)$ to $(\boldsymbol{G}_0, \boldsymbol{g})$. The result is

$$-\tfrac{4}{9}\sigma^4 \chi n^2 (\pi m k T)^{\frac{1}{2}} (\tfrac{6}{5}\mathbf{S} + (\boldsymbol{\nabla} \cdot \boldsymbol{v})\mathbf{I}),$$

where, as before, $\mathbf{S}$ is the symmetric traceless part of $\boldsymbol{\nabla v}$. Thus we have shown that the integral $\mathbf{P}_\Phi$ is equal to

$$\mathbf{P}_\Phi = \tfrac{2}{3}\pi n \sigma^3 \chi n^2 kT \mathbf{I} - \tfrac{8}{15}\pi n \sigma^3 (1 + \tfrac{4}{15}\pi n \sigma^3 \chi) \eta^{(0)} \mathbf{S} - \tfrac{4}{9}\sigma^4 \chi n^2 (\pi m k T)^{\frac{1}{2}} (\tfrac{6}{5}\mathbf{S} + (\boldsymbol{\nabla} \cdot \boldsymbol{v})\mathbf{I}).$$

D.6. *The integral $\boldsymbol{q}_\Phi$*

From the definition (12.5-20) we have

$$\boldsymbol{q}_\Phi = \tfrac{1}{2}\sigma^3\chi \iiint \tfrac{1}{2}m(C'^2 - C^2) f f_1 (\boldsymbol{g} \cdot \boldsymbol{k})\boldsymbol{k} \, \mathrm{d}^2 k \, \mathrm{d}^3 c \, \mathrm{d}^3 c_1 + \tfrac{1}{4}\sigma^4\chi \iiint \tfrac{1}{2}m(C'^2 - C^2) f^{(0)} f_1^{(0)} \left(\boldsymbol{k} \cdot \boldsymbol{\nabla}_r \log \frac{f^{(0)}}{f_1^{(0)}}\right) (\boldsymbol{g} \cdot \boldsymbol{k})\boldsymbol{k} \, \mathrm{d}^2 k \, \mathrm{d}^3 c \, \mathrm{d}^3 c_1 .$$

The first term is treated in the following manner. First, $C'^2 - C^2$ is replaced by $2(\boldsymbol{G}_0 \cdot \boldsymbol{k})(\boldsymbol{g} \cdot \boldsymbol{k})$, then the integration over $\boldsymbol{k}$ is performed. Next $\boldsymbol{G}_0$ and $\boldsymbol{g}$ are replaced by $\tfrac{1}{2}(\boldsymbol{C} + \boldsymbol{C}_1)$ and $\boldsymbol{C}_1 - \boldsymbol{C}$, respectively, and finally the integrations over $\boldsymbol{c}$ and $\boldsymbol{c}_1$ are performed. Again, the first three steps are easy and we illustrate only the final step. One has

$$\begin{aligned}
\tfrac{1}{2}\sigma^3\chi \iiint \tfrac{1}{2}m(C'^2 - C^2) f f_1 (\boldsymbol{g} \cdot \boldsymbol{k})\boldsymbol{k} \, \mathrm{d}^2 k \, \mathrm{d}^3 c \, \mathrm{d}^3 c_1 \\
&= \tfrac{1}{15}\pi\sigma^3\chi m \iint f f_1 [(C_1^2 - C^2)(\boldsymbol{C}_1 - \boldsymbol{C}) \\
&\qquad + \tfrac{1}{2}(C_1^2 - 2\boldsymbol{C}_1 \cdot \boldsymbol{C} + C^2)(\boldsymbol{C}_1 + \boldsymbol{C})] \, \mathrm{d}^3 c \, \mathrm{d}^3 c_1 \\
&= \tfrac{1}{15}\pi\sigma^3\chi m \iint f f_1 (\tfrac{3}{2}C^2 \boldsymbol{C} + \tfrac{3}{2}C_1^2 \boldsymbol{C}_1) \, \mathrm{d}^3 c \, \mathrm{d}^3 c_1 \\
&= \tfrac{1}{5}\pi\sigma^3\chi m \iint f f_1 \, C^2 \boldsymbol{C} \, \mathrm{d}^3 c \, \mathrm{d}^3 c_1 \\
&= \tfrac{1}{5}\pi\sigma^3\chi n m \int f C^2 \boldsymbol{C} \, \mathrm{d}^3 c.
\end{aligned}$$

Now, we substitute the expression for f, given under D.5. This time, the term proportional to $\nabla \boldsymbol{v}$ is odd in the components of $\boldsymbol{C}$ and, hence, its integral over $\boldsymbol{c}$ vanishes. Thus, the first integral in the definition of $\boldsymbol{q}_\Phi$ is equal to

$$\tfrac{1}{5}\pi\sigma^3\chi nm \int f^{(0)} \left\{1 - \frac{1}{n\chi}(1+\tfrac{2}{5}\pi n\sigma^3\chi)\boldsymbol{A} \cdot \nabla \log T\right\} C^2\boldsymbol{C}\, \mathrm{d}^3c$$

$$= -\tfrac{2}{5}\pi n\sigma^3(1+\tfrac{2}{5}\pi n\sigma^3\chi)\lambda^{(0)}\nabla T,$$

where $\lambda^{(0)} = \tfrac{1}{3}k[\boldsymbol{A}, \boldsymbol{A}]$, i.e., $\lambda^{(0)}$ is the coefficient of thermal conductivity for a gas at low density, cf. Eq. (5.4-42).

In the second integral, one replaces $C'^2 - C^2$ by $2\,(\boldsymbol{G}_0 \cdot \boldsymbol{k})(\boldsymbol{g} \cdot \boldsymbol{k})$ and performs the integration over $\boldsymbol{k}$. Thus,

$$\tfrac{1}{4}\sigma^4\chi \iiint \tfrac{1}{2}m(C'^2 - C^2)f^{(0)}f_1^{(0)} \left(\boldsymbol{k} \cdot \nabla_r \log \frac{f^{(0)}}{f_1^{(0)}}\right) (\boldsymbol{g} \cdot \boldsymbol{k})\boldsymbol{k}\, \mathrm{d}^2k\, \mathrm{d}^3c\, \mathrm{d}^3c_1$$

$$= \tfrac{1}{48}\pi\sigma^4\chi m \iint f^{(0)}f_1^{(0)}[(\boldsymbol{G}_0 \cdot \boldsymbol{g})(\boldsymbol{g}\boldsymbol{g}+g^2\mathsf{I})/g$$

$$+g(\boldsymbol{G}_0\boldsymbol{g}+\boldsymbol{g}\boldsymbol{G}_0)] \cdot \left(\nabla_r \log \frac{f^{(0)}}{f_1^{(0)}}\right) \mathrm{d}^3c\, \mathrm{d}^3c_1 .$$

Now, substituting the expression for $\nabla_r \log (f^{(0)}/f_1^{(0)})$ given under D.5 we see that the terms involving $\nabla \boldsymbol{v}$ are odd in the components of $\boldsymbol{C}$ and $\boldsymbol{C}_1$ and, therefore, their integrals vanish. Thus, the second integral in $\boldsymbol{q}_\Phi$ is equal to

$$-\tfrac{1}{48}\pi\sigma^4\chi \frac{m^2}{kT} \iint f^{(0)}f_1^{(0)}(\boldsymbol{G}_0 \cdot \boldsymbol{g})[(\boldsymbol{G}_0 \cdot \boldsymbol{g})(\boldsymbol{g}\boldsymbol{g}+g^2\mathsf{I})/g$$

$$+g(\boldsymbol{G}_0\boldsymbol{g}+\boldsymbol{g}\boldsymbol{G}_0)] \cdot \nabla \log T\, \mathrm{d}^3c\, \mathrm{d}^3c_1 .$$

These integrals are most easily evaluated after a transformation of variables from $(\boldsymbol{c}, \boldsymbol{c}_1)$ to $(\boldsymbol{G}_0, \boldsymbol{g})$. The result is

$$-\tfrac{2}{3}n^2\sigma^4\chi(\pi k^3T/m)^{\frac{1}{2}}\nabla T.$$

Thus we have shown that the integral $\boldsymbol{q}_\Phi$ is equal to

$$\boldsymbol{q}_\Phi = -\left[\tfrac{2}{5}\pi n\sigma^3(1+\tfrac{2}{5}\pi n\sigma^3\chi)\lambda^{(0)} + \tfrac{2}{3}n^2\sigma^4\chi\left(\frac{\pi k^3 T}{m}\right)^{\frac{1}{2}}\right]\nabla T.$$

Appendix E
Evaluation of the flux vectors in dense gases

In this appendix we evaluate the pressure tensor and heat flow vector for a dense gas in the first-order Chapman-Enskog approximation. The notation is the same as in Chapter 13.

E.1. *The pressure tensor*

The pressure tensor $\mathbf{P}$ consists of two parts, viz., a kinetic part $\mathbf{P}_K$ and a potential part $\mathbf{P}_\Phi$, defined by Eq. (13.1-25). The evaluation of the kinetic part is straight-forward,

$$\mathbf{P}_K = \int m\boldsymbol{C}\boldsymbol{C} f_1^{(0)}(1+\phi)\,\mathrm{d}^3c$$

$$= \int m\boldsymbol{C}\boldsymbol{C} f_1^{(0)} \left[1 - \frac{1}{n} A(\mathscr{C})\boldsymbol{C}\cdot\boldsymbol{\nabla}\log\theta\right.$$

$$\left. - \frac{1}{n} B(\mathscr{C})(\mathscr{C}\mathscr{C} - \tfrac{1}{3}\mathscr{C}^2\boldsymbol{I}) : \boldsymbol{\nabla v} - \frac{1}{n}\Gamma(\mathscr{C})\boldsymbol{\nabla}\cdot\boldsymbol{v}\right]\mathrm{d}^3c$$

$$= n\theta\boldsymbol{I} - \frac{4\theta}{15n}\mathbf{S}\int f_1^{(0)}B(\mathscr{C})\mathscr{C}^4\,\mathrm{d}^3c - \frac{2\theta}{3n}(\boldsymbol{\nabla}\cdot\boldsymbol{v})\boldsymbol{I}\int f_1^{(0)}\Gamma(\mathscr{C})\mathscr{C}^2\,\mathrm{d}^3c, \qquad (1)$$

where $\mathbf{S}$ is the symmetric traceless part of $\boldsymbol{\nabla v}$. The potential part consists of three terms, each term corresponding to a single term in the expression (13.2-59) of f_2. The first term is trivial,

$$\mathsf{P}_{\Phi,0} = -\tfrac{1}{2}\int \boldsymbol{r}_{12}\boldsymbol{r}_{12}\frac{\varphi'(r_{12})}{r_{12}}\left\{\iint f_2(x_1 x_2|f_1^{(0)})\mathrm{d}^3c_1\,\mathrm{d}^3c_2\right\}\mathrm{d}^3r_2|_{\boldsymbol{r}_1=\boldsymbol{r}}$$

$$= -\tfrac{1}{2}n^2\int \boldsymbol{r}_{12}\boldsymbol{r}_{12}\frac{\varphi'(r_{12})}{r_{12}}g(r_{12})\mathrm{d}^3r_2|_{\boldsymbol{r}_1=\boldsymbol{r}}$$

$$= -\tfrac{1}{6}n^2\mathbf{I}\int r\varphi'(r)g(r)\,\mathrm{d}^3r. \tag{2}$$

The second term is more complicated. Its definition is

$$\mathsf{P}_{\Phi,1} = -\tfrac{1}{2}\int \boldsymbol{r}_{12}\boldsymbol{r}_{12}\frac{\varphi'(r_{12})}{r_{12}}\iiint f_2'(x_1 x_2|f_1^{(0)};\boldsymbol{c}')f_1^{(0)}(\boldsymbol{c}')$$

$$\times\left[-\frac{1}{n}A(\mathscr{C}')\boldsymbol{C}'\cdot\boldsymbol{\nabla}\log\theta - \frac{1}{n}B(\mathscr{C}')(\mathscr{C}'\mathscr{C}'-\tfrac{1}{3}\mathscr{C}'^2\mathbf{I}):\boldsymbol{\nabla v}\right.$$

$$\left.-\frac{1}{n}\Gamma(\mathscr{C}')\boldsymbol{\nabla}\cdot\boldsymbol{v}\right]\mathrm{d}^3c'\,\mathrm{d}^3c_1\,\mathrm{d}^3c_2\,\mathrm{d}^3r_2|_{\boldsymbol{r}_1=\boldsymbol{r}}.$$

The term involving $\boldsymbol{\nabla}\log\theta$ does not contribute, since its integrand is odd in the components of $\boldsymbol{C}'$. In the term involving $\boldsymbol{\nabla v}$ we decompose the tensor $\boldsymbol{r}_{12}\boldsymbol{r}_{12}$ into a diagonal part $(\tfrac{1}{3}r_{12}^2\mathbf{I})$ and a traceless part $(\boldsymbol{r}_{12}\boldsymbol{r}_{12}-\tfrac{1}{3}r_{12}^2\mathbf{I})$. The former leads to a contribution

$$-\frac{1}{6n}\mathbf{I}\int \mathrm{d}^3r_2\,r_{12}\varphi'(r_{12})\iiint f_2'(x_1 x_2|f_1^{(0)};\boldsymbol{c}')\,f_1^{(0)}(\boldsymbol{c}')$$

$$\times B(\mathscr{C}')(\mathscr{C}'\mathscr{C}'-\tfrac{1}{3}\mathscr{C}'^2\mathbf{I})\mathrm{d}^3c'\,\mathrm{d}^3c_1\,\mathrm{d}^3c_2|_{\boldsymbol{r}_1=\boldsymbol{r}}:\boldsymbol{\nabla v},$$

which is zero because $\mathscr{C}'\mathscr{C}'-\tfrac{1}{3}\mathscr{C}'^2\mathbf{I}$ is traceless. (We recall that the integral must yield an isotropic tensor of rank four.) Thus, the contribution from the term involving $\boldsymbol{\nabla v}$ is reduced to

$$\frac{1}{2n}\int \mathrm{d}^3r_2\int \mathrm{d}^3c'\,\mathrm{d}^3c_1\,\mathrm{d}^3c_2\,w(\boldsymbol{r}_{12}\boldsymbol{r}_{12}-\tfrac{1}{3}r_{12}^2\mathbf{I})(\mathscr{C}'\mathscr{C}'-\tfrac{1}{3}\mathscr{C}^2\mathbf{I})|_{\boldsymbol{r}_1=\boldsymbol{r}}:\boldsymbol{\nabla v},$$

where we have defined the weight w,

$$w = \frac{\varphi'(r_{12})}{r_{12}}f_2'(x_1 x_2|f_1^{(0)};\boldsymbol{c}')\,f_1^{(0)}(\boldsymbol{c}')\,B(\mathscr{C}').$$

As the integrand involves only symmetric traceless tensors, the integral yields an isotropic tensor of rank four. Therefore, the general element of the tensor must be of the form

$$\frac{1}{2n}\int \mathrm{d}^3 r_2 \int \mathrm{d}^3 c' \,\mathrm{d}^3 c_1 \,\mathrm{d}^3 c_2 \, w(\boldsymbol{r}_{12}\boldsymbol{r}_{12} - \tfrac{1}{3} r_{12}^2 \mathsf{I})_{ij} (\mathscr{C}'\mathscr{C}' - \tfrac{1}{3}\mathscr{C}'^2 \mathsf{I})_{kl}$$
$$= \alpha \delta_{ij}\delta_{kl} + \beta(\delta_{ik}\delta_{jl} + \delta_{il}\delta_{jk})$$

The constants α and β are determined in the following manner. First, one keeps k and l fixed and puts $i = j$. Using the Einstein summation convention one obtains the result $\alpha = -\frac{2}{3}\beta$, which is then substituted in the right member of the equation. Then one multiplies both members by $(-\frac{2}{3}\delta_{ij}\delta_{kl} + \delta_{ik}\delta_{jl} + \delta_{il}\delta_{jk})$. Again using the Einstein summation convention one finds $\beta = \frac{1}{10}$ times the expression in the left member with the index k replaced by i and the index l replaced by j. Thus,

$$\frac{1}{2n}\int \mathrm{d}^3 r_2 \int \mathrm{d}^3 c' \,\mathrm{d}^3 c_1 \,\mathrm{d}^3 c_2 \, w(\boldsymbol{r}_{12}\boldsymbol{r}_{12} - \tfrac{1}{3} r_{12}^2 \mathsf{I})_{ij} (\mathscr{C}'\mathscr{C}' - \tfrac{1}{3}\mathscr{C}'^2 \mathsf{I})_{kl}$$
$$= \frac{1}{20n}(-\tfrac{2}{3}\delta_{ij}\delta_{kl} + \delta_{ik}\delta_{jl} + \delta_{il}\delta_{jk})$$
$$\times \int \mathrm{d}^3 r_2 \int \mathrm{d}^3 c' \,\mathrm{d}^3 c_1 \,\mathrm{d}^3 c_2 \, w(\boldsymbol{r}_{12}\boldsymbol{r}_{12} - \tfrac{1}{3} r_{12}^2 \mathsf{I}) : (\mathscr{C}'\mathscr{C}' - \tfrac{1}{3}\mathscr{C}'^2 \mathsf{I}).$$

Consequently, the contribution to $\mathsf{P}_{\Phi,1}$ due to the term involving $\nabla \boldsymbol{v}$ is equal to

$$\frac{1}{10n}\mathsf{S}\int \mathrm{d}^3 r_2 \int \mathrm{d}^3 c' \,\mathrm{d}^3 c_1 \,\mathrm{d}^3 c_2 \, w(\boldsymbol{r}_{12}\boldsymbol{r}_{12} - \tfrac{1}{3} r_{12}^2 \mathsf{I}) : (\mathscr{C}'\mathscr{C}' - \tfrac{1}{3}\mathscr{C}'^2 \mathsf{I})|_{\boldsymbol{r}_1 = \boldsymbol{r}}.$$

Finally, the contribution to $\mathsf{P}_{\Phi,1}$ due to the term involving $\nabla \cdot \boldsymbol{v}$ is trivially evaluated and we find

$$\mathsf{P}_{\Phi,1} = \frac{1}{10n}\mathsf{S}\int \mathrm{d}^3 r_2 \frac{\varphi'(r_{12})}{r_{12}} \iiint f_2'(x_1 x_2 | f_1^{(0)}; \boldsymbol{c}') f_1^{(0)}(\boldsymbol{c}') B(\mathscr{C}')$$
$$\times [(\boldsymbol{r}_{12}\cdot\mathscr{C}')^2 - \tfrac{1}{3} r_{12}^2 \mathscr{C}'^2] \,\mathrm{d}^3 c' \,\mathrm{d}^3 c_1 \,\mathrm{d}^3 c_2 |_{\boldsymbol{r}_1 = \boldsymbol{r}}$$
$$+ \frac{1}{6n}(\nabla \cdot \boldsymbol{v})\mathsf{I} \int \mathrm{d}^3 r_2 \, r_{12}\varphi'(r_{12})$$
$$\times \iiint f_2'(x_1 x_2 | f_1^{(0)}; \boldsymbol{c}') f_1^{(0)}(\boldsymbol{c}') \Gamma(\mathscr{C}') \,\mathrm{d}^3 c' \,\mathrm{d}^3 c_1 \,\mathrm{d}^3 c_2 |_{\boldsymbol{r}_1 = \boldsymbol{r}}. \tag{3}$$

We now evaluate the last term in the expression for the potential part of the pressure tensor,

$$\boldsymbol{P}_{\Phi,2} = -\tfrac{1}{2}\int \boldsymbol{r}_{12}\boldsymbol{r}_{12}\,\frac{\varphi'(r_{12})}{r_{12}}\iiint f_2'(x_1\,x_2|f_1^{(0)};x')$$
$$\times(\boldsymbol{\nabla}_{r'}f_1^{(0)}(x'))_{\boldsymbol{r}'=\boldsymbol{r}_1}\cdot(\boldsymbol{r}'-\boldsymbol{r}_1)\,\mathrm{d}^6x'\,\mathrm{d}^3c_1\,\mathrm{d}^3c_2\,\mathrm{d}^3r_2|_{\boldsymbol{r}_1=\boldsymbol{r}}.$$

Here we substitute

$$(\boldsymbol{\nabla}_{r'}f_1^{(0)}(x'))_{\boldsymbol{r}'=\boldsymbol{r}_1} = f_1^{(0)}(\boldsymbol{c}')\left[\boldsymbol{\nabla}\log n+(\mathscr{C}'^2-\tfrac{3}{2})\boldsymbol{\nabla}\log\theta+\frac{m}{\theta}(\boldsymbol{\nabla v})\cdot\boldsymbol{C}'\right],$$

where all quantities in the right member are evaluated at $\boldsymbol{r}_1$. The integrals arising from the coefficients of $\boldsymbol{\nabla}\log n$ and $\boldsymbol{\nabla}\log\theta$ are zero, as the result must be an isotropic tensor of rank three and the average of $\boldsymbol{r}_{12}\boldsymbol{r}_{12}(\boldsymbol{r}'-\boldsymbol{r}_1)$ is generally not isotropic. Thus

$$\boldsymbol{P}_{\Phi,2} = -\left(\frac{m}{2\theta}\right)^{\frac{1}{2}}\int \mathrm{d}^3r_2\iiint \mathrm{d}^6x'\,\mathrm{d}^3c_1\,\mathrm{d}^3c_2\,w\boldsymbol{r}_{12}\boldsymbol{r}_{12}\mathscr{C}'(\boldsymbol{r}'-\boldsymbol{r}_1)|_{\boldsymbol{r}_1=\boldsymbol{r}} : \boldsymbol{\nabla v},$$

where

$$w = \frac{\varphi'(r_{12})}{r_{12}}f_2'(x_1\,x_2|f_1^{(0)};x')f_1^{(0)}(\boldsymbol{c}').$$

Decomposing both tensors $\boldsymbol{r}_{12}\boldsymbol{r}_{12}$ and $\mathscr{C}'(\boldsymbol{r}'-\boldsymbol{r}_1)$ in their symmetric traceless, antisymmetric and diagonal parts and retaining only those tensor products which have the proper tensorial character (i.e., isotropic tensors of rank four) we find

$$\boldsymbol{P}_{\Phi,2} = -\left(\frac{m}{2\theta}\right)^{\frac{1}{2}}\int \mathrm{d}^3r_2\iiint \mathrm{d}^6x'\,\mathrm{d}^3c_1\,\mathrm{d}^3c_2\,w[(\boldsymbol{r}_{12}\boldsymbol{r}_{12}-\tfrac{1}{3}r_{12}^2\boldsymbol{I})\boldsymbol{T}$$
$$+\tfrac{1}{9}r_{12}^2\mathscr{C}'\cdot(\boldsymbol{r}'-\boldsymbol{r}_1)\boldsymbol{II}]|_{\boldsymbol{r}_1=\boldsymbol{r}} : \boldsymbol{\nabla v},$$

where $\boldsymbol{T}$ is the symmetric traceless part of $\mathscr{C}'(\boldsymbol{r}'-\boldsymbol{r}_1)$. This expression is easily seen to be equivalent to

$$\boldsymbol{P}_{\Phi,2} = -\frac{1}{5}\left(\frac{m}{2\theta}\right)^{\frac{1}{2}}\boldsymbol{S}\int \mathrm{d}^3r_2\,\frac{\varphi'(r_{12})}{r_{12}}\iiint f_2'(x_1\,x_2|f_1^{(0)};x')f_1^{(0)}(\boldsymbol{c}')$$
$$\times[(\boldsymbol{r}_{12}\cdot\mathscr{C}')(\boldsymbol{r}_{12}\cdot(\boldsymbol{r}'-\boldsymbol{r}_1))-\tfrac{1}{3}r_{12}^2\mathscr{C}'\cdot(\boldsymbol{r}'-\boldsymbol{r}_1)]\,\mathrm{d}^6x'\,\mathrm{d}^3c_1\,\mathrm{d}^3c_2|_{\boldsymbol{r}_1=\boldsymbol{r}}$$
$$-\frac{1}{9}\left(\frac{m}{2\theta}\right)^{\frac{1}{2}}(\boldsymbol{\nabla}\cdot\boldsymbol{v})\boldsymbol{I}\int \mathrm{d}^3r_2\,r_{12}\,\varphi'(r_{12})\iiint f_2'(x_1\,x_2|f_1^{(0)};x')\,f_1^{(0)}(\boldsymbol{c}')$$
$$\times\mathscr{C}'\cdot(\boldsymbol{r}'-\boldsymbol{r}_1)\,\mathrm{d}^6x'\,\mathrm{d}^3c_1\,\mathrm{d}^3c_2|_{\boldsymbol{r}_1=\boldsymbol{r}}. \tag{4}$$

The combined results (1)–(4) correspond to the results (13.2-62) through (13.2-64).

E.2. *The heat flow vector*

The heat flow vector $\boldsymbol{q}$ is computed in much the same way as the pressure tensor. The vector $\boldsymbol{q}$ consists of a kinetic part, $\boldsymbol{q}_K$, and two potential parts, $\boldsymbol{q}_{\Phi 1}$ and $\boldsymbol{q}_{\Phi 2}$, defined by Eq. (13.1-30). The kinetic part is given by

$$\boldsymbol{q}_K = \int \tfrac{1}{2}mC^2\boldsymbol{C}f_1^{(0)}(1+\phi)\mathrm{d}^3c$$

$$= -\frac{2\theta}{3mn}\int f_1^{(0)}A(\mathscr{C})\mathscr{C}^4\,\mathrm{d}^3c\,\nabla\theta. \tag{5}$$

$\boldsymbol{q}_{\Phi 1}$ and $\boldsymbol{q}_{\Phi 2}$ each consist of three terms, each term corresponding to a single term in the expression (13.2-59) of f_2. The first term is zero in both cases as the integrands are odd in the components of $\boldsymbol{C}$. The second term in $\boldsymbol{q}_{\Phi 1}$ is

$$\boldsymbol{q}_{\Phi 1,1} = \tfrac{1}{2}\int \varphi(r_{12})\iiint \boldsymbol{C}_1\, f_2'(x_1\, x_2|f_1^{(0)};\, \boldsymbol{c}')\, f_1^{(0)}(\boldsymbol{c}')$$

$$\times\left[-\frac{1}{n}A(\mathscr{C}')\boldsymbol{C}'\cdot\nabla\log\theta - \frac{1}{n}B(\mathscr{C}')(\mathscr{C}'\mathscr{C}'-\tfrac{1}{3}\mathscr{C}'^2\boldsymbol{I}):\nabla\boldsymbol{v}\right.$$

$$\left.-\frac{1}{n}\Gamma(\mathscr{C}')\nabla\cdot\boldsymbol{v}\right]\mathrm{d}^3c'\,\mathrm{d}^3c_1\,\mathrm{d}^3c_2\,\mathrm{d}^3r_2|_{\boldsymbol{r}_1=\boldsymbol{r}}$$

$$= -\frac{1}{mn}\int \mathrm{d}^3r_2\iiint \mathrm{d}^3c'\,\mathrm{d}^3c_1\,\mathrm{d}^3c_2\, w\mathscr{C}_1\,\mathscr{C}'|_{\boldsymbol{r}_1=\boldsymbol{r}}\cdot\nabla\theta$$

with

$$w = \varphi(r_{12})\, f_2'(x_1\, x_2|f_1^{(0)};\, \boldsymbol{c}')\, f_1^{(0)}(\boldsymbol{c}')\, A(\mathscr{C}').$$

Decomposing the tensor $\mathscr{C}_1\mathscr{C}'$ into its symmetric traceless, antisymmetric and diagonal parts one may verify that only the latter part contributes, as the coefficient of $\nabla\theta$ must be an isotropic tensor of rank two. Thus we find

$$\boldsymbol{q}_{\Phi 1,1} = -\frac{1}{3mn}\int \mathrm{d}^3r_2\iiint \mathrm{d}^3c'\,\mathrm{d}^3c_1\,\mathrm{d}^3c_2\,\varphi(r_{12})\, f_2'(x_1\, x_2|f_1^{(0)};\, \boldsymbol{c}')$$

$$\times f_1^{(0)}(\boldsymbol{c}')\, A(\mathscr{C}')\,\mathscr{C}_1\cdot\mathscr{C}'|_{\boldsymbol{r}_1=\boldsymbol{r}}\nabla\theta. \tag{6}$$

The second term in $\boldsymbol{q}_{\Phi 2}$ is evaluated in an analogous manner,

$$\boldsymbol{q}_{\Phi 2,1} = -\tfrac{1}{4}\int \boldsymbol{r}_{12}\boldsymbol{r}_{12}\frac{\varphi'(r_{12})}{r_{12}}\cdot\iiint (\boldsymbol{C}_1+\boldsymbol{C}_2)\,f_2'(x_1\,x_2|f_1^{(0)};\boldsymbol{c}')\,f_1^{(0)}(\boldsymbol{c}')$$
$$\times\left[-\frac{1}{n}A(\mathscr{C}')\boldsymbol{C}'\cdot\nabla\log\theta-\frac{1}{n}B(\mathscr{C}')(\mathscr{C}'\mathscr{C}'-\tfrac{1}{3}\mathscr{C}'^2\mathsf{I}):\nabla\boldsymbol{v}\right.$$
$$\left.-\frac{1}{n}\Gamma(\mathscr{C}')\nabla\cdot\boldsymbol{v}\right]\mathrm{d}^3c'\,\mathrm{d}^3c_1\,\mathrm{d}^3c_2\,\mathrm{d}^3r_2|_{\boldsymbol{r}_1=\boldsymbol{r}}$$
$$=\frac{1}{2mn}\int\mathrm{d}^3r_2\iiint\mathrm{d}^3c'\,\mathrm{d}^3c_1\,\mathrm{d}^3c_2\,w\boldsymbol{r}_{12}\mathscr{C}'|_{\boldsymbol{r}_1=\boldsymbol{r}}\cdot\nabla\theta,$$

with

$$w=\frac{\varphi'(r_{12})}{r_{12}}\boldsymbol{r}_{12}\cdot(\mathscr{C}_1+\mathscr{C}_2)\,f_2'(x_1\,x_2|f_1^{(0)};\boldsymbol{c}')\,f_1^{(0)}(\boldsymbol{c}')A(\mathscr{C}').$$

Again, only the diagonal part of the tensor $\boldsymbol{r}_{12}\mathscr{C}'$ gives a nonzero contribution. One finds

$$\boldsymbol{q}_{\Phi 2,1}=\frac{1}{6mn}\int\mathrm{d}^3r_2\frac{\varphi'(r_{12})}{r_{12}}\boldsymbol{r}_{12}\cdot\iiint\mathrm{d}^3c'\,\mathrm{d}^3c_1\,\mathrm{d}^3c_2\,(\mathscr{C}_1+\mathscr{C}_2)$$
$$\times f_2'(x_1\,x_2|f_1^{(0)};\boldsymbol{c}')\,f_1^{(0)}(\boldsymbol{c}')A(\mathscr{C}')(\boldsymbol{r}_{12}\cdot\mathscr{C}')|_{\boldsymbol{r}_1=\boldsymbol{r}}\cdot\nabla\theta. \qquad (7)$$

We now evaluate the last terms in the expressions for $\boldsymbol{q}_{\Phi 1}$ and $\boldsymbol{q}_{\Phi 2}$,

$$\boldsymbol{q}_{\Phi 1,2}=\tfrac{1}{2}\int\varphi(r_{12})\iiint\boldsymbol{C}_1\,f_2'(x_1\,x_2|f_1^{(0)};x')$$
$$\times(\nabla_{\boldsymbol{r}'}f_1^{(0)}(x'))_{\boldsymbol{r}'=\boldsymbol{r}_1}\cdot(\boldsymbol{r}'-\boldsymbol{r}_1)\,\mathrm{d}^6x'\,\mathrm{d}^3c_1\,\mathrm{d}^3c_2\,\mathrm{d}^3r_2|_{\boldsymbol{r}_1=\boldsymbol{r}}.$$

The only nonzero contribution comes from the term involving $\mathscr{C}'^2\nabla\log\theta$ in the expression for $(\nabla_{\boldsymbol{r}'}f_1^{(0)}(x'))_{\boldsymbol{r}'=\boldsymbol{r}_1}$. It yields

$$\boldsymbol{q}_{\Phi 1,2}=\frac{1}{2\theta}\left(\frac{2\theta}{m}\right)^{\frac{1}{2}}\int\mathrm{d}^3r_2\iiint\mathrm{d}^6x'\,\mathrm{d}^3c_1\,\mathrm{d}^3c_2\,w\mathscr{C}_1(\boldsymbol{r}'-\boldsymbol{r}_1)|_{\boldsymbol{r}_1=\boldsymbol{r}}\cdot\nabla\theta,$$

where

$$w=\varphi(r_{12})\,f_2'(x_1\,x_2|f_1^{(0)};x')\,f_1^{(0)}(\boldsymbol{c}')\mathscr{C}'^2.$$

Again, only the diagonal part of $\mathscr{C}_1\,(\boldsymbol{r}'-\boldsymbol{r}_1)$ contributes, so

$$\boldsymbol{q}_{\Phi 1,2} = \frac{1}{6\theta}\left(\frac{2\theta}{m}\right)^{\frac{1}{2}} \int \mathrm{d}^3 r_2 \iiint \mathrm{d}^6 x'\, \mathrm{d}^3 c_1\, \mathrm{d}^3 c_2\, \varphi(r_{12})\, f_2'(x_1\, x_2 | f_1^{(0)}; x')$$
$$\times f_1^{(0)}(\boldsymbol{c}')\mathscr{C}'^2(\boldsymbol{\mathscr{C}}_1 \cdot (\boldsymbol{r}' - \boldsymbol{r}_1))|_{\boldsymbol{r}_1=\boldsymbol{r}} \cdot \boldsymbol{\nabla}\theta. \tag{8}$$

Finally, the vector $\boldsymbol{q}_{\Phi 2,2}$ becomes

$$\boldsymbol{q}_{\Phi 2,2} = -\tfrac{1}{4} \int \boldsymbol{r}_{12}\boldsymbol{r}_{12} \frac{\varphi'(r_{12})}{r_{12}} \cdot \iiint (\boldsymbol{C}_1 + \boldsymbol{C}_2)\, f_2'(x_1\, x_2 | f_1^{(0)}; x')$$
$$\times (\boldsymbol{\nabla}_{\boldsymbol{r}'} f_1^{(0)}(x'))_{\boldsymbol{r}'=\boldsymbol{r}_1} \cdot (\boldsymbol{r}' - \boldsymbol{r}_1)\, \mathrm{d}^6 x'\, \mathrm{d}^3 c_1\, \mathrm{d}^3 c_2\, \mathrm{d}^3 r_2 |_{\boldsymbol{r}_1=\boldsymbol{r}}$$
$$= -\frac{1}{4\theta}\left(\frac{2\theta}{m}\right)^{\frac{1}{2}} \int \mathrm{d}^3 r_2 \iiint \mathrm{d}^6 x'\, \mathrm{d}^3 c_1\, \mathrm{d}^3 c_2 \frac{\varphi'(r_{12})}{r_{12}} \boldsymbol{r}_{12} \cdot (\boldsymbol{\mathscr{C}}_1 + \boldsymbol{\mathscr{C}}_2)$$
$$\times f_2'(x_1\, x_2 | f_1^{(0)}; x') f_1^{(0)}(\boldsymbol{c}')\mathscr{C}'^2 \boldsymbol{r}_{12}(\boldsymbol{r}' - \boldsymbol{r}_1)|_{\boldsymbol{r}_1=\boldsymbol{r}} \cdot \boldsymbol{\nabla}\theta$$
$$= -\frac{1}{12\theta}\left(\frac{2\theta}{m}\right)^{\frac{1}{2}} \int \mathrm{d}^3 r_2 \iiint \mathrm{d}^6 x'\, \mathrm{d}^3 c_1\, \mathrm{d}^3 c_2 \frac{\varphi'(r_{12})}{r_{12}} \boldsymbol{r}_{12} \cdot (\boldsymbol{\mathscr{C}}_1 + \boldsymbol{\mathscr{C}}_2)$$
$$\times f_2'(x_1\, x_2 | f_1^{(0)}; x')\, f_1^{(0)}(\boldsymbol{c}')\mathscr{C}'^2(\boldsymbol{r}_{12} \cdot (\boldsymbol{r}' - \boldsymbol{r}_1))|_{\boldsymbol{r}_1=\boldsymbol{r}} \boldsymbol{\nabla}\theta. \tag{9}$$

The combined results (5)–(9) correspond to the results (13.2-60) through (13.2-61).

References

Anderson, D. G., 1966, Numerical solution of the Krook kinetic equation, J. Fluid Mech. **25**, 271–287.

Barua, A. K., A. Saran and Y. Singh, 1967, On the representation of the interaction energy between two polar molecules, Appl. Sci. Res. **18**, 43–49.

Beenakker, J. J. M., 1969, The influence of electric and magnetic fields on the transport properties of polyatomic dilute gases, in: *Festkörperprobleme* **8** (F. Vieweg & Sohn, Braunschweig).

Bhatnagar, P. L., E. P. Gross and M. Krook, 1954, A model for collision processes in gases, I. Small amplitude processes in charged and neutral one-component systems, Phys. Rev. **94**, 511–525.

Bird, G. A., 1968, The structure of normal shock waves in a binary gas mixture, J. Fluid Mech. **31**, 657–668.

Bogoliubov, N. N., 1946, Problems of a dynamical theory in statistical physics (Russian), (Moscow); English transl. in: *Studies in statistical mechanics* **1**, eds. J. de Boer and G. E. Uhlenbeck (North-Holland Publishing Company, Amsterdam).

Boltzmann, L., 1872, Weitere Studien über das Wärmegleichgewicht unter Gasmolekülen, Sitzungs Berichte Kaiserl. Akad. der Wissenschaften **66** (2), 275–370.

Boltzmann, L., 1875a, Ueber das Wärmegleichgewicht von Gasen, auf welche äussere Kräfte wirken, Sitzungs Berichte der Oesterr. Akad. der Wissenschaften, Mathem.-Naturwiss. Klasse **72** (2), 427–457.

Boltzmann, L., 1875b, Bemerkungen über der Wärmeleitung der Gase, Sitzungs Berichte der Oesterr. Akad. der Wissenschaften, Mathem.-Naturwiss. Klasse **72** (2), 458–483.

Born, M. and H. S. Green, 1949, *A general kinetic theory of liquids* (Cambridge University Press).

Brokaw, R. S., 1958, Approximate formulas for the viscosity and thermal conductivity of gas mixtures, I, J. Chem. Phys. **29**, 391–397.

Brokaw, R. S., 1965, Approximate formulas for the viscosity and thermal conductivity of gas mixtures, II, J. Chem. Phys. **42**, 1140–1146.

Bruch, L. W. and I. J. McGee, 1967, Semiempirical potential and bound state of the helium-4 diatom, J. Chem. Phys. **46**, 2959–2967.

Brush, S. G., 1966, *Kinetic theory*, I. The nature of gases and of heat; II. Irreversible processes (Pergamon Press, Oxford).

Buddenberg, J. W. and C. R. Wilke, 1949, Calculation of gas mixture viscosity, Industrial and Eng. Chemistry **41**, 1345–1347.

Burgers, J. M., 1969, *Flow equations for composite gases* (Academic Press, New York).

Burnett, D., 1935a, The distribution of velocities in a slightly non-uniform gas, Proc. London Math. Soc. **39**, 385-430.

Burnett, D., 1935b, The distribution of molecular velocities and the mean motion in a non-uniform gas, Proc. London Math. Soc. **40**, 382–435.

Carleman, T., 1933, Sur la théorie de l'équation intégro-différentielle de Boltzmann, Acta Mathematica **60**, 91–146.

Carleman, T., 1957, *Problèmes mathématiques dans la théorie cinétique des gaz*, Publ. Sci. de l'Inst. Mittag-Leffler **2** (Almquist & Wiksells, Uppsala).

Case, K. M., 1960, Elementary solutions of the transport equation and their applications, Ann. Phys. (N.Y.) **9**, 1–23.

Case, K. M. and P. F. Zweifel, 1967, *Linear transport theory* (Addison Wesley Publishing Company).

Cercignani, C., 1962, Elementary solutions of the linearized gas-dynamics Boltzmann equation and their applications to the slip-flow problem, Ann. Phys. (N. Y.) **20**, 219–233.

Cercignani, C., 1967, Existence and uniqueness in the large for boundary value problems in kinetic theory, J. Math. Phys. **8**, 1653–1656.

Cercignani, C., 1969a, *Mathematical methods in kinetic theory* (Plenum Press, New York).

Cercignani, C., 1969b, Boundary value problems in linearized kinetic theory, in: *Transport theory* **1**, SIAM-AMS Proc. (American Mathematical Society, Providence, R. I).

Cercignani, C. and H. Lampis, 1970, Kinetic models for gas-surface interactions, in: *Rarefied gas dynamics*, ed. C. Cercignani (Academic Press, New York).

Chapman, S., 1916, On the law of distribution of molecular velocities, and on the theory of viscosity and thermal conduction, in a non-uniform simple monatomic gas, Phil. Trans. Roy. Soc. London **216**, 279–341.

Chapman, S., 1917, On the kinetic theory of a gas; Part II, A composite monatomic gas, diffusion, viscosity and thermal conduction, Phil. Trans. Roy. Soc. London **217**, 118–192.

Chapman, S. and F. W. Dootson, 1917, A note on thermal diffusion, Phil. Mag. **33**, 248–253.

Chapman, S. and T. G. Cowling, 1939, 1952, *The mathematical theory of non-uniform gases* (Cambridge University Press).

Chmieleski, R. M., 1966, Transport properties of a nonequilibrium partially ionized gas, Ph. D. Diss., Stanford University.

Chmieleski, R. M. and J. H. Ferziger, 1967a, Transport properties of a nonequilibrium partially ionized gas, Phys. of Fluids **10**, 364–371.

Chmieleski, R. M. and J. H. Ferziger, 1967b, Transport properties of a nonequilibrium partially ionized gas in a magnetic field, Phys. of Fluids **10**, 2520–2530.

Choh, S. T. (and G. E. Uhlenbeck), 1958, The kinetic theory of dense gases, Ph. D. Diss., University of Michigan.

Cohen, E. G. D., 1961, On the connection between various derivations of the Boltzmann equation, Physica **27**, 163–184.

Cohen, E. G. D., 1962a, Generalization of the Boltzmann equation, Physica **28**, 1025–1044.

Cohen, E. G. D., 1962b, Cluster expansions and the hierarchy, Physica **28**, 1045–1059, 1060–1073.

Cohen, E. G. D., 1966a, On the statistical mechanics of moderately dense gases not in equilibrium, in: *Lectures in theoretical physics* **8A**, ed. W. E. Brittin (University of Colorado Press, Boulder, Colorado).

Cohen, E.G.D., 1966b, Kinetic theory of dense gases, in: *Cargèse lectures in theoretical physics, Statistical mechanics*, ed. B. Jancovici (Gordon & Breach, New York).

Cohen, E. G. D., 1967, Kinetic theory of dense gases, in: *Lectures in theoretical physics* **9C**, ed. W. E. Brittin (Gordon & Breach, New York).

Cohen, E. G. D., 1968, The kinetic theory of dense gases, in: *Fundamental problems in statistical mechanics* **2**, ed. E. G. D. Cohen (North-Holland Publishing Company, Amsterdam).

Coope, J. A. R., R. F. Snider and F. R. McCourt, 1965, Irreducible Cartesian tensors, J. Chem. Phys. **43**, 2269–2275.

Coremans, J. M. J. and J. J. M. Beenakker, 1966, The influence of the density on the viscosity coefficient of gases, Physica **26**, 653–663.

Cotter, J. R., 1952, Conduction of heat in a monatomic gas, Proc. Roy. Irish Acad. A **55**, 1–28.

Courant, R. and D. Hilbert, 1953, *Methods of mathematical physics*, I (Interscience, New York).

Cowling, T. G., P. Gray and P. G. Wright, 1963, The physical significance of formulae for the thermal conductivity and viscosity of gaseous mixtures, Proc. Roy. Soc. A **276**, 69–82.

Curtiss, C. F., 1956, Kinetic theory of nonspherical molecules, J. Chem. Phys. **24**, 225–241.

Curtiss, C. F. and Ch. Muckenfuss, 1957, Kinetic theory of nonspherical molecules, II, J. Chem. Phys. **26**, 1619–1636.

Curtiss, J., 1968, Symmetric gaseous diffusion coefficients, J. Chem. Phys. **49**, 2917–2919.

Dahler, J. S., 1959, Transport phenomena in a fluid composed of diatomic molecules, J. Chem. Phys. **30**, 1447–1475.

Dahler, J. S., 1965, High density phenomena, in: *Research frontiers in fluid dynamics*, eds. R. J. Seeger and G. Temple (Interscience, New York).

Darrozès, J. S., 1968a, Approximate solutions of the Boltzmann equation for flows past bodies of moderate curvature, in: *Proc. Intern. Symp. on Rarefied gas dynamics* **6**, eds. L. Trilling and H. Y. Wachman (Academic Press).

Darrozès, J. S., 1968b, On the approximate solutions of the Boltzmann equations for a binary gas mixture, 12th Intern. Congress of Applied mechanics, Stanford University, 1968.

Darrozès, J. S. and J.-P. Guiraud, 1966, Généralisation formelle du théorème *H* en présence de parois, C. R. de l'Acad. des Sciences, Paris, **262** A, 1368–1371.

Davies, T. V. and E. M. James, 1966, *Nonlinear differential equations* (Addison Wesley Publishing Company).

De Boer, J. and J. Van Kranendonk, 1948, The viscosity and heat conductivity of gases with central intermolecular forces, Physica **14**, 442–452.

De Groot, S. R. and P. Mazur, 1962, *Non-equilibrium thermodynamics* (North-Holland Publishing Company, Amsterdam).

Dorfman, R., 1963, Note on the linearized Boltzmann integral equation for rigid sphere molecules, Proc. Nat. Acad. Sci. USA **50**, 804–806.

Dorfman, J. R., 1967, The binary collision expansion method in kinetic theory, in: *Lectures in theoretical physics* **9C**, ed. W. E. Brittin (Gordon & Breach, New York).

Dorfman, J. R. and E. G. D. Cohen, 1967, Difficulties in the kinetic theory of dense gases, J. Math. Phys. **8**, 282–297.

Dymond, J. H., M. Rigby and E. B. Smith, 1964, Intermolecular potential energy functions for simple molecules, Nature **204**, 678

Enskog, D., 1911, Bermerkungen zu einer Fundamentalgleichung in der kinetischen Gastheorie, Physik. Zeitschr. **12**, 533–539.

Enskog, D., 1917, Kinetische Theorie der Vorgänge in mässig verdünnten Gasen, Diss., Uppsala.

Enskog, D., 1922a, Die numerische Berechnung der Vorgänge in mässig verdünnten Gasen, Ark. Mat. Astron. Fys. **16**, no. 16, 1–60.

Enskog, D., 1922b, Kinetische Theorie der Wärmeleitung, Reibung und Selbstdiffusion in gewissen verdichteten Gasen und Flüssigkeiten, Kungl. Svenska Vet.-Ak. Handl. **63**, no. 4.

Enskog, D., 1928, Ueber die Grundgleichungen in der kinetischen Theorie der Flüssigkeiten und der Gase, Ark. Mat. Astron. Fysik **21** A, no. 13, 1–28.

Epstein, M., 1967, A model of the wall boundary condition in kinetic theory, AIAA Journal **5**, 1797–1800.

Ernst, M. H., 1966, Transport coefficients and temperature definition, Physica **32**, 252–272.

Eucken, E., 1913, Ueber das Wärmeleitvermogen, die spezifische Wärme und die innere Reibung der Gase, Physik. Zeitschr. **14**, 324–332.

Ferziger, J. H., 1965, Completeness property of solutions of the relaxation problem in kinetic theory, Phys. of Fluids **8**, 426–431.

Ferziger, J. H., 1969, Behavior of the H-function at a local Maxwellian state, Physica **43**, 29–32.

Frieman, E. A., 1963, On a new method in the theory of irreversible processes, J. Math. Phys. **4**, 410–418.

Frieman, E. A. and R. Goldman, 1966, Propagation of correlations in a Boltzmann gas, J. Math. Phys. **7**, 2153–2170.

Furry, W. H., R. Clark Jones and L. Onsager, 1939, On the theory of isotope separation by thermal diffusion, Phys. Rev. **55**, 1083–1095.

Gakhov, F. D., 1966, *Boundary value problems* (Pergamon Press, Addison Wesley Publishing Company).

Gandhi, J. M. and S. C. Saxena, 1966, Thermal conductivity of multicomponent mixtures of inert gases, Indian J. Pure and Appl. Phys. **4**, 461–466.

García-Colín, L. S., M. S. Green and F. Chaos, 1966, The Chapman-Enskog solution of the generalized Boltzmann equation, Physica **32**, 450–478.

Goldman, R. and E. A. Frieman, 1967, Logarithmic density behavior of a nonequilibrium Boltzmann gas, J. Math. Phys. **8**, 1410–1426.

Goldstein, H., 1950, *Classical mechanics* (Addison Wesley Publishing Company).

Gorter, C. J., 1938, Zur Interpretierung des Senftleben-Effektes, Naturwiss. **26**, 140.

Grad, H., 1949, On the kinetic theory of rarefied gases, Comm. Pure and Appl. Math. **2**, 311–407.

Grad, H., 1958, Principles of the kinetic theory of gases, in: *Handbuch der Physik* **12**, ed. S. Flügge (Springer Verlag).

Grad, H., 1963a, Asymptotic theory of the Boltzmann equation, Phys. of Fluids **6**, 147–181.

Grad, H., 1963b, Asymptotic theory of the Boltzmann equation, II, in: *Rarefied Gas Dynamics* **1**, ed. J. A. Laurmann (Academic Press, New York).

Grad, H., 1965a, On Boltzmann's H-theorem, J. Soc. for Ind. and Appl. Math. **13**, 259–277.

Grad, H., 1965b, Solution of the Boltzmann equation in an unbounded domain, Comm. Pure and Appl. Math. **18**, 345–354.

Grad, H., 1965c, Asymptotic equivalence of the Navier-Stokes and nonlinear Boltzmann equations, in: *Proc. Symp. Appl. Math.* **17** (American Mathematical Society, Providence, R.I.).

Green, M. S., 1956, Boltzmann equation from the statistical mechanical point of view, J. Chem. Phys. **25**, 836–855.

Green, M. S., 1958, The non-equilibrium pair distribution function at low densities, Physica **24**, 393–403.

Grew, K. E. and J. N. Mundy, 1961, Thermal diffusion in some mixtures of inert gases, Phys. of Fluids **4**, 1325–1332.

Gross, E. P. and E. A. Jackson, 1959, Kinetic models and the linearized Boltzmann equation, Phys. of Fluids **2**, 432–441.

Guernsey, R., 1960, Ph. D. Diss., University of Michigan (unpublished).
Guiraud, J. P., 1968a, Sur le problème de Couette et l'équation de Boltzmann, J. de Mécanique **7**, 171–203.
Guiraud, J. P., 1968b, Théorie mathématique de l'équation de Boltzmann, ONERA, T. P. no. 597, Chatillon (France).
Guiraud, J. P., 1968c, Problème aux limites intérieur pour l'équation de Boltzmann linéaire, C. R. Acad. Sci. Paris **266A**, 671–673.
Haines, L. K., J. R. Dorfman and M. H. Ernst, 1966, Divergent transport coefficients and the binary collision expansion, Phys. Rev. **144**, A 207–215.
Hecke, E., 1922, Ueber die Integralgleichung der kinetischen Gastheorie, Mathem. Zeitschr. **12**, 274–286.
Hess, S., 1967, Verallgemeinerte Boltzmanngleichung für mehratomige Gase, Diss., Erlangen; Z. Naturf. **22a**, 1871–1889.
Hilbert, D., 1912, *Grundzüge einer allgemeinen Theorie der linearen Integralgleichungen* (Teubner, Leipzig, 1912; Chelsea Publishing Company, New York, 1953).
Hille, E. and R. S. Phillips, 1957, *Functional analysis and semi-groups* (American Mathematical Society, Providence, R. I.).
Hirschfelder, J. O., 1957, Heat conductivity in polyatomic or electrically excited gases, II, J. Chem. Phys. **26**, 282–285.
Hirschfelder, J. O., C. F. Curtiss and R. B. Bird, 1954, *The molecular theory of gases and liquids* (J. Wiley & Sons, New York).
Holway, L. H., 1963, Approximation procedure for kinetic theory, Ph. D. Diss., Harvard University.
Ikenberry, E., 1955, A system of homogeneous spherical harmonics, Am. Math. Monthly **62**, 719–721.
Ikenberry, E. and C. Truesdell, 1956, On the pressures and the flux of energy in a gas according to Maxwell's kinetic theory, J. of Rat. Mech. and Analysis **5**, 1–54, 55–128.
Jancel, R. and Th. Kahan, 1962a, Relations entre les diverses solutions de l'équation intégro-différentielle de Boltzmann, C. R. Acad. Sci. Paris **254**, 1929–1931.
Jancel, R. and Th. Kahan, 1962b, Rôle des fonctions propres de l'opérateur linéaire Maxwellien dans la résolution de l'équation de Boltzmann, C. R. Acad. Sci. Paris **254**, 2292–2294.
Jancel, R. and Th. Kahan, 1966, *Electrodynamics of plasmas* (J. Wiley & Sons Ltd., London).
Jaynes, E. T., 1967, Foundations of probability theory and statistical mechanics, in: *Delaware Seminar in the foundations of physics*, ed. M. Bunge (Springer Verlag, Berlin).
Jeans, J. H., 1901, The distribution of molecular energy, Phil. Trans. Roy. Soc. A **196**, 397–431.
Jeffreys, H., 1931, *Cartesian tensors* (Cambridge University Press).
Kagan, Yu. and A. M. Afanas'ev, 1961, On the kinetic theory of gases with rotational degrees of freedom (Russian), Zh. Eks. i Teor. Fiz. **41**, 1536–1545; English transl. in: Soviet Physics–JETP **14** (1962) 1096–1101.
Kahn, B., 1938, On the theory of the equation of state, Diss., Utrecht; also published in: *Studies in statistical mechanics* **3**, eds. J. de Boer and G. E. Uhlenbeck (North-Holland Publishing Company, Amsterdam, 1965).
Katz, A., 1967, *Principles of statistical mechanics – The information theory approach* (W. H. Freeman and Co., San Francisco).
Kawasaki, K. and I. Oppenheim, 1965, Logarithmic terms in the density expansion of transport coefficients, Phys. Rev. **139**, A 1763–1768.
Keller, J. M. and W. L. Taylor, 1969, Evaluation of transport properties of gases: The Bruch-McGee potential for helium, J. Chem. Phys. **51**, 4829–4837.

Kestin, J., 1964, The viscosity and the thermal conductivity of gases, in: Proc. Intern. seminar on the transport properties of gases, Brown University.

Kihara, T., 1949, *Imperfect gases* (Japanese), (Asakusa Bookstore, Tokyo); available in English transl.: U.S. Office of Air Research, Wright-Patterson Air Force Base.

Kim, S. K. and J. Ross, 1965, The viscosity of moderately dense gases, J. Chem. Phys. **42**, 263–271.

Kirkwood, J. G., 1946, 1947, The statistical mechanical theory of transport processes, J. Chem. Phys. **14**, 180–201, 347; **15**, 72–76.

Kogan, M. N., 1969, *Rarefied gas dynamics* (transl. from Russian), (Plenum Press, New York).

Kohler, M., 1948, Behandlung von Nichtgleichgewichtsvorgängen mit Hilfe eines Extremalprinzipes, Zeitschr. Physik **124**, 772–789.

Kumar, K., 1966, Polynomial expansions in kinetic theory of gases, Ann. Phys. (N.Y.) **37**, 113–141.

Kuščer, I. and M. M. R. Williams, 1967, Relaxation constants of a uniform hard-sphere gas, Phys. of Fluids **10**, 1922–1927.

Ladyzhenskaia, O. A., 1969, *The mathematical theory of viscous incompressible flow* (transl. from Russian), (2nd ed., Gordon & Breach, New York).

Landau, L. D. and E. M. Lifshitz, 1958, *Quantum mechanics – Nonrelativistic theory* (transl. from Russian), (Pergamon Press, Oxford).

Landoldt-Börnstein, 1951, *Zahlenwerte und Funktionen* (6 Aufl., Springer Verlag, Berlin).

Laranjeira, M. F., 1960, An elementary theory of thermal and pressure diffusion in gaseous binary and complex mixtures, Physica **26**, 409–416.

Lenard, A., 1960, On Bogoliubov's kinetic equation for a spatially homogeneous plasma, Ann. Phys. (N.Y.) **10**, 390–400.

Lennard-Jones, J. E. (formally J. E. Jones), 1924, On the determination of molecular fields, II. From the equation of state of a gas, Proc. Roy. Soc. A **106**, 473–477.

Liepmann, H. W., R. Narasimha and M. T. Chahine, 1962, Structure of a plane shock layer, Phys. of Fluids **5**, 1313–1324.

London, F., 1930, Ueber einige Eigenschaften und Anwendungen der Molekularität, Zeitschr. Phys. Chem. **11**, 222–251.

London, F., 1937, The general theory of molecular forces, Trans. Faraday Soc. **33**, 8–26.

Longmire, C. L., 1963, *Elementary plasma physics* (Interscience Publ., New York).

Lorentz, H. A., 1887, Ueber das Gleichgewicht der lebendigen Kraft unter Gasmolekülen, Sitz. Ber. Kaiserl. Akad. der Wiss, Mathem.-Naturwiss. Klasse, **95**, 115–152.

Lorentz, H. A., 1905, The motion of electrons in metallic bodies, Proc. Sect. Sci. Kon. Akad. Wet., Amsterdam **7**, 438–453, 585–593, 684–691.

Loyalka, S. K., 1967, Momentum and energy slip in rarefied gases, Ph. D. Diss., Stanford University.

Loyalka, S. K. and J. H. Ferziger, Model dependence of the slip coefficient, Phys. of Fluids **10**, 1833–1839.

Lunn, A. C., 1913, Integral equations in the kinetic theory of gases (abstract), Bull. Am. Math. Soc. **19**, 455.

Marshall, W., 1960, Kinetic theory of an ionized gas, AERE T/R 2247, 2352, 2419, Atomic Energy Research Establishment, Harwell, Berks.

Mason, E. A., 1954, Transport properties of gases obeying a modified Buckingham (exp-6) potential, J. Chem. Phys. **22**, 169–186.

Mason, E. A., 1957a, Higher approximations for the transport properties of binary gas mixtures, I. General formulas, J. Chem. Phys. **27**, 75–84.

Mason, E. A., 1957b, Higher approximations for the transport properties of binary gas mixtures, II. Applications, J. Chem. Phys. **27**, 782–790.

Mason, E. A. and L. Monchick, 1962a, Heat conductivity of polyatomic and polar gases, J. Chem. Phys. **36**, 1622–1639.

Mason, E. A. and L. Monchick, 1962b, Transport properties of polar-gas mixtures, J. Chem. Phys. **36**, 2746–2757.

Mason, E. A., R. J. Munn and F. J. Smith, 1966, Thermal diffusion in gases, in: *Advances in atomic and molecular physics* **2**, ed. D. R. Bates (Academic Press, New York).

Mason, E. A. and S. C. Saxena, 1958, Approximate formulae for the thermal conductivity of gas mixtures, Phys. of Fluids **1**, 361–369.

Mason, E. A., S. Weissman and R. P. Wendt, 1964, Composition dependence of gaseous thermal diffusion factors and mutual diffusion coefficients, Phys. of Fluids **7**, 174–179.

Maxwell, J. C., 1860, Illustrations of the dynamical theory of gases, I. On the motions and collisions of perfectly elastic spheres; II. On the process of diffusion of two or more kinds of moving particles among one another; III. On the collision of perfectly elastic bodies of any form, Philos. Magazine **19**, 19–32; **20**, 21–32; **20**, 33–36.

Maxwell, J. C., 1867, On the dynamical theory of gases, Phil. Trans. Roy. Soc. London **157**, 49–88.

McCourt, F. R. W., 1966, Transport properties of gases with rotational states, Ph. D. Diss., University of British Columbia.

McCourt, F. R. W. and R. F. Snider, 1964, 1965, Thermal conductivity of a gas with rotational states, J. Chem. Phys. **41**, 3185–3194; **43**, 2276–2283.

McCoy, B. J., S. I. Sandler and J. S. Dahler, 1966, Transport properties of polyatomic fluids, IV. The kinetic theory of a dense gas of perfectly rough spheres, J. Chem. Phys. **45**, 3485–3512.

McCune, J. E., T. F. Morse and G. Sandri, 1963, On the relaxation of gases toward continuum flow, in: *Rarefied gas dynamics* **1**, ed. J. A. Laurmann (Academic Press).

Michels, A. and R. O. Gibson, 1931, The measurement of the viscosity of gases at high pressures. The viscosity of N_2 to 1000 atm., Proc. Roy. Soc. A **134**, 288–307.

Monchick, L., 1962, Free flight theory of gases, Phys. of Fluids **5**, 1393–1398.

Monchick, L. and E. A. Mason, 1961, Transport properties of polar gases, J. Chem. Phys. **35**, 1676–1697.

Monchick, L. and E. A. Mason, 1967, Free-flight theory of gas mixtures, Phys. of Fluids **10**, 1377–1390.

Montgomery, D., 1967, The foundations of classical kinetic theory, in: *Lectures in theoretical physics* **9C**, ed. W. E. Brittin (Gordon & Breach, New York).

Morgenstern, D., 1954, General existence and uniqueness proof for spatially homogeneous solution of maximal Boltzmann equation in the case of Maxwellian molecules, Proc. Nat. Acad. Sci. USA **40**, 719–721.

Mott, N. F. and H. S. W. Massey, 1965, *The theory of atomic collisions* (3rd ed., Oxford).

Muckenfuss, C. and C. F. Curtiss, 1958a, Kinetic theory of nonspherical molecules, III, J. Chem. Phys. **29**, 1257–1272.

Muckenfuss, C. and C. F. Curtiss, 1958b, Thermal conductivity of multicomponent gas mixtures, J. Chem. Phys. **29**, 1273–1277.

Munn, R. J., 1964, Interaction potential of the inert gases, I, J. Chem. Phys. **40**, 1439–1446.

Munn, R. J. and F. J. Smith, 1965, Interaction potential of the inert gases, II, J. Chem. Phys. **43**, 3998–4002.

Munn, R. J., F. J. Smith and E. A. Mason, 1965, Transport collision integrals for quantum gases obeying a 12–6 potential, J. Chem. Phys. **42**, 537–539.

Muskhelishvili, N. I., 1953, *Singular integral equations* (transl. from Russian), (Noordhoff, Groningen).

Navier, C. L. M. H., 1822, Mémoire sur les lois du mouvement des fluides, Mém. Acad. Sci. **6**, 389.

O'Neal, C. and R. S. Brokaw, 1962, Relation between thermal conductivity and viscosity for some nonpolar gases, Phys. of Fluids **5**, 567–574.

Pauling, L. and E. B. Wilson, 1935, *Introduction to quantum-mechanics* (McGraw-Hill, New York).

Pekeris, C. L., 1963, Note on the square-integrability of the kernel of the linearized Boltzmann integral equation for rigid sphere molecules, Proc. Nat. Acad. Sci. USA **49**, 38–40.

Pidduck, F. B., 1922, The kinetic theory of a special type of rigid molecules, Proc. Roy. Soc. A **101**, 101–112.

Povzner, A. Ja., 1962, On the Boltzmann equation in the kinetic theory of gases (Russian), Mat. Sbornik **58 (100)**, 65–86.

Rostoker, N. and M. N. Rosenbluth, 1960, Test particles in a completely ionized plasma, Phys. of Fluids **3**, 1–14.

Sandri, G., 1963, The foundations of nonequilibrium statistical mechanics, Ann. Phys. (N.Y.) **24**, 332–379, 380–418.

Scharf, G., 1967, Functional-analytic discussion of the linearized Boltzmann equation, Helv. Phys. Acta **40**, 929–945.

Schiff, L. I., 1949, *Quantum mechanics* (McGraw-Hill, New York).

Segal, B. M., 1971, Kinetic theory of shock structure, Ph.D. Diss., Stanford University.

Segal, B. M. and J. H. Ferziger, 1970, Shock structure according to some new models of the Boltzmann equation, in: *Proc. 2nd Intern. Conf. Numerical methods in fluid mechanics* (Springer Verlag).

Senftleben, H., 1930, Einfluss eines Magnetfeldes auf das Wärmeleitungsvermögen von paramagnetischen Gasen, Physik. Zeitschr. **31**, 822, 961–963.

Sengers, J. V., 1965a, Thermal conductivity and viscosity of simple fluids, Int. J. Heat and Mass Transfer **8**, 1103–1116.

Sengers, J. V., 1965b, Density expansion of the viscosity of a moderately dense gas, Phys. Rev. Letters **15**, 515–517.

Sengers, J. V., 1966a, Triple collision contribution to the transport coefficients of a rigid sphere gas, Phys. of Fluids **9**, 1333–1347.

Sengers, J. V., 1966b, Divergence in the density expansion of the transport coefficients of a two-dimensional gas, Phys. of Fluids **9**, 1685–1696.

Sengers, J. V., 1967, Triple collision contribution to the transport coefficients of gases, in: *Lectures in theoretical physics* **9C**, ed. W. E. Brittin (Gordon & Breach, New York).

Shapiro, A., 1953, *The dynamics and thermodynamics of compressible fluid flow*, 2 vols. (Ronald Press Co., New York).

Shapiro, C. S. and N. Corngold, 1965, Approach to equilibrium of a neutron gas, Phys. Rev. **137**, A 1686–1696.

Simon, A. and E. G. Harris, 1960, Kinetic equations for plasma and radiation, Phys. of Fluids **3**, 245–254.

Smith, F. J. and R. J. Munn, 1964, Automatic calculation of the transport collision integrals with tables for the Morse potential, J. Chem. Phys. **41**, 3560–3568.

Smith, F. J., R. J. Munn and E. A. Mason, 1967, Transport properties of quadrupolar gases, J. Chem. Phys. **46**, 317–321.

Snider, R. F., 1960, Quantum-mechanical modified Boltzmann equation for degenerate internal states, J. Chem. Phys. **32**, 1051–1060.

Snider, R. F., 1964a, Variational methods for solving the Boltzmann equation, J. Chem. Phys. **41**, 591–595.

Snider, R. F., 1964b, Perturbation variation methods for a quantum Boltzmann equation, J. Math. Phys. **5**, 1580–1587.

Spitzer, L., 1962, *Physics of fully ionized gases* (2nd ed., Interscience Publishers, New York).

Stokes, G. G., 1845, On the friction of fluids in motion and the equilibrium and motion of elastic solids, Trans. Cambridge Philos. Soc. **8**, 287 (Papers **1**, 75).

Storvick, T. S. and E. A. Mason, 1966, Determination of diffusion coefficients from viscosity measurements: Effect of higher Chapman-Enskog approximations, J. Chem. Phys. **45**, 3752–3759.

Struminskii, V. V., 1964, Hilbert's method of solving the Boltzmann equation (Russian), Dokl. Akad. Nauk SSSR **158**, 70–73; English transl. in: Sov. Phys. Doklady **9** (1965) 733–735.

Su, C. H. and Young-ping Pao, 1969, Boltzmann equation with power-law interparticle potentials, Phys. of Fluids **12**, 552–556.

Sutherland, W., 1893, The viscosity of gases and molecular force, Philos. Magazine **36** (5th Ser.), 507–531.

Sutherland, W., 1909, Molecular diameters, Philos. Magazine **17** (6th Ser.), 320–321.

Taxman, N., 1958, Classical theory of transport phenomena in dilute polyatomic gases, Phys. Rev. **110**, 1235–1239.

Uhlenbeck, G. E. and G. W. Ford, 1962, The theory of linear graphs with applications to the theory of the virial development of the properties of gases, in: *Studies in statistical mechanics* **1**, eds. J. de Boer and G. E. Uhlenbeck (North-Holland Publishing Company, Amsterdam).

Uhlenbeck, G. E. and G. W. Ford, 1963, *Lectures in statistical mechanics* (American Mathematical Society, Providence, R. I.).

Van de Ree, J., 1967, On the definition of the diffusion coefficient in reacting gases, Physica **36**, 118–126.

Van Heijningen, R. J. J., 1967, Diffusion in binary gaseous mixtures, Diss., Leiden.

Van Kampen, N. G., 1955, On the theory of stationary waves in plasmas, Physica **21**, 949–963.

Van Leeuwen, J. M. J. and A. Weyland, 1967a, Non-analytic density behavior of the diffusion coefficient of a Lorentz gas, Physica **36**, 457–490.

Van Leeuwen, J. M. J. and A. Weyland, 1967b, The density expansion of the diffusion coefficients of a Lorentz gas, in: *Statistical mechanics, foundations and application*, ed. Th. A. Bak (Benjamin, New York).

Waldmann, L., 1947, Der Diffusionsthermoeffekt, II, Zeitschr. Physik **124**, 175–195.

Waldmann, L., 1957, Die Boltzmann Gleichung für Gase mit rotierenden Molekülen, Zeitschr. Naturf. **12a**, 660–662.

Waldmann, L., 1958, Transporterscheinungen in Gasen von mittlerem Druck, in: *Handbuch der Physik* **12**, ed. S. Flügge (Springer Verlag).

Waldmann, L., 1965, Quantum theoretical transport equations for polyatomic gases, in: *Statistical mechanics of equilibrium and non-equilibrium*, ed. J. Meixner (North-Holland Publishing Company, Amsterdam).

Waldmann, L., 1968, Kinetic theory of dilute gases with internal degrees of freedom, in: *Fundamental problems in statistical mechanics* **2**, ed. E. G. D. Cohen (North-Holland Publishing Company, Amsterdam).

Wang Chang, C. S. and G. E. Uhlenbeck, 1951, Transport phenomena in polyatomic gases, Univ. of Michigan Eng. Res. Inst. Report CM-681.

Wang Chang, C. S. and G. E. Uhlenbeck, 1952, On the propagation of sound in monatomic gases, Univ. of Michigan Eng. Res. Inst. Report.

Wang Chang, C. S., G. E. Uhlenbeck and J. de Boer, 1964, The heat conductivity and viscosity of polyatomic gases, in: *Studies in statistical mechanics* **2**, ed. J. de Boer and G. E. Uhlenbeck (North-Holland Publishing Company, Amsterdam).

Wassiljewa, A., 1904, Wärmeleitung in Gasgemischen, Physik. Zeitschr. **5**, 737–742.

Weinstock, J., 1965, Nonanalyticity of transport coefficients and the complete density expansion of momentum correlation functions, Phys. Rev. **140**, A 460–465.

Weissman, S. and E. A. Mason, 1962, Determination of gaseous diffusion coefficients from viscosity measurements, J. Chem. Phys. **37**, 1289–1300.

Whittaker, E. T. and G. N. Watson, 1915, *A course of modern analysis* (Cambridge University Press).

Wild, E., 1951, On Boltzmann's equation in the kinetic theory of gases, Proc. Cambridge Philos. Soc. **47**, 602–609.

Wilke, C. R., 1950, A viscosity equation for gas mixtures, J. Chem. Phys. **18**, 517–519.

Williams, M. M. R., 1971, *Mathematical methods in particle transport theory* (Butterworths, London).

Wu, Ta-You, 1966, *Kinetic equations of gases and plasmas* (Addison Wesley Publishing Company).

Yan, C. C., 1969, Relaxation rate spectrum of the linearized Boltzmann equation for hard spheres, Phys. of Fluids **12**, 2306–2312.

Yosida, K., 1968, *Functional analysis* (2nd ed., Springer Verlag, Berlin).

Yvon, J., 1935, *La théorie statistique des fluides et l'équation d'état*, Actualités Scientifiques et Industrielles no. 235 (Hermann et Cie, Paris).

Ziman, J., 1956, The general variational principle of transport theory, Can. J. Phys. **34**, 1256–1273.

Zygmund, A., 1959, *Trigonometric series* (Cambridge University Press).

List of symbols

In general, symbols which occur only in a few consecutive pages are not included in this list. The number following the short description of the meaning of a symbol indicates the page on which the symbol is first introduced.

$[\,,\,]$	bracket integral, 94–96
$\{\,,\,\}$	Poisson bracket, 35
$[\ ]_n$	*n*th order Chapman-Cowling approximation, 138
$[\ ]_n^K$	*n*th order Kihara approximation, 140
$[\ ]_{\mathrm{r.s.}}$	value for rigid sphere model potential, 207
b	impact parameter, 28
b^*	reduced impact parameter, 252
b	covolume per unit mass, 370
c_v	specific heat, 198
c	speed of light, 429
$\mathbf{c}$	molecular velocity, 10
$\bar{c}$	average molecular velocity, 17
d	Debye length, 428
$\mathbf{d}$	diffusion driving force, 171
e	charge of an electron, 427
f	velocity distribution function, 11
f	Eucken ratio, 216
$\mathbf{g}$	relative velocity, 26

g^{*2}	reduced relative kinetic energy, 252
$\mathscr{g}$	dimensionless relative velocity, 203
h	Planck's constant, 259
i	imaginary unit, 438
$\boldsymbol{j}$	current density, 432
k	Boltzmann's constant, 12
k_T	binary thermal diffusion ratio, 189
k_{Ti}	thermal diffusion ratio, 176
l	mean free path, 19
m	molecular mass, 10
m_{ij}	reduced mass, 205
n	number density, 11
n_2	pair distribution function, 381
p	hydrostatic pressure, 15
$\boldsymbol{p}$	molecular momentum, 33
$\boldsymbol{q}$	heat flow vector, 16
r^*	reduced intermolecular distance, 252
r_0	range of intermolecular potential, 50
r_c	gyromagnetic radius, 429
$\boldsymbol{r}$	molecular position vector, 10
t	time, 10
t_coll	mean collision time, 51
u	internal energy per unit mass, 12
$\mathfrak{v}$	hydrodynamic velocity, 12
x	molecular coordinates in phase space, 33
x_i	molecular fraction, 219
A^*	ratio of reduced Ω-integrals, 207
$\boldsymbol{A}$	vector function in Chapman-Enskog theory, 126
$\mathfrak{A}$	vector function in Chapman-Enskog theory, 438
B^*	ratio of reduced Ω-integrals, 207
$\mathbf{B}$	tensor function in Chapman-Enskog theory, 126
$\mathfrak{B}$	tensor function in Chapman-Enskog theory, 448
C^*	ratio of reduced Ω-integrals, 207
$\boldsymbol{C}$	peculiar velocity, 12
$\mathscr{C}$	dimensionless peculiar velocity, 135
D_{ij}	diffusion coefficient, 174
D_N	distribution function in Γ-space, 33

D_T binary thermal diffusion coefficient, 189
D_{Ti} thermal diffusion coefficient, 174
$\boldsymbol{D}^j$ vector function in Chapman-Enskog theory, 172
$\boldsymbol{\mathfrak{D}}^j$ vector function in Chapman-Enskog theory, 439
$\mathscr{D}$ self-diffusion coefficient, 190
$\mathscr{D}_{12}$ binary diffusion coefficient, 189
$\mathfrak{D}$ streaming operator, 62
$\mathrm{E}^\star$ ratio of reduced Ω-integrals, 207
$\boldsymbol{E}$ electric field, 232
E tensor function in Chapman-Enskog theory, 339
$\mathscr{E}$ dimensionless energy variable, 317
$\mathrm{F}^\star$ ratio of reduced Ω-integrals, 207
F_s reduced distribution function, 36
$\boldsymbol{F}$ force per unit mass, 24
$\boldsymbol{\mathfrak{F}}$ external force acting on a molecule, 35
$\boldsymbol{G}$ center-of-mass velocity, 26
$\mathscr{G}_0$ dimensionless center-of-mass velocity, 203
H Boltzmann's H-function, 71
H_N Hamiltonian of a system of N particles, 33
$\boldsymbol{H}$ magnetic field, 429
$\mathscr{H}$ Hilbert space, 100
$\mathfrak{H}_N$ Hamiltonian operator, 35
I linearized collision operator, 87
$I_2(I_3)$ linearized binary (ternary) collision operator, 409, 410
I unit tensor of rank two, 15
J collision operator, 62
$\boldsymbol{J}$ internal angular momentum, 335
K number of components of a gas mixture, 17
Kn Knudsen number, 464
$\mathfrak{K}$ operator describing gas-surface interaction, 78
L linearized collision operator, 465
L angular momentum flux tensor, 336
$\mathfrak{L}$ phase mixing operator, 38
$\boldsymbol{M}$ angular momentum operator, 335
N Avogadro's number, 16
P pressure tensor, 15
Q partition function, 336
Q charge density, 431

$Q^{(l)}$	transport cross section, 205
$Q_{ij}^{(l)}$	transport cross section, 206
R	universal gas constant, 16
S	entropy, 75
$S_\nu^{(n)}$	Sonine polynomial, 132
$S_t^{(s)}$	streaming operator, 46
$\boldsymbol{S}$	rate-of-shear tensor, 129
$\mathfrak{S}_t^{(s)}$	composite streaming operator, 50
$\tilde{\mathfrak{S}}_t^{(s)}$	composite streaming operator, 47
T	temperature, 12
T^*	reduced temperature, 252
$T(t)$	semigroup of operators, 100
$\mathfrak{T}_t^{(s)}$	composite streaming operator, 407
$\tilde{\mathfrak{T}}_t^{(s)}$	composite streaming operator, 48
$\boldsymbol{V}_i$	diffusion velocity, 18
W	integral kernel describing gas-surface interaction, 78
Z_{int}	partition function, 316
α_T	binary thermal diffusion factor, 190
α_{ij}	polynary thermal diffusion factor, 224
β	thermal expansion coefficient, 393
$\boldsymbol{\beta}$	vector of macroscopic observables, 113
ε	geometrical collision variable, 26
ε	parameter in Hilbert and Chapman-Enskog methods, 109
$\in$	strength parameter for intermolecular potential, 240
$\boldsymbol{\epsilon}$	alternating tensor, 340
η	viscosity, 130
η_{ij}	parameter in multicomponent transport properties, 217
θ	energy variable in theory of dense gases, 390
κ	bulk viscosity
κ	isothermal compressibility, 393
λ	thermal conductivity, 130
λ_{ij}	parameter in multicomponent transport properties, 217
λ'	partial coefficient of thermal conductivity, 178
μ	dimensionless mass ratio, 26
μ	uniformity parameter in theory of dense gases, 388
$\boldsymbol{\mu}$	dipole moment, 232
ν	collision frequency, 19

ν	index for intermolecular potential, 239
ρ	mass density, 11
σ	differential collision cross section, 31
σ	range parameter for intermolecular potential, 240
σ	electrical conductivity, 443
τ	mean free time, 19
τ_κ	relaxation time, 323
φ	intermolecular potential, 33
φ	electro-thermal coefficient, 443
φ^*	reduced intermolecular potential, 252
χ	geometrical collision variable, 26
χ	density correction factor in Enskog's theory of dense gases, 358
ψ	collisional invariant, 64
ω	gyrofrequency, 351
Γ	scalar function in Chapman-Enskog theory, 318
H^{qr}	linear combination of bracket integrals, 137
H_{ij}^{qr}	linear combination of bracket integrals, 186
Θ	molecular interaction operator, 35
Λ^{qr}	linear combination of bracket integrals, 136
Λ_{ij}^{qr}	linear combination of bracket integrals, 182
$\boldsymbol{\Phi}$	flux vector, 14
$\Omega^{(l,r)}$	effective cross section (Ω-integral), 205
$\Omega_{ij}^{(l,r)}$	effective cross section (Ω-integral), 205
$\Omega^{(l,r)\star}$	reduced Ω-integral, 207
$\Omega_{ij}^{(l,r)\star}$	reduced Ω-integral, 207

Author index

Numbers in italics refer to the pages on which the complete references are listed.

E

F

G

H

I

J

K

L

M

Subject index